T0140131

Advances in Intelligent Systems and Computing

Volume 1034

The series "Advances in Intelligent Systems and Computing" contains publications on theory, applications, and design methods of Intelligent Systems and Intelligent Computing. Virtually all disciplines such as engineering, natural sciences, computer and information science, ICT, economics, business, e-commerce, environment, healthcare, life science are covered. The list of topics spans all the areas of modern intelligent systems and computing such as: computational intelligence, soft computing including neural networks, fuzzy systems, evolutionary computing and the fusion of these paradigms, social intelligence, ambient intelligence, computational neuroscience, artificial life, virtual worlds and society, cognitive science and systems, Perception and Vision, DNA and immune based systems, self-organizing and adaptive systems, e-Learning and teaching, human-centered and human-centric computing, recommender systems, intelligent control, robotics and mechatronics including human-machine teaming, knowledge-based paradigms, learning paradigms, machine ethics, intelligent data analysis, knowledge management, intelligent agents, intelligent decision making and support, intelligent network security, trust management, interactive entertainment, Web intelligence and multimedia.

The publications within "Advances in Intelligent Systems and Computing" are primarily proceedings of important conferences, symposia and congresses. They cover significant recent developments in the field, both of a foundational and applicable character. An important characteristic feature of the series is the short publication time and world-wide distribution. This permits a rapid and broad dissemination of research results.

** Indexing: The books of this series are submitted to ISI Proceedings, EI-Compendex, DBLP, SCOPUS, Google Scholar and Springerlink **

More information about this series at http://www.springer.com/series/11156

Vikrant Bhateja · Suresh Chandra Satapathy ·
Yu-Dong Zhang · V. N. Manjunath Aradhya
Editors

Intelligent Computing and Communication

Proceedings of 3rd ICICC 2019, Bangalore

 Springer

Editors
Vikrant Bhateja
Department of Electronics and
Communication Engineering
Shri Ramswaroop Memorial Group
of Professional Colleges (SRMGPC)
Lucknow, Uttar Pradesh, India

Dr. A.P.J. Abdul Kalam
Technical University
Lucknow, Uttar Pradesh, India

Yu-Dong Zhang
Department of Informatics
University of Leicester
Leicester, UK

Suresh Chandra Satapathy
School of Computer Engineering
Kalinga Institute of Industrial
Technology (KIIT)
Bhubaneswar, Odisha, India

V. N. Manjunath Aradhya
Department of MCA
J. S. S. Science and Technology University
Mysuru, India

ISSN 2194-5357 ISSN 2194-5365 (electronic)
Advances in Intelligent Systems and Computing
ISBN 978-981-15-1083-0 ISBN 978-981-15-1084-7 (eBook)
https://doi.org/10.1007/978-981-15-1084-7

This Springer imprint is published by the registered company Springer Nature Singapore Pte Ltd.
The registered company address is: 152 Beach Road, #21-01/04 Gateway East, Singapore 189721, Singapore

Conference Organization Committees

Chief Patrons

Dr. D. Hemachandra Sagar, Chancellor, DSI
Dr. D. Premachandra Sagar, Pro Chancellor, DSI

Patrons

Shri. Galiswamy, Secretary, DSI
Dr. A. N. N. Murthy, Vice Chancellor, DSU
Shri. R. Janardhan, Pro-Vice Chancellor, DSU
Dr. Puttamadappa C., Registrar, DSU

Honorary Chairs

H. P. Khincha, Evangelist, DSU
K. Jairaj, Advisor, DSU

General Chair

Dr. S. C. Satapathy, KIIT Bhubaneswar, Odisha, India

Publication Chair

Dr. Vikrant Bhateja, SRMGPC, Lucknow, U.P., India

Organizing Chairs

Dr. M. K. Banga, Chairman CSE, Dean Research, DSU
Dr. A. Srinivas, Dean SoE, DSU

Technical Program Chair

Dr. V. N. Manjunatha Aradhya, SJCE, Mysore, Karnataka, India

Organizing Secretaries

Dr. Rajesh T. M., Department of CSE, DSU
Dr. Mallanagouda Patil, Department of CSE, DSU

Program Secretaries

Dr. Shaila S. G., Department of CSE, DSU
Dr. Jasma Balasangameshwara, Department of CT, DSU

Advisory Committee

Dr. Aimé Lay-Ekuakille, University of Salento, Lecce, Italy
Dr. Amira Ashour, Tanta University, Egypt
Dr. Aynur Unal, Stanford University, USA
Dr. Bansidhar Majhi, IIIT Kancheepuram, Tamil Nadu, India
Dr. Dilip Kumar Sharma, Vice-Chairman, IEEE UP Section
Dr. Ganpati Panda, IIT Bhubaneswar, Odisha, India
Dr. Jagdish Chand Bansal, South Asian University, New Delhi, India
Dr. João Manuel R. S. Tavares, Faculdade de Engenharia da Universidade do Porto (FEUP), Porto, Portugal

Dr. Jyotsana Kumar Mandal, University of Kalyani, West Bengal, India
Dr. K. C. Santosh, University of South Dakota, USA
Dr. Le Hoang Son, Vietnam National University, Hanoi, Vietnam
Dr. Naeem Hanoon, Multimedia University, Cyberjaya, Malaysia
Dr. Nilanjan Dey, TIET, Kolkata, India
Dr. Noor Zaman, Universiti Tecknologi Malaysia; PETRONAS, Malaysia
Dr. Roman Senkerik, Tomas Bata University, Zlin, Czech Republic
Dr. Swagatam Das, Indian Statistical Institute, Kolkata, India
Dr. Siba K. Udgata, University of Hyderabad, Telangana, India
Dr. Shri Nivas Singh, MMMUT, Gorakhpur, U.P., India
Dr. Steven L. Fernandez, University of Alabama at Birmingham, USA
Dr. Tai Kang, Nanyang Technological University, Singapore
Dr. Valentina Balas, Aurel Vlaicu University of Arad, Romania

Technical Program Committee

Prof. H. R. Vishwakarma, VIT, Vellore, India
Prof. Dan Boneh, Stanford University, California, USA
Prof. Alexander Christea, University of Warwick, London, UK
Prof. Ahmad Al- Khasawneh, Hashemite University, Jordan
Dr. Bharat Singh Deora, JRNRV University, India
Prof. Jean Michel Bruel, Departement Informatique, IUT de Blagnac, Blagnac, France
Prof. Ngai-Man Cheung, Singapore University of Technology and Design, Singapore
Prof. Yun-Bae Kim, Sungkyunkwan University, South Korea
Prof. Ting-Peng Liang, National Chengchi University, Taipei, Taiwan
Prof. Sami Mnasri, IRIT Laboratory, Toulouse, France
Prof. Lorne Olfman, CISAT, Claremont Graduate University, California, USA
Prof. Brent Waters, The University of Texas at Austin, TX, USA
Prof. Philip Yang, PricewaterhouseCoopers, Beijing, China
Prof. R. K. Bayal, Rajasthan Technical University, Kota, Rajasthan, India
Prof. Martin Everett, The University of Manchester, England
Prof. Feng Jiang, Harbin Institute of Technology, China
Prof. Prasun Sinha, The Ohio State University, Columbus, OH, USA
Dr. Savita Gandhi, Gujarat University, Ahmedabad, India
Prof. Xiaoyi Yu, Chinese Academy of Sciences, Beijing, China
Dr. Mukesh Shrimali, Pacific University, Udaipur, India
Prof. Komal Bhatia, YMCA University of Science and Technology, Faridabad, Haryana, India
Prof. S. R. Biradar, SDM College of Engineering and Technology, Dharwad, Karnataka

Dr. Soura Dasgupta, SRM University, Chennai, India
Dr. Sushil Kumar, Jawaharlal Nehru University, New Delhi, India
Dr. Amioy Kumar, IIT Delhi, India

Preface

This book is a collection of high-quality peer-reviewed research papers presented at the '3rd International Conference on Intelligent Computing and Communication (ICICC 2019)' held at School of Engineering, Dayananda Sagar University, Bengaluru, India, during June 7–8, 2019.

After the success of past two editions of ICICC held in 2016 at University of Kalyani, West Bengal, India, and in 2017 at MAEER's MIT College of Engineering, Pune, India, the 3rd International Conference on Intelligent Computing and Communication is organized by School of Engineering, Dayananda Sagar University, Bengaluru, India. All papers of past ICICC editions are published by Springer AISC Series. Presently, ICICC 2019 provided a platform for academicians, researchers, scientists, professionals and students to share their knowledge and expertise in the diverse domain of intelligent computing and communication.

ICICC 2019 had received a number of submissions from the fields of ICT, intelligent computing and its prospective applications in different spheres of engineering. The papers received have undergone a rigorous peer review process with the help of the technical program committee members of the conference from the various parts of country as well as abroad. The review process has been crucial with minimum two–three reviews each along with due checks on similarity and content overlap. This conference has featured eleven exclusive theme-based special sessions as enlisted below along with main track and hence witnessed more than 300 submissions in total.

S1: Internet of Things (IoT) and Smart Computing in Modern Technical Arena
S2: Knowledge Discovery from Imprecise and Vague Data
S3: Next-Gen of Healthcare: Revolutionizing Clinical Decision Making with Predictive Analytics
S4: Intelligent System for Internet of Things and its Application
S5: Image and Video Processing for Smart Surveillance Systems using Deep Learning
S6: Big Data and e-Health
S7: Intelligent Computational Systems for a Smart Planet

S8: Computational Intelligence Optimization Techniques for Engineering Applications
S9: Application of Machine Learning: Multidisciplinary Domains
S10: Visual Recognition and Biometrics Security
S11: Multimedia Security

Out of the above pool, only 80+ quality papers were finally presented in seven parallel tracks (of oral presentation sessions chaired by esteemed professors from premier institutes of the country) and compiled in this volume for publication.

The conference featured many distinguished keynote addresses by eminent speakers like Dr. Rabi N. Mahapatra (Texas A&M University, USA) delivering a talk on 'IoT's Bold Vision: Computing & Security Concerns'; Mr. Manoj Wagle (Head of HPE/Aruba Bangalore WLAN Business & Site) delivering a talk titled 'Enterprise Wireless (WiFi) Network'; and Mr. Aninda Bose (Senior Publishing Editor, Springer Nature) delivering an interesting and informative lecture on 'Importance of Ethics in Research Publishing.' These keynote lectures/talks embraced a huge toll of audience of students, faculties, budding researchers as well as delegates. The editors thank the general chair, TPC chair and the organizing chair of the conference for providing valuable guidelines and inspirations to overcome various difficulties in the process of organizing this conference. The editors also thank School of Engineering, Dayananda Sagar University, Bengaluru, for their wholehearted support in organizing the third edition of ICICC series.

The editorial board take this opportunity to thank the authors of all the submitted papers for their hard work, adherence to the deadlines and patience during the review process. The quality of a refereed volume depends mainly on the expertise and dedication of the reviewers. We are indebted to the TPC members who not only produced excellent reviews but also did these in short time frames.

Lucknow, India   Dr. Vikrant Bhateja
Bhubaneswar, India   Dr. Suresh Chandra Satapathy
Leicester, UK   Dr. Yu-Dong Zhang
Mysuru, India   Dr. V. N. Manjunath Aradhya

About This Book

This book covers proceedings of the 3rd International Conference on Intelligent Computing and Communication (ICICC 2019) that provided a forum for presenting original research findings, innovations, practical experiences—algorithms, applications as well as current issues from the domains of intelligent computing, communication and other related areas of ICT. This conference was organized by the Department of CSE, School of Engineering, Dayananda Sagar University, Bengaluru, India, during June 7–8, 2019.

This volume covers broad areas of intelligent computing and communication focussing on data engineering, signal and image processing, geo-informatics, sensor networks, Web security, privacy and e-commerce. The presented works have shown implementation of various computational intelligence techniques like swarm intelligence, evolutionary algorithms, bio-inspired algorithms, etc. The volume will serve as knowledge center for students of postgraduate level and research scholars in various engineering disciplines.

Contents

About the Editors

Dr. Vikrant Bhateja is an Associate Professor at the ECE Department, SRMGPC, Lucknow (U.P.), India; where he also serves as the Head of Academics & Quality Control. He holds a doctorate in biomedical imaging and has 16 years of academic teaching experience, with over 145 publications in reputed international conference proceedings and journals to his credit. His areas of research include digital image and video processing, computer vision, medical imaging, and machine learning. Dr Vikrant has edited 20 books with Springer Nature. He is Editor-in-Chief of IGI Global—International Journal of Natural Computing and Research (IJNCR); an Associate Editor of the International Journal of Ambient Computing and Intelligence (IJACI); and a Guest Editor for journals including Evolutionary Intelligence and the Arabian Journal of Science and Engineering.

Prof. (Dr.) Suresh Chandra Satapathy is currently working as a Professor, School of Computer Engineering, KIIT Deemed to be University, Bhubaneswar, Odisha, India. He obtained his Ph.D. in CSE from JNTU Hyderabad and M. Tech. in CSE from NIT, Rourkela, Odisha, India. He has 27 years of teaching experience. His research interests are data mining, machine intelligence and swarm intelligence. He has acted as program chair of many international conferences and edited over 25 volumes of proceedings from Springer series from LNCS, AISC, LNNS, LNEE, SIST, etc. He is also in the Editorial board of few international Journals and has over 130 research publications in International journals and conference proceedings.

Prof. (Dr.) Yu-Dong Zhang received his Ph. D. degree from Southeast University, China in 2010. He worked as post-doc from 2010 to 2012 and a research scientist from 2012 to 2013 at Columbia University, USA. He served as Professor from 2013 to 2017 in Nanjing Normal University, where he was the director and founder of Advanced Medical Image Processing Group in NJNU. From 2017, he served as Full Professor in Department of Informatics, University of Leicester, UK. His research interests are deep learning in communication and signal processing, medical image processing. He is now the editor of Scientific Reports, Journal of Alzheimer's Disease, International Journal of Information Management, etc. He has conducted and joined many successful academic grants and industrial projects, such as NSFC, NIH, EPSRC, etc.

Dr. V. N. Manjunath Aradhya is currently working as an Associate Professor in the Dept. of MCA, Sri Jayachamarajendra College of Engineering, Mysuru, India. He received the M.S. and Ph. D degrees in Computer Science from University of Mysore, Mysuru, India, in 2004 and 2007 respectively. He is a recipient of "Young Indian Research Scientist" from Italian Ministry of Education, University and Research, Italy during 2009–2010. His professional recognition includes as a Technical Editor for Journal of Convergence Information Technology (JCIT), Editor Board in Journal of Intelligent Systems. His research interest includes, Pattern Recognition, Image Processing, Document Image Analysis, Computer Vision and Machine Intelligence. He has published more than 135 research papers in reputed international journals, conference proceedings, and edited books.

Task Scheduling in Heterogeneous Cloud Environment—A Survey

Roshni Pradhan and Suresh Chandra Satapathy

Abstract Cloud computing environment provides various computing services along with provision of resources in pay per use basis. Task scheduling and resource scheduling in cloud environment mostly take the help of Infrastructure as a Service (IaaS). In the current computing era, allocation of task to a machine is one of the most important issues. It is a NP–Hard problem. Many task scheduling algorithms are available to resolve such problems in a heterogeneous cloud environment. In this paper various task scheduling methods are compared and described. The performance of the different scheduling techniques are measure with the help of metrics like makespan, cloud utilization, etc.

Keywords Cloud · Infrastructure as a service (IaaS) · NP–Hard · Heterogeneous cloud · Makespan · Cloud utilization

1 Introduction

Task scheduling in cloud environment is carried out, based on different factors so that it enhances the cloud performance all together. It is based on various parameters in order that to increase the overall Cloud Performance. A group of task can be used for data processing, code accessing or to store data using some function. Tasks are characterized as per the Service Level Agreement (SLA) and asked for services. Each task is processed using one of the available machines in heterogeneous cloud. After completion of task processing the overall result is reflected to the user side [5, 9]. Cloud scheduler collects all the tasks or jobs after task processing and schedules them using some task scheduling techniques. During this process cloud scheduler ask for cloud information services for available resources. On the basis of accessible

R. Pradhan · S. C. Satapathy (✉)
Kalinga Institute of Industrial Technology, Bhubaneswar, India
e-mail: suresh.satapathyfcs@kiit.ac.in

R. Pradhan
e-mail: roshni.pradhanfcs@kiit.ac.in

© Springer Nature Singapore Pte Ltd. 2020
V. Bhateja et al. (eds.), *Intelligent Computing and Communication*,
Advances in Intelligent Systems and Computing 1034,
https://doi.org/10.1007/978-981-15-1084-7_1

1

resources and their features, tasks are allotted to the machines or servers. Cloud schedulers manage and allocate numerous tasks to various virtual machines. The task scheduling will be efficient if and only if it gives an optimal result. It will increase the total throughput along with CPU utilization and turnaround time [20]. Tasks can also be statistically allotted to available machine resources at compilation time and can be at run-time dynamically [4, 16].

2 Classification of Task Scheduling

Scheduling in cloud environment can be classified in to three types. Those are Resource Scheduling, Workflow Scheduling, and Task scheduling. The overall classification is clearly given in Fig. 1. Resource Scheduling is nothing but allocation of jobs or tasks to the physical machine or virtual machines. Workflow scheduling aimed to complete the execution of workflow by considering some factors like deadline or budget. Task scheduling is to distribute the arriving tasks or jobs to the available machines [24]. It may be centralized or distributed. Tasks can also be independent and dependent. Tasks are allocated to either homogeneous or heterogeneous cloud environment. In centralized scheduling, there is only one scheduler is available which map all the tasks but in distributed environment, bag of tasks are distributed among various scheduler [14, 19]. It creates high implementation complexity and processor cycles are spared. Centralized scheduling is difficult to achieve from scalability and fault tolerance point of view [18]. It is prone to single point failure. The hierarchical classification of cloud task scheduling is shown in Fig. 1.

In distributed computing environment, scheduling is categorized into two methods. First one is heuristics. Heuristics algorithms are again two types, i.e., Static and Dynamic scheduling. Dynamic scheduling is again of two types, one Online Mode and other one is Batch Mode. In Static scheduling, arriving task is known to the scheduler and they are allocated to the different virtual machines to avail the resources. In Dynamic scheduling, tasks are allocated immediately as soon as arrived. In comparison to static scheduling, dynamic scheduling outperforms and gives better results. But it results a high overhead in dynamic scheduling as the tasks are scheduled immediately and the information for the task scheduling gets updated periodically.

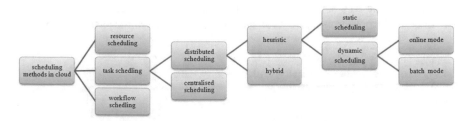

Fig. 1 Classification of task scheduling methods

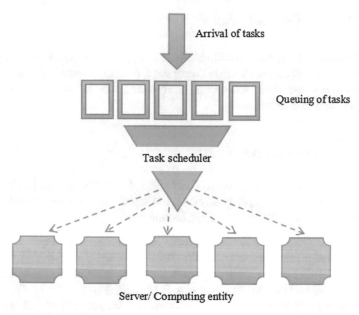

Fig. 2 Task scheduling components

3 Scheduling Model in the Cloud Datacenters

Task scheduling in the cloud environment has various functional components as shown in Fig. 2. The functional components are described as follows.

- **Computing entity**: Virtualization is one of the methods used in cloud computing environment. Virtual machines deliver facilities like operating system. Many software are available to execute the submitted task by task scheduler. Computing capacity of the software is calculated as number of instructions process per second.
- **Job scheduler**: The vital segment of the task scheduling in the cloud environment is job scheduler. It decides task execution sequence. It holds the tasks in to a queue until the current executing task is finished.
- **Job waiting queue**: Specific machines are chosen for task allocation in a waiting queue.
- **Job arriving process**: Jobs are arranged according to a scheduling algorithm.

4 Existing Task Scheduling Algorithms

Existing task scheduling methods are categorized into two distinct classes. They are Heuristic task scheduling and Hybrid task scheduling. Each of this is clarified quickly in the coming subsegments with their subcategorization.

4.1 Heuristic Task Scheduling

Heuristics scheduling approaches [2, 4, 21, 25, 26] give an ideal solution in which it utilizes the knowledge bases for generating the scheduling order. Heuristic methodologies can be either static scheduling or dynamic scheduling. Static scheduling algorithm is depicted below.

4.1.1 Static Scheduling Approaches

In this method of scheduling, all the jobs or tasks arrive at the same time. Those tasks are free from resources. Simple strategies like First Come First Service (FCFS) and Round Robin (RR) are Static scheduling techniques.

- **FCFS**: In this method of scheduling, tasks are get gathered in a queue. The tasks wait until the resources are available. Then the tasks are allocated to distinct machines keeping on the view of availability. This method of scheduling is less complex in comparison to other techniques [13, 22, 23].
- **RR**: This strategy utilizes a similar FIFO method for scheduling [13, 22, 23]. A specific time quantum is assigned to each task which plays an important role in task scheduling. The current task will be preempted by the upcoming tasks after the time quantum gets over. Opportunistic load balancing algorithm in cloud does not consider the current load of virtual machine. The machines are kept busy as they are loaded with queue of tasks. It does not give a good result from makespan calculation point of view.
- **Minimum Execution Time (MET)**: MET [4, 11–13] scheduling heuristics allocate the tasks into the machine based on which machine it requires less execution time. It chooses the efficient machine to execute the incoming tasks. Sometimes it results in load imbalance as it does not consider the resource availability.
- **Minimum Completion Time (MCT)**: Minimum completion time of task is taken into consideration in MCT. According to the algorithm the tasks are chosen and assigned to machines for scheduling. The existing load of the machines is also consider while task allocation. Completion time of a task is the sum total of execution time of the task on the machine and the ready time of the specific machine. It is the real time required for task execution [3].
- **Min-Min**: Min-Min [6, 9, 16] task scheduling method is used to chose the smallest task from the available tasks group. The task having minimum completion time is chosen and allocated to the machine. The disadvantage of this is, it increases the minimum completion time of the process and hence increases the makespan. Previously, allocated machine load is not considered in this type of scheduling technique. It simply allocates the task to faster machine. The longer task takes much time for waiting then the smaller task. This technique of scheduling enhances the throughput of the system.
- **Max-Min**: This type of scheduling is almost similar to Min-Min [3, 12]. Here the longest task (with most maximum completion time) is chosen from the bag

of tasks. The task is allocated to the machine having minimum completion time. Here load balancing is not considered. The tasks having smaller completion time starves for a period of time. The makespan of the process increases as compare to Min-Min as the longer task takes much time to get execute in machine taking less time. Genetic Algorithms and Simulated Annealing are two methods which give near optimal solutions.

- **Genetic Algorithm (GA)**: This approach of scheduling [1, 7], there are four sub-operations, i.e., evaluation, selection, mutation, is there. The initial population is the probable mappings of the tasks to the available machines. In this kind of scheduling, vector position is represented through a task. The vector is chosen from task list. The value of each position determines the machine allocation process. It will decide which task to map on which machines. Here each job represents a chromosome. There is fitness value associated with each chromosome. The fitness value is considered as the execution time (makespan) of the tasks. It is calculated as the time taken to complete the execution of task in a particular machine. This technique takes the past result and generates an improved result. It follows the rule of survival of fittest for the scheduling technique. Another research addresses the resource allocation and task scheduling in multi-cloud environment. Genetic Algorithm-based Customer-Conscious Resource Allocation and shortest task first scheduling concept are taken into consideration [8, 15].
- **Simulated Annealing (SA)**: It is an iterative method of scheduling. It is very similar to genetic algorithm where the mapping of tasks begins with a random distribution. In each iteration, SA gives better results. Mutation is another part of SA which is used to evaluate makespan. The updated result is replaced after completion of each iteration. But the disadvantage is that, it gives low results in comparison to genetic algorithm. The features used in GA and SA can be merged to get a better result, i.e., Makespan [7, 10, 26].

4.1.2 Dynamic Scheduling Approaches

Dynamic scheduling heuristic [7, 22] task arrives at different time slots and it depends on the current state of machine. It is broadly categorized in to two groups. First one is Online mode, where tasks are assigned to the available virtual machines immediately. Most-fit, MCT, MET, and OLB algorithms are examples of dynamic scheduling [1, 22]. Second type scheduling algorithm is batch mode scheduling algorithms. Here tasks are collected in a group and schedule at predefined time.

- **Switching Algorithm**: It is another method in which, switching occurs between MET and MCT as per the load of the system.
- **K-Percent Best**: In this heuristic, a subset of k computationally higher ranking machines is first chosen during the scheduling. The efficiency of the algorithm depends on the value of k and according to it tasks are assigned to the machines. It performs well in achieving a good makespan as compared to MCT [17, 23].

- **Sufferage Heuristic**: In Sufferage heuristic, the tasks are scheduled based on a sufferage value. Sufferage value is calculated from the first and second earliest completion time of a scheduling heuristics. It is calculated for all the tasks. Task having highest sufferage value is selected and allocated to a machine [3, 18].

5 Scheduling Parameters

Various scheduling parameters are considered in the previously mentioned methods are listed below:

- **Makespan**: The total time taken by all the tasks waiting in a job queue is the makespan time in task scheduling. The lower the makespan, the more efficient the algorithm. The goal of the scheduling algorithm is to reduce the makespan of the system.
- **Cloud utilization**: The goal of cloud or machine utilization is to maximize the use of all the machines to avoid idle time of machine.
- **Deadline**: It is the whole time period from start time. Deadline constraint is always used by a good scheduling algorithm to increase the efficiency of overall process.
- **Execution Time:** Actual time taken by a task is the execution time. Scheduler always tries to minimize the total execution time of an algorithm to get an optimal result.
- **Completion Time**: Completion time is the time taken to complete the entire execution of a task. It includes the execution time and delay caused by the cloud system. Many existing scheduling algorithms considered minimization of completion time as a scheduling constraint.
- **Performance**: This parameter is related to the overall efficiency of the scheduling algorithm. Performance is considered by the algorithm at the end user and along at the cloud service provider end.
- **Quality of Service (QoS)**: Execution cost, deadline, performance, makespan, etc. are the constraints in Quality of service. All the QoS are defined in Service Level Agreement (SLA). It is a contract between cloud user and cloud service provider.

6 Comparison of Existing Scheduling Algorithm

In cloud environment, many scheduling algorithms are performed. A comparison study is shown in Table 1.

Table 1 Comparison of cloud scheduling algorithm

Scheduling method	Parameters considered	Advantages	Disadvantages
First come first serve [13, 22, 23]	Arrival time	Simple in implementation	Does not consider any other criteria for scheduling
Round robin [13, 22, 23]	Arrival time, time quantum	Less complexity and load is balanced more fairly	Preemption is required
Minimum execution time algorithm [4, 11–13]	Expected execution time	Selects the fastest machine for scheduling	Load imbalanced
Minimum completion time algorithm	Expected completion time, load balancing	Load balancing is considered	Optimization in selection of best resource is not there
Min-Min, Max-Min [1, 6, 7, 9, 16]	Makespan, expected completion time	Better makespan compared to other algorithms	Poor load balancing and QoS factors are not considered
Genetic algorithm [1, 7, 8, 15]	Makespan, efficiency, performance, optimization	Better performance and efficiency in terms of makespan	Complexity and long time consumption
Simulated annealing [7, 10, 26]	Makespan, optimization	Finds more poorer solutions in large solution space, better makespan	QoS factors an heterogeneous environments can be considered
Switching algorithm	Makespan, load balancing, performance	Schedules as per load of the system, better makespan	Cost and time consumption in switching as per load
K-percent best [17, 23]	Makespan, performance	Selects the best machine for scheduling	Resource is selected based on the completion time only
Sufferage heuristic [3, 18]	Minimum completion time, reliability	Better makespan along with load balancing	Scheduling is done based on a sufferage value

7 Conclusion

Efficient scheduling algorithms dependably assume a noteworthy part in many performance metrics point of view provided by a cloud environment. This paper presented a study of scheduling algorithms having many scheduling parameters. An analytical study is done for each scheduling algorithm. One or two-parameter is chosen for comparative study between algorithms. A better and efficient heuristic approach can be developed by adding more metrics. The result or outcome can be useful to deploy in a cloud environment. An efficient scheduling technique is required to satisfy the SLA prepared by the cloud provider. It also fulfill the user requirements by the same time. It is beneficial to the cloud providers. The combination of different parameters to compute an efficient scheduling algorithm will promote to improve the overall cloud performance. It will be considered as an enhancement for cloud service providers.

References

1. Armstrong, R., Hensgen, D., Kidd, T.: The relative performance of various mapping algorithms is independent of sizable variances in run-time predictions. In: Proceedings Seventh Heterogeneous Computing Workshop (HCW98) (n.d.). https://doi.org/10.1109/hcw.1998.666547
2. Beloglazov, A., Abawajy, J., Buyya, R.: Energy-aware resource allocation heuristics for efficient management of data centers for cloud computing. Future Gener. Comput. Syst. **28**(5), 755–768 (2012). https://doi.org/10.1016/j.future.2011.04.017
3. Braun, T.D., Siegel, H.J., Beck, N., Blni, L.L., Maheswaran, M., Reuther, A.I., Freund, R.F.: A comparison of eleven static heuristics for mapping a class of independent tasks onto heterogeneous distributed computing systems. J. Parallel Distrib. Comput. **61**(6), 810–837 (2001). https://doi.org/10.1006/jpdc.2000.1714
4. Buyya, R., Yeo, C.S., Venugopal, S., Broberg, J., Brandic, I.: Cloud computing and emerging IT platforms: vision, hype, and reality for delivering computing as the 5th utility. Future Gener. Comput. Syst. **25**(6), 599–616 (2009). https://doi.org/10.1016/j.future.2008.12.001
5. Buyya, R., Broberg, J., Goscinski, A.: Cloud Computing: Principles and Paradigms. Wiley (2011)
6. Chen, H., Flann, N., Watson, D.: Parallel genetic simulated annealing: a massively parallel SIMD algorithm. IEEE Trans. Parallel Distrib. Syst. **9**(2), 126–136 (1998). https://doi.org/10.1109/71.663870
7. Coli, M., Palazzari, P.: Real time pipelined system design through simulated annealing. J. Syst. Archit. **42**(6–7), 465–475 (1996). https://doi.org/10.1016/s1383-7621(96)00034-3
8. Falco, I.D., Balio, R.D., Tarantino, E., Vaccaro, R.: Improving search by incorporating evolution principles in parallel Tabu Search. In: Proceedings of the First IEEE Conference on Evolutionary Computation, IEEE World Congress on Computational Intelligence (n.d.). https://doi.org/10.1109/icec.1994.349949
9. Fernandez-Baca, D.: Allocating modules to processors in a distributed system. IEEE Trans. Softw. Eng. **15**(11), 1427–1436 (1989). https://doi.org/10.1109/32.41334
10. Forell, T., Milojicic, D., Talwar, V.: Cloud Management: Challenges and Opportunities (2011). Retrieved from http://ieeexplore.ieee.org/stamp/stamp.jsp?tp=&arnumber=6008934&isnumber=6008799. https://doi.org/10.1109/IPDPS.2011.233
11. Freund, R.F., Sunderam, V.: Special Issue on Heterogeneous Processing. Academic Press, San Diego (1994)
12. Freund, R.F.: Scheduling resources in multi-user heterogeneous computing environment with smart net. In: 7th IEEE HCW, pp. 184–199. Orlando, FL, USA (1998).https://doi.org/10.1109/HCW.1998.666558
13. Huang, C.J., Guan, C.T., Chen, H.M., Wang, Y.W., Chang, H.C., Li, C.Y., Weng, C.H.: An adaptive resource management scheme in cloud computing. Eng. Appl. Artif. Intell. **26**(1), 382–389. Retrieved from https://doi.org/10.1016/j.engappai.2012.10.004 (2013)
14. Jadeja, Y., Modi, K.: Cloud computing - concepts, architecture and challenges. In: International Conference on Computing, Electronics and Electrical Technologies, pp. 877–80 (2012) https://doi.org/10.1109/ICCEET.2012.6203873
15. Jena, T., Mohanty, J.R.: GA-based customer-conscious resource allocation and task scheduling in multi-cloud computing. Arab. J. Sci. Eng. **43**(8), 4115–4130 (2017). https://doi.org/10.1007/s13369-017-2766-x

16. Li, J., Qiu, M., Ming, Z., Quan, G., Qin, X., Gu, Z.: Online optimization for scheduling preemptable tasks on IaaS cloud systems. J. Parallel Distrib. Comput. **72**(5), 666–677 (2012). https://doi.org/10.1016/j.jpdc.2012.02.002
17. Maheswaran, M., Ali, S., Siegal, H., Hensgen, D., Freund, R.: Dynamic matching and scheduling of a class of independent tasks onto heterogeneous computing systems. In: Proceedings Eighth Heterogeneous Computing Workshop (HCW99) (n.d.). https://doi.org/10.1109/hcw.1999.765094
18. Nagadevi, S., Satyapriya, K., Malathy, D.: A survey on economic cloud schedulers for optimized task scheduling. Int. J. Adv. Eng. Technol. **4**(1), 58–62 (2013)
19. Nathani, A., Chaudhary, S., Somani, G.: Policy based resource allocation in IaaS cloud. Future Gener. Comput. Syst. **28**(1), 94–103 (2012). https://doi.org/10.1016/j.future.2011.05.016
20. Panda, S.K., Jana, P.: An efficient task scheduling algorithm for heterogeneous multicloud environment. In: 3rd IEEE International Conference on Advances in Computing, Communications Informatics (2014). Retrieved from http://ieeexplore.ieee.org/stamp/stamp.jsp?tp=&arnumber=6968253&isnumber=6968191. https://doi.org/10.1109/ICACCI.2014.6968253
21. Sherihan, A.E., Kitakami, M.: Integrating trust into scheduling algorithms in grid system. In: 2009 2nd International Conference on Computer Science and Its Applications (2009). https://doi.org/10.1109/csa.2009.5404219
22. Shrivastava, V., Bhilare, D.: Algorithms to improve resource utilization and request acceptance rate in iaas cloud scheduling. Int. J. Adv. Netw. Appl. **3**(5), 1367–1374 (2012)
23. Velte, A.T., Velte, T.J., Elsenpeter, R.: Cloud Computing a Practical Approach. McGraw-Hill, New York, NY (2010)
24. Xhafa, F., Barolli, L., Durresi, A.: Batch mode scheduling in grid systems. Int. J. Web Grid Serv. **3**(1), 19 (2007). https://doi.org/10.1504/ijwgs.2007.012635
25. Xhafa, F., Carretero, J., Barolli, L., Durresi, A.: Immediate mode scheduling in grid systems. Int. J. Web Grid Serv. **3**(2), 219 (2007). https://doi.org/10.1504/ijwgs.2007.014075
26. Yang, X., He, H., Sun, Y.: Data dependence graph directed scheduling for clustered VLIW architectures. Tsinghua Sci. Technol. **15**(3), 299–306 (2010). https://doi.org/10.1016/S1007-0214(10)70065-1

Improving the Efficiency of Ensemble Classifier Adaptive Random Forest with Meta Level Learning for Real-Time Data Streams

Monika Arya and Chaitali Choudhary

Abstract New challenges have emerged in data mining as the traditional techniques have floundered with real-time data streams. The traditional technique needs refurbishing so as to acclimatize with concept drifting data streams. Thus dealing with the concept changes is the most imperative task of stream data mining. Ensemble classifiers have the ability to automatically adapt with the incoming drifts and, therefore, it is the most interesting research area in data stream mining. Bagging, Boosting and Random forest generation are the common ensemble techniques and are the most popular machine learning approaches in the current scenario for static data (Gomes HM, Bifet A, Read J, Barddal JP, Enembreck F, Pfharinger B, Abdessalem T (2017) Adaptive random forests for evolving data stream classification. Mach Learn 106(9–10):469–1495, [1]). A large number of base classifiers in an ensemble can cause computational overhead. Data mining classifiers for real-time data streams, therefore, need to be updated constantly and retrained with the labeled instances of the newly arrived novel classes in data streams and to cope with concept drift; otherwise, the mining models will become less and less accurate as time passes by. However, for data streams, adaptive random forest algorithms have been widely used for ensemble generation due to its competence to handle different types of drifts. This paper proposes a modified adaptive random forest with meta level learner algorithm and concept adaptive very fast decision tree to overcome the concept drift problem in real-time data streams. The proposed algorithm is experimentally compared with state-of-the-art adaptive random forest algorithm on several real synthetic datasets. Results indicate its efficiency in terms of accuracy and processing time.

Keywords Data stream mining · Random forests · Ensemble · Concept drift · Pruning · Forest · Adaptive random forest · Data streams

M. Arya · C. Choudhary (✉)
University of Petroleum and Energy Studies, Dehradun, India
e-mail: chaitali.choudhary@gmail.com

M. Arya
e-mail: arya.akshara@gmail.com

© Springer Nature Singapore Pte Ltd. 2020
V. Bhateja et al. (eds.), *Intelligent Computing and Communication*,
Advances in Intelligent Systems and Computing 1034,
https://doi.org/10.1007/978-981-15-1084-7_2

1   Introduction

Technological advancement have propelled machine learning to become more prevalent in real-world applications like business, where several machine learning algorithms adopted to analyze user's interest, product recommendations, predictions and many more task which are aided to make crucial decisions that can judge the fate of the business. Exclusive of the business application machine learning is also gaining popularity for weather forecast, face recognition, credit card fraud detection, spam filtration, sentimental analysis, network monitoring, etc., covering nearly all aspects of life. The source data of these applications are non-static and new instances and new class labels can appear in this streaming data in the form of concept drift at any time. The existing data mining techniques are trained in instances where the numbers of labels are fixed. They cannot detect and classify these novel classes and thus can misclassify these new instances [2, 3]. Therefore data streams environment demand for real-time data processing. Data mining classifiers for real-time data streams, therefore, need to be updated constantly and retrained with the labeled instances of the newly arrived novel classes in data streams; otherwise, the mining models will become less and less accurate as time passes by. The machine learning algorithm is expected to learn from these evolving data streams as soon as they are made available due to limited memory and storage capacity. This context has become more challenging due to the concept drifting nature of data streams. For classification problems, concept drift is formally defined as the change of joint distribution of data, i.e., $p(x, y)$, where x is the feature vector and y is the class label [4]. Ensemble classifiers have the ability to automatically adapt with the incoming drifts and therefore it is the most interesting research area in data stream mining. In ensemble technique, multiple learning algorithms are employed. Usage of multiple algorithms results in better predictive performance rather than using an individual algorithm. Classification accuracy significantly improves by generating an ensemble of trees, where each tree votes for the most popular class.

In this paper, we propose an ensemble model based on the random forest for classification. This significantly extends previous work [1] on adapting the random forest ensemble to adapt with concept drifting data and accurately classify the incoming new instances in streaming environment.

The remainder of this work is organized as follows. In Sect. 2 we study various aspects that affect the performance of the ensemble model for data stream classification. In Sect. 3 we have discussed the literature being studied related to data stream classification and experimentally compare various decision tree based ensemble approaches. In Sect. 4 we have discussed our newly suggested algorithm, i.e., adaptive random forests with Meta level learning for data stream classification (ARFML). In Sect. 5, we have discussed the required experimental setting required and data sets used to implement the newly designed algorithm. In Sect. 6, the results of the experiments are presented and thoroughly discussed. Finally, Sect. 6 concludes this work.

2 Problem Statements

In data stream classification, instances are not readily available for training as part of a large static data set, instead, they are provided as a continuous stream of data in a fast-paced way. Prediction requests are expected to arrive at any time and the classifier must use its current model to make predictions. On top of that, it is assumed that concept drifts may occur (evolving data streams), which damage (or completely invalidate) the current learned model. Concept drifts might be interleaved with stable periods that vary in length, and as a consequence, besides learning new concepts it is also expected that the classifier retains previously learned knowledge. The ability to learn new concepts (*plasticity*) while retaining knowledge (*stability*) is known as the stability–plasticity dilemma [5, 6]. In other words, a data stream learner must be prepared to process a possibly infinite amount of instances, such that storage for further processing is possible as long as the algorithm can keep processing instances at least as fast as they arrive. Also, the algorithm must incorporate mechanisms to adapt its model to concept drifts, while selectively maintaining previously acquired knowledge. Random forests are currently one of the most used machine learning algorithms in the non-streaming (batch) setting [1]. The attributed properties of the random forest like its robustness to outlier and noise, its speed as compared to bagging or boosting, its usefulness in giving internal estimates of error, strength, correlation, and variable importance and its simplicity along with easily parallelized nature favor's its popularity [7]. However, RF's are used for static data in offline applications. This fact limits its applicability for real-world problems where data is in the form of a continuous stream which is subjected to concept drift. To overcome the limitations and challenges of evolving data streams the Random forest classifier needs to be adaptive so as to work efficiently under the strict constraint of storage space and computational time. Many types of research proceeded in this direction.

This research emphasizes on developing an efficient ensemble-based classification algorithm for the data stream. Following is the problem statement:

(a) Is it possible to improve the efficiency of the adaptive random forest ensemble based classifier in terms of accuracy, computational time, memory usage, and precision?

There are various aspects that need to be considered before building an ensemble as they can directly or indirectly affect its overall performance. Following questions should be answered before building an ensemble:

- How are subsets of the training data chosen for each individual learner? Training subsets can be chosen by random selection, by examining which training patterns are difficult to classify and focusing on those, or by other means.
- How are classifications made by the different individual learners combined to form the final prediction? They can be combined by averaging, majority vote, weighted majority vote, etc.
- What types of learners are used to form the ensemble? Do all the learners use the same basic learning mechanism or are there differences? If the learners use the

same learning algorithm, then do they use the same initialization parameters? For handling, concept drift many learning algorithms were used as base models. One of the most popular and heavily studied is the decision tree.

• What should be the size of the ensemble, i.e., how many component classifiers should be used to design the ensemble model?

3   Related Works

A Bifet et al. [8] proposed the Hoeffding Adaptive Tree, which is an adaptive extension to the Hoeffding Tree that uses ADWIN as a change detector and error estimator. Abdulsalam et al. [9] in their work extended the standard Random Forest algorithm so that it can be applied to streaming data. Marron et al. [10] gave the idea of Random Forests of Very Fast Decision Trees for data streams. They explored the usage of Graphics Processing Units (GPUs) to applications of data stream mining in order to overcome the increasing need for more computational power to process a large amount of data in real time. Hulten et al. [11] presented the concept-adapting very fast decision tree (CVFDT) algorithm as an extension of VFDT to deal with concept drift, maintaining a model that is consistent with the instances stored in a sliding window. Zhukov et al. [12] proposed a novel approach to classification to adapt to concept drifts called Proximity Driven Streaming Random Forest (PDSRF). Kumar et al. [13] in their work compared Hoeffding tree, Streaming Random forest and CVFDT (Concept-Adapting Very Fast Decision Tree) which are used for stream data classification and concluded that CVFDT handles 'concept drift' very efficiently by creating alternative subtree to find the best attribute at the root node. Jankowski et al. [14] proposed algorithm, named Concept-adapting Evolutionary Algorithm For Decision Tree. It does not require any knowledge of the environment such as numbers and rates of drifts. The novelty of the approach is combining tree learners and evolutionary algorithms. Gomes et al. [1] in their work proposed an effective resampling method in adaptive random forest (ARF) algorithm for the classification of evolving data streams (ARF). The proposed resampling method and adaptive operators can cope with different types of concept drifts without complex optimizations for different data sets.

An experimental comparison of Random forest based ensemble classifier algorithms is done. The Algorithms are compared in terms of accuracy, kappa, kappa temp, memory and time elapsed. All the algorithms are compared using synthetic datasets generated in the MOA framework. Ten thousand learning instances of real stream data having concept drift are used. The ensemble size is 10. The drift detection method used is ADWIN change detector. The classifiers which are compared are the following:

1. Adaptive random forest (ARF) evaluates a classifier on a stream by testing than training with each example in sequence.
2. Random Hoeffding tree: These are random decision trees for data streams.

3. Hoeffding tree (VFDT): A very fast decision tree algorithm for streaming data, where instead of reusing instances, we wait for new instances to arrive. The most interesting feature of the Hoeffding Tree is that it builds a tree that provably converges to the tree built by a batch learner with sufficiently large data.
4. Hoeffding Adaptive tree: The *Hoeffding Adaptive Tree* is an adaptive extension to the Hoeffding Tree that uses ADWIN as a change detector and error estimator.
5. (CVFDT)-concept-adapting very fast decision tree (CVFDT) algorithm as an extension of VFDT to deal with concept drift, maintaining a model that is consistent with the instances stored in a sliding window.
6. ARF Hoeffding tree: ARF Hoeffding tree for data streams. These are base learners for the adaptive random forest classifiers.
7. ASHoeffding tree: Adaptive size Hoeffding tree uses trees of different sizes.

4   The Proposed Modified Adaptive Random Forest Algorithm

Random forests are currently one of the most used machine learning algorithms in the non-streaming (batch) setting [1]. Random forest ensemble generation approach allows many trees to grow and at the same time prevents overfitting. Overfitting is prevented by decorrelating the base classifiers using bootstrap aggregation and randomly selecting the features for node splitting. In static data environment, Random forest algorithm requires many passes over input data for bootstrap creation of a tree while in the data stream environment it is not possible to perform multiple passes. The random forest algorithm for static data thus needs to be adapted for the streaming environment. Many researchers proceeded in this direction suggesting adaptations of random forest algorithm. The experimental comparisons of these algorithms are done in Sect. 2. In this work we present modified Adaptive random forest algorithm with the following modifications:

1. Original adaptive random forest algorithm uses a bagging approach while in the modified algorithm Meta Level Learner is used.
2. The adaptive random forest uses VDFT while the modified algorithm uses CVFDT.
3. Splitting criteria in original ARF is Hoeffding bound while in modified approach misclassification error is used.

Justification for first modification: The popular approaches of combining the base classifiers to form an ensemble using are voting and stacking. The voting approach combines different machine learning classifiers, and allows the classifier to a vote on what is the predicted class label(s) are for a record. The class which is predicted by the majority of the base classifier is selected. The voting approach is further categorized as bagging & boosting. Bagging technique employs building multiple models (typically same type) from different subsamples of the training dataset. The training set is divided into the number of subsets and each subset is used to train their decision tree. As a result, we end up with an ensemble of different models. Now an average of all the predictions is used for final prediction. In this way, the ensemble model is more robust than the single decision tree. In Boosting the classifiers learn sequentially by learning from errors made by the previous model. So boosting employs building multiple models (typically of the same type) each of which learns to fix the prediction errors of a prior model in the chain. By combining the whole set at the end converts a weak classifier into a better performing model. Stacking is another ensemble model, where a new model is trained from the combined predictions of two (or more) previous model. The predictions from the models are used as inputs for each sequential layer, and combined to form a new set of predictions.

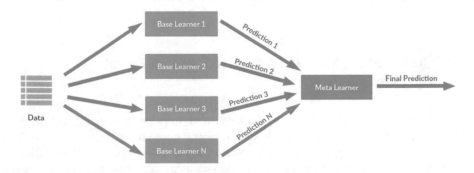

The basic difference between stacking and voting is that in voting the final class is decided by the majority of votes cast by base-level classifiers and no learning takes place at the Meta level, whereas in stacking learning takes place at the meta level. This adds an extra layer of the learner and can improve the accuracy of the ensemble classifiers. Mathematically the stacking can be explained as follows: let there be an ensemble model with n number of base classifiers and each classifier having an error rate ER < 1/2. The ER for each base classifier is independent. Then the probability that the ensemble makes an error will be more than N/2. The estimated error added with the output of the base classifiers act as an input to the Meta level classifier and thus the Meta level classifiers are trained to learn from the errors of the base-level classifiers. Therefore it can provide improved final classification prediction.

Justification for the second modification: CVFDT (Concept-Adapting Very Fast Decision Tree) is the modification of VFDT to solve the concept drift problem in data streams. In CVFDT, a sliding window approach is used but it does not construct an alternate subtree each time from the beginning. It can update the statistics at each node by incrementing the counter with every new example and decrementing the counter with an old example. CVFDT has better accuracy than VDFT and the size of tree is also smaller than VFDT.

Justification for third modification: For constructing decision trees splitting criteria has a crucial role. Splitting criteria is the measure based on which the decision is taken that which attributes to choose as the root node and which attribute will be the child node. Splitting criteria should be chosen such that it should increase the accuracy and decrease the errors of the classifier. The different splitting criteria are entropy or information gain, Gini index, misclassification error, etc.

Among the decision tree based classifiers for the data stream, VFDT uses information gain or Gini index as splitting measure. While the CVFDT used for mining data streams with concept drifts employs splitting criterion based on Hoeffding bound. In data stream mining recent data are taken for further process. As the data stream arrives continuously, therefore, a time window is maintained which keeps the entire example in a bunch that arrives in that particular time window. Whenever a new example arrives, the window is checked whether it has sufficient examples or not. If the example exceeds and the window overflows then the oldest example is forgotten. The decision of forgetting a particular attribute depends upon certain statistics maintained at every node. As in the data streams, the concept is changing so the split attribute can become outdated. To overcome this drawback CVFDT grows an alternate subtree with a new attribute having a higher gain as its root node. Using Hoeffding bound as the splitting criteria may lead to less effective tree generation. Jadhav and Kosbatwar [15] in their work proposed to use misclassification error as splitting criteria. Misclassification error is another measure that is used rarely in the literature for decision tree construction and this criterion has strong mathematical foundations and allows determining the optimal attribute to split the node of the tree.

Therefore, the main novelty of MARF is the introduction of meta level learner and the usage of misclassification error in CVFDT to deal with evolving data streams. The algorithm uses drift monitor and warning methods as in the previous adaptive random forest algorithm.

In original ARF algorithm, the random forest tree training module is based on VFDT with some modifications. In our approach, the random forest tree training algorithm is based on CVFDT with misclassification error. In the original CVFDT, the splitting criterion is based on Hoeffding bound. In contrast to VFDT, CVFDT grows an alternate subtree with the new best attribute as its root node. It is based on the fact that in data streams the concept keeps on changing, therefore, the selected root attribute may get outdated with time. Thus new root attribute needs to be selected. The best attribute is selected on the basis of the misclassification error. And based on this the decision to replace the old subtree with alternate new subtree is taken.

5 Experimental Settings

In this section, we have given the details of the experimental settings. We have evaluated the experiments in terms of memory, time and classification performance. We have performed a tenfold cross-validation prequential evaluation. Along with classification accuracy, we have also reported Kappa M and Kappa Temporal. We have used 3 synthetics in our experiments. Synthetic data sets include LED Generator, SEA generator, concept drift streams. Table 1 gives the details of the data set used. The

Table 1 Comparison of decision tree based ensemble classifiers for real-time data streams

Classifier	Adaptive random forest (ARF)		Random Hoeffding tree		Hoeffding tree (VFDT)		Hoeffding adaptive tree (CVFDT)		ARF Hoeffding tree		AS Hoeffding tree	
Comparison criteria	Current	Mean	Current	Mean	Current	Mean	Current	Mean	Current	Mean	Current	Mean
Accuracy	87.5	85.8	83	80.17	99.1	97.45	99.1	97.8	73.8	71.42	99.1	97.45
Kappa	74.04	70.76	65.37	59.01	98.18	94.77	98.18	95.48	45.28	39.49	98.18	94.77
Kappa temp	72.77	69.79	65.93	59.33	98.2	94.75	98.2	95.46	46.96	41.38	98.2	94.75
Time elapsed	4 m 2 s		1 m 17 s		2 m 19 s		4 m 24 s		50.48 s		2 m 18 s	
Memory	132.42	102.86	7.5	4.12	14.46	8.79	22.04	13.59	2.89	1.59	14.46	8.79
Time	143.43	104.66	72.93	31.1	130.78	57.65	254.41	118.3	49.67	21.63	129.55	57.14

Table 2 Comparison of ARF and MARF with split criteria as info gain and data set LED generator

Comparison criteria	ARF		MARF	
	Current	Mean	Current	Mean
Accuracy	73.80	73.22	89.00	87.00
Kappa	45.46	43.62	77.26	73.32
Kappa temp	44.73	45.16	76.69	72.54
Time	186.87	74.74	115.43	84.79
Memory	6.27	3.41	115.77	90.66

Table 3 Comparison of ARF and MARF with split criteria as Gini index and data set LED generator

Comparison criteria	ARF		MARF	
	Current	Mean	Current	Mean
Accuracy	84.70	82.81	84.00	82.78
Kappa	68.96	64.5	65.67	63.48
Kappa Temp	69.58	64.76	64.84	63.58
Time	224.64	93.84	174.21	95.89
Memory	3.97	2.10	34.99	21.98

Table 4 Comparison of ARF and MARF with split criteria as info gain and data set SEA generator

Comparison criteria	ARF		MARF	
	Current	Mean	Current	Mean
Accuracy	89.70	87.70	92.10	91.25
Kappa	77.33	72.47	82.51	80.36
Kappa temp	77.06	72.98	82.33	79.95
Time	55.0	22.20	52.87	38.52
Memory	10.35	5.30	33.86	25.70

evaluator used is a window classification performance evaluator. For drift detection and drift warning, Adwin change detector has been employed. We have compared the adaptive random forest having meta level learner with the state-of-the-art adaptive random forest under two splitting criterion info gain and Gini index (Tables 2, 3, 4, 5, 6 and 7).

Table 5 Comparison of ARF and MARF with split criteria as Gini index and data set SEA generator

Comparison criteria	ARF		MARF	
	Current	Mean	Current	Mean
Accuracy	88.60	87.46	92.10	91.25
Kappa	74.08	71.96	82.51	80.36
Kappa Temp	73.85	72.44	82.33	79.95
Time	27.27	11.52	52.98	38.48
Memory	6.33	3.31	33.91	25.75

Table 6 Comparison of ARF and MARF with split criteria as info gain and data set concept drift stream generator

Comparison criteria	ARF		MARF	
	Current	Mean	Current	Mean
Accuracy	79.80	77.19	90.70	90.70
Kappa	57.58	52.45	81.18	81.18
Kappa temp	57.52	53.28	80.42	80.42
Time	80.20	33.69	42.06	42.06
Memory	7.11	3.99	52.30	52.36

Table 7 Comparison of ARF and MARF with split criteria as Gini index and data set concept drift stream generator

Comparison criteria	ARF		MARF	
	Current	Mean	Current	Mean
Accuracy	79.10	79.00	90.70	90.70
Kappa	56.62	56.61	81.18	81.18
Kappa temp	58.20	56.98	80.42	80.42
Time	48.66	21.25	41.64	41.64
Memory	2.08	1.09	52.33	52.33

6 Results and Conclusion

We have proposed a modified adaptive random forest with meta level learner algorithm and concept adaptive very fast decision tree to overcome the concept drift problem in real-time data streams. The proposed algorithm is experimentally compared with state-of-the-art adaptive random forest algorithm on several synthetic datasets like LED generator, SEA generator and concept drift real stream. Results indicate that the meta level learner algorithm is efficient than the traditional random forest in terms of accuracy, Kappa, Kappa temp time, and memory.

References

1. Gomes, H.M., Bifet, A., Read, J., Barddal, J.P., Enembreck, F., Pfharinger, B., Abdessalem, T.: Adaptive random forests for evolving data stream classification. Mach. Learn. **106**(9–10), 1469–1495 (2017)
2. Biswas, A., Farid, D.M., Rahman, C.M.: A new decision tree learning approach for novel class detection in concept drifting data stream classification. J. Comput. Sci. Eng. **14**(1), 1–8 (2012)
3. Masud, M., Gao, J., Khan, L., Han, J., Thuraisingham, B.M.: Classification and novel class detection in concept-drifting data streams under time constraints. IEEE Trans. Knowl. Data Eng. **23**(6), 859–874 (2011)
4. Sun, Y., Tang, K., Minku, L.L., Wang, S., Yao, X.: Online ensemble learning of data streams with gradually evolved classes. IEEE Trans. Knowl. Data Eng. **28**(6), 1532–1545 (2016)
5. Harrison, R.M., Tilling, R., Romero, M.S.C., Harrad, S., Jarvis, K.: A study of trace metals and polycyclic aromatic hydrocarbons in the roadside environment. Atmos. Environ. **37**(17), 2391–2402 (2003)
6. Gama, J., Žliobaitė, I., Bifet, A., Pechenizkiy, M., Bouchachia, A.: A survey on concept drift adaptation. ACM Comput. Surv. (CSUR) **46**(4), 44 (2014)
7. Breiman, L.: Random forests. Mach. Learn. **45**(1), 5–32 (2001)
8. Bifet, A., Gavaldà, R. Adaptive learning from evolving data streams. In: International Symposium on Intelligent Data Analysis, pp. 249–260. Springer, Berlin, Heidelberg (2009)
9. Abdulsalam, H., Skillicorn, D.B., Martin, P.: Streaming random forests. In: 11th International Database Engineering and Applications Symposium, 2007 (IDEAS 2007), pp. 225–232. IEEE (2007)
10. Marron, D., Bifet, A., Morales, G.D.F.: Random forests of very fast decision trees on GPU for mining evolving big data streams. In: ECAI, vol. 14, pp. 615–620 (2014)
11. Hulten, G., Spencer, L., Domingos, P.: Mining time-changing data streams. In: Proceedings of the Seventh ACM SIGKDD International Conference on Knowledge Discovery and Data Mining, pp. 97–106. ACM (2001)
12. Zhukov, A.V., Sidorov, D.N., Foley, A.M.: Random forest based approach for concept drift handling. In: International Conference on Analysis of Images, Social Networks and Texts, pp. 69–77. Springer, Cham (2016)
13. Kumar, A., Kaur, P., Sharma, P.: A survey on Hoeffding tree stream data classification algorithms. CPUH-Res. J. **1**(2) (2015)
14. Jankowski, D., Jackowski, K., Cyganek, B.: Learning decision trees from data streams with concept drift. Proc. Comput. Sci. **80**, 1682–1691 (2016)
15. Jadhav, S.A., Kosbatwar, S.P.: Concept-adapting very fast decision tree with misclassification error

Plain Text Encryption Using Sudoku Cipher

Kashi Nath Dey, Souvik Golui, Neha Dutta, Arnab Kumar Maji
and Rajat Kumar Pal

Abstract During the transmission of secret messages through any network, encryption is a necessity. The message must be encrypted in such a manner that only authorized user(s) can decrypt it using suitable key(s). Different ciphers are used for this purpose, which follows some particular sequence of steps. In this paper, a new cipher is proposed for the encryption of plaintext using Sudoku puzzles. The advantage of using Sudoku puzzles for cipher generation is that it can easily take care of the integrity of a message. It is known that Sudoku is usually a 9×9 matrix, wherein each of the digits, 1–9, is present in each row, each column, and in each minigrid. Therefore, during transmission, if the encrypted message gets modified, it can easily be detected, as in due course it may violate the uniqueness constraints in each row, column, or minigrid of the Sudoku puzzle. Moreover, in our proposed scheme, we have employed different random permutations of the digits 1–9 to add more security in the ciphertext. Implementations have also been carried out to support the proposed approach.

Keywords Cipher · Decryption · Encryption · Minigrid · Permutation · Sudoku

K. N. Dey
Department of Computer Science and Engineering, Dr. Sudhir Chandra Sur Degree Engineering College, Kolkata 700074, India
e-mail: kndey@jisgroup.org

S. Golui · N. Dutta · R. K. Pal
Department of Computer Science and Engineering, University of Calcutta, Kolkata 700106, India
e-mail: souvikgolui88@gmail.com

N. Dutta
e-mail: nehadutta1994@gmail.com

R. K. Pal
e-mail: pal.rajatk@gmail.com

A. K. Maji (✉)
Department of Information Technology, North-Eastern Hill University, Shillong 793022, India
e-mail: arnab.maji@gmail.com

© Springer Nature Singapore Pte Ltd. 2020
V. Bhateja et al. (eds.), *Intelligent Computing and Communication*,
Advances in Intelligent Systems and Computing 1034,
https://doi.org/10.1007/978-981-15-1084-7_3

1 Introduction

In this digital era, there are needs for transmitting messages through the communication medium. When the message is transmitted through some communication medium, there is a chance that intruder may catch hold of the secret message which is intended for a particular receiver. There are chances that intruders may modify the secret message. To prevent that, several cryptographic techniques are used. The written text is basically known as "plaintext". The plaintext can be converted into some other form using some keys. A cipher can be defined as the method for hiding words or text with encryption by replacing original letters, words or text with other letters, numbers, and symbols [1]. There are different types of cipher exist namely plaintext cipher, block cipher, substitution cipher, stream cipher, permutation cipher, transposition cipher [2], etc.

In the proposed scheme, we have used a Sudoku matrix as a key to encrypt the plaintext. Sudoku is a Japanese puzzle game having a structure of matrix $n \times n$. n values can be 4, 9, 16, 25 or even more [3]. But the most common value of n is 9. In the Sudoku puzzle instance, some values are given as clues. There are 9 rows, 9 columns and 9 minigrids of size 3×3. The numbers 1 to 9 is needed to be placed in the blank cell in such a way that in each row, column and minigrid there is only one instance of them.

The benefit of using Sudoku Puzzle as the key is that, if an intruder got the encrypted secret message, there will be very less possibility of modification because of the inherent structure of the Sudoku matrix, i.e., in each row, column, and minigrid each digit presents only once. Moreover, a blank Sudoku puzzle of 9×9 can be solved in 6.67×10^{21} [4] in different ways. If we treat each solution as a different permutation, then it can be easily found that a huge number of keys can be generated.

2 Techniques for Solving Sudoku Puzzle

Sudoku is an NP (Non-deterministic Polynomial) class of problems [5]. It means solution to this kind of problem is needed to be guessed, but the correctness of the solution can be verified in polynomial time. The most common solution strategy for solving the Sudoku puzzle is backtracking [6] based methodologies. In this method, the value of any of the blank cell of the Sudoku puzzle instance is guesses based on the clues given. Then without violating the Sudoku uniqueness constraints for row, column or minigrid the other values for the other blank cells are also guessed. If at some point by time, no more options are available, then the program needs to backtrack to its earlier step. This method is also known as a good guessing strategy. As this strategy includes several backtracking to an earlier step, it is comparatively slower than the other algorithm. But in this method, there is a guarantee for finding out the solution of the Sudoku puzzle instance.

Other methods include elimination-based strategies [6] and soft computing based methodologies [7]. In elimination-based strategies, a probable list of candidates is created for each blank cell of the Sudoku puzzle. Then from the probable list, candidates are eliminated based on certain patterns. Elimination-based strategies includes Unique Missing Candidates [6], Naked Singles [8], Hidden Singles [8], Lone Ranger [8], Locked Candidate [8], Twin [8], Triplet [8], Quad [8], X-Wing [8], XY-Wing [8], XYZ-Wing [6], Forced Chain [6] etc. Although these methods are relatively faster than the backtracking method, it does not guarantee the solution of the Sudoku puzzle [6].

Soft computing based methodologies are also can be used for finding the solution of the Sudoku puzzle. Perez et al. [9] solved the Sudoku puzzle using Cultural Genetic Algorithm (CGA) [9], Repulsive Particle Swarm Optimization (RPSO) [9], Quantum Simulated Annealing (QSA) [9], Hybrid Genetic Algorithm and Simulated Annealing (HGASA) [9], Bee Colony Optimization [9], Artificial Immune System (AIS) based Optimization [9]. Although these techniques guarantee a unique solution to Sudoku puzzle, these methods are very much computational extensive.

Maji et al. [10] propose a unique permutation-based methodologies for solving the Sudoku puzzle. In this strategy, permutations are created for each minigrid based on the given clues. In this algorithm, first of all, each minigrid is numbered as shown in Fig. 1.

In Figure 1, the minigrid in the top-left corner is numbered as 1. Then other minigrids in the top are numbered as 2, 3. Similarly, all other minigrids are numbered in a sequential manner. Then a unique permutation tree is generated. The leaf nodes will indicate the valid permutations of the minigrid. All of these kinds of permutations will be generated for each of the minigrids. Then these permutations are checked for compatibility amongst each other so that the final solutions can be generated. In our proposed algorithm for plaintext encryption, we have followed these permutation tree based solution strategies to solve the Sudoku puzzle, as we have used a minigrid-based strategies. Moreover instead of each individual blank cell, which is 81 in numbers, as it considered minigrids (only 9 in numbers) the computational complexity may be lesser.

Fig. 1 Numbering of minigrid for a 9 × 9 Sudoku puzzle

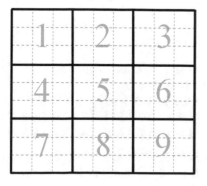

3   Proposed Algorithm for Plaintext Encryption Using Sudoku Puzzle

In this paper, we have proposed a new 9×9 sudoku matrix based cipher for encrypting and decrypting the plaintext. Many algorithms has been proposed so far to encrypt the image data [11, 12] using Sudoku puzzle, as well as Sudoku puzzle have a huge application in the domain of Image Steganography and Watermarking [13]. But this kind of encryption using Sudoku based cipher is completely new in this domain. To date, no attempt has been made to encrypt the plaintext using Sudoku based cipher. The main reason for choosing Sudoku because a blank Sudoku puzzle produces 6.67 $\times 10^{21}$ different Sudoku Matrix as a solution [4]. Any of these solutions can be used for the key generation for our proposed method. So if we choose Sudoku as key, there will be less chance for the attacker to guess the key. Moreover, Sudoku is a type of Latin square where each row, column, and minigrid having each digit uniquely. So any ciphertext which is encrypted using Sudoku Matrix is less vulnerable to modification by the attacker as the uniqueness constraint of this Sudoku matrix may be violated.

The proposed algorithm is discussed as follows:

Input: A $n \times n$ solution instance, $P'_{n \times n}$ of Sudoku. For encryption, the binary plaintext block of $n \times n$ where each character has been converted into n–bits each. For decryption, the binary ciphertext block of $n \times n$ in binary.

Output: For encryption, a binary ciphertext block of $n \times n$ of the corresponding binary plaintext block. For decryption, a binary plaintext block of $n \times n$ of the corresponding binary ciphertext block.

Step 1: Take the sudoku problem instance, P, and calculate number of blanks in each minigrid. The blank cells are replaced with 0. The Sudoku instance matrix of the problem (P) instance is as shown.
[['0' '1' '0' '0' '2' '0' '3' '4' '0'] ['4' '0' '0' '1' '0' '0' '5' '0' '0'] ['0' '0' '6' '7'
'0' '8' '0' '0' '0']
['0' '2' '1' '4' '0' '0' '7' '5' '0'] ['5' '7' '0' '0' '1' '0' '0' '8' '9'] ['0' '3' '9' '0'
'0' '6' '4' '1' '0']
['0' '0' '0' '9' '0' '7' '8' '0' '0'] ['0' '0' '5' '0' '0' '1' '0' '0' '7'] ['0' '9' '8' '0'
'3' '0' '0' '6' '0']]

Step 2: Find the solution grid of P. The solution of the Sudoku matrix shown in step 1 is as shown:
[[9 1 7 6 2 5 3 4 8] [4 8 2 1 9 3 5 7 6] [3 5 6 7 4 8 2 9 1]
[6 2 1 4 8 9 7 5 3] [5 7 4 3 1 2 6 8 9] [8 3 9 5 7 6 4 1 2]
[1 6 3 9 5 7 8 2 4] [2 4 5 8 6 1 9 3 7] [7 9 8 2 3 4 1 6 5]]

Step 3: While number of permutation tables generated in the current iteration is less than 16 and number of epochs less than 100 do steps 4 to 11.

Step 4: Generate a random permutation of the numbers 1, 2, …, n.

Step 5: Initialize *count* as 0. For each cell in minigrid, m_i, corresponding to the number present in random permutation $p_1 \, p_2 \, \ldots \, p_i \, \ldots \, p_n$, do steps 6 to 7.

Step 6: $P'_{i,j}(\text{new}) = P'_{i,j}(\text{old}) + ((count \times n) - 1)$.

Step 7: Incremented *count* by 1.

Let us elaborate on the steps with the solution grid shown in step 2. Take a random permutation say [1–8]. Each element of the permutation points to the corresponding minigrid numbers as shown in Fig. 1. So we will move to the corresponding minigrids and perform the operation as discussed in step 6 and 7. For example in this Sudoku matrix, we need to move to minigrid number 5 first. This minigrid 5 is a [3 × 3] matrix having the values [[4 8 9] [3 1 2] [5 7 6]].

As initially *count* = 0 and $n = 9$, after the after the operation as mentioned in step 6, corresponding values will be [[3 7 8] [2 0 1] [4 6 5]].

So applying all the operation discussed in step 6 and 7, we get the modified Sudoku matrix as follows:

[[44 36 42 50 46 49 20 21 25] [39 43 37 45 53 47 22 24 23] [38 40 41 51 48 52 19 26 18] [14 10 9 3 7 8 60 58 56] [13 15 12 2 0 1 59 61 62] [16 11 17 4 6 5 57 54 55] [27 32 29 80 76 78 70 64 66] [28 30 31 79 77 72 71 65 69] [33 35 34 73 74 75 63 68 67]]

Step 8: For generating the list of unique permutation tables,

Do the Steps 9–11, while permutation table generated is different from any of the previously generated permutation tables and number of permutation tables generated less than 100.

Step 9: Apply transpose to P' in the previous iteration.

Step 10: Apply permutation operation on the transposed minigrid P' using the minigrid P' in the previous iteration.

Step 11: Store the permutations.

For an example let's consider the Sudoku matrix as shown in step 7. So after transpose the matrix will be

[[44 39 38 14 13 16 27 28 33] [36 43 40 10 15 11 32 30 35] [42 37 41 9 12 17 29 31 34] [50 45 51 3 2 4 80 79 73] [46 53 48 7 0 6 76 77 74] [49 47 52 8 1 5 78 72 75] [20 22 19 60 59 57 70 71 63] [21 24 26 58 61 54 64 65 68] [25 23 18 56 62 55 66 69 67]]

The permutation operation is described as follows:

Result = [Block to be permuted] o [Permutation Table]

The symbol "o" denotes the permutation operation.

For example, if [3 1 2] is the block to be permuted and [1 3 2] is the permutation table, then the result will be [3 2 1]. Actually the 1, 3, 2 in the permutation table correspondingly showing the position of 3, 1, 2 in the block.

So, if we apply the permutation operation between the modified matrix discussed in step 7 and step 11 it will be.

[[74 46 76 5 47 1 41 9 31] [7 77 53 49 75 52 12 29 17] [48 0 6 78 8 72 37 34 42]
[11 43 36 14 28 33 70 59 19] [15 32 10 38 44 39 57 71 63] [30 40 35 13 27 16 60
20 22] [50 4 51 67 62 66 65 24 58] [45 3 2 69 55 25 68 26 64] [80 73 79 23 18 56
21 54 61]]]
Apply the steps recursively until we get the sufficient numbers of matrix.

Step 12: **Encryption**: For encryptions, all the permutation tables generated from
the Sudoku Matrix using earlier steps are used as keys. These keys of $n \times n$ will
be first converted into binary: Then the plaintext (to be encrypted) will be converted
into binary and each character of the plaintext will be represented by n bits. Then
we divide the binarized plaintext into block multiple of n. Next, we perform XOR
operation between the keys and the block of text. Finally, after the Ex-OR operation
of all blocks, we permute the resultant binary text using the method discussed in Step
11.

Step 13: **Decryption**: Decryption is the opposite method of encryption. As it is a
kind of symmetric cryptographic method, the same permutation tables will be used
as a key. First, the inverse of all permutation tables of size $n \times n$ are generated. Then
in a similar way, as mentioned earlier the binarized cipher is permuted. Then we
perform XOR between the permutation table and permuted cipher block to generate
the original text.

4 Experimental Results

We have computed the experimental results in intel core i5 machines with 8 GB of
RAM. A sample run is of our algorithm is as shown. We have considered blocks of
texts for these run. The text block is as shown:

> Alice was beginning to get very tired of sitting by her sister on the bank and of having
> nothing to do once or twice she had peeped into the book her sister was reading but it had no
> pictures or conversations in it and what is the use of a book thought Alice without pictures
> or conversation So she was considering in her own mind as well as she could for the hot day
> made her feel very sleepy and stupid whether the pleasure of making a daisy chain would be
> worth the trouble of getting up and picking the daisies when suddenly a White Rabbit with
> pink eyes ran close by her There was nothing so VERY remarkable in that nor did Alice
> think it so VERY much out of the way to hear the Rabbit say to itself Oh dear Oh dear I
> shall be late when she thought it over afterwards it occurred to her that she ought to have
> wondered at this but at the time it all seemed quite natural but when the Rabbit actually
> TOOK A WATCH OUT OF ITS WAISTCOAT POCKET and looked at it and then hurried
> on Alice started to her feet for it flashed across her mind that she had never before seen a
> rabbit with either a waistcoat pocket or a watch to take out of it and burning with curiosity
> she ran across the field after it and fortunately was just in time to see it pop down a large
> rabbit hole under the hedge.

The Sodoku matrix used as a key is as shown:

[[9 1 7 6 2 5 3 4 8]
[4 8 2 1 9 3 5 7 6]
[3 5 6 7 4 8 2 9 1]
[6 2 1 4 8 9 7 5 3]
[5 7 4 3 1 2 6 8 9]
[8 3 9 5 7 6 4 1 2]
[1 6 3 9 5 7 8 2 4]
[2 4 5 8 6 1 9 3 7]
[7 9 8 2 3 4 1 6 5]]

After encryption the ciphertext is

"ǓŏↃŵĞ□oßfī1CŲ·BⱢĪTYꞫUeU9ꞱYŃUƷƏsGĪbꞱĞg⁻ÍĒŪefŲö□ī|ÆŬzķUwğNĒĪ
dDŽꞰ·□țâꞱ\$šɥ·□ǫ}ÒĪↃœƷ□Ⱡ□^ŪdŖŻY¥û9¢ŪTŶƷoNÄĪ@GşözH□ƊeǫŶğ□ď',^GD
hûǔĞīƷ?Ū`ǏŰDz×ǿ\²Ǐ@ƨŵĞÕ/ý²Ṱ7ṣŷĞ□Ǐ□ĊꞱdꝐŲơ□ǫ»2ĪbŃHɥU□.Ṱ\$dżŮç¥⁻XdžG
PğŹò□p□ꞱłœẑłΘŝÄĮubtDzÕˉ>|4DŽꞰꝐ□p¹®ŰDYŰDz÷ŏŬ6Ʇ
ƏŰöⱢ^ꞱŪ@ʐŲ£ÕtĶÒꞱ\$PŶU³3[2Ǐ%CŘH□ǏỷŲyĞ7yŽHķƏ□ĆHPьŒæ□oůłôꞱŚó·
;ƀÄGьů¶#/áꞱTьŖ··?□ṰꞱ\$Ğŵãhɥž}□GhɥꞱ²Ǥ¡*łŠ%sŷzđůĒꞱtdžś·□Ļ.û2ŮpTŲơ□ǏªʜꞱ
\$GŷYʉ□ûĞ¢ŪebŲzÕK»²Ǐ\$ЬŲz÷ŋ
IJꞱDŽşⱭµş¹vʜ\$ЬꞲĞ¡/źĖ|tDžœƷ□į»¦Ủ\$hɥŬòÕŇrꞱQYŜöêû·ꞱePŷYëûƒŠbꞲp⁻□2ĪtGꞰ
·ěůěÆʜ4gꞰꝐůNŪƎŶHьβǏžVĞGDZÆ□ï□ꝐH4ЬꞱz□□2źŪ7LJœꝐbpØ□ŪdYŴDzÕǿŮ
6ꞱuↃłzbğŻRG!hɥž|ꝯↃↃ:6H4džėꝐⱢkꞱĴьỷƷ·ěŜIJŪ7DŽœƷbƏ□zŠaṣŷHьɥdºꞱHYtöbp¿NJ
ŮtÄǨò□®¾ĴG4LJœHь□ěę"|ƀŵDzǏ^RꞱAhɥLJDz□CŽĤꞱ5Cɥ·□śøRŨ
CŽĞGo□□%CtøꝐ/ú&Ū@VŶ³Dž3□zĊYħ|çŋpĶꞱYŰğŨį□ꞰŠƎŲæ□û®ŰGdG§PhɥŮ^r
ŮłŴDzŎÔŨṰ5ŲĞhɥį□ʐĪePꞰ¶Ʞꞅ§ÖĮŠDŽŷHьƷ□ŨǏŠDŽŷHьƷ□ŨŪꞱlŷUŰį□ŋŨdↃUŖÕ̃ë»:
ŠꞱŲĞ□/□ƎG1hɥŲ§hɥǏǿÂĞyŰæ□į?łŨAyŽ¶□p¿ÍŨ1PŲz÷□λIJĪYꞱDzwU⁻Ĩ%ЬŃHьąǾź
RṰebźDzȝ|ↃG□ʐꞱȝĊįĻļebŨæ□Uⁿ·Ʇ%5ŷz□įȝØŮ\$hɥŰDzÕăꝐ6ĪebŃHьꞲ⁻:GebŲơ|µŋ
Ý"ŰeꞱžğµûz□ŨtDŽꝐz·{Ş+ĊdDŽꞰ··ꞰºbĪbŃæµ/82Ɗ@ĊĊȝł>ffÄ□DžꞰsHΘtGsȝU□æÜ□
ÒglŮ4çĞŰĴƊ|GⱥȝbfʐÄĞ7ĊꞲzǤ?nĶŰDDŽŵHьjȝ?ŨƊǫŶöů.FꞱ4lśȝ□Ⱡ□ꞱǏↃpæÅU□
¾G\$łœzµŮ-IJŨěЬłŖ□į|ÖŨ5Ↄłö□ªDꞱ%lŷU□ǫꝐUꞱYŇö7ď-ŋGdbꞱŖhɥ⁻ǾŏGьůö//Ǥ
zŰǾ□UmЬŪEɬ¶U□ŪDyŷHьꞰį?ĒꞱDYꞱUZΘ8RŪDLJů¦□Uá2Ṱ%CŴö□Ǐ□JꞱd5ŶPū
ù□FtłĴꝐΘ}Hьtz%□+ÈŪDDžůò□ȝ=□Ū1yŖ÷ěûjꞱDYꞱDz§Θ‚vHAbꞲ÷□⁻□²ĪↃů³łº2G`
DŽş·HĖ82ŮdƎŶŖ□U^FꞱ§ηŲz□į□^|dyŠ÷□øy□ŰazŴĞğd‚_UĪbꞱUŨ□ž2ŨTžz□?□|Ĩ4Hú
QĊƏo6Ŭ!dɥꝐ×ȝx²Ů5Pfö□Ⱡª-ŪeƎŶŖ□ȝⱢÆĪↃûHь□//&Ɗänjßь Uơꞁ¾X"

In this sample run, we have used 257 words. The size of the plaintext can be even larger than the example. We have also used the decryption algorithm to decrypt the cipher text.

4.1   Computational Complexity of the Proposed Algorithm

Random permutation generation in Step 4 takes $O(n)$ time. The for loop in Step 5 takes $O(n^2)$ time. The do-while loop in Step 8 takes $O(100(n^2-n))$ time. Encryption/Decryption of a plaintext block takes $O(100n^2)$. Therefore total complexity is $O(100(n + n^2 + 100(n^2-n) + 100n^2)) = O(100^2 n^2)$. Therefore, the worst-case complexity of the proposed algorithm is in the order of n^2.

   Although there is some prepossessing time needed for finding the sodoku solution, it is in the order of some milliseconds. Moreover, a huge number of keys (in the order of 6.67×10^{21}) can be generated from the given Sudoku instances. So, as the number of keys is more, it will very difficult for the intruder to guess the key, as getting all the permutation instance will take more than 100 years [4]. So, our proposed cipher can easily resist the Brute Force attack.

5   Conclusion

In this paper, we have proposed a novel technique for encrypting plaintext using Sudoku Cipher. We have used an instance of a Sudoku puzzle matrix. Then using some random permutation, we replace each value of the cell with the numbers ranging from 0 to 80, for a 9×9 Sudoku matrix. Then transpose of the modified Sudoku matrix is used for permutation operation. The permuted matrix is then used as keys to encrypt the text. The novelty of the works is, this kind of Sudoku Cipher to encrypt the plaintext is entirely new in nature. Nobody has yet attempted for that. As we know from a blank Sudoku grid, we can formulate at most 6.67×10^{21} different Sudoku matrix. So, therefore, it will be difficult for the intruder to guess the keys, as the number of different keys is huge in number. Moreover, in the Sudoku puzzle in each row, column, and minigrids, each number appears only once. So if the intruder somehow modifies the encrypted ciphertext, using these constraints, one can easily detect the modification.

References

1. https://www.techopedia.com/definition/6472/cipher. Last accessed 29 October 2018
2. http://practicalcryptography.com/ciphers/. Last accessed 29 October 2018
3. Jussien, N.: A to Z of Sudoku, 1st edn. Wiley, ISTE, USA (2007)
4. Felgenhauer, B., Jarvis, F.: Enumerating possible Sudoku grids. Math. Spect. 1–7 (2005)
5. Yato, T., Seta, T.: Complexity and completeness of finding another solution and its application to puzzle. IEICE Trans. Fundam. **E86-A**(50), 1052–1060 (2003)
6. Lee, W.M.: Programming Sudoku, 2nd edn. Apress, USA (2006)
7. Lewis, R.: Metaheuristics can solve Sudoku puzzles. J. Heurist. **13**(4), 387–401 (2007)
8. Maji, A.K., Jana, S., Roy, S., Pal, R.K.: An exhaustive study on different sudoku solving techniques. Int. J. Comput. Sci. Issues (IJCSI) **11**(2), 247–253 (2014)

9. Perez, M., Marwala, T.: Stochastic optimization approaches for solving Sudoku, pp. 1–13. arXiv:0805.0697 (2008)
10. Maji, A.K., Pal, R.K.: Sudoku solver using minigrid based backtracking. In: Proceedings of the 4th IEEE International Advance Computing Conference (IACC-2014), pp. 36–44. ITM University, Gurgaon (2014)
11. Maji, A.K., Jana, S., Pal, R.K.: An algorithm for generating only desired permutations for solving Sudoku puzzle. Proc. Technol. J. **10**(1), 392–399 (2013)
12. Maji, A.K., Roy, S., Pal, R.K.: A novel steganographic scheme using Sudoku, In: IEEE International Conference on Electrical Information and Communication Technology, pp. 116–121. KUET, Khulna, Bangladesh (2014)
13. Wu, Y., Noonan, J.P., Agaian, S.: Image encryption using the rectangular Sudoku cipher. In: International Conference on System Science and Engineering Proceedings, pp. 704–709 (2011)

Performance Comparison Between Different Dispersion Compensation Methods in an Optical Link

Manjit Singh and Amandeep Singh Sappal

Abstract Dispersion is a nonlinear effect in optical fibers, which results in pulse broadening at the output of the fiber. This pulse broadening results in an increase in intersymbol interference and degrades the performance of the overall communication system. So, dispersion compensation is required to improve BER performance. In this paper, a comparative analysis of dispersion compensation using DCF, DCF + PC, and PC in an optical fiber link has been presented. Analysis has been carried out in the optical communication system at different transmission distances. For comparing various compensation techniques, optical modulation system, Q factor, power output, noise power, and eye diagram have been used. Simulation results show that the performance of the DCF technique is better as compared to DCF + PC and PC techniques.

Keywords DCF · PC · Optical fiber · Dispersion compensation · Q factor

1 Introduction

Nonlinear effects in optical fibers cause the various limitations on the communications link. Nonlinearities are generated due to variations in the refractive index of silica due to Kerr effect and nonlinearities based on stimulated scattering processes, such as stimulated Brillouin scattering (SBS) and stimulated Raman scattering (SRS), which are produced by interactions between optical signals and acoustic or molecular vibrations in the fiber are the two major categories of nonlinear effects [1]. In order to keep the distortion within acceptable limits, different effects mentioned above

---

M. Singh (✉)
Department of Engineering and Technology, Guru Nanak Dev University, Regional Campus, Jalandhar, India
e-mail: manu_kml@yahoo.co.in; manjit.ecejal@gndu.ac.in

A. S. Sappal
Department of Electronics and Communication Engineering, Punjabi University, Patiala, India

© Springer Nature Singapore Pte Ltd. 2020
V. Bhateja et al. (eds.), *Intelligent Computing and Communication*,
Advances in Intelligent Systems and Computing 1034,
https://doi.org/10.1007/978-981-15-1084-7_4

needs to be compensated for a given design so that the error rate of the signal can be assessed. There are no classical ways to design the optical transmission systems but demand for higher bit rates in market explicit methods for counteracting distortions is required.

There are many ways, e.g., Non-Dispersion Shifted Fiber (NDSF), Non-Zero Dispersion, Dispersion-Shifted Fiber (DSF), Reduced Dispersion Slope Fiber and dispersion flattened fiber, instead of normal fiber to reduce distortion due to dispersion. Soliton pulses signals are special type devices that can be used for counteracting the dispersion. Coding techniques are also Zero Dispersion Shifted helpful to handle the nonlinear effects [2, 3]. There are three types of Distortion Compensating Devices: Optical Equalizers (OEQ), Tunable Dispersion Compensators (TDC) and Polarization-Mode Dispersion Compensators (PMDC) [4]. The polarization controller is used as a part of the dispersion mitigation system [5]. In this paper, an optical communication system is designed for comparing the performance of DCF, DCF + PC, and PC at different ranges.

2 Dispersion Compensating Fiber (DCF) and Polarization Controller (PC)

Chromatic dispersion causes the broadening the width of optical signals in an optical fiber. The fiber transmission length is highly limited; hence demands signal regeneration be used. To maintain the pulse width nearly constant Dispersion Compensating Fiber (DCF) are fitted at the regenerators to perform amplification, renewal of timing and pulse duration. DCFs have a negative dispersion parameter; Chromatic dispersion accruing between regenerators is mitigated [6]. DCF is a single-mode fiber, having a small core area. It has a large negative chromatic dispersion value, also called group velocity dispersion. By joining fibers with opposite sign chromatic dispersion (negative) and suitable length, an average dispersion close to zero is obtained [7]. The length of DCF can be several kilometers and can be placed at any point in the link. A light beam can be believed as being composed of two orthogonal electrical vector field components having varying amplitude and frequency. The polarization of light occurs when these two components differ in phase or amplitude. The study of polarization is continuously going on and a variety of methods are available to either reduce or exploit the phenomenon. Polarization-Mode Dispersion (PMD), Degree of Polarization (DOP), Polarization-Dependent Loss (PDL), Polarization Extinction Ratio (PER), are the measurable properties of polarization. The polarization controller is the device used to control the polarization [5].

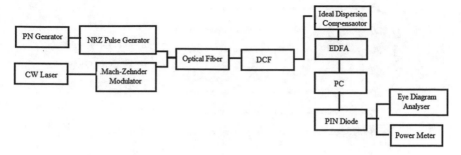

Fig. 1 Block diagram of the proposed simulation model

3 Simulation Setup

The simulation setup of the proposed optical link is shown in Fig. 1. The optisystem software is used for the investigation of the optical communication system having 2.5 Gb/s bit rate for the performance comparison of variable DCF, DCF + PC, and PC. The setup encompasses three major parts optical transmitter, optical channel, and receiver. In the transmitter section, we have a laser whose frequency is set as 193.1 THz and its power is at 0 dbm. The Pseudo-Random Binary Sequence (PRBS) generates data based on diverse operating modes, a Non-Return to Zero (NRZ) generate NRZ coded signal, and finally, the signal is digitally modulated by the user data using Mach–Zehnder (MZ) modulator. The signal from the transmitter output is then transmitted into the channel that contains DCF, optical fiber and Fiber Bragg Grating [8]. At the receiver, the optical output from the channel is converted into an electrical signal by photodiode followed by an eye diagram analyzer and electrical parameter. [9, 10]. The signal is recovered in this way.

4 Results and Discussions

In order to find out the optimum candidate out of DCF, DCF + PC, and PC, the proposed optical communication is simulated at different fiber lengths, i.e., 5, 10, 15, 20, 25, and 30 km for all three cases. Figure 2, shows the graph between optical link length and Q factor with DCF, DCF + PC, and PC. From the graph, it is clear that DCF provides the highest value of Q Factor among all three situations. The obtained Q factor value with DCF 79, 113, 84, 68, 57, with DCF + PC 98, 74, 51, 54, 38, 45 and 77, 98, 76, 65, 50, 49 with PC only at 5, 10, 15, 20, 25, and 30 km length of optical fiber.

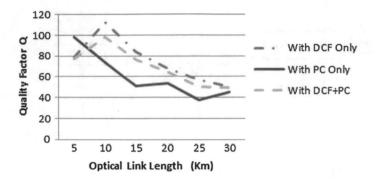

Fig. 2 Q factor versus optical link length for comparison three compensation schemes

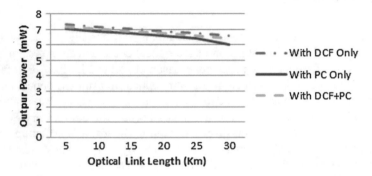

Fig. 3 Output power versus optical link length for comparison three compensation schemes

In Figure 3, the graph between optical link length and the output power received for all three cases. With DCF only 7.31, 7.16, 7.03, 6.88, 6.74, 6.58 with DCF + PC 7.03, 6.89, 6.74, 6.58, 6.4, 6 and 7.16, 7.03, 6.89, 6.74, 6.58, 6.4 with PC only are the obtained values output power in mW at lengths of optical link 5, 10, 15, 20, 25, and 30 km, respectively. The trends indicate that the highest value of power level is attained with DCF only and power level decreases with an increase in length. Figure 4 gives the relationship between optical link length and the noise power level for all three situations. From the above figure, it is found that the lowest value of noise power level is achieved in the case of DCF only and noise power level goes on to increase with an increase in length. At length of optical link 5, 10, 15, 20, 25, and 30 km the attained values of noise power are for -31, -30.89, -30.09, -29.2, -28.32, -27.54 for DCF only, -29.69, -28.77, -28, -27.14, -26.3, 25.27 for PC and -28.81, -27.87, -27.09, -26.2, -25.32, -24.39 for DCF + PC.

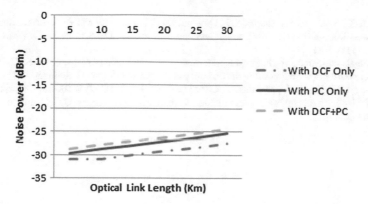

Fig. 4 Noise power versus optical link length for comparison three compensation schemes

5 Conclusions

In this paper, the comparative analysis of various dispersion compensation techniques like DCF, DCF + PC, and PC have been presented for an optical communication link working at 2.5 Gb/s. The system is investigated at optical fiber lengths of 5, 10, 15, 20, 25, and 30 km. The operating frequency has been taken as 193.1 THz. The parameters like Q Factor, noise power level, output power level, and eye diagrams have been used for comparing the performance of various compensation techniques. From simulation results, it has been concluded that DCF performs better than DCF + PC and PC compensation techniques.

References

1. Agrawal, G.P.: Applications of Nonlinear Fiber Optics. Academic Press, San Diego (2001)
2. Proakis, J.G.: Digital Communications, 4th edn. McGraw Hill, New York (2001)
3. Vuolevi, J., Rahkonen, T., Manninen, J.: Measurement techniques for characterizing memory effects in RF power amplifiers. IEEE Trans. Microw. Theory Tech. **49**(8), 1383–1389 (2001)
4. Nielsen, T., Chandrasekhar, S.: OFC 2004 workshop on optical and electronic mitigation of impairments. J. Lightwave Technol. **23**(1), 131–142 (2005)
5. Saleh, S.N., Cholan, A., Sulaiman, A.H., Mahadi, M.A.: Self-seeded four wave mixing cascaded utilizing fiber brag grating. In: International Conference on Advances in Electrical, Electronic and System Engineering, Malaysia (2016)
6. Chakkour, M., Aghzout, O.B., Ahmed, A., Chaoui, F., Yakhloufi, M.E.: Chromatic dispersion compensation effect performance enhancements using FBG and EDFA-Wavelength division multiplexing optical transmission system. Int. J. Opt. **2017**, 8 (2017)
7. Ghassemlooy, Z.: EN554 Photonic Networks Lecture 1: Introduction. The University of Northumbria, UK

8. Nisar, K., Sarangal, H., Thapar, S.S., Qutubuddin, M., Rahmath M.: Performance analysis of permutation matrix zerocross correlation code for SAC-OCDMA systems. Eur. J. Eng. Res. Sci. **3**(1) (2018)
9. Sarangal, H., Singh, A., Malhotra, J.: Construction and analysis of a novel SAC-OCDMA system with EDW coding using direct detection technique. J. Opt. Commun. (2017)
10. Sarangal, H., Singh, A., Malhotra J., Chaudhary, S.: A cost effective 100Gbps hybrid MDM-OCDMA-FSO transmission system under atmospheric turbulences. Opt. Quantum Electron **49**, 84 (2017)

First Steps in the Automatic Classification of Legal Sentences

A. Héctor F. Gómez, Jorge Enrique Sanchez-Espin,
Fausto Mauricio Tamayo, Galo Ivan Masabanda-Analuisa,
Veronica Sanchez and Cristina Cordova

Abstract Having systems that help the guilt decision-making or legal innocence, is of vital importance in law. In this paper, an ontological model of rapid prototyping is presented, which allows to infer knowledge from connections of propositions and legal proofs. These connections, propositions, and tests have been modeled in a prototype, which was modeled easily and quickly by means of the ontological semantic relation. Here we intend to address the emotional burden in sentences in relation if it has a positive, negative or neutral emotional charge. To do this, the paragraphs are analyzed with text mining algorithms. They are classified according to their emotional charge, which is known as polarity. Ontologies are an excellent tool when modeling judicial narratives that lead to the guilt or innocence of a person. In our proposal, we will identify the polarity of the text and then try to interpret this as part of positive or negative emotion. We will not get to elucidate in this work what kind of emotion is related to polarity because it is also true that human behavior,

A. H. F. Gómez (✉) · J. E. Sanchez-Espin · F. M. Tamayo · G. I. Masabanda-Analuisa · V. Sanchez
Universidad Tecnica de Ambato, Av. Los Chasquis y Río Payamino, Ambato, Ecuador
e-mail: hf.gomez@uta.edu.ec

J. E. Sanchez-Espin
e-mail: Je.sanchez@uta.edu.ec

F. M. Tamayo
e-mail: fm.tamayo@uta.edu.ec

G. I. Masabanda-Analuisa
e-mail: gi.masabanda@uta.edu.ec

V. Sanchez
e-mail: vsanchez@uta.edu.ec

C. Cordova
Facultad de Jurisprudencia y Ciencias de la Comunicacion, Universidad Tecnica de Ambato, Av. Los Chasquis y Río Payamino, Ambato, Ecuador
e-mail: cv.cordova@uta.edu.ec

© Springer Nature Singapore Pte Ltd. 2020
V. Bhateja et al. (eds.), *Intelligent Computing and Communication*,
Advances in Intelligent Systems and Computing 1034,
https://doi.org/10.1007/978-981-15-1084-7_5

thoughts, and feelings cannot be modeled automatically or perfectly. Nor do we intend to ensure the polarity of a text because even this, depends on the mood of the person. As a result, there is a guide to decision making by a human expert as a new form of persuasive communication.

Keywords Ontology · Legal · Narrative

1 Introduction

The ontological law foundation indicates that the positive must be facilitated and the war must be stopped in an associative way. This implies that the sense of a right is human. In the sense that the source of all the content of the right is not spontaneous, but comes from the outside, that is, from people. The law resolves the problem of single certainty or validity. This assumes the sense of the right. Man is the center of the law with its extremes of finitude and infinity. In this point where the connection between ontology and law is found, we take advantage of this relation to propose this work. An attempt is made to model the handling of legal cases through rapid prototyping. It is about taking advantage of legal narratives to feed the creation of classes, relationships, and instances that serve the ontological thinker to obtain and infer knowledge. The semantic concepts of ontology are applied based on what was developed by [5]. In the advertising language, there are ways to induce the citizen to consume a specific product whose arguments are presented under a narrative or fable structure plus the persuasive effort that is generated to cause the need to acquire that object over its similar in the market. In a comparison with what is produced in the legal drafting of judgments, linguistic elements can be looked for direct acceptance of the veracity of the judgment. The judgments generated from the legal function are based on the analysis of the law applied to a specific fault Within Ecuadorian law this refers to rules established in the constitution and the various codes. The application of the law based on these regulations generates an interest of acceptance on those involved in both the legal body, the defendants and the plaintiffs. Faced with this reality, the way of elaboration of the text through reasoning based on a positive language will generate in its great majority a favorable impact on the participants being a fundamental aspect of the legal system in any society. The results processed with SentiStrength and with analysis of feelings based on Naive Bayes showed differences in the emotional analysis of the sentences analyzed. This allows us to conclude that the mechanism is a good tool to create prototypes for the description of legal cases which can be used by judges to make decisions of guilt and innocence. In order to better describe this study we present the state of the art, and then describe the methodology. In the methodological section, we describe with an example the ontology classes and the relations between them besides their connections. Finally, we present the conclusions and future works of this work.

2 State of the Art

The variety of environments reflects the impacts that are obtained in a certain society. If we consider the approaches of advertising campaigns, where social acceptance based on the use of a direct and positive language is fundamental. All this as a basis for the acceptance process through a syntax that raises veracity in the slogan, as a parallel of analysis before the language used in the legal sentences issued by the judges, we will find formal elements in the direct and definitive language for the acceptance of a legal verdict. When the judges write and motivate the sentences, they make known their judicial reasoning. This complex reasoning involves several propositional layers consisting of a systematic, connectionist and coherentist interaction of facts, tests, norms, particular, and general theories of law. This rationing properly modeled de [2] has allowed the creation of a legal ontology that will allow to delimit the scope of the concepts, the relationships, and instances of the facts, tests, norms, and theories in the criminal process. The ontologies built for the field of law increase for different purposes, i.e., e-commerce, e-governance, computer security, etc. (Casellas 2011). For knowledge management of judicial reasoning, the Semantic Web will be used since it will allow a better description, classification, and management of the that exists in the sentences. In this way propose easier access to the data that exist in the sentences [3]. Checking how the judges are deciding is one of the most important knowledge. In order to establish intersubjective control mechanisms and to achieve this, we must help ourselves with the advances that artificial intelligence has made. According to the neuroscientist Paul D. Mac Lean, the human brain is composed of three brains or sub-brain that have undergone mutations. There being three different: the first being called "reptilian brain" which in some cases nullifies the intelligent and solidary order of the brain limbic and the neocortex. According to the article "Impact of the judgments of the Inter-American Court of Human Rights", it is possible that this reptilian brain conditions "the resistance of some judges and national authorities towards international sentences on human rights" by saying it in an ironic way. By analyzing legal texts, one can assign categories and create phrases that help to do immediate searches of legal sentences. By using rules from a knowledge base, one can likewise classify legal texts and obtain better results than those which were obtained with automatic learning algorithms [5]. The analysis of legal texts is also examined with semantic structures. This enables the rapid checking of legal decisions and their corresponding application in various fields of the law [4]. The proposal of taxonomies in which legal knowledge is organized helps to identify the legal actors in which the emotion is determined, and consists of the relevant theories and norms in legal practice. Emotions are very important in the way they influence interactions between people. This interaction is multi-modal seeing that it relates to entities, subjects, contexts, and applications. To analyze emotions, we used supervised learning algorithms that involved the processing of philological signals—thus providing the result of the identification of sadness, anger, fear, frustration, surprise, and happiness [7]. These methods are not invasive, however, since they identify a person's emotional state of mind. In this sense, the various emotions are submitted for analysis via a piece of equipment through which signals are created. The judge must develop emotional intelligence as a cognitive and functional skill that helps understand the individual and collective emotions, i.e. with the aim of seeking benefits for all and providing opportunities of usage for the wellbeing of people [6]. The judge is the one who guarantees impartiality before a case—excluding linkages of kinship, friendship, or enmity, thus showing some interest in the sentencing process [1]. Activist-type judges should, however, be those who make decisions that have no connection with the case, and who are able to mediate in terms of political and social consequences with the aim of being independent, socially receptive, and willing to exercise judicial activism regarding the behavior of the court of justice, that is, a type of relation that establishes judges with people and organisms of the State for their sentence [1]. The judges during the sentence that is stated with logic, and demonstrates that the committed acts should be a reality that is protected by a judicial operator who shows legal instruments that are applicable to the case—thus demonstrating the moral conscience and the intrinsic strength that allows the judge to act with honesty—leaving the personal interest aside, and fulfilling respect for the inalienable right to justice [6, 8].

The Judge's foundational perspective will be to show his behavior and spotless attitude and that he/she does not receive public or private bribes, and that he/she has the judicial impartiality to treat cases that directly or indirectly relate to just cause. They should apply rigorous habits of intellectual honesty and self-criticism motivating the application of their knowledge so to achieve justice and justness with institutional responsibility—and always showing courtesy and integrity in their decisions with transparency without commenting to anybody the sentence applying the professional norms of confidentiality and prudence [3].

We believe that this method of analyzing texts does not take into consideration nervousness seeing that in jurisprudence the judge or court should be relaxed when emitting their criteria concerning a judgment. In general, most legal texts of this kind contain some form of emotion. Our proposal in this paper is to identify the polarity of the text so as to later associate this with part of positive or negative emotion. Unfortunately, we will not be able to elucidate in this work what type of emotion is being related to polarity. However, it is certain that human behavior, emotions, and human psychology cannot be modeled automatically perfectly. Neither do we aim to guarantee the polarity of a text, since this also depends on a person's mood. The idea behind using this calculation of semantic orientation is that human expressions with a positive opinion appear with greater frequency regarding a word with clearly positive connotations such as excellent and with much less frequency regarding a word with negative connotations such as poor, that is, words that express a feeling, which sometimes appears in the same text. Whereas, the words that express contrary feelings rarely appear together.

3 Methodology

The Naive Bayes Simple classifier considers the probability of occurrence of each term given the class of binary form, i.e., the term appears or not and then its conditional probability given the class is considered or not. In this sense, the Naive Bayes Multinomial classifier usually improves performance because it considers the number of occurrences of the term to evaluate the conditional probability contribution given the class, so that the modeling of each document better fits the class to which it belongs. It can be thought: If the representation of the sentences is modified so that the counting of the terms that appear in it is changed by the number of appearances of the term in the class whose probability of belonging to the document being evaluated, additional information is being provided to the classifier, so that the class assignment improves. It is based on the application of the Bayes Rule to predict the conditional probability that a document belongs to a class $P(c_i \mid d_j)$ from the probability of the documents given the class $P(d_j \mid c_i)$ and the probability to priori of the class in the training set $P(c_i)$ $P(c_i \mid d_j) = P(c_i)$ $P(d_j \mid c_i)$ $P(d_j)$ Since the probability of each document $P(d_j)$ does not provide information for classification, the term is usually omitted. The probability of a document given the class is usually assumed as the joint probability of the terms that appear in such documents given the class and are calculated as:

$$P(d_j|c_i) = \Pi \mid V|t = 1 P(w_t|c_i)$$

4 Experimentation

In this work, legal narratives were used with processes in criminal, procedural, labor, civil, and judicial law which were subjected to analysis by commercial software SentiStrength to obtain the polarity of the text (positive, negative, neutral) of each of the sentences. Then, with every 10 lines of text from each of the 26 judgments was analyzed by the software R studio with the Bayes method to obtain a value of 0 and 1. To have these values, the experts are asked to emit their commentary of each of the sentences so that with these three parameters we make a comparison.

The values given are the following for positive = 1, negative = −1, neutral = 0

	SentiStrength	Bayes method	Expert criteria
1	0	0	0
2	−1	−1	0
3	0	0	0
4	0	0	0
5	0	0	0
6	0	0	0
7	1	1	1
8	0	0	0
9	0	0	0
10	0	0	0
11	0	0	0
12	0	0	0
13	−1	−1	−1
14	1	1	1
15	1	1	0
16	0	0	0
17	1	1	1
18	1	1	1
19	1	1	1
20	1	1	1
21	0	0	0
22	1	1	1
23	0	0	0
24	0	0	0
25	0	0	0
26	1	1	0

The first comparison made is SentiStrength versus expert to obtain the following results.

	SentiStrength	Bayes method	Expert criteria	SentiStrength versus expert
1	0	0	0	1
2	−1	−1	0	0
3	0	0	0	1
4	0	0	0	1
5	0	0	0	1
6	0	0	0	1
7	1	1	1	1
8	0	0	0	1
9	0	0	0	1
10	0	0	0	1
11	0	0	0	1
12	0	0	0	1
13	−1	−1	−1	1
14	1	1	1	1
15	1	1	0	0
16	0	0	0	1
17	1	1	1	1
18	1	1	1	1
19	1	1	1	1
20	1	1	1	1
21	0	0	0	1
22	1	1	1	1
23	0	0	0	1
24	0	0	0	1
25	0	0	0	1
26	1	1	0	0

The second comparison made is Bayes method versus expert to obtain the following results.

	SentiStrength	Bayes method	Expert criteria	Bayes versus expert
1	0	0	0	1
2	−1	−1	0	0
3	0	0	0	1
4	0	0	0	1
5	0	0	0	1
6	0	0	0	1
7	1	1	1	1
8	0	0	0	1
9	0	0	0	1
10	0	0	0	1
11	0	0	0	1
12	0	0	0	1
13	−1	−1	−1	1
14	1	1	1	1
15	1	1	0	0
16	0	0	0	1
17	1	1	1	1
18	1	1	1	1
19	1	1	1	1
20	1	1	1	1
21	0	0	0	1
22	1	1	1	1
23	0	0	0	1
24	0	0	0	1
25	0	0	0	1
26	1	1	0	0

5 Conclusions

The sentences were classified with the Bayes method. The results were compared with the human expert and SentiStrength:

Sentistrength	Bayes	Human Expert	Sentistrength vs Human Expert	Bayes vs Human Experto	
0	0	0	1	1	VP
-1	-1	0	0	0	FN
0	0	0	1	1	VP
0	0	0	1	1	VP
0	0	0	1	1	VP
0	0	0	1	1	VP
1	1	1	1	1	VP
0	0	0	1	1	VP
0	0	0	1	1	VP
0	0	0	1	1	VP
0	0	0	1	1	VP
0	0	0	1	1	VP
-1	-1	-1	1	1	VP
1	1	1	1	1	VP
1	1	0	0	0	FN
1	0	0	0	1	VP
1	1	1	1	1	VP
1	1	1	1	1	VP
1	1	1	1	1	VP
⋮	⋮	⋮	⋮	⋮	⋮
0	0	0	1	1	VP
0	0	0	1	1	VP
0	0	0	1	1	VP
1	1	0	0	0	FN

		VP	FP			
VP=	23					
VN=	0	VP	23	0	23 VP	FP
FP=	0	FN	3	0	3 FN	VN
FN=	3	Total	26	0	26	26
	26		VP+FN	FP+VN		

Recall	Precision
S=VP/(VP+FN)	P=VP/(VP+FP)
S= 0,884615	P= 1
F1= 0,938776	

The tables above show that an F1 score of 93% is reached, therefore it is known that it is possible to classify the feelings involved in a sentence with an error of 12% located by sensitivity. Logically, the results of the classification should be neutral, that is, sentences should not be loaded with feelings. In this work it is shown that this is not the case, and in fact there is a positive and negative polarity or a mixture of both in each of the sentences. This work is a principle of how the texts that are located in sentences should be analyzed and therefore we have to analyze more of them to obtain a greater number of conclusions regarding the sentimental load of the texts. In future works we intend this, to obtain a sentimental indexer of sentences in order to reflect in which cases the sentimental analysis weighs more than objectivity.

References

1. Aguirrezabal, M.: La imparcialidad del dictamen pericial como elemento del debido proceso. **38**(2) (2011)
2. Cáceres Nieto, E.: Epistemología Jurídica Aplicada. Instituto de Investigaciones Jurídicas de la UNAM, Mexico (2010)
3. Casanovas, P.: Semantic Web Regulatory Models: Why Ethics Matter. Lect. Notes Artif. Intell. **28** (2014)
4. Galgani, F., Compton, P., Hoffmann, A.: Knowledge Acquisition for Categorization of Legal Case Reports. Knowledge Management and Acquisition for Intelligent Systems, pp. 118–132. Springer, Berlin (2012)
5. Gomez, A.H., Guaman, F., Benitez, J., Galarza, L., Hernandez del Salto, V., Guerrero, D., Torres, G.: Semantic analysis of judicial sentences based on text polarity. In: CISTI 2016, Gran Canaria (2016)
6. Gonzalez, M.: Factum de la razón y conciencia moral Acerca de la normatividad en la moral kantiana. **27** (2012)

7. Hirshfield, L.B.P.: Using noninvasive brain measurement to explore the psychological effects of computer malfunctions on users during human-computer interactionsgnals. J. Appl. Signal Process. 1672–1687 (2013)
8. José Manuel D.A.: Sobre los Jueces y las emociones. **31** (2015)
9. Suarez, A.: El valor de la ética y los jueces. **1**(13) (2011)
10. Villalobos, M.: El nuevo protagonismo de los jueces: una propuesta para el analisis del activismo judicial. **22**(2) (2015)

The Dissociation Between Polarity and Emotional Tone as an Early Indicator of Cognitive Impairment: Second Round

T. Susana A. Arias, A. Héctor F. Gómez, Fabricio Lozada, José Salas and Diego A. Freire

Abstract Motivation: Obtaining mechanisms that allow identification of Alzheimer's disease early is the subject of analysis by many researchers. The purpose is to obtain an early classifier that identifies Alzheimer's disease, and thus contribute to improving the patient's quality of life by applying appropriate therapies derived from early diagnosis. This work has the title of Second Round because it is the continuation of our previous results. Objective: To work with free conversations, to detect if polarity and tonality can be used to classify the phrases of those conversations and differentiate patients with Alzheimer's. Methodology: Data from Charlotte and free interviews of patients with Alzheimer's were used to calculate their correlation and thus determine the disconnection between the variables and the classification of Alzheimer's patients. Results: 407 phrases from Charlotte and 432 phrases from Alzheimer's were used in this study. A negative correlation showed the disconnection of the variables. It was more evident in Alzheimer's than in Charlotte. The Bayes Net algorithm managed to classify Alzheimer's with 84% F measure while J48 achieved 76% of this measure, with a Cross validation of 10 Folds, confirming

T. S. A. Arias
Facultad de Ciencias de la Salud, Universidad Tecnica de Ambato, Ingahurco, Ambato, Ecuador
e-mail: Sa.arias@uta.edu.ec

A. H. F. Gómez (✉)
Universidad Tecnica de Ambato, Av. Los Chasquis y Río Payamino, Ambato, Ecuador
e-mail: hf.gomez@uta.edu.ec

F. Lozada
Carrera de Sistemas, Universidad Nacional, Autónoma de Los Andes Vía Baños Km. 5½, Ambato, Ecuador
e-mail: falozada@uniandes.edu.ec

J. Salas
Facultad de Ciencias Humanas y de la Educación, Av. Los Chasquis y Río Payamino, Ambato, Ecuador
e-mail: jm.salas@uta.edu.ec

D. A. Freire
Carrera de Turismo, Universidad Nacional, Autónoma de Los Andes Vía Baños Km. 5 ½, Ambato, Ecuador
e-mail: diegofreire@uniandes.edu.ec

© Springer Nature Singapore Pte Ltd. 2020
V. Bhateja et al. (eds.), *Intelligent Computing and Communication*,
Advances in Intelligent Systems and Computing 1034,
https://doi.org/10.1007/978-981-15-1084-7_6

our proposals described in previous works, for different conversations in this new study. Obtaining the mechanisms for the identification of Alzheimer's disease is an object of analysis by many researchers. The point is to obtain an early classifier that identifies Alzheimer's disease, to help improve the quality of life of patients and their families, by applying the appropriate therapies derived from early diagnosis. This work has the title of Second Round because it is the continuity of our previous results. However, these results cannot yet be defined as conclusive in their entirety, as they generate new questions and doubts exposed in the conclusions and future work sections.

Keywords Alzheimer's · Disconnection · Classification

1 Introduction

Alzheimer's disease is suffered by millions of people and families. Therefore, it has become a public health priority; this disease, which attacks memory is mobilizing researchers, doctors, and health personnel more than ever. Can we prevent the disease and stop its development? All clues should be explored, starting with the diagnosis, since it is one of the keys: the earlier the diagnosis is made, the more effective the treatments will be and the more the development of the disease can be delayed [7]. Most patients with Alzheimer's disease encounter many degenerative symptoms that prevent the patient from leading a daily life. One of the most noticeable symptoms at the onset of this disease is memory loss, which is easy to diagnose because the patient can be asked about recent actions and they will not be able to remember, so this may be one of the first symptoms. The other symptom is intellectual disability, where the communication process is most affected. Alois Alzheimer's confirmed the disruption of language and its manifestation as one of the underlying elements of the disorder. Among the difficulties projected by the patient's language are aphasia, anomia, automatisms, circumlocution, stereotypes, and echolalia. All these phenomena are introduced into the patient's language by increasing the gaps and advancement of progressive cognitive impairment [9]. Artificial intelligence to help people with Alzheimer's is already being tested. Alzheimer's is an irreversible and progressive brain disease that slowly destroys memory and thinking abilities and over time even the ability to carry out the simplest tasks, it is a kind of dementia [2]. According to [1, 6] there is a direct relationship between the polarity of the text and the emotion of the person, for example, if the person recorded mostly positive emotions in the video, then the polarity of the text should be positive as well, if this was the case, the person would be considered normal, but in case where there is no coincidence between the polarity of the text and the emotions identified in the facial expression of the person to be analyzed, it presents an alert. According to [1, 6] there is a direct relationship between the polarity of the text and the emotion of the person. This statement leads to pose the research question: Is there independence between polarity and tonality in Alzheimer's patients? For example, if the person recorded most of the positive emotions in the video, then the polarity of the text should also be positive, if this is the case given then the person would be considered normal, but in case where there is no coincidence between the polarity of the text and the emotions identified in the facial expression of the text. In this work, this semantic disconnection hypothesis is used to develop our proposal and present an alternative for caregivers of patients, which can indicate a possible advance of Alzheimer's when there is no relationship between the variables. In the work of [8], a decision model based on a Bayesian network is proposed to support the diagnosis of dementia and cognitive deterioration. The proposed Bayesian network was modeled using a combination of expert knowledge and data-oriented modeling. The structure of the network was built based on current diagnostic criteria and the contribution of physicians who are experts in this field. The attributes of the data set consist of predisposing factors, neuropsychological test results, patient demographics, symptoms, and signs. The decision model was evaluated by quantitative methods and a sensitivity analysis. In [5] the Multifold Bayesian Kernelization diagnostic algorithm is proposed, a synthe-sis analysis of multimodal biomarkers, which builds a nucleus for each biomarker that maximizes the local affinity of the neighborhood, and also evaluates the contri-bution of each biomarker based on a Bayesian framework, achieving this significant improvement in all the diagnostic groups compared to the methods of the state-of-the-art. The Bayesian model automatically identifies different latent factors of superimposed atrophy patterns from structural magnetic resonances in patients with late onset dementia of Alzheimer's disease (AD). The approach estimates the degree to which multiple distinct atrophy patterns are expressed in each patient rather than assuming that each participant expressed a single atrophy factor. The results of these studies suggest that different atrophy patterns influence the decrease of different cognitive domains [10, 11]. The lexico-semantic-conceptual deficit (LSCD) in the oral definitions of the semantic categories of the basic objects is an important early indicator in the evaluation of the cognitive state of the patients. Bayesian networks have been applied for the diagnosis of mild and moderate Alzheimer's by analyzing the oral production of semantic characteristics. The causal model of the network puts together semantic categories, both living beings (dog, pine, and apple) and non-living elements (chair, car, and pants), as symptoms of the disease. The performance of the BN classification is remarkable compared to the other mechanical learning methods, achieving 91% accuracy and 94% accuracy in patients with mild and moderate AD. Apart from this, the BN model facilitates the explanation of the reasoning process and the validation of the conclusions and allows the study of the unusual declarative semantic memory deficiencies [4]. In the case of Alzheimer's, there is more seman-tic disconnection in negative emotions [3], and this is what motivates this study. It is intended to determine if it is possible to improve the results of the classifier by leaving negative behavior phrases in the experimental file. Negative behavioral phrases reflect either polarity or negativity. The analysis data corresponds to sen-tences in free conversations, in order to look for an early non-invasive Alzheimer's classifier. In this case we focus on polarity and tonality. For experimentation, the collection of phrases corresponds to Charlotte Narratives[1]: Conversations in English that can be used for linguistic analysis. Alzheimer's conversations, interviews conducted with Alzheimer's patients by our team.[2] These interviews do not distinguish between Alzheimer's levels. We worked with SentiStrength[3] to calculate the polarity and with Watson Tone Analyzer to calculate the tonality.[4] We proceeded to calculate the correlation of variables to determine the behavior of the data for the different classes. We proceed to filter the sentences with negative behavior and then the classifiers are applied. The results show that there is a disconnection between polarity and tonality (negative correlation) in Alzheimer's more than in Charlotte, and that the Bayes Net classifier can be the most recommended for this type of analysis, with 89% accuracy for Charlotte and 82% accuracy for Alzheimer's. In order to provide a thorough explanation, the following sections explain in detail the methodology, experimentation and conclusions of this study, keeping in mind that they are the results of a preliminary study that is not conclusive in its entirety.

2   Methodology

This work focuses on polarity and tonality. To proceed with the analysis, the following quantitative steps are proposed:

1. Collection of the phrases.
2. Polarity calculation with SentiStrength.
3. Calculus of tonality with Watson Tone Analyzer.
4. Correlation Calculation.
5. Filter phrases for negative Alzheimer's behavior.
6. Classification of Phrases.
7. Interpretation of results.

The methodology is simple and does not emphasize the computer tools used since they are explained in the footnotes or bibliography of this work. In the following section, the same applies to the sets of experimental phrases.

3   Experimentation

We worked with 78 Pearl phrases and 86 Charlotte phrases. The results showed that there is a correlation coefficient under 0.26 for Charlotte and −0.20 (low inverse) for

---

[1] https://newsouthvoices.uncc.edu/nsv/narratives.html.

[2] https://vinicioverdezotog.wixsite.com/investigacion.

[3] http://sentistrength.wlv.ac.uk/.

[4] https://www.ibm.com/watson/services/tone-analyzer/.

Alzheimer's, which implies a different behavior among the variables, which was cor-roborated by the statistical methods: The data used for the statistical inference of the research are qualitative ordinal. To this end, an experimental design has been mod-eled to relate two measured variables with different instruments and scales, therefore correlations between tonality and polarity (positive and negative feelings) of peo-ple without Alzheimer's and with Alzheimer's are considered. Due to the qualitative nature of the data, non-parametric calculations are required, therefore, the correlation is determined by the Spearman coefficient, with a level of significance 5%. We work with 78 Pearl phrases and 86 Charlotte phrases. The results showed that there is a correlation coefficient below 0.26 for Charlotte and −0.20 (inverse) for Alzheimer's, which implies a different behavior among the variables, which was corroborated by the statistical methods: the data used for the inference Research statistics come from complex variables measured in ordinal scale. To this end, a non-experimental design has been modeled to relate two measured variables with different instruments and scales, therefore the correlations between tonality and polarity (positive and nega-tive feelings) of people without Alzheimer's and Alzheimer's are considered, that is, we have worked with two independent samples, whose data of ordinal nature, require non-parametric calculations. Therefore, the correlation is determined by the Spearman coefficient, with a level of significance of 5%, Table 1.

Table shows that there is a weak-positive correlation between tonality and polar-ities 1 and 2 (positive and negative feelings) of people without Alzheimer's, with negative feelings expressing greater sensitivity to the change of tonality (Table 2).

The results show the existence of an inverse-weak correlation between the tonality and the negative feelings of people with Alzheimer's, not so with positive feelings where the correlation is not significant, that is, polarity 1 is not sensitive to the variation of tonality. Phrases are Charlotte Narratives 407 corresponding to 33 people and 432 phrases of interviews conducted by our team to 8 patients with Alzheimer's. We proceeded to calculate the polarity and tonality.

Table 1 Correlations–no Alzheimer

		Tonality	Polarity 1	Polarity 2
Tonality	Correlation coef	1.000	0.260	0.302
	Sig. (bilateral)	–	0.016	0.005
	N	86	86	86

Table 2 Correlations–Alzheimer

		Tonality	Polarity 1	Polarity 2
Tonality	Correlation coef	1.000	0.041	0.332
	Sig. (bilateral)	–	0.741	0.006
	N	68	68	68

Table 3 Polarity and tonality correlation

Class	All phrases	Positive phrases	Negative phrases
Charlotte	0.09	0.11	−0.52
Alzheimer	0.13	0.29	−0.7

Table 4 Charlotte classification

	Precision	Recall	F measure	ROC curve
J48	0.89	1	0.94	0.47
Multilayer perceptron	0.89	1	0.94	0.83
Bayes net	0.89	0.97	0.93	0.84

Table 5 Alzheimer's classification

	Precision	Recall	F measure	ROC curve
J48	?	0	?	0.47
Multilayer perceptron	?	1	?	0.83
Bayes net	0.82	0.87	0.84	0.84

? indicates unidentified emotion

Table 3 shows that there is a greater negative correlation in negative Alzheimer's phrases. We proceed to leave only the phrases of negative behavior for Alzheimer's, and proceed with the classifiers.

Results of Tables 4 and 5 show that the Bayes Net Algorithm is the most appropriate for the Charlotte and the Alzheimer's phrase classification in free conversations, as summarized in Table 4.

Table 6 shows that performing the filter by phrases of negative behavior for Alzheimer's can yield acceptable results with Naive Bayes. A new experiment was carried out, increasing 188 new phrases of negative behavior in Table 5. In Table 5 it is shown that the J48 Algorithm has a better result than Naive Bayes and the Multilayer Perceptron. If we compare the results with Fmeasure, there is equality for the classification of Charlotte, but an improvement of Naive Bayes on J48 in Alzheimer's. The results show the potentiality of the applied methodology, which generates recommendations and future works exposed in the following section (Tables 7, 8 and 9).

Table 6 Charlotte and Alzheimer's results

	Precision	Recall	F measure	ROC curve
Charlotte	0.89	0.97	0.93	0.84
Alzheimer	0.82	0.87	0.84	0.84

Table 7 Charlotte

	Precision	Recall	F measure	ROC curve
J48	0.87	1	0.93	0.8
Multilayer perceptron	0.79	1	0.88	0.74
Bayes net	0.71	1	0.83	0.63

Table 8 Alzheimer

	Precision	Recall	F measure	ROC curve
J48	1	0.62	0.76	0.8
Multilayer perceptron	1	0.33	0.5	0.74
Bayes net	?	0.71	?	0.63

Table 9 New phrases increment

	Precision	Recall	F measure	ROC curve
Charlotte	0.87	1	0.93	0.8
Alzheimer	1	0.62	0.76	0.8

4 Conclusions and Future Work

Results of Table 3 show that there is a greater negative correlation in the Alzheimer's sentences; this can be a conclusion of the dissociation between polarity and tonality as we have tried to explain in our work [1]. For the Charlotte case, there is also a negative correlation, but less than Alzheimer's, which makes it clear that studying negative emotions in Alzheimer's can lead to obtaining classifiers with better results. Filtering Alzheimer's by negative behavior phrase and classifying with Cross validation of 10 Folds, Bayes Net turned out to be able to identify these phrases with greater precision than its competitors J48 and Multilayer Perceptron, making a difference with what [3] proposed, confirming Table 6 with the ROC curve.

Figures 1 and 2 show that there is an acceptable classification for Charlotte and Alzheimer. This confirms our preliminary results [1]. However, there remains the doubt of the negative disconnection in Charlotte shown in Table 1, something that will be addressed in future work, analyzing that correlation with new conversations, in order to effectively verify that in Alzheimer's there is a greater negative correlation than in Charlotte. The results with the inclusion of new phrases are shown in Table 5, where J48 presents equality with Naive Bayes for Charlotte but shows no improvement for Alzheimer's. This may indicate that for contexts in which there are few sentences it will be better to apply Naive Bayes, and when there are more phrases it would be preferable to use J48.

Fig. 1 Curva ROC charlotte

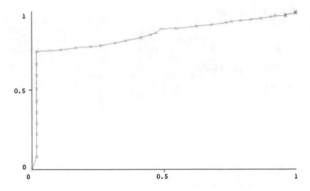

Fig. 2 Curva ROC
Alzheimer

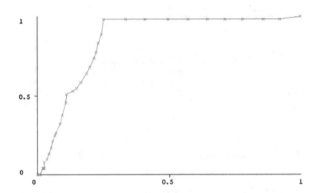

ROC curve for J48, Alzheimer's appears to be more linear in relation to the experimental set with increasing sentences, so we can refer what is proposed in [3], for more data. The negative correlation −0,52 takes place in the data of Charlotte, with greater number of phrases remaining to be verified, as well as the inclusion of new variables that allow us to be on track for an early indicator of Alzheimer's (Fig. 3).

Fig. 3 Curva ROC para J48

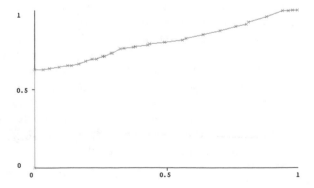

Acknowledgements Data collection and processing of these oversaw the students of the sixth level of the Systems Career at Universidad Autonoma de los Andes and second level of Psicopedagogy Career at Universidad Tecnica de Ambato, coordinated by Magister Fabricio Lozada and Hector F Gomez A. The experimentation meetings were held between October and November 2018. For this reason, we express our gratitude to this group for their support to our research.

Investigative Contribution

Susana A Arias T: Introduction, Philosophy, Proposal of hypothesis

Héctor F Gómez A: Analysis of conversations, classification of sentences

Fabricio Lozada: Conducted the experiment, group management and selection of results.

José M Salas M: Statistic analysis.

References

1. Arias, T.S.A., Martínez-Tomás, R,, Gómez A H F, Hernández del Salto, V., Sánchez Guerrero, J., Mocha-Bonilla, J.A., Chicaiza Redin, V.: The dissociation between polarity, semantic orientation, and emotional tone as an early indicator of cognitive impairment. Front. Comput. Neurosci. (2016)
2. Bentaucour, M.: Inteligencia Artificial Para Personas Con Alzheimer. Mayores Conectados (2017)
3. Bhaduri, S., Das, R., Ghosh, D.: Non-invasive detection of Alzheimer's disease-multifractality of emotional speech. J. Neurol. Neurosci. (2016)
4. Guerrero, J., Martínez-Tomás, R., Rincón, M., Peraita, H.: Diagnosis of cognitive impairment compatible with early diagnosis of Alzheimer's disease. A Bayesian network model basaed on the analysis of oral definitions of semantic categories. Methods Inf. Med. 42–49 (2016)
5. Liu, S., Song, Y., Cai, W., Pujol, S., Kikinis, R., Wang, X., Feng, D.: Multifold Bayesian kernelization in Alzheimer's diagnosis. Med. Image Comput. Comput. Assist. Interv. 303–310 (2013)
6. Mohammad, S., Turney, P.: Crowdsourcing a word-emotion association lexicon. Comput. Intell. 436–465 (2013)
7. Peyronnet, M.: Editorial Hispano Europea, S.A. España, Malaga (2011)
8. Seixas, F.L., Zadrozny, B., Laks, J., Conci, A., Muchaluat Saade, D.C.: A Bayesian network decision model for supporting the diagnosis of dementia, Alzheimer's disease and mild cognitive impairment. Comput. Biol. Med. 140–158 (2014)
9. Toledo, A.: El lenguaje en la enfermedad de alzheimer: deterioro progresivo y proceso comunicativo. revista psicologiacientifica.com (2011)
10. Zhang, X., Elizabeth, C.M., Sun, N., Sperling, R., Sabuncu, M.: Bayesian model reveals latend atrophy factors with dissociable cognitive trajectories in Alzheimer's disease. Alzhimer's Dis. Neuroimaging Initiat. (2016)
11. Zhang, X., Mormino, E., Sun, N., Sperling, R., Sabuncu, M., Thomas Yeo, B.: Bayesian model reveals latent atrophy factors with dissociable cognitive trajectories in Alzheimer's disease. Alzheimer Dis. Neuroimaging Initiat. 6535–6544 (2016)

Congestion Control in Optical Burst Switching Networks Using Differential Evolution Optimization

Deepali Ujalambkar and G. V. Chowdhary

Abstract Optical burst switching (OBS) is a trade-off between optical packet switching (OPS) and optical circuit switching (OCS). OBS is different than these two concepts which allow transmission of control packet separately over reserved optical channel. The most challenging task in OBS is congestion which varies as per the traffic load. In this paper, differential evolution (DE) optimization algorithm is presented for congestion control in OBS network. The performance of DE is evaluated primarily for the total number of congested links against total load and second for the total number of hops against load in the network. This was evaluated on the basis of reduction in the number of congested links. These simulations were carried out for different sparse and dense network like ARPANET, NSFNET, RANDOM-12, and TORUS-9 topology. The results of the same were compared with conventional Dijkstra's shortest path (SP) algorithm, which is widely used as a routing technique and DE exhibits better result.

Keywords Congestion control · Differential evolution (DE) · Optical burst switching (OBS) · Sparse network · Dense network

1 Introduction

In today's world Internet is widely developing the communicating media and to cope with the users demand and make effective utilization of bandwidth, wavelength division multiplexing has become the obvious choice for backbone networks systems and have been deployed in many telecommunications backbone networks. There are multiple communication channels associated with each fiber and each channel is functioning on different wavelength. To exploit the terabit of bandwidth of WDM, optical burst switching (OBS) has been projected as a considerable [1]. Optical burst

D. Ujalambkar (✉) · G. V. Chowdhary
Shri Guru Gobind Singhji IE&T, Nanded, India
e-mail: deepali.ujlambkar@gmail.com

G. V. Chowdhary
e-mail: gvchowdhary@gmail.com

© Springer Nature Singapore Pte Ltd. 2020
V. Bhateja et al. (eds.), *Intelligent Computing and Communication*,
Advances in Intelligent Systems and Computing 1034,
https://doi.org/10.1007/978-981-15-1084-7_7

switching network is analogous to tell-n-go one-way reservation protocol in which control packet bits are sent first to establish connection, then without waiting for the acknowledgment about connection establishment, a burst is sent. Burst is a unit of transmission in OBS [2].

In OBS, the control packets are transmitted electronically and then data bursts are sent optically. These control signals are processed electronically to set the optical light path for the data burst transmission. OBS provides more flexibility in bandwidth utilization than wavelength routing but it requires faster switching and control technology [2, 3]. Thus OBS provides bandwidth effectiveness and implementation simplicity. The data bursts are transmitted just after the offset time. The total time taken by the control packet for processing at every intermediate node in the network while routing is the offset time. The wavelength on a link utilized by the burst will be freed when the burst goes through the link, either naturally as per the reservation made or by using explicit release packet. Thus, effective utilization of bandwidth on a link can be done by the burst from different sources to different sink which allows statistical multiplexing and time sharing [4]. Several IP packets with the same destination address are grouped together at the ingress node and then are processed through the optical channel. The source and destination nodes can be a personal computer, mainframe, telephone, or any other communication device [4, 5]. There are many optimization algorithms which fail to give optimal solution for the real time problem having objective function which is noncontinuous, noisy, non-differentiable, nonlinear, multidimensional, etc. Differential evolution algorithm is an optimization algorithm which continuously tries to improve the performance of such problems. DE algorithm was invented by Storn and Price in 1997 [6]. DE is population-based algorithm that can be used for multidimensional real-valued problems. DE algorithm optimizes the solution by creating new population based on fitness function and by selecting the best set from the population [6, 7].

2 Related Works

The congestion control approach has a significant effect on network performance in terms of loss rate, average delay, and network utilization. This is necessary to enhance the performance of the network. Many methods have been carried for congestion control in OBS network. Barakat et al. [8] have presented an analytical model for estimation of minimum average latency and control plane loss reduction in OBS network. Wang [9] have proposed a modified TCP decoupling approach for the reduction of congestion by regulating the burst sending timing, Triay et al. [10] proposed an ant colony optimization (ACO) algorithm for the distributed routing and wavelength assignment (RWA) in the OBS network. But parameter tuning in ACO was a complicated task. Du et al. [11] presented a load balancing deflection routing method for congestion control in OBS network. Load balancing deflection perform better for less load but fails to give optimum solution in heavy load. Jin et al. [12] discussed a heuristic congestion control method by adjusting the burst interval.

This method was simple to implement but resulted in very high delay. Zhani et al. [13] proposed a feedback control approach to improve the quality of service in terms of burst loss using feedback control approach. A. Abid, et al. proposed flow control and reservation bits (FCRB) of the staged reservation scheme for the congestion control in OBS and can be used effectively. Argos et al. [14] proposed multipath routing schemes for congestion control in OBS network which provided lower burst loss probability (BLP) and improvement in the load balancing.

None of the author have emphasized on multi objective optimization technique for OBS. This paper emphasizes on multi objective optimization, which is designed as integer linear programming problem to minimize the congested links and number of hops traversed by the burst in the network by using differential evolution optimization algorithm. Rest of the paper is organized as follows: the second section describes the system methodology for congestion control in OBS network using DE algorithm. In the third section, experimental results and their analysis are discussed. The final section discusses about the conclusion followed by references.

3 System Methodology

Consider a network topology $G(N, L)$ where N is the set of nodes in the network graph and L is the set of link joining two nodes. Let $T(Si, Dj)$ be the demand vector which is the set of source and destination node pair which are ready for the transmission.

For DE, mutation factor (*MFact*), Population size (*PSize*), and crossover probability (*CrPr*) are kept constant throughout the simulation of the program. *MFact* and *CrPr* are considered in the range of [0, 1].

(a) **Initial Population Generation**

First step in DE algorithm is initial population generation. In this work, initial population is generated by using individual encoding method. In this, for the demand set T for every pair of source (*Si*) and destination node (*Dj*) shortest path is calculated using Dijkstra's shortest path algorithm which calculates the number of hops to the egress node and is used later to compare with DE. To find the minimum distance one of the link is disabled at a time and new shortest path is calculated if it exists [4].

(b) **Fitness Calculation**

Each individual of the population which is selected for the fitness calculation is called as target vector. For every target vector, total number of hops and total number of congested hops are calculated. The fitness function to be minimized for the individual target vector (Fitness(indv)) is calculated using Eq. (1).

$$\text{Fitness(indv)} = w_1 * \frac{\text{Cong(indv)}}{|T|} + w_2 * \frac{\text{total\_hop(indv)}}{(V(G) - 1) * |T|} \tag{1}$$

Here, cong(indv) is the number of congested link present in the target vector and total_hop(indv) represents total number of hops present in the target vector. The total number of nodes in the network is referred as $N(G)$ and the total numbers of source-destination node pairs in the demand vector are $|T|$.

From the sample space (population) generated from the individual encoding, new generations of population are selected for the mutation and crossover operation. The new vector having lesser fitness value is added in new population.

(c) Mutation

For every target vector bi in population, mutant children are created using Eq. (2). For the implementation mutation factor $MFact$ is selected as 0.2.

$$\text{Mutant}_{childi} = r1 + Mfact * (r2 - r3) \tag{2}$$

where $i = 1, 2, 3, ...,$ popsize and r_1, r_2, r_3 are randomly selected from the population such that $r_1 \neq r_2 \neq r_3 \neq$ target vector [4].

(d) Crossover

To increase the diversity in the mutation process, crossover is used. The trail vector is created if a random number is less than crossover probability which has set to 0.5. The trial vector created using crossover is given by (3),

$$\text{trial}_{vectori} = \begin{cases} \text{Mutant\_Child}_i & \text{if } randT < CrPr \\ b_i & \text{Otherwise} \end{cases} \tag{3}$$

The infeasible solutions in mutation and crossover population generation are avoided using boundary constraints. If the new population goes beyond the boundary constraint then those are limited using Eq. (4).

$$\text{trial}_{vectori} = lz + \text{round}(u_Z - l_Z) * \text{rand}(i) \tag{4}$$

where the reference number available in the demand vector is given by u_Z and l_Z which are the upper and lower bound, respectively.

(e) Selection

Finally, if the fitness of trail vector is less than fitness of the target vector then trail vector is selected as new optimized population member.

4 Experimental Results

This section provides the analytical results to show better performance of DE algorithm. The work is compared with shortest path algorithm. All the simulations are carried in MATLAB. The networks considered for the analysis are shown in Fig. 1a–d.

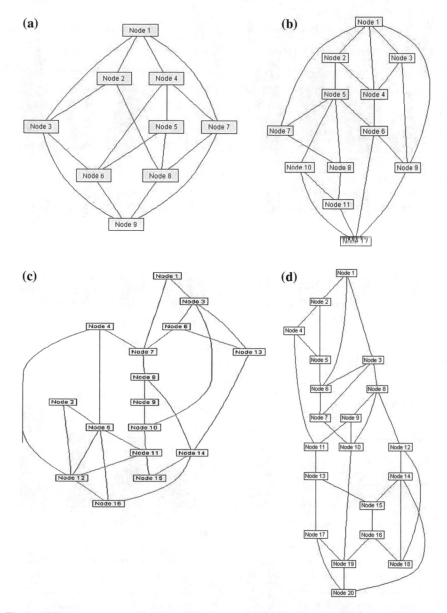

Fig. 1 Different network topologies **a** TORUS-9. **b** RANDOM-12. **c** NFSNET. **d** ARPANET

The results for total number of congested link in the network topology verses load are shown in following Fig. 2a–d. From the results it has been observed that though DE algorithm reduces the total congestion in the path, total number of hops may increase or remain the same.

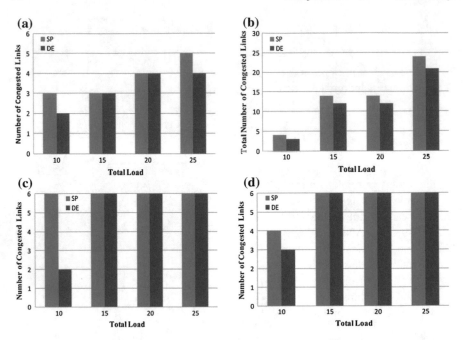

Fig. 2 Total number of congested links versus total load for **a** TORUS-9. **b** RANDOM-12. **c** NFSNET. **d** ARPANET network

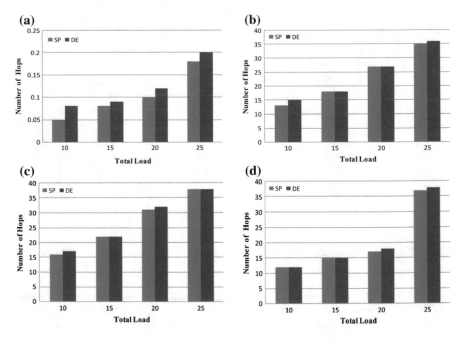

Fig. 3 Total hops versus load for **a** TORUS-9. **b** ANDOM-12. **c** NFSNET. **d** ARPANET

The results for total hop count verses load are shown in Fig. 3a–d. From the simulation result, it has been found that DE algorithm exhibits better results for the sparse network and for the network with higher traffic load. As TORUS-9 is a dense network, DE algorithm performance for TORUS-9 is poor compared to ARPANET, NSFNET, and Random-12, etc.

5 Conclusion

In this paper, we implemented the system for reduction of the total number of congested links and total number of hops using differential evolution algorithm. Experiment is carried out on the various network topologies like ARPANET, NSFNET, TORUS-9, and RANDOM-12 for a different load. The results are compared with the conventional shortest path algorithm on the basis of total hop count and congested hop count, and it has been observed that DE algorithm performs better than the SP algorithm. DE performs better for the sparse and highly loaded network than dense network. As a future work, the system can be extended to check the performance evaluation burst loss probability for multi-objective function using DE.

References

1. Jue, J.P., Vokkarane, V.M.: Optical Burst Switched Networks, 1st ed. Reading, Springer (2005)
2. Chen, Y., Qiao, C., Yu, X.: Optical burst switching: a new area in optical networking research. IEEE Netw. **18**(3), 16–23 (2004)
3. Xiong, Y., Vandenhoute, M., Cankaya, H.C.: Control architecture in optical burst-switched WDM Networks. IEEE J. Sel. Areas Commun. **18**(10), 1838–1851 (2000)
4. Barpanda, R.S., Turuk, A.K., Sahoo, B.: A review of optical burst switching in wavelength division multiplexing Networks. In: International Conference on Information Technology (ICIT), pp. 142–147, Bhubaneswar, India, December 2014
5. Rosberg, Z., Le Vu, H., Zukerman, M., White, J.: Performance analyses of optical burst-switching Networks. IEEE J. Sel. Areas Commun. **21**(7):1187–1197 (2003)
6. Storn, R., Price, K.: Differential evolution—a simple and efficient heuristic for global optimization over continuous spaces. J Glob. Optim. **11**, 341–359 (1997)
7. Mohamed, A.W., Sabry, H.Z.: Constrained optimization based on modified differential evolution algorithm. Inform. Sci. **194**, 171–208 (2012)
8. Barakat, N., Darcie, T.E.: Control-plane congestion and provisioning guidelines for OBS networks. In: IEEE INFOCOM Workshops 2008, Phoenix, AZ, pp. 1–6 (2008)
9. Wang, S.Y.: Using TCP congestion control to improve the performances of optical burst switched networks. In: 2003 IEEE International Conference on Communications, ICC '03, Anchorage, AK, vol. 2, pp. 1438–1442 (2003)
10. Triay, J., Cervello-Pastor, C.: An ant-based algorithm for distributed RWA in optical burst switching. In: 11th International Conference on Transparent Optical Networks, Azores, pp. 1–4 (2009)
11. Du, Y., Zhu, C., Zheng, X., Guo, Y., Zhang, H.: A novel load balancing deflection routing strategy in optical burst Switching Networks. In: OFC/NFOEC 2007-2007 Conference on Optical Fiber Communication and the National Fiber Optic Engineers Conference, Anaheim, CA, pp. 1–3 (2007)

12. Jin, M., Yang, O.W.W.: A traffic shaping heuristic for congestion control in optical burst-switched Networks. In: Proceedings 2006 31st IEEE Conference on Local Computer Networks, Tampa, FL, pp. 529–530 (2006)
13. Zhani, M.F., Aly, W.H.F., Elbiaze, H.: On providing QoS in optical burst switched networks using feedback control. In: 2009 IEEE 34th Conference on Local Computer Networks, Zurich, pp. 309–312 (2009)
14. Abid, A., Abbou, F.M., Ewe, H.T.: On the congestion control in optical burst switching networks. In: IEEE International Conference on Telecommunications and Malaysia International Conference on Communications, Penang, pp. 299–302 (2007)
15. Argos, C.G., de Dios, O.G., Aracil, J.: Adaptive multi-path routing for OBS Networks. In: 9th International Conference on Transparent Optical Networks, Rome, pp. 299–302 (2007)
16. Jewett, J., Shrago, J., Yomtov, B.: Designing Optimal Voice Networks for Businesses, Government And Telephone Companies. Telephony Pub. Corp. (1980)

Locality Parameters for Privacy Preserving Protocol and Detection of Malicious Third-Party Auditors in Cloud Computing

Akheel Mohammed and D. Vasumathi

Abstract Cloud computing is an anticipated and inevitable solution for large storage capacities, the virtual presence of ownership and universal accessibility which extends the boundaries of a user today. These benefits also open up a challenge for being too exposed to welcoming attackers to steal potential information from those cloud environments. Irrespective of types of data, information stored on these cloud environments varies from databases to coding of software and raw and processed information of various organizations including governments. To ensure confidentiality and integrity of data input from various users, cloud service providers employ third-party auditors to verify the safety measures are intact and maintained. As the name predicts, an auditor may not possess a responsibility to protect the data after their intended verification is over. Third-Party Auditors have the function of confirming that the saved data has the same integrity as of original data, does not seek any information in near future to disturb the integrity and finally does not introduce any loopholes for new attacking attempts from external sources. This proposal intends to derive a mechanism to limit the resources provided to third-party auditors in the architecture. The location of data and users is protected in this approach along with locking of data segments to enhance the security of the architecture. Locality information of data storage space and the data information is protected from third-party service providers to enhance the privacy level. The simulation results compare the technique and prove that this method outperforms the previous techniques when addressing the sturdiness, communication complexity and time consumption concerns.

Keywords Public auditing · Location-based privacy preserving · Third-party auditor (TPA) · Cloud client (CC) · Cloud service provider (CSP) · Security and privacy of storage (SPS) · Panda public auditing (PPA)

A. Mohammed (✉)
Shadan Women's College of Engineering and Technology, Hyderabad, India
e-mail: alikhancrm@gmail.com

D. Vasumathi
Jawaharlal Nehru Technological University Hyderabad, Hyderabad, India
e-mail: rochan44@gmail.com

© Springer Nature Singapore Pte Ltd. 2020   67
V. Bhateja et al. (eds.), *Intelligent Computing and Communication*,
Advances in Intelligent Systems and Computing 1034,
https://doi.org/10.1007/978-981-15-1084-7_8

1 Introduction

Enhancing the technological services in storage and computation, many resources are now available to the common man for cheaper and faster rates. These services are preferred more than ever before and technologies are advancing the second we speak. Limitations of handheld devices with minimal storage capacity, processing speed, and computation complexity were answered in cloud computing. The concept of master and slave computing has evolved to a great new extent which was never imagined. New features [1] are added to existing cloud computing technologies every day by various cloud service providers (CSP). These resourced technologies have opened a universal access theory with endless benefits. The space offered by public or private data centres is virtually segmented, assigned and dedicated to users. The space is utilized for storage through request and automatic back up from registered devices. Cloud computing demands innovations for several automated processes like allocation of memory spaces, extension and reuse of resources, billing and protection plays a significant role in defining a standard for versatility [2, 3].

An Internet service is a background and fundamental component of cloud computing. This establishes and completes the connection between the user and remote data centres for data storage, processing and retrieval [4]. Based on the functionality of these cloud servers, the service is categorized into an environment that acts either as Software, Platform, or Infrastructure. Software as a Service is a component designed for simplicity to both sides of users being end users or administrators. The next category will encourage users to implement carefully designed software into the cloud environment. An Infrastructure defines how the components of the cloud, users, and data are supposed to function in a given virtual space [4, 5].

Despite these advantages, the openness of such a cloud is susceptible as the same leniency applies to attackers present in the same domain. The challenge comes when the data owners have the belief that their data is safe and sound in a very open sphere of influence. Protective measures are defined and imposed in spite of the growing attacking mechanism, and this proposal intends to give a promising solution for identifying malicious attacks from so-called third-party auditors. The solutions to these problems should be confined to consume less energy and the cost of a cloud service provider. Third-party auditors gained the confidence of data owners and cloud service providers as they charge a lesser cost delivering a trustworthy comment on existing security standards [6]. The rate of increasing users is exponential and thus imposes an additional challenge for a cloud service provider to audit and periodically report to its users.

The challenges in protecting the data in a cloud are categorized based on the following factors. Access to clients with potential information, the partiality of service to different clients, location of data and data owners, uniform service through a standard administration, technical support, retrieval, and overall maintenance are the factors which need to be balanced when security implementations are enforced.

2 Related Works

Third-Party Auditor is believed to be a common implementer of both cloud service providers and data owners where they test how the mechanism works and if the data are really secured. The question is whether the third-party auditor has the intention to steal potential information as they possess the credibility to access the same [7]. Provided with better benefits than auditing alone, bribed auditors may also lead to loss of information and degrading the integrity of data. The following section illustrates how a system is defined to check on such malicious attacks from trusted third-party auditors. This system gives relief to either benefactor of the cloud service and ensures that only trusted auditors are appointed to ensure the protective schemes. Adding to the time and computation complexity of overall processes, this mechanism should utilize minimal resources for defining a strategy for identifying dishonest third-party auditors. A protocol was defined for certifying security and conserving the privacy details of data owners when they use cloud storage [8]. The protocol has limited the accessibility of third-party auditors using RSA (Rivest, Shamir, and Adelman) encryption and preventing the exposure of owners' data but with details required for auditing. This technique enabled users to modify, add, and delete data whenever needed. Participants of this model were the data owners, cloud service provider, and third-party auditors. The data owners will compare the services of different cloud service providers and prefer one based on economic norms. The cloud service provider will offer the space required and provide mechanisms to protect the data. The data owners will decide to use the service as a platform or service as mentioned in previous sections. The benefits of one CSP will be justified by a third-party auditor but the decision depends on how trustworthy they are. The TPA cannot be blindly believed to be reliable and honest as they are no way in connection with either of the other participants. Both parties cannot confirm that their data is protected when a third-party auditor is involved. This led to the absolute necessity of an algorithm for securing the data, commencing the encryption through RSA.

(1) A public and private key is generated by the data owners for encrypting the data. The public key is shared with the CSP for outsourcing for the auditing process to complete. The audit process is initiated from the TPA side and sent to CSP for providing sufficient resources.
(2) The encrypted data is transferred to TPA when this request is initiated from Audit.
(3) The encrypted data will be verified by TPA once the public key is received from the data owners through the CSP.

Evaluation of this technique is done in two different analyses. One for estimating the time taken for the transfer of encrypted data between the cloud space and the third-party auditor is communication time. The next is the computation cost, which is consumed by TPA to audit the given data and confirm that information integral.

The next methodology implements the usage of a modern ciphertext [9] for encrypting the data before it is sent to TPA. This scheme does not require a copy of the data which removes redundancy cost and concentrates more on the storage of integrated data. Five steps are included in the procedure, namely data owners, CSP, TPA and algorithms for encryption and sharing. A function KeyGen is used by TPA and data owners to generate private and public keys like other encryption schemes. After verifying the data in cloud space, TPA generates metadata about verification done on data in cloud space using the function SigGen. CSP then generates the proof that the data stored is at the right place and time by the function GenProof, which is later used by TPA to verify and authenticate that data is integral by the function VerifyProof.

The public key is generated for sharing between the data owners, cloud service providers, and third-party owners once the owned data is shared. Whereas the private key is used to protect data that will not be revealed to third-party auditors. The cloud service provider will establish a communication medium between TPA and data owners to share the keys and data for verifying the correctness. The data requested by TPA will be verified and a Proof will be generated. This proof will be forwarded back to data owners through CSP. The verified proof should match with the key generated by the data owners. This confirms that the data is not disturbed by any factors internally and externally [9]. Performance is evaluated in terms of communication cost, computation cost and storage/retrieval cost. The aim of this study is to impose a stronger encryption standard with a lesser computational cost. Requests and responses are made shorter to reduce the communication cost.

The foundation of this privacy preservation commenced with a homomorphic linear authenticator scheme [2]. Original content is marked by an arbitrary masking technique that prevents the need for a local copy when the verification process is done by third-party auditors. The operations are dependent on the exchange of keys without the need of preparing a local copy for verification processes. Privacy is preserved with the same set of algorithms used in previously mentioned schemes [10].

The next scheme introduced the scheme of auditing the data in terms of batches without exposing all the available data. This is modified in our previous paper to lock certain modules based on the importance of data. The importance is defined by the data owners themselves. The method implements a bilinear map for the encryption process [11]. The background study explains the common terms used for encryption and transferring of data to and fro. KeyGen is used for producing two sets of keys for public and private usage. The public key will be transferred to third-party auditors who are authorized by CSPs. A signature is generated by cloud service providers before outsourcing the data files for auditing. After the auditing process, proof for integral data and outsourced files is compared for ensuring that the verification process is complete [12]. A challenge-based authentication process is also provided in the literature survey [13]. The schemes' complexity is constant even with the different approaches are implemented and investigated [14].

Rao et al. [15] introduced the method, namely, Social group optimization (SGO) for performing the job scheduling task. The main goal of this research method is to allocate the resources optimally with enhanced customer satisfaction rate and the reduced makespan time. Das et al. [16] adapted the pay as you go metric of cloud computing to provide the services to the customers in order to balance the cost and service quality. This method effectively achieves trade off balance between the cost and service quality by introducing the load balancing method.

3 Proposed Algorithm

The technique used in this proposal is dependent on the algorithm and its phases. Key exchange is regulated between the cloud service providers, data owners, and third-party auditors.

Algorithm I: Authentication of Keys The cloud service provider is asked to prepare a list of random numbers along with legitimate secret keys. Data owners will be asked to segregate the data based on the importance and save the credential information into the locked modules [4]. The testing process is initiated once these keys and random numbers are shared with the data owners. Using these secret keys, data is encrypted. Now the request for auditing is initiated from the TPA side. Unless the keys exist in both lists obtained from CSP and TPA, malicious activity is detected. The data owners will prevent the transfer of encrypted data to TPAs who exist without the right keys.

Initialization: (kk:known key; sk: Secret key; TPAT: Third-Party Auditor Test; TPATR: Third-Party Auditor Test Result; CSPR: Cloud Service Provider Result; DO: Data Owner; pn: prime number, rn: random number, and V: validation) Input: Dc
Input: (kk; sk;TPAT; rn; p)/* Other than Locked modules
Output: (TPATR; CSPR; V)
Select rn within range $1 < rn < p$
Compute Srn/* Set of random numbers
DO initiates TPAT
Declare TPAT = Srn
DO computes CSPR = Srn.K
SPres to CC: Srn.K
DO computes TPAT = (kk)rn
DO checks if TPAT = Srn.K = CSPR then
DO confirms successful V
else if CC determines the malicious activity of TPA
End if
End else if

Algorithm II: Trust Helmet The trust helmet is a common party that maintains the common keys to be shared within all the participants. The trust helmet will issue security checks over the keys and participants, making sure that keys are confined within the group. When the keys are shared with the TPA by the CSP, TOA also checks for the presence of those keys in Trust Helmet. If the keys are verified to be present, then the process continues. Similarly, the same process is also confirmed with Data owners. At any given time, the keys should be commonly present in all participants,

Initialization: (Srk: Set of Random keys to be shared; Sk: Hk: Hidden key; p: prime number; rk: Random key; TPA: Third-Party Auditor; and TH: Trust Helmet)
Input: (Srk; CSP; Sk; TH)
Output: (TPA; HK)
DO & CSP use rk
Set $1 < rk < p$ && TPA knows Srk && Srk belongs to T
CSP to TPA: HK = Sk + rk mod p
TPA examines SHK if Srk Ssk = Ssk + rk mod p then
Set SHK = Srk + rk
Else if Srk rk not equal to Srk + rk then
Set CSP to SHK
End if
End else-if
TPA to CS belongs to HK
CC computes HK - rk = Srk mod p if DO = HK then
Set TPA = DO
End if

The CSP will assign a trusted key to the Data owners for encryption following which the auditors will request for verifying the integrity. Now if the auditors will ensure that data owners possess the same set of keys as given by the cloud service providers or else, the presence of manipulated entries is present either in the auditing or cloud service providers. The following section implements these concepts into a model and examines the possible outcomes. Previous strategies are also compared to provide a detailed report on the advantages of this proposed system.

4 Results and Discussions

Conditions	Statistics
Number of Chassis Switches in L4	1980
Packet Size	1260 KB
Line cards at L4	1630

(continued)

(continued)

Conditions	Statistics
Ports at L4	72
Number of racks at L4	16
Number of Chassis Switches t L3	432
Line Cards at L3	164
Ports at L3	48
Number of racks at L3	128
Used virtual machines	1800
Number of Servers	64
Maximum number of Cloud Service Users	18000
Hosts in each rack	132
Each Host supports	16 processors
Memory with each processor	256 GB
Storage Memory	512 GB
Virtual Disk Memory	430 GB
Bandwidth for L4	256 GB/sec
Bandwidth for L3	128 GB/sec
Bandwidth for L2	64 GB/sec
Bandwidth for L1	16 GB/sec
Queue delay	0.005 s
Burst time	0.0056 s
Idle time	0.0032 s

Green Cloud simulator is utilized for investigating the performance of privacy-preserving third-party auditing using locality parameters. This is an open-source simulation tool that can be implemented with C++ and results demonstrate the computational cost and energy consumption. Simulated results were obtained as the scenarios will portray the real-time occurrences. The conditions that were initialized are portrayed in the next table (Figs. 1, 2).

These investigations are tested with dishonesty rates of 0–5% of Third-Party Auditors. The proposed algorithms have shown that this method has outperformed the previous strategies and the results are shown in the following graphs for malicious attacks of 2%. The model was tested to express how the trustworthy third-party auditors are identified in an environment. The next set of results demonstrates the communication cost of this system in spite of a new scheme introduced. The first figure provides sufficient evidence that when the number of cloud users, i.e., data owners increase, the durability of locality-based privacy-preserving scheme remained constant at 100%, where the other models produced results of 99% approximately. This simulates the real-time environment where the actions and users are dynamic in nature. Subsequently, the system was tested for time taken for auditing actions by the third-party auditor. The proposed scheme also demonstrated reduced false positives

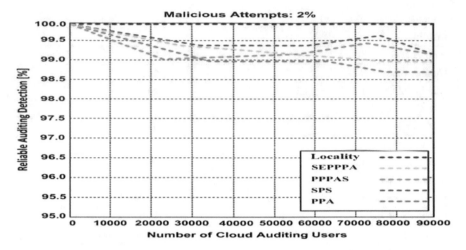

Fig. 1 Reliable auditing detection

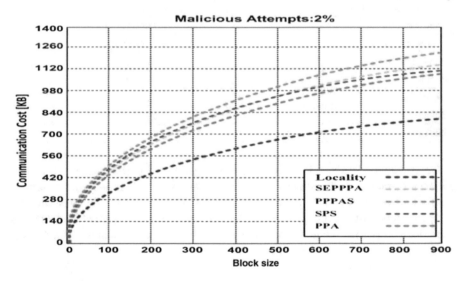

Fig. 2 Communication cost

than the other methods. Communication cost was also estimated with increasing number of blocks, ranging from 0 to 900. The simulation results show that the proposed model took 830 KB while the previous methods cost around 1100 KB. These evidences provide enough justification to prove that this model is outperforming the existing standards.

The approach has been an adept technique to identify the presence of malicious third-party auditors even before the encrypted data is shared among the participants in the domain. The Cloud Service Provider, Data Owners, and Third-Party Auditors can rely on this methodology for efficient and secured communication for auditing and privacy preserving. The proposed method has shown 100% reliability even in the presence of 90,000 users at the same time. The rate of false positives was addressed and this approach has been more efficient than existing methods. The third simulation proves that computation cost also turned up lesser for blocks of size 900 KB. The processing and communication time was considerably reduced after enhancing the security mechanism.

5 Conclusion

Organizations including governmental operations are deploying their services through cloud environments. The data owners believe that along with unlimited benefits of accessing a cloud environment, security will also be achieved. Yet the system is susceptible to a number of cases. This proposal introduced a scheme for authenticating the third-party auditor to ensure that data is secured when verification is one. The CSPs rather than being an economical need to be more secure in this virtual world with no boundaries. Having these objectives in mind, computational and processing overhead cannot be compromised when security is added. This proposal intends to address all these defects in a mechanism to authenticate the third-party auditors based on their locations. The method was to signify that auditors are trusted users who participate just to verify the integrity of data stored. Schemes were introduced to regulate the exchange of keys between the typical participants. An external trust helmet was maintained and referred to whenever there is a need to transfer data in an encrypted form. Instead of passing data, the keys were shared and compared before the original data is at stake. Results from simulation have shown promising results at an affordable computation and processing time. In the future, tracing back to the location where the system failed to preserve the privacy of data owners is to be derived with betterments to this scheme.

References

1. Razaque, A., Rizvi, S.S.: Privacy-preserving model: a new scheme for auditing cloud stakeholders. J. Cloud Comput. 6(1), 7 (2017)
2. Wang, C., Chow, S.S.M., Wang, Q., Ren, K., Lou, W.: Privacy-preserving public auditing for secure cloud storage. IEEE Trans. Comput. 62(2), 362–375 (2013)
3. Razaque, Abdul, Rizvi, Syed S.: Triangular data privacy-preserving model for authenticating all key stakeholders in a cloud environment. Comput. Secur. 62, 328–347 (2016)

4. Mohammed, A., Vasumathi, D.: Restrictive ambiguity and add-on architecture prototype for privacy preservation in cloud auditing. Int. J. Comput. Theor. Nanosci. (American Scientific Publishers, SCOPUS)
5. Das, M.S., Govardhan, A., Lakshmi, D.V.: Cost minimization through load balancing and effective resource utilization in cloud-based web services. Int. J. Nat. Comput. Res. (IJNCR) 8(2), 51–74 (2019)
6. Lee, W.-B., Chang, C.-C.: Efficient group signature scheme based on the discrete logarithm. In: IEE Proc. Comput. Digit. Tech. 145(1), 15–18 (1998)
7. Monika, V., Parwekar, P.: Survey on cloud data storage security techniques. Indian J. Res. Pharm. Biotechnol. 2321–5674 ISSN: 2320-3471 (Online) (2015)
8. Wei, L., Zhu, H., Cao, Z., Dong, X., Jia, W., Chen, Y., Vasilakos, A.V.: Security and privacy for storage and computation in cloud computing. Inform. Sci. 258, 371–386 (2014)
9. Hussien, Z.A., et al.: Public auditing for secure data storage in cloud through a third party auditor using modern ciphertext. In: 2015 11th International Conference on Information Assurance and Security (IAS). IEEE (2015)
10. Wang, Boyang, Li, Baochun, Li, Hui: Panda: public auditing for shared data with efficient user revocation in the cloud. IEEE Trans. Serv. Comput. 8(1), 92–106 (2015)
11. Rizvi, S., Razaque, A., Cover, K.: Third-party auditor (TPA): a potential solution for securing a cloud environment. In: 2015 IEEE 2nd International Conference on Cyber Security and Cloud Computing (CSCloud), pp. 31–36. IEEE (2015)
12. Worku, S.G., Xu, C, Zhao, J., He, X.: Secure and efficient privacy-preserving public auditing scheme for cloud storage. Comput. Electr. Eng. 40(5), 1703–1713 (2014)
13. Moghaddam, F.F., Karimi, O., Alrashdan, M.T: A comparative study of applying real-time encryption in cloud computing environments. In: 2013 IEEE 2nd International Conference on Cloud Networking (CloudNet), pp. 185–189. IEEE (2013)
14. Yang, K., Jia, X.: An efficient and secure dynamic auditing protocol for data storage in cloud computing. IEEE Trans. Parallel Distrib. Syst. 24(9), 1717–1726 (2013)
15. Parwekar, P., Kumar, P., Saxena, M., Saxena, S.: Public auditing: cloud data storage. In: 2014 5th International Conference on Confluence the Next Generation Information Technology Summit (Confluence), 169–173. https://doi.org/10.1109/confluence.2014.6949366publication
16. Rao, K.T.: Client-awareness resource allotment and job scheduling in heterogeneous cloud by using social group optimization. Int. J. Nat. Comput. Res. (IJNCR) 7(1), 15–31 (2018)

Analysis of NaCl Electrolyte Based SOI-ISFET pH Sensor

Narendra Yadava and R. K. Chauhan

Abstract In this paper, the analysis on the characteristic of the Silicon-on-Insulator Ion-Sensitive Field Effect Transistor (SOI-ISFET) pH Sensor has been carried out. NaCl is used as an electrolyte to sense the change in pH value. The analysis includes DC analysis which is modeled using Poisson/Boltzmann equations and frequency-dependent AC analysis which includes Poisson/Nernst/Planck–Poisson/Drift/Drift equations together with site binding charge model equations at the electrolyte and insulator interfaces. The simulation is carried out using ENBIOS-2D Lab tool provided by nanohub group. From the results, it is observed that with an increase in the pH value, the slope of the transfer curve decreases and the threshold voltage of the device increases. Also, the frequency-dependent performance of the device at pH value of 7 is analyzed and it is observed that the device becomes less sensitive at higher frequencies.

Keywords SOI-MOSFET · pH sensor · Electrolytes · ISFET · Poisson equation · Nernst equation

1 Introduction

As per the growing need for portable electronic gadgets, MOS technologies are pushed into deep sub-micrometers range and by today the technology reached in the tens of the nanometer range [1]. Also, it is well known that with the diminishing of the MOS technology, several small geometry effects arises such as gate controllability decreases, leakage current increases, etc. [2]. To overcome these adverse effects, researchers are making very great effort and even various designs already have been reported [3–5]. Out of the several proposed designs, Silicon on Insulator (SOI) technology is found to be a very promising technology to overcome the adverse effect

N. Yadava (✉) · R. K. Chauhan
Department of Electronics and Communication Engineering, MMMUT, Gorakhpur, India
e-mail: narendrayadava5@gmail.com

R. K. Chauhan
e-mail: rkchauhan27@gmail.com

© Springer Nature Singapore Pte Ltd. 2020
V. Bhateja et al. (eds.), *Intelligent Computing and Communication*,
Advances in Intelligent Systems and Computing 1034,
https://doi.org/10.1007/978-981-15-1084-7_9

of the small geometry devices [2]. Also, due to cost-effective and simple fabrication process, SOI technology is widely used.

In the field of pH measurement technique until now, several electrochemical and non-electrochemical methods are in practice. However, in the year 1970 Bergveld [6] has proposed an idea to sense the pH value using a FET-based sensor called Ion-Sensitive FET (ISFET) sensor. The main purpose of introducing FET-based sensor is due to some of its advantageous properties like portability, low cost, high speed, low noise, low power dissipation, high sensitivity, etc.

In the field of ISFET technology until now several designs have been proposed by the researcher [7, 8]. However, due to several advantages of SOI technology which are stated above made it a leading candidate to be used in ISFET sensors.

In this paper, the characteristics of the SOI-based ISFET sensor is analyzed, in which Na^+Cl^- is used as an electrolyte to sense the change in the pH value. Here both DC, as well as AC performances, have been investigated in which DC analysis is modeled using Poisson/Boltzmann equations while AC analysis is modelled using Poisson/Nernst/Planck–Poisson/Drift/Diffusion equations together with site binding charge model equations at the electrolyte and insulator interface.

All the analysis has been carried out using ENBIOS-2D Lab tool provided by nanohub group [9].

2 ISFET Overview

The detailed overview of the ISFET architecture over which DC and AC analysis have been carried out is explained below.

The structure represented here is a two-dimensional structure in which the sample under test is made in contact with the Floating gate electrode. The complete structure is in the hundreds of nanometers range which is comparatively smaller in comparison to the other electrochemical sensors. Also, the FET is an Si material based device in which Na^+Cl^- is used as an electrolyte to sense the pH value of the sample.

2.1 ISFET Geometry

The schematic representation of the SOI-based ISFET sensor is shown in Fig. 1. The structure is n-type SOI-based ISFET device in which the thickness of Buried oxide (BOX) layer is nearly 30 times to that of the gate oxide layer. The passivation is provided over both the Source as well as Drain regions in order to avoid any conduction due to ions present in the floating electrolyte material over the device. The complete structure is designed using ENBIOS 2D Lab tool.

Fig. 1 Schematic representation of SOI-based ISFET pH sensor

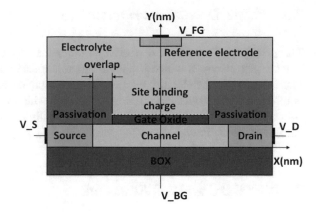

2.2 ISFET Parameters

The ISFET device is made of Si material in which the Source and Drain regions are n-type diffusion regions which are doped with a doping concentration of 1e19/cm^3. The channel doping is p-type having doping concentration of 1e16/cm^3. The gate oxide material plays a crucial role in sensing the change in the voltage of the electrolyte material. In this work, SiO$_2$ is used as a gate oxide material over which the site binding charge accumulates and results in a change in the drain current of the device. The passivation which is an insulator made up of SiO$_2$ is provided over both the Source as well as Drain diffusion regions in order to make them isolated from the site binding charges. The electrolyte material used in this work is Na$^+$Cl$^-$ with a concentration of 0.001 mol/L. All the analysis is carried out at an environmental temperature of 298.15 K with a constant drain to source DC voltage of 0.05 V having a substrate body electrode grounded (Table 1).

Table 1 SOI-based ISFET device structural parameters

S. No.	Parameters	Value (nm)
1	Channel length	200
2	Source/drain length	50
3	Passivation overlap	50
4	Semiconductor film thickness	20
5	Gate oxide thickness	3
6	BOX thickness	100
7	Passivation oxide thickness	400
8	Height of electrolyte above passivation	600
9	Width of the reference electrode	20

2.3   ISFET Operating Principle

The principle of operation of ISFET is the same as that of the MOSFET except that the gate electrode which is known as a reference gate electrode in the ISFET is made floating in the electrolyte material. The sample of having a certain pH value is made in contact with the reference electrode which leads to a change in the ion concentration of the electrolyte material and finally results in a change in the drain current of the device.

3   ISFET DC and AC Equations

The DC analysis is carried out using the *Poisson–Boltzmann* (PB) equation at equilibrium so that no ion quasi-potential gradient exists in the electrolyte and it can be written as [10]:

$$\nabla \cdot (\varepsilon \nabla V_0) = -\left(\rho_f + \sum_{m=1}^{N_{sp}} Z_m q n_m^\infty \exp\left(\frac{Z_m q}{k_B T}(\varnothing_{0m} - V_0)\right)\right)$$

where V_0 is the DC potential, ρ_f is the fixed charge density(volume or surface charge density which depends whether the model is 2D or 3D), N_{sp} is the number ion species, Z_m represents the sign of the valence and n_m is the mth ion concentration which is given by

$$n_m = n_m^\infty \exp\left(\frac{Z_m q}{k_B T}(\varnothing_m - V)\right)$$

Where $\varnothing_m$ denotes the quasi-potential and n_m^∞ is the ion concentration at $\varnothing_m = V$.

To perform AC analysis, the linearized Poisson/Nernst/Plank–Poisson/Drift/Diffusion equation is used. Assume all the physical quantities are time-harmonic functions of the form [10]:

$$\varnothing_m = \varnothing_{0m} + \Re[\widetilde{\varnothing}_m \exp(j\omega t)]$$

$$V = V_0 + \Re[\widetilde{V} \exp(j\omega t)]$$

We get

$$\nabla \cdot \left(\varepsilon \nabla \tilde{V}\right) + \sum_{m=1}^{N_{sp}} \frac{Z_m^2 q^2}{k_B T} n_{0m}\left(\widetilde{\varnothing}_m - \tilde{V}\right) = 0$$

$$Z_m q \mu_m \nabla \cdot \left(n_{0m} \left(\frac{Z_m q}{k_B T} \left(\widetilde{\emptyset}_m - \tilde{v} \right) \nabla \emptyset_{0m} + \nabla \widetilde{\emptyset}_m \right) \right) - j\omega n_{0m} \frac{Z_m q}{k_B T} \left(\widetilde{\emptyset}_m - \tilde{v} \right) = 0$$

and

$$\widetilde{n}_m = n_{0m} \frac{Z_m q}{k_B T} \left(\widetilde{\emptyset}_m - \tilde{v} \right)$$

4 Results and Discussion

The DC and AC performances of the ISFET device are analyzed investigated in which DC analysis is modeled using Poisson/Boltzmann equations while AC analysis is modeled using Poisson/Nernst/Plank–Poisson/Drift/Diffusion equations together with site binding charge model equations at the electrolyte and insulator interface. Figure 2 represents the current voltage relationship of floating gate at two different pH values. It is observed that when pH changes from 3 to 7 the drain current get shifted towards right side, i.e., the slope of the curve decreases. Also the threshold voltage of the device gets altered with alteration of the pH value. The shift in the slope as well as in the threshold voltage of the device is due to decrease in the ion concentration of the electrolyte which in results decreases the number of free carriers available and finally decreases the drain current with increased threshold voltage.

The DC potential distribution of the device shown in Fig. 3 represents that as the DC bias voltage decreases, the potential distribution shifts upward in the channel region which implies that the concentration of the electrolyte decreases.

Figure 4 shows that the concentration of electrons decreases and correspondingly the concentration of holes increases across the channel region of the device when the DC voltage increases from 0 to 1 V across the floating gate electrode.

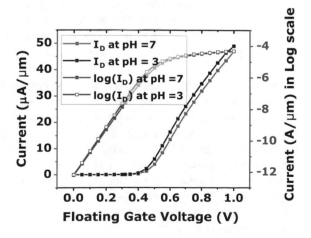

Fig. 2 Current voltage relation of floating gate electrode at pH value of 3 and 7

Fig. 3 DC potential
distribution at different DC
bias voltages

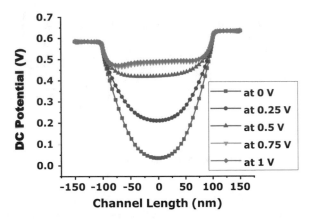

Fig. 4 Electrons and holes
concentration across the
channel length

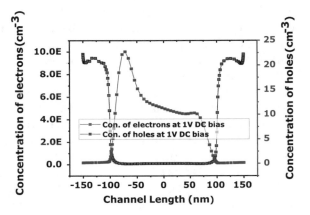

The frequency-dependent performance of the device at a pH value of 7 is analyzed. The magnitude and the phase responses of AC potential at frequency change from 1 kHz to 1 GHz are shown in Fig. 5a, b, respectively. In Fig. 5 as the frequency increases from 1 kHz to 1 GHz, the magnitude of AC potential increases from about 300 to 800 μV and in the phase response, the phase of the AC potential changes from 0° to nearly 1.9°.

Figure 6 represents the variation in the impedance (Z) of the device when the AC signal is applied. In Fig. 6a the real value of impedance is plotted with respect to the imaginary value of impedance in the log scale (both the axes). From the imaginary impedance value of 67 kΩ × μm to 0.3 GΩ × μm, the real impedance value increases gradually, i.e., from 2.1 MΩ × μm to 0.7 GΩ × μm but as the imaginary impedance value increases beyond 0.3 GΩ × μm the increase is exponential in nature and it reaches up to tera-ohms range. Figure 6b represents the magnitude and the phase response of the AC impedance. The magnitude of the AC impedance decreases from 800 MΩ to nearly 0.7 MΩ as the frequency increases from 1 to 100 kHz and beyond 100 kHz the decrease in the impedance becomes very slow. In the phase response of

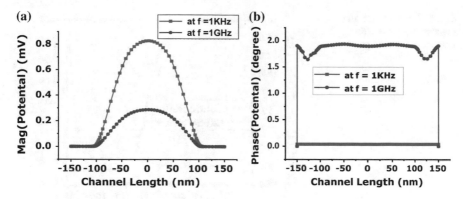

Fig. 5 a Magnitude plot of AC potential at different frequencies and **b** Phase plot of AC potential at different frequencies

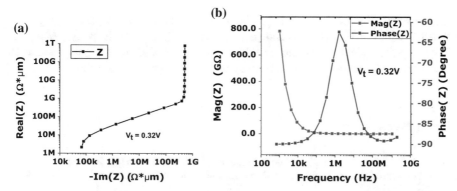

Fig. 6 a Plot of real impedance value with respect to the imaginary impedance value of the device. **b** Magnitude and phase response of the impedance

the AC impedance, as the frequency increase from 1 to 1.4 MHz the phase changes from -90Ω to $-62°$ and finally, it is around $-88°$ at 1 GHz.

The conductance and c plots are shown in Fig. 7. The conductance is almost constant up to 100 MHz and beyond this, it increases sharply. The value of capacitance is constant up to 100 kHz and then it makes a sharp transition up to 5 MHz and finally, it gets saturated.

5 Conclusion

The effect of pH change on the DC characteristics of the ISFET sensor is analyzed and it is observed that with an increase in pH value, the slope of the transfer curve decreases and the threshold voltage of the device increases. At pH value of 3, the

Fig. 7 Conductance and capacitance plots with respect to frequency

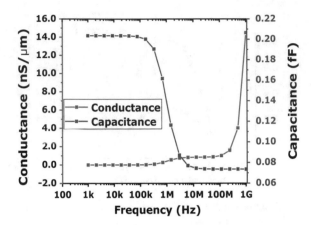

threshold voltage obtained is equal to 0.319 V while at pH value of 7 (the pH value of freshwater), the threshold voltage obtained is equal to 0.320 V. The I_{on}/I_{off} ratio is nearly equal to 10^5 which is enough to provide a fast output response. Also, the frequency-dependent performance of the device at pH value of 7 is analyzed and it is observed that the device becomes less sensitive at higher frequencies.

References

1. Carballo, J.A., Chan, W.T.J., Gargini, P.A., et al.: ITRS 2.0: toward a re-framing of the semi-conductor technology roadmap. In: 32nd IEEE International Conference on Computer Design, ICCD (2014)
2. Yan, R.H., Ourmazd, A., Lee, K.F.: Scaling the Si MOSFET: from bulk to SOI to bulk. IEEE Trans. Electron Devices **39**(7), 1704–1710 (1992)
3. Yamada, T., Nakajima, Y., Hanajiri, T., Sugano, T.: Suppression of drain-induced barrier lowering in silicon-on-insulator MOSFETs through source/drain engineering for low-operating-power system-on-chip applications. IEEE Trans. Electron Devices **60**, 260–267 (2013)
4. Mishra, V.K., Chauhan, R.K.: Performance analysis of fully-depleted ultra-thin-body (FD UTB SOI) MOSFET based CMOS inverter circuit for low power digital applications. In: Springer AISC Series, vol. 434, pp. 375–382 (2016)
5. Mishra, V.K., Chauhan, R.K.: Performance analysis of modified source and tunnel diode body contact based fully-depleted silicon-on-insulator MOSFET for low power digital applications. J. Electron. Optoelectron. **12**(1), 59–66 (2017). America Scientific Publisher
6. Bergveld, P.: Development of ion-sensitive solid-state device for neurophysiological measurements. IEEE Trans. Biomed. Eng. **17**(1), 70–71 (1970)
7. Kaisti, M., Zhang, Q., Prabhu, A., Lehmusvuori, A., Rahman, A., Levon, K.: An ion-sensitive floating gate FET model: operating principles and electrofluidic gating. IEEE Trans. Electron Devices **62**(8), 2628–2635 (2015)
8. Abdolkader, T.M.: A numerical simulation tool for nanoscale ion-sensitive field-effect transistors. Int. J. Numer. Model. Eletron. Netw. Devices Fields **29**(6), 1118–1128 (2016)

9. Hoxha, A., Scarbolo, P., Cossettini, A., Pittino, F., Selmi, L.: ENBIOS-2D Lab. http://nanohub. org/resources/biolabisfet (https://doi.org/10.4231/d3v11vm7d)
10. Pittino, F., Selmi, L.: Use and comparative assessment of the CVFEM method for Poisson-Boltzmann and Poisson–Nernst–Planck three dimensional simulations of impedimetric nano-biosensors operated in the DC and AC small signal regimes. Comput. Methods Appl. Mech. Eng. **278**, 902–923 (2014)

Quantitative Modeling and Simulation for Stator Inter-turn Fault Detection in Industrial Machine

Amar Kumar Verma, P. Spandana, S. V. Padmanabhan and Sudha Radhika

Abstract This paper deals with quantitative modeling and simulation for stator inter-turn fault detection in industrial machine. Quantitative modeling and simulation for both healthy and faulty induction machine at different loading conditions are studied and validated with experimental results under same operating condition. Fault identification techniques are mainly categorized into signature extraction-based, model-based, and knowledge-based approach. Introduction of statistical feature extraction, feature selection, and fault classification using knowledge-based approach are intelligent methodology for stator inter-turn fault identification and diagnosis for industrial machine.

Keywords Fault detection and diagnosis (FDD) · Motor current signature analysis (MCSA) · Artificial intelligent · Condition monitoring (CM)

1 Introduction

Three-phase squirrel cage induction machine (SCIM) is widely used in industries due to their reliability, ease of construction, high efficiency, and high overloading capability. In spite of being highly efficient and reliable in operation, induction machine has to go through various breakdowns due to mechanical, electrical, and thermal stress as shown in Fig. 1a. According to electric power research institute (EPRI) and IEEE, industry applications society (IEEE-IAS) survey standard, short circuit in stator winding of SCIM accounts for 37% of overall faults as shown in Fig. 1b.

A. K. Verma (✉) · P. Spandana · S. V. Padmanabhan · S. Radhika
Department of Electrical and Electronics Engineering, BITS Pilani, Hyderabad Campus,
Hyderabad 500078, India
e-mail: amarverma710@gmail.com
URL: https://www.bits-pilani.ac.in/hyderabad/EEE/PhDs

S. Radhika
e-mail: sradhika@hyderabad.bits-pilani.ac.in
URL: https://www.bits-pilani.ac.in/hyderabad/sudharadhika/Profile

© Springer Nature Singapore Pte Ltd. 2020   87
V. Bhateja et al. (eds.), *Intelligent Computing and Communication*,
Advances in Intelligent Systems and Computing 1034,
https://doi.org/10.1007/978-981-15-1084-7_10

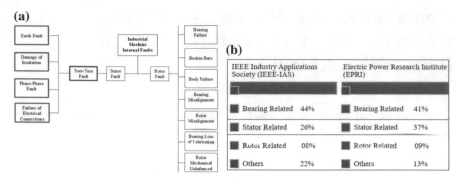

Fig. 1 **a** Fault classifications. **b** % component of machine failure

Turn-turn, phase-phase, phase-neutral, open circuit are an example of stator winding faults whereas turn-turn is considered as the major fault in induction machine and if not predicted at incipient stage, it can be propagated into above-mentioned faults. Thus, fault identification at incipient stage can prevent severe damage in due course of time to improve the machine operational reliability and cost of energy savings [1].

Earlier detection of growing stator inter-turn fault can prevent damage of stator windings and eventually increases lifetime by preventing machine unplanned downtime. To understand the behavior of stator winding faults; dynamic modeling of induction machine (IM) with and without fault have been derived and validated using simulated stator windings. The performance parameter of induction machine such as stator current, angular speed of rotor, and developed electromagnetic torque are analyzed and plotted, respectively.

Various fault identification and diagnosis techniques have been developed. The conventional fault detection and diagnosis uses motor current signature analysis [2], flux [3], artificial intelligence [4], vibration monitoring [5], temperature measurements, acoustic noise measurements, radio frequency (RF) emissions, infrared recognition, and NN-based techniques are commonly categorized under signature extraction-based, model-based, and knowledge-based approach [6]. Artificial intelligence tools with signal processing methods together have captured the attention of many researchers in last few years. Artificial intelligence-based approaches comprise fuzzy logic, support vector machine (SVM), and artificial neural networks among others [7]. The fault identification and diagnosis process based on multiple signatures are more reliable in identifying multistage fault and severity evaluation [3].

The paper is organized as follows: In Sect. 2 methodology includes experimental setup and pattern recognition problem. Dynamic modeling and simulation of squirrel cage induction machine (SCIM) with and without stator inter-turn fault have been studied in Sect. 3. As the next step, simulation results have been validated with experimental under same operating condition as explained in Sect. 4. The overall work studied in this paper is elaborately summarized in Sect. 5 followed by the future scope of work in the last Sect. 6.

2 Methodology

Fault identification techniques are mainly categorized into signature extraction, model, and knowledge-based approach. Motor current signature analysis (MCSA) of stator winding short-circuit fault is used as a pattern recognition problem which constituent of feature extraction, selection and fault classification. Initially by conventional methods, statistical features have been extracted and the best prominent ones can be selected by one of the methods using ANOVA, F-test, or Forward/Backward greedy algorithm. Finally, selected features have been used to classify fault using supervised pattern classification techniques, namely, artificial neural network (ANN). MCSA with and without stator inter-turn is used for fault identification as shown in Fig. 2.

2.1 Experimental Setup

Y-Connected induction machine with and without stator inter-turn fault along with brake drum arrangement has been set up as shown in Fig. 3a. Various machine performance parameters such as angular speed of the rotor, developed electromagnetic torque, slip, power factor, efficiency, etc. have been studied under loading condition before creating stator inter-turn fault. The current signal from both the machine has been sensed by transducer (hall effect) as shown in Fig. 3b and then it has been given to data acquisition and LabVIEW software for further data processing.

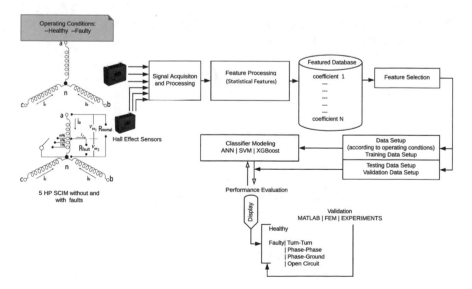

Fig. 2 Intelligent fault diagnosis methodology for stator winding faults

(a)

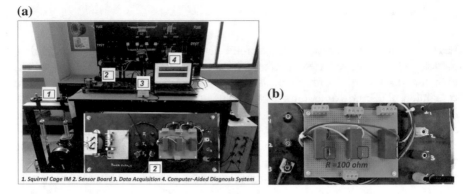

(b)

1. Squirrel Cage IM 2. Sensor Board 3. Data Acquisition 4. Computer-Aided Diagnosis System

Fig. 3 **a** Experimental setup. **b** Hall effect sensors

Table 1 Extracted statistical features under different machine loading conditions

Loading (%)	Mean	Median	Mode	Sample variance	Kurtosis	Skewness	Count
0	0.02218	0.02238	−0.0980	0.00721	−1.5029	−0.00039	10,000
25	0.00214	0.02235	0.26351	0.02962	−1.4811	0.00146	10,000
50	0.02245	0.02238	0.44047	0.09095	−1.4955	1.84E-05	10,000

2.2 Feature Extraction

Various time statistical features of acquired data signal under different loading condition are shown in Table 1 and some of the mathematical expression of statistical features are represented in Eqs. 1–6.

$$\text{mean} = \bar{x} = \frac{1}{N}\left(\sum_{i=1}^{N} x_i\right) \tag{1}$$

$$\text{median} = ((N+1)/2)\text{th value} \tag{2}$$

$$\text{standard deviation}(s_N) = \sqrt{\frac{1}{N}\sum_{i=1}^{N}(x_i - \bar{x})^2} \tag{3}$$

$$\text{Variance}(Var(X)) = \frac{1}{N}\sum_{i=1}^{n}(x_i - \mu)^2 \tag{4}$$

Table 2 ANN model description

Hidden layer	Activation function	No. of dense layer
1	relu	1500
2	sigmoid	500
3	softmax	5

$$\text{Kurtosis}(Kurt[X]) = \text{E}\left[\left(\frac{X - \mu}{\sigma}\right)^4\right] \tag{5}$$

$$\text{Skewness}(\gamma_1) = \text{E}\left[\left(\frac{X - \mu}{\sigma}\right)^3\right] \tag{6}$$

2.3 Feature Selection

Initially by the conventional method, various statistical features such as mean, median, standard deviation, variance, kurtosis, and skewness have been extracted and then the best prominent features have been selected using ANOVA for fault classification.

2.4 Fault Classification

As the next step, selected features have been used to classify fault using supervised pattern classification techniques, namely, artificial neural network (ANN). The model description of neural network for given set of input and output is listed in Table 2. This ANN model classifies 76% of test data correctly with given set of input parameters.

3 Modeling and Simulation

A dynamic model for both healthy and faulty induction machine has been studied and derived as represented in Eqs. 7–15.

3.1 Mathematical Modeling of Induction Machine

Healthy squirrel cage induction machine. The graphical representation of induction machine without stator inter-turn fault, i.e., healthy machine is shown in Fig. 4a and its dynamic modeling is represented in Eqs. 7–12.

Stator equation:

$$\frac{d\lambda_{qs}}{dt} = v_{qs} - r_s i_{qs} - \omega \lambda_{ds} \tag{7}$$

$$\frac{d\lambda_{ds}}{dt} = v_{ds} - r_s i_{ds} + \omega \lambda_{qs} \tag{8}$$

Rotor equation:

$$\frac{d\lambda_{qr}}{dt} = -r_r i_{qr} - (\omega - \omega_r)\lambda_{dr} \tag{9}$$

$$\frac{d\lambda_{dr}}{dt} = -r_r i_{dr} + (\omega - \omega_r)\lambda_{qr} \tag{10}$$

Speed equation:

$$\frac{d\omega_r}{dt} = \frac{T_e - B_m \omega_r - T_{load}}{J} \tag{11}$$

Torque equation:

$$T_e = \frac{3}{2}\frac{p}{2}(\lambda_{ds} i_{qs} - \lambda_{qs} i_{ds}) \tag{12}$$

Faulty squirrel cage induction machine. The graphical representation of induction machine with stator inter-turn fault is shown in Fig. 4b and its dynamic modeling is represented in Eqs. 13–15.

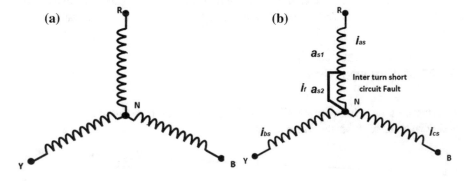

Fig. 4 **a** Healthy stator winding. **b** Stator inter-turn fault

Stator equation:

$$\frac{d\lambda_{qs}}{dt} = v_{qs} - r_s i_{qs} - \omega \lambda_{ds} + \frac{2}{3} \mu\, r_s i_f \cos\theta \tag{13}$$

$$\frac{d\lambda_{ds}}{dt} = v_{ds} - r_s i_{ds} + \omega \lambda_{qs} + \frac{2}{3} \mu\, r_s i_f \sin\theta \tag{14}$$

Torque equation:

$$T_e = \frac{3}{2}\frac{p}{2} L_m(i_{qs}i_{dr} - i_{ds}i_{qr}) + \frac{p}{2}\mu L_m i_f(i_{qr}\sin\theta - i_{dr}\cos\theta) \tag{15}$$

3.2　Simulation Results

Simulation setup for 400 V, 50 Hz, 5.4 HP, 1430 RPM healthy induction machine under different loading conditions is simulated as shown in Fig. 5 and corresponding stator current, angular speed of the rotor, and electromagnetic torque are shown in Fig. 6a, b, respectively, and also the same have been tabulated in Table 3. Induction machine is driven by universal bridge and load torque is applied for determining the performance of the SCIM under different loading conditions such as no load, 25, 50, 75%, and at full load.

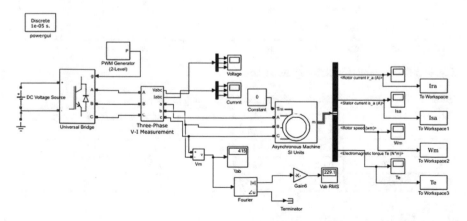

Fig. 5 Simulation setup

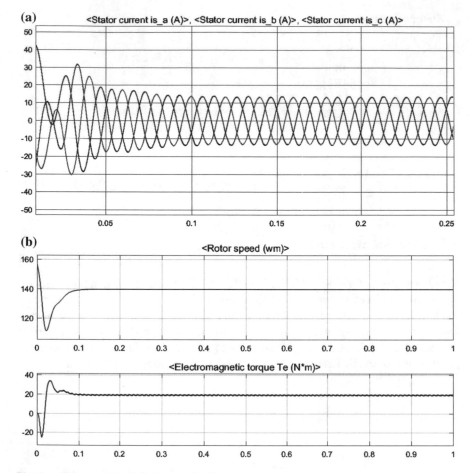

Fig. 6 a Stator currents. **b** Rotor speed and torque

Table 3 Simulation results under different machine loading condition

Load (%)	Slip (s)	Load torque (N-m)	Rotor speed (rad/s)	Electromagnetic torque (N-m)	Stator current (Amp)
50	0.065	12.0	146.9	15.0	4.20
25	0.031	6.00	152.1	9.00	2.82
0	0.000	0.00	157.0	3.00	1.34

4 Validation

In three-phase healthy SCIM, stator currents in all three phases are equal but have a different amplitude in case of faulty machine. Figure 7a shows the stator current for healthy SCIM under different loading conditions whereas Fig. 7b shows the variation

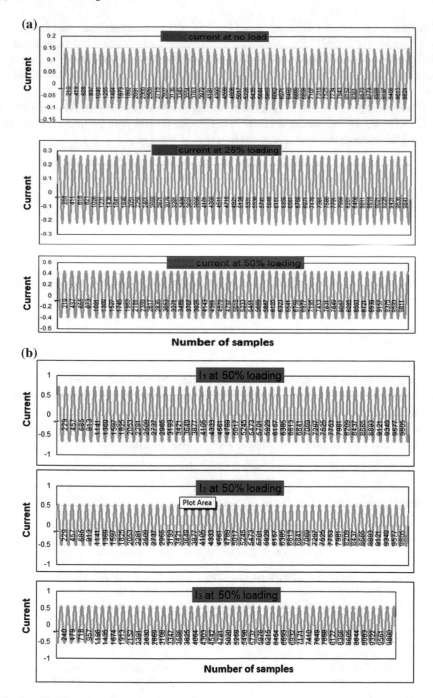

Fig. 7 a Healthy SCIM. **b** Faulty SCIM

of current amplitude for different severities of stator winding short-circuit fault such as 1, 5, and 10%. This work validates experimental result obtained from healthy SCIM under different loading conditions using Simulink environment. Further, this work will be extended to validate experimental result obtained from faulty SCIM as well.

Note: The conversion ratio (K_N) of sensor is 1:1000. The total number of sample is 10,000 over 10 kS/s sampling rate. A resistance of 100 ohm is connected across each of the sensors to give current out in terms of voltage.

5 Conclusion

Mathematical modeling of a 3-ϕ SCIM with and without stator inter-turn fault have been studied and similar experimental setup has been installed. The results obtained from the experimental setup under different loading conditions have been validated with results obtained from the mathematical model using Simulink environment.

6 Future Scope

The proposed work uses statistical features to classify machine operating condition, i.e., healthy or faulty; as the next step, time-frequency-based wavelet transform will be used to improve the resolution of current signal to enhance classification accuracy for a multistage fault such as stator inter-turn fault, phase-phase, phase-ground, and phase-neutral fault at incipient stage to prevent machine downtime loss.

References

1. Radhika, S., Sabareesh, G.R., Jagadanand, G., Sugumaran, V.: Precise wavelet for current signature in 3ϕ IM. Expert Syst. Appl. **37**, 450–455. https://doi.org/10.1016/j.eswa.2009.05.046 (2010)
2. Verma, A.K., Radhika, S., Padmanabhan, S.V.: Wavelet based fault detection and diagnosis using online MCSA of stator winding faults due to insulation failure in industrial induction machine. In: IEEE Recent Advances in Intelligent Computational Systems (RAICS), pp. 204–208. https://doi.org/10.1109/RAICS.2018.8635058 (2018)
3. Haroun, S., Seghir, A.N., Touati, S.: Multiple features extraction and selection for detection and classification of stator winding faults. IET Electr. Power Appl. **12**(3), 339–346. https://doi.org/10.1049/iet-epa.2017.0457 (2017)
4. Seera, M., Lim, C.P., Nahavandi, S., Loo, C.K.: Condition monitoring of induction motors: a review and an application of an ensemble of hybrid intelligent models. Expert Syst. Appl. **41**(10), 4891–4903. https://doi.org/10.1016/j.eswa.2014.02.028 (2014)
5. Alameh, K., Cité, N., Hoblos, G., Barakat, G.: Vibration-based fault diagnosis approach for permanent magnet synchronous motors. IFAC-PapersOnLine **48**(21), 1444–1450. https://doi.org/10.1016/j.ifacol.2015.09.728 (2015)

6. Devi, N.R., Sarma, D.V.S., Rao, P.V.R.: Detection of stator incipient faults and identification of faulty phase in three-phase induction motor-simulation and experimental verification. IET Electr. Power Appl. **9**(8), 540–548. https://doi.org/10.1049/iet-epa.2015.0024 (2015)
7. Seera, M., Lim, C.P., Nahavandi, S., Loo, C.K.: Condition monitoring of induction motors: a review and an application of an ensemble of hybrid intelligent models. Expert Syst. Appl. **41**(10), 4891–4903. https://doi.org/10.1016/j.eswa.2014.02.028 (2014)
8. Vamsi, I.V., Abhinav, N., Verma, A.K., Radhika, S.: Random forest based real time fault monitoring system for industries. In: IEEE 4th International Conference on Computing Communication and Automation (ICCCA), pp. 1–6. (2018, December)

CMOS-Based XOR Gate Design for Full Swing Output Voltage and Minimum Power Delay Product (PDP)

Mangaldeep Gupta, B. P. Pandey and R. K. Chauhan

Abstract In this paper, the performance of the different structures of XOR/XNOR circuit has been evaluated. The non-full swing output voltage level and power delay product (PDP) of the earlier structures has been improved in the proposed XOR/XNOR circuit. The performance of the proposed circuit has been investigated in terms of full swing output voltage, total power dissipation and computational delay using Pyxis Schematics Tool of Mentor Graphics. The Simulation is based on TSMC018 CMOS technology model file.

Keywords XOR/XNOR gate · VLSI · Output voltage swing · Transistor sizing · Power dissipation · Propagation delay

1 Introduction

Digital circuits comprise a large part of electronic circuits and every electronic circuit are made by fundamental gates such as AND, NAND, OR, NOR, XOR, XNOR, etc. In these circuits, there are various designing constraints such as reduction of power dissipation and area of such circuits while preserving their speed. But the primary designing constraint is the full swing output voltage of every node in the entire circuit which means that strong logic "1" voltage is equal to supply voltage (V_{DD}) and strong logic "0' voltage is equal to the ground for all possible input conditions. In such a case node voltage swing is from 0 to V_{DD} that is known as full swing output voltage. However, for some cases node voltage swing in circuit may be degraded due to the threshold voltage (V_T) drop problem [1–4] or due to sneak path between V_{DD} and

M. Gupta (✉) · B. P. Pandey · R. K. Chauhan
Department of Electronics and Communication Engineering, MMMUT, Gorakhpur 273010, UP, India
e-mail: mangaldeepgcet@gmail.com

B. P. Pandey
e-mail: panday.bramha@mmmut.ac.in

R. K. Chauhan
e-mail: rkchauhan27@gmail.com

© Springer Nature Singapore Pte Ltd. 2020
V. Bhateja et al. (eds.), *Intelligent Computing and Communication*,
Advances in Intelligent Systems and Computing 1034,
https://doi.org/10.1007/978-981-15-1084-7_11

ground in which all transistors between sneak path have a finite resistance and act as a voltage divider [5, 6], and finally results in non-full swing node voltage. The output and some internal nodes have a non-full swing problem which leads to a contradiction between logic "0" and "1" and other drawbacks like low driving and long propagation delay.

In Digital circuits like microprocessors, digital signal processor, digital communication devices, etc., have a large part of electronics modules. Due to scaling of CMOS technology, its on-chip integration increases. And also due to tradeoff between area, computational time and power dissipation, it is very challenging to meet the required design constraints. Hence, with the increasing demand for battery operating devices such as laptops, cell phone, and other electronics devices, power dissipation is a critical parameter, so the main objective to minimize power consumption and area while maintain the speed of systems. Power delay product (PDP) is an important parameter that is used to measuring performance and quality of CMOS circuits design [7, 8]. PDP can be explicating as the energy requirement of the circuit for its output voltage is switched from high to low and low to high. In complementary CMOS logic gates, power is dissipated in a pull up network during charging load capacitance from 0 to V_{DD}. Similarly, power is dissipated in pull-down network during discharging the load capacitance from V_{DD} to 0. The energy is nothing but heat dissipation in NMOS and PMOS transistor due to conduction of current during switching. Thus, for designing the point of view, PDP should be minimum as possible.

The performance of various digital applications such as microprocessors, digital signal processor, digital communication devices, etc. relates to the performance of the arithmetic unit such that adder, multiplier, and divider. Because of the fundamental role of addition in every arithmetic module, several efforts are created to enhance the effective adder circuits, example carry select adder, carry skip adder, and look ahead carry adder [8–10]. In various circuits such as adder, comparators, parity checker, Galois field modular arithmetic, compressors, etc., XOR and XNOR circuits play a vital role. Hence their behavior will be affect circuits' performance. So the main objective of this paper is to design XOR and XNOR circuits with minimum PDP and full swing output voltage.

1.1 Review of XOR/XNOR Gate

As we know that XOR/XNOR circuits is the main body of the adder circuit and it has also many applications in digital electronics circuits design. Various circuits for XOR/XNOR gate have been proposed earlier, in [6] XOR-XNOR circuit has been implemented simultaneously by double pass transistor (DPL) logic in which required 2-static inverter, 2-PMOS and 2-NMOS pass transistors, it is used for low power application but problem in this circuit is not getting strong logic voltage for all possible input condition. In recent [6, 9, 11], the simultaneous XOR/XNOR circuits are used widely in full adder structure, in simultaneous XOR/XNOR circuit used the concept of complementary pass transistor logic which has pass transistor network with inverse pass transistor network. In this structure, output is driven by NMOS transistors and two PMOS transistor are cross-coupled with the output of XOR and XNOR gate, the problem of this topology increases the delay and short circuits power dissipation due to feedback structure.

In [5], 7T XOR and XNOR circuits with full swing except single input logic combination (AB = 01) and minimum power delay product (PDP) have been proposed. Figure 1a shows that XOR circuit in which a combination of transistors (MN1, MN2, MP4, and MP5) is also XOR circuit but the problem in this circuit for input combination (AB = 00) is output voltage $|Vtp|$ because PMOS transistor pass a weak 0. Hence in this situation output swing is from V_{DD} to $|Vtp|$. To overcome this problem, added one more NMOS transistor (MN4) in which gate terminal is driven by input $\bar{A}$. by adding transistor (MN4), non-swing output problem has been solved but still adding one more problem i.e. sneak path for input combination (AB = 01), so that output node in Fig. 1a behave like a potential divider from V_{DD} to ground, in this situation transistors (MN1, MN2, MN3 and MP4) are in off state and output node voltage will be non-zero potential that is some finite voltage which is depend on ON resistance of NMOS transistor (MN4) and PMOS transistors (MP5) between input B and ground terminal. which is shown in Fig. 2. Hence non-swing output problem of this circuit is still present.

Similarly Fig. 1b show that XNOR circuit in which combination of transistors (MN1, MN2, MP1 and MP2) is also XNOR circuit but problem in this circuit for input combination (AB = 11) is output voltage ($V_{DD}-Vtn$) because NMOS transistor pass a weak 1, hence in this situation output swing is from 0 to ($V_{DD}-Vtn$). To overcome this problem added one more PMOS transistor (MP4) which gate terminal is driven by input $\bar{A}$ by adding transistor (MP4), non-swing output problem have been solved but still adding one more problem i.e. for input combination (AB = 01), output node in Fig. 1b behave like a potential divider from V_{DD} to ground in this

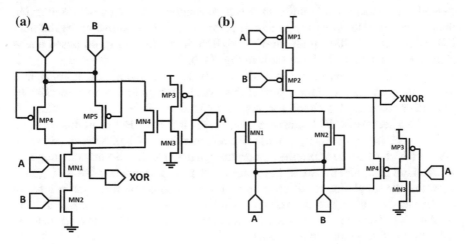

Fig. 1 Non-full Swing XOR/XNOR circuit [5]

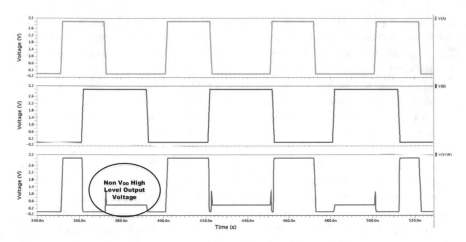

Fig. 2 Non full-swing performance of circuit Fig. 1a

situation transistors (MP1, MP2, MN3 and MN2) are in off state and output node voltage will be non-zero potential that is some finite voltage which depends on ON resistance of NMOS transistor (MN1) and PMOS transistor (MP4) between input B and ground terminal. Hence, output swing voltage is the primary designing constraint of any VLSI circuits.

2 Proposed XOR/XNOR Circuits

As we know the full-swing node voltage of any electronic circuits is a primary designing constraint. And minimum power consumption and area with preserving their speed are also necessary to design parameters. Hence to overcome the non-full swing output voltage problem of XOR/XNOR circuits in [5], proposed a new XOR/XNOR circuits which are show in Fig. 3a, b.

Figure 3a, show that proposed XOR circuits with full swing output voltage, in this circuit combination of transistors (MN4, MN5, MP5 and MP6) make an XNOR gate but problem in this circuit for input combination AB = 11, output voltage is ($V_{DD}-Vtn$) because NMOS transistor pass a weak logic '1'. Hence in this situation output swing is from 0 to ($V_{DD}-Vtn$). To overcome this problem, we can add static CMOS inverter with an appropriate noise margin high (NM_H). So that the proposed XOR circuit gives the strong output voltage for all possible input conditions. Figure 4 show that the simulation results of proposed XOR circuits, in this output waveform clearly show the strong output voltage for logic '1' and '0' at every possible input combination. The role of static CMOS inverter gives the strong output voltage, using the concept of noise margin in inverter circuits.

Similarly, Fig. 3b shows that the proposed XNOR circuit with full swing output voltage, in this circuit, which is a combination of transistors (MN9, MN8, MP9, and

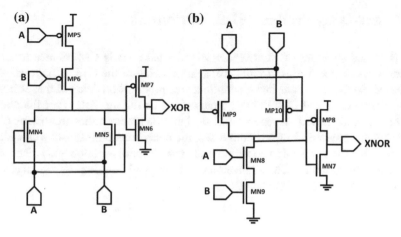

Fig. 3 A proposed full-swing XOR XNOR circuits

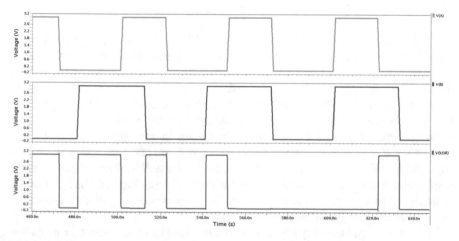

Fig. 4 A full-swing output waveform of circuit Fig. 3a

MP10) make an XOR circuit but problem in this circuit for input combination (AB = 00) is output voltage $|Vtp|$ because PMOS transistor pass a weak logic '0'. Hence in this situation output swing is from $|Vtp|$ to V_{DD}. To overcome this problem, we can add static inverter with an appropriate noise margin low (NM_L). Hence circuit in Fig. 3b becomes an XNOR gate with a strong output voltage for all possible input conditions.

3 Power-Delay Product (PDP) Optimization

The proposed structure of XOR/XNOR circuits has a static CMOS inverter at the output node. Hence in this circuits propagation delay of the critical path has been increased. So the concern is how to minimize the propagation delay of these circuits.

In general design of CMOS logic circuits base on timing (delay) specification is one of the fundamental issues in VLSI design which determines the performance of the complex system. In maximum cases, the delay constraint must be considered together with other designing constraints such as power dissipation and silicon area. The delay expression for CMOS inverter can be applied for complex digital circuits which are shown in Eqs. (1) and (2):

$$\frac{W_N}{L_N} \propto \frac{C_{\text{load}}}{\tau_{PHL} \cdot \mu_n \cdot C_{ox}} \text{ and } \frac{W_P}{L_P} \propto \frac{C_{\text{load}}}{\tau_{PLH} \cdot \mu_P \cdot C_{ox}} \tag{1}$$

$$\text{Delay} \propto \frac{C_{\text{load}}}{C_{\text{gate}} \text{ or } C_{\text{input}}} \tag{2}$$

To determine the channel dimensions (W_N, W_P) of NMOS and PMOS transistors which is satisfied the timing requirement. In the above formula, propagation delay (rise time and fall time) is directly proportional to the load capacitance of the circuit. The load capacitance (C_{load}) consists of two components which are intrinsic capacitances (parasitic capacitances or drain capacitances which is depend on transistor dimension) and extrinsic capacitances (fan-out capacitances and wiring or interconnect capacitances) which are independent on transistor dimensions. Load capacitance (C_{load}) mainly consist of extrinsic component of capacitances hence load capacitance will be independent of the transistors dimension. Finally, propagation delay (input to output signal delay during the high to low and low to high) is directly proportional to load capacitance and inversely proportional to gate capacitance or input capacitance of the corresponding circuit.

To minimize power delay product (PDP), we have to try to reduce propagation delay as well as power consumption. In CMOS circuits there are three major source of power dissipation which are switching power dissipation (power consumed by charging and discharging of node capacitances or load capacitance), short circuit power dissipation (power dissipation due to current flow from supply voltage to ground) and static power dissipation (due to flow of reverse leakages current, subthreshold current and tunneling current). Total power dissipation of any circuit mainly consists of total switching power dissipation of all node ($P_{\text{switching}} = \sum C_{\text{load}} \cdot V_{DD}^2 \cdot f$). where f is the transition frequency at particular node and C_{load} is node capacitances.

Consider the XOR circuit in Fig. 1a, in which static CMOS inverter with suitable noise margin high (MN_H) is used, so that critical delay of XOR circuit has been increased. To minimize the delay of the proposed circuit, the above methodology have been used. As we know that propagation delay is directly proportional to load capacitance at particular node and inversely proportional to gate capacitance or input capacitance of circuit. If capacitance at the internal node of XOR circuit act as a load capacitance which is nothing but gate capacitance of Static CMOS inverter, one way to reduce the propagation delay, decrease the gate capacitance or channel width (W) of NMOS (MN6) and PMOS (MP7) transistor but at the same time decrease the noise margin of CMOS inverter. Hence we have to used an alternate way to reduce delay which is increase the input Capacitance or width of gate input transistors (MP5 and MP6). Due to increasing the width of gate input transistors, increase the current driving capability causes increase in the power dissipation of the circuit. Hence propagation delay and dissipation power have a tradeoff between them.

4 Simulation and Performance Analysis of XOR Circuit

In this simulation analysis, all circuits are simulated using Pyxis schematics tool by Menton Graphics with TSMCUI8 CMOS model technology. Figure 2 is the transient or AC analysis of XOR circuit [5] which is show that non-swing performance of XOR circuit for input condition AB = 01 due to high level node voltage is not to equal V_{DD} because transistors MN$_4$ and MP$_5$ is ON between V_{DD} and ground, hence these transistors have a finite resistance act like a voltage divider. To overcome this non full swing voltage problem proposed a new XOR circuit in Fig. 3a, transient analysis of these circuits is shown in Fig. 4 which gives the strong output voltage for all input conditions. Table 1 shows that the Simulation Results of XOR circuits with optimum aspect ratio for two worst case (AB = 10 to 00) and (AB = 01 to 11) with an optimum aspect ratio of each transistor [5] corresponding aspect ratio PDP have been shown. And Table 2 show that the Simulation Results of proposed XOR circuit with an optimum aspect ratio of each transistor for minimum PDP for two worst case (AB = 10 to 00) and (AB = 01 to 11). According to Tables 1 and 2, the worst-case PDP is a minimum for the proposed circuit as compare to Table 1.

Table 1 Simulation results for worst case (AB = 10 to 00) and (AB = 01 to 11) with optimum aspect ratio of XOR circuit [5]

AB = 10 ⟶ 00									
W/L							Power (pW)	Delay (ps)	PDP (10^{-24} J)
N1	N2	N3	P3	N4	P4	P5			
2.61	5.07	2.00	2.00	2.00	5.29	3.79	74.32	40.380	3001.04
AB = 01 ⟶ 11									
W/L							Power (μW)	Delay (ps)	PDP (10^{-18} J)
N1	N2	N3	P3	N4	P4	P5			
2.61	5.07	2.00	2.00	2.00	5.29	3.79	**920.91**	63.578	**58549.6**

Table 2 Simulation results for worst case (AB = 10 to 00) and (AB = 01 to 11) with optimum aspect ratio of proposed XOR circuit

AB = 10 ⟶ 00								
W/L						Power (pW)	Delay (ps)	PDP (10^{-24} J)
P5	P6	N4	N5	N6	P7			
11.43	11.43	2.00	2.00	4.00	4.00	**122.1396**	67.8005	**8281.126**
AB = 01 ⟶ 11								
W/L						Power (pW)	Delay (ps)	PDP (10^{-24} J)
P5	P6	N4	N5	N6	P7			
11.43	11.43	2.86	2.86	4.00	4.00	88.1382	93.804	8267.716

5 Conclusion

In this paper, the performance of previously reported structures of XOR/XNOR circuits has been evaluated. The evaluation revealed that circuits have the output dependency on the ON resistance between the sneak path and leads to the non-full swing output voltage level. Finally proposed XOR/XNOR circuits with 6T that offer full swing output voltage. In simulation, the result shows that worst-case performance of the XOR circuit for input transitions AB = 01 to 11 and 10 to 00. Based on simulation find the optimum aspect ratio of transistors to minimize the PDP. Finally, it is concluded that the proposed XOR circuits are superior in terms of performance such that full swing output voltage and minimum PDP for all input combinations. And also show the optimum dimension of all transistors which are used in XOR circuit, for minimum PDP and our result have been compared with previous work. Future work includes designing of full adder using proposed circuits and comparison with previous work.

References

1. Bui, H.T., Wang, Y., Jiang, Y.: Design and analysis of low-power 10-transistor full adders using novel XOR-XNOR gates. IEEE Trans. Circuits Syst. II Analog Digit. Signal Process. **49**(1), 25–30 (2002)
2. Wairya, S., Nagaria, R.K., Tiwari, S.: New design methodologies for high-speed low-voltage 1 bit CMOS full adder circuits. Int. J. Comput. Technol. Appl. **2**(2), 190–198 (2011)
3. Chowdhury, S.R., Banerjee, A., Roy, A., Saha, H.: A high speed 8 transistor full adder design using novel 3 transistor XOR gates. Int. J. Electron. Circuits Syst. **2**(4), 217–223 (2008)
4. Wang, J.-M., Fang, S.-C., Feng, W.-S.: New efficient designs for XOR and XNOR functions on the transistor level. IEEE J. Solid-State Circuits **29**(7), 780–786 (1994)
5. Naseri, H., Timarchi, S.: Low-power and fast full adder by exploring new XOR and XNOR gates. in IEEE Trans. Very Large Scale Integr. (VLSI) Syst. **26**(8), 1481–1493 (2018)
6. Aguirre-Hernandez, M., Linares-Aranda, M.: CMOS full-adders for energy-efficient arithmetic applications. IEEE Trans. Very Large Scale Integr. (VLSI) Syst. **19**(4), 718–721 (2011)
7. Kim, N.S., et al.: Leakage current: Moore's law meets static power. Computer **36**(12), 68–75 (2003)

8. Weste, N.H.E., Harris, D.M., Design, C.M.O.S.V.L.S.I.: A Circuits and Systems Perspective, 4th edn. Addison-Wesley, Boston, MA, USA (2010)
9. Goel, S., Kumar, A., Bayoumi, M.: Design of robust, energy-efficient full adders for deep-submicrometer design using hybrid-CMOS logic style. IEEE Trans. Very Large Scale Integr. (VLSI) Syst. **14**(12), 1309–1321 (2006)
10. Timarchi, S., Navi, K.: Arithmetic circuits of redundant SUT-RNS. IEEE Trans. Instrum. Meas. **58**(9), 2959–2968 (2009)
11. Radhakrishnan, D.: Low-voltage low-power CMOS full adder. IEE Proc. Circuits Devices Syst. **148**(1), 19–24 (2001)

Potential Threat Detection from Industrial Accident Reports Using Text Mining

Sohan Ghosh, Arnab Mondal, Kritika Singh, J. Maiti and Pabitra Mitra

Abstract Analysis of industrial incident reports may aid in identifying potential danger and means to avoid them. We aim to automate the potential threat detection process as much as possible using text mining techniques such as multiclass classification and information extraction. While most of the related works concentrate mostly on either the cause of the incident or its effect and consider only a limited number of possible categories, we aim to predict both hazardous elements and accidents while not considering any fixed number of classes. Further, we also predict the potential target and the potential threat the incident poses.

Keywords Industrial accidents · Text mining · Support vector machines · Information extraction

1 Introduction

Accident refers to an unfortunate incident that occurs unexpectedly and unintentionally, typically resulting in damage or an injury. With advancements in technologies and increasing socio-technical interaction in the workplace, the number of incidents

S. Ghosh (✉) · A. Mondal
Heritage Institute of Technology, Kolkata, India
e-mail: sohanghosh29@gmail.com

A. Mondal
e-mail: arnab.mondal473@gmail.com

K. Singh · J. Maiti · P. Mitra
Indian Institute of Technology, Kharagpur, India
e-mail: kritika.swati@gmail.com

J. Maiti
e-mail: jhareswar.maiti@gmail.com

P. Mitra
e-mail: pabitra@gmail.com

© Springer Nature Singapore Pte Ltd. 2020
V. Bhateja et al. (eds.), *Intelligent Computing and Communication*,
Advances in Intelligent Systems and Computing 1034,
https://doi.org/10.1007/978-981-15-1084-7_12

109

reported at the workplace has seen a proportional increase. If the cause and effect of these accidents can be predicted, then they can be controlled [1].

Certain statistical analysis methods derived from Binomial distribution and Poisson distribution show how information interpreted from the summaries of these accidents can be helpful to predict future events [2]. In safety-critical systems, the analysis of significant accidents focuses on investigating the causal factors of system failures to prevent similar occurrences in the future or minimize their consequences [3]; hence accident analysis is extremely important.

Incident reports indicate the presence of a problem in the system which, if remained unresolved, can be the cause of a high- risk accident in the industry. The narrative obtained from the incident reporter remains the best available source of information explaining the cause of the accident [4]. Bulzacchelli et al. [5] studied on fatal injuries in the maintenance and service department of a particular US manufacturing industry by using narrative text from 1984–1997, the typical causes being electrocution, or struck by an object, or being caught in parts of equipment. Similarly, in another study, the implementation of the fuzzy Bayesian model to examine the accuracy of a data mining algorithm to classify the injury narratives was done [6].

Classification of narratives into the Bureau of Labor Statistics occupational injury and illness categories utilizing four machine learning algorithms has been done in [7]. Chang et al. [8] used classification and Regression Trees (CART) and the Negative Binomial Regression model for analyzing traffic accident behavior. Similar algorithms have also been used in this domain in another study [9]. Here, an approach has been presented for incident risk factor identification and analysis using data from the Aviation Administration.

Some of the ever-present challenges in extracting essential information from the voluminous unstructured text are technical data summarization, complaint extraction and analysis, suggestions, feedback, and failure analysis. Some approaches to this study are presented in [10–12]. Atkinson-Abutridy et al. [13] proposed a report on the information extraction approach involving a genetic algorithm along with the use of semantic and generic heuristic rules to optimize the extraction process. However, the use of semantic and rhetorical information is limited in the information extraction process. Han et al. proposed SVM classification based method for extracting metadata from the header of research papers [14].

One common agreement among these studies is the importance of the narrative in the incident report and the need for more research to explore the unstructured raw data for better identification of incident/accident causes. The information about "the initiating mechanism, causing a hazardous element to actuate into an accident causing a threat to target" is often implicit in these incident narratives.

In this article, we propose an integrated information extraction system that predicts not only the hazardous element but also the potential target and potential threat associated with the accident, using text mining algorithms. Further, we attempt to predict the correct specific incident/accident using information extraction. While finding out the hazardous elements and incident/accident we do not consider any fixed number of classes.

The rest of the paper is organized as follows: The methodology adopted is explained in Sect. 2; a case study including results and discussions is provided in Sect. 3 and conclusion of the work is given in Sect. 4.

2 Overview of Methodology

The overall methodology adopted has been demonstrated in Fig. 1. Before applying text mining techniques to the incident narrative, it was preprocessed to obtain term-document matrix (TDM) which then served as input to different text mining algorithms (discussed later) to identify key components of accident path: hazardous element (HE), accident, potential target, and potential threat. The specific hazardous element is classified using a two-level classification technique involving linear SVM and Word2Vec/WordNet/keyword matching. Next, information extraction was used for identifying accidents caused by HE with the narrative and HE as input to the algorithm. Using an accident, TDM and HE, we then extracted the Potential Target using linear SVM. Finally, potential target and accident were used to predict potential threats to target after the accident by applying linear SVM.

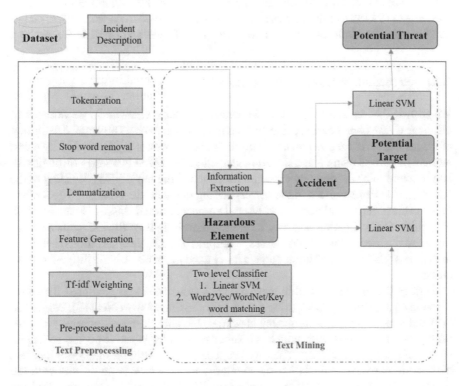

Fig. 1 Block diagram of the overall methodology

2.1 Data Preprocessing

The raw input in the form of incident description was first tokenized using Python NLTK (Natural Language Tool Kit) library [15] following which stop words were removed using the "English" stopwords list. After that lemmatization was performed. Term frequency-inverse document frequency (tf-idf) weighting was then applied to extract the words as features that are comparatively more important than the others, thus forming the TDM.

2.2 Text Mining

Text Mining extracts meaningful information from the text, and thus makes the information present in the accident reports available for various statistical and machine learning algorithms. Hence, text mining served as a feasible solution to extract the information about HEs involved and potential threat an accident may cause, from existing manually labeled data, to ease the process of analysis.

Text Classification. SVM is a supervised machine learning algorithm and serves the purposes of both classification and regression analysis [16]. In this study, we have used linear SVM as a multiclass classifier as it was observed to perform better on our dataset than other well-known classifiers like Naïve Bayes and Logistic Regression.

2.2.1 Prediction of Hazardous Elements

We treated this as a multiclass classification problem, where a large number of different classes are possible, and this number is not fixed. The brief description of the incident is treated as the data to be classified. However, trying to classify a vast number of classes using a classification algorithm like SVM is a fairly difficult task and is bound to be detrimental to the performance of the model. So according to the hazardous elements present in the corresponding industry, we propose to generalize the possible hazardous elements into a small number of broad classes. So the first task is to solve a classification problem where the descriptions of the incidents serve as input and the broad generalized classes as outputs. The Linear SVM model available in Python's "Scikit-Learn" library does a fairly good job in classifying the generalized classes.

Further, the subclasses within these broad classes need to be predicted. To do so, we adopted a hybrid approach using a model comprising of WordNet [17], Word2Vec [18] and simple keyword/ frequency checking. We chose a representative word for each generalized class with which the subclasses under it are expected to have a strong similarity in meaning. Then we propose to scan through the entire incident description narrative, calculating the similarity of each word one by one with the representative word using either WordNet or Word2Vec. In this way, we find the word which is most similar in meaning to the representative word. We argue that this word, in most cases is the subclass or is a part of the required subclass. For example, one can pick "chemical" as the representative word for the broad class "Chemical" or "metal" as the representative word for the broad class "Hot Metal and Steel". Now subclasses under "Chemical" category can include chemicals such as "Chlorine", "Sulfur", etc. If the narrative contains these terms, then they are bound to show higher similarity with the word "chemical" than other terms in the narrative which are not actual chemicals. To address situations where the word found using this method was not the correct subclass, we propose to consider a threshold similarity value such that if the word with maximum similarity shows lower similarity value than this threshold we simply output the broader class and do not predict the subclass for that particular incident narrative. Specific subclassification in such cases can be done by a human manually.

We now need to judge whether to use WordNet or Word2Vec for this purpose, as both of these models can evaluate the similarity between two words. It is observed that out of the broader classes some show better results using WordNet while the remaining using Word2Vec. So we propose a hybrid approach to improve the performance altogether, where we select either WordNet or Word2Vec simply on the basis of which performs better for the selected representative word corresponding to the broad class.

For certain generalized classes, it is observed that no such representative word can be suitably picked to find a good enough similarity with the subclasses; however, the total number of subclasses is limited and a list of those generally found in the industry can be prepared. In such cases, we do not follow the similarity-based approach and, a keyword and frequency checking approach can be followed to match the tokens present in the incident description with the prepared list to predict the subclasses. For example, for the generalized class "Gas", a list of commonly present harmful gases in the industry are prepared (the list can contain gases like CO, BF, Oxygen, etc.). Then the tokens in the incident narrative are matched with the already prepared list. The frequency of matched terms is then checked, with the term having the highest frequency predicted as the subclass.

Algorithm 1. Hazardous element prediction

Variables used :
broad_class: the predicted generalized class of hazardous element
t1: Threshold for instances under the broad class
max_similar_word: the word in the incident description with the maximum similarity value
 t2: Threshold similarity value
Knowledge_Base: A list of hazardous elements usually found in the industry

---

Functions used:
Classifier(x): Uses Linear SVM to predict the generalized classes
Find_Representative(x): Returns TRUE if it is possible to select a suitable representative word for the generalized class
Find_Most_Similar(x,y): Finds the word in text y with the highest similarity to the representative word of generalized class x
Keyword_Frequency_Check(x,y): Predicts the sub-class within broad class x using keyword and frequency checking rules over narrative y

Input: description - the brief incident description
Output: hazard - the hazardous element responsible for the accident

function Classify_Hazardous_Element (description)
broad_class = Classifier (description)
if instances of broad_class < t1 **then**
    hazard = broad_class
    return hazard
else if Find_Representative(broad_class) = = TRUE **then**
    max_similar_word=Find_Most_Similar (broad_class, description)
    hazard = Keyword_Frequency_Check (max_similar_word, description)
    if max_similar_word.similarity > t2 and hazard ∈ Knowledge_Base **then**
        return hazard
    else
        hazard = broad_class
        return hazard
    end if
else
    hazard = Keyword_Frequency_Check (broad_class, description)
    return hazard
end if

At times certain subclasses appear as bigrams like "CO gas" or "sulfuric acid". In such situations, we take a rule-based approach. For example, if the representative word is "acid" for the broad class "Acid", when we find the similarity of the words in the incident description with the word "acid" the most similar word will obviously be "acid" itself if it is present in the narrative. So we can always check the word that appears before "acid" in the description. If the word exists in the knowledge base, then we can derive the required subclass easily using that word. Like if "sulfuric" exists just before "acid", then we can conclude that our subclass is "sulfuric acid" since it exists in our knowledge base. Algorithm 1 depicts the algorithm for our proposed approach.

2.2.2 Information Extraction of Accident Type

Information extraction [19] plays an important role in extracting structured information from unstructured sources such as raw text. The incident description provided most of the time contains the actual accident/incident in some form or other which needs to be derived. We first tokenize the incident descriptions into sentences, following which we tokenize them further into word tokens using NLTK's "sent_tokenizer" and "word_tokenizer" modules, respectively. After the process of tokenization, we apply Part-Of-Speech (POS) tagging [15] to the tokens. This way every word has its own POS tag. For example ("acid", "NN"; "splashed", "VBD"), meaning that "acid" is a singular noun and "splashed" is a verb of past tense. We then propose to extract phrases present in the incident description by the process of "chunking" using appropriate Regular Expression rules. Chunking is a task that follows Part-Of-Speech Tagging that adds structure to the sentence by using a regular expression. The result is a set of a group of the words in "chunks". For example, one can define a simple regular expression rule like {<JJ.?>*<NN.?><VB.?><IN><NN.?>}. This rule can extract phrases of the form: any number of adjectives followed by any kind of noun, then a verb of any kind followed by a preposition and finally another noun of any kind. A sample phrase/chunk that can be extracted through this rule can be "hot acid fell on track". Of course, in practice, a lot of such simple rules need to be combined to generate the required regular expression rule which will suffice our purpose. We claim that one of the chunks extracted in this way is the required actual incident/accident.

A dictionary or list of tokens is created using frequency distribution of words (tokens having higher frequency will be given higher priority) collected from the accident column present in the training dataset. Another list is created by collecting tokens present in the hazardous element corresponding to a particular incident description. We then go on to check the presence of tokens of these two lists in the chunks obtained from the incident description, with the tokens in the hazardous element collection given higher priority than those formed from incidents of the training set. The chunk which gets a matching token first is predicted as the actual incident/accident. For example, suppose the corresponding hazardous element is "CO_2 gas". And suppose our rule extracts the chunks "one CO_2 gas cylinder exploded," "was kept in stand" and "an upright position" from the incident description. Now out of the chunks extracted, the first chunk "one CO_2 gas cylinder exploded" contains tokens from the hazardous element "CO_2" and also "gas". So we output this chunk as our Incident/Accident. If none of the tokens from either list can be matched, then it is notified that human intervention is needed to find the incident/accident manually. Our proposed approach is described in Algorithm 2.

Algorithm 2. Accident extraction

Variables used:
text_tokens: tokens present in the text narrative
part_of_speech: text_tokens with their part of speech tags
all_chunks: a list of all the chunks formed based on the regular expression rule
hazardous_element: the predicted hazardous element for the text_narrative
hazard_tokens: a list of tokens present in hazardous_element
accident_tokens: tokens collected from the accident descriptions in the training set

---

Functions used:
Tokenize(x): Tokenizes x into sentences and then word tokens
POS_Tagging(x): assigns part of speech tags to the tokens x
Chunking (x, y): creates chunks from x based on the rule y
Classify_Hazardous_Element(x): predicts the hazardous element from x (See Algorithm 1)
Sort_By_Frequency(x): Sorts the list x according to the frequency of the tokens in x

---

Input: text_narrative - the provided brief incident description

Output: accident - the actual accident

function Classify_Accident (text_narrative)
text_token = Tokenize (text_narratives)
part_of_speech = POS_Tagging (text_token)
all_chunks = Chunking (part_of_speech, regular expression rule)
hazardous_element = Classify_Hazardous_Element (text_narrative)
hazard_tokens = Tokenize (hazardous_element)
for each chunk ∈ all_chunks
    for each token ∈ hazard_tokens
        if token ∈ chunk **then**
            accident = chunk
            return accident
        end if
    end for
end for
Sort_By_Frequency (tokens ∈ accident_tokens)
for each token ∈ accident_tokens
    for each chunk ∈ all_chunks
        if token ∈ chunk **then**
            accident = chunk
            return accident
        end if
    end for
end for
return 'do manually'

The predicted results are verified by domain experts to correct any grammatical errors arising during the process of chunking. However, note that only inspecting the predicted chunk is sufficient to correct these errors; one does not necessarily have to take up the tedious job of manually going through the incident description.

2.2.3 Potential Target Classification

When an accident occurs in an industry, there is always a high chance that a worker might sustain an injury or property might get damaged. In this paper, we predict this potential target of an accident from the accident report. For that first, we use the algorithm for hazardous element classification (Algorithm 1) to predict the hazardous element associated with the incident. Then we use the algorithm for accident extraction (Algorithm 2) to predict the actual accident related to the incident. Then the TDM and the already predicted hazardous element involved and accident/incident are used to train a Linear SVM classifier. The predicted class is the required Potential Target. Algorithm 3 shows the algorithm for our approach.

Algorithm 3. Potential target classification

Variables used:
hazard: hazardous element associated with the incident
accident: the actual accident

Functions used:
Classify_Hazardous_Element(x): predicts the hazardous element from x (See Algorithm 1)
Classify_Accident(x): predicts the accident from x (Described in Algorithm 2)
Classifier (x, y, z): Uses Linear SVM to predict the potential target using x, y and z as input

---

Input: description – the brief incident description
Output: potential_target – the Potential Target of the incident

---

function Classify_Potential_Target (description)
hazard = Classify_Hazardous_Element (description)
accident = Classify_Accident (description)
potential_target = Classifier (description, hazard, accident)
return potential_target

2.2.4   Potential Threat Classification

After finding out the potential target related to an accident report, our final task is finding out the potential threat as the final output. At first, we use the algorithm for accident extraction (Algorithm 2) to predict the actual accident related to the incident. Then the algorithm to predict the potential target of an incident (Algorithm 3) is used to predict the potential target. The Linear SVM classifier then predicts the potential threat treating the already predicted accident and potential target attributes as the input data. Algorithm 4 describes our proposed approach.

Algorithm 4. Potential threat classification

Variables used:
accident: the actual accident
potential_target: the potential target of the incident

Functions used:
Classify_Accident(x): predicts the accident from x (See Algorithm 2)
Classify_Potential_Target(x): predicts the Potential Target from x (See Algorithm 3)
Classifier(x, y): Uses Linear SVM to predict the potential threat using x and y as input

---

Input: description – the brief incident description
Output: potential_threat – the Potential Threat of the incident

---

function Classify_Potential_Threat (description)
accident = Classify_Accident (description)
potential_target = Classify_Potential_Target (description)
potential_threat = Classifier (accident, potential_target)
return potential_threat

---

3 Case Study

3.1 Dataset

The dataset consists of details of 640 brief incident descriptions of events for the years 2015–2016 in a steel plant in Eastern India. There are 6 primary attributes of the dataset on which we have worked on—"Brief Description" of the incident, "Hazardous Element Broad Category", "Specific Hazardous Element", "Incident/Accident", "Potential Target", and "Potential Threat". The narratives were given in "Brief Description" column, often describing a series of events leading to the incident. Around 130 different classes are present for the "Specific Hazardous Element". Since the classification of 130 classes using 640 samples is infeasible, we did not consider any fixed number of classes while predicting the hazardous element, thus predicting newer elements beyond the ones in the dataset possible. The "Hazardous Element Broad Category" attribute generalizes the specific hazardous elements into eight broad classes. Also, no fixed number of classes were considered for the incident/accident as the number of distinct classes possible was too large. The Potential Target consists of a limited number of possible classes viz. "human"; "property"; "human, property"; "human, property, environment"; "environment", "property, environment".

Similarly, Potential Threat also consists of only 8 classes viz. "property damage", "injury", "fatality" to name a few. In the last two cases, the limitation on the number of classes allows traditional multiclass classification techniques to be used to predict the required class. From the total dataset, 70% of the available data is used for training and rest is used for testing. Few sample instances of the dataset have been shown in Table 1.

Table 1 Sample instances of the dataset

Brief description	HE broad category	Specific HE	Incident/accident	Potential target	Potential threat
Alkali solution overflew from tank and spilled out and some liquid went in trench	Chemical	Alkali	Alkali spillage	Property	Property loss
Hot metal and slag was overflowing from trough #1 and immediately cast closed in not dry condition	Hot metal and steel	Hot metal and slag	Overflowing of hot metal and slag	Human, property	Injury, property loss

3.2 Sample Output

The brief description of an incident is treated as the input in our model, and the potential threat is the final output. In the process, we derive the hazardous element, the incident/accident and the potential target as intermediate outputs.

An example of a typical Input and Output is given as follows:

Input	Output
"There was a hydraulic leakage in DC#2 chute plate due to hose burst resulting fire"	*Hazardous element*: hydraulic oil *Accident*: fire accident *Potential target*: property *Potential threat*: property loss

3.3 Performance Measures

After the implementation of the model and getting the required attributes as outputs in the form of a probability or a class, we must find out how effective our model is based on some metric. In our study we have used Accuracy (A), Precision (P), Recall (R), and *F*-measure (F) [20] to evaluate our model. These are calculated as follows:

$$\text{Accuracy} = \frac{\text{number of correct predictions}}{\text{total number of predictions}} \tag{1}$$

$$\text{Precision} = \frac{\text{true positives}}{\text{true positives} + \text{false positives}} \tag{2}$$

$$\text{Recall} = \frac{\text{true positives}}{\text{true positives} + \text{false negatives}} \tag{3}$$

$$F - \text{measure} = 2 * \frac{\text{precision} * \text{recall}}{\text{precision} + \text{recall}} \tag{4}$$

3.4 Results and Discussions

In this work, we have proposed an integrated text mining model that is used to detect Potential Threat for an incident with the Hazardous Element, actual Accident and Potential Target as intermediate outputs. The summarized performance of our approach has been shown in Table 2.

Table 2 Summarized performance of the proposed system

Predicted element	Accuracy	Precision	Recall	F-measure
Hazardous element	0.71	–	–	–
Accident	0.66	–	–	–
Potential target	0.39	0.35	0.35	0.34
Potential threat	0.33	0.30	0.34	0.30

3.4.1 For Hazardous Elements

There were a total of about 130 distinct specific hazardous elements present in our dataset, and there is scope for newer ones as well. Although our primary aim was to predict the specific hazardous elements, 8 generalized, broad classes of Hazardous Elements viz. "Hot Metal and Steel", "Acid", "Chemical", "Gas", "Dust and Steam", "High-Pressure Material", "Activity Related", and "Others" were also provided in the dataset. TDM from the incident narratives was used as input to Linear SVM function available in the "Scikit-learn" library to predict these broad classes. Out of these we did not try to find subclasses within "Activity Related" and "Others" as their instances are few, and they are of less importance.

For "Gas" and "High-Pressure Material" finding a suitable representative word to calculate similarity was not possible, and it was found that simple keyword and frequency matching rules were enough to predict the subclasses. For the remaining broad categories, a hybrid model of WordNet and Word2Vec was used to find similarity between the selected representative word and the tokens of the narrative. With "acid" as the representative word for "Acid", WordNet was found to perform better than Word2Vec, while for the rest of the classes Word2Vec was observed to perform better. A threshold similarity value of 0.31 was set such that if the similarity value of the word with maximum similarity is found to be below this threshold then the subclassification is not performed and simply the broad class is predicted as the output.

A summary of our generalized classes and techniques used to sub-classify them have been depicted in Table 3. The accuracy of our model was 0.71 (see Table 2). Considering the small size of the training dataset, our proposed algorithm was observed to perform quite well. Due to the nature of the dataset, it was not possible to calculate precision, recall, and f-measure.

3.4.2 For Incident/Accident

We have not considered any fixed number of classes for predicting the incident/accident either. A regular expression rule was developed to generate the phrases or "chunks" using the process of chunking. Python's NLTK library was used for this purpose. The chunks were then matched with the token lists prepared from the accidents present in the training set and the hazardous element of the corresponding

Table 3 Generalized classes and techniques used to find sub-classes within them

Broad category	Sub-classification method	Representative word
Hot metal and steel	Similarity matching (Word2Vec), keyword matching rules	Metal
Acid	Similarity matching (WordNet), keyword matching rules	Acid
Gas	Keyword matching rules	N/A
Chemical	Similarity matching (Word2Vec), keyword matching rules	Chemical
Dust and steam	Similarity matching (Word2Vec), keyword matching rules	Dust, steam
High-pressure material	Keyword matching rules	N/A
Activity related hazard	N/A	N/A
Others	N/A	N/A

accident. Our model showed an accuracy of 0.66 (see Table 2), while the calculation of the other performance measures was not possible due to the nature of the dataset. Considering our small dataset and the fact that we have not taken any fixed classes into account, the performance achieved is considerably good. The predicted chunks were then verified by domain experts for grammatical errors and exactness of the prediction.

3.4.3   For Potential Target

Since only a limited number of classes were possible for this attribute, the Linear SVM function of the "Scikit-learn" library was used to solve this multiclass classification problem using TDM, already predicted hazardous element and accident as the input data. Accuracy, precision, recall, and f-measure of the model were found to be 0.39, 0.35, 0.35, and 0.34, respectively (see Table 2). The reduction in the performance level of this stage can be attributed to the small size and nature of our dataset. Also since the results of the previous stages of the pipeline are being used as inputs in this stage, the error of the previous stages affects the performance of later stages. However, for a considerably large dataset, the performance of our model is expected to improve by a large margin.

3.4.4   For Potential Threat

There were only 8 distinct classes for this attribute. So it was treated as a multiclass classification problem and the Linear SVM of the "Scikit-learn" library was used for classification. The already predicted accident and potential target were used as the input data with the output being the required Potential Threat. Accuracy, precision, recall, and f-measure of the model were found to be 0.33, 0.30, 0.34, and 0.30, respectively (see Table 2). The low performance of this stage can be attributed to the same reasons as in the case of prediction of Potential Target.

4 Conclusion

This study proposed an efficient method of constructing a text mining based model aimed to reduce manual effort as much as possible by automating the entire threat detection process in a steel plant. Noteworthy is the fact that we have not considered any fixed number of classes while predicting the specific hazardous element and actual accident, thus outperforming most of the works in this field which aim to predict only a few generalized classes. As future work, advanced methods for removing noise more efficiently from the dataset can be adopted to improve the performance. Experiments can also be conducted on much larger datasets to see the improvement in performance with the increase in the size of the dataset. Further study can also be done to extract chunks that are grammatically more correct during the accident extraction process of our approach, which would help to reduce manual labor even further.

References

1. Sarkar, S., Pateshwari, V., Maiti, J.: Predictive model for incident occurrences in steel plant in India. In: 2017 8th International Conference on Computing, Communication, Networking and Technology ICCCNT, pp. 1–5 (2017)
2. Collins, R.L.: Heinrich's fourth dimension. Open J. Saf. Sci. Technol. **1**, 19–29 (2011)
3. Kontogiannis, T., Leopoulos, V., Marmaras, N.: A comparison of accident analysis techniques for safety-critical man-machine systems. Int. J. Ind. Ergon. **25**(4), 327–347 (2000)
4. Posse, C., Matzke, B., Anderson, C., Brothers, A., Matzke, M., Ferryman, T.: Extracting information from narratives: an application to aviation safety reports. In: 2005 IEEE Aerospace Conference Proceedings, pp. 3678–3690 (2005)
5. Bulzacchelli, M.T., Vernick, J.S., Sorock, G.S., Webster, D.W., Lees, P.S.J.: Circumstances of fatal lockout/tagout-related injuries in manufacturing. Am. J. Ind. Med. **51**(10), 728–734 (2008)
6. Marucci-Wellman, H., Lehto, M., Corns, H.: A combined fuzzy and Naive Bayesian strategy can be used to assign event codes to injury narratives. Inj. Prev. **17**, 407–414 (2011)
7. Marucci-Wellman, H.R., Corns, H.L., Lehto, M.R.: Classifying injury narratives of large administrative databases for surveillance—a practical approach combining machine learning ensembles and human review. Accid. Anal. Prev. **98**, 359–371 (2017)
8. Chang, L.Y., Chen, W.C.: Data mining of tree-based models to analyze freeway accident frequency. J. Saf. Res. **36**(4), 365–375 (2005)
9. Shi, D., Guan, J., Zurada, J., Manikas, A.: A data-mining approach to identification of risk factors in safety management systems. J. Manag. Inf. Syst. **34**(4), 1054–1081 (2017)
10. Srivastava, S., Haroon, M., Bajaj, A.: Web document information extraction using class attribute approach. In: 2013 4th International Conference on Computer, Communication and Technology, pp. 17–22 (2013)

11. Rocha, O.R., Vagliano, I., Figueroa, C., Torino, P., Licciardi, C.A., Marengo, M., Italia, T.: Semantic annotation and classification in practice. IT Prof. **17**(2), 33–39 (2015)
12. Sleiman, H.A., Corchuelo, R.: A survey on region extractors from web documents. IEEE Trans. Knowl. Data Eng. **25**(9), 1960–1981 (2013)
13. Atkinson-abutridy, J., Mellish, C., Aitken, S.: Combining information extraction with genetic algorithms for text mining. IEEE Intell. Syst. **19**(3), 22–30 (2004)
14. Han, H., Giles, C.L., Manavoglu, E., Zha, H., Zhang, Z., Fox, E.A.: Automatic document metadata extraction using support vector machines. In: Proceedings of 3rd ACM/IEEE-CS Joint Conference on Digital Libraries, pp. 37–48 (2003)
15. Bird, S., Klein, E., Loper, E.: Natural Language Processing with Python. O'Reilly Media, Sebastopol, CA (2009)
16. Cortes, C., Vapnik, V.: Support-vector Networks. Mach. Learn. **20**(3), 273–297 (1995)
17. Miller, G.A., Beckwith, R., Fellbaum, C., Gross, D., Miller, K.J.: Introduction to wordnet: An on-line lexical database. Int. J. Lexicogr. **3**(4), 235–244 (1990)
18. Mikolov, T., Chen, K., Corrado, G., Dean, J.: Efficient estimation of word representations in vector space. In: Proceedings of ICLR Workshop (2013)
19. Pande, V.C., Khandelwal, A.S.: A survey of different text mining techniques. IBMRD's J. Manag. Res. **3**(1), 125–133 (2014)
20. Euzenat, J., Shvaiko, P.: Ontology Matching. Springer, Heidelberg (2007)

Bagging for Improving Accuracy of Diabetes Classification

Prakash V. Parande and M. K. Banga

Abstract The quality of human life is improved by detecting diseases effectively based on the rapid development of digital image processing, internet of things and effective deep learning processing. We propose a novel application of Bootstrap Aggregation, i.e., Bagging for improving the accuracy of diabetes classification in this paper. The model of bagged logistic regression is designed to classify diabetes effectively. The ROC is used to visualize performance of the algorithm based on False_Positive_Rate (FPR) and True_Positive_Rate (TPR). The parameters performances of the proposed method are compared with the existing techniques. It is concluded that the performance of the current method with bagging is better compared to traditional techniques that do not apply any additional measures.

Keywords Bagging · Multiple classifier systems · Diabetes

1 Introduction

In recent advances, many researchers have explored the technique of combining multiple classifiers to produce an ensemble classifier. Researchers have clearly demonstrated that the ensemble classifier is an applicably good individual classifier that is accurate and the errors that are made are always on different parts of the input space. Basically, ensemble classifiers are statistical learning algorithms that have achieved an enhanced performance by constructing a set of classifiers like decision trees. Ensemble methods combine multiple models to fit the training data in a certain

P. V. Parande (✉)
Research Scholar, School of Computing and Information Technology, Reva University, Bengaluru 560064, Karnataka, India
e-mail: prakashvp2010@gmail.com

M. K. Banga
Ex-Director of School of Computing & Information Technology, Reva University, Bengaluru 560064, Karnataka, India
e-mail: banga.mkrishna@gmail.com

Department of CSE, Dayanandsagar University, Bengaluru 560068, Karnataka, India

© Springer Nature Singapore Pte Ltd. 2020   125
V. Bhateja et al. (eds.), *Intelligent Computing and Communication*,
Advances in Intelligent Systems and Computing 1034,
https://doi.org/10.1007/978-981-15-1084-7_13

way like Bagging and Boosting. Bagging is a Bootstrap Aggregation method which is an algorithm of ensemble machine learning. Bootstrap fits separate decision trees to multiple copies of the earliest training set. It merges all such decision trees to generate a single predictive model. It is a powerful statistical method designed to reduce variances for better accuracy, machine learning stability and the problem of overfitting. It is widely used in statistical classification and as well as in regression.

Diabetes is also called Diabetes Mellitus (DM). It is a disorder of metabolism, where high blood sugars levels are found for a longer period. The high blood sugar levels are because of which the body is unable to produce adequate insulin for its own needs. This may be due to either impaired insulin secretion or impaired insulin action or maybe both. Worldwide 300 million people have been affected by Diabetes and the count is increasing day by day. Diabetes Mellitus can be diagnosed by traditional methods like blood test and urine test to check the glucose levels.

Contribution: In this paper, the mathematical model of the bagged logistic regression is developed to identify diabetes samples efficiently. The performance of the present bagging technique is compared to the current technique is presented with results.

Organization: The other sections are considered in the paper as; Sect. 2 is the literature survey of existing techniques. Section 3 describes the Mathematical model for the Bagged Prediction Method. The Algorithm to train a bagged logistic regression for improving predictions is elaborated in Sect. 4. The results of work are provided in Sects. 5, and 6 delivers the conclusion.

2 Literature Survey

In this section, literature reviews of existing techniques on boosting algorithms for classification are explained.

Abellán and Masegosa [1] describe the employment of bagging ensembles of credal sets. Credal sets are inaccurate probability decision trees with uncertainty measures. Continuously featured data sets and unavailability of data is prone to noise in class labels or wrong stint. Their work showed that decision trees with bagging credal would reduce error in classification for data sets with significant class noise.

Huang et al. [2] have shown an algorithm using the classification tree of bagging to envision the amino acid composite receptor type. The proposed G-protein coupled receptors (GPCRs) described the correction rate identity is 91.1% in tenfold cross-validation for subfamily classification.

Chi and Bruzzone [3] developed a novel sample-driven semi-labeled bootstrap aggregating (bagging) method for co-inference process framework to address classification issue of ill-possession. The architecture ensembles of classifiers are hybrid training sets of bootstrap which are of labeled and semi-labeled patterns with a balanced number. It develops basic classifier with Multi-layer Perceptron neural networks (MLPs).

Zhao et al. [4] proposed bagging incorporation using Bagging Evolutionary Feature Extraction (BEFE). It overcomes the limitations of the Revised a Direct Evolutionary Feature Extraction algorithm (DEFE) algorithm. It uses Whitened Principal Component Analysis (WPCA) for minimum training sets and the maximum number of classes along with weighted fitness. It used Yale and ORL face databases on face recognition applications. Cord´on et al. [5] illustrated FRBCS which is design for classifiers ensembles in bagging, accuracy-complexity trade-off random subspace and machine learning field. The main objective is to increase the accuracy and multi-criteria Genetic Algorithm (GA) is used to reduce the complexity of classifier for lexicographic order. Yu and Wong [6] proposed a repertory dictionary to store a set of classifiers representing distinguish concepts with similar characteristics. The proposed Bagging-Adaboost Ensemble (BAE) improves the accuracy, stability, and robustness using techniques of bagging and Adaboost for image data.

Cord´on et al. [7] proposed work with a training error-guided multi-criteria GA used greedy, random subspace, and GRASP-based Battiti's methods. It designed FRBCS ensembles to deal with the problem of classification having a large number of features in the final multi-classifier. The work demonstrated the UCI machine learning repository for different characteristics with and without the genetic selection stage on four data sets. Tu et al. [8] presented bagging which identifies the patient heart disease using C4.5 algorithm of Naïve Bayes algorithm and decision tree. This bagging method computes the confusion matrix of each model using precision, recall, F-measure, tenfold cross-validation and ROC space to evaluate the performance. The data set for coronary artery disease collected from the UCI KDD Archive. It shows that clinical application provided great advances in healing CAD. Zhao et al. [9] proposed bagging on the random sampling training set with an ensemble learning framework. The approach classifies the marine plankton images got from Evaluation Recorder and Shadowed Image Particle Profiling. It repeatedly perturbs the training set of origin to get nondependent bootstrap replicates. The multiple classifiers are constructed which are compatible with one another, which have no considerable effects on the whole training set trained with a single classifier. More than 93% accuracy for improved tenfold cross-validation is achieved by fusing such classifiers with major votes.

Kotsiantis and Pintelas [10] proposed manifold learning and bagging using nonlinear dimensionality reduction Isomap procedure to reduce high dimension of text feature space. The method applied to the text pretreatment avoids dimension curse to achieve a reduction in vector dimension and merges Bagging C4.5 algorithms to categorize complete text. It demonstrated a manifold learning technique that reduces the effect of text dimension in the pretreatment of text classification. Ghosh and Afroge [11] proposed a comparison of the Bangla handwritten characters with recognition of optical characters. The technique used the Hog algorithm and Binary pixel algorithm for the feature extraction method. Aggregation of Bootstrap and Support Vector Machine (SVM) are considered to recognize a character. It has shown that the classifier of SVM and aggregation of Bootstrap for trained characters gets 100% accuracy. It gets 89.8% and 93% accuracy for random untrained characters.

Ameri et al. [12] proposed a bagged regression tree application for multiple degrees of freedom (DOFs) in simultaneous myoelectric estimation. It investigated 10 subjects that participated in able-bodies with flexion-extension of wrist, pronation-supination, and abduction-adduction. The baseline ANNs in abduction-adduction of the wrist ($p < 0.05$) are outperformed by bagged trees, there is no considerable difference get in flexion-extension of wrist and pronation-supination ($p > 0.1$). It concluded that myoelectric prostheses with simultaneous control can use bagged trees potentially. Kong et al. [13] described a bootstrap filtering with a weight updating process to solve the Non-Intrusive Load Monitoring (NILM) problem using smart meter data. The proposed system solver design preserves high accuracy in runtime to update weights, while solving real-life NILM problems. It evaluated with multiple common metrics of REDD dataset. It illustrated that some commonly evaluated metrics are more suitable and additional merit of bootstrap filtering gives a confidence level of estimation. Zhang and Fisher [14] proposed reliable multi-objective online optimization control using bootstrap-aggregated neural networks of a fed-batch fermentation process. It developed copies of bootstrap re-samples of the actual training data with high prediction confidence bounds. Online re-optimization reduces model plant mismatch effects and optimization results of unknown disturbances. To produce baker's yeast, it is used in an imitated fermentation process of fed-batch.

Riana et al. [15] considered similarities to classify conventional and nonconventional classes in cervical cancer. The seven data classes are classified by Bagging (NvB + BG), Naïve Bayes model (NvB), Naïve Bayes, and Greedy Forward Selection (NvB + GFS). The model uses NvB, BG, and GFS to manage the issue of disparity class and Pap smear single image classification to select relevant features. The high accuracy value for the classification of normal and abnormal classes using NvB + GFS is 92.15%. The Naïve Bayes method and Bagging technique with Greedy Forward Selection of seven classification classes is 63.25%. Kim and Lim [16] proposed a classification scheme of the vehicle for multi-view surveillance camera images. Augmented Data increases the efficiency of the model in bagging using deep learning. The result has shown 97.84% classification accuracy on classification challenge dataset of 103,833 images. Spanakis et al. [17] described the Ecological Momentary Assessment (EMA) data organization to deal with decision trees. The proposed method combined boosted trees and bagging to create bootstrap samples using data correlation with respect to the subject. The boosting method is manipulated to develop a bagging function with minimum trees. The BBT (Bagged Boosted Trees) with over/under sampling method estimates conditional class probability function. The BBT on imitation of the system and live EMA data demonstrates high accuracy prediction than classic decision tree algorithms. It also estimated for conditional class probability function.

Rong et al. [18] divided dataset into several subspaces according to the location of samples in input space using Location Bagging-based Under Sampling (LBUS). Each individual base classifier constructs the training dataset by selecting samples from different subspaces. The input space is divided into various small spaces using the ITQ hashtag technique and under_sampling hash buckets with the location. The work proved that LBUS outperforms the Random-based Under Sampling (RUS) and Inverse RUS (IRUS). Shah et al. [19] combines bagging methods with random forest,

Reduced Error Pruning Tree and J48. It is to found pathological and apprehensive foetus from conventional ones. The UCI repository CTG data is taken to carry out this work. The optimized classifiers parameters were applied to the data sets. A proposed bagging method with decision trees analyzed the full feature space. The feature space is reduced by correlation feature selection (cfs) which is a subset evaluation technique. The results show considerable improvement in healthy and pathological subjects classification with values 0.90 for full and reduced feature space. Prasetio and Riana [20] proposed a genetic algorithm (bagging) to improve spinal disorder class classification accuracy. Genetic techniques with a selection of feature and bagging are solving the issue of imbalance class. It is applied to neural networks, naïve Bayes, and k-nearest neighbor which achieved accuracy of 88.06%, 86.13%, and 89.03% respectively. It also shows signified improvement for most classifier algorithms in the disorders of spine classification.

3 Mathematical Model of the Bagged Prediction Method

Let there be a probability distribution p from which we draw m datasets:

$$(X_1, Y_1)^1, \ldots (X_n, Y_n)^1 \tag{1}$$

$$(X_1, Y_1)^m, \ldots (X_n, Y_n)^m \tag{2}$$

Let there be a new point x for which prediction Y is required.
Let the true value for x be y.
We estimate y by training a model on each of these subsets and get Y_i for the ith dataset.
The final prediction is $Z = (\frac{1}{m})(Y_1 + Y_2 \ldots Y_m)$.
Risk function for one prediction model:

$$E((Y - y)^2) = E((Y - E(Y))^2) = Var(Y) \tag{3}$$

Risk function for aggregated predictions:

$$E((Z - y)^2) = E((Z - E(Z)^2) \tag{4}$$

because y is the expected value of Z.

$$
\begin{aligned}
&= Var(Z) \\
&= Var((1/m)(Y^1 + Y^2 \ldots Y^m)) \\
&= 1/m^2 (Var(Y^1 + Y^2 \ldots Y^m)) \\
&= 1/m^2 (Var(Y^1) + Var(Y^2) \cdots + Var(Y^m))
\end{aligned} \tag{5}
$$

Assuming we draw independent datasets and hence independent

$$Y_i = 1/m(Var(Y))$$ (6)

This shows that the variance in aggregated prediction is reduced by a factor of m.
Since we cannot draw m datasets from the population, we approximate m bootstrap
samples from the training dataset.

4   Algorithm to Train a Bagged Logistic Regression for Improving Predictions

(a) Input the dataset (Pima Indians Diabetes Database) from UCI repository
     in R. (http://archive.ics.uci.edu/ml/machinelearningdatabases/pima-indians-
     diabetes/)
(b) Create samples with replacement from the diabetes dataset for training and
     testing the models each.
(c) Instruct a classification of logistic regression on the training data set with Class
     Variable as the response variable and other variables as predictors and generate
     predictions for the test dataset.
(d) Take 50 samples with the replacement of size one-third of the training dataset.
(e) Train a logistic regression classifier on each sample with Class Variable as the
     response variable and other variables as predictors.
(f) Compute predictions for testing dataset by the 50 models and average the m to
     create a final prediction.
(g) Compare the predictions from the logistic regression model trained in step c
     with the predictions by the bagged model in step f.

Input: Pima Indians Diabetes Database in R from UCI repository.
Output: Compared Prediction for the logistic regression model and bagged
model.
Create_samples_replacement (diabetes dataset n).
If (Class Variable = the response variable and dataset size $n = 50$)).
Train (logistic regression-based classifier).
Create_samples_replacement (diabetes dataset $n/3$)// Bagging.
Train (classifiers of logistic regression).
Compare (classifier of logistic regression results with bagged logistic regression-
based classifier).
Output (result).

5 Results

5.1 Receiver Operating Characteristics (ROC)

The ROC graphs are often used for and visualization of classifier performance and their ranking. The focus of most of the available classification systems is to minimize the classification error rate. An effective criterion in this case to measure the quality of a decision rule would be the ROC curve area. The ranking of the classifier also plays a major role when being chosen for real-time applications.

The models such as logistic regression and bagged logistic regression for classification prediction are used in our model. The output Y is a probability of the outcome of diabetic classification.

The accuracy of logistic and bagging models are compared by setting the threshold of probability at $c = 0$ $i.e. P > = 0.5$ the correct output prediction of the model is assumed to be $y = 1$ and for $P < 0.5$ the wrong prediction is taken as $y = 0$. ROC curve is a better way to compare the predictions using plots of True Positive Rate (TPR) and False-Positive Rate (FPR) at all possible manifested magnitude levels.

$$\text{TPR} = \text{Percentage of diabetic samples predicted as diabetic}$$
$$= P[p >= c | y = 1]$$

$$\text{FPR} = \text{Percentage of non-diabetic samples predicted as diabetic}$$
$$= P[p >= c | y = 0]$$

and the

ROC curve is plotting the entire set of possible true- and false-positive fractions obtained by varying c from $(0, 1)$

$$ROC = \{(FPR, TPR), c \in (0, 1)\} \tag{7}$$

The ROC curve which is a combination of TPR and FPR is as shown in Fig. 1 for bagging and logistic regression. The area under the ROC curve has been used as a measure of the accuracy of the model as it compares the model predictions without any assumptions. It is observed that the bagging is better than the logistic regression.

5.2 Comparison

The performance parameters such as Positive and negative prediction rates, the accuracy, sensitivity, detection rate, others... are compared for the existing Logistic Regression technique and proposed Bagged Logistic Regression in Table 1.

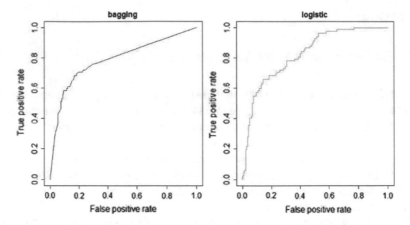

Fig. 1 The ROC curve which is a combination of TPR and FPR for bagging and logistic regression

Table 1 Comparison between logistic regression technique and bagged logistic regression

Performance parameters	Logistic regression	Bagged logistic regression
Accuracy	0.7826	0.8043
Kappa	0.5087	0.5504
Sensitivity	0.8377	0.8701
Specificity	0.6711	0.6711
Positive Pred. value	0.8377	0.8428
Negative Pred. value	0.6711	0.7183
Prevalence	0.6696	0.6696
Detection rate	0.5609	0.5826
Detection prevalence	0.6696	0.6913
Balanced accuracy	0.7544	0.7706

From Table 1. It is evident that the performance parameters of the proposed method, i.e., with bagging show improvement compared to the existing technique.

6 Conclusions

The consideration and importance given to a variant of multi classifier have seen some rise in the recent years as it promises higher accuracy and improved response time. The method proposed in this research work is to apply bagged logistic regression instead of the logistic regression model to identify diabetes in a dataset of samples. From the experimental results and the comparison shown in Table 1, it is apparent that the proposed method fares well against the available methods considerably.

To conclude; using ROC of TPR, FPR, and other performance parameters, it can be abundantly assured that the proposed model is better compared to the existing techniques of logistic regression for prediction of diabetes in patients.

References

1. Abellan, J., Masegosa, A.R.: Bagging schemes on the presence of class noise in classification. In: Elsevier Publications, Expert Systems with Applications, vol. 39, pp. 6827–6837 (2012)
2. Huang, Y., Cai, J., Ji, L., Li, Y.: Classifying G-protein coupled receptors with bagging classification tree. In: Elsevier Publications, Computational Biology and Chemistry, vol. 28, pp. 275–280 (2004)
3. Chi, M., Bruzzone, L.: A semilabeled-sample-driven bagging technique for ill-posed classification problems. IEEE Geosci. Remote Sens. Lett. 2(1), 69–73 (2005)
4. Zhao, T., Zhao, Q., Lu, H., Zhang, D.: Bagging evolutionary feature extraction algorithm for classification. In: IEEE Third International Conference on Natural Computation (ICNC), pp. 1–6 (2007)
5. Cord´on, O., Quirin, A., S´anchez, L.: A First Study on Bagging Fuzzy Rule-based Classification Systems with Multicriteria Genetic Selection of the Component Classifiers. In: IEEE 3rd International Workshop on Genetic and Evolving Fuzzy Systems, pp 1–6, (2008)
6. Yu, Z., Wong, H.-S.: Mage classification based on the bagging-adaboost ensemble. In: IEEE ICME, pp. 1481–1484 (2008)
7. Cord´on, O., Quirin, A., Sanchez, L.: On the use of bagging, mutual information-based feature selection and multi criteria genetic algorithms to design fuzzy rule-based classification ensembles. In: IEEE Eighth International Conference on Hybrid Intelligent Systems, pp. 549–554 (2008)
8. Tu, M.C., Shin, D., Shin, D.: A comparative study of medical data classification methods based on decision tree and bagging algorithms. In: Eighth IEEE International Conference on Dependable, Autonomic and Secure Computing, pp. 183–187 (2009)
9. Zhao, F., Feng, Lin., Hock, Soon, Seah.: Bagging based plankton image classification. In: IEEE ICEP, pp. 2081–2084 (2009)
10. Kotsiantis, S.B., Pintelas, P.E.: A method based on manifold learning and bagging for text classification. In: IEEE 2nd International Conference on Artificial Intelligence, Management Science and Electronic Commerce (AIMSEC), pp. 2713–2716 (2011)
11. Ghosh, A.K., Afroge, S.: A comparison between support vector machine (svm) and bootstrap aggregating technique for recognizing bangla handwritten characters. In: International Conference of Computer and Information Technology (ICCIT2017), pp. 1–5 (2017)
12. Ameri, A., Scheme, E.J., Englehart, K.B., Parker, P.A.: Bagged regression trees for simultaneous myoelectric force estimation. In: The 22nd Iranian Conference on Electrical Engineering (ICEE), pp. 2000–2003 (2014)
13. Kong, W., Dong, Z., Xu, Y., Hill, D.: An enhanced bootstrap filtering method for non-intrusive load monitoring. In: The IEEE Power and Energy Society General Meeting (PESGM), pp. 1–5 (2016)
14. Zhang, J., Fisher, R.: Reliable multi-objective on-line re-optimization control of a fed-batch fermentation process using bootstrap aggregated neural networks. In: The International Symposium on Computer Science and Intelligent Controls (ISCSIC), pp. 49–56 (2017)
15. Riana, D., Hidayanto, A.N., Fitriyani.: Integration of bagging and greedy forward selection on image pap smear classification using naïve bayes. In: The 5th International Conference on Cyber and IT Service Management (CITSM), pp. 1–7 (2017)
16. Kim, P.-K., Lim, K.-T.: Vehicle type classification using bagging and convolutional neural network on multi view surveillance image. In: IEEE Conference on Computer Vision and Pattern Recognition Workshops (CVPRW), pp. 914–919 (2017)

17. Spanakis, G., Weiss, G., Roefs, A.: Enhancing classification of ecological momentary assessment data using bagging and boosting. In: IEEE International Conference on Tools with Artificial Intelligence (ICTAI), pp. 388–395 (2016)
18. Rong, T., Tian, X., Ng, W.W.Y.: Location bagging-based under sampling for imbalanced classification problems. In: The International Conference on Wavelet Analysis and Pattern Recognition (ICWAPR), pp. 72–77 (2016)
19. Shah, S.A.A., Aziz, W., Arif, M., Nadeem, M.S.A.: Decision trees based classification of cardiotocograms using bagging approach. In: The International Conference on Frontiers of Information Technology (FIT), pp. 12–17 (2015)
20. Prasetio, R.T., Riana, D: A comparison of classification methods in vertebral column disorder with the application of genetic algorithm and bagging. In: The Fourth International Conference on Instrumentation, Communications, Information Technology, and Biomedical Engineering (ICICI-BME), pp. 163–168 (2015)

Two-Way Handshake User Authentication Scheme for e-Banking System

B. Prasanalakshmi and Ganesh Kumar Pugalendhi

Abstract The e-commerce system that involves many fraudulent activities as of today needs enormous research to be done to raise the level of security. The security even if provided in many aspects needs it to be reviewed and authenticated. In most of the existing systems like SSL, OTP generation to the mobile number or mail id, or any such systems proves that there are some drawbacks. Such drawbacks can be overcome by using multiple biometric entities for authentication even failing one authentication proves to be a spooler. Such a foolproof system is proposed that involves three biometric traits and the security measures are analyzed.

Keywords E-commerce · Banking · Security · Biometric · Authentication

1 Introduction

Security concerns are of great importance in the evolving world of human authentication. Many techniques are in use for building security in any system. The foremost used technique for identification and authentication is the use of photograph face image or fingerprint. Hence, biometrics has played an important role in this scenario. To communicate or store data, much security is required to hide such data. Some of the data hiding techniques include steganography, watermarking, etc. In the emerging world of digitalization of banking communications, it becomes vital in securing the identity of the individual to avoid fraudulent systems. To bring out security and enhance its impact in the electronic transactions much more effort has to be depicted and the related security measures are to be analyzed.

B. Prasanalakshmi (✉)
King Khalid University, Abha, Saudi Arabia
e-mail: drsanaksa@gmail.com

G. K. Pugalendhi
Anna University, Coimbatore, India
e-mail: ganesh23508@gmail.com

© Springer Nature Singapore Pte Ltd. 2020
V. Bhateja et al. (eds.), *Intelligent Computing and Communication*,
Advances in Intelligent Systems and Computing 1034,
https://doi.org/10.1007/978-981-15-1084-7_14

2 Previous Study

A lot of such research contributions have been done by several researches proving
the weakness of the existing system and their genuinity. To point out some Dhillon
et al. [1] introduced an approach to secure the stages online based on encryption and
compression of data. Even though many upcoming researches exist in cybersecurity
specifically with e-commerce transactions, many fraudulent systems also develop
to break the new systems. Hence a day to day development is needed to bring out
foolproof systems. The use of Pretty Good Privacy has been handled by Slamy [2] to
provide confidentiality, authentication, compression and segmentation services for
E-commerce security. An advanced secure payment system(ASEP) as introduced
by Lee and Lee [3] includes the combination of three cryptosystems Elliptic Curve
Cryptosystem, Secure Hash Algorithm, and Block Byte Bit Cipher, replacing RSA
and DES in order to improve its performance. An implementation of the new payment process to assure atomicity, and to protect the sensitive information of the
cardholder and the merchant was proposed by Zhang [4], and Craft and Kakar [5].
Another commonly used security measure over web transactions is the SSL(Secure
Sockets Layer). In order to solve the weaknesses of SSL, SET [6] (Secure Electronic
Transaction) has evolved. VISA introduced a protocol that is 3D secure [7] based on
additional control on online transactions using data of the customer to be validated
by the user himself.

3 Proposed System

Traditionally, the user identity like pin, date of birth and many can be utilized for
proving the authenticity of the user making transactions.

Even then, fraudulent systems may exist hence comes a solution with biometric
authentication. The proposed system includes three biometric real-time entity to be
verified by the user himself. Authenticating each biometric entity is detailed so that
the performance of such existing biometric authentication can be increased. The
proposed system includes four individual steps contributing equally for two phases.

The overall design of the proposed scheme is divided into four submodules which
fall into two phases of the biometric authentication system. The enrollment phase
as in Fig. 1. includes Image acquisition and processing and Data embedding. In the
proposed system three biometric traits, such as Finger vein, Ear and Voice are used.
Finger vein key generation is a published work [8] proving the genuinity in results.

Key generation from the finger vein is a part of the proposed technique. Ear
feature vector is generated by means of combined 5 level DWT and 1 level DCT and
finally Voice feature is extracted by means of combined LPC and MFCC algorithm
and processed by DFT. The verification phase includes Data Extraction and check
module where the biometric template stored in the database is used in the verification
phase, which is a one to one comparison. The live samples of the traits involved in

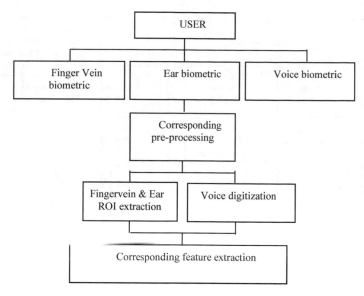

Fig. 1 Enrollment phase—overall

the verification process are obtained through their respective scanners. The output images, which are raw, are processed to convert them to their corresponding templates using the similar procedures adopted in the embedding phase. These templates are also compared to the templates stored in the database. This, step-by-step process, indicates whether the person claiming his identity is authorized or unauthorized. The technical steps as shown in Fig. 2. involved are given as watermark extraction and checking phase.

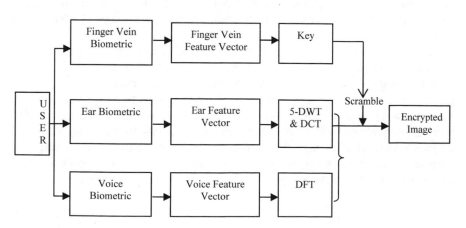

Fig. 2 Enrollment phase—detailed structure

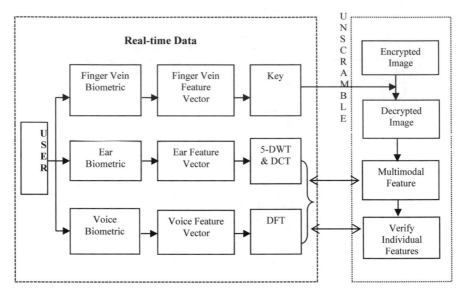

Fig. 3 Verification phase

The verification phase as discussed in Fig. 3 involves checks at three stages. If the system identifies failure at even one stage the entire authentication scheme assumes the person to be fraudulent or unauthorized. The checks performed are:

a. The ear image obtained as a result of the extraction algorithm is verified against the ear template stored in the database during enrollment. If this matching move in a positive manner, key is generated from the finger vein image of the person, else the person who provided the data for authentication is rejected to be unauthorized.

The key generated from the live scan image of the finger vein is compared with the key stored in the database at the time of enrollment. If this moves well the fingerprint which is the private key is used to unscramble the encrypted image, else the person is unauthorized.

As discussed already with a multimodal biometric authentication system which could be applicable for any domain of interest on authentication, here is presented an authentication scenario in the banking enterprise which includes finger vein, voice, and ear involving the customer of a bank to authenticate himself with the three biometric traits over the phone. When the customer enters the customer id in the browser an automated call is raised to his mobile (authentication device) which has a finger vein sensor. The user is authenticated by his voice received as a response for a set of authentication questions. The second authentication lies with the front camera captured ear image and the finger vein included in the mobile device.

The enrollment phase which includes enrolling one's finger vein, ear, and voice data at the spot of enrollment are taken into consideration and the technical steps involved are discussed in the previous figures. The technical strategy for the verification phase is as given the details of the technical steps involved are not discussed since they are basic to the reader and could involve any known algorithm.

The entire process taken place in the banking sector authentication is shown in Fig. 4 of a customer over phone is given as a brief figure to represent the actual handshake flow of the entire process. Instead of OTPs generated to mail or mail, this would provide much more security.

The proposed system as in Fig. 4 explains the case study taken over on the proposed security model for e-banking transactions. Initially, the customer of the bank registers himself/herself with their biometric identities as requested at the instance of registering themselves for enabling e-commerce transactions. The Handshake model

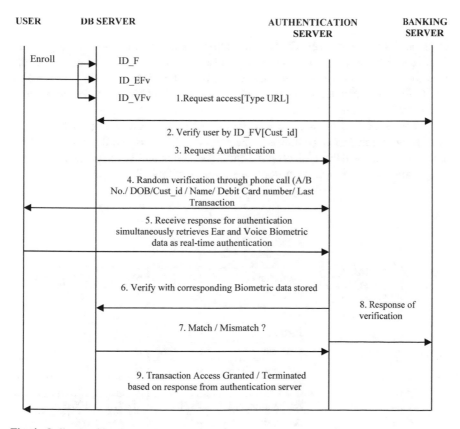

Fig. 4 Online banking transaction system

depicted shows the flow of communication from the time of initiating the transaction until the authentication of the transaction. Initially while the customer requests for authentication initiation, the bank website retrieves the finger vein data through the mobile inbuilt biometric sensor to detect the finger vein image of the customer that can be used as preliminary authentication, as a second stage of authentication the customer is asked to answer any of the IVR question from which his voice is recognized and the biometric sensor placed near the speaker of the mobile detects the ear biometric image. This system assumes the mobile to have these inbuilt biometric sensors.

4 Results and Conclusion

The proposed scheme on the e-banking system utilizes nine solid steps where each and every step takes into account several technical specifications and calculations to prove the identity of the user. The feature vector generation of Finger vein as given in [8] is done. The ear features vector extraction involved to point to the middle of the ear and crop the portion as required, remove noise and the results of it as produced by MATLAB. The voice authentication mechanism also involves a combined method of LPC and MFCC as discussed in Table 1.

This section also discusses the security considerations of any cryptographic system. The proposed system promises to meet the security considerations of Confidentiality, Integrity, Availability, Accountability. The advantages of the proposed system as compared with the existing systems are figured in Table 2.

Table 1 Comparative analysis of LPC, MFCC, and concatenated LPC + MFCC performance

Parameters	LPC	MFCC	LPC + MFCC
Epoch	30	50	12
Time	0	0.0026	0
Performance	0.43	0.527	0.0356
Gradient	0.000106	8.26	0.00481
Validation checks	6	6	6
Neurons	20	10	5
Recognition rate	93.30%	100%	100%
Error rate	6.70%	0%	0%

Table 2 Method comparison of security

	SSL	ASEP	OTP	Proposed
Confidentiality	Yes	Yes	Yes	Yes
Integrity	Yes	Yes	Yes	Yes
Anonymity	No	Yes	Yes	Yes
Non–revocation	No	No	No	Yes
Protection of sensitive data on sending	No	No	No	Yes
Protection of receiving data	No	No	No	Yes
Complexity	Low	High	Low	Low
Cost	Low	High	Low	Low
Possibility for man in the middle attack	Yes	Yes	Yes	No

References

1. Dhillon, G., Ohri, J.: Optimizing security in E-commerce through implementation of hybrid technologies, CSECS'06. In: Proceedings of the 5th WSEAS International Conference on Circuits, Systems, Electronics, Control and Signal Processing, pp. 165–170
2. Slamy, A.A.: E-Commerce security. IJCSNS Int. J. Comput. Sci. Netw. Secur. **8**(5) (2008)
3. Lee, B., Lee, T.: An ASEP (Advanced Secure Electronic Payment) protocol design using 3BC and ECC(F2 m) algorithm,e-technology, e-commerce and e-service. In: EEE 2004. IEEE International Conference on, pp. 341–346 (2004)
4. Zhang, X.: Implementation of a suggested e-commerce model based on set protocol, software engineering research, management and applications (SERA). In: 2010 Eighth ACIS International Conference on, pp. 67–73
5. Craft, A., Kakar, R.: E-commere Security, Conference on Information Systems Applied Research 2009, v2 Washington
6. Houmani, H., Mejri, M.: Formal analysis of SET and NSL protocols using the interpretation functions-based method. J. Comput. Netw. Commun. 18, Article ID 254942 (2012)
7. Jarupunphol, P., Mitchell, C.: Measuring 3-D secure and 3D SET against e-commerce end-user requirements. In: Proceedings of the 8th Collaborative Electronic Commerce Technology and Research Conference (CollECTeR)
8. Prasanalakshmi, B., Kannammal, A.: A secure cryptosystem from palm vein biometrics. In: ACM International Conference on Interaction Sciences: Information Technology, Culture and Human, pp. 1401–1405 (2009)

Periocular Biometrics for Non-ideal Images Using Deep Convolutional Neural Networks

Punam Kumari◉ and Seeja K. R.◉

Abstract The objective of this research is to study the effect of eyeglasses and the masking of the eye portion on the recognition accuracy of the periocular biometric authentication system. In this paper, six different off-the-shelf deep Convolutional Neural Networks (CNN) are implemented. Experimental results show that in both the cases VGG 19 CNN model outperforms others on the UBIPr database.

Keywords Periocular biometrics · Deep convolutional neural network · Transfer learning

1 Introduction

Periocular region-based biometric authentication system considers eyelashes, eye shape, skin texture, eyebrow, eye fold, tear duct eye socket, etc., as key features for identification. Eyeglasses on the periocular region could hide important features that make the system less reliable [1]. Similarly, masking of eye region will hide sclera, iris and pupil features which can also degrade the recognition accuracy [2]. In this proposed work, we have implemented deep CNN via transfer learning to analyze the usability of periocular authentication for the two non-ideal scenarios (1) when the subject is wearing glasses and (2) when the eye portion is hidden.

Deep neural network is a field that is highly inspired by biological neuron system. CNN are one of the most popular deep neural networks. For recognition via deep CNN, we need to train a model first. In this paper, instead of training our model from scratch, we have used the Transfer Learning approach. Transfer Learning is a method in which a model which has already been trained for one purpose can be used as an initial point to train a model for a similar type of application, by replacing the final classification layer accordingly [3]. For transfer learning, lots of pretrained models already exist. In this paper, we have used 6 different pretrained

P. Kumari (✉) · S. K. R.
Computer Science Department, Indira Gandhi Delhi Technical University for Women, Delhi, India
e-mail: punam_taurus@hotmail.com

© Springer Nature Singapore Pte Ltd. 2020   143
V. Bhateja et al. (eds.), *Intelligent Computing and Communication*,
Advances in Intelligent Systems and Computing 1034,
https://doi.org/10.1007/978-981-15-1084-7_15

models AlexNet, GoogLeNet, ResNet50, ResNet101, VGG16, and VGG19, for the matching of non-ideal images of the periocular region.

2 Related Work

Park et al. [2] were the first researchers who analyzed the feasibility of the periocular region in the field of biometric authentication systems. Later, many of the researchers consider the periocular region as a standalone modality and use it for authentication [4, 5] as well as for soft biometric classification [6, 7].

Some of the researchers considered the concept of periocular region and its fusion with iris [8] and found that fusion of both of the modality can dramatically increase the recognition accuracy of iris-based biometric authentication system.

Researchers also implemented the concept of deep learning in the field of periocular biometrics. Such as in [9] researchers implemented the concept of visual attention mechanism using deep neural network and make focus on critical components such as eyebrow and eye for feature extraction and obtained better recognition accuracy.

Alahmadi et al. [10] used two different pretrained models—VGG-Face and VGG-Net for discriminative feature extraction. Whereas Tapia et al. [11] implemented a deep learning-based CNN model, i.e., Lenet-5 for gender prediction. They implemented a fusion of two different Lenet-5 CNN models (for left and right eye both) and obtained 87.26% recognition accuracy.

The authors in [12] analyzed the recognition accuracy of the images captured from the subjects before and after cataract surgery. For feature extraction, they used three different approaches—average Gabor filter response, deep neural network-based CNN ScatNet, and DSIFT. Cosine similarity was used for matching and after fusion of matching score, they obtained 69% Rank-10 recognition accuracy and 24% Genuine Acceptance Rate at 1% False Acceptance Rate.

Zhao and Kumar [13] proposed a novel concept of semantic assisted convolutional neural network. They added an additional convolutional neural network branch to already existing CNN. The attached CNN was trained with gender-based semantic information Their model was obtained remarkable recognition accuracy of ~92%.

3 Methodology

Methodology used in the proposed work is show in Fig.1.

Step 1: Input database: UBIPr database [14] which contains a total of 10252 images (344 subjects and 30images/subject) with different degradation factors such as pose variation, hair occlusion, subject with wearing glasses, etc.

Step 2: ROI extraction: Canthus points are used as the reference point (see Fig. 2) for extracting the ROI. The reason is that they are least affected whether the eyes are

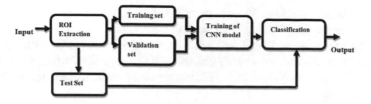

Fig. 1 Methodology used in the proposed work

Fig. 2 Eye Image with inner and outer canthus point

open or closed. In UBIPr database coordinates of Inner and Outer Canthus points are provided in its metadata. For ROI extraction, we have used the algorithm proposed by Liu. P. et al. [15]. Steps of the algorithm are as follows:

1. (x1, y1): Coordinate for inner Canthus points and
   (x2, y2): Coordinate for outer canthus points.
2. Euclidean distance between inner and outer canthus points is

$$D((x1, y1), (x2, y2)) = \sqrt{(x1 - x2)2 + (y1 - y2)2} \qquad (1)$$

3. Calculate points Lp = (Lpx, Lpy)

$$Lpx = (x1 + x2)/2 \qquad (2)$$

$$Lpy = (y1 + y2)/2 \qquad (3)$$

4. Calculate top left (x3, y3)and bottom right (x4, y4) coordinate points of rectangular ROI

$$(x3, y3) = (Lpx - 1.2 \times D, Lpy - 0.8 \times D) \qquad (4)$$

$$(x4, y4) = (Lpx + 1.2 \times D, Lpy + 0.8 \times D) \qquad (5)$$

5. Extract rectangular ROI with the above-calculated points and resize all ROI images to the size of 64 X 128 X 3.

Step 3: Dataset Creation: We have partitioned our experimental dataset into three subparts: Training set, Validation set, and Test set.

Step 4: Training: We have used the Transfer learning approach [16]. In this approach, first, select a model which is already trained on some large dataset then we have fine-tuned these models for classification. For our proposed work, we have used 6 different pretrained models:

1. **AlexNet** [17] was developed in 2012 and contains 62.4 million parameters. It was one of the first convolutional neural networks which is 8 layers deep and trained on the ImageNet database.
2. **Google Net** [18] is also known as Inception V1. It is a convolutional neural network, which is 22 layers deep and has 9 inception modules. The inception module uses different sizes of filters at the same level. Because of the inception module, this network looks a bit wider instead of deeper.
3. **ResNet** [19] is made of equal "residual" blocks. The primary idea behind residual blocks is to re-route the input, which facilitates the next layer to learn the concepts of the previous layer plus the input of that previous layer. Residual Net 50 is a convolutional neural network which 50 layers deep and contains 25.6 million parameters whereas Residual Net 101 is a convolutional neural network which is 101 layers deep.
4. **VGG** [20] proposed an improvement on AlexNet by replacing its large 11 X11 and 5X5 filters with multiple 3X3 kernel size filters one after another. With this improvement VGG16 which is 16 layers deep network is able to learn more complex features as compare to AlexNet because small size kernels arranged in multiple nonlinear layers increase the depth of the network. VGG19 is 19 layers deep network.

Step 5: Validation Validate all of the model using validation set.

Step 6: Classification Classify the test data using the trained CNN model and then calculated its recognition accuracy.

4 Experiments and Results

The proposed work is implemented on MATLAB R2018a. The selected CNN models follow two different architectures—Series network and Directed acyclic graph. Networks are fine-tuned with a deeper knowledge of parameters and several hit and trial attempts. The network specification for both Series and DAG is given below.

Series network (AlexNet, VGG 16, and VGG19)

Image Input Layer Size:	64X128X3 with 'Zerocenter' normalization
Weight learn rate factor:	10 Bias learn rate factor: 10
Classification o/p type:	Crossentropyex.
Training Option:	Stochastic gradient descent with momentum(sgdm)

Max Epochs: 20 Initial Learning Rate:1e^{-4} Shuffle:'every epoch'
Validation Frequency: 3 Min. batch size for training: 10

Directed Acyclic Graph (DAG) network (Google Net, ResNet50 and ResNet 101)

Image Input Layer Size: 64X128X3 with 'Zerocenter' normalization
Weight learn rate factor: 10 Bias learn rate factor: 10
Classification output type: Crossentropyex.
Training Option: Stochastic gradient descent with momentum(sgdm)
Max Epochs: 6 Initial Learning Rate: 3e^{-4} Shuffle:'every epoch'
Validation Frequency: 3 Min. batch size for training: 10

Experiment No. 1: Subjects are wearing glasses

In this experiment, dataset contains only those images in which subjects are wearing glasses (see Fig. 3.) Recognition accuracy obtained by 6 different CNNs is shown in Table 1.

\# Training images = 7527   \# Validation images = 1720   \# Testing images = 1005

Experiment 2: Eye portion of the images are hidden

In this experiment, we manually mask the eye portion of the images and hide features like eye shape, iris, sclera, etc. The example image is shown in Fig. 3. Recognition accuracy obtained by 6 different CNNs is shown in Table 2.

\# Training images = 5025   \# Validation images = 2152   \# Testing images = 3075

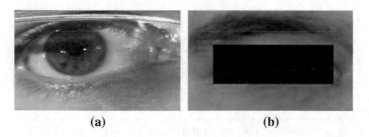

(a)   (b)

Fig. 3 Example images of periocular region **a** subject wearing glasses **b** masked eye portion

Table 1 Recognition accuracy on the dataset with subjects wearing glasses

CNN	Validation accuracy	Testing accuracy	CNN	Validation accuracy	Testing accuracy
AlexNet	98	97.22	ResNet101	97.22	95
GoogLeNet	96.3	91.11	VGG16	93.52	88.33
ResNet50	99	98.89	VGG19	100	98.89

Table 2 Recognition accuracy on the dataset with masked eye portion

CNN	Validation accuracy	Testing accuracy	CNN	Validation accuracy	Testing accuracy
AlexNet	82.80	62	ResNet101	93	71.2
GoogLeNet	87	67	VGG16	92.67	72.6
ResNet50	93.67	69	VGG19	94.83	78

5 Discussion

5.1 Comparison with Already Existing Work in Literature

Experiment 1: (When the subject is wearing glasses): According to the literature, eyeglasses could degrade the recognition accuracy [1] and make the system less reliable. But in the proposed work, VGG19 CNN model obtained remarkable 98.89% recognition accuracy.

Experiment 2: (with masked eye portion): After masking of eye portion VGG19 outperform other techniques already existing in the literature (Table 3).

To analyze the performance of the proposed fine-tuned Series and DAG network on the original UBIPr dataset we have evaluated the recognition accuracy of all deep CNN models (as shown in Table 4) and compare it with existing work in literature which have used UBIPr database (Table 5).

\# Training images = 5025   \# Validation images = 2152   \# Testing images = 3075

Table 3 Comparison of recognition accuracy of datasets with masked eye portion

Reference	Database	Method/Feature extraction technique[a]	Result (%)
Park et al. [2]	FRGC	Combination of HoG, LBP, and SIFT	69.63
Miller et al. [21]	FERET	LBP	74.07
Proposed approach	UBIPr	CNN (VGG 19)	78

[a]**HoG**: Histogram of Oriented Gradients, **LBP**: Local Binary Pattern, **SIFT**: Scale Invariant Feature Transform, **CNN**: Convolutional Neural Network

Table 4 Recognition accuracy of fine-tuned off-the-shelf CNN models on whole UBIPr dataset

CNN	Validation accuracy	Testing accuracy	CNN	Validation accuracy	Testing accuracy
AlexNet	82.80	79.80	ResNet101	93	86
GoogLeNet	87	87.20	VGG16	92.67	89.20
ResNet50	93.67	88	VGG19	94.83	89.50

Table 5 Comparison of recognition accuracy obtained in Table 4 with existing methods in literature

S.No.	References	Method	Results (%)
1	Proposed approach	Convolutional neural network VGG19	89.50
2	Zhao and Kumar [13]	Semantic assisted convolutional neural network	82.43
3	Nie et al. [22]	Convolutional restricted Boltzman machine (CRBM) with dense SIFT	50.1

6 Conclusion and Future Work

The primary objective of this research work is to analyze the effect of eyeglasses and masking of the eye portion of the images on recognition accuracy using deep CNN on the UBIPr dataset. For the first issue, Deep CNN models with transfer learning approach ResNet 50 and VGG 19 obtained remarkable recognition accuracy of 98.89% and for the second issue, we found that VGG 19 obtained maximum recognition accuracy of 78%. Means, masking of eye region (hiding of features eye shape, sclera, and eye socket) in periocular region images degrade the recognition accuracy by 35–40%. So, they must be considered as a critical component for periocular Biometrics.

We have also compared the maximum accuracy obtained with our fine-tuned network to other reported research in literature and found that the proposed model outperforms others.

7 Future Work

Future work includes implementation of the concept of attention mechanism and integration of semantic information such as gender information with basic features (edge, corners, key points, etc.) in a single learning model.

References

1. Mahalingam, G., Ricanek, K.: LBP-based periocular recognition on challenging face datasets. EURASIP J. Image Video Process. **1**, 1–36 (2013). https://doi.org/10.1186/1687-5281-2013-36
2. Park, U., Jillela, R.R., Ross, A., Jain, A.K.: Periocular biometrics in the visible spectrum. IEEE Trans. Inf. Forensics Secur. **6**(1), 96–106 (2011). https://doi.org/10.1109/TIFS.2010.2096810

3. A comprehensive hands-on guide to transfer learning with real-world applications in deep learning. https://towardsdatascience.com/a-comprehensive-hands-on-guide-to-transfer-learning-with-real-world-applications-in-deep-learning-212bf3b2f27a

4. Oh, B.S., Oh, K., Toh, K.A.: On projection-based methods for periocular identity verification In: International Conference on Industrial Electronics and Applications, pp. 871–876. Singapore (2012). https://doi.org/10.1109/iciea.2012.6360847

5. Uzair, M., Mahmood, A., Mian, A. et al.: Periocular biometric recognition using image sets. In: IEEE Workshop on Applications of Computer Vision, pp. 246– 251. Tampa, FL, USA (2013). https://doi.org/10.1109/wacv.2013.6475025

6. Chen, H., Gao, M., Ricanek, K., Xu, W., et al.: A novel race classification method based on periocular features fusion. Int. J. Pattern Recognit Artif Intell. 31(8), 1–21 (2017). https://doi.org/10.1142/S0218001417500264

7. Lyle, J., Miller, P., Pundlik, S. et al.: Soft biometric classification using periocular region features. In: IEEE International Conference on Biometrics: Theory Applications and Systems, pp. 1–7. Washington, DC, USA (2010). https://doi.org/10.1109/btas.2010.5634537

8. Boddeti, V., Smereka, J., Kumar, B.: A comparative evaluation of iris and ocular recognition methods on challenging ocular images. In: IEEE International Joint Conference on Biometrics, pp. 10–18. Washington, DC, USA (2011). https://doi.org/10.1109/ijcb.2011.6117500

9. Zhao, Z., Kumar, A.: Improving periocular recognition by explicit attention to critical regions in deep neural network. IEEE Trans. Inform. Frensics Secur. 13(12), 1–15 (2018). https://doi.org/10.1109/TIFS.2018.2833018

10. Alahmadi, M., Hussain, H., Aboalsamh, et al.: Convsrc: Smartphone based peri ocular recognition using deep convolutional neural network and sparsity augmented collaborative representation (2018). https://doi.org/10.1016/j.patcog.2016.12.017, arXiv:1801.05449

11. Tapia, J., Aravena Carlos C.: Gender Classification from periocular NIR images using fusion of CNNs models. In: IEEE International Conference on Identity, Security, and Behavior Analysis, pp. 1–6. Singapore (2018). https://doi.org/10.1109/isba.2018.8311465

12. Keshari, R., Ghosh, S., Agarwal, A., et al.: Mobile periocular matching with pre- Post cataract surgery. In: IEEE International Conference on Image Processing, pp. 1–6. Phoenix, AZ, USA (2016). https://doi.org/10.1109/icip.2016.7532933

13. Zhao, Z., Kumar, A.: Accurate periocular recognition under less constrained environment using semantics-assisted convolutional neural network. IEEE Trans. Inf. Forensics Secur. 12(5), 1017–1030 (2017). https://doi.org/10.1109/TIFS.2016.2636093

14. UBIPr http://socia-lab.di.ubi.pt/~ubipr/. Last accessed 19 May 2019

15. Liu, P., Jing-Ming Guo, J.-M., et al.: Ocular recognition for blinking eyes. IEEE Trans. Image Process. 26(10), 5070–5081 (2017). https://doi.org/10.1109/TIP.2017.2713041

16. Hussain, M., Bird, J.J., Faria, D.R.: A study on CNN transfer learning for im age classification. In: Workshop on Computational Intelligence, pp. 191–202. Nottingham, Springer, Berlin (2018). https://doi.org/10.1007/978-3-319-97982-3_16

17. Krizhevsky, A., Sutskever, I., Hinton, G.E.: ImageNet classification with deep convolutional neural networks. Commun. ACM 60(6), 84–90 (2012). https://doi.org/10.1145/3065386

18. Szegedy, C., Liu, W., Jia, Y., et al.: Going deeper with convolutions. In: IEEE Conference on Computer Vision and Pattern Recognition, pp. 1–9. Boston, MA, USA (2015). https://doi.org/10.1109/cvpr.2015.7298594

19. He, K., Zhang, X., Ren, S. et al.: Deep residual learning for image recognition. In: IEEE Conference on Computer Vision and Pattern Recognition, pp. 770– 778. Las Vegas, NV, USA (2016). https://doi.org/10.1109/cvpr.2016.90

20. Simonyan, K., Zisserman, A.: Very deep convolutional networks for large-scale image recognition (2014). arXiv:1409.1556
21. Miller P.E., Rawls, A.W., Pundlik, S.J., Woodard, D.L.: Personal identification using periocular skin texture. In: ACM Symposium on Applied Computing, pp. 1496–1500 ACM (2010). https://doi.org/10.1145/1774088.1774408
22. Nie, L., Kumar, A., Zhan, S.: Periocular recognition using unsupervised convolutional RBM feature learning. In: IEEE International Conference on Pattern Recognition, pp. 399–404. Stockholm, Sweden (2014). https://doi.org/10.1109/icpr.2014.77

An Elliptic Curve Cryptography-Based Multi-Server Authentication Scheme Using Cancelable Biometrics

Subhas Barman, Abhisek Chaudhuri, Atanu Chatterjee and Md. Ramiz Raza

Abstract Authentication is a very important element in any secure cryptographic mechanism. For secure authentication between two parties in an open network, various schemes have been proposed. Previously many of them involved communication between clients and a single server and biometrics were not used. Later, to increase the level of security, three-factor authentication schemes involving passwords, biometrics, and smart cards were introduced for secure communication in a multi-server environment. But, majority of these schemes do not provide facility for biometrics update phase. Additionally, these existing schemes are vulnerable to different forms of adverse attacks. A three-factor-based authentication scheme in a multi-server environment using Elliptic Curve Cryptography (ECC) has been introduced in our paper to remove these kinds of security flaws. In order to prove the validity of the secrecy provided in the proposed scheme, the widely accepted ProVerif tool has been used. We have also discussed informal security analysis of the proposed scheme. Finally, the proposed scheme is compared with existing related schemes based on different performance and security parameters.

Keywords Authentication · Multi-server · ECC · ProVerif

S. Barman (✉) · A. Chaudhuri · A. Chatterjee · Md. Ramiz Raza
Jalpaiguri Government Engineering College, Jalpaiguri, India
e-mail: subhas.barman@gmail.com

A. Chaudhuri
e-mail: abhisek.chaudhuri.1959@gmail.com

A. Chatterjee
e-mail: atanuc13@gmail.com

Md. Ramiz Raza
e-mail: raza00120@gmail.com

© Springer Nature Singapore Pte Ltd. 2020
V. Bhateja et al. (eds.), *Intelligent Computing and Communication*,
Advances in Intelligent Systems and Computing 1034,
https://doi.org/10.1007/978-981-15-1084-7_16

153

1 Introduction

In this modern era of information technology, secret and secure communication is of utmost importance. Due to advancement of the technologies, most of the services are now provided using a multi-server architecture. As more of these services are being provided, safe, secure and authenticated access to data is becoming an issue of utmost importance. The first work in this field was done by Lamport [1] in 1981 where he used password verifier table. But since it stored the passwords, it could not withstand stolen-verifier attack [2]. Several two-factor-based authentication schemes [3, 4] were proposed to overcome this flaw. In single-server-based authentication schemes, a user needs to individually register to each server in order to access them securely. But, this causes inconvenience since user has to register repeatedly. Moreover, in a multi-server environment, the protocols designed for authentication in single-server environments, are impossible to be applied directly. To obviate these problems, several multi-server authentication schemes have been proposed [5–7].

In 2009, an authentication scheme [8] using dynamic ID was introduced by Liao and Wang. Later, it was shown that this scheme could not withstand server spoofing attack, insider attack, and masquerade attack. Later, Hsiang and Shih [5] made some modifications to this scheme and proposed an improvement. But, it could not withstand server spoofing attack, stolen smart card attack, and masquerade attack. So Sood et al. proposed a new dynamic ID-based scheme [6], claiming prevention of attacks and achieving user anonymity.

A significant drawback of two-factor-based authentication schemes is that passwords might be divulged and smart cards may be stolen or lost. So to increase the level of security, an additional layer of security was introduced in the form of biometrics. Several three-factor-based multi-server schemes were proposed [9, 10]. A strong biometric-based scheme [9] was proposed by He and Wang. However, it was discovered that this could not prevent replay attack, impersonation and known session temporary information attack by Odelu et al. and proposed a newer and improved scheme [10].

Mishra proposed a secure multi-server authentication scheme [11] in 2014. But, it was found out that it was vulnerable to replay, masquerading and denial of service attacks by Lu et al. [12]. However, Lu et al.'s scheme [12] were vulnerable to lack of user anonymity attack, impersonation attack, man in middle and perfect forward secrecy. So Reddy et al. [13] proposed a robust and improved multi-server based authentication scheme using ECC. Ali and Pal [14] proposed a confidentiality preserving authentication scheme. However, Barman et al [15] found out that Ali and Pal's scheme [14] suffered from privileged insider attack. Majority of the proposed schemes are susceptible to different network attacks. They do not offer strong user anonymity and biometrics change phase. We have put forward a secure authentication scheme in multi-server environment in this paper. Here, we have used ECC for session key exchange and revocable template generations [15, 16] for biometrics data.

2 Proposed Scheme

Our scheme consists of six major phases: (1) Server Registration, (2) User Registration, (3) Login, (4) Mutual Authentication and Key Exchange,(5) Password and Biometric Update Phase, and (6) Smart Card Revocation Phase. Also, this scheme has three participants: (1) User, (2) Server, and (3) Registration Center (RC). Used notations and their descriptions are given in Table 1

2.1 Server Registration Phase

Each server, say S_j, sends its identity SID_j and x_{s_j} via a secure channel to the Registration Center RC. On receiving them, RC calculates $x_j = h(PSK \| SID_j)$ and via a secure channel, sends it to RC. Also RC creates a table T_{SK} in which it stores the pair $\{SID_j, H_j = x_{s_j}P\}$ for every server S_j.

2.2 User Registration Phase

User U_i inputs ID_i, PW_i, and BIO_i. User also selects T_{P_i}. User then generates $C_{T_i} = f(BIO_i, T_{P_i})$. Also, user calculates $RPW_i = h(PW_i \| C_{T_i})$ and $F_i = h(ID_i)$. Via a secure channel, user then sends F_i and RPW_i to RC. On receiving these, RC creates a

Table 1 Used notations

Symbol	Description
U_i, ID_i	ith user and User ID of user U_i
PW_i, BIO_i, SC_i	Password, biometrics, and Smart Card of user U_i
S_j, SID_j	jth server and identity of server S_j
RC, PSK	Registration Center and pre-shared key of RC
P, E	Generating point P of elliptic curve E
x_{s_j}, x_{u_i}	Private keys of server S_j and user U_i, respectively
$f(.)$	One-way biometric transformation function
C_{T_i}	Cancelable Biometric Template of user U_i
T_{P_i}	Transformation parameter for cancelable template generation
N_1, N_2	Random nonce generated by U_i and S_j, respectively
SK_{ij}	Session Key between U_i and S_j
$h(.)$	One-way collision resistant cryptographic hash function
ΔT	Acceptable threshold for transmission delay.
$\oplus, \|$	XOR and string concatenation operator, respectively

table T_R and stores a pair $\{F_i, T_c\}$ into it, where T_c indicates current timestamp. RC calculates $A_i = h(F_i||PSK)$, $B_i = h(A_i||h(RPW_i))$ and $C_i = A_i \oplus RPW_i$. RC then selects a random nonce n_1. Then RC calculates $D_i = n_1 \oplus h(RPW_i||F_i||A_i)$. RC then stores B_i, C_i, D_i, T_{SK} into smart card SC_i. Via a secure channel, SC_i is issued to U_i by RC. User then stores T_{P_i} and $h(.)$ into SC_i. Finally, SC_i contains $\{B_i, C_i, D_i, T_{SK}, T_{P_i}, \text{ and } h(.)\}$.

2.3   Login Phase

User U_i inputs ID_i, PW_i and BIO_i^*. User then calculates $C_{T_i}^* = f(BIO_i^*, T_{P_i})$, $RPW_i = h(PW_i||C_{T_i}^*)$, $A_i^* = C_i \oplus RPW_i$, $B_i^* = h(A_i^*||h(RPW_i))$. If B_i^* equals B_i, then the user is authentic and is allowed to login. Then U_i further calculates $F_i = h(ID_i), n_1 = D_i \oplus h(RPW_i||F_i||A_i^*)$. U_i then chooses a random nonce N_1. U_i then calculates $K_u = x_{u_i}P$, $K = x_{u_i}H_j$, $RPW_{ij} = h(RPW_i||SID_j)$, $M_1 = N_1 \oplus K$, $M_2 = N_1 \oplus RPW_{ij}$, $M_3 = n_1 \oplus RPW_{ij} \oplus N_1$, $M_4 = F_i \oplus n_1 \oplus N_1$ and $Z_1 = h(F_i||N_1||n_1||RPW_{ij}||SID_j||T_1)$. Then U_i sends $K_u, M_1, M_2, M_3, M_4, Z_1, T_1$ via a login secure channel to server as login message T_1 is the current time at U_i.

2.4   Mutual Authentication and Key Exchange Phase

Upon receiving the login messages, server S_j checks if $|T_1 - T_c| > \Delta T$, where T_c is the current server time. In case yes, the session is immediately terminated. Otherwise, it goes ahead and calculates $K = x_{s_j}K_u$, $N_1 = M_1 \oplus K$, $RPW_{ij} = M_2 \oplus N_1$, $n_1 = M_3 \oplus N_1 \oplus RPW_{ij}$ and $F_i = M_4 \oplus n_1 \oplus N_1$. Then S_j checks if $Z_1 = h(F_i||N_1||n_1||RPW_{ij}||SID_j||T_1)$. If true, authenticity of the user U_i is established, otherwise session is immediately terminated. Then server chooses a random nonce N_2. Then S_j calculates $M_5 = x_j \oplus h(RPW_{ij}||N_1||n_1)$, $M_6 = N_2 \oplus h(RPW_{ij}||x_j)$, $SK_{ji} = h(N_1||N_2||K||n_1||F_i)$, $Z_2 = h(SK_{ji}||N_2||x_j||RPW_{ij}||T_2)$. S_j then sends M_5, M_6, Z_2, T_2 via a public channel where T_2 is current time in S_j, to U_i. After receiving these, U_i checks if $|T_2 - T_c| > \Delta T$. If true, then it terminates the session immediately. Otherwise, it calculates $x_j = M_5 \oplus h(RPW_{ij}||N_1||n_1)$, $N_2 = M_6 \oplus h(RPW_{ij}||x_j)$ and $SK_{ij} = h(N_1||N_2||K||n_1||F_i)$. U_i then checks if $Z_2 = h(SK_{ij}||N_2||x_j||RPW_{ij}||T_2)$. If true, it is established that the server is authentic. This entire process is graphically explained in Fig. 1.

U_i	S_j
Input ID_i, PW_i, BIO_i^*	

$C_{T_i}^* = f(BIO_i^*, T_{P_i})$
$RPW_i = h(PW_i||C_{T_i}^*)$
$A_i^* = C_i \oplus RPW_i$
$B_i^* = h(A_i^*||h(RPW_i))$
if ($B_i^* \neq B_i$) {reject}
$F_i = h(ID_i)$
$n_1 = D_i \oplus h(RPW_i||F_i||A_i^*)$
User chooses random nonce N_1
$K_u = x_{u_i}P$
$K = x_{u_i}H_j$
$RPW_{ij} = h(RPW_i||SID_j)$
$M_1 = N_1 \oplus K$
$M_2 = N_1 \oplus RPW_{ij}$
$M_3 = n_1 \oplus RPW_{ij} \oplus N_1$
$M_4 = F_i \oplus n_1 \oplus N_1$
$Z_1 = h(F_i||N_1||n_1||RPW_{ij}||SID_j||T_1)$

$$\xrightarrow{\quad K_u, M_1, M_2, M_3, M_4, Z_1, T_1 \quad}$$
(Public Channel)

if ($|T_1 - T_c| > \Delta T$) {reject}
$K = x_{s_j} K_u$
$N_1 = M_1 \oplus K$
$RPW_{ij} = M_2 \oplus N_1$
$n_1 = M_3 \oplus N_1 \oplus RPW_{ij}$
$F_i = M_4 \oplus n_1 \oplus N_1$
Check $Z_1 \stackrel{?}{=} h(F_i||N_1||n_1||RPW_{ij}||SID_j||T_1)$
Choose random nonce N_2
$M_5 = x_j \oplus h(RPW_{ij}||N_1||n_1)$
$M_6 = N_2 \oplus h(RPW_{ij}||x_j)$
$SK_{ji} = h(N_1||N_2||K||n_1||F_i)$
$Z_2 = h(SK_{ji}||N_2||x_j||RPW_{ij}||T_2)$

$$\xleftarrow{\quad M_5, M_6, Z_2, T_2 \quad}$$
(Public Channel)

if ($|T_2 - T_c| > \Delta T$) {reject}
$x_j = M_5 \oplus h(RPW_{ij}||N_1||n_1)$
$N_2 = M_6 \oplus h(RPW_{ij}||x_j)$
$SK_{ij} = h(N_1||N_2||K||n_1||F_i)$
Check $Z_2 \stackrel{?}{=} h(SK_{ij}||N_2||x_j||RPW_{ij}||T_2)$ {Server Authentic}

Fig. 1 Mutual authentication and key exchange phase

2.5 Password and Biometric Change Phase

User inputs ID_i, PW_i^* and BIO_i^*. SC_i calculates $C_{T_i}^* = f(BIO_i^*, T_{P_i})$, $RPW_i^* = h(PW_i^*||C_{T_i}^*)$, $A_i^* = C_i \oplus RPW_i^*$, $B_i^* = h(A_i^*||h(RPW_i^*))$ $F_i = h(ID_i)$, $n_1 = D_i \oplus h(RPW_i^*||F_i||A_i^*)$. It then checks if $B_i = B_i^*$. If true, user is allowed to enter new password and biometric data. User inputs new PW_i^{new}, BIO_i^{new} and $T_{P_i}^{new}$, $C_{T_i}^{new} = f(BIO_i^{new}, T_{P_i}^{new})$, $RPW_i^{new} = h(PW_i^{new}||C_{T_i}^{new})$, $B_i^{new} = h(A_i^*||h(RPW_i^{new}))$, $C_i^{new} = A_i^* \oplus RPW_i^{new}$, $D_i^{new} = n_1 \oplus h(RPW_i^{new}||F_i||A_i^*)$. Finally, B_i^{new}, C_i^{new}, D_i^{new}, $T_{P_i}^{new}$ are replaced in SC_i.

2.6   Smart Card Revocation Phase

In case SC_i is damaged, lost or stolen, U_i can request a new one from RC. U_i inputs ID_i, PW_i and imprints biometric BIO_i. User calculates $F_i = h(ID_i)$ $C_{T_i} = f(BIO_i, T_{P_i})$ and $RPW_i = h(PW_i||C_{T_i})$. U_i sends F_i and RPW_i via a secure channel to the RC. On receiving F_i and RPW_i, RC checks if an entry associated with F_i is present in the table T_R. If it is present then it indicates U_i is a valid user and then, the steps in Sect. 2.2 are executed by RC and entry $\{F_i, T_c\}$ in T_R is replaced with $\{F_i, T_c^*\}$ where T_c^* is current timestamp.

3   Security Analysis

Here, we have analyzed our scheme from different aspects of security. For formal security analysis, we have used ProVerif. We also have discussed the informal security analysis of this scheme and lastly, a comparison is made between our scheme and some other multi-server authentication schemes.

3.1   Formal Security Analysis

For the verification of robustness of any cryptographic protocol against any kind of active and passive adversaries, various verification tools are used. Among them, ProVerif is widely accepted as a protocol verification tool [15]. For the verification our proposed scheme, we have used ProVerif, for its ease of use automated testing given some cryptographic parameters. It uses applied π-calculus for all the verification purposes. In our implementation, we have used the latest version (2.00). All the communications take place in channels. So, we have taken a public channel.

```
free Ch_pub:channel.
```

The variables and functions declared in our implementation are show in Table 2. We have taken two processes, $User U_i$ and $Server S_j$, shown in Table 3. The events are defined as follows:

```
event begin_UserUi(bitstring).
event end_UserUi(bitstring).
event begin_ServerSj(bitstring).
event end_ServerSj(bitstring).
```

The main process is defined as follows:

```
process ( (!ServerSj) | (!UserUi) )
```

The security of the random nonces and session key are verified using a few queries. The corresponding results are summarized in Table 4

Table 2 Functions and variables

Variables	Functions
free IDi:bitstring [private].	
free PWi:bitstring [private].	
free BIOi:bitstring [private].	fun f(bitstring,bitstring): bitstring.
free SIDj:bitstring.	fun h(bitstring): bitstring.
free Hj:bitstring [private].	fun mult(bitstring,bitstring): bitstring.
const P:bitstring.	fun concat(bitstring,bitstring):
free xsj:bitstring [private].	bitstring.
free xj:bitstring [private].	fun xor(bitstring,bitstring): bitstring.
free Bi:bitstring [private].	equation forall a:bitstring,b: bitstring;
free Ci:bitstring [private].	xor(xor(a,b),b)=a.
free Di:bitstring [private].	
free TPi:bitstring [private].	

Table 3 Processes

Process User U_i	Process Server S_j
let UserUi=	
let CTi=f(BIOi,TPi) in	
let RPWi=h(concat(PWi,CTi)) in	
let Ai'=xor(Ci,RPWi) in	
let Bi'=h(concat(Ai',h(RPWi))) in	
if(Bi'=Bi) then	
let Fi=h(IDi) in	let ServerSj=
let n1=xor(Di,h(concat(RPWi,	in (Ch_pub,(xKu:bitstring,
concat(Fi,Ai')))) in	xM1:bitstring,xM2:bitstring,
new N1:bitstring;	xM3:bitstring,xM4:bitstring,
new xui:bitstring;	xZ1:bitstring,xT1:bitstring));
let Ku=mult(xui,P) in	let K=mult(xsj,xKu) in
let K=mult(xui,Hj) in	let N1=xor(xM1,K) in
let RPWij=h(concat(RPWi,SIDj)) in	let RPWij=xor(xM2,N1) in
let M1=xor(N1,K) in	let n1=xor(xM3,xor(N1,RPWij)) in
let M2=xor(N1,RPWij) in	let Fi=xor(xM4,xor(n1,N1)) in
let M3=xor(n1,xor(RPWij,N1)) in	if(xZ1=h(concat(Fi,concat(N1,
new T1:bitstring;	concat(n1,concat(RPWij,
let M4=xor(Fi,xor(n1,N1)) in	concat(SIDj,xT1))))))) then
let Z1=h(concat(Fi,concat(N1,	new N2:bitstring;
concat(n1,concat(RPWij,	let M5=xor(xj,h(concat(RPWij,
concat(SIDj,T1)))))) in	concat(N1,n1)))) in
out (Ch_pub, (Ku,M1,M2,M3,M4,Z1,T1));	let M6=xor(N2,h(concat(RPWij, xj))) in
in (Ch_pub,(xM5:bitstring,	let SKji=h(concat(N1,concat(N2,
xM6:bitstring,xZ2:bitstring,	concat(K,concat(n1,Fi))))) in
xT2:bitstring)); let	new T2:bitstring;
xj=xor(xM5,concat(RPWij, concat(N1,n1)))	let Z2=h(concat(SKji,concat(N2,
in	concat(xj,concat(RPWij,T2))))) in
let N2=xor(xM6, h(concat(RPWij,xj))) in	out (Ch_pub,(M5,M6,Z2,T2))
let SKij=h(concat(N1,concat(N2,	else 0.
concat(K,concat(n1,Fi))))) in	
if(xZ2=h(concat(SKij,concat(N2,	
concat(xj,concat(RPWij,xT2))))) then	
0.	

Table 4 Results of proverif code

Queries	Result
query attacker(SKij).	not attacker(SKij[]) is true
query attacker(SKji).	not attacker(SKji[]) is true
query attacker(n1).	not attacker(n1[]) is true
query attacker(N1).	not attacker(N1[]) is true
query attacker(N2).	not attacker(N2[]) is true
query id:bitstring; event(end_UserUi(id)) ==> event(begin_UserUi(id)).	event(end_UserUi(id)) ==> event(begin_UserUi(id)) is true
query id:bitstring; event(end ServerSj(id)) ==> event(begin_ServerSj(id)).	event(end_ServerSj(id_38)) ==> event(begin_ServerSj(id_38)) is true

3.2 Informal Security Analysis

Here, we have analyzed our scheme for informal security and also shown this scheme withstands several attacks. Here, we denote adversary as $\mathcal{A}$.

1. **Anonymity and Untraceability**: If $\mathcal{A}$ intercepts communication messages $M_4 = F_i \oplus n_1 \oplus N_1$ through public channel, it is not computationally possible to guess n_1 and N_1 within polynomial time. It is also not possible to calculate ID_i from F_i due to one-way hash function. Due to involvement of random nonces N_1 and N_2 transmitted messages, $K_u, M_1, M_2, M_3, M_4, Z_1, M_5, M_6$ and Z_2 between user and server are dynamic and different for an user in each sessions. So, over different sessions, $\mathcal{A}$ will not be able to trace an user. This shows preservation of anonymity and untraceability property.

2. **Stolen Smart Card Attack**: Suppose $\mathcal{A}$ extracts all secret informations $B_i, C_i, D_i, T_{SK}, T_{P_i}$ which are in SC_i, he/she cannot find PW_i and C_{T_i} from secret information because of the hashing functions used. So, it can withstand stolen smart card attack.

3. **Replay Attack**: If $\mathcal{A}$ tries to send the messages $K_u, M_1, M_2, M_3, M_4, Z_1, M_5, M_6$ and Z_2 later again after intercepting these messages from public channel to gain access of secret credentials of an user, the checks $|T_1 - T_c| > \Delta T$ at user side and $|T_2 - T_c| > \Delta T$ at server side would fail as ΔT is used for verification and moreover time stamp is included in message Z_1 and Z_2. So, it can resist replay attacks.

4. **Man-in-the-Middle Attack**: Here $\mathcal{A}$ intercepts all messages between valid U_i and S_j and it tries to modify these message to pretend as a valid user. $\mathcal{A}$ needs the knowledge of K, RPW_{ij}, F_i, n_1 and SID_j, which are hashed so cannot be found. This shows our scheme resists man-in-the-middle attacks.

5. **Privileged Insider Attack**: User sends F_i and RPW_i to the RC. But $\mathcal{A}$ from RC cannot compute U_i's identity ID_i, password PW_i, and biometrics template C_{T_i} with the knowledge of F_i and RPW_i due to one-way hash function. $\mathcal{A}$ from server S_j is also unable to compute ID_i, PW_i and C_{T_i} from the messages M_1, M_2, M_3, M_4 and Z_1. Therefore, it is not vulnerable against privileged insider attacks.

6. **Known Session Key Temporary Information Attack**: Let $\mathcal{A}$ has information about random nonces N_1, N_2 and n_1 and tries to compute session key $SK_{ij} = h(N_1||N_2||K||n_1||F_i)$, but SK_{ij} does not depend only on N_1 and N_2 but it also depends on F_i and K. For $\mathcal{A}$, it is computationally infeasible to compute F_i and K. So, $\mathcal{A}$ cannot find the session key.

7. **Impersonation attack**: Suppose $\mathcal{A}$ intercepted all the communication messages and imitates as valid U_i or S_j after doing some modifications.

   (a) $\mathcal{A}$ selects a random Nonce N_1^* and tries to compute $M_1 = N_1 \oplus K$, $M_2 = N_1 \oplus RPW_{ij}$, $M_3 = n_1 \oplus RPW_{ij} \oplus N_1$, $M_4 = F_i \oplus n_1 \oplus N_1$ and $Z_1 = h(F_i||N_1||n_1||RPW_{ij}||SID_j||T_1)$ where $F_i = h(ID_i)$, $RPW_{ij} = h(RPW_i||SID_j)$ and $RPW_i = h(PW_i||C_{T_i})$. However, $\mathcal{A}$ has no knowledge about user's secret credentials ID_i, PW_i, and C_{T_i}. So, user impersonation attacks can be foiled by this scheme.

   (b) $\mathcal{A}$ generates a random nonce N_2^* and tries to compute $M_5 = x_j \oplus h(RPW_{ij}||N_1||n_1)$, $M_6 = N_2 \oplus h(RPW_{ij}||x_j)$, $[x_j = h(PSK||SID_j)]$. Also, $Z_2 = h(SK_{ji}||N_2||x_j||RPW_{ij}||T_2)$, $SK_{ji} = h(N_1||N_2||K||n_1||F_i)$ is calculated. But $\mathcal{A}$ needs the knowledge about K, x_j, n_1 and F_i at the same time which is not computationally feasible. Therefore, it can resist server impersonation attacks.

8. **Forward Secrecy**: If $\mathcal{A}$ knows the secret keys K, n_i and $F_i = h(ID_i)$ still it is not possible to calculate $SK_{ij} = h(N_1||N_2||K||n_1||F_i)$ as N_1 and N_2 are randomly generated and not repeated. So, forward secrecy is preserved in this scheme.

9. **Efficient and Secure Password/Biometric Template Update**: User U_i enters ID_i, existing password PW_i and biometric C_{T_i} into the smart card SC_i. If everything is correct, user is allowed to enter new PW_i^* and $C_{T_i}^*$. SC_i update the information stored into it according to new PW_i^* and $C_{T_i}^*$ without any involvement of RC. Thus, efficient and secure password/biometric template update is ensured in this scheme.

3.3 Performance Analysis

We have compared our protocol with a few existing schemes based on the security provided by our proposed scheme. It is summarized in the Table 5. The notations used are Th: one-way cryptographic hash computing time, Tecc: time taken for point multiplication over an elliptic curve, Tsym: symmetric decryption/encryption time, Texp: modular exponentiation time, Tbio: time taken to create cancelable biometric template. Based on the experimental results used in [17], we have used the times for above computations as follows: Th: 0.0023 ms, Tecc: 2.226 ms, Tsym: 0.0046 ms, Texp: 0.0046 ms, and Tbio: 2.226 ms. Also, the communication costs have been calculated on the basis of the results in [17] and are as follows: U_i, S_j, $h(.)$, and $f(.)$ are 160bits, respectively, timestamp is of 32bits, ECC point is of $160 + 160 = 320$ bits.

Table 5 Comparison of security features

Security Features	Ref. [9]	Ref. [12]	Ref. [14]	Ref. [18]	Ref. [19]	Proposed scheme
Anonymity and Untraceability	No	No	No	No	No	Yes
Stolen smart card attack	No	Yes	Yes	Yes	Yes	Yes
Replay attack	Yes	Yes	Yes	Yes	Yes	Yes
Efficient and Secure password and biometric template update	No	No	No	No	No	Yes
Privileged insider attack	Yes	No	No	Yes	Yes	Yes
Known session key/Temporary information attack	Yes	Yes	Yes	Yes	No	Yes
Impersonation attack	No	No	Yes	No	Yes	Yes
Forward secrecy	Yes	No	Yes	Yes	Yes	Yes

Table 6 Comparison of computation time and communication cost

PP	CCRP	CCLAP	CCPBCP	TCC	TCT	CC
Ref. [9]	3Th	23Th + 8Tecc	2Th	28Th + 8Tecc	17.8724	3520
Ref. [12]	4Th	16Th	4Th	24Th	0.0552	1216
Ref. [14]	4Th	16Th + 2Texp	10Th	30Th + 2Texp	0.0782	1664
Ref. [18]	8Th	13Th + 4Texp	6Th	27Th + 4Texp	0.0805	2688
Ref. [19]	4Th	12Th + 10Tsym	4Th	20Th + 10Tsym	0.092	4032
Proposed scheme	6Th + 1Tbio	15Th + 3Tecc + 1Tbio	8Th + 2Tbio	29Th + 3Tecc + 4Tbio	15.6487	2304

Abbreviations used in Table 6 are PP: performance parameters, CCRP: registration phase's computation cost, CCLAP: login and authentication phase's computation cost, CCPBCP: computation cost for password and biometrics change phase, TCC: total computation cost, TCT: total computation time (in milliseconds), CC: communication cost (in bits).

Here, we can see marginal increase in computation cost when compared to other existing protocols. However, that is an acceptable tradeoff given the enhanced security provided by our proposed scheme.

4 Conclusion

In this paper, we have proposed a new three-factor-based secure key exchange scheme in a multi-server environment. We have verified the security of the proposed protocol and also have compared with a few existing schemes on the basis of security provided and computation time. We have shown that when compared to existing schemes, our proposed scheme provides better security. Also, efficient and secure mechanism to update biometric data of users is provided by our scheme. Although, when compared to existing schemes, computation time taken in our scheme is more, but given the enhanced security and features provided in our scheme, it is an acceptable tradeoff. The communications cost in our scheme is similar to existing schemes.

References

1. Lamport, L.: Password authentication with insecure communication. Commun. ACM **24**(11), 770–772 (1981)
2. Hwang, M.-S., Li, H.L.: A new remote user authentication scheme using smart cards. IEEE Trans. Consum. Electron. **46**(1), 28–30 (2000). https://doi.org/10.1109/30.826377
3. Juang, W.S., Chen, S.T., Liaw, H.T.: Robust and efficient password-authenticated key agreement using smart cards. IEEE Trans. Ind. Electron. **55**(6), 2551–2556 (2008)
4. Sun, D.Z., Huai, J.P., Sun, J.Z., Li, J.X., Zhang, J.W., Feng, Z.Y.: Improvements of juang's password-authenticated key agreement scheme using smart cards. IEEE Trans. Ind. Electron. **56**(6), 2284–2291 (2009)
5. Hsiang, H.C., Shih, W.K.: Improvement of the secure dynamic id based remote user authentication scheme for multi-server environment. Comput. Stand. Interfaces **31**(6), 1118–1123 (2009)
6. Sood, S.K., Sarje, A.K., Singh, K.: A secure dynamic identity based authentication protocol for multi-server architecture. J. Netw. Comput. Appl. **34**(2), 609–618 (2011)
7. Satapathy, S.C., Bhateja, V., Raju, K.S., Janakiramaiah, B.: Computer communication, networking and internet security. In: Proceedings of IC3T, vol. 5. Springer, Berlin (2017)
8. Liao, Y.P., Wang, S.S.: A secure dynamic id based remote user authentication scheme for multi-server environment. Comput. Stand. Interfaces **31**(1), 24–29 (2009)
9. He, D., Wang, D.: Robust biometrics-based authentication scheme for multiserver environment. IEEE Syst. J. **9**(3), 816–823 (2015)
10. Odelu, V., Das, A.K., Goswami, A.: A secure biometrics-based multi-server authentication protocol using smart cards. IEEE Trans. Inform. Forensics Secur. **10**(9), 1953–1966 (2015)
11. Mishra, D., Das, A.K., Mukhopadhyay, S.: A secure user anonymity-preserving biometric-based multi-server authenticated key agreement scheme using smart cards. Exp. Syst. Appl. **41**(18), 8129–8143 (2014)
12. Lu, Y., Li, L., Yang, X., Yang, Y.: Robust biometrics based authentication and key agreement scheme for multi-server environments using smart cards. PLoS One **10**(5), e0126323 (2015)
13. Reddy, A.G., Das, A.K., Odelu, V., Yoo, K.Y.: An enhanced biometric based authentication with key-agreement protocol for multi-server architecture based on elliptic curve cryptography. PloS One **11**(5), e0154308 (2016)
14. Ali, R., Pal, A.K.: Three-factor-based confidentiality-preserving remote user authentication scheme in multi-server environment. Arabian J. Sci. Eng. **42**(8), 3655–3672 (2017)
15. Barman, S., Guha, P., Saha, R., Ghosh, S.: Cryptanalysis and improvement of three-factor-based confidentiality-preserving remote user authentication scheme in multi-server environment. In: Proceedings of International Ethical Hacking Conference, pp. 75–87. Springer, Berlin (2019)
16. Barman, S., Shum, H.P.H., Chattopadhyay, S., Samanta, D.: A secure authentication protocol for multi-server-based e-healthcare using a fuzzy commitment scheme. IEEE Access **7**, 12557–12574 (2019). https://doi.org/10.1109/ACCESS.2019.2893185
17. Barman, S., Das, A.K., Samanta, D., Chattopadhyay, S., Rodrigues, J.J., Park, Y.: Provably secure multi-server authentication protocol using fuzzy commitment. IEEE Access **6**, 38578–38594 (2018)
18. Li, X., Niu, J., Kumari, S., Liao, J., Liang, W.: An enhancement of a smart card authentication scheme for multi-server architecture. Wirel. Pers. Commun. **80**(1), 175–192 (2015)
19. Wen, F., Susilo, W., Yang, G.: Analysis and improvement on a biometric-based remote user authentication scheme using smart cards. Wirel. Pers. Commun. **80**(4), 1747–1760 (2015)

Dynamic TDMA Scheduling for Topology Controlled Clustering in Wireless Sensor Networks

Hemantaraj M. Kelagadi, B. Shrinidhi and Priyatamkumar

Abstract The need for interconnection among multiple sensor nodes in Wireless Sensor Networks (WSN) leads to a higher collision rate and correlated contention based on the incremental node participation in the network. Limited power resources and variation in the topological structures are the major constraints for centralized timeslot assigning algorithm. Though the topology variation gets addressed by distributed timeslot management algorithm, they fail to reduce the energy depletion. Incorporation of Time Division Multiple Access (TDMA) for scheduling the slot is found to eliminate the collision by guaranteeing the energy conservation. Efficient distribution of TDMA slots to minimize the energy and topology factors has been addressed by several algorithms. In this paper, a technique for energy minimization is proposed for controlling the topology of the WSN. We propose a dynamic TDMA scheduling and integrate within TL-LEACH protocol. The proposed algorithm is compared with LEACH protocol (Low Energy Adaptive Clustering Hierarchy) and TL-LEACH (two-level LEACH) for analyzing the performance and is found to perform better by minimizing the energy depletion and help in enhancing the lifetime of the network.

Keywords Wireless sensor networks · Topology control · Dynamic TDMA scheduling · Energy consumption

H. M. Kelagadi · B. Shrinidhi (✉) · Priyatamkumar
School of Electronics and Communication, K. L. E Technological University,
Huballi 580031, India
e-mail: shrinidhibhat13@gmail.com

H. M. Kelagadi
e-mail: hmkelagadi@bvb.edu

Priyatamkumar
e-mail: priyatam@kletech.ac.in

© Springer Nature Singapore Pte Ltd. 2020
V. Bhateja et al. (eds.), *Intelligent Computing and Communication*,
Advances in Intelligent Systems and Computing 1034,
https://doi.org/10.1007/978-981-15-1084-7_17

165

1   Introduction

The advancements in the field of microelectronics, sensor, and wireless communications in recent years have led to the rise of wireless sensor networks. They are set up with wireless sensor devices that are battery powered, self-configured, and well equipped with sensors and transceivers and thus, form an organized communication network. Every sensor node does the job of data collection, data processing, and communicating it to a particular base station [1, 2]. The sensor nodes are deployed over a large area and are used to gather the information of the area with respect to a particular feature. The significant applications of wireless sensor networks incorporate monitoring of the temperature, pressure, humidity, vehicular movements, noise levels, soil makeup, the presence or absence of certain obstacles, etc [3].

The wireless sensor nodes have restricted power supply and are often deployed in a rough physical environment. Hence, power efficiency becomes one of the main challenges in these networks [4]. In recent years, many topology control algorithms have been developed to reorganize various parameters of the node and its mode of operation [5]. Each of them is aimed to decrease the overall energy consumption and in this way increment the system's lifetime.

Topology control mainly includes topology construction and topology maintenance. One of the important topology construction techniques is building hierarchical topologies. In hierarchical based topology, the entire network is partitioned into clusters or grids aiming to enhance the efficiency in terms of energy and scalability of the network. One of the first and basic hierarchical protocols is the LEACH (Low Energy Adaptive Clustering Hierarchy) protocol [6]. Many other hierarchical protocols have been built up based on LEACH [7], one among them being TL-LEACH (two-level LEACH). In this work, a solution has been proposed to modify the TDMA schedule of TL-LEACH thereby reducing energy consumption.

The remainder of the paper is organized as follows: Sect. 2 introduces TL-LEACH protocol in detail. Section 3 presents the proposed protocol and the detailed algorithm is given in Sect. 4. The analysis of simulation results is carried out in Sects. 5 and 6 concludes the work.

2   TL-LEACH: A Hierarchical Routing Protocol

TL-LEACH is the modification of LEACH, one of the most popular hierarchical, cluster-based, energy-efficient protocols aimed at minimum energy consumption and increase in network lifetime [8]. TL-LEACH protocol mainly divides the nodes into two types, each with a different task to perform [9]. First ones are the Cluster Heads (CH) which are further classified into primary (PCH) and secondary cluster heads (SCH). The task of SCH is to collect the data from all the other nodes belonging to the same cluster and aggregate them. It also has to generate a TDMA schedule [10] for each node to send its data at a specific TDMA slot. The PCH handles the responsibility of transmitting the data, which is sent by SCH to Base Station (BS).

The second type is the node, part of a particular cluster which sends the data from the sensors to the CH in a particular timeslot.

The communication process includes a number of rounds each consisting of two phases, one being the setup phase and other is the steady phase. Formation of clusters and selection of cluster heads are carried out in the setup phase. In the steady-state main focus is given on the data collection, aggregation, and transmission. In the first round, the PCH is selected as follows: A random number between zero and one is chosen by every sensor node. The sensor node becomes a PCH if this randomly generated value falls below threshold T(n), which is given by Eq. 1 [11].

$$T(n) = \begin{cases} \frac{p}{1-p*[r \ mod(1/p)]} & \text{if } n \in G \\ 0 & \text{if } n \notin G \end{cases} \tag{1}$$

where p is the probability of the number of cluster heads to be selected in the rth round, n is the number of nodes, and G represents the nodes which were not clustered heads in last $1/p$ rounds.

In the steady phase, the environmental information is sensed and gathered by the sensor nodes, and this data is transmitted to SCH in the specified TDMA slot. After schedule completion, data aggregation is done by the SCH which then sends it to PCH. The latter transmits the data to the base station. The steady-state lasts for a specific period after which setup phase begins again and new cluster heads are picked. The energy spent during transmission, E_{tx} is given in Eq. 2.

$$E_{tx} = \begin{cases} kE_{elec} + k\varepsilon_{fs}d^2 & \text{if } d \leq d_0 \\ kE_{elec} + k\varepsilon_{mp}d^4 & \text{if } d > d_0 \end{cases} \tag{2}$$

where k is packet length, E_{elec} is energy consumed by electronics per bit, ε_{fs} is the free space factor, ε_{mp} is the multipath factor, d is the distance of transmission, and d_0 is the threshold distance [12].

The threshold distance d_0 can be calculated using Eq. 3.

$$d_0 = \sqrt{\frac{\varepsilon_{fs}}{\varepsilon_{mp}}} \tag{3}$$

The energy consumed for the reception, E_{rx} is given by Eq. 4.

$$E_{rx} = kE_{elec} \tag{4}$$

3 Proposed Solution

The TL-LEACH protocol includes multi-hop, which reduces the energy consumption to a certain extent. But, it does not explore another component in the protocol, TDMA schedule. In TL-LEACH, the time slots are fixed for every sensor with a fixed number of packets to be sent every time. This raises two main concerns. First is, suppose a sensor node has fewer packets in the buffer to that of the time slot, the sensor node is activated till the end of timeslot even though there are no packets to be sent. Alternatively, if a sensor node has more packets in the buffer to that of the timeslot, many packets might get dropped. This problem becomes critical when there is an emergency. Hence, we propose a solution to have a dynamic time slot in the TDMA schedule based on the number of packets sent by each sensor node.

In this paper, we enable the nodes to create a dynamic TDMA schedule which adapts to the changes in the number of packets in the buffer of the sensor nodes and in turn modifies the sleep schedule of the sensors. Figure 1 shows a TDMA schedule. Every frame comprises M timeslots and each slot can deliver k packets.

Considering the empty slots in each frame, the algorithm decides whether the timeslot must be increased or decreased or continue the same. The timeslot is adjusted according to the requirement and thus, energy consumption is monitored.

The changes in the timeslot are done using harmonic mean. Equation 5 for the harmonic mean (H) is given as

$$H = \frac{n}{\sum_{i=1}^{n} \frac{1}{x_i}} \tag{5}$$

where n is the total quantity of sensor nodes and x_i represents the vacant spaces in the timeslot due to the absence of the desired number of packets. Thus, the sum of x_i represents the total vacant spaces when all timeslots are taken into account. The rules to be followed are

1. If none of the timeslots are empty, i.e., if the packets are equal to the size of the timeslots, then no changes are needed in the next frame.
2. There must be a maximum and minimum limit for the number of packets in the buffer to avoid it to reach infinity.
3. If the value of $H > 1$, the length of the timeslot has to be increased by the integer value of H. It allows the other sensor nodes to be in sleep mode for more time. Higher the value of H more is stability.

Fig. 1 TDMA scheduling model

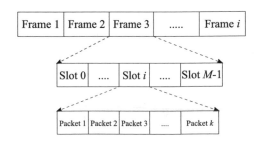

4. If the value of $0 < H \leq 1$, the timeslot has to be decreased by the first value after the decimal. It represents that the data needs to be sent frequently. Thus, the data sent to the SCH is increased.
5. If $H \leq 0$, there is no variation in the timeslot.

The timeslot modifications are reflected on all nodes and not just on a particular node. The harmonic mean is calculated and the timeslot is adjusted in every round. The dynamic scheduling decreases the active time when only a few packets are to be sent and optimizes the TDMA schedule.

4 Algorithm

The important step in the proposed protocol is the calculation of the harmonic mean to vary the size of the timeslot. The steps to calculate the harmonic mean is given in algorithm 1.

Algorithm 1 Timeslot size variation

1: **initialize** PCH, SCH, nodes(*Node*), rounds(*rmax*), energy of nodes(E_r), T(n), and distance(*d*)
2: **for** *round* $= 0 : 1 : rmax$ **do**
3:    **for** $i = 1 : n$ **do**
4:       Find the number of packets in the buffer of *Node*[*i*].
5:       $x_i = Timeslot - Packets[i]$ {Find the number of empty slots in a node}
6:    **end for**
7:    Find the total number of empty timeslots.
8:    Calculate harmonic mean using Eq. 5.
9:    **if** $H > 1$ **then**
10:       *Offset* $= H$
11:       *Timeslot* $= Timeslot + Offset$
12:    **else if** $0 < H \leq 1$ **then**
13:       *Offset* $= int((H - int(H)) * 10))$
14:       *Timeslot* $= Timeslot - Offset$
15:    **else**
16:       *Offset* $= 0$
17:    **end if**
18: **end for**

The steps carried out in operation of TL-LEACH with Dynamic TDMA scheduling is given in algorithm 2. Algorithm 1 is a part of the implementation.

Algorithm 2 Dynamic TDMA scheduling in TL-LEACH

1: **initialize** PCH, SCH, nodes(*Node*), rounds(*rmax*), energy of nodes(E_r), T(n), and distance(d)
2: **for** *round* = 0 : 1 : *rmax* **do**
3:   **if** *round* = 0 **then**
4:      PCH is elected based on threshold value T(n) as given in Eq. 1.
5:      Clusters are created based on distance proximity.
6:      SCH is selected based on the distance to BS and residual energy.
7:      The size of timeslot is calculated using algorithm 1.
8:   **end if**
9:   **if** *round* > 0 **then**
10:      PCH and SCH are selected based the distance to BS and residual energy.
11:      The size of timeslot is calculated using algorithm 1.
12:   **end if**
13: **end for**

---

5 Simulation and Result Analysis

The simulation done in MATLAB platform provides the results of the tested solution. The parameters used are depicted in Table 1. The quantity of nodes in the network is assumed to be 100 as LEACH protocol is very efficient with the said number of nodes. Based on the literature survey and considering the average energy in the primary batteries which power the sensor nodes, each node has initial energy of 2 J.

The proposed protocol accomplishes the task of modifying the timeslots of the nodes in accordance with the packets in the buffer of the nodes. An example of a timeslot for a node is shown in Fig. 2.

The first six rounds of the operation are described in Fig. 2. In the first round, each node has a timeslot to deliver 8 packets. But, since some of the slots are empty, the value of H is less than 1. Hence in the next round, the size of timeslot reduces to 6.

Table 1 Parameters of the network

Parameter	Value
Network area	300 m * 300 m
Initial energy	2 J
BS selection	(150, 150) m
Number of nodes	100
Packet size	512 bytes
Percentage of cluster head	5%
E_{elec}	50 nJ/bit
ε_{mp}	0.0013 pJ/bit/m^4
ε_{fs}	10 pJ/bit/m^2
E_{DA}	5 nJ/bit/signal

Fig. 2 Timeslot of a node

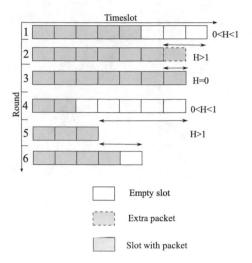

In the second round, the sensor nodes have more packets in the buffer than the size of timeslots. The value of H is more than 1 and to rectify this, in the third round the timeslot is increased to accommodate 7 packets. In the third round, all the slots are filled and there are no less or more packets to be sent. Hence, in the next round, the timeslot remains the same as the value of H is zero. In the fourth round, the number of packets sent is very less compared to the timeslot size. Hence in round five, the timeslot is modified to accumulate fewer packets. The sixth round has an increase of timeslot slot since the timeslot in the previous round cannot deliver all packets in the buffer. The same process continues for subsequent rounds.

The simulation compares three protocols: LEACH, TL-LEACH, and TL-LEACH with dynamic TDMA. The results show the live nodes and the residual energy in the network at the end of 200 rounds.

Figure 3 shows the active nodes present in all three protocols at the end of all rounds. The TL-LEACH with dynamic TDMA has maximum live nodes followed by LEACH and TL-LEACH. At around 100th round all three protocols have nearly lost half of their nodes but in the end, TL-LEACH with dynamic TDMA manages to maintain more live nodes than the other two. The number of active nodes in LEACH has reduced almost to zero while TL-LEACH with dynamic TDMA has more than 15 active nodes.

Figure 4 shows the total energy remaining in all the nodes at the end of 200 rounds. The LEACH protocol consumes maximum energy while TL-LEACH with dynamic TDMA consumes minimum energy at the end of all rounds. The average residual energy decreases gradually in all three protocols and at the end of all rounds, the residual energy of both TL-LEACH and LEACH differs by a small quantity. The TL-LEACH with dynamic TDMA protocol manages to conserve more energy as compared to the other two protocols.

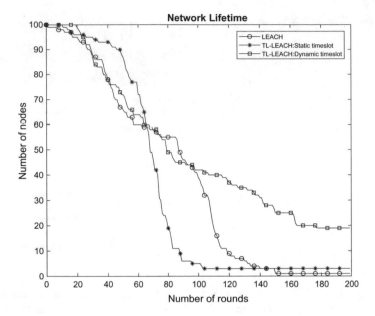

Fig. 3 Network lifetime for all three protocols

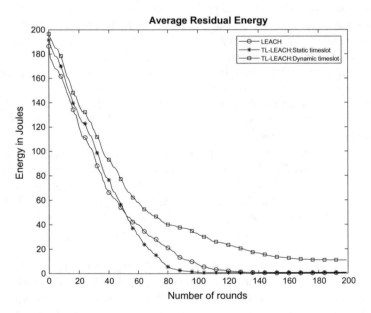

Fig. 4 Residual energy in the network for all three protocols

6 Conclusion and Future Scope

In this paper, a scheduling algorithm is designed by varying the size of TDMA slots dynamically. It, in turn, helps in proper manipulation of the active state and sleep state of the sensor nodes deployed in the network. This is achieved by altering the size of the timeslot based on the number of packets sent by the node in every next frame to minimize the energy consumption of the nodes.

While the timeslot size changes in every round, the change is based on the empty slots in the previous round. Any sudden fluctuation in the buffer size of the sensor nodes cannot be immediately regulated, and hence some data might be lost.

The protocols discussed so far mainly focus on energy reduction through sleep schedule and multi-hop. There is a need for an algorithm for data fusion to diminish the amount of data to be sent by means of the cluster heads. Time scheduling algorithm with dynamic timeslot management for topology control in WSN though reduces the energy consumption, inclusion of this mechanism in data aggregation further helps in minimizing the energy depletion. In future works, the main focus will be to achieve energy efficiency in terms of data collection, aggregation, and transmission through scheduling.

References

1. Pantazis, N.A., Nikolidakis, S.A., Vergados, D.D.: Energy-efficient routing protocols in wireless sensor networks: a survey. IEEE Commun. Surv. Tutor. **15**(2), 551–591 (2013). https://doi.org/10.1109/SURV.2012.062612.00084
2. Akyildiz, I., Su, W., Sankarasubramaniam, Y., Cayirci, E.: Wireless sensor networks: a survey. Comput. Netw. **38**(4), 393–422 (2002). https://doi.org/10.1016/S1389-1286(01)00302-4, http://www.sciencedirect.com/science/article/pii/S1389128601003024
3. Arampatzis, T., Lygeros, J., Manesis, S.: A survey of applications of wireless sensors and wireless sensor networks. In: Proceedings of the 2005 IEEE International Symposium on, Mediterrean Conference on Control and Automation Intelligent Control, pp. 719–724 (2005). https://doi.org/10.1109/.2005.1467103
4. Sharma, S., Bansal, R.K., Bansal, S.: Issues and challenges in wireless sensor networks. In: 2013 International Conference on Machine Intelligence and Research Advancement, pp 58–62 (2013). https://doi.org/10.1109/ICMIRA.2013.18
5. Kelagadi, H.M., Priyatamkumar, R.S.: Clustering techniques and need for computational intelligence for topology control in wireless sensor network: an investigation. Int. J. Eng. Technol. **7**, 220–224 (2018). https://doi.org/10.14419/ijet.v7i4.41.25386
6. Polkade, J.W., Baradkar, H.M.: Leach algorithm based on clustering for enhancement of wireless sensor network. In: Int. J. Eng. Trends Technol. (IJETT) **42**(3), 142–145 (2016)
7. Singh, K.: Wsn leach based protocols: A structural analysis. In: 2015 International Conference and Workshop on Computing and Communication (IEMCON), pp 1–7 (2015). https://doi.org/10.1109/IEMCON.2015.7344478
8. Kodali R.K., Sarma, N.: Energy efficient routing protocols for wsn's. In: 2013 International Conference on Computer Communication and Informatics, pp 1–4 (2013). https://doi.org/10.1109/ICCCI.2013.6466285
9. Sibahee M.A.A., Lu, S., Masoud, M.Z., Hussien, Z.A., Hussain, M.A., Abduljabbar, Z.A.: LEACH-T: LEACH clustering protocol based on three layers. In: 2016 International Conference

on Network and Information Systems for Computers (ICNISC), pp 36–40 (2016). https://doi.org/10.1109/ICNISC.2016.018

10. Biazi, A., Marcon, C., Shubeita, F., Poehls, L., Webber, T., Vargas, F.: A dynamic TDMA-based sleep scheduling to minimize WSN energy consumption. In: 2016 IEEE 13th International Conference on Networking, Sensing, and Control (ICNSC), pp 1–6 (2016). https://doi.org/10.1109/ICNSC.2016.7478994

11. Fahimi, M., Abad, K., Jabraeil Jamali, M., Jamalip, J.: Modify LEACH algorithm for wireless sensor network. Int. J. Comput. Sci. Issues **8**, 219–224 (2015)

12. Comeau, F., Aslam, N.: Analysis of LEACH energy parameters. Procedia Comput. Sci. **5**, 933–938 (2011). https://doi.org/10.1016/j.procs.2011.07.131, http://www.sciencedirect.com/science/article/pii/S187705091100456X

Deep Learning-Based Music Chord Family Identification

Himadri Mukherjee, Ankita Dhar, Bachchu Paul, Sk. Md. Obaidullah,
K. C. Santosh, Santanu Phadikar and Kaushik Roy

Abstract Research in the field of audio signal processing has developed considerably and music signal processing has not been an exception to this. Musicians from all over the globe have benefited tremendously with different technological advancements thereby leading music industry on to the next level. Music composers and DJs are always interested in the background music (BGM) of a song which is extremely critical in setting the mood. It is also very important for automatic music transcription and track composition for stage performers. Chords are one of the fundamental entities of BGM which are constituted with the aid of two or more musical notes. Identification of chords is thus a very important task which becomes challenging when the audio clips are short or not of studio quality. In this paper, a system is presented which can aid in distinguishing chords based on their type/family. We have experimented with two of the most fundamental type of chords major and minor at the outset and obtained a highest accuracy of 99.28% for more than 6000 very short clips of one-second duration with a deep learning-based approach.

H. Mukherjee (✉) · A. Dhar · K. Roy
Department of Computer Science, West Bengal State University,
Kolkata, India
e-mail: himadrim027@gmail.com

B. Paul
Department of Computer Science, Vidyasagar University,
Midnapore, India

Sk. Md. Obaidullah
Department of Computer Science and Engineering,
Aliah University, Kolkata, India

K. C. Santosh
Department of Computer Science,
The University of South Dakota, SD, USA
e-mail: santosh.kc@ieee.org

S. Phadikar
Department of Computer Science and Engineering, Maulana Abul Kalam Azad
University of Technology, Kolkata, India

© Springer Nature Singapore Pte Ltd. 2020   175
V. Bhateja et al. (eds.), *Intelligent Computing and Communication*,
Advances in Intelligent Systems and Computing 1034,
https://doi.org/10.1007/978-981-15-1084-7_18

Keywords Chord type identification · LSF · Deep learning

1   Introduction

Research in audio signal processing has highly progressed over the years. Scientists have demonstrated an interest in analyzing musical signals which have led to the development of different useful tools for the musicians. Music production has been highly simplified due to multifarious developments which has aided the musicians greatly. A music piece, be it a song or an instrumental, has two components. The former is the foreground or the leading melody and the latter is termed as background or background music. The BGM is very much important as the foreground, because it not only makes a piece complete but also is very much responsible for bringing out the mood of a composition as thought of by an artist. The background melody involves playing of chords.

A basic chord is composed of three notes which are termed as triad. It consists of a root note, the third and the fifth note. The name of a chord is based on the root note while the fifth note completes a chord. It is the third note which determines whether a chord is major or minor thereby making the two very close in terms of the constituent notes. Every music piece has a chord chart or set of chord progressions associated with it. A music piece without background chords often sounds empty at times or appear to be missing texture. Identification of the chord chart of a piece is one of the primitive tasks for a musician prior to either experimenting with it or understanding a composition. Identifying the same is also very important for transcribing a piece completely and for the production of background tracks for artists to perform on the stage. There are disparate type of chords, major and minor being two of the most popular and widely used ones. Almost every song has a major or minor or both of them associated with it. Identification of the type of chord is important prior to the attempt of recognizing the exact chord. Guerrero-Turrubiates et al. [1] attempted the distinction of four types of chords in the thick of major, minor, major seventh, and minor seventh. They used a neural network-based classification technique on a database composed with the aid of an electric and acoustic guitar and reported an accuracy of 93%. Rajpurkar et al. [2] presented an online system for recognition of chords in real time. They used hidden markov model (HMM) in addition to Gaussian discriminant analysis for classification along with chroma-based features on clean audio and reported an accuracy of 99.19%. Zhou and Lerch presented a deep learning-based approach for identifying chords. They experimented on a set of 317 pieces and reported a best weighted chord symbol recall value of 0.916 with max pooling. A detailed account of their experiments is presented in [3].

Cheng et al. [4] presented a system recognizing chords for classifying music and retrieving them. They used a N-gram-based approach along with HMM. They used several chord level features in thick of chord histogram and longest common chord subsequence. They obtained a best overall accuracy of 67.3% using a frame wise evaluation scheme. Muludi et al. [5] attempted chord identification using Fourier transformation-based features along with pitch class profile. They worked on a dataset of 432 guitar chords which produced an accuracy of 70.06%. Osmalskyj et al. used a neural network-based approach along with pitch class profile values. They worked on a dataset of chords recorded with guitar. They experimented with a smaller dataset put together with the aid of a few other instruments as well in the thick of piano, violin, and accordion. The reported an error rate of 6.5% for chord identification. They also presented results for instrument identification using their technique which is detailed in [6].

Oudre et al. [7] presented a chord transcription scheme with the aid of disparate parameters in the thick of chroma vectors, chord templates, etc. They experimented on the 13 Beatles album and reported average overlap scores of 0.711 by considering major-minor and dominant seventh chords and 0.698 by considering only major and minor chords. Costantini and Casali [8] attempted to recognize chords with frequency analysis-based technique. They experimented with 2, 3, and 4 note chords from several instruments like clarinet, flute, violin, and saxophone. They reported highest accuracies of 98, 97, and 95% for the 2, 3, and 4 note chords using clarinet.

In this paper, we have presented a system to distinguish chord families from short audio clips. Such a system has the potential of aiding in automatic music transcription as well as background generation. Experiments have been performed on major and minor chords with LSF-based features and deep learning-based classification. The proposed system is pictorially illustrated in Fig. 1.

In the rest of the paper, the details about the dataset are presented in Sect. 2 followed by the proposed methodology in Sect. 3. The results are presented in Sect. 5 and finally, we have concluded in Sect. 6.

2 Dataset

Data is very important for any experiment. The quality of data is also very critical for the same. The data needs to possess real world characteristics to aid in the development of robust systems. The dataset of the present experiment was put together with the aid of a Hertz acoustic guitar (HZR3801E). The guitar was tuned at 440 Hz,

Fig. 1 Pictorial representation of the proposed system

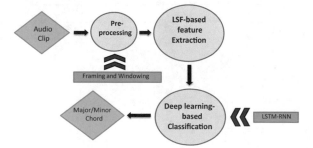

which is the industry standard. The clips were recorded using the ordinary soundcard supplied with the motherboard (Gigabyte B150M-D3H) of our PC. We did not use an audio interface and neither an electric guitar in order to maintain a nonstudio data scenario. Handling such data is critical due to low sensitivity because of the absence of preamplifiers and low-quality pickup of the semi acoustic guitar. Musicians were asked to play two major chords (C and G) and two minor chords (Em and Am). The following chords were chosen based on the fact that these are the most popular major and minor keys for musicians [9]. Another factor which inspired us to choose this chords is the fact that Am is the relative minor key of C and Em is the relative minor key of G. Two scales are said to be relatives of each other if they share the common notes. The notes for the scales of (Am-C) pair and (Em-G) pair are presented in Table 1 along with the notes of the chords. It can be seen that the chords differ only in a single note. We wanted to test our system's capability of handling such minute differences which further inspired us to pick the aforementioned chords.

The volunteers used different plectrums including Jim Dunlop XH, Jim Dunlop H, and Jim Dunlop Riffs 0.46 mm. Different plectrums were used to uphold the minute tonal variations in our dataset which occur for different plectrums. Volunteers played the chords both in nut as well as barre shape to incorporate different variations. The collected clips were used to put together 2 datasets (D_1 and D_2) having clips of lengths 1 and 2 seconds, respectively. We wanted to observe our system's performance for very short length clips and thus we engendered such datasets. The details of the engendered datasets are presented in Table 2.

Table 1 Notes involved in the chords

Scale	Notes	Chord notes	Similar notes
C	C, D, E, F, G, A, B	C, E, G	C, E
Am	A, B, C, D, E, F, G	A, C, E	
G	G, A, B, C, D, E, $F^\sharp$	G, B, D	G, B
Em	E, $F^\sharp$, G, A, B, C, D	E, G, B	

Table 2 Details of the engendered datasets

Dataset (Clip length in second)	Major chord	Minor chord	Total clips
D_1 (1)	3107	3114	6221
D_2 (2)	1549	1552	3101

3 Proposed Method

3.1 Preprocessing

The spectral contents of an audio clip show high deviations which pose difficulty in analysis. In order to deal with this, clips are partitioned into small sections which are known as frames and analyzed independently. Audio clips are split in an overlapping mode in order to ensure continuity in between two consecutive frames. However, splitting audio clips into frames often produces jitters which interfere with the frequency domain analysis. In order to handle this, the frames are subjected to a windowing function. In the present experiment, the audio clips were split into 256 sample frames with 100 sample overlap which were then subjected to hamming window [10] which is mathematically presented as under where N represents the frame size and n is a point within the frame boundary. Noise removal was not performed to ensure real-world scenario.

$$w(n) = 0.54 - 0.46\cos(\{2\pi n\}/\{N - 1\}) \tag{1}$$

4 Feature Extraction

We had extracted standard Line spectral frequency (LSF) [11] features from the clips in frame level also has the strength of performing high interpolation. This feature considers a signal as the output of an all pole filter H(z). Its inverse I(z) is shown below where $I_{1...n}$ are the predictive coefficients

$$I(z) = 1 + i_1 z^{-1} + \cdots + i_n z^{-n} \tag{2}$$

The LSF representation is obtained by decomposing I(z) into $I_x(z)$ and $I_y(z)$ which are detailed below

$$I_x(z) = I(z) + z^{-(n+1)} I(z^{-1}) \tag{3}$$

$$I_y(z) = I(z) - z^{-(n+1)} I(z^{-1}) \tag{4}$$

We extracted 5, 10, 15, 20, and 25-dimensional features for every dataset. In real world, audio clips are of different sizes which produce different number of frames. Since features are extracted in a frame wise manner, thus a disparity in the feature dimension is observed. In order to handle this situation, sum of the energy values for each dimension (band) was calculated. Then the ratio of distribution of the energy values in the bands was calculated which was used as a feature. Along with this, the bands were also graded based on total energy content. Thus, we obtained features of dimension 10 (F_1), 20 (F_2), 30 (F_3), 40 (F_4), and 50 (F_5) dimensions which were

Fig. 2 Trend of the feature values for F_4

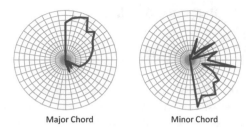

Major Chord			Minor Chord

independent of the length of the clips. The trend of the feature values for F_4 (best result) is presented in Fig. 2.

4.1 Deep Learning-Based Classification

Deep learning has recently attracted the attention of researchers across the globe and has outperformed several standard machine learning techniques. In this experiment, we have used a recurrent neural network-based classifier with long short-term memory for segregating the feature sequences. It can preserve states in contrary to standard neural networks [12] which helps in handling sequential data. It is a generalized deep learning technique with an ability to handle long sequences [13].

Long-short Term Memory-RNNs (LSTM-RNN) solves the vanishing gradient problem as observed in simple RNNs [13]. It also has advantage over simple RNNs for complex extended cases [14]. A LSTM block has long-term memory known as the cell state. It has three gates: forget gate, input gate, and output gate.

The input gate (i_n) is involved in determination of the values associated in deduction of the new state shown in Eq. (5) where Wt_i is the associated weight.

$$i_n = \sigma(Wt_i S_{n-1} + Wt_i X_n) \tag{5}$$

The forget gate is involved in determination of the values which are discarded or forgotten from the previous state in the present state. It is mathematically shown below where Wt_f is the associated weight.

$$f_n = \sigma(Wt_f S_{n-1} + Wt_f X_n) \tag{6}$$

The output gate helps to determine the values from the present state which determines the next state. It is shown as under where Wt_o is the associated weight.

$$o_n = \sigma(Wt_o S_{n-1} + Wt_o X_n) \tag{7}$$

The intermediate cell state c_n is shown as follows

$$c_n = \tanh(Wt_c S_{n-1} + Wt_c X_n) \tag{8}$$

which generates cell state C_n as

$$C_n = (i_n * c_n) + (f_n * C_{n-1}) \tag{9}$$

The new state h_n thus produced is

$$h_n = o_n * \tanh(C_n) \tag{10}$$

In the current experiment, we used a 100-dimensional LSTM layer which was carried forward to a 100-dimensional fully connected layer. The output of this layer was passed to a 50-dimensional fully connected layer. Both these layers had a relu activation function. Finally, the output was passed to a 2-dimensional layer for determination of the classes with a softmax activation. The training epochs were initially set to 100 with a batch size of 50. The network structure along with the activations and the dimensions of the layers were set based on experimental trials. We had used a cross validation technique for the evaluation of our system.

5 Result and Analysis

Each of the feature sets F_1–F_5 for the datasets D_1 and D_2 were subjected to the LSTM-RNN-based classifier whose results are presented in Table 3.

It is seen from the Table that we obtained better results the 1 s clips as compared to the 2 s clips. This is mostly due to the greater effect of noise within the played rhythms which crept in because of the ordinary guitar pickup. Since the best result was obtained for D_1 using F_4, so we carried it forward for further tests.

We experimented by varying the batch size during training whose results are presented in Table 4. It is seen from the Table that the best result was obtained for 150 instances per batch. Increasing the batch size beyond this led to overfitting and thus the batch size was set to 150 for the next phase.

Table 3 Obtained accuracies using the different feature dimensions for the datasets

Dataset	Features				
	F_1	F_2	F_3	F_4	F_5
D_1	45.20	88.94	94.76	97.17	91.95
D_2	50.05	71.24	50.02	81.33	89.71

Table 4 Obtained accuracies for different batch sizes on D_1

Batch size	50	100	**150**	200
Accuracy (%)	98.07	98.26	**98.59**	96.82

Table 5 Obtained accuracies for different folds on D_1

Fold	5	**10**	15	20
Accuracy (%)	98.59	**99.26**	99.23	99.13

Table 6 Obtained accuracies for different training iterations on D_1

Epochs	100	150	**200**	250
Accuracy (%)	99.26	99.18	**99.28**	99.16

Table 7 Confusion matrix for the best performance settings

	Major	Minor
Major	3104	3
Minor	42	3072

The folds of cross-validation were varied to find the best size of training and testing sets. The obtained results are presented in Table 5 where it is seen that the best result was obtained for 10-fold which led to a perfect distribution of the different variations of playing in our dataset.

The training epochs were also experimented with and it was found that the best result was obtained for 200 training epochs. Increasing the training epochs beyond this led to overfitting. The obtained results for the different training epochs is presented in Table 6. The confusion matrix for our best result of 99.28% in our experiment is presented in Table 7. This result was obtained for a batch size of 150, with 10-fold cross-validation and 200 training epochs. It is seen from the Table that the misclassification was greater for the minor chords as compared to the major chords. The dataset was manually inspected and it was found that many instances of the minor chord clips often had higher noise as compared to the major chords mostly due to the fingering positions. We had also subjected this dataset to some other popular classifiers [15] whose results are presented in Fig. 3.

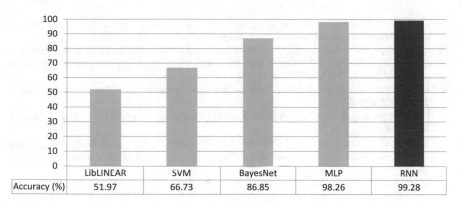

Accuracy (%)	LibLINEAR	SVM	BayesNet	MLP	RNN
	51.97	66.73	86.85	98.26	99.28

Fig. 3 Performance of different classifiers on D_1 with 40-dimensional features

6 Conclusion

In this paper, a system is presented to distinguish chord types/families with a deep learning-based approach. The system has been tested with the two most popular chord types and a misclassification rate of only 0.72 has been obtained which is encouraging considering the low quality pick up of the instrument as well as the extremely short clip lengths. Such scenario was upheld mainly to test the robustness of our system. In future, we will experiment with other chord families and instruments along with exact chord identification. We plan to use deep learning-based feature extraction as well as experiment with other acoustic features. We also plan to experiment with other machine learning techniques to test the performance of our system. The system will also be subjected to artificial noise to test its robustness.

Acknowledgements The authors would like to thank Mr. Soukhin Bhattacherjee for his help during data collection.

References

1. Guerrero-Turrubiates, J., Ledesma, S., Gonzalez-Reyna, S., Avina-Cervantes, G.: Guitar chords classification using uncertainty measurements of frequency bins. Math. Probl. Eng. **2015** (2015)
2. Rajpurkar, P., Girardeau, B., Migimatsu, T.: A Supervised Approach To Musical Chord Recognition (2015)
3. Zhou, X., Lerch, A.: Chord detection using deep learning. In: Proceedings of the 16th ISMIR Conference, vol. 53 (2015)
4. Cheng, H.T., Yang, Y.H., Lin, Y.C., Liao, I.B., Chen, H.H.: Automatic chord recognition for music classification and retrieval. In: 2008 IEEE International Conference on Multimedia and Expo, pp. 1505–1508. IEEE (2008)
5. Muludi, K., Loupatty, A.F.S.: Chord identification using pitch class profile method with fast fourier transform feature extraction. Int. J. Comput. Sci. Issues (IJCSI) **11**(3), 139 (2014)

6. Osmalsky, J., Embrechts, J.J., Van Droogenbroeck, M., Pierard, S.: Neural networks for musical chords recognition. In: Journees d'informatique Musicale, pp. 39–46 (2012)
7. Oudre, L., Grenier, Y., Févotte, C.: Chord recognition using measures of fit, chord templates and filtering methods. In: 2009 IEEE Workshop on Applications of Signal Processing to Audio and Acoustics, pp. 9–12. IEEE (2009)
8. Costantini, G., Casali, D.: Recognition of musical chord notes. WSEAS Trans. Acoustics Music 1(1), 17–20 (2004)
9. https://insights.spotify.com/us/2015/05/06/most-popular-keys-on-spotify/. Visited on 1 March 2019
10. Mukherjee, H., Obaidullah, S.M., Santosh, K.C., Phadikar, S., Roy, K.: A lazy learning-based language identification from speech using MFCC-2 features. Int. J. Mach. Learn. Cybern. 1–14 (2019)
11. Mukherjee, H., Obaidullah, S.M., Santosh, K.C., Phadikar, S., Roy, K.: Line spectral frequency-based features and extreme learning machine for voice activity detection from audio signal. Int. J. Speech Technol. 21(4), 753–760 (2018)
12. Lipton, Z.C., Berkowitz, J., Elkan, C.: A critical review of recurrent neural networks for sequence learning (2015). arXiv:1506.00019
13. Li, J., Mohamed, A., Zweig, G., Gong, Y.: LSTM time and frequency recurrence for automatic speech recognition. In: 2015 IEEE Workshop on Automatic Speech Recognition and Understanding (ASRU), pp. 187–191. IEEE (2015)
14. Hochreiter, S., Schmidhuber, J.: Long short-term memory. Neural Comput. 9(8), 1735–1780 (1997)
15. Hall, M., Frank, E., Holmes, G., Pfahringer, B., Reutemann, P., Witten, I.H.: The WEKA data mining software: an update. ACM SIGKDD Explor. Newslett. 11(1), 10–18 (2009)

A Multilevel CNN Architecture for Character Recognition from Palm Leaf Images

R. L. Jyothi and M. Abdul Rahiman

Abstract Deep Learning networks have proven its significance in almost all the areas related to object recognition. With the advent of deep learning concepts, there is a drastic improvement for object recognition problems in various machine learning domain. Convolutional Neural Networks (ConvNets or CNNs) are a special category of Neural Networks that have proven very effective in areas such as image recognition and classification. ConvNets have been successful in most of the pattern recognition problems. Due to the high impact produced by deep learning networks in machine learning applications, different variations of CNN have been developing in a competing manner overriding the performance of the just previous version. In this work, a modified version of CNN is proposed constituting multilevel layers of CNN. Here the input image entrenched in the form of a pyramid is given as input to the system. The performance of the proposed work is tested with degraded grantha characters extracted from palm leaf documents and characters from MNIST dataset. Grantha characters from palm leaves documents are chosen to test the performance of the system with very small sized and highly degraded dataset.

Keywords Multilevel convolutional neural network · Grantha dataset · Small dataset · Scale invariant · Palm leaves

1 Introduction

Pattern Recognition Systems are being developed with the aim of reaching human performance in recognition of objects. But none of the traditional methods on pattern recognition has not reached the required performance on tasks such as the recognition

R. L. Jyothi (✉) · M. Abdul Rahiman
University of Kerala, Thiruvananthapuram, India
e-mail: jyothianil@gmail.com

M. Abdul Rahiman
e-mail: rahimapaika@yahoo.com

R. L. Jyothi
College of Engineering Chengannur, Chengannur, India

© Springer Nature Singapore Pte Ltd. 2020
V. Bhateja et al. (eds.), *Intelligent Computing and Communication*,
Advances in Intelligent Systems and Computing 1034,
https://doi.org/10.1007/978-981-15-1084-7_19

of handwritten digits. Convolutional neural network had overrided the performance
of state-of-the-art methods in almost most benchmark datasets [1–4]. CNN is a
artificial neural network created based on the proposal of first artificial neuron in [5].
CNN is closely related to Hubel and Wiesels experimental analysis in simple cells
and complex cells in the primary visual cortex [6, 7].

The training of these deep hierarchical neural networks has greatly improved
supervised pattern classification [8, 9]. Deep Convolutional neural networks(DNN)
was introduced by [10] modified and simplified by [11–14]. Deep Convolutional
networks (DNN) have already proved their efficiency on different types of datasets
[14–17]. The recognition efficiency of DNN increases with the increase in the number of layers and maps per layers in the network [4]. Training these networks with
CPU requires weeks, months, and even years. But through parallel processing using
graphics processing unit (GPU) the training time of the network has considerably
been reduced. In this work, a multilevel CNN architecture is proposed where the input
images are scaled into multiple levels (creating a pyramid of scale) before applying to the CNN architecture. Moreover the CNN architecture consists of multilevel
CNN's each working on a different scale of the input image pyramid. The network
is trained on MNIST dataset and newly created dataset of ancient grantha characters
extracted from palm leaves. Section 2 depicts literature survey on some of the recent
efficient CNN networks and their performance in MNIST dataset. Section 3 explains
the proposed system architecture and Sect. 4 depicts the experimental analysis of the
proposed system on MNIST and grantha dataset.

2   Related Work

Fukushima [10] proposed a hierarchical model called Neocognitron, consisting of
stacked pairs of simple unit layer and complex unit layer based on Hubel and Wiesels
findings in cat [6, 7]. The first CNN was proposed in [18, 19]. The main difference
between Neocognitron and CNN architecture is that in CNN a backpropagation (BP)
algorithm for learning is incorporated. Over the past years, many techniques have
been developed for improving the performance of CNN. The first successful GPU
implementation of CNN was AlexNet [20] that won the Large Scale Visual Recognition Challenge (ILSVRC) 2012 on Imagenet. Since then, most of the submissions
to this competition were based on GPU implemented CNN. As this work concentrated on the recognition of handwritten characters, some of the existing efficient
CNN networks experimented on MNIST dataset is analyzed here. Experimental
analysis of CNN network on MNIST dataset was first introduced in [11]. In this
work, several versions of LeNet-5 were trained on the regular MNIST database.
After 10 passes the test error rate stabilizes to 0.95% and training error rate reaches
0.35% after 19 passes. The LeNet-5 is compared which LeNet-1, LeNet-4, and
Boosted LeNet-4. The small CNN LeNet-1 produced an error rate of 1.7% while
Lenet-4 produced an error rate of 1.1% and Boosted LeNet-4 produced an error rate
of 0.7% when tested on MNIST dataset. A Multicolumn Deep Neural Network was

introduced in [21]. Here, several CNN's are combined to form multicolumn CNN. In this work, a comparative analysis of proposed version of CNN with previous versions are carried out and found the multicolumn CNN produced the lower error rate of 0.23%. In [21] a modified CNN network is proposed where regularization done with dropout module was replaced with dropconnect. Dropconnect is the modification of dropout where each connection is dropped with a probability of '1−p' compared to dropout where each random output is dropped. The network was able to reduce the recognition of MNIST dataset to a considerable amount. In [22] a modified CNN is proposed by incorporating recurrent connections in convolution layer. Recurrent convolutional neural network contains stack of recurrent convolutional layers interleaved with max-pooling layers. The network was able to reduce the error rate in recognition of MNIST dataset to 0.31%. In [23], CNN is analyzed using different types of pooling strategies. By combining max pooling and average pooling different types of pooling methods are created. The different types of pooling functions compared in the work are mixed max-average pooling, gated max-average pooling, and tree pooling. From the analyzed result in the work, it can be seen that CNN network based on Tree pooling produces the lowest error rate on MNIST dataset. In [24] CNN is modified by adopting a fractional pooling strategy. Here as stochastic pooling, a degree of randomness is introduced in the pooling region selection. Fractional pooling reduces the size of the image by a factor α. CNN based on fractional pooling reduced the error rate on CNN to 0.34%. In [25] a multiscale convolution based CNN is proposed. Here, a patch of image centered at some spatial position 'i' is convoluted with filters of varying scale followed by average pooling and softmax activation function. In this network different modules are formed using various filters. The output of each module is subjected to maxout activation function followed by max pooling and dropout. The output of the final block is subjected to average pooling and softmax activation function. Competitive CNN when tested on MNIST dataset produced an average error rate of 0.33%.

In [26] a multiscale CNN network is proposed comprising of two parts a multiscale object proposal network and multiscale object detection. In proposal network, detection step is performed at multiple layers. The proposed network when analyzed with KITTI dataset produced the highest recognition efficiency of 89.02 for car object detection, 83.92 for pedestrians, and 84.06 on cyclist dataset. In [27] a multiscale sequential convolutional neural network was designed to detect eye optic disc and fovea center. The experiments were carried out in MESSIDOR and Kaggle datasets. The proposed method was able to produce an average successful detection rate of 97% for optic disc detection and 96.6% for fovea center detection.

3 Multilevel CNN Architecture

The architecture of multilevel CNN is divided into five parts-image pyramid layer, CNN layers, concatenation layer, and two sets of fully connected layers with batch normalization and activation layers.

3.1 Image Pyramid Layer

A pyramid of input image is created by re-sizing the images in the form of a pyramid. The input image is re-sized to eight different resolutions to create a pyramid of height of 8. The different resolutions chosen for re-sizing the input images are 1024×1024, 512×512, 256×256, 128×128, 64×64, 32×32, 16×16, 8×8. The images are converted in the form of pyramid to make the system scale invariant (Fig.1).

3.2 CNN Layers

The sub-image of each resolution of the pyramid is subjected to eight different CNN layers. Each set of CNN layers has two set of convolution layers, batch processing layer, activation layers, and one set of max pooling and flattening layer. First convolution layer consists of 16 ($3 \times 3 \times 3$) filters initialized with Gabor filter values of 16 orientations. The Gabor filter values [28] are calculated using the equation given below:

$$G(x, y) = \exp\left(\left(-\frac{(M^2 + \gamma^2 N^2)}{2\sigma^2}\right)\cos\left(\frac{2\pi}{\lambda}M\right)\right) \tag{1}$$

Here $M = x\cos\theta - y\sin\theta$ and $N = x\cos\theta + y\sin\theta$

x and y determines the size of Gabor filter. γ represent aspect ratio taken as 5.2, λ determines wavelength(chosen as 0.3), σ determines the scale (effective width-chosen as 4.1) and θ depicts orientation of Gabor kernels. Here 16 θ values are

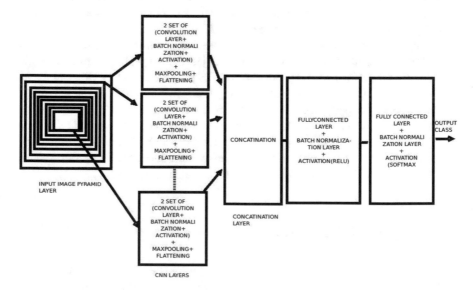

Fig. 1 System architecture

chosen. The θ values chosen are 0, 12, 24, 36, 48, 60, 72, 84, 96, 108, 120, 132, 144, 156, 168, 180.

The second convolution layer consists of 8 (3 × 3)filters of eight different orientations initialized with Gaussian random values. The input to second convolution layer is a size of image w × l × 16. Here the width 'w' and length 'l' of the input image to second convolution layer varies on each CNN layer based on the size of the input image on each CNN layer. The image of size 1024 × 1024 × 3 applied as input to the first convolution layer produces an output of 1022 × 1022 × 16 sized image as output. The 1022 × 1022 × 16 image when applied to second convolution layer produces an image of size 1020 × 1020 × 128 as output. The output image from previous layer is subjected to batch normalization based on mean and variance values. A batch normalization is performed in a column order by selecting a batch of the size 1020 × 1 × 128 when the input sub-image to the CNN layer is of size 1024 × 1024 × 3. The size of the batch selected for batch normalization varies from one CNN layer to another depending upon the input size of applied. In the next stage, the image is applied with ReLU activation function. Finally, max pooling is applied to the previous layer image at a stride of 2 × 2 × 2. Finally, a flattening operation is applied to the resultant image thereby producing a one-dimensional vector of size 1 × v. Where the value of 'v' varies from one CNN layer to another based on the different in size of input applied to different CNN layers. When the input image size is 1024 × 1024 × 3, the value of v is 16257044.

3.3 Concatenation Layer

Input vector from each CNN layer is re-sized to a vector size 'lm' by padding with zero values. The size lm is the size of the image resulted from the first CNN layer which is the largest of all vectors produced from the CNN layers. In this case, the value of lm is 16257044. The re-sized image from all CNN layers is concatenated to form a matrix of size lm × 8. The concatenated matrix is subjected to flattening operation to convert into a single one-dimensional vector. The resultant vector will be of size 130056192.

3.4 Fully Connected Layers

The last part of the architecture constitute of two sets of fully connected layer, batch normalization layer, and activation layer. Here entire vector wise batch normalization is performed based on mean and variance values. The activation function used in the first part is ReLU and in the second part is softmax. The final output from the fully connected layer is the recognition accuracy percentage corresponding to each class in the domain.

4 Experimental Analysis

Experimental analysis was carried out in two different sets of dataset MNIST dataset and grantha dataset. MNIST is one of the most frequent datasets used in the machine learning community. It consists of handwritten digits of 0–9. There are 60000 training images and 10000 testing images. The images are in grayscale with size 28 × 28 pixels. The second dataset is the newly created data set of Grantha characters extracted from palm leaves. These Palm leaves were obtained from Oriental research Institute in Kerala (a small state in the Southern part of India). Figure 2 shows the cropped segment of a grantha palm leaf. Grantha characters were extracted from 4015 folios of ancient palm leaves. The palm leaves used in this work are scanned with resolution 400 dpi. 41 Grantha characters of 290 samples each were extracted from the palm leaves. Sufficient samples of only 41 characters were obtained from palm leaves therefore only 41 characters out of the 50 Grantha characters were used in this work. Figure 3 shows the grantha characters list.

The cropped grantha characters without applying any preprocessing like thinning and binarization are applied to the proposed system. The system is trained with 200 samples of each grantha character and about 50,000 images in case of MNIST dataset. The rest of the images in both the dataset were used for testing purposes. Table 1 shows the recognition efficiency produced by the proposed system in both the dataset. From the tabulated result, it can be affirmed that the proposed system works effectively in degraded dataset of small number of images. Moreover, the system produces very low error rate in case of MNIST images. Comparative analysis of the proposed work with state-of-the-art methods are shown in Table 2. From the analyzed result it can be confirmed that proposed system work efficiently compared to the state-of-the-art methods.

Fig. 2 Grantha palmleaf

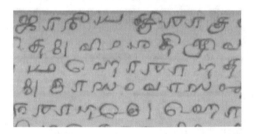

Fig. 3 Grantha characters

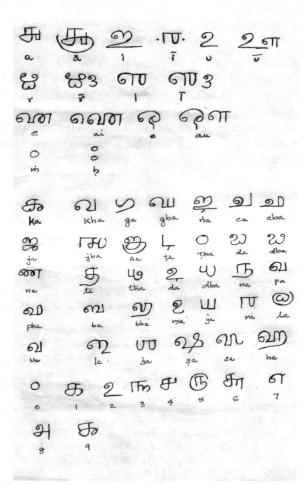

Table 1 Experimental analysis result of the proposed method

Dataset	Recognition efficiency
MNIST	99.79
Grantha	96.3

Table 2 Comparative analysis of proposed system compared to other CNN networks

Method	Error rate
CNN [13]	0.40
CNN [29]	0.39
MLP [15]	0.35
CNN committe [16]	0.27
Proposed work	0.21

5 Conclusion

A multilevel CNN architecture is proposed in this work. Here input image is scale to multilevel to form a pyramid. This is done to make the system scale invariant. Multiple levels of CNN is created with each CNN working on a single scale in the pyramid. The proposed work is experimented on MNIST dataset and a very small sized dataset of degraded characters extracted from palm leaves. From experimental analysis, it can be concluded that proposed system produces less error rate compared to the existing state-of-the-art methods.

References

1. Szegedy, C., Liu, W., Jia, Y., Sermanet, P., Reed, S., Anguelov, D., Erhan, D., Vanhoucke, V., Rabi-novich, A.: Going deeper with convolutions (2014). arXiv preprint arXiv:1409.4842
2. Chatfield, K., Simonyan, K., Vedaldi, A., Zisserman, A.: Return of the devil in the details: delving deep into convo- lutional nets. In: British Machine Vision Conference (2014)
3. Lin, M., Chen, Q., Yan, S.: Network in network. In: International Conference on Learning Representations (ICLR) (2014)
4. Krizhevsky, A.: Learning multiple layers of features from tiny images. Masters thesis. Computer Science Department, University of Toronto (2009)
5. McCulloch, W., Pitts, W.: A logical calculus of the ideas immanent in nervous activity. Bull. Math. Bio-Phys. 115–133 (1943)
6. Hubel, D.H., Wiesel, T.N.: Receptive fields of single neurones in the cats striate cortex. J. Physiol. **148**(3), 574–591 (1959)
7. Hubel, D.H., Wiesel, T.N.: Receptive fields, binocular interaction and functional architecture in the cats visual cortex. J. Physiol. **160**(1), 106–154 (1962)
8. Bengio, Y., Lamblin, P., Popovici, D., Larochelle, H.: Greedy layer-wise training of deep networks. In: Neural In formation Processing Systems (2007)
9. Courville, A., Manzagol, P.A., Vincent, P., Erhan, D., Bengio, Y., Bengio, S.: Why does unsupervised pre-training help deep learning?. J. Mach. Learn. Res. 625–660 (2010)
10. Fukushima, K.: Neocognitron: a self-organizing neural network for a mechanism of pattern recognition unaffected by shift in position. Biolo. Cybern. 193–202 (1980)
11. LeCun, Y., Bottou, L., Bengio, Y., Haffner, P.: Gradient-based learning applied to document recognition. In: Proceedings of the IEEE, pp. 2278–2324 (1998)
12. Behnke, S.: Hierarchical neural networks for image interpretation. Lecture Notes in Computer Science, vol. 2766. Springer (2003)
13. Simard, P.Y., Steinkraus, D., Platt, J.C.: Best practices for convolutional neural networks applied to visual document analysis. In: Seventh International Conference on Document Analysis and Recognition, pp. 958–963 (2003)
14. Ciresan, D.C., Meier, U., Masci, J., Gambardella, L.M., Schmidhuber, J.: Flexible, high performance convolutional neural networks for image classification. In: International Joint Conference on Artificial Intelligence, pp. 1237–1242 (2011)
15. Ciresan, D.C., Meier, U., Gambardella, L.M., Schmidhuber, J.: Deep, big, simple neural nets for handwritten digit recognition. Neural Comput. 3207–3220 (2010)
16. Ciresan, D.C., Meier, U., Gambardella, L.M., Schmidhuber, J.: Convolutional neural network committees for handwritten character classification. In: International Conference on Document Analysis and Recognition, pp. 1250–1254 (2011)
17. Uetz, R., Behnke, S.: Large-scale object recognition with CUDA-accelerated hierarchical neural networks. In: IEEE International Conference on Intelligent Computing and Intelligent Systems (ICIS) (2009)

18. LeCun, Y., Boser, B., Denker, J., Henderson, D., Howard, R., Hubbard, W., Jackel, L.: Handwritten digit recognition with a back-propagation network. In: Advances in Neural Information Processing Systems (NIPS), pp. 396–404 (1990)
19. LeCun, Y., Boser, B., Denker, J.S., Henderson, D., Howard, R.E., Hubbard, W., Jackel, L.D.: Backpropagation applied to handwritten zip code recognition. Neural Comput. 1(4), 541–551 (1989)
20. Krizhevsky, A., Sutskever, I., Hinton, G.E.: Imagenet classification with deep convolutional neural networks. In: Advances in Neural Information Processing Systems (NIPS), pp. 1097–1105 (2012)
21. Cireşan, D., Meier, U., Schmidhuber, J.: Multi-column deep neural networks for image classification. In: IEEE Conference on Computer Vision and Pattern Recognition, pp. 3642–3649 (2012)
22. Liang, M., Hu, X.: Recurrent convolutional neural network for object recognition. In: International Conference on Pattern Recognition and Computer Vision, pp. 3337–3375 (2015)
23. Lee, C.-Y., Gallagher, P.W., Tu, Z.: Pooling functions in convolutional neural networks: mixed, gated, and tree (2015). arXiv:1509.08985v2 [stat.ML]
24. Graham, B.: Fractional max-pooling (2015). arXiv:1412.6071v4 [cs.CV]
25. Lee, C.-Y., Carneiro, G.: Multi-scale convolution (2015). arXiv:1511.05635v1 [cs.CV]
26. Cai, Z., Fan, Q., Feris, R., Vasconcelos, N., San Diego, U.C.: A unified multi-scale deep convolutionalneural network for fast object detection. arXiv:1607.07155 [cs.CV]
27. Al-Bander, B., Al-Nuaimy, W., Williams, B.M., Zheng, Y.: Multiscale sequential convolutional neural networks for simultaneous detection of fovea and optic disc. Biomed. Signal Process. Control 91–11 (2018)
28. Jyothi, R.L., Rahiman, A., Anilkumar, A.: Unadorned Gabor based convolutional neural network overrides transfer learning concept. Int. J. Appl. Eng. Res. 11012–11017 (2018)
29. Ranzato, M.A., Poultney, C., Chopra, S., Lecun, Y.: Efficient learning of sparse representations with an energy-based model. In: Advances in Neural Information Processing Systems (NIPS 2006) (2006)

Novel Multimodel Approach for Marathi Speech Emotion Detection

Vaijanath V. Yerigeri and L. K. Ragha

Abstract Speech is a human vocal communication using the vocal tract. While speaking, the speaker may perform verities of intentional speech act. Voice tones of different emotions are easily understood even by primitive. Voice is the first and foremost reactive action by homosapiens to express intended emotions. In today's fast world every professional human being is under an enormous amount of stress. Stress is a serious threat to human health because it is proving as a **silent killer**. The remedy to counter this problem is early detection of symptoms which will reduce these cases. This paper focuses on stress-emotion detection in the regional language, i.e., Marathi and presents a multimodel approach. Different features like Mel-Frequency Cepstral Coefficient (MFCC), energy, pitch, and vocal tract frequency. These features are extracted from the given speech signal and the same has been trained using Artificial Neural Network (ANN). The implementation is done in MATLAB. The combination of a multi-model feature set with ANN gives satisfactory results up to 83–84% for identifying negative emotion, i.e., sadness and anger in speech.

Keywords Speech emotion recognition system · Feature extraction ·
Mel-frequency Cepstral coefficient · Vocal tract frequency · Pitch · Energy ·
Artificial neural network

1 Introduction

Charles Darwin monographed on emotion expression by homosapiens [1]. This milestone work generated a new wave of interest in psychologists and they gradually started collecting information and increased the knowledge-base. Human emotions have a long evolutionary purpose for our survival as a species. Psychologists observed that homosapiens use lots of nonverbal cues like facial expressions, gesture, body

V. V. Yerigeri (✉)
M.B.E.S. College of Engineering, Ambajogai 431 517, M.S, India

L. K. Ragha
Terna Engineering College, Nerul, Navi-Mumbai, M.S, India

© Springer Nature Singapore Pte Ltd. 2020
V. Bhateja et al. (eds.), *Intelligent Computing and Communication*,
Advances in Intelligent Systems and Computing 1034,
https://doi.org/10.1007/978-981-15-1084-7_20

language or tone of voice, to convey their emotions. Emotions are expressions of an internal thought process or maybe a reaction to an external stimulus.

Cognitive science is the interdisciplinary branch which examines human cognition. This science deals with emotion, reasoning, attention, memory, perception, and language. To understand these functions technological support from Artificial Intelligence (AI), linguistic and neuroscience is required. Cognitive science works on the fundamental concept that "process of thinking may be best understood by representational structures of the mind and the computational procedures that can be applied to those structures". Identifying emotions during social interaction is paramount and is done subconsciously by humans. Building and maintaining human relationships depends upon a person's ability to understand emotions communicated by other person(s). Misidentification of emotion(s) may result in breaking up of relationships and hence the social detachment.

Voice emotions are clearly recognized by all the people [2–4]. Even the most primitive can recognize the tones of different emotions. Animals can also recognize the meaning of the human voice. Thus the tone language is universal for all means of communication and can convey a physical state that includes personality, appearance, intelligence [5] age group, and gender of a speaker [6].

The author is working on Speech Emotion Recognition (SER) because, in the case of face emotion recognition, both the speakers should meet personally or have to be on a video call in case of distant communication. Even though the technology has advanced where video call is feasible; still it has its own limitation due to video streaming speed and quality. So voice call is preferred way of communication which does not put the binding on speakers to meet physically. A phone call or Voice over Internet Protocol (VoIP) is the best and convenient way.

SER is a sub-branch of speech perception research. It uses AI technology so it is therefore also referred to as Emotion AI (EAI) or Artificial Emotion Recognition (AER) or Affective computing (AC) [7]. AC deals with recognizing, interpreting, understanding and replicating human emotions [8]. The success of Human–Computer Interface (HCI) crucially depends upon the accuracy of SER system [9, 10].

Cowie and Cornelius [11] proposed a 2D model for empirical analysis which is shown below.

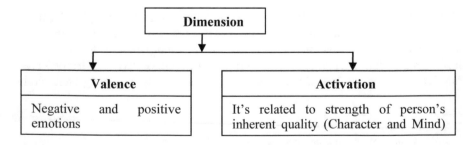

Valence indicates emotion is positive (happy, joy, trust) or negative (sad, disgust, fear, anger). Activation reflects the inherent qualities of persons like character and mindset. It represents the strength of a person to react in a particular situation. Speech emotions are verified statistically as well empirically. Two-dimensional model is necessary as well as useful in deriving empirical results. Hearer can rate the emotions presented by the speaker in concern with activation and valence dimension. After that statistical operations will be performed on the speech signal to determine emotion(s). Empirical results will be compared with the statistical to calculate the ground truth [12], presented a model having different emotions states, e.g., joy, trust, fear, surprise, sadness, anger, disgust, and anticipation.

SER being multimodal, it is a complex system. But due to application in medical diagnosis, it has obtained more attention. SER can discern mental disorder [13]. It can also diagnose Alzheimer and Parkinson [14]. Increasing suicide rates, catastrophic weather, nasty political climate, everyday annoyance like nasty colleagues, train delay, traffic, etc., are outside factors of stress. Looking for a job, managing own chronic condition, parenting an anxious, depressed, autistic, disabled, learning child; caring for loved one, etc. are home stressors. People working in the military and Information Technology (IT) sector are highly stressed due to workload. Children are stressed due to cut-throat competition in the education field. Thus in modern world stress is *constant*. Stress will be chronic if it is paired with uncontrolled circumstances. It may lead to the physical problem (alcoholism, asthma, fatigue) or even impulsive behavior like committing suicide. Therefore, stress is defined as *Silent Killer* (*SK*) by medical experts [15]. Solution to this problem is Computer Voice Stress Analysis (CVSA). SER is sub-branch of CVSA. SER gives realistic and natural HCI [16].

The objective of the research paper is to detect stress by analyzing speech of a person, spoken in Marathi language. Escalating growth of Information Technology (IT) sector in Maharashtra is motivational factor for selecting this language. Late hour working and immense mental stress due to cut-throat competition, approximately 68% of Indian IT professionals are heavily stressed or depressed [17]. Over-stress leads to health issues and attracts verities of diseases like acid peptic, asthma, fatigue, diabetics [18–20]. Even the intake of alcohol increases substantially. The situation is vicious in itself, i.e., stress leads to physical problems and physical issues leads to more depression. Human emotions have high correlation with Fatigue and depression. One can figure out emotional prosody i.e. speech tone of individual is conveyed through changes in speech rate, timber, loudness, pitch and pauses which is different from semantic and linguistic information. Thus speech is normal outlet to demonstrate stress.

AbdelWahab and Busso [21] used artificial intelligence. Different databases were trained and tested using the multi-corpus framework. The adaptive Supervised Model (ASM) was considered to improve system performance. To evaluate the ruggedness of a system they used mismatched training and testing conditions. The concept was Adaptive Supervised Learning (ASL). Jin et al. [22] worked on low-level features like jitter, shimmer, special contour, fundamental frequency (F_0), intensity, etc. Based on low-level features they generated acoustic feature presentations those are divergent. It includes statistical features as well. A new representation was a combination of Gaussian Super vectors and a set of low-level acoustic code words. Gammatone Cepstral Coefficients of speech signal were proposed by Garg and Bahl [23]. Variance

Modeling and Discriminant Scale Frequency Maps (VM-DSFM) was proposed by Wang et al. [24].

2 Database Creation

There are six basic emotions, namely *happy, sad, disgust, anger, surprise, and fear.* Emotions may be detected using facial expressions and physiological methods, but it has drawbacks [25]. Databases for different languages are available like Danish, KISMET, BabyEars, SUSAS, MPEG-4, etc. [26]. CMU INDIC is the only source which provides Marathi database. Therefore, Vishal et al. recorded 25 audio of male as well as female speaker of age group ranging from 21 to 41 years [27]. He considered three emotions and each speaker recorded 24 words per emotion. This database is not made available to public. No benchmark database is available for Marathi language [28]. Therefore, creating a database was a desideratum.

Spoken language depends upon gender, age group, accent, and geographical regions. While recording audio clip parameters like sampling rate, duration, type of recording (mono/stereo), number of bits, etc. should be set in accordance to get the maximum result out of it. As the system is supposed to be used by common people, one should avoid recording by professionals. Recording by professionals may add up superlative emotions [29, 30]. Based on these criteria, the author created **Marathi database** considering variations, shown as follows.

1. Age group: young one (7–10 years), adolescent (15–18), youngsters (18–35), middle age (35–55) and old age (above 60).
2. Gender: Male/Female.
3. Emotions: happy, anger, sad, surprise fear, and neutral.
4. Sentence size: 5–7 words.
5. Repetition rate −3.
6. Types of statement per speaker: 10.
7. Geographical zones covered: Pune, Ambajogai, Jalgaon, Kolhapur, Nagpur

Spoken Marathi has variations based on district, locality in its accent, and tone. The author considered points 1–7 and created 6000 recordings in .WAV format. The recording was done on PC, with a sampling rate of 8000 Hz, 8 bits and mono, using sound recorder of Windows 7 operating system. Normally sound recorder provides .WMA format. The same was then converted to .WAV format.

3 Formulating Problem Statement: Block Diagram of SER

Emotions are exceptionally associated to speech features, e.g., energy, pitch, jitter, shimmer, eloquent, timing, etc. [20, 31]. Varieties of models of emotions in different dimensions are presented by the researcher [30]. Arousal is correlated with intensity or energy with which the speaker conveys his/her emotion. High activation is related to anger and happiness emotion. Thus, for a robust system, one should consider number of features.

Refer Fig. 1. In SER two blocks are major, i.e. feature extraction and classifications. Phonetic and prosodic two major feature sets. Features extracted for the SER is described in Sect. 5. SER system uses a supervised learning system to classify features extracted from speech signal. There will be two phases, training and testing.

MATLAB is used for Monte carlo simulation. The software can handle audio file formats like .wav/.wma/.mp3. The input speech signal is first preprocessed. Speech signal should have a mono track, but still, if input given by the user is stereo track then it will be converted to a mono track. Normalizing speech signal, removal of noise and segregating unvoiced-voiced sound are the next stages in preprocessing [32].

The author has proposed features to be extracted from audio wave are energy, vocal tract, Mel-frequency Cepstrum Coefficient (MFCC) and pitch. These features are extracted for varieties of audio signals described in Sect. 2. After operating upon a certain percentage of dataset, the feature vector is generated. The generated vector is trained by the Supervised Machine Learning (SML) algorithm [33]. SML produces a network file ('.*net*') based on the number of datasets. The file will be later used in the testing phase.

In the testing phase, real-time audio or the audio file which was not the part of the training phase will be input to SER. The next two stages, i.e., preprocessing and feature extraction, are the same as that of the training phase. New feature set generated with new input file will be input to the classifier and the same will be juxtaposed with the trained model. Euclidean distance will be computed. The minimum distance will show the correlation of input audio with a particular emotion. The confusion matrix is presented in Sect. 7. It depicts the accuracy of a system.

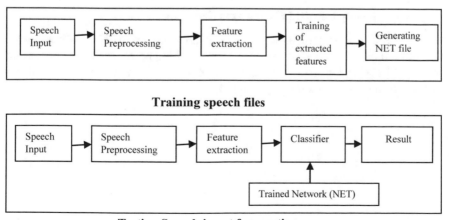

Training speech files

Testing Speech input for emotion

Fig. 1 Block diagram of SER system

4 Proposed Work—Preprocessing

For stereo recorded audio files conversion to mono track is necessary. This can be done by performing element-wise addition and averaging out the speech value. If the number of recording channels is N and the length of recording is M then, $y(m) = \frac{\sum_{n=0}^{N} S(n)}{N}$ $m = 0, 1, \ldots, M$. The speech signal value varies from 0 to 255. By normalization operation, the same is brought to within the range of 0–1. This makes the algorithm independent of the basic value of speech signal. Noise removal is achieved using the threshold technique. The signal value below a certain level will be considered as noise signal and is given as follows.

Algorithm for separating voiced/unvoiced sound:
1. Frame the signal into short non overlapped frames.
2. Apply hamming window to all the frames.
3. Calculate short time energy of the output generated from step – 2.
4. Calculate zero crossing rates of the frames generated in step -1.
5. If (zero crossing rate < Threshold # short time energy >Threshold)
   $TRUE \longrightarrow$ Voiced Frame segment
   $FALSE \longrightarrow$ Frame is Non-voiced segment

---

5 Proposed Work—Feature Extraction

Proposed work extracts four features of the speech signal and uses it for training purpose, thus presents a multi-model approach for emotion detection.

a. **Mel-Frequency Cepstral Coefficients (MFCC):**

MFCC provides short term power spectrum presentation. MFCC uses nonlinear Mel scale. It calculates cosine transform (linear) of log power spectrum. Advantage of using MFCC is for better representation of speech wave that matches with the response of the human auditory system [34].

MFCC Algorithm:
1.   Frame the signal into short frames.
2.   Split speech buffer into separate segments.
3.   Power spectrum is calculated by periodogram estimation of each and every segments.
4.   To calculate mel frequency use Equation-1[16].
5. $mel(f) = 1127 \ln\left(1 + \frac{f}{700}\right)$ ....(1)
6.   Mel filter bank is applied to power spectrum.
7.   Energy of each filter is calculated and summed up.
8.   Calculate log of energy of all the filter bank.
9.   Output of step 6 is used to calculate Discrete Cosine Transform (DCT).
10.   Retain 2 to 13 coefficients of DCT.

In the proposed scheme, MFCC feature of all different speakers with varieties of emotions is extracted.

b. **Vocal Tract Frequency**:

Vocal tract (VT) of humans is typically 17–18 cm, considering VT as a closed cylinder, the typical frequency generated is 500 Hz. This prediction leads to set of formant frequencies, i.e., 0.5, 1.5, and 2.5 kHz. But these frequencies may change as the articulator introduces vowel sounds. Male voice formant frequencies are first formant 0.15–0.85 kHz, second formant 0.5–2.5 kHz, third formant 1.5–3.5 kHz, and fourth formant 2.5–4.8 kHz.

c. **Pitch**:

Pitch is the main acoustic correlate of tone and intonation. Pitch is the number of vibrations produced by the vocal tract per second. Ear detects pitch by highness or lowness of tone.

d. **Energy**

Teager energy operator (Eq. 2) [30] is used to calculate the energy of speech signal.

$$\Psi[S(n)] = S^2(n) - S(n-1) * S(n+1) \tag{2}$$

Algorithm:
1. Read speech file
2. Detect voiced part.
3. Perform operation $y(n) = \lvert S(n)\rvert^2 - S(n-1) * \mathbb{C}\, S(n+1)$, where $\mathbb{C}$ is complex conjugate.
4. Calculate cross correlation of $y(n)$
5. Evaluate $z(n) = \sum \lvert y(n)\rvert^2$
6. Finally $energy = 20 \log_{10} z(n)$

Table 1 ANN classifier pseudocode

Setup network hidden layer
Set Input, required output, learning rate and epochs (iterations)
Weights initialised with random number within the range from -1 to 1
Do
Pattern presented to training network
Each layer and every node in layered network
    1.   Summation of all the inputs to node(s)
    2.   Threshold is applied to summation result
    3.   Calculating activation to node
// Propagating errors backward to Network
for all nodes in output layer
Perform Error calculation
end
for hidden layers in the network
for each node in the layer
    1.   Calculating Nodes' signal error
    2.   Updating weights of the nodes'
end
end
// Global error
Error function calculation
Repeat
While ((maximum number of epochs < specified AND Error function > specified)

6 Proposed Work—Artificial Neural Network (ANN)

After the feature extraction matrix is created for different types of emotions expressed by different speakers. These features are trained using Artificial Neural Network (ANN)—a pattern recognition technique. ANN is also a machine learning algorithm [35]. The base of neural network learning is on a similar line to biological neurons. Structure of ANN is input, hidden layers, and output. Each stage has a number of nodes that are interconnected. The training phase generates *.net*structure. The same should be used for testing system performance. Pseudocode for ANN is shown in Table 1.

7 Results

For implementation purpose, MATLAB (2014b) is used. Figure 2 shows the actual speech signal plot. Figure 3 shows output after removing the unvoiced or silence part. Only voiced part considered for the feature extraction process.

Feature extraction is the base for the classifier. The multi-model approach is used to make the system robust and reliable. 30% part of the database is used for training

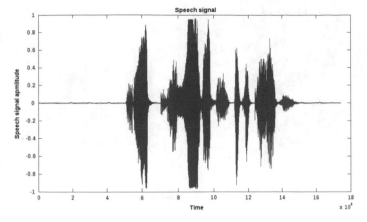

Fig. 2 Speech signal

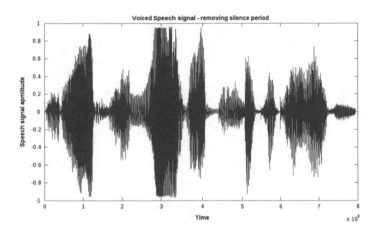

Fig. 3 Voiced part of speech signal

purpose. Before initiating training it has been observed that extracted feature provides variations based on emotions. The following diagram represents a 3D graph. The X-axis represents the average number of speakers. Y-axis represents different features like MFCC, vocal tract, pitch, and energy. Z-axis represents different emotions Fig. 4.

Feature database set was trained using '*nftool*' in MATLAB. Out of 6000 speech files, 30%, i.e., 1800 speech files (360 files per emotions) were trained. Trained model is used for testing of the database. Figure 5 depicts performance graph.

Refer Fig. 4. The plot specifies that MFCC features of emotions are having very less overlap, hence, they are distinguishable. Stress related emotion viz. Sad and angry are distinct as compared to happy, neutral and surprise, in all features. In sad condition, pitch and energy both are low. In angry emotion VT, pitch and energy all are maximum level.

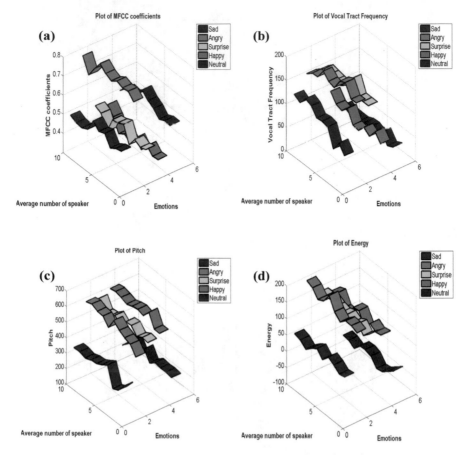

Fig. 4 **a** MFCC plot. **b** Vocal tract frequency plot. **c** Pitch plot. **d** Energy plot

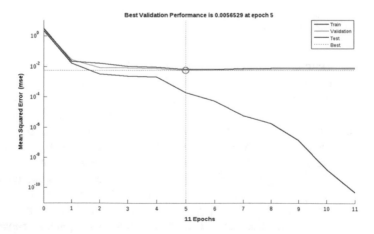

Fig. 5 Performance plot

Table 2 ANN classifier performance analysis

Emotions	Emotion Recognition %				
	Sad	Angry	Surprise	Happy	neutral
Sad	87	0	0	0	13
Angry	0	80	12	8	0
Surprise	0	6	85	9	0
Happy	0	7	10	75	8
Neutral	6	0	0	4	90

As shown, the validation and test graph are near to best. Accuracy is calculated based on the following equation.

$$\text{Accuracy} = \frac{\text{Number of correct emotions detected}}{\text{Total number of samples of that emotion}} \times 100$$

Table 2 depicts overall results:

8 Conclusion

Database creation is the base of the Marathi language emotion detection system. Diversity in database creation is the need of the day. In India, for the same language pronunciations differs with the zone. Covering all these zones was a great challenge. Multi-model approach for extracting features is suggested in this research work with ANN as a classifier. Stress typically introduces negative emotions in human mind. Depression and anxiety may lead to sad/angry emotions. So if the average of performance accuracy of these both emotions are considered then it is around 83–84%.

Present multi-model approach deals with four sets of features. The same may be further extended by adding quality features like jitter and shimmer.

References

1. Darwin, C.: The Expression of Emotion in Man and Animals. JohnMurray, London (1872)
2. Blanton, S.: The voice and the emotions. Q. J. Speech **1**(2), 154–172 (1915)
3. Fairbanks, G., Pronovost, W.: Vocal pitch during simulated emotion. Science **88**(2286), 382–383 (1938)
4. Soskin, W.F., Kauffman, P.E.: Judgment of emotion in word-free voice samples. J. Commun. **11**(2), 73–80 (1961)
5. Kramer, E.: Elimination of verbal cues in judgments of emotion from voice. J. Abnorm. Soc. Psychol. **68**(4), 390 (1964)

6. Bozkurt, O.O., Taygi, Z.C.: Audio-based gender and age identification. In: 22nd Signal Processing and Communications (2014). https://doi.org/10.1109/siu.2014.6830493
7. McCarthy, J.: What is artificial intelligence (2007). http://wwwformal.stanford.edu/jmc/whatisai.html
8. Higginbotham, A.: Welcome to Rosalind Picard's touchy-feelyworld of empathictech (2011). http://www.wired.co.uklmagazine/archive/2012/11(features/emotion-machines)
9. Calvo, R.A., D'Mello, S.: J. IEEE Trans. Affect. Comput. Arch. (IEEE Computer Society Press Los Alamitos, CA, USA) (1), 18–37 (2010)
10. El Ayadi, M., Kamel, M.S., Karray, F.: Survey on speech emotion recognition: features, classification schemes, and databases. Pattern Recognit. (2011). https://doi.org/10.1016/j.patcog.2010.09.020
11. Cowie, R., Cornelius, R.R.: Describing the emotional states that are expressed in speech. Speech Commun. (Elsevier Science) 40, 5–32 (2003)
12. Ekman, P.: An argument for basic emotions. pp. 169–200. 07 Jan 2008. https://doi.org/10.1080/02699939208411068
13. Kostoulas.: Emotion recognition from speech using digital signal (2012)
14. López-de-Ipiña, K., Alonso, J.-B., Travieso, C.M., Solé-Casals, J., Egiraun, H., Faundez-Zanuy, M., Ezeiza, A., Barroso, N., Ecay-Torres, M., Martinez-Lage, P., Martinez, U.: On the selection of non-invasive methods based on speech analysis oriented to automatic alzheimer disease diagnosis. Sensors. ISSN 1424-8220. www.mdpi.com/journal/sensors
15. https://www.everydayhealth.com/wellness/united-states-of-stress/
16. Guo, B., Hershey, P.A.: Creating Personal, Social, and Urban Awareness Through Pervasive Computing. Information Science Reference (2014)
17. Schuller, B., Rigoll, G., Lang, M.: Speech emotion recognition combining acoustic features and Linguistic information in a hybrid support vector machine belief network architecture. In: Proceedings of the ICASSP 2004, vol. 1 (2004)
18. Williams and Stevens.: Robust Emotion Recognition using Spectral and Prosodic Features. Springer (2013)
19. Benesty, J., Sondhi, M.M., Huang, Y.: Handbook of Speech Processing. Springer(2008)
20. Rao, K.S., Koolagudi, S.G.: Emotion Recognition Using Speech Features. Springer (2013)
21. Abdelwahab, M., Busso, C.: Supervised domain adaptation for emotion recognition from speech. IEEE (2015). ISSN. 1520-6149
22. Jin, Q., Li, C., Chen, S., Wu, H.: Speech emotion recognition with acoustic and lexical features. IEEE (2015). ISSN. 1520-6149
23. Garg, E., Bahl, M.: Emotion recognition in speech using gammatone cepstral coefficients. Int. J. Appl. Innov. Eng. Manag. (IJAIEM) 3(10) (2014). ISSN 2319-4847
24. Wang, J.-C., Chin, Y.-H., Chen, B.-W., Lin, C.-H., Chung-Hsien, W.: Speech emotion verification using emotion variance modeling and discriminant scale frequency maps. IEEE Trans. Speech Lang. Process. 23(10), 1552–1562 (2015)
25. Hasrul, M.: Human affective (Emotion) behaviour analysis using speech signals: a review. In: 2012 International Conference on Biomedical Engineering ICoBE 2012, pp. 27–28 (2012)
26. Shrishrimal, P.P., Deshmukh, R.R., Waghmare, V.B.: Indian language speech database: a review. Int. J. Comput. Appl. 47(5), 17–21 (2012)
27. Waghmare, V.B., Deshmukh, R.R., Shrishrimal, P.P., Janvale, G.B.: Development of isolated Marathi words emotional speech database. IJCA 94(4), 0975–8887, 19–22 (2014)
28. Pahune, S., Mishra, N.: Emotion recognition through combination of speech and image processing. Int. J. Recent Innov. Trends Comput. Commun. (2015). ISSN: 2321-8169
29. Nayak, B., Madhusmita, M., Sahu, D.K.: Speech emotion recognition using different centred GMM, 3(9) (2013). ISSN: 2277 128X
30. El Ayadi, M., Kamel, M.S. Karray, F.: Survey on speech emotion recognition: features, classification schemes, and databases. Pattern Recogn. 44, 572–587 (2011)
31. http://www.phon.ucl.ac.uk/courses/spsci/expphon/week9.php
32. Elissa, K.: Input processing for cross language information access (1991–92). ISSN No. 0972-645

33. Eyben, F., Scherer, K., Schuller, B., Sundberg, J., Andre, E., Busso, C., Devillers, L., Epps, J., Laukka, P., Narayanan, S., Truong, K.: The Geneva minimalistic acoustic parameter set (GeMAPS) for voice research and affective computing. IEEE Trans. (2015)
34. Waghmare, V.B., Deshmukh, R.R., Shrishrimal, P.P., Janvale, G.B.: Emotion recognition system from artificial marathi speech using MFCC and LDA techniques. Elsevier (2014)
35. http://machinelearningmastery.com/a-tour-of-machine-learning-algorithms/

Assessment of Autistic Disorder Using Machine Learning Approach

Camellia Ray, Hrudaya Kumar Tripathy and Sushruta Mishra

Abstract Autistic Spectrum Disorder is a neurobehavioral disorder with multiple limitations including social interaction, communication, and restrictive behaviors. Nowadays, ASD has become one of the quick spreading diseases all over the world. Therefore, there is a huge need to provide a time-consuming and easily accessible diagnostic tool to detect autism at an early stage that helps the clinicians to provide prior medications. Though there is no proper curability of autism, still easy detection helps to provide better therapy sessions and supports the autistic child to lead a comfort independent life. Our paper deals with the classification of datasets of autistic suspected patients using various classifiers in the WEKA tool to find out the one with higher accuracy that can predict the child to be truly suffering from autism.

Keywords Autistic spectrum disorder · Diagnostic tool · Random forest algorithm

1 Introduction

Autism Spectrum Disorder is a neurodevelopmental disorder having five of its types including Pervasive Developmental Disorder, Asperger Syndrome, Childhood Disintegrative Disorder, Rett syndrome, and Autistic Disorder. The five types of disorders are combined to have their own limitations in proper communication, social interaction, and attention problems [1]. The disease lasts through the entire life of an individual with its symptoms arising in early childhood [2].

The real cause for autism is still under research but the possibility of its occurrence can be through genetically related issues, advanced aging of parents, severe

C. Ray (✉) · H. Kumar Tripathy · S. Mishra
School of Computer Engineering, Kalinga Institute of Industrial Technology, Deemed to be University, Bhubaneswar 751024, Odisha, India
e-mail: camellia.jhilik@gmail.com

H. Kumar Tripathy
e-mail: hktripathyfcs@kiit.ac.in

S. Mishra
e-mail: sushruta.mishrafcs@kiit.ac.in

© Springer Nature Singapore Pte Ltd. 2020
V. Bhateja et al. (eds.), *Intelligent Computing and Communication*,
Advances in Intelligent Systems and Computing 1034,
https://doi.org/10.1007/978-981-15-1084-7_21

infections or complications during pregnancies like the crisis of oxygen supply in the brain while birth or premature babies. The dysfunction in the fusiform gyrus, amygdala, cerebellum, cerebral cortex in the brain can be a major cause for autism. The environmental factors like vaccines (MMR), pollutants, and dietary additives can be the causes after birth [1].

There is an increased notice of the autism all over the world, where a huge number of children are already affected by the autistic disorder. Therefore, the autistic disorder needs to be identified earlier so that it can create a roadmap for early treatment. Various diagnostic tools have been invented to conduct a clinical assessment of the patients like Social Communication Disorder Checklist (SCDC), Social Communication Questionnaire (SCQ), Autism Spectrum Quotient (ASQ), Autism Diagnostic Observation Schedule (ADOS). But all of those processes are lengthy and not cost-effective. Therefore, an urgent need for the development of an easily implemented and effective screening method is truly a matter of concern to diagnose whether the individuals are suffering from ASD or not [2].

2 Literature Survey

Rudra et al. [3] has described some screening and diagnostic toolkits like SCQ, SCDC, AQC, and ADOS consisting of a set of questionnaires which were further translated to Hindi and Bengali languages since the majority of the population in India are not proficient in English. The participants were provided with a certain checklist and their respective cutoff. The participants meeting the cut-off score of a certain tool proceeds for further analysis. Later the ADOS positive children were referred to the Mental Health Foundation of India.

Rudra et al. [4] described the extension of paper [2] where the teachers and parents are provided with the hard copy of SCDC in English, Hindi and Bengali languages. Some of the positive children do not respond to other tools or the children do not possess some of the questionnaires present in the following spreadsheets of respective tools. There are some formulae to find out whether the non-responders possess autism to get the overall weighted estimation of the broader autism spectrum. Osman Altay in [5] has analyzed the two machine learning algorithms which have been utilized for the identification of autism through classification. The 292 examples are pre handled and characterized later on through the WEKA apparatus where 70% of the information was taken for preparing and 30% for testing. In spite of the fact that LDA gives the most astounding precision of 90.8% though KNN with the exactness of 88.5%, still different assessment measures to test the arrangement like affectability that signifies the ASD patients are more in KNN than that of LDA.

Fadi Tabtah in [6] has established an autism screening method, where the dataset of cases and controls were gathered from a few indicative instruments like ADOS and ADI-R which were further preprocessed and prepared to utilize AI calculations to develop a model for determination of ASD to diminish the screening time and improving the affectability and explicitness.

3 Data Collection and Description

The dataset utilized in the examination was gotten from the National Institute for Health Research (NHS). NHS is the biggest national clinical research funder in Europe that supports fantastic research to improve wellbeing. The reserve prepares and underpins wellbeing scientists by giving world-class examine offices. The following dataset was also used in a study conducted by [2].

The dataset consists of 704 samples with their 21 attributes. Out of those 21 attributes in Table 1, there are 10 questions related to the behavioral traits of autism that help to detect whether the child is suffering from autism or not [7].

Figure 1 consists of the details of all questionnaires related to behavioral traits to understand the conditions of respective individuals [7].

The output values consist of two classes Yes and No to symbolize if the suspected individuals are suffering from autism or not [8].

3.1 WEKA

WEKA (Waikato Environment for Knowledge Analysis) is an open-source software named after a flightless bird WEKA found in New Zealand. WEKA consists of a collection of machine learning algorithms that help in data mining tasks like classification, clustering, regression, association rule mining, and attribute selection for data analysis and predictive modeling [9].

WEKA has several graphical interfaces, out of them one is Explorer to deal with the smaller dataset. The explorer consists of different panels like **Preprocess** that have filters to load and transfer the data through pre-processing tools. **Classify** helps

Table 1 Attributes related to characteristics of individuals with their respective values [7]

Attribute name	Values
Age	4–11
Gender	Male, Female
Ethnicity	White European, South Asian, Asian, Middle Eastern, Pasifika, Hispanic, Turkish, Latino, Black, Others, Unknown
Jaundice	Yes, No
Autism	Yes, No
Country of res	52 different countries
Used app before	Yes, No
Result	0–10
Age description	Range
Relation	Parent, Relative, Health care professionals, Unknown
Class	Yes, No

Please tick one option per question only:	Definitely Agree	Slightly Agree	Slightly Disagree	Definitely Disagree
1	S/he often notices small sounds when others do not				
2	S/he usually concentrates more on the whole picture, rather than the small details				
3	In a social group, s/he can easily keep track of several different people's conversations				
4	S/he finds it easy to go back and forth between different activities				
5	S/he doesn't know how to keep a conversation going with his/her peers				
6	S/he is good at social chit-chat				
7	When s/he is read a story, s/he finds it difficult to work out the character's intentions or feelings				
8	When s/he was in preschool, s/he used to enjoy playing games involving pretending with other children				
9	S/he finds it easy to work out what someone is thinking or feeling just by looking at their face				
10	S/he finds it hard to make new friends				

Fig. 1 Ten questionnaires

in the classification of instances through different classifiers for predictions. The test options within Classify are used to divide the dataset for training and testing during classification. In the Training set, researchers have their own training instances to make the model and the same training set is used in testing purposes as well. In the case of a Supplied test set, a model is developed with researchers' own training sets and applied on a particular supplied test set for testing. Thirdly during cross-validation, 90% of the data are used in training purpose and 10% into testing purposes with every time leaving a different fold for testing. Lastly, in percentage split, the respective percentage is used for training the dataset and the remaining for testing. **Clustering** is used to group similar instances and ungroup the dissimilar ones with the same test options for training and testing. **Associate** uses the association and a priori algorithm to find out the probability of co-occurrence of items. **Select Attributes** choose the necessary attribute through Attribute Evaluator and the best search option to reduce the dimensionality for better prediction. **Visualization** helps in visualizing each attribute with respect to their values by selecting individual plots in the matrix to ensure whether the models are performing as expected [9].

4 Methodology

The dataset consisting of 21 attributes including 10 questionnaires related to behavioral traits are preprocessed using several filters. The values of all the attributes are converted to numerical values in order to make the classification easier. The missing values present in the dataset are removed using the Interquartile range (IQR). Later the dataset has been sampled randomly to balance the positive and negative instances for smooth classification.

After preprocessing, the dataset is classified using a different algorithm to discover a proper model for predicting the ASD children that differ from the normal ones, based on the value of other attributes. The test option committed cross-validation of 10 folds, which is the most usable test set nowadays and generates a different fold for testing every time. The cross-fold of 10 clears a meaning of ninefold for training purpose and onefold for testing purpose.

(i) *Naive Bayes*

Naive Bayes algorithm is a technique used for classification [10]. The classifier dependent on Bayes theorem expects the estimation of a specific component to be autonomous of the estimation of whatever another element when the class is given. The class with the most extreme likelihood is chosen to be the new class of the new occurrence [11]. They are really easy to implement and often obtain good results while classifications. But due to their class conditional independencies, they lose their accuracy.

$$P(A_1, A_2, \ldots, A_n | C) = P(A_1 | C_j) P(A_2 | C_j) \ldots, P(A_n | C_j). \tag{1}$$

where, $A_1, A_2, \ldots, A_n$ is the attribute and C is the class given.

(ii) *Radial Basis Function Network*

RBFN is an artificial neural network that uses a radial basis function as activation functions for classification. It comprises of 3 layers namely the input layer, a hidden layer with nonlinear RBF activation function, and a linear output layer [12]. Each neuron in the RBFN stores a prototype of the examples in the training set. With the income of new inputs, the Euclidian distances are calculated between the new input with each example of the training set. The one that closely resembles one of the training set possesses the same class label with the particular training set.

(iii) *K-star*

K-star is an instance-based classifier, where the class of a test case depends on the class of a preparation occurrence like it [13]. The entropic measure is used as a distance function to determine the similarity. K-star provides high performance to deal with missing values and smoothness problems [14]. The k-star distance can find out by summing up the probability of all possible transformation between two instances a and b where transformation starts from a and ends at 'b' being motivated through Information theory. The probability of all paths from instance a and b can be defined by

$$P * (b|a) = \sum_{\bar{t} \in P : \bar{t}(a) = b}^{n} p(\bar{t}) \tag{2}$$

where P^* is the probability of all paths from instance a to b, represents each transformation from a to b

(iv) *Random Trees*

An irregular tree is a gathering of tree pointers called a forest. They are sorted out with a progression of inquiry sets and conditions in a tree-like structure. The test conditions are connected to the record and are caught up with the proper branch dependent on the result of the test. That point either drives us to an interior hub to apply with another test again or straightforwardly to a leaf hub classifier takes the information highlight vector and gatherings with each tree in the forest. The class name of the trees with larger part votes will be the class names of the updated one [15].

(v) *Random Forest*

Random forest is an accumulation of multiple decision trees that are organized with a series of question sets and conditions in a tree-like structure. The test conditions are applied to the record and are followed up with the appropriate branch based on the outcome of the test [16]. It then either leads us to an internal node to apply with a new test again or directly to a leaf node. Random forest merges the decision trees with a random subset of features to get a more accurate and stable prediction [17]. The class that comes up with the majority voting will be the new class of the respective instances.

5 Result and Discussions

After classifying through five classifiers as discussed above in the WEKA tool, the classifier possesses different accuracy rates depending on the respective algorithm. After the detailed analysis discussed below, the random forest appears to be with the highest accuracy than any classifier (Figs. 2, 3, 4, 5, 6 and 7).

From the above graph, it is crystal clear that Random Forest is providing the highest accuracy of 100% than any other classifier discussed earlier.

The TP or true positive rate that represents the number of positive events correctly categorized as the positive rate was found out to be 1, whereas the FP or false positive rate that wrongly categorized the negative events as positive was 0 for both Yes and No classes. Likewise, the Precision which is the fraction of relevant instances among the retrieved instances was 1 and the recall among the total relevant instances was

```
Scheme:weka.classifiers.bayes.NaiveBayes
Relation:       adult-weka.filters.unsupervised.attribute.NumericToNominal-Rfirst-10
Instances:      704
Attributes:     21

Time taken to build model: 0.02 seconds

--- Stratified cross-validation ---
--- Summary ---

Correctly Classified Instances        683              97.017  %
Incorrectly Classified Instances       21               2.983  %
Kappa statistic                         0.9262
Mean absolute error                     0.0326
Root mean squared error                 0.136
Relative absolute error                 8.2965 %
Root relative squared error            30.6985 %
Total Number of Instances             704

--- Detailed Accuracy By Class ---

              TP Rate   FP Rate   Precision   Recall   F-Measure   ROC Area   Class
              0.963     0.011     0.996       0.963    0.979       0.999      NO
              0.989     0.037     0.908       0.989    0.947       0.999      YES
Weighted Avg. 0.97      0.018     0.972       0.97     0.971       0.999

--- Confusion Matrix ---

   a   b   <-- classified as
 496  19 |  a = NO
   2 187 |  b = YES
```

Fig. 2 Confusion matrix on results using Naive Bayes

```
Scheme:weka.classifiers.functions.RBFNetwork -B 2 -S 1 -R 1.0E-8 -M -1 -W 0.1
Relation:       adult-weka.filters.unsupervised.attribute.NumericToNominal-Rfirst-10
Instances:      704
Attributes:     21

Correctly Classified Instances        695              98.7216 %
Incorrectly Classified Instances        9               1.2784 %
Kappa statistic                         0.9676
Mean absolute error                     0.0143
Root mean squared error                 0.1075
Relative absolute error                 3.631  %
Root relative squared error            24.2529 %
Total Number of Instances             704

--- Detailed Accuracy By Class ---

              TP Rate   FP Rate   Precision   Recall   F-Measure   ROC Area   Class
              0.988     0.016     0.994       0.988    0.991       0.991      NO
              0.984     0.012     0.969       0.984    0.976       0.991      YES
Weighted Avg. 0.987     0.015     0.987       0.987    0.987       0.991

--- Confusion Matrix ---

   a   b   <-- classified as
 509   6 |  a = NO
   3 186 |  b = YES
```

Fig. 3 Confusion matrix on results using RBF network

Fig. 4 Graphical
representation of the
accuracy rate of five
algorithms

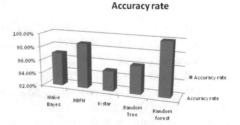

```
KStar options : -B 20 -M a

Time taken to build model: 0 seconds

  ---- Stratified cross-validation ----
  ---- Summary ----

Correctly Classified Instances          672              95.4545 %
Incorrectly Classified Instances         32               4.5455 %
Kappa statistic                          0.885
Mean absolute error                      0.0543
Root mean squared error                  0.1802
Relative absolute error                 13.8047 %
Root relative squared error             40.6673 %
Total Number of Instances                704

  ---- Detailed Accuracy By Class ----

              TP Rate   FP Rate   Precision  Recall  F-Measure  ROC Area  Class
               0.965     0.074     0.973     0.965     0.969      0.991     NO
               0.926     0.035     0.907     0.926     0.916      0.991     YES
Weighted Avg.  0.955     0.064     0.955     0.955     0.955      0.991

  ---- Confusion Matrix ----

   a    b    <-- classified as
  497   18 |  a = NO
   14  175 |  b = YES
```

Fig. 5 Confusion matrix on results using *K*-star

```
Scheme:weka.classifiers.trees.RandomTree -K 0 -M 1.0 -S 1
Relation:      adult-weka.filters.unsupervised.attribute.NumericToNominal-Rfirst-10
Instances:     704
Attributes:    21

Correctly Classified Instances          677              96.1648 %
Incorrectly Classified Instances         27               3.8352 %
Kappa statistic                          0.9022
Mean absolute error                      0.0403
Root mean squared error                  0.1854
Relative absolute error                 10.2564 %
Root relative squared error             41.8422 %
Total Number of Instances                704

  --- Detailed Accuracy By Class ---

              TP Rate   FP Rate   Precision  Recall  F-Measure  ROC Area  Class
               0.975     0.074     0.973     0.975     0.974      0.96      NO
               0.926     0.025     0.931     0.926     0.928      0.96      YES
Weighted Avg.  0.962     0.061     0.962     0.962     0.962      0.96

  --- Confusion Matrix ---

   a    b    <-- classified as
  502   13 |  a = NO
   14  175 |  b = YES
```

Fig. 6 Confusion matrix on results using random tree

1 as well for both the classes. Both the F-measure that measures the test's accuracy and ROC Area to measure the classification accuracy yields to 1 in both the labels.

The major part of the result is the confusion matrix that summarizes the performance of a classification algorithm. In the confusion matrix of random forest, 515 samples were classified as true negative and remaining 129 as true positive which means none of them are misclassified.

In Table 2, the total analysis represented the best performance of the random forest Algorithm than any other one. Thus, it can be clearly concluded that in case of diagnosis of autism even after enlarging the dataset, Random Forest will firmly predict the ASD -affected children than any other classifier for their early recovery.

```
Random forest of 100 trees, each constructed while considering 5 random features.
Out of bag error: 0


Time taken to build model: 0.25 seconds

--- Stratified cross-validation ---
=== Summary ===

Correctly Classified Instances        704              100      %
Incorrectly Classified Instances        0                0      %
Kappa statistic                         1
Mean absolute error                     0.056
Root mean squared error                 0.0951
Relative absolute error                14.2421 %
Root relative squared error            21.463  %
Total Number of Instances             704

=== Detailed Accuracy By Class ===

               TP Rate   FP Rate   Precision   Recall   F-Measure   ROC Area   Class
                  1         0          1          1         1           1       NO
                  1         0          1          1         1           1       YES
Weighted Avg.     1         0          1          1         1           1

=== Confusion Matrix ===

   a    b    <-- classified as
 515    0  |   a = NO
   0  189  |   b = YES
```

Fig. 7 Confusion matrix on results using random forest

Table 2 Accuracy rates of different classifiers while classifying through WEKA tool	Classifier	Accuracy (%)
	Naïve Bayes	97.01
	RBFN	98.72
	K-star	95
	Random tree	96.16
	Random forest	100

6 Conclusion

The paper deals with a pilot search of the dataset that contains 10 questionnaires and other individual characters of 706 samples to predict whether they fall in the category of ASD or not. The search has been processed through the WEKA tool where the Random Forest algorithm responds with the highest accuracy during classification. The vast analysis of TP rate, FP rate, Precision, Recall, and other important features proved it in the same way. There is a need to focus on future research initiatives in the area of ASD diagnosis that are as follows:

- Enhancing the dataset by collecting real-time data from different autism centers.
- A classifier with the highest accuracy can later classify better with the enhanced dataset and will help to provide a healthier model for prediction.

Acknowledgements The paper is one of the pieces of my execution to diagnosing the medically introverted kid. I need to express my most significant gratefulness to every single one of the people, who has given me the probability to finish the paper. I am really thankful to my guide and co-guide in empowering suggestions and backing and helping to arrange my points especially in forming this paper.

References

1. Howlin, P.: Autism spectrum disorders. Psychiatry **5**(9), 320–324 (Elsevier) (2006)
2. Thabtah, F., Kamalov, F., Rajab, K.: A new computational intelligence approach to detect autistic features for autism screening. Int. J. Med. Inform. **117**, 112–124 (Elsevier) (2018)
3. Rudra, A., Belmonte, S., Chakrabarti, B.M.: Translation and usability of autism screening and diagnostic tools for autism spectrum conditions in India
4. Rudra, A., Belmonte, M.K., Soni, P.K., Mukerji, S., Chakrabarti, B.: Prevalence of autism spectrum disorder and autistic symptoms in a school-based cohort of children in Kolkata, India. Autism Res. **10**(10), 1597–1605 (Wiley Online Library) (2017)
5. Altay, O., Ulas, M.: Prediction of the autism spectrum disorder diagnosis with Linear discriminant analysis classifier and K-nearest neighbor in children. In: International Symposium on Digital Forensic and Security (ISDFS), vol. 4, pp. 748–753. IEEE (2018)
6. Thabtah, F.: Autism spectrum disorder screening: machine learning adaptation and DSM-5 fulfillment. In: Proceedings of the 1st International Conference on Medical and Health Informatics 2017 (ICMHI), vol. 5, pp. 1–6. IEEE (2017)
7. Altay, O., Ulas, M.: Prediction of the autism spectrum disorder diagnosis with linear discriminant analysis classifier and K-nearest neighbor in children. In: 6th International Symposium on Digital Forensic and Security (ISDFS), pp. 1–4. IEEE, Antalya (2018)
8. Hall, M., Frank, E., Holmes, G., Pfahringer, B., Reutemann, P., Witten, I.H.: The WEKA data mining software: an update: ACM SIGKDD explorations newsletter. ACM Digit. Libr. **11**(1), 10–18 (2009)
9. Mukherjee, S., Sharma, N.: Intrusion detection using Naive Bayes classifier with feature reduction. Procedia Technol. **4**, 119–128 (Elsevier) (2012)
10. Barone, R., Alaimo, S., Messina, M., Pulvirenti, A., Bastin, J., Ferro, A., Frye, R.E., Rizzo, R.: A subset of patients with autism spectrum disorders show a distinctive metabolic profile by dried blood spot analyses. Front. Psychiatry **9**, 1–11 (2018)
11. Radha, V.: Neural network based face recognition using RBFN classifier. In: Proceedings of the World Congress on Engineering and Computer Science, vol. 1, pp. 555–560. IEEE (2011)
12. Dayana, Hernandez: An experimental study of $K*$ algorithm. Int. J. Inf. Eng. Electron. Bus. (IJIEEB) MECS **2**, 14–19 (2015)
13. Cleary, J.G., Trigg, L.E.: K*: an instance-based learner using an entropic distance measure. In: 12th International Conference on Machine Learning, pp. 108–114. ACM Digital Library, California (1995)
14. Li, M., Song, Q., Zhao, Q.: A fuzzy adaptive rapid-exploring random tree algorithm. In: 3rd International Conference on Materials Science and Mechanical Engineering (ICMSME), pp. 167–171. DEStech, Pennsylvania (2016)
15. Belgiu, M., Dragut, L.: Random forest in remote sensing: a review of applications and future directions. ISPRS J. Photogramm. Remote Sens. **114**, 24–31 (Elsevier) (2016)

16. Feczko, E., Balba, N.M., Dominguez, O., Cordova, M., Karalunas, S.L., Irwin, L., Demeter, D.V., Hill, A.P., Langhorst, B.H., Painter, J.G., Santen, J., Fombonne, Nigg, Fair: Subtyping cognitive profiles in autism spectrum disorder using a functional random forest algorithm. NeuroImage **172**, 674–688 (Elsevier) (2018)
17. Abbas, H., Garberson, F., Glover, E., Wall, D.P.: Machine learning approach for early detection of autism by combining questionnaire and home video screening. J. Inform. Health Biomed. **25**(8), 1000–1007 (Oxford University Press) (2018)

Use of Data Mining Techniques for Data Balancing and Fraud Detection in Automobile Insurance Claims

Slokashree Padhi and Suvasini Panigrahi

Abstract A novel hybrid data balancing method based on both undersampling and oversampling with ensemble technique has been presented in this paper for efficiently detecting the auto insurance frauds. Initially, the skewness from the original imbalance dataset is removed by excluding outliers from the majority class samples using Box and Whisker plot and synthetic samples are generated from the minority class samples by using synthetic minority oversampling (SMOTE) technique. We employed three supervised classifiers, namely, support vector machine, multilayer perceptron, and K-nearest neighbors for classification purpose. The final classification results are obtained by aggregating the results obtained from these classifiers using the majority voting ensemble technique. Our model has been experimentally evaluated with a real-world automobile insurance dataset.

Keywords Automobile insurance · Outliers · Box and whisker plot · Synthetic minority oversampling · Supervised classifier

1 Introduction

An automobile insurance is an agreement between vehicle's owner and insurance organization which gives financial protection in case of any auto accident or vehicular theft. Automobile insurance fraud occurs upon submitting forged credentials to the insurance organization by showing break down of the automobile in a fake accident or exaggerated claims [1].

The missing or inappropriate depiction of the information regarding a claim makes the detection of fraud more challenging [1]. Secondly, it is noticed that the count of

S. Padhi · S. Panigrahi (✉)
Department of CSE, Veer Surendra Sai University of Technology, Burla, Sambalpur 768018, Odisha, India
e-mail: spanigrahi_cse@vssut.ac.in

S. Padhi
e-mail: slokashree.padhi@gmail.com

© Springer Nature Singapore Pte Ltd. 2020
V. Bhateja et al. (eds.), *Intelligent Computing and Communication*,
Advances in Intelligent Systems and Computing 1034,
https://doi.org/10.1007/978-981-15-1084-7_22

fraudulent claims are very less as compared to the total number of cases submitted. The unequal distribution or skewness in the dataset makes the fraud investigators work much more difficult. By using unbalanced dataset almost all of the supervised classifiers give inefficacious results because they are bias toward the majority class samples and evade the minority class samples. For this reason, reducing skewness from the dataset is very much important.

It is observed from the study that only undersampling or oversampling data balancing methods were unable to generate a balanced class distribution. This paper presents a novel hybrid data balancing method that initially applies both undersampling and oversampling techniques on majority (non-fraud) and minority (fraud) class instances, respectively, and combine their results to obtain a modified balanced dataset. Thereafter, three different classifiers, namely, MLP, KNN, and SVM are applied on the modified balanced dataset for classification of the instances as legitimate or fraud. The final decision is made by aggregating the results obtained from the classifiers using the majority voting technique.

The remaining paper is structured as follows: Sect. 2 focuses on a review of the related work done in automobile domain. Section 3 presents the brief overviews of the algorithms used in our approach and Sect. 4 describes our proposed approach. Results of our approach are discussed in Sects. 5 and 6 draws the conclusion from the research.

2   Related Work

This section reviews various methods that have been proposed for data balancing and fraud detection in this domain.

A novel approach named FC-ANN based on fuzzy clustering along with artificial neural network has been presented in [2]. K-RNN- and OCSVM-based undersampling approach which used K-RNN to remove the outliers from the genuine samples after that used OCSVM to extract the support vectors has been presented in [3]. Chun and Li [4] proposed a nearest neighbor method with pruning rules of linear time complexity to detect the outliers. For detecting fraudulent claims, Nia et al. [5] proposed a supervised method named spectral ranking anomaly. A novel genetic-algorithm-based FCM approach has been presented in [6] for identifying the fraudulent claims. The work in [7] shows the importance of SMOTE oversampling on an unbalanced dataset. A detailed overview of different ensemble techniques for decision-making has been presented in [8]. Paper [9] focuses on a privacy-preserving data mining technique by hiding the sensitive patterns from the public before sharing the dataset. For the processing of large dataset, paper [10] presented a similarity-based K-medoids clustering method which is used for reducing the clustering overhead. For the detection of malware in the dataset, paper [11] proposed a tree-based supervised classification algorithm.

3 Background Study

For better understanding of our proposed AIFDS, the techniques applied have been described in this section. The classifiers support vector machine, multilayer perceptron, and KNN are too popular for an introduction. Hence, we present a brief overview of fuzzy C-means clustering (FCM), Box and Whisker plot, synthetic minority oversampling (SMOTE), and majority voting.

3.1 Fuzzy C-Means Clustering (FCM)

By using FCM, all the data points of the dataset are assigned with some membership values and all the meaningful clusters present in the dataset are found [6]. The objective function of FCM is expressed by the equation as follows:

$$j_m(U, C; D) = \sum_{k=1}^{n} \sum_{i=1}^{c} (\mu_{ik}^m) B_{ik}(c_i, d_k) \qquad (1)$$

subject to $\sum_{i=1}^{c} \mu_{ik} = 1 \, \forall k$ and $0 \leq u_{ik} \leq 1$. The objective function is represented as J_m, the value of $m > 1$, and U is the membership matrix. The dataset with n points is represented as $D = \{d_1, d_2, \ldots, d_n\}$ on which FCM is applied. The set of cluster center is represented as $C = \{c_1, c_2, \ldots, c_n\}$. The distance between data point and cluster center is represented as $B_{ik}(c_i, d_k)$. After obtaining the cluster center, the Euclidian distance can be calculated between a cluster center (c_i) and a data point (d_i) with n number of instances by the formula as follows:

$$e_u = \sqrt{\sum_{i=1}^{n} c_i - d_i} \qquad (2)$$

3.2 Box and Whisker Plot

This technique was proposed by Tukey [12] which is used to remove outliers [12]. Here, the interquartile range (IRQ) is calculated using the following formula:

$$IRQ = |Q_3 - Q_1| \qquad (3)$$

where Q_1 denotes the lower quartile of the data and Q_3 denotes the upper quartile of the data. The 75th and 25th percentile of the samples represented by Q_3 and Q_1, respectively, and Q_2 is the median of the data points. The threshold value is calculated by the formula expressed below:

$$U_{th} = Q_3 + 3 \times IRQ \tag{4}$$

3.3 Synthetic Minority Oversampling (SMOTE)

SMOTE [7] is one of the most common oversampling techniques for data balancing. To oversample the minority class, take a sample from the minority class and identify its k-nearest neighbors in feature space. Among the k-nearest neighbors, we have to randomly choose j neighbors. The value of j denotes the amount of oversampling needed. Take the difference between the chosen minority class samples and one of its nearest neighbors and multiply an arbitrary number between 0 and 1 with that difference value. By adding this value to the chosen minority class sample, we can generate new synthetic data point. Continue with the next nearest neighbors up to user-defined j neighbors.

The above steps can be represented by the equation:

$$s_{new} = s_j + \left(s_j - s_i\right)\Delta \tag{5}$$

s_{new} denotes the newly generated synthetic minority class sample. s_j is the randomly chosen minority class sample. s_i is the chosen nearest neighbor among the k-nearest neighbors. Δ is a random number between 0 and 1.

3.4 Majority Voting

Majority voting is one of the most perceptive ensemble techniques that used multiple learning models to give better results than single learning model. The correct decision is made by the ensemble if it obtains votes by majority of the classifiers [8]. Let us consider the decision of the nth classifier D_n as $d_{nj} \in \{0, 1\}$, $n = 1,..., N$ and $j = 1,..., c$, where N and c denote the number of classifier and number of classes, respectively. If class j is chosen by nth classifier, then $d_{n,j} = 1$ and zero, otherwise [13]. An ensemble decision for class k is made by vote if

$$\sum_{n=1}^{N} d_{n,k} = \max_{j=1}^{c} \sum_{n=1}^{N} d_{n,j} \tag{6}$$

4 Proposed Approach

Initially, in our proposed AIFDS, both undersampling and oversampling are used to reduce the skewness from the original imbalanced dataset. After that, we are using three different classifiers, namely, MLP, KNN, and SVM and combined the outcomes of those classifiers by majority voting technique for achieving our goal. The flow of events of the entire process is depicted as shown in Fig. 1.

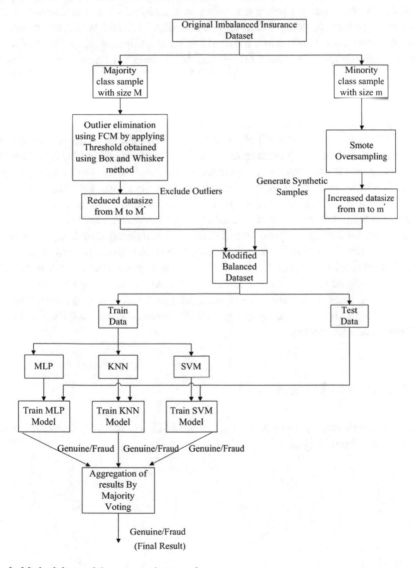

Fig. 1 Methodology of the proposed approach

We have first divided the available dataset into majority (non-fraudulent) class and minority class (fraudulent) samples. In this method, first we have employed the fuzzy c-means clustering algorithm on the majority class samples by giving appropriate number of cluster as input. For determining the required number of cluster, we have used two fuzzy validity indices [6] that are partition coefficient (PC) and partition entropy (PE). It is examined that by using two number of clusters PC gives maximum value (0.5) and PE gives minimum value (0.69) which give best results. We compute the Euclidian distance (e_u) between the data points of majority class sample and cluster centers according to Eq. (2) and sort the distance values in ascending order. To detect and remove the outlier, Box and Whisker technique is applied as discussed above and a threshold value is computed according to Eq. (4). We calculate the quartile values $Q1 = 0.6619$, $Q3 = 0.7217$ which are needed for threshold computation. We get the threshold value $U_{th} = 0.9011$. If the Euclidian distance e_u of any data point is greater than the threshold value, then it is marked as an outlier. The SMOTE technique is applied on the minority class samples which increased the number of instances of minority class samples as described in Sect. 3.3.

After completion of undersampling and oversampling approach, we have merged both minority and majority class samples and obtained a modified balanced dataset. For training and testing purpose, we are using 80% and 20% of the dataset, respectively. The training samples are fed into three classifiers, namely, MLP, KNN, and SVM individually. We have used the 10-fold cross-validation technique for efficient results. The train dataset is partitioned into ten equal partitions or folds. We are using the first fold for testing purpose and the union of remaining nine folds for training purpose to train three classifiers individually. Repeat the above procedure ten times using different folds as testing set every time. The average testing results are used as final prediction. Finally, we validated our proposed model using the rest 20% of the dataset. A decision (genuine/fraud) regarding each test record is made. The final decision is made by aggregating the results obtained from the classifiers using the majority voting technique.

5   Results and Discussions

Our proposed model has been implemented in MATLAB R2016a environment on a Windows 8.1, Intel i3 system.

5.1 Explanation and Preprocessing of Insurance Fraud Dataset

We experimentally evaluated our proposed approach with an automobile insurance dataset [14] known as "carclaims.txt". The data preprocessing steps are done before we have applied the dataset in our proposed model as mentioned in [6].

5.2 Performance Analysis

Initially, out of 15,420 samples, 14,497 are legitimate and 923 are malicious in the original unbalanced insurance dataset. After applying the Box and whisker technique for outlier removal, the number of legitimate samples has been decreased from 14,497 to 13,450 and after applying SMOTE, the malicious class samples have been increased from 923 to 11,999. Hence, by applying the above two techniques, we have generated an almost balanced dataset. We are using sensitivity, specificity, and accuracy as the parameter for measuring the performance of our model which are defined as follows:

$$Sensitivity = TP/(TP + FN) \tag{7}$$

$$Specificity = TN/(TN + FP) \tag{8}$$

$$Accuracy = TP + TN/(TP + TN + FP + FN) \tag{9}$$

where TP, TN, FP, and FN denote true positive, true negative, false positive, and false negative, respectively.

In Table 1, we have analyzed the performance of unbalanced dataset with the modified dataset after removal of outliers and also compared the performance of unbalanced dataset with the modified dataset after applying SMOTE technique on minority class samples. Table 2 shows the comparative performance of original unbalanced dataset with the modified balanced dataset obtained after undersampling and oversampling. It is observed from the table that the modified balanced dataset produces better results than the original unbalanced dataset. Figure 2 compares the results of each individual classifier after data balancing with the aggregation results after applying majority voting which shows that after applying majority voting all the parameters get improved.

Table 1 Comparison of average results on the insurance dataset

Classifiers	Performances metrics (in %)											
	Original unbalanced dataset			Modified dataset after outlier removal			Modified dataset after applying SMOTE		
	Sensitivity	Specificity	Accuracy	Sensitivity	Specificity	Accuracy	Sensitivity	Specificity	Accuracy
MLP	0	100	94.03	32.20	81.00	82.00	49.98	28.75	74.98
KNN	9.22	99.10	93.00	42.33	89.87	84.44	71.50	41.26	69.02
SVM	70.76	63.2	63.65	74.21	62.95	64.28	90.53	38.92	58.41

Table 2 Comparative performance of supervised classifiers on original unbalanced dataset with the modified balanced dataset

Classifiers	Performances metrics (in %)					
	Original unbalanced dataset			Modified balanced dataset after outlier removal + SMOTE		
	Sensitivity	Specificity	Accuracy	Sensitivity	Specificity	Accuracy
MLP	0	100	94.03	58.21	52.80	79.30
KNN	9.22	99.10	93.00	82.80	53.20	75.00
SVM	70.76	63.2	63.65	**92.30**	65.30	77.74

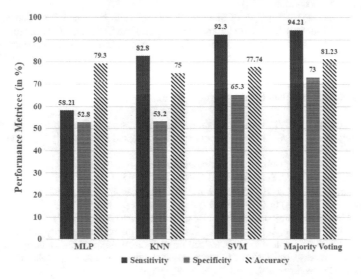

Fig. 2 Comparative results of each individual classifier with majority voting

6 Conclusions

In this paper, we have proposed a hybrid data balancing method with ensemble technique for efficiently detecting the auto insurance frauds. It is experimentally evaluated that our proposed model using MLP, KNN, and SVM classifiers gives better sensitivity compared to the original unbalanced dataset. We achieved 92.30% and 82.80% fraud detection rate (sensitivity) on a balanced dataset by applying SVM and KNN, respectively. After aggregating the results obtained from different classifiers by majority voting, it yielded 94.21% fraudulent claim detection rate. Experimental results verify that the balanced dataset provides better result than the unbalanced skewed dataset with respect to all the performance metrics. It is also noticed that the majority voting ensemble technique gives better performance than the individual classifiers.

References

1. Ngai, E.W., Hu, Y., Wong, Y.H., Chen, Y., Sun, X.: The application of data mining techniques in financial fraud detection: a classification framework and an academic review of literature. Decis. Support Syst. **50**(3), 559–569 (2011)
2. Wang, G., Hao, J., Ma, J., Huang, L.: A new approach to intrusion detection using artificial neural networks and fuzzy clustering. Expert Syst. Appl. **37**(9), 6225–6232 (2010)
3. Sundarkumar, G.G., Ravi, V.: A novel hybrid undersampling method for mining unbalanced datasets in banking and insurance. Eng. Appl. Artif. Intell. **37**, 368–377 (2015)
4. Li, Y., Yan, C., Liu, W., Li, M.: A principle component analysis-based random forest with the potential nearest neighbor method for automobile insurance fraud identification. Appl. Soft Comput. **70**, 1000–1009 (2018)
5. Nian, K., Zhang, H., Tayal, A., Coleman, T., Li, Y.: Auto insurance fraud detection using unsupervised spectral ranking for anomaly. J. Financ. Data Sci. **2**(1), 58–75 (2016)
6. Subudhi, S., Panigrahi, S.: Use of optimized fuzzy C-means clustering and supervised classifiers for automobile insurance fraud detection. J. King Saud Univ.-Comput. Inf. Sci. (2017)
7. Chawla, N.V., Bowyer, K.W., Hall, L.O., Kegelmeyer, W.P.: SMOTE: synthetic minority oversampling technique. J. Artif. Intell. Res. **16**, 321–357 (2002)
8. Polikar, R.: Ensemble based systems in decision making. IEEE Circuits Syst. Mag. **6**(3), 21–45 (2006)
9. Kalyani, G., Rao, M.C.S., Janakiramaiah, B.: Privacy-preserving classification rule mining for balancing data utility and knowledge privacy using adapted binary firefly algorithm. Arab. J. Sci. Eng. 1–23 (2017)
10. Narayana, G.S., Vasumathi, D.: An attributes similarity-based K-medoids clustering technique in data mining. Arab. J. Sci. Eng. **43**(8), 3979–3992 (2018)
11. Dasgupta, S., Saha, S., Das, S.K.: Malware detection in android using data mining. Int. J. Nat. Comput. Res. (IJNCR) **6**(2), 1–17 (2017)
12. Tukey, J.W.: Exploratory data analysis. Reading. Addison-Wesley (1977)
13. Stepenosky, N., Green, D., Kounios, J., Clark, C.M., Polikar, R.: Majority vote and decision template based ensemble classifiers trained on event related potentials for early diagnosis of Alzheimer's disease. In: 2006 IEEE International Conference on Acoustics Speech and Signal Processing Proceedings, vol. 5, pp. V–V. IEEE (2006)
14. Phua, C., Alahakoon, D., Lee, V.: Minority report in fraud detection: classification of skewed data. ACM SIGKDD Explor. Newsl. **6**(1), 50–59 (2004)

Application of Generalized Possibilistic Fuzzy C-Means Clustering for User Profiling in Mobile Networks

K. Ashwini and Suvasini Panigrahi

Abstract User profiling is the process of constructing a normal profile by accumulating the past calling behavior of a user. The technique of clustering focusses on outcome of a structure or an intrinsic grouping in unlabeled data collection. In this paper, our main intention is on building appropriate user profile by applying generalized possibilistic fuzzy c-means (GPFCM) clustering technique. All the call features required to build a user profile is collected from the call detail record of the individual users. The behavioral profile modeling of users is prepared by implementing the clustering on two relevant calling features from the reality-mining dataset. The labels are not present in the dataset and thus we have applied clustering which is an unsupervised approach. Before applying the clustering algorithm, a proper cluster validity analysis has to be done for finding the best cluster value and then the cluster analysis is done using some performance parameters.

Keywords Call detail records · Unsupervised learning · GPFCM · User profiling · Cluster validity index

1 Introduction

Day by day, the world is experiencing new things as the technology is moving to a direction which has brought so much advancement. Focusing on the telecom industry it had seen many changes, the use of landlines has decreased due to the mobile communication since 2005 both in developed and developing countries. According to ICT Facts and Figures-International Telecommunication Union (2014), there has been an increase of mobile phone subscriptions from 12% of the population in 2000 up to 96% in 2014 [1], overall it can be said that there has been significant increase of the mobile phone usage. Apart from the advantages offered, the telecommunication

K. Ashwini · S. Panigrahi (✉)
Department of CSE, Veer Surendra Sai University of Technology, Burla, Sambalpur 768018, India
e-mail: spanigrahi_cse@vssut.ac.in

K. Ashwini
e-mail: ashwinikamakshi@gmail.com

© Springer Nature Singapore Pte Ltd. 2020
V. Bhateja et al. (eds.), *Intelligent Computing and Communication*,
Advances in Intelligent Systems and Computing 1034,
https://doi.org/10.1007/978-981-15-1084-7_23

sector is badly affected by frauds. The main purpose is to get benefits either free of cost or at a lessened rate with the likelihood of selling it to a third party. The effective solution for telecom fraud is to depreciate the damages and cut down the losses by detecting it as soon as possible since it cannot be completely eradicated [2].

According to the Communications Fraud Controls Association (CFCA), the amount of revenue loss is nearly $30 billion globally last year. Fraud has become a global and most damaging issue, and it has turned out to become a cause of revenue loss and terrible debts to the telecommunication industry and in the upcoming years as the revenue will increase so as the fraud will increase proportionately. The frauds will continue to exist, what will change are the different methods to commit fraud and the rate of detection of fraud. The survey which was done worldwide reported 750 fraudsters in the middle of March 2013 and August 2015. It was also found that 79% of fraudsters were men and also the proportion of women has moved up from 13 to 17% in 2010.

The superimposed fraud has seen the highest growth rate because of the fact that it has become easy to gain access to the already existing account. This fraud remains undetected in some cases due to the reason that the number of fraudulent calls is comparatively less as compared to the overall call volume [3]. There is a need to develop methods, planning, and guidelines for the detection and prevention of fraud. The objective is to detect illegal activities and to minimize the false alarms. In this work, we have implemented GPFCM on the subscriber's call detail records for accurate user-profile generation, so that a proper classification technique can be applied afterward for detecting fraud. We have shown the proposed work on the reality-mining dataset. As far as we know, this is the first attempt to implement GPFCM on the reality-mining dataset for detecting superimposed fraud.

This paper proceeds in the following order: Sect. 2 reviews the erstwhile work done in case of superimposed mobile fraud detection. A brief presentation of the clustering algorithm for user profiling is described in Sect. 3. The next section defines the work done for carrying out our proposed approach. After that in Sect. 5, we have discussed the different results obtained along with the implementation environment and the cluster validity metrics used for evaluation of the proposed system. Lastly, Chap. 6 gives the conclusion of our work.

2 Related Work

In the current section, we have discussed some of the established works related to the mobile fraud detection system. The authors of paper [4] proposed two anomaly detection methods which are dedicated to the concept of signatures in which the initial method is based on a signature-deviation-based approach, whereas the succeeding one is based on the method of dynamic cluster analysis. Multilayer perceptron and hierarchical agglomerative clustering have been proposed in paper [5] for fraud detection using five unalike profiles for each individual subscriber. In paper [6], five different user profiles are selected and then a genetic-programming-based classifier is applied for the detection of fraudulent patterns. In a recent work [2], the problem to come across the fraudulent behavior in mobiles has been addressed by analyzing the user's calling behavior and implementing fuzzy clustering along with support vector machine (SVM). The fuzzy clustering technique has been applied for construction of user profile in order to further proceed for the detection of fraud. The recent work in [7] suggests the usage of possibilistic fuzzy c-means (PFCM) for the construction of user-behavior representation. After the user profiling has been done, the testing dataset has been implemented on the clustering technique for the identification of any specious behavior, and this is done by calculating the Euclidean distance of the incoming call record with each of the cluster centroid and relating it by means of the threshold value.

Even though several techniques have been applied for developing an appropriate user profile, the recent work in which PFCM clustering is used has some disadvantages which are overcome by applying generalized possibilistic fuzzy c-means clustering (GPFCM) in our present work. The applied technique is better than other fuzzy clustering algorithms—fuzzy c-means (FCM), possibilistic c-means (PCM), and possibilistic fuzzy c-means (PFCM).

3 Background Study

This section illustrates the GPFCM algorithm [8] in detail which is used for carrying out our work. GPFCM clustering is the combination of commonly used clustering algorithms FCM and PFCM. The PFCM algorithm overcomes the drawback of FCM by handling the noisy instances and GPFCM improves the PFCM by finding accurate cluster centers. GPFCM takes unlabeled instances of a dataset and performs the clustering process by first performing FCM and then by taking the output of FCM it loads PFCM and then the results which are achieved from PFCM initialize GPFCM to give as output the required cluster centers. A particular membership value along with typicality value is allocated to each point of the cluster. The objective function of the GPFCM which needs to be minimized to obtain the results is stated below:

$$
\begin{aligned}
F(U, T, V; X) = \sum_{k=1}^{D} \sum_{j=1}^{c} & (c_{\text{FCM}} u_{jk}^{m} f_{j,\,\text{FCM}} \left(||x_k - v_j||_A^2 \right) \\
& + c_{\text{PCM}} t_{jk}^{\eta} f_{j,\,\text{PCM}} \left(\left(||x_k - v_j||_A^2 \right) \right) \\
& + \sum_{j=1}^{c} \gamma_j \sum_{k=1}^{D} (1 - t_{jk})^{\eta} + \sum_{k=1}^{D} \lambda_k \left(\sum_{j=1}^{c} u_{jk} - f_k \right) \quad (1)
\end{aligned}
$$

where the objective function is symbolized by F, $U = [u_{jk}]$ and $T = [t_{jk}]$ are the obtained membership matrix and typicality matrix, respectively, and V is a matrix of c cluster centers. $X = \{x_1, x_2, x_3, \ldots, x_n\}$ is the dataset in which the GPFCM is implemented. In GPFCM, the functions $f_{j,\text{FCM}}\left(\left\|x_k - v_j\right\|_A^2\right)$ and $f_{j,PCM}\left(\left\|x_k - v_j\right\|_A^2\right)$ are used for fuzzy and possibilistic terms, respectively. The parameters c_{FCM} and c_{PCM} are the user defined which is taken as 1 for both the variables in GPFCM, m represents the fuzzifier weighting exponent, whereas η terms the scale parameter, $m > 1$ and $\eta > 1$.

The other fuzzy clustering techniques FCM, PCM, and PFCM are applicable in case of Euclidean norm, whereas GPFCM is able to operate with covariance norm. Covariance norm has the advantage that it generates ellipsoidal clusters with finer alignment with the configurations of the data. GPFCM can find the accurate cluster centers as compared to FCM, PCM, and PFCM despite the data if it is noisy as it is less sensitive to noise.

4 Proposed Approach

The proposed approach examines the past behavioral patterns of the users for identifying fraudulent cases. The certain call detail records are not accessible for the use of public from any service provider because if dataset is revealed then it may cause privacy policy disruption. Thus, the features which are used for our proposed approach are the data extracted from the reality-mining dataset [9] to categorize appropriate user-behavior representations (profiles) inside the dataset, in order to spot fraudulent usage. The GPFCM technique is used in our approach for analyzing the dataset and creating appropriate clusters, and other fuzzy clustering techniques are compared with it to find the best results.

We have taken some of the mentioned features from the reality-mining dataset for the profile making of the user: user ID (user_id), date and time of the call (date/time), call duration (dur), and type of the call (call_type). The user_id represents the device MAC ID of the callee or the caller, date/time signifies the date and time of the calls, dur indicates the duration of the call in seconds, and call_type denotes the type of call formed: 0 indicates local call, 1 is for national calls, and 2 indicates international calls.

The GPFCM clustering technique is then implemented on the dataset using covariance norm matrix to obtain the user behavioral profiles, and the inputs on which the clustering is applied are user_id, date/time, dur, and call_type. After applying the clustering techniques, Fukuyama and Sugeno CVI (cluster validity index) is applied in the dataset to find the precise number of the clusters and parameters like compactness and separation are used to analyze the best clustering technique. Figure 1 represents the outline of our proposed approach. To get the final cluster center matrix of GPFCM step by step, we have to proceed. In the first step, we have to apply the fuzzy c-means algorithm to get the cluster center matrix and partition matrix. Then, the outputs obtained are used for initializing the possibilistic fuzzy c-means clustering to get the cluster centers of PFCM along with partition matrix and typicality

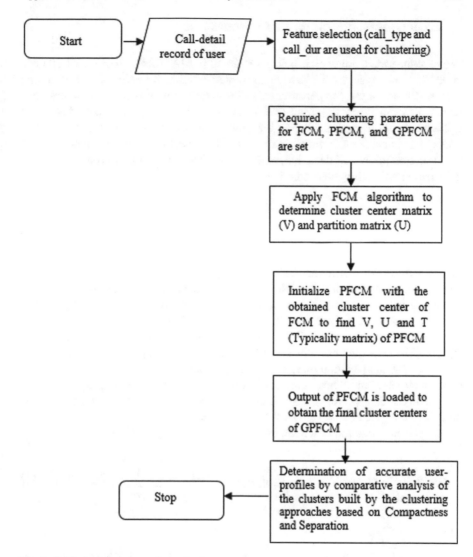

Fig. 1 Outline of the proposed approach

matrix. The process stops in all these algorithms when the maximum number of iterations has reached. Now the cluster centers of PFCM are used to obtain the results of generalized possibilistic fuzzy c-means technique. The various results are computed using the different equations of the GPFCM algorithm. Thus, the final cluster centers, partition matrix, and typicality matrix are used for further evaluation of the clustering algorithms.

5   Experimental Results and Discussions

The recommended approach is being performed in MATLAB 2016a on a 1.70 GHz i5-4210U CPU system. We have done implementation by using a reality-mining dataset [9] to analyze the performance. This dataset contains data which has been gathered over a 9-month period (2004–2005) of 106 users and it contains phone call logs, location logs, messaging logs, unlike applications usage logs, individual status of a phone like charging, switch off or idle state, etc. As our main objective is based on fraudulent call detection, attributes similar to user activity, device-specific information are mainly considered.

5.1   Cluster Validation

The various fuzzy algorithms demand prerequisitely to determine the number of clusters but it is not always possible to decide the accurate number of clusters. Since the results obtained from the fuzzy algorithms are dependent on c, i.e., the number of clusters, hence, there is a need to evaluate it which is called cluster validity. The two most commonly used validity indexes are partition coefficient (PC) and partition entropy (PE). The major disadvantage of these indexes is that both these indexes are dependent on the partition matrix, not on the dataset and they are sensitive to the fuzzy parameter "m". Thus, some other cluster validity indexes are proposed which depend on the data as well as involve the membership values [10]. Among them, the most common one is Fukuyama and Sugeno CVI.

The validity function is defined by Fukuyama and Sugeno CVI:

$$V_{\text{FS}} = J_m(u, v) - K_m(u, v) = \sum_{i=1}^{c} \sum_{j=1}^{d} u_{ij}^m ||x_j - v_i||^2 - \sum_{i=1}^{c} \sum_{j=1}^{d} u_{ij}^m ||v_i - v_A||^2 \quad (2)$$

where $v_A = \sum_{i-1}^{c} v_i / c$.

After finding V_{FS} for different numbers of clusters, the best value of c is chosen to produce the best clustering result:

$$c^* = \min_{2 \le c \le d-1} V_{\text{FS}} \quad (3)$$

Table 1 depicts the Fukuyama and Sugeno CVI values for finding the optimal number of clusters in case of FCM, PFCM, and GPFCM. The CVI is calculated using Eqs. (2) and (3) which gives the minimal value when cluster number is 2 in all the three cases. Thus, the optimal value of c is chosen as 2. After choosing the cluster number, we have implemented GPFCM on the dataset and other fuzzy clustering techniques related to the GPFCM are also applied for comparative analysis.

Table 1 Determination of ideal number of clusters by applying Fukuyama and Sugeno CVI

Value of cluster number (c)	V_{FS} (FCM)	V_{FS} (PFCM)	V_{FS} (GPFCM)
2	3.7750	1.1913	1.0228
3	7.0704	1.9391	7.4613
4	3.9532	3.8534	5.4774

5.2　Cluster Analysis

Compactness and separation are two basic estimation measures taken to evaluate the clustering algorithms [11]. When $\sum_{j=1}^{d} u_{jk}^{m}$ is higher, it means the cluster is more compact and when the volume of the clusters $\sum_{j=1}^{c} \text{Vol}_j$ is smaller then the cluster is more compact. Thus, the ratio of $\sum_{j=1}^{c} \text{Vol}_j$ to $\sum_{j=1}^{d} u_{jk}^{m}$ is defined as the measure of compactness which should be minimalized. The separation parameter which needs to be maximized is defined as the total difference between the values of the membership matrix or the typicality values or a combination of them per cluster so as to get the maximum separation of the clusters.

Table 2 shows the results obtained when the various fuzzy clustering algorithms are analyzed using the parameters compactness and separation. The minimum value of the compactness measure has been found in the case of GPFCM, similarly the maximum value of separation measure is seen in the case of GPFCM. Thus, this hybrid technique which is combination of FCM and PFCM gives accurate results as compared to others.

Table 3 shows the elapsed time for carrying out the different clustering techniques. It clearly shows that GPFCM takes the largest amount of time as compared to others because more computations are involved in calculating the result using nonlinear updating equations of the cluster centers but faster convergence is not a priority here since the clusters would be inaccurate. Thus, it is clearly evident from the results that although GPFCM takes more time, it gives accurate outcomes as compared to other clustering approaches.

Table 2 Cluster analysis of FCM, PCM, PFCM, and GPFCM

Clustering technique	Compactness	Separation
FCM	1.4370	1.2207
PCM	0.0486	1.6424
PFCM	0.0158	1.3144
GPFCM	0.0114	1.7922

Table 3 Time required for
the reality-mining dataset to
perform FCM, PCM, and
PFCM

Clustering technique	Elapsed time (in seconds)
FCM	369.8396
PCM	709.1831
PFCM	2016.8955
GPFCM	2761.9914

6 Conclusions

In our proposed work, GPFCM technique has been suggested for behavioral profile construction of the users present in the reality-mining dataset. After the profile
building, a proper classification technique needs to be applied in the dataset for the
effective fraud detection. Here, GPFCM has been applied to cluster noisy data and
it is more accurate as compared to other fuzzy algorithms. The fuzzy clustering
algorithms have their application in many areas which include image processing,
fault diagnosis, fuzzy time-series forecasting, etc. FCM, PCM, and PFCM fail to
give accurate cluster centers when executed using covariance norm matrix while the
GPFCM algorithm is insensitive to the covariance norm matrix. The first index we
have experimented considers the cluster's compactness within them and the second
index considers the separation of the clusters. Both these measures prove GPFCM
better than other fuzzy clustering algorithms. The computational costs of GPFCM
and time taken by it for execution are more as compared to FCM and PFCM because
it involves nonlinear updating equations, but all we need are the accurate cluster
centers. Based upon the results, we conclude that by employing GPFCM clustering
technique, this kind of real-world problems can be resolved effectually.

References

1. Blondel, V.D., Decuyper, A., Krings, G.: A survey of results on mobile phone datasets analysis.
   EPJ Data Sci. **4**(1), 10 (2015)
2. Subudhi, S., Panigrahi, S.: Use of fuzzy clustering and support vector machine for detecting
   fraud in mobile telecommunication networks. IJSN **11**(1/2), 3–11 (2016)
3. Cox, K.C., Eick, S.G., Wills, G.J., Brachman, R.J.: Brief application description; visual data
   mining: recognizing telephone calling fraud. Data Min. Knowl. Disc. **1**(2), 225–231 (1997)
4. Alves, R., Ferreira, P., Belo, O., Lopes, J., Ribeiro, J., Cortesão, L., Martins, F.: Discovering
   telecom fraud situations through mining anomalous behavior patterns. In: Proceedings of the
   DMBA Workshop, on the 12th ACM SIGKDD (2006)
5. Hilas, C.S., Mastorocostas, P.A.: An application of supervised and unsupervised learning
   approaches to telecommunications fraud detection. Knowl.-Based Syst. **21**(7), 721–726 (2008)
6. Hilas, C.S., Kazarlis, S.A., Rekanos, I.T., Mastorocostas, P.A.: A genetic programming
   approach to telecommunications fraud detection and classification. In: Proceedings of the 2014
   International Conference on Circuits, Systems Signal Processing Communication Computer,
   pp. 77–83 (2014)

7. Subudhi, S., Panigrahi, S.: Use of possibilistic fuzzy C-means clustering for telecom fraud detection. In: Computational Intelligence in Data Mining, pp. 633–641. Springer, Singapore (2017)
8. Askari, S., Montazerin, N., Zarandi, M.F.: Generalized possibilistic fuzzy C-means with novel cluster validity indices for clustering noisy data. Appl. Soft Comput. **53**, 262–283 (2017)
9. Eagle, N., Pentland, A.S.: Reality mining: sensing complex social systems. Pers. Ubiquit. Comput. **10**(4), 255–268 (2006)
10. Wang, W., Zhang, Y.: On fuzzy cluster validity indices. Fuzzy Sets Syst. **158**(19), 2095–2117 (2007)
11. Deborah, L.J., Baskaran, R., Kannan, A.: A survey on internal validity measure for cluster validation. Int. J. Comput. Sci. Eng. Survey **1**(2), 85–102 (2010)

A Method to Detect Blink from the EEG Signal

Narayan Panigrahi, Amarnath De and Somnath Roy

Abstract Electroencephalography (EEG) and its application in the development of brain-computer interfacing (BCI) is a rapidly growing topic of research. In this paper, we propose a method to analyze EEG data to detect the blink artifacts. Initially, blink samples were taken manually and analyzed visually. The moving average technique was applied to smoothen the signal. Eventually, the gradient of the signal was calculated and a modified threshold technique was applied to detect the blink artifacts.

Keywords Electroencephalography · Electrooculogram · Brain-computer interface · Event-related potential

1 Introduction

Electroencephalogram (EEG) is a measure of the electrical signals of the brain of a human being. It provides evidence of how the brain functions over time. Brain-computer interface (BCI) is the coordination of a brain and compatible devices that enable EEG signals from the brain to manipulate some external activity, such as control of a cursor or a prosthetic limb [1]. The interface allows an instantaneous communications pathway between the brain and also the object to be controlled. Electrooculography (EOG) is a technique that measures the standing potential that is caused due to the polar nature of the eyes. The resulting signal is called the

N. Panigrahi
Center for AI and Robotics(CAIR), Bangalore, India
e-mail: npanigrahi7@gmail.com

A. De (✉) · S. Roy
Jadavpur University, Kolkata, India
e-mail: amarnathde1992work@gmail.com

S. Roy
e-mail: somroymail@gmail.com

© Springer Nature Singapore Pte Ltd. 2020
V. Bhateja et al. (eds.), *Intelligent Computing and Communication*,
Advances in Intelligent Systems and Computing 1034,
https://doi.org/10.1007/978-981-15-1084-7_24

(a)

(b)

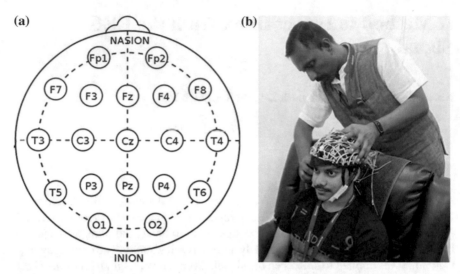

Fig. 1 **a** Cross sectional view of 10–20 electrode placement diagram. **b** Subject wearing EEG cap while recording of EEG and EOG signal. (Courtesy Axxonet Pvt. Ltd., Bangalore, India)

electrooculogram. Tracking the movement of the eye through sensors enables us to fix the position where one's eyes are focused [2]. The study of EOG can determine presence, attention, focus, drowsiness, consciousness, or other mental states of the subject [3–5]. Event-Related Potential (ERP) is a small voltage generated in the brain due to the occurrence of a specific event or stimulus. ERPs can be reliably measured from EEG. In this research, we characterize the EEG signal and propose a method to detect and process the EOG signal. When active, i.e., not in the state of sleep the external functioning of the human eye is characterized by three distinct functions, viz., saccadic, fix, and blink. Saccadic and fix are voluntary actions or actions controlled by a human. Whereas, the blink is an involuntary action that is associated with randomness and high fluctuation of EEG voltage [6]. We discuss how to capture the EEG and EOG signal and how to filter the EEG channel to delineate the EOG and related signals from other channels. Then we discuss the process to characterize and delineate the blinks from the rest of the EEG signals so that saccadic and fix are delineated (Fig.1).

2 Data Acquisition

2.1 Process of EEG Data Acquisition

Before starting the recording session for the acquisition of the EEG, the subject was asked to stay as calm as possible during the test. First, a scene containing the picture of an object was shown to the user. Then next, five slides containing several instances of the object in different locations in the scene were displayed. The subject was instructed to read the scene and identify the objects in the scene. During this process, the EEG of the subject was acquired and recorded. This experiment was conducted on different subjects and in the same scene. The process is depicted in Fig. 2. The EEG acquired from the subject is plotted in Fig. 3. While recording the EEG the distance from the monitor to the subject is kept at a distance of 93 cm. The monitor size is (30 cm × 60 cm). The distance from the camera to the subject is 27 cm. The camera feed is used for gaze estimation and provide approximate coordinates of the eye fix on the monitor screen.

2.2 EOG Data Acquisition

Before starting the recording session for acquisition of the EEG, the subject was asked to stay as calm as possible during the test. First, a scene containing the picture of an object was shown to the user. Then next, 5 slides containing several instance of the object in different locations in the scene was displayed. The subject was instructed to read the scene and identify the objects in the scene. During this process the EEG of the subject was acquired and recorded. This experiment was conducted on different subject and on same scene. The process is depicted in the Fig. 2. The EEG acquired from the subject is plotted in Fig. 3. While recording the EEG the distance from the monitor to the subject is kept at a distance of 93 cm. Monitor size is (30 cm × 60 cm). The distance from the camera to subject is 27 cm. The camera feed is used for gaze estimation and provide approximate coordinates of the eye fix on the monitor screen.

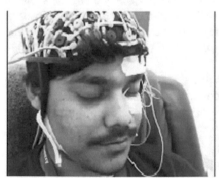

Fig. 2 EOG electrode placement on subject

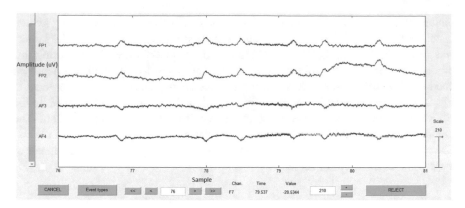

Fig. 3 4 channel EEG data plot using EEGLAB Toolbox (Version 14.1.2b)

3 Blink Detection

The EEG signal is highly varying both in voltage and frequency. Also, the baseband of the signal varies frequently. Blink in EEG & EOG signals is considered as artifacts in the EEG. To filter and detect the blink from the EEG we propose a novel technique of blink detection in the paper:

– EEG data acquired in .edf format is loaded in MATLAB using the EEGLAB toolbox (version 14.1.2b).
– Plot the EEG acquired in step-1 to visualize the recorded data for all channels. The sampling rate was set to 1024 Hz.
– The application was mostly concerned with frequencies in the range 0–30 Hz. We filtered the overall channel signals into 1–30 Hz frequency using the standard filtering technique available in the EEGLAB Toolbox. After a visual analysis of the plot obtained, it was found that the impact of the blinks was mostly affecting FP1, FP2, A1(EOG1), A2(EOG2). Hence, these four channels are used for further processing of blink detection.
– After filtering, the signal contained too many fluctuations over a very short time period. In order to get rid of unwanted fluctuations and make the signals smoother, the signals were subjected to a moving average technique. Figure 4 illustrates the results obtained for two individual blink occurrences.

As shown in Fig. 5 the DC value for each individual blink sample is highly varying, therefore standard threshold technique will produce very poor results if applied. Thus, the calculation of Gradient for the signal will result in better detection of the activity of the peaks caused by blinks.

Blinks were detected using the following method:

1. The signal from channel EOG1 was selected and the moving average technique was applied on it with lead parameter value 10 and lag parameter value 20. The result is shown in Fig. 6, the first subplot.

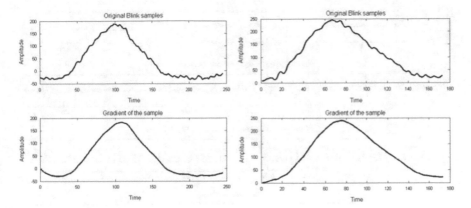

Fig. 4 Two blink samples and the resulting signal after applying moving average technique

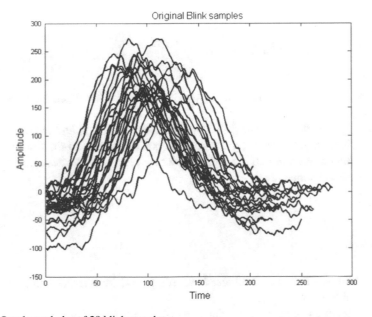

Fig. 5 Overlapped plot of 29 blink samples

2. The gradient of the signal was obtained and shown in the second subplot.
3. A modified threshold technique was applied on the gradient to obtain the duration of each individual blinks. The threshold function is defined as below:

$$f(x) = \begin{cases} 2T, & x \geq T \\ T, & 0 \leq x < T \\ -T, & 0 > x > -T \\ -2T, & x \geq -T \end{cases} \tag{1}$$

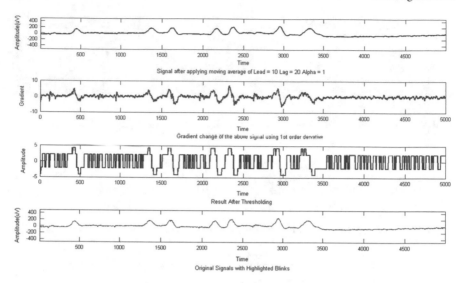

Fig. 6 Detection of blinks in EEG data stream

C is a constant that needs to be adjusted for proper blink detection generally in range (0.1–0.5).

4. After applying the threshold function, the resultant starting and ending points of the blink are highlighted on the original signal and shown in the fourth subplot. Hence, this technique of blink detection was applied on the whole signal and for all selected channels.

4 Results

The algorithm was tested on 2 subjects' recordings and the corresponding results are shown in Table 1.

Due to the huge length of the data, we divided the data in a range of 30,000 data points (29.3 s as the sampling rate is 1024 Hz) and analyzed them one by one. The constant C was adjusted for each data point range but it observed to be the same throughout the course of the experiment. Similarly, the Lag and the Lead for the moving average applied to the signal were also found to be the same. These parameters were found to be the best for detection. Too high values of Lead and Lag led to the omission of the blinks and too low values of the Lead and Lag led to noisy blink signatures. Taking subject 1 into consideration it was observed that subject 1 blinked frequently during the course of the experiment. A total of 130 blinks were recorded manually out of which 85 were correctly detected which gives an accuracy of 65%. The subject was tired while performing this experiment with signs of drowsiness. Consecutively such high blink rate was observed as it was difficult

Table 1 Algorithm accuracy for 2 subjects' EEG data

Subject	Data point range (×1000)	C	Lag	Lead	No of blinks present	True positives	False negatives	Accuracy (%)
1	5–30	0.3	5	10	11	8	0	65.38
	30–60	0.3	5	10	14	8	0	
	60–90	0.3	5	10	31	19	1	
	90–120	0.3	5	10	25	14	1	
	120–160	0.3	5	10	27	23	2	
	160–180	0.3	5	10	13	9	0	
	180–198.656	0.3	5	10	9	4	2	
2	5–35	0.2	5	10	6	5	1	90
	35–65	0.3	5	10	1	1	0	
	65–95	0.3	5	10	1	1	0	
	95 125	0.3	5	10	–	–	–	
	125–185	0.3	5	10	1	1	0	
	185–204.8	0.3	5	10	1	1	0	

to concentrate on the task. On the other hand, subject 2 was well rested and relaxed and throughout the experiment, a total of 10 blinks were recorded manually out of which 9 were detected successfully. Thus, the detection accuracy was 90%. Subjects 1 and 2 were given the same set of tasks to perform. As the subject 2 was well rested, it can be concluded easily that the subject performed the task with a very low blink rate as the concentration level was high.

5 Applications

Applications of blink detect are many. One of them is the determination of one's concentration level and drowsiness. It was found out according to the latest research [7] that the mean blink rate at rest was 17 blinks per min and during a conversation, it increased to 26 and it was as low as 4.5 while reading. Based on the statistics we can infer that a subject's concentration level is high if it has a blink rate of at most 6 per min. Otherwise, the subject is either low in his concentration level or distracted. However, if it is observed that the blink duration of the subject is above a certain level, then the subject is feeling drowsy. Blink is considered as an artifact during EEG analysis. Therefore, the detection of blinks is used to remove them from the original signal for further analysis.

6   Conclusions

The methodology used in the experiment includes gradient calculation and a modified threshold function to detect the blinks in the EEG signal. As a result, this method for detecting blinks is computationally cost-effective compared to other traditional methods [8–10]. Blinks can also be used to measure alertness, drowsiness and other cognitive states. Future research can be extended towards the detection of different mind states by using the blink detection technique. Further pattern recognition techniques using a Deep Neural Network with a multiclass classifier can be implemented to recognize cognitive states.

Declaration

We have taken permission from competent authorities to use the images/data as given in the paper. We also took the consent of the subject for recording of data. In case of any dispute in the future, we shall be wholly responsible.

Acknowledgements We are thankful to Axxonet Pvt. Ltd for providing us the facility to record EEG data. We give special thanks to Dr. Arun Sasidhran, Mr. Chetan S M, Mr. Sumit Sharma, Ms. Deepa D for their valuable help and guidance. We are highly indebted to our guide Dr. Narayan Panigrahi for the constant support and motivation.

References

1. Roy, R., Konar, A., Tibarewala, D.N.: Control of artificial limb using EEG and EMG-a review (2011)
2. Panigrahi, N., Lavu, K., Gorijala, S.K., Corcoran, P., Mohanty, S.P.: A method for localizing the eye pupil for point-of-Gaze estimation. In: IEEE Potentials **38**(1), 37–42 (2019). https://doi.org/10.1109/MPOT.2018.2850540
3. Gimeno, P.T., Cerezuela, G.P., Montanes, M.C.: On the concept and measurement of driver drowsiness, fatigue and inattention: implications for countermeasures. Int. J. Veh. Des. **42**, 67–86 (2006). https://doi.org/10.1504/IJVD.2006.010178
4. Liu, N.-H., Chiang, C.-Y., Chu, Hsuan-Chin: Recognizing the degree of human attention using EEG signals from mobile sensors. Sensors (Basel, Switzerland) **13**, 10273–86 (2013). https://doi.org/10.3390/s130810273
5. Lin, C.-T., et al.: EEG-based drowsiness estimation for safety driving using independent component analysis. IEEE Trans. Circuits Syst. I: Regular Paper **52**(12), 2726-2738 (2005). https://doi.org/10.1109/TCSI.2005.857555
6. Landau, A.N., Esterman, M., Robertson, L.C., Bentin, S., Prinzm, W., et al.: Different effects of voluntary and involuntary attention on EEG activity in the gamma band. https://doi.org/10.1523/JNEUROSCI.3092-07.2007
7. Bentivoglio AR, Bressman SB, Cassetta E, Carretta D, Tonali P, Albanese A. Analysis of blink rate patterns in normal subjects. Mov Disord. **12**, 1028 (1997). https://doi.org/10.1002/mds.870120629

8. Sawant, H.K., Jalali, Z.: Detection and classification of EEG waves. Orient. J. Comput. Sci. Technol. **3**(1), 207–213 (2010). https://bit.ly/2M36Jub
9. Amin, H.U., Mumtaz, W., Subhani, A.R., Saad, M.N.M., Malik, A.S.: Classification of EEG signals based on pattern recognition approach. Front. Comput. Neurosci. 1662–5188 (2017). https://doi.org/10.3389/fncom.2017.00103
10. Lotte, F., et al.: A review of classification algorithms for EEG-based brain computer interfaces. J. Neural Eng. **4**(2), R1 (2007). https://doi.org/10.1088/1741-2560/4/2/R01

Power-Efficient Reversible Logic Design of S-Box for N-Bit AES

Rohini Hongal, Rohit Kolhar and R. B. Shettar

Abstract S-Box is the main core of AES. The S-Box provides confusion capability for AES. The S-Box used in standard AES is static throughout the encryption process. But nowadays, cyberattacks are continuously developing and power attack is one among them. The main aim of this work is to provide higher security for several applications over the Internet by enhancing the overall strength of the existing Advanced Encryption Standard algorithm (AES). Paper explores the usage of new logic family called reversible logic gates as a part of quantum technology to mitigate power attacks. As S-Box involves heavy mathematical calculations, it dissipates more power which may lead to power attacks. To improve the security of the conventional AES, reversible logic is used to build S-Box, as reversible gates offer ideally zero internal power dissipation. This work discusses the implementation details of S-Box using reversible logic and does the performance analysis w.r.t hardware requirements and power.

Keywords Reversible logic · AES · S-Box · Affine transform (AT)

1 Introduction

S-Box is the main core of AES. The confusion capability for AES is provided by this S-Box. The S-Box used in standard AES is static throughout the encryption process. The AES algorithm operates on state of a matrix array of bytes. The state matrix consists of four rows of N bytes each, where N is the block length divided by 32. Each byte of data in state matrix is replaced by byte of data from the lookup table.

R. Hongal (✉) · R. B. Shettar
ECE, B.V.B.C.E.T, Hubli, India
e-mail: rohini_sh@bvb.edu

R. B. Shettar
e-mail: raj@bvb.edu

R. Kolhar
ECE, KLETU, Hubli, India
e-mail: rohitkolhar@gmail.com

© Springer Nature Singapore Pte Ltd. 2020
V. Bhateja et al. (eds.), *Intelligent Computing and Communication*,
Advances in Intelligent Systems and Computing 1034,
https://doi.org/10.1007/978-981-15-1084-7_25

Due to digitalization and new emerging technologies, there are advantages as well as disadvantages. Therefore, need of data security becomes utmost important. We need to secure our sensitive data from interceptors. Therefore, it is important to have design techniques that give secure, reliable, and practical transmission. One of the important techniques to make our data secure is the cryptography. It is the concept of encrypting or hiding the data and making it complex for an interceptor to decrypt it. In order to restrict the interceptors from knowing the original messages, classical cryptography involves mathematical techniques.

The cryptography which we use is referred to as classical cryptography, which involves mathematical equations. Classical cryptosystems consist of three types, namely, symmetric systems, asymmetric systems, and hash functions. When we consider the quantum computers, there are threats toward the security of classical cryptography. The quantum computers can efficiently compute the hard-mathematical problems faster. Quantum cryptography involves the laws of quantum mechanics, which mainly relies on two important principles. They are Heisenberg uncertainty and photon polarization [1–3].

1.1 Reversible Logic

Reversible logic has gained greater attention due to its applications in low-power VLSI and also has drawn the attention of researchers nowadays because of the emerging quantum technology [4–6]. The special feature of reversible logic gate is that input and outputs are uniquely related and recovered from each other (that is n-input and n-output device). We can implement one or more applications using single reversible logic gate. One such example for 2×2 matrix of reversible gate is Feynman gate (FG) shown in Fig. 1 gives XOR operation. Quantum cost of this gate is 1 and other output is known as garbage which is used to maintain the reversibility. Feynman gate is also called as copying gate or duplicating gate. The inputs are A, B and the outputs are X, Y, where $X = A$, $Y = A \oplus B$. Such few other reversible gates used in this work are shown in Figs. 2, 3, 4, 5 as BVF gate, Peres gate, Toffoli gate, and TS-3 gate, respectively.

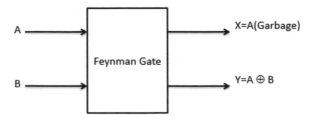

Fig. 1 Feynman gate

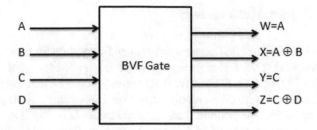

Fig. 2 BVF gate

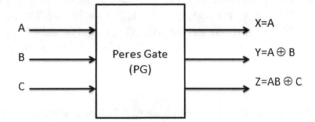

Fig. 3 Peres gate (PG)

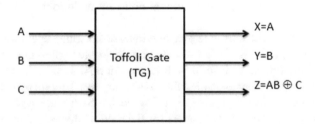

Fig. 4 Toffoli gate

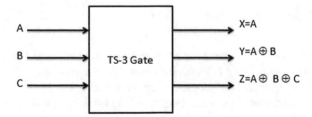

Fig. 5 TS-3 gate

2   S-Box Design Methodology

We know that in 2001, the NIST approved Advance Encryption Standard (AES) after
rigorous evaluation [7]. AES uses a message of particular block size of 128 bits and
key of 128 bits, 192 bits, and 256 bits for its operation. The AES uses AddRoundKey,
SubByte, ShiftRow, and MixColumn operations on 4×4 array of bytes called state
[8–10]. Among all these operations, nonlinear operation is on SubByte where each
byte in the state array is replaced with values from predefined table called as S-Box
and computationally heavy calculations are used in this phase [6, 11, 12].

The multiplicative inverse in GF (2^8) and an affine transformation (AT) are applied
for computation of SubByte. For opposite calculation, the InvSubByte transformation
and the inverse affine transformation are applied in Galois Field (2^8) [13]

One simple method used in executions of the S-Box for the SubByte is by using
pre-calculated values stored in a read-only memory based reference table. But as
ROMs have fixed access time for its read/write operation, it leads to unbreakable
delay [12]. Further, such operation is costly in terms of hardware. Thus, we can use
combinational-logic-based design to implement S-Box as discussed in Refs. [12–14].
This type of design helps to save area occupied. Also, pipelined operation can be
used to improve performance in given clock frequency. The construction of S-Box
is presented here using both combinational and reversible logic to make it power
efficient.

As both the designs of substitute byte and inverse substitute operation involve the
AT and its inverse, only the design of the substitute byte is presented in this work.
The design of multiplicative inverse is discussed in next section. Later, AT will be
discussed with steps for implementing the S-Box for the substitute byte operation. In
this paper, we have referred papers [13, 15] to design S-Box using reversible logic.
Readers are advised to refer these papers for detailed mathematical analysis.

2.1   Multiplicative Inverse

The GF(2^8) is used to represent discrete bits in a byte, viewed as coefficients of power
term in GF polynomial. From [15], $bx + c$ is used to represent any arbitrary polynomial, given an complicated polynomial of $x^2 + Ax + B$. From [13], the complicated
polynomial selected was $x^2 + x + \lambda$. Since $A = 1$ and $B = \lambda$, then simplified equation
can be written as

$$(bx + c)^{-1} = b\left(b^2\lambda + c(b + c)\right)^{-1}x + (c + b)\left(b^2\lambda + c(b + c)\right)^{-1} \quad (1)$$

Equation (1) indicates that multiplication, addition, squaring, and multiplication
inversion operations in GF(2^4) are used. From this, the multiplicative inverse circuit
GF(2^8) is shown in Fig. 6.

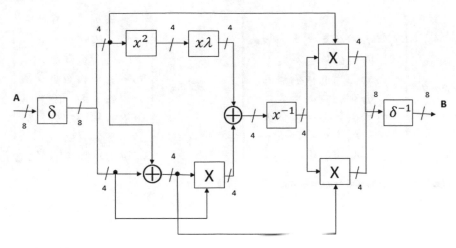

Fig. 6 Multiplicative inverse circuit

where

δ   Isomorphic mapping,
δ^{-1} Inverse isomorphic mapping in GF(2^8),
x^2 Squarer in GF(2^4), X^{-1}—Multiplicative inverse in GF(2^4),
X   Multiplication in GF(2^4), xλ—Multiplication with constant in GF(2^4),
⊕   Addition operation in GF(2^4) (Fig. 7).

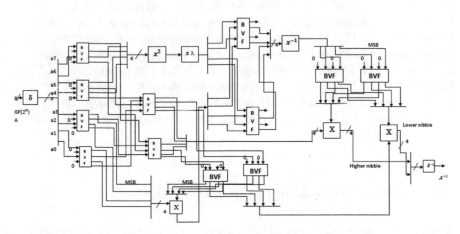

Fig. 7 Reversible multiplicative inverse circuit

$$\delta \times q = \begin{pmatrix} q_7 \oplus q_5 \\ q_7 \oplus q_6 \oplus q_4 \oplus q_3 \oplus q_2 \oplus q_1 \\ q_7 \oplus q_5 \oplus q_3 \oplus q_2 \\ q_7 \oplus q_5 \oplus q_3 \oplus q_2 \oplus q_1 \\ q_7 \oplus q_6 \oplus q_2 \oplus q_1 \\ q_7 \oplus q_4 \oplus q_3 \oplus q_2 \oplus q_1 \\ q_6 \oplus q_4 \oplus q_1 \\ q_6 \oplus q_1 \oplus q_0 \end{pmatrix} \qquad \delta^{-1} \times q = \begin{pmatrix} q_7 \oplus q_6 \oplus q_5 \oplus q_1 \\ q_6 \oplus q_2 \\ q_6 \oplus q_5 \oplus q_1 \\ q_6 \oplus q_5 \oplus q_4 \oplus q_2 \oplus q_1 \\ q_5 \oplus q_4 \oplus q_3 \oplus q_2 \oplus q_1 \\ q_7 \oplus q_4 \oplus q_3 \oplus q_2 \oplus q_1 \\ q_5 \oplus q_4 \\ q_6 \oplus q_5 \oplus q_4 \oplus q_2 \oplus q_0 \end{pmatrix}$$

Fig. 8 Isomorphic and inverse isomorphic mapping

2.2 Isomorphic Mapping and Inverse Isomorphic Mapping

Lower order fields of $GF(2^1)$, $GF(2^2)$, and $GF((2^2)^2)$ are used to compute multiplicative inverse. In order to achieve this, the following polynomials are used [13]:

$$\begin{aligned} GF(2^2) &\to GF(2) &&: x^2 + x + 1 \\ GF\left((2^2)^2\right) &\to GF(2^2) &&: x^2 + x + \varphi \\ GF\left(((2^2)^2)^2\right) &\to GF\left((2^2)^2\right) &&: x^2 + x + \lambda \end{aligned} \qquad (2)$$

where $\varphi = \{10\}_2$ and $\lambda = \{1100\}_2$. An element based on $GF(2^8)$ in composite field cannot be directly applied to calculate multiplicative inverse. Thus, isomorphic function (δ) is used for mapping this element. After implementing the multiplicative inverse, the result also has to be mapped back with inverse isomorphic function, δ^{-1}. 8×8 matrix can be used to denote both δ and δ^{-1}. Let q be the element in $GF(2^8)$, then the isomorphic mappings and its inverse can be written as $\delta*q$ and $\delta^{-1}*q$, as shown in Fig. 8 [13]. The matrices shown in Fig. 8 are altered using reversible logic gates to implement isomorphic and inverse isomorphic mapping hardware as shown in Figs. 9 and 10, respectively.

2.3 Arithmetic Operations in Composite Field

Again q can be spilt to $q_H x + q_L$ in GF using $bx + c$, from [14, 15]. For example, if $q = \{1101\}_2$, it can be represented as $\{11\}_2 x + \{01\}_2$, where q_H is $\{11\}_2$ and $q_L = \{01\}_2$. Using this, the logical expressions for the arithmetic operations like addition, multiplication, and inverse can be calculated.

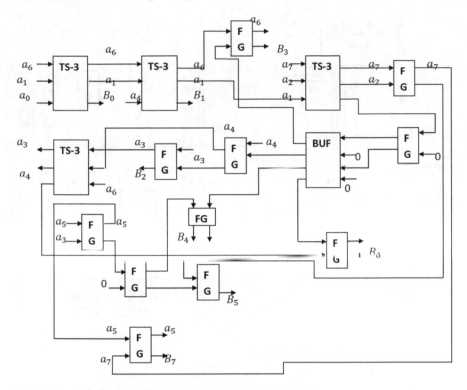

Fig. 9 Reversible hardware diagram for isomorphic mapping

2.3.1 GF(2^4) Addition

XOR operation is used for adding two elements in Galois field.

2.3.2 GF(2^4) Squaring

Equations for calculating the squarer operation in GF(2^4) are studied and presented as shown in Fig. 11 and reversible equivalent in Fig. 12 with equations used as follows (3):

$$
\begin{aligned}
k_3 &= q_3 \\
k_2 &= q_3 \oplus q_2 \\
k_1 &= q_2 \oplus q_1 \\
k_0 &= q_3 \oplus q_1 \oplus q_0
\end{aligned}
\tag{3}
$$

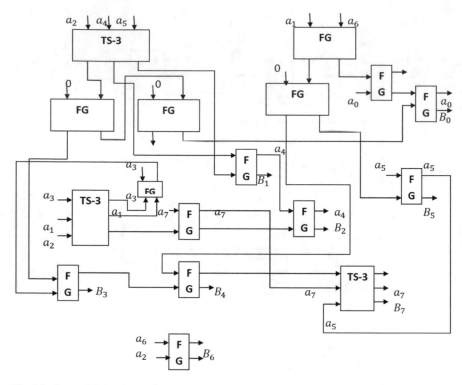

Fig. 10 Reversible hardware diagram for inverse isomorphic mapping

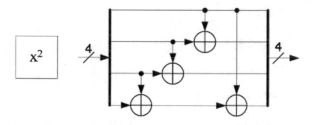

Fig. 11 Squarer in GF(2^4) [12]

2.4 Multiplication with Constant λ

The equations to calculate multiplication with constant is depicted in Fig. 13 and reversible implementation in Fig. 14 with following Eq. (4):

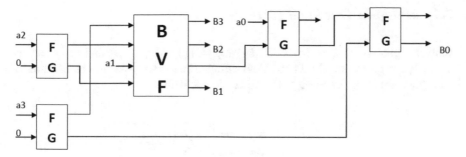

Fig. 12 Reversible equivalent of squarer in GF(2^4)

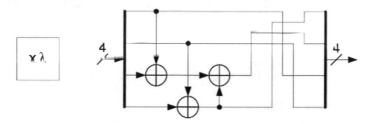

Fig. 13 Multiplication with constant λ [12]

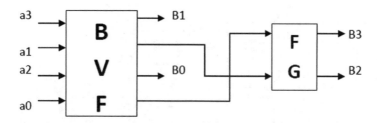

Fig. 14 Reversible multiplication with constant λ

$$k_3 = q_2 \oplus q_0$$
$$k_2 = q_3 \oplus q_2 \oplus q_1 \oplus q_0$$
$$k_1 = q_3$$
$$k_0 = q_2$$

(4)

2.5 *Multiplication in GF(2^4)*

With $k = qw$, where $k = \{k_3\ k_2\ k_1\ k_0\}_2$, $q = \{q_3\ q_2\ q_1\ q_0\}_2$, and $w = \{w_3\ w_2\ w_1\ w_0\}_2$ are elements of GF(2^4). Conventional and reversible implementation of multiplication is shown in Figs. 15 and 16, respectively.

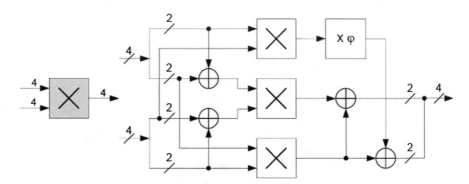

Fig. 15 Multiplication in GF(2^4) [12]

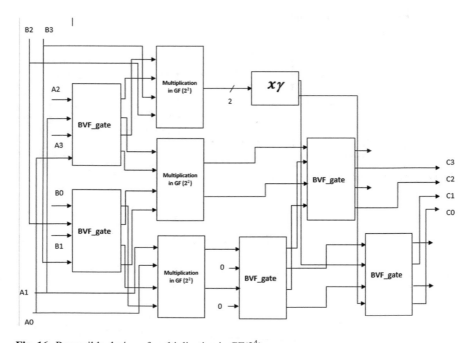

Fig. 16 Reversible design of multiplication in GF(2^4)

2.6 Multiplication in GF(2^2)

The logic used for calculating multiplication in GF(2) is shown in Fig. 17 and its reversible logic is shown in Fig. 18 with $k_1 = q_1w_1 \oplus q_0w_1 \oplus q_1w_0$, $k_0 = q_1w_1 \oplus q_0w_0$.

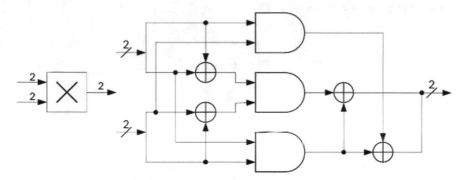

Fig. 17 Multiplication in GF(2) [12]

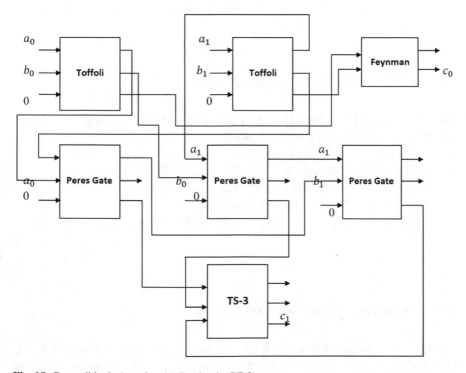

Fig. 18 Reversible design of multiplication in GF(2)

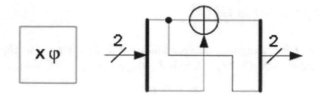

Fig. 19 Multiplication with constant φ [12]

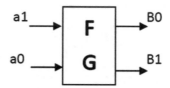

Fig. 20 Reversible multiplication with constant φ

2.7 Multiplication with Constant φ

Figure 19 shows the formulation for calculating multiplication with constant φ with $k_1 = q_1 \oplus q_0$ and $k_0 = q_1$ (Fig. 20).

2.8 Multiplicative Inversion in GF(2^4)

The authors of [12] have provided formulae to find multiplicative inverse of q from equations [12] and reversible design is shown in Fig. 21.

$$
\begin{aligned}
q_3^{-1} &= q_3 \oplus q_3 q_2 q_1 \oplus q_3 q_0 \oplus q_2 \\
q_2^{-1} &= q_3 q_2 q_1 \oplus q_3 q_2 q_0 \oplus q_3 q_0 \oplus q_2 \oplus q_2 q_1 \\
q_1^{-1} &= q_3 \oplus q_3 q_2 q_1 \oplus q_3 q_1 q_0 \oplus q_2 \oplus q_2 q_0 \oplus q_1 \\
q_0^{-1} &= q_3 q_2 q_1 \oplus q_3 q_2 q_0 \oplus q_3 q_1 \oplus q_3 q_1 q_0 \oplus q_3 q_0 \oplus q_2 \oplus q_2 q_1 \oplus q_2 q_1 q_0 \oplus q_1 \oplus q_0
\end{aligned}
$$

$$(5)$$

2.9 Affine Transformation and Its Inverse

The SubByte or InvSubByte is calculated using both multiplicative inverse in GF(2^8). Figures 22 and 23 show the reversible implementation of both affine transformation and inverse affine transformation, respectively.

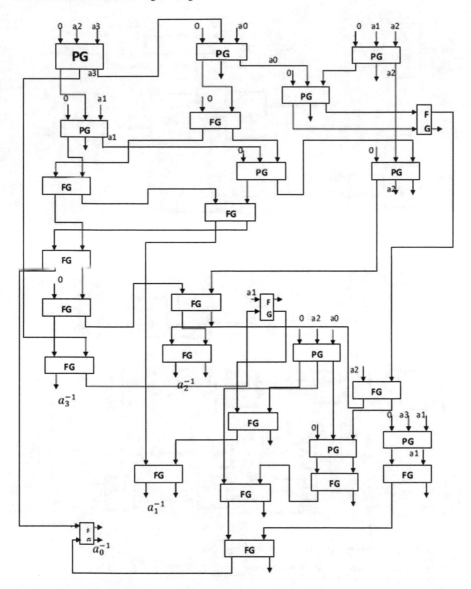

Fig. 21 Reversible hardware for multiplicative inverse of 4 bits

3 Testing Environment

The design of S-Box is done using reversible logic. Library of reversible gate is created using Xilinx tool and Verilog coding is used to verify the functionality of S-Box. 256 bytes generated are verified with conventional S-Box values as shown in Fig. 24. Inverse S-Box is designed using reversible gates and values are shown in Fig. 25.

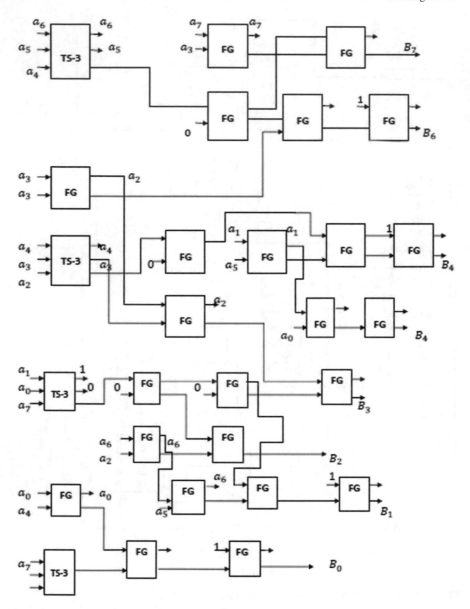

Fig. 22 Reversible affine transformation

Genus tool from cadence is used to perform power analysis. It can be seen from Table 1 that power consumed by design using reversible logic is less in comparison with conventional. Hardware required to implement S-Box using reversible logic is shown in Table 2.

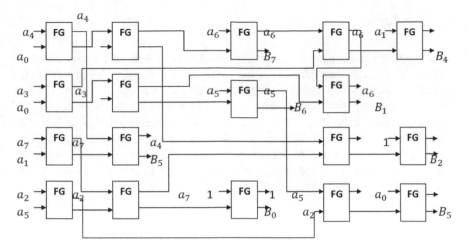

Fig. 23 Reversible inverse affine transformation

Fig. 24 Simulation result of S-Box

Fig. 25 Result of inverse S-Box

Table 1 Power analysis

Instance	Leakage power (nW)	Dynamic power (nW)	Total power (nW)
S-Box conventional	17.306	62168.520	62185.826
S-Box reversible	11.596	44553.652	44565.221

Table 2 Hardware requirements for the design of reversible S-Box and inverse S-Box

S. no.	Gate	No. of gates	Constants	Garbage
1	BVF	20	20	4
2	FG	93	15	46
3	TS_3	12	0	10
4	PG	13	13	12
5	Toffoli	2	2	0

4 Conclusion

AES algorithm is stronger against many attacks and has a satisfactory speed of data encryption and decryption. But upcoming future technologies like quantum technology may cause a threat to existing security algorithms. As S-Box used in AES is static throughout encryption process, power attacks are very prominent which may cause threat in practice. First attempt made in reversible design of S-Box is proposed in order to make AES algorithm more secure. Hardware requirements in terms of gate count and performance in terms of garbage outputs and ancilla constants are presented. Further work can be extended to use this power-efficient S-Box to implement security algorithms.

References

1. Soeken, M., Roetteler, M., Wiebe, N., De Micheli, G.: Logic synthesis for quantum computing, arXiv preprint arXiv:1706.02721 (2017)
2. Farik, M., Ali, S.: The need for quantum-resistant cryptography in classical computers. In: 2016 3rd Asia-Pacific World Congress on Computer Science and Engineering (APWC on CSE), Nadi, pp. 98–105 (2016)
3. Li, J., et al.: A survey on quantum cryptography. Chin. J. Electron. **27**(2), 223–228 (2018)
4. Datta, K., Shrivastav, V., Sengupta, I., Rahaman, H.: Reversible logic implementation of AES algorithm. In: 2013 8th International Conference on Design and Technology of Integrated Systems in Nanoscale Era (DTIS), pp. 140–144. IEEE (2013)
5. Berent: Advanced encryption standard by example. http://www.networkdls.com/Articles/AESbyExample.pdf (2007). Accessed June 2013
6. Rohini, H., Rajashekar, S.: Design of reversible logic based combinational circuits. Commun. Appl. Electron. (CAE) **5**, 38–43 (2016)
7. The Advanced Encryption Standard, http://en.wikipedia.org/wiki/Advanced_Encryption_Standard
8. Advanced Encryption Standard (AES), Federal Information Processing Standards Publication 197, 26th November 2001
9. Rohini, S.H., Nikhita, M., Pooja, A., Rajashekar, B.S.: Performance analysis of AES-128 bits, 192 bits & 256 bits using reversible logic. In: Seventh International Conference on Advanced electrica Measurement and Instrumentation Engineering (EMIE), 13, 14 July 2018, pp. 165–170 (2018)
10. Rohini, H., Jyoti, H., Rajashekar, S.: An approach towards design of N-bit AES to enhance security using reversible logic. Commun. Appl. Electron. (CAE) **7**(22) (2018)
11. Rohini, S.H., Jyoti, R.H., Rajashekar, B.S.: Reversible logic based modified design of AES-CBC mode. In: Seventh International Conference on Advanced Electrical Measurement and Instrumentation Engineering (EMIE), 13, 14 July 2018, pp. 171–176 (2018)
12. Zhang, X., Parhi, K.K.: High-speed VLSI architectures for the AES algorithm. IEEE Trans. Very Large Scale Integration (VLSI) Syst. **12**(9) (2004)
13. Satoh, A., Morioka, S., Takano, K., Munetoh, S.: A Compact Rijndael Hardware Architecture with S-box Optimization. Springer, Berlin Heidelberg (2001)
14. Good, T., Benaissa, M.: Very small FPGA application-specific instruction processor for AES. IEEE Trans. Circuits Syst I Regular Papers **53**(7) (2006)
15. Rijmen, V.: Efficient Implementation of the Rijndael S-Box. Katholieke Universiteit Leuven, Dept. ESAT, Belgium

16. Agrawal, M., Mishra, P.: A comparative survey on symmetric key encryption techniques. Int. J. Comput. Sci. Eng. **4**(5), 877 (2012)
17. Nag, K.S., Bhuvaneswari, H., Nuthan, A.: Implementation of advanced encryption standard-192 bit using multiple keys. In: IET Conference Proceedings. The Institution of Engineering & Technology (2013)
18. Mahmoud, E.M., Abd, A., Hafez, E., Elgarf, T.A., et al.: Dynamic aes-128 with key-dependent S-box (2013)
19. Prasad, H., Kandpal, J., Sharma, D., Verma, G.: Design of low power and secure implementation of sbox for aes. In: 2016 3rd International Conference on Computing for Sustainable Global Development (INDIACom), pp. 2092–2097. IEEE (2016)
20. Meghanathan, N.: A tutorial on network security: attacks and controls, arXiv preprint arXiv: 1412.6017 (2014)
21. Jacob, G., Murugan, A., Viola, I.: Towards the generation of a dynamic key-dependent S-box to enhance security. IACR Cryptol. ePrint Arch. **2015**, 92 (2015)
22. Ariffin, S., Mahmod, R., Jaafar, A., Ariffin, M.R.K.: Symmetric encryption algorithm inspired by randomness and non-linearity of immune systems. Int. J. Nat. Comput. Res. (IJNCR) **3**(1), 56–72 (2012)
23. Satapathy, S.C., et al.: Information systems design and intelligent applications. Adv. Intell. Syst. Comput. **2**, 219–223 (2016)

Salient Edge(s) and Region(s) Extraction for RGB-D Image

Nikhila Ratakonda, Navya Kondaveeti, Anunitya Alla and O. K. Sikha

Abstract In human visual systems, detection of the salient region plays an important role, as it ensures effective allocation of resources and fast processing. Though depth is an essential cue for visual saliency, it has not been well explored. This introduces a salient edge-based, region extraction model for RGB-D images. Most of the computational saliency models, reported in, produce smooth saliency maps; however, the edge information is not preserved. This paper presents a simple framework for the detection of salient edges by the preservation of the edges with high saliency scores. The final saliency map is obtained by fusion of the generated salient edge map and the RGB saliency map. Experiments are conducted over the publicly available RGB-D-2 dataset. The stability of the proposed RGB-D saliency model was assessed against standard evaluation metrics such as precision, recall, F-measure, MAE, NSS and AUC scores.

Keywords Salient edge · RGB-D saliency · Saliency map

1 Introduction

Image saliency is the distinctive trait of an image, which makes a particular section of the image more prominent to the human eye than the rest of the image [1]. In simple words, saliency is what that stands bent on you and the way you are ready to target the first relevant items of what you see quickly. Detection of the salient region

N. Ratakonda · N. Kondaveeti · A. Alla · O. K. Sikha (✉)
Amrita School of Engineering, Amrita Vishwa Vidyapeetam, Coimbatore, India
e-mail: ok_sikha@cb.amrita.edu

N. Ratakonda
e-mail: cb.en.u4cse15040@cb.students.amrita.edu

N. Kondaveeti
e-mail: cb.en.u4cse15543@cb.students.amrita.edu

A. Alla
e-mail: cb.en.u4cse15003@cb.students.amrita.edu

© Springer Nature Singapore Pte Ltd. 2020
V. Bhateja et al. (eds.), *Intelligent Computing and Communication*,
Advances in Intelligent Systems and Computing 1034,
https://doi.org/10.1007/978-981-15-1084-7_26

is an emerging research topic in the area of computer vision, since it affords effective allocation of computational resources, along with diverse applications such as object detection and localization [2], image segmentation [3], image retrieval [4], etc. A saliency map is an image that shows the unique quality of individual pixels in the image, as compared with other pixels in the image.

In the literature, extensive studies have been reported for extraction of salient regions of a two-dimensional (2D) image. Computational saliency models for 2D images can be mainly grouped into three classes: local contrast-based methods, global contrast-based methods and statistical learning-based methods. Local contrast-based models [5] use low-level image features, from a small image area to produce a saliency map. The saliency maps, generated by this approach, are accurate, albeit computationally complex as they use features from a smaller segment of an image. Global contrast-based models use the colour contrast of an image region relative to the whole image, as a measure to compute the saliency map [1]. Saliency maps derived from this model were able to generate uniform saliency maps, with less computational complexity. Statistical learning-based models [6] utilize machine learning algorithms to generate the saliency map.

Recently, a few studies were conducted to examine the significance of incorporating depth, in the extraction of saliency maps. A depth map is an image in which the value of a pixel represents the distance between the image plane and the corresponding object in the scene. An RGB-D image can be considered as an integration of an RGB image and the respective depth map. A saliency model for stereoscopic images was put forward by Cong et al. [7], in which they have used the depth map to generate a graph which was then used to generate the saliency map. Bao et al. [8] addressed the problem of detecting unknown objects, in an indoor environment by using the visual saliency scheme incorporating the depth information. Karthik et al. [9] confirmed the role of depth in the salient region extraction of an image. They adopted depth-induced blur, centre bias and competing saliencies, to generate the saliency map for an RGB-D image. Ran Ju et al. [10] suggested an anisotropic, centre-surround difference-based algorithm, where the minimum depth value of an image segment was considered as the background pixel for that segment. The input depth map was oversegmented using SLIC algorithm. For each superpixel, the difference between background (segment's minimum pixel value) and the centre pixel was computed to gauge the extent of its contrast, from its surroundings. Albeit the futuristic computational saliency models (RGB-D), reported in the literature, were able to generate smooth saliency maps, and they failed to conserve sharp edges of the salient object under consideration. The edges become blurred or smoothened in the output saliency map, as an explicit remedy for the preservation of the edge information, which has not been revealed in the literature [11].

An *edge* can be defined as a boundary of an object within an image or as a sudden change in the pixel intensity. Edge is an important visual cue, which facilitates the differentiation of the image regions. A salient edge can be defined as the edges with high a saliency score, which represents the boundary of a salient object in the image [24].

The primary objective of this investigation was to propose a salient edge detection model using the depth map and the salient derived edge to produce a saliency map, with high resolution. The remainder of this paper is structured as follows: the proposed RGB-D saliency model is explained in Sect. 2. Section 3 describes the results and discussions and Sect. 4 gives the conclusions.

2 Proposed System

This research, predominantly examined the efficacy of the interplay of the colour and the depth attributes to deduce a high-resolution saliency map by the preservation of the salient edge features. In this study, colour and depth information were processed separately to obtain the final saliency map. Depth information was used to extract the salient edge of an image since it supplied additional layering details and well-defined edges. Figure 1 shows a framework of the proposed system.

2.1 Salient Edge Extraction from Depth Map

1. The smoothened depth map is divided into N superpixels using the SLIC over-segmentation algorithm, as shown in Fig. 1c. A superpixel is an irregular image patch, which is better aligned with intensity edges, rather than a rectangular patch [10].

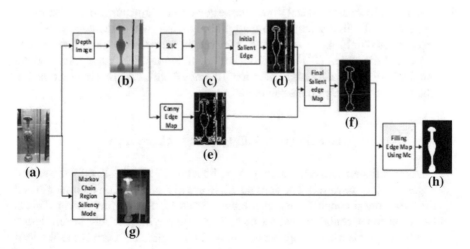

Fig. 1 Overview of the proposed RGB-D saliency model

2. For each superpixel, the pixel with a minimum depth value is considered as the background pixel for that segment. The distances between individual pixels and the background are calculated for every superpixel to generate Depth_Sal1, as shown in Eq. 1.

$$\text{Depth\_Sall}(x_i, y_i) = \left\| D(x_i, y_i) - \min(S_i) \right\|^2 \tag{1}$$

In Eq. 1, $D(x_i, y_i)$ represents the individual pixels from the depth map D and Si represents the ith superpixel, after SLIC segmentation. The mean value of the superpixel is subtracted from the individual pixels of that segment to generate Depth_Sal2 as in Eq. 2.

$$\text{Depth\_Sal}(x_i y_i) = \left\| D(x_i, y_i) - \text{mean}(S_i) \right\|^2 \tag{2}$$

3. An initial salient edge map is generated by combining *Depth_Sal1* and *Depth_Sal2* as shown below:

$$\text{SalEdge\_init} = \text{Depth\_Sall} + \text{Depth\_Sal2}$$

Figure 1d shows the initial salient edge obtained from depth map using the proffered method. It is clear from the figure that few edges, other than the boundary of salient object, are also appearing in the resultant map and the edges seem broken (no connectivity).

4. The output from the classical 'Canny edge detection' algorithm is used to enhance the generated initial salient edge map, SalEdge_init, where the edge pixels are disconnected or broken, we have used the output from the classical 'Canny edge detection' algorithm [12]. The initial salient edge map, (SalEdge_init), and the output from the 'Canny edge detection', (CannyEdge), Fig. 1e, are combined to obtain the commonly connected edges.

$$\text{SalcanEdge} = \text{SalEdge\_init} * \text{CannyEdge}$$

As mentioned above, not all edges are beneficial for the task; only the salient edges are to be retrieved. The derived edges are ranked based on their length and the two longest connected edges are considered for further processing. This is based on the speculation that a true salient object will generally have a larger number of pixels when compared to a non-salient region. For each of the obtained edges, average saliency scores are calculated and the edge, with the highest average saliency score, is retained in the final, salient edge map, as shown in Fig. 1f.

2.2 Colour Saliency Map Extraction from the RGB Image

Markov's chain-based salient region extraction algorithm, suggested by Jiang et al. in [13], was used to extract the salient region for the colour image Fig. 1g. A Markov chain is arithmetic and statistical procedure that changes from state to state, within a limited number of possible states. As it is a group of various states and likelihoods of a variable, its futuristic condition or state, in the future, is determined by its nearest preceding state. The authors have used spatial distribution of the background, appearance divergence and the salient object for the generation of the saliency map. Figure 2 shows the results, obtained from MC, for five random images, from the dataset proposed by Peng et al. [14].

2.3 Fusion of Salient Edge and Colour Saliency Map Resume

This section describes the generation of the final saliency map, by integrating the generated colour saliency map and the salient edge, as explained in the previous sections. Since the salient edge delineates the boundary of the salient object, the region bounded by the edges should be filled up to generate the final saliency map. A horizontal traversing algorithm was used to cover up the area enclosed by the salient edge, with maximum saliency score obtained from the colour saliency map, as follows:

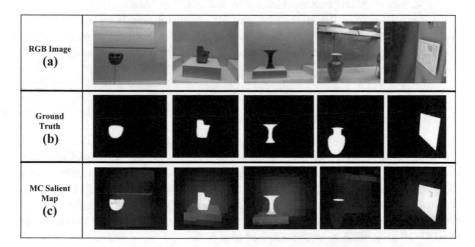

Fig. 2 Saliency map obtained for RGB image (two dimensional) using MC [13]

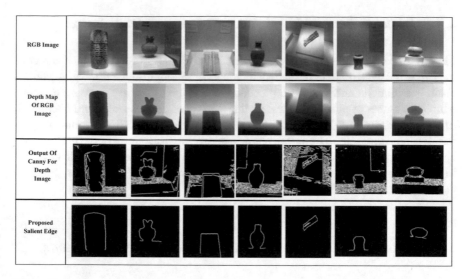

Fig. 3 Comparison of edge map obtained from canny edge detection and the proposed salient edge detection

- Traverse horizontally from both right and left of the final salient edge map.
- When the scan line comes across a white pixel from the left, it will be considered as the first pixel, and when it comes across the white pixel from the right, it will be taken as the last pixel.
- The stretch of pixels between those two end pixels is filled with the maximum value obtained from the saliency map of MC, as in Fig. 4e. Both the boundaries and the information of the salient object are preserved from the proposed salient edge Fig. 1h. It suppresses the background and highlights the salient edge Fig. 3e.

3 Experimental Results and Analysis

This section presents an evaluation of the proposed RGB-D saliency model, quantitatively, using six benchmark evaluation metrics on the well-known, publicly available, RGB-D2 dataset [14]. The RGB-D2 dataset is constituted of 1000 images of various indoor and outdoor scenes along with a map of their depth and ground truth. The efficiency of the proposed saliency model was compared against five state-of-the-art RGB-D saliency models, from published literature.

3.1 Evaluation Metrics

The proposed method was appraised against the following metrics: F-measure rate, precision–recall curve (PR), normalized scanpath saliency [NSS], mean absolute error (MAE) proposed by Achanta et al. in [15], Pearson's correlation coefficient (CC) [16] and area under ROC curve (AUC) as follows:

1. **F-measure**: It is the weighted mean of precision and average recall. It is calculated as follows:

$$F_\beta = \frac{(1 + \beta^2) \bullet \text{Precision} \bullet \text{Recall}}{\beta^2 \bullet \text{Precison} + \text{Recall}}$$

As suggested by Achanta et al. [17], β^2 were set to 0.3 to foreground precision rather than recall.

2. **Precision–recall [PR]**: It is the fraction of total number of relevant pixels (ground truth) to the number of retrieved pixels (salient region) from the salient pixels in the map, where B is the binary image and G is the ground truth.

$$\text{Precision} = \frac{|B \cap G|}{B}$$

The fraction of retrieved pixels (salient region) to the total number of relevant pixels (ground truth) is known as recall or sensitivity.

$$\text{Recall} = \frac{|B \cap G|}{G}$$

3. **Normalized scanpath saliency [NSS]**: The correlation between the saliency map and the ground truth, as the average normalized saliency at fixed locations [18], is measured by normalized scanpath saliency (NSS).

4. **Mean absolute error [MAE]**: The average difference between the resultant saliency map and the ground truth is recorded by MAE.

$$AE = \frac{1}{W \times H} \sum_{x=1}^{W} \sum_{Y=1}^{H} ||S(x, y) - G(x, y)||$$

where W is the width, H is the height of the saliency map S and G is the binary ground truth mask.

5. **Pearson's correlation coefficient (CC)**: It computes the measure of the association, or the reliance, between two variables, i.e. it quantifies the direct proportionality relationship between the saliency map and the ground truth, where

the ground truth and the saliency map have same magnitude, and the correlation coefficient will have large values in those sections.

6. **Area under ROC curve [AUC]**: The AUC score is computed by considering the derived saliency map as a classifier, which classifies the foreground and the background of the images. Receiver operating characteristic (ROC) is a plot of the true positive rate (TPR) versus false positive rate (FPR), by testing all possible thresholds. The prevalent appraise for analysing the produced saliency map is the area under ROC curve (AUC). The AUC considers arbitrary sample of values of pixels as negatives.

Figure 3 describes the edge map obtained from the classical Canny edge detection algorithm and the salient edge map produced by the proposed RGB-D method. It is conspicuous from the figure that the depth map-based, salient edge detection model is capable of retaining edges that belong to the salient object. Figure 4 shows the final salient edge and map obtained from the RGB-D model. It is clear from the figure that the RGB-D saliency model is capable of preserving sharp edges and produces high-quality saliency maps. Visual comparison of saliency maps, obtained from the proposed RGB-D model, and conventional saliency models is shown in Fig. 5. The output procured from the RGB-D model was compared to the following five state-of-the-art models: CD [19], LB [20], FT [15], SUN [21] and DES [22]. In the figure, the first row shows the original colour image from the RGB-D2 dataset, and from the second row onwards, the saliency map from various state-of-the-art models, as well as the proposed model, is shown. The saliency maps, obtained from SUN, FES and

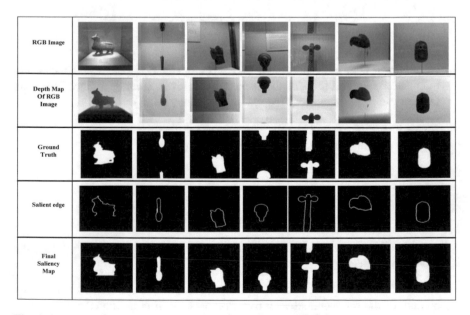

Fig. 4 Salient edges and maps obtained from the proposed RGB-D model

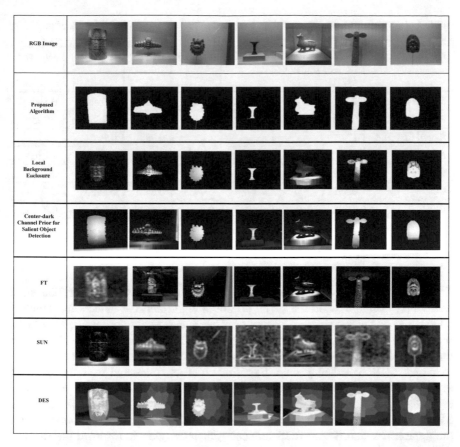

Fig. 5 Perceptible differentiation of saliency map procured from the proposed algorithm and the reference algorithms

DES, capture most of the salient pixels, but few non-salient pixels were also present in the saliency map. The figure also shows that false positive detection rate is notably less in the RGB-D saliency model compared to other methods. Table 1 tabulates the results obtained from various performance measures discussed in the previous section and the highlighted values obtained for the proposed RGB-D model. As seen in the above table, by comparing the evaluation metrics, it is perceived that the suggested algorithm's performance is more desirable. This division presents an evaluation of the proposed RGB-D saliency model, quantitatively, using six benchmark evaluation metrics on the well-known, publicly available, RGB-D2 dataset [14]. The RGB-D2 dataset is constituted of 1000 images, of various indoor and outdoor scenes, along with a map of their depth and ground truth. The efficiency of the proposed saliency model was compared against five state-of-the-art RGB-D saliency models, from published literature.

Table 1 Comparison of the six algorithms in regard to mean F-measure, maximum of precision, maximum of recall, NSS, MAE, CC and AUC

Metric	Proposed RGB-D saliency (SER)	CD [19]	LB [20]	FT [15]	SUN [21]	DES [22]
Mean F-measure	0.821	0.825	0.898	0.824	0.596	0.895
Precision (max)	0.803	0.833	0.9110	0.725	0.445	0.925
Recall (max)	1	1	1	0.988	0.995	1
NSS	1.5649	0.9117	2.5198	0.7890	1.3764	2.1105
MAE	0.0729	0.0855	0.0435	0.1900	0.2701	0.2828
CC	0.3912	0.4780	0.0435	0.2427	0.3485	0.6652
AUC	0.6950	0.6235	0.8135	0.5596	0.6367	0.8344

4 Conclusion

This paper has proposed a new structure to improvise the detection of the salient edge of an image using a depth map. The proposed algorithm first detects the saliency edge based on the respective depth maps. The obtained edges are ranked based on the average saliency score and the edge with the highest saliency was retained in the output. Final saliency map was obtained by fusion of the generated salient edge map and the RGB saliency map. Experiments were conducted over the publicly available RGB-D-2 dataset. Stability of the proposed RGB-D saliency model was evaluated against the standard evaluation metrics such as precision, recall, F-measure, MAE, NSS and AUC score. If project is scaled in bigger applications, we can join our approach with different RGB-induced saliency models to discover the RGB-D saliency models on various applications, e.g. video saliency detection and co-saliency detection. These outcomes reaffirmed that the submitted algorithm outperforms other algorithms, in both robustness and accuracy, for various structures and scenarios.

Declaration We have taken permission from competent authorities to use the images/data as given in the paper. In case of any dispute in the future, we shall be solely responsible.

References

1. Itti, L.: Visual salience. Scholarpedia **2**(9), 3327 (2007)
2. Vidhya, N., Itti, L.: An integrated model of top-down and bottom-up attention for optimizing detection speed. In: 2006 IEEE Computer Society Conference on Computer Vision and Pattern Recognition (CVPR'06), vol. 2. IEEE (2006)
3. Liu, Z., Shi, R., Shen, L., Xue, Y., Ngan, K.N., Zhang, Z.: Unsupervised salient object segmentation based on kernel density estimation and two-phase graph cut. IEEE Trans. Multimedia **14**(4), 1275–1289 (2012)
4. Aarthi, R., Amudha, J.: Saliency based modified chamfers matching method for sketchbased image retrieval. In: 2015 International Conference on Innovations in Information, Embedded

and Communication Systems (ICIIECS), pp. 1–4. IEEE (2015, March)

5. Itti, L., et al. Attentional modulation of human pattern discrimination psychophysics reproduced by a quantitative model. In: Advances in Neural Information Processing Systems (1999)
6. Wang, L., et al.: Automatic salient object extraction with contextual cue. In: 2011 International Conference on Computer Vision. IEEE (2011)
7. Cong, R., et al.: Saliency detection for stereoscopic images based on depth confidence analysis and multiple cues fusion. IEEE Signal Proc. Lett. **23**(6), 819–823 (2016)
8. Bao, J., et al.: Saliency-guided detection of unknown objects in RGB-D indoor scenes. Sensors **15**(9), 21054–21074 (2015)
9. Desingh, K., et al.: Depth Really Matters: Improving Visual Salient Region Detection with Depth. BMVC (2013)
10. Achanta, R., Shaji, A., Smith, K., Lucchi, A., Fua, P., Süsstrunk, S.: SLIC superpixels compared to state-of-the-art superpixel methods. IEEE Trans. Pattern Anal. Mach. Intell. **34**(11), 2274–2282 (2012)
11. Bhateja, V., Devi, S., Urooj, S.: An evaluation of edge detection algorithms for mammographic calcifications. In: Proceedings of the Fourth International Conference on Signal and Image Processing 2012 (ICSIP 2012). Springer, India (2013)
12. Ding, L., Goshtasby, A.: On the canny edge detector. Pattern Recognit. **34**(3), 721–725 (2001)
13. Jiang, B., Zhang, L., Lu, H., Yang, C., Yang, M.H.: Saliency detection via absorbing markov chain. In: Proceedings of the IEEE International Conference on Computer Vision, pp. 1665–1672 (2013)
14. Peng, H., et al.: Rgbd salient object detection: a benchmark and algorithms. In: European Conference on Computer Vision. Springer, Cham (2014)
15. Achanta, R., Hemami, S., Estrada, F., Susstrunk, S.: Frequency-Tuned Salient Region Detection (2009)
16. Bylinskii, Z., Judd, T., Oliva, A., Torralba, A., Durand, F.: What do different evaluation metrics tell us about saliency models? IEEE Trans. Pattern Anal. Mach. Intell. **41**(3), 740–757 (2019)
17. Gupta, A., et al.: A novel color edge detection technique using hilbert transform. In: Proceedings of the International Conference on Frontiers of Intelligent Computing: Theory and Applications (FICTA), pp. 725–732. Springer, Berlin, Heidelberg (2013)
18. Peters, R.J., Iyer, A., Itti, L., Koch, C.: Components of bottom-up gaze allocation in natural images. Vis. Res. **45**(18), 2397–2416 (2005)
19. Zhu, C., et al.: Exploiting the Value of the Center-dark Channel Prior for Salient Object Detection. arXiv preprint arXiv:1805.05132 (2018)
20. Feng, D., Barnes, N., You, S., McCarthy, C.: Local background enclosure for RGB-D salient object detection. In: Proceedings of the IEEE Conference on Computer Vision and Pattern Recognition, pp. 2343–2350 (2016)
21. Zhang, L., Tong, M.H., Marks, T.K., Shan, H., Cottrell, G.W.: SUN: a Bayesian framework for saliency using natural statistics. J. Vis. **8**(7), 32–32 (2008)
22. Cheng, Y., Fu, H., Wei, X., Xiao, J., Cao, X.: Depth enhanced saliency detection method. In: Proceedings of International Conference on Internet Multimedia Computing and Service, p. 23. ACM (2014, July)

Web Text Categorization: A LSTM-RNN Approach

Ankita Dhar, Himadri Mukherjee, Sk. Md. Obaidullah, K. C. Santosh, Niladri Sekhar Dash and Kaushik Roy

Abstract Categorization of text documents has become a prime task with plenty of applications. It is a process that assigns the text documents to their individual categories based on the contents. In this paper, a long short term memory (LSTM) recurrent neural network (RNN) based system along with handcrafted graph-based features is proposed for text categorization. In comparison to the traditional bag of words model for representation of documents, this graph-based model represents each document by a weighted graph for determining the relationships among the tokens. The significance of a token to a text document is determined by the graph-based feature. Experiments were tested on 14,373 Bangla text documents (around 57,22,569 tokens) and a maximum accuracy of 99.21% has been obtained.

Keywords Text categorization · Weighted graph · Recurrent neural network

A. Dhar (✉) · H. Mukherjee · K. Roy
Department of Computer Science, West Bengal State University, Kolkata, India
e-mail: ankita.ankie@gmail.com

H. Mukherjee
e-mail: himadrim027@gmail.com

K. Roy
e-mail: kaushik.mrg@gmail.com

Sk. Md. Obaidullah
Department of Computer Science & Engineering, Aliah University, Kolkata, India
e-mail: sk.obaidullah@gmail.com

K. C. Santosh
Department of Computer Science, The University of South Dakota, Vermillion, USA
e-mail: santosh.kc@ieee.org

N. S. Dash
Linguistic Research Unit, Indian Statistical Institute, Kolkata, India
e-mail: ns_dash@yahoo.com

© Springer Nature Singapore Pte Ltd. 2020 281
V. Bhateja et al. (eds.), *Intelligent Computing and Communication*,
Advances in Intelligent Systems and Computing 1034,
https://doi.org/10.1007/978-981-15-1084-7_27

1   Introduction

The volume of digital text data has expanded considerably with expeditious growth of the Internet in recent years. Hence, there is a pressing need for developing an automatic and efficient text categorization system for managing and sorting such large and unstructured text data. Categorization of text has become an important problem in text mining. It is a process that assigns the text documents to their individual categories based on the contents. Various researches have been performed in this field in languages, like English, Arabic, Chinese, Japanese, etc., but very few studies have been done in Indian languages especially in Bangla. Since Bangla is one of the most popular languages in the world with approximately having 243 million first-language speakers [3] and also the official language of Bangladesh and regional language of West Bengal, India, thus, a text categorization system in Bangla can help users to retrieve domain-specific information efficiently according to their need. These facts motivated us to involve our time and knowledge in this field of interest. This paper presents the recurrent neural network-based approach with long-short-term memory (LSTM) along with handcrafted graph-based features for categorization of Bangla text documents. The present experiment was performed on 14,373 Bangla text documents having around 57,22,569 tokens from nine text categories and obtained an accuracy of 99.21%.

In the remaining sections of the paper, a brief survey on some of the recent works in text classification is provided in Sect. 2, followed by Sect. 3 illustrating the proposed methodology that involves information about data, features and deep learning-based classification; in Sect. 4, the experimental results have been analyzed and comparative study with the state-of-the-art methods have been provided; and Sect. 5 concludes the paper demonstrating some future works in this field.

2   Related Works

Currently, deep neural network models proposed by various researchers in English have led to new ideas for solving several NLP problems. Du and Huang [2] proposed a technique that works based on the attention mechanism for assigning weights to each token. The domain-specific tokens will be having more weightage compare to the general tokens. Thus, representing texts not only deals with entire token set but also pays attention to domain-specific tokens which were produced to a softmax classifier. The experiment was tested on two news classification databases published by NLPCC2014 and Reuters and obtained F measure of 88.5% and 51.8%, respectively. Lai et al. [10] introduced a recurrent convolutional neural network without involving handcrafted features. They implemented a recurrent network for capturing contextual information that introduces less noise compared to traditional neural networks. They employed a max-pooling layer which determines the relevancy of a token in text categorization task. The experiments were conducted on four standard databases and reported an accuracy of 96.79% on 20Newsgroup dataset.

However, deep learning approaches have not been explored much in Indian languages but few researches performed using traditional methods are mentioned here. Gupta and Gupta [6] experimented with a hybrid method by coupling NB and ontology-based techniques together for categorizing Punjabi text documents. The experiment was conducted on 184 text documents from seven subdomains of sports category and reported that the hybrid method outperformed compared to the traditional methods. Swamy and Thappa [14] worked on 100 documents each from Tamil, Kannada, and Telugu languages using Zipf's law, VSM, and TF-IDF along with DT, Knn, and NB classifiers. Patil and Bogiri [12] worked based on user's profile using LINGO (Label Induction Grouping) model along with VSM. The systems efficacy were tested on 200 documents from 20 text domains. Rakholia and Saini [13] used NB classification algorithm for classifying Gujarati text documents on a dataset of 280 documents from six domains and obtained an accuracy of 75.74% and 88.96% without and with feature selection approach respectively. In case of Bangla, in spite of various shortfalls, some efforts have been made for developing text categorization system. Mandal and Sen [11] tested the performance of their system using four supervised machine learning approaches on 1000 labeled news text documents consists of 22,218 tokens from five text classes. They reported maximum accuracy of 89.14% using SVM algorithm. On the other hand, Kabir et al. [9] used SGD algorithm for categorizing Bangla texts from nine distinct text domains. The performance of their system was slightly better than previous work as they obtained an accuracy of 93.85%. Islam et al. [8] reported an accuracy of 92.57% by carrying out the experiment on Bangla text documents from 12 text domains using TF-IDF and SVM as feature selection method and classifier, respectively.

From the literature review, it can be observed that deep learning approaches have not been explored much in Indian languages including Bangla because of the unavailability of standard datasets, resources, and tools. Thus, this paper aims to develop a system using LSTM-RNN approach to help users for retrieving useful information in Bangla.

3 Proposed Methodology

The text documents were first tokenized and then stop words were removed followed by graph-based feature and thereafter LSTM-RNN-based classification algorithm was used to determine the text categories of the given text documents. The working methodology is diagrammatically demonstrated in Fig. 1.

Fig. 1 Graphical illustration of the proposed methodology

Table 1 Text documents distribution for each domain

Domain (distribution)	Bs (2168)	En (2166)	Fd (1051)	Li (1103)	Md (1047)
	Sa (2122)	Sp (2134)	St (1194)	Tl (1388)	

3.1 Dataset

Data is considered as an important step while performing any experiment. The data in terms of quality and quantity plays an important role in the outcome of any experiment. For the experiment, total of 14,373 Bangla text documents were obtained from nine text categories: Business (Bs), Entertainment (En), Food (Fd), Literature (Li), Medical (Md), State affairs (Sa), Sports (Sp), Science & Technology (St), and Travel (Tl). The detailed distribution of the text documents is presented in Table 1. The sources from which the documents were obtained is provided in [1].

3.2 Preprocessing

While dealing with text documents, before processing further with the data, the raw texts need to be splitted into 'token'. The process of splitting the sentences into tokens is called 'tokenization'. In our experiment, the tokens were retrieved based on a 'space' delimiter and the total tokens counts to be 57,22,569. Since each and every token does not hold domain-specific information essential in categorizing the documents, they need to be discarded from the feature vector and such tokens are termed as stop words. The selection of stop words is problem specific, therefore in the present work, punctuation marks, pronouns, English and Bangla digits, English equivalent words, postpositions, conjunctions, interjections, all the articles, some adjectives, and adverbs were treated as stop words and after removal, the number of tokens results to be 44,47,689.

3.3 Feature

The present work proposes a graph-based feature extraction scheme using a weighted graph algorithm as follows:

A graph Gr was determined by 3-tuple: Gr = (Vc, Ed, Wt), where Vc determines a set of vertices, Ed denotes the weighted edges that connect the Vc and Wt denotes the weight vector of Ed.

1. Node: Unique tokens obtained from the text documents being considered.
2. Edge: Generated based on the frequency of token in text documents. If two tokens arrive in a particular text document then an edge was assigned between two tokens.

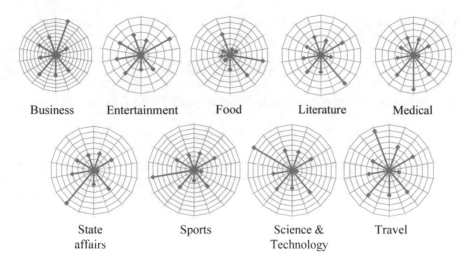

Business   Entertainment   Food   Literature   Medical

State   Sports   Science &   Travel
affairs   Technology

Fig. 2 Trend of the feature

3. Weight Vector: It was determined using Eq. 1, where Wt(l,m) represents the weight assigned to tokens l and m, f(l,m) represents the frequency of tokens l and m within a text document, f(l) denotes the frequency of token l and f(m) denotes the frequency of token m separately in that particular text document. The higher weight Wt(l,m) confirms the stronger edge between two vertices.

$$Wt(l, m) = f(l, m)/\{f(l) + f(m) - f(l, m)\} \tag{1}$$

The trend of the feature for each nine categories is depicted through Fig. 2.

3.4 Deep Learning-Based Classification

Deep learning has recently gained attention from researchers across the globe and has outperformed various standard machine learning algorithms. Recurrent neural networks (RNN) [4, 5] are a particular kind of neural network applied in classifying sequences. Comparing to other traditional neural networks, RNNs have the capability of retaining states during sequential data categorization. It is more generalized compare to DNNs and can be stretched for longer pattern.

The present state S_t is represented below where S_{t-1}, X_t, and F_w determine the previous state, input at tth state and a recursive function.

$$S_t = F_w(S_{t-1}, X_t) \tag{2}$$

Each of the weights was multiplied with corresponding weights and the recursive function (tanh function) is shown below.

$$S_t = tanh(W_S S_{t-1} + W_X X_t) \tag{3}$$

The final output Y_t was obtained by initiating a weight W_y to S_t as presented below.

$$Y_t = W_y * S_t \tag{4}$$

Normally, RNNs experiences vanishing gradient problem that is solved using Long-Short-Term Memory-RNNs (LSTM-RNN). It has the potential to resolve varying composite as well as elongated problems where simple RNN faces lots of challenges. Each LSTM block comprises long-term memory known as the cell state involves three gates (forget gate, input gate, and output gate). The input gate (i_t) determines which values are involved in establishment of the new state presented below where W_i denotes the associated weight.

$$i_t = (W_i S_{t-1} + W_i X_t) \tag{5}$$

The forget gate represents the values, which requires to be eliminated from the previous state while establishing values of present state. The mathematical expression is given below where W_f denotes the associated weight.

$$f_t = (W_f S_{t-1} + W_f X_t) \tag{6}$$

The output gate represents the values to be transferred from the present state while establishing the next state. The mathematical expression is given below where W_o denotes the associated weight.

$$o_t = (W_o S_{t-1} + W_o X_t) \tag{7}$$

The intermediate cell state c_t can be represented as below producing cell state C_t as shown in Eq. 9.

$$c_t = tanh(W_c S_{t-1} + W_c X_t) \tag{8}$$

$$C_t = (i_t * c_t) + (f_t * c_{t-1}) \tag{9}$$

The new state h_t thus obtained is

$$h_t = o_t * tanh(C_t) \tag{10}$$

In the present scenario, an embedding layer was used and the matrix was initialized with a value of 300 and a 100-dimensional LSTM layer. The output of this layer was passed to a 100-dimensional layer with a relu activation function whose outcome was fed to the second 50-dimensional dense layers with the same function. Finally, the results were produced to the final 8-dimensional dense layers with a softmax activation. The training epochs were initially set to 100 with a batch size of 50. The dimensionality of the layers as well as the number of layers and activations were chosen based on trial. We had used a cross-validation folds for evaluation of the system.

4 Results and Discussion

The graph-based features extracted from 14,373 Bangla text documents from nine different text categories were subjected to the RNN-LSTM classifier. The experiments were conducted for 100, 150, and 200 iterations whose results are presented in Table 2. It can be seen from Table 2, that the best outcome is for 150 iterations. The confusion matrix for these iterations are also presented in Tables 3, 4 and 5 as well. This shows the capability of the system while working with text documents from different text categories which are often required in real-world scenario.

A few other popular machine learning algorithms from WEKA [7] in the thick of Naïve Bayes Multinomial (NBM), Multilayer Perceptron (MLP), Rule-based (PART), Decision Tree (J48), and K Nearest Neighbor (KNN) were also applied to the same whose results are presented in Fig. 3.

Table 2 Obtained accuracies using the different iterations

Iterations	100	150	200
Accuracy (in %)	99.15	**99.21**	99.04

Table 3 The confusion matrix obtained using 100 iterations

	Bs	En	Fd	Li	Md	Sa	Sp	St	Tl
Bs	**2144**	1	0	2	1	4	1	15	0
En	0	**2161**	2	3	0	0	0	0	0
Fd	0	0	**1050**	1	0	0	0	0	0
Li	4	5	1	**1074**	1	3	0	5	10
Md	0	0	0	0	**1043**	3	0	0	1
Sa	8	0	0	2	2	**2109**	1	0	0
Sp	1	1	0	2	0	3	**2127**	0	0
St	7	0	0	2	1	0	1	**1183**	0
Tl	1	0	0	20	1	1	0	5	**1360**

Table 4 The confusion matrix obtained using 150 iterations

	Bs	En	Fd	Li	Md	Sa	Sp	St	Tl
Bs	**2144**	0	0	2	2	6	1	13	0
En	1	**2163**	0	1	0	1	0	0	0
Fd	0	0	**1048**	3	0	0	0	0	0
Li	3	3	0	**1080**	5	2	4	1	5
Md	2	0	0	3	**1038**	1	0	3	0
Sa	12	0	0	1	4	**2104**	1	0	0
Sp	0	2	0	1	0	1	**2130**	0	0
St	4	0	0	2	1	0	1	**1186**	0
Tl	1	0	0	20	0	1	0	0	**1366**

Table 5 The confusion matrix obtained using 200 iterations

	Bs	En	Fd	Li	Md	Sa	Sp	St	Tl
Bs	**2144**	0	0	5	1	4	1	12	1
En	0	**2159**	2	3	0	1	1	0	0
Fd	0	0	**1048**	3	0	0	0	0	0
Li	3	5	4	**1073**	0	1	4	0	13
Md	5	0	0	2	**1039**	0	0	0	1
Sa	11	1	0	2	0	**2106**	1	1	0
Sp	1	5	0	1	0	1	**2126**	0	0
St	5	1	0	1	1	1	0	**1183**	2
Tl	2	0	0	21	3	2	1	2	**1357**

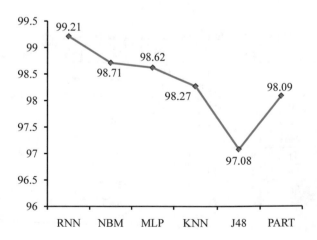

Fig. 3 Performance of different classifiers

Table 6 Comparison of the working methodology with other existing approaches

Reference	Used feature	Accuracy (in %)
Mandal and Sen [11]	TF-IDF	96.79
Kabir et al. [9]	TF-IDF	97.29
Islam et al. [8]	TF-IDF	98.02
Proposed work	**graph-based**	**99.21**

4.1 Comparison with Existing Methods

The working methodology has been compared with some of the existing methods following the feature extraction schemes and experimented on our dataset using RNN-LSTM algorithm for 150 iterations. The obtained accuracies for all the approaches are demonstrated in Table 6 where it can be observed that the working methodology outperformed all other existing methods for Bangla in terms of recognition accuracy.

5 Conclusion

In this paper, a LSTM-RNN categorization system is presented for categorizing Bangla text documents. The system was tested on 14,373 Bangla text documents from various web sources and obtained maximum accuracy of 99.21% using graph-based features for 150 iterations. In the future, exploring various feature extraction schemes are in our plan for dealing with the text documents belonging to more than one text category. We also plan to implement other text representation schemes and machine learning techniques in a combined form for further improvement in the performance of the system. Finally, we plan to explore deep learning-based approaches in the feature extraction stage and observe the system performance.

Acknowledgements One of the authors thank DST for support in the form of INSPIRE fellowship.

References

1. Dhar, A., Dash, N.S., Roy, K.: Classification of bangla text documents based on inverse class frequency. In: Proceedings of IoT-SIU, pp. 1–6 (2018)
2. Du, C., Huang, L.: Text classification research with attention-based recurrent neural networks. Int. J. Comput. Commun. Control **13**(1), 50–61 (2018)
3. Ethnologue.: https://www.ethnologue.com/language/ben (2019)
4. Fan, S., Huang, B.: Recurrent collective classification. Knowl. Inf. Syst. 1–15 (2018)
5. Gou, C., Shen, H., Du, P., Wu, D., Liu, Y., Cheng, X.: Learning sequential features for cascade outbreak prediction. Knowl. Inf. Syst. **57**, 721–739 (2018)

6. Gupta, N., Gupta, V.: Punjabi text classification using naive bayes , centroid and hybrid approach. In: Proceedings of WSSANLP, pp. 109–122 (2012)
7. Hall, M., Frank, E., Holmes, G., Pfahringer, B., Reutemann, P., Witten, I.H.: The weka data mining software: an update. SIGKDD Explor. **11**, 10–18 (2009)
8. Islam, M.S., Jubayer, F.E.M., Ahmed, S.I.: A support vector machine mixed with tf-idf algorithm to categorize bengali document. In: Proceedings of ICECCE, pp. 191–196 (2017)
9. Kabir, F., Siddique, S., Kotwal, M.R.A., Huda, M.N.: Bangla text document categorization using stochastic gradient descent (sgd) classifier. In: Proceedings of CCIP, pp. 1–4 (2015)
10. Lai, S., Xu, L., Liu, K., Zhao, J.: Recurrent convolutional neural networks for text classification. In: Proceedings of AAAICAI, pp. 2267–2273 (2015)
11. Mandal, A.K., Sen, R.: Supervised learning methods for bangla web document categorization. Int. J. Artif. Intell. Appl. **05**, 93–105 (2014)
12. Patil, J.J., Bogiri, N.: Automatic text categorization: Marathi documents. In: Proceedings of ICESA, pp. 689–694 (2015)
13. Rakholia, R.M., Saini, J.R.: Classification of Gujarati documents using naive bayes classifier. Indian J. Sci. Technol. **10**(5), 1–9 (2017)
14. Swamy, M., Thappa, M.: Indian language text representation and categorization using supervised learning algorithm. Int. J. Data Min. Tech. Appl. **02**, 251–257 (2013)

Closed-Set Device-Independent Speaker Identification Using CNN

Tapas Chakraborty, Bidhan Barai, Bikshan Chatterjee, Nibaran Das,
Subhadip Basu and Mita Nasipuri

Abstract Speaker Identification(SI) has numerous applications in real world.
Traditional classifiers like Gaussian Mixture Models (GMM), Support Vector
Machine (SVM), and Hidden Markov Models (HMM) were used earlier for SI. Fea-
tures like Mel Frequency Cepstral Coofficient (MFCC), and Gammatone Frequency
Cepstral Coefficients (GFCC) need to be generated first. But these approaches do
not perform well when audio data captured through multiple devices and recorded
in different environments, i.e., in mismatch condition. Whereas Machine Learning
(ML) algorithms usually provide better accuracy, and hence became more popular. Restricted Boltzmann Machine(RBM), Long-Short-Term Memory (LSTM), and
Convolutional neural network (CNN) are some of the ML approaches applied on SI.
In this paper, CNN is used for automatic feature extraction and speaker classification
on IITG-MV noisy dataset. CNN performs better than GMM, specially for device
mismatch case.

Keywords SI · MFCC · GFCC · GMM · SVM · HMM · RBM · LSTM · CNN

T. Chakraborty (✉) · B. Barai · B. Chatterjee · N. Das · S. Basu · M. Nasipuri
Jadavpur University, Kolkata, India
e-mail: ju.tapas@gmail.com

B. Barai
e-mail: bidhan.barai@jadavpuruniversity.in

B. Chatterjee
e-mail: bchatterjee7980@gmail.com

N. Das
e-mail: nibaran.das@jadavpuruniversity.in

S. Basu
e-mail: subhadip.basu@jadavpuruniversity.in

M. Nasipuri
e-mail: mita.nasipuri@jadavpuruniversity.in

© Springer Nature Singapore Pte Ltd. 2020   291
V. Bhateja et al. (eds.), *Intelligent Computing and Communication*,
Advances in Intelligent Systems and Computing 1034,
https://doi.org/10.1007/978-981-15-1084-7_28

1   Introduction

Closed-set Speaker Identification(SI) is the process to identify a speaker from his voice (audio signal) by comparing it against a set of known voices. SI systems perform well for clean speech, when audio is recorded without any noise. However, noise can not be avoided in real-life scenario. As a result unwanted frequencies may present in audio data along with speaker-specific frequencies, for which identification becomes challenging. Moreover, audio capturing devices have their own propertics. Difference in sampling rate, frequency range, and bandwidth will be there if same audio speech is recorded with different devices. So device mismatch makes identification process more challenging.

Before the invention of neural networks, traditional classifiers like Gaussian Mixture Models (GMM) [1], Support Vector Machine (SVM) [3], and Hidden Markov Models (HMM) [9] were used as classifiers. Those classifiers require prior knowledge and human effort in feature design. Mel Frequency Cepstral Coefficient (MFCC) and Gammatone Frequency Cepstral Coefficients (GFCC) [2] features, extracted from audio signals, were used as input to traditional classifiers. In recent years, i-Vector became state-of-the-art technique, and RBM has been used for Speaker Recognition using i-Vector [5]. With the advance of Neural Networks, Long-Short-Term Memory (LSTM) and Convolutional neural network (CNN) were applied in speaker recognition domain. Convolutional Neural Network [8] does not require any human effort in feature extraction as it automatically extracts features from the data.

In this work, CNN model is used to extract useful features of audio signal so that voice recognition becomes more accurate. Initially, audio data has been preprocessed to remove noise and silent parts. Log-spectrograms are then generated from those processed audio signals. Python signal analysis library **Librosa** [10] is used for generating Log-spectrogram. Log-spectrograms are then used as input to CNN. Initially, CNN is trained using audio files of known speakers. Then it is tested with unknown speaker audio files.

Below is the block diagram of overall process (Fig. 1).

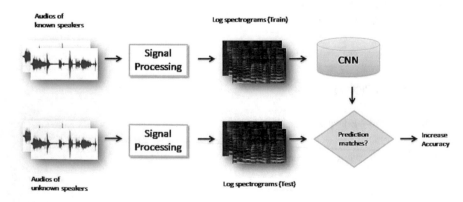

Fig. 1   A block diagram of overall process

CNN is implemented on IITG-MV dataset [7]. Indian Institute of Technology Guwahati Multi-variability Speaker Recognition (IITG-MV SR) database has five phases. In this work, only phase-1 data is used. It contains speech data in *.wav* format from hundred speakers. Each speaker spoke in multiple styles (reading and conversation); in multiple languages (one in English and another in their favorite language); in various sessions and environments. Audio was recorded using five different devices (tablet, phone, etc.) which makes this dataset exceptional.

Both device matched and device mismatched situations are tested and CNN performed better than traditional GMM-based approach in both of these cases. Rest of the paper is organized as follows: Preprocessing of audio signal is described in Sect. sec:2. Third section describes CNN architecture. Experiments, results, and conclusions are discussed in the remaining sections.

2 Audio Preprocessing

First step of the experiment is to decide what should be the input for CNN. Processed data like MFCC should not be used as input to CNN, as CNN needs to extract features by itself. Rather raw audio signal should be used as input. However, Dieleman et al. [4] showed that CNN performs better when Log-spectrogram is given as input rather than raw audio signals. Audio signals are preprocessed to remove noise and silent parts.

2.1 Preemphasis

Audio speech signal is passed through a High Pass Filter (HPF) to increase the amplitude of higher frequencies. It also removes low-frequency noises. If $\mathscr{S}(n)$ is the audio signal, preemphasis can be done using below equation [1].

$$\mathscr{S}(n) = \mathscr{S}(n) - \alpha * \mathscr{S}(n-1) \tag{1}$$

Here α is a parameter whose value is chosen as 0.97.

2.2 Silence Frame Removal

To remove silence regions of input audio signal, Short-Time Signal Analysis (STSA) is performed. Output of HPF, i.e., preemphasized signal is divided into several time frames of short duration (window of 20 ms with 10 ms overlap). To identify the silence frames, energy of those frames is compared with average frame energy of audio signal [1].

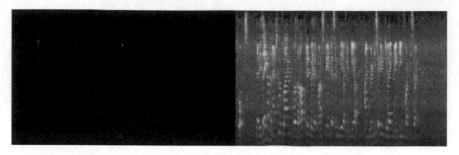

Fig. 2 Diagram of Mel-Spectrogram (left) and Log-spectrogram(right) generated from audio data

$$E_{avg}(\mathbf{x}) = \Sigma_i \left(\frac{|A(\mathbf{x})^2|}{N} \right) \tag{2}$$

If $E_i > k * E_{avg}$ for a specific frame, then that frame is considered as voiced or active frame otherwise considered as silence frame. Here, k is a parameter whose value was experimentally chosen as 0.2.

2.3 Log-Spectrogram

Mel-Spectrograms (left) are then generated on preprocessed audio data. Mel-Spectrogram indicates that samples from the dataset are not significant enough to be fed into CNN network without audio processing. That is the reason why Log-spectrogram (right) is used to overcome this problem. Log-spectrogram indicates that it can efficiently extract voice characteristics of each speaker from audio signals. If s is the audio signal, Log-spectrogram is computed using the below formula.

$$L(s) = \log(1 + \beta * s) \tag{3}$$

Parameter β was experimentally chosen as 10000 (Fig. 2).

3 CNN Architecture

Convolutional neural network (CNN) is a class of deep neural networks which uses a variation of multilayer perceptron. CNN is also known as shift and space invariant neural networks, due to their shared-weights architecture and translation invariant characteristics. CNN requires less preprocessing compared to other algorithms. This indicates that CNN has the capability to learn key features automatically. Figure 3 shows architecture of custom CNN model that we have used in this paper.

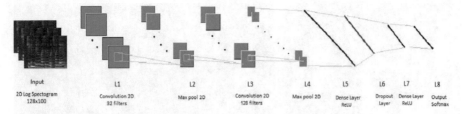

Fig. 3 Block diagram of CNN Architecture

Apart from the custom CNN model, standard CNN model, Resnet is also implemented. Resnet gave similar accuracy. Custom CNN model performs faster than Resnet, hence, chosen for this experiment.

Parameter tuning is a crucial step for any deep learning experiments. Several approaches mentioned by Jiuxiang Gu et al. [6] has been taken into consideration. Kernel of size 23 gave best performance, padding set to half of kernel size and stride is set to 1. Max-pooling is used as pooling. Dropout layer is used to prevent overfitting with a rate of $n = 0.5$.

4 Experiment, Results, and Discussion

4.1 Data Preparation

IITG-MV database is the source for this experiment, which has hundred speakers. Each speaker has two audio files, one in conversation style and the other one in reading style. CNN requires balanced training data so that each class has equal contribution in overall loss calculation. However, IITG-MV database has unbalanced data. Some speaker has extremely short audio (around one-tenth of other speakers). Lowest duration is noted and signals of only that duration are taken from all speakers, rest ignored.

In earlier GMM-based experiment, conversation style data was used as training while reading style data was used for testing. However, that approach cannot be followed in CNN, as conversation style data is significantly less compared to corresponding reading style data for some speakers. These two issues make the situation challenging as deep learning-based approaches require considerable amount of data to train model properly. So we have mixed conversation and reading style data, and then performed cross-validation on entire dataset. Figure 4 shows detail flow diagram of this experiment.

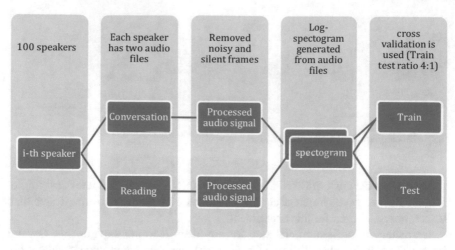

Fig. 4 A block diagram of data preparation

4.2 CNN Model Training

CNN is implemented using pytorch. During training, data was picked randomly to train the model. Training data was of the form (X_i, Y_i) where X_i is input data for ith speaker of shape 128*100*3 and Y_i is input label for ith speaker. Objective of the training is to minimize overall training loss with respect to all speakers.

4.3 Identification Using CNN

Let there are n speakers $\mathscr{S} = \{1, 2, 3, \ldots\ldots, n\}$. Output layer of CNN has n nodes, one for each speaker. When a test speaker data is given into CNN model, a vector with n scores will be returned as output. ith scores indicate probability of that unknown speaker to become to ith speaker. Maximum score is considered in this case. Decision rule of speaker identification is given below

$$\hat{S} = \arg\max_{k \in \mathscr{S}} (p(\mathbf{x}_i)) \tag{4}$$

Here $\hat{S}$ is the identified speaker and ith speaker's score is given by $p(\mathbf{x}_i)$, formula given below. The identified speaker $\hat{S}$ has the maximum score.

$$p(\mathbf{x}_i) = \frac{e_i^x}{\sum_1^n e_k^x} \tag{5}$$

4.4 Performance Measure

The accuracy is measured by the percentage of correct identification, equation given below:

$$Accuracy(\%) = \Sigma_i \left(\frac{Number\ of\ speakers\ correctly\ classified}{Total\ number\ of\ Speakers} \right) * 100 \quad (6)$$

Performance of this system can be measured by confusion matrix. Confusion matrix is nothing but a table with two dimensions "actual" and "predicted", and similar sets of "classes" in both dimensions.

Two popular performance metrics, precision, and recall have also been used to measure the performance of this system. Precision is the fraction of events where speaker i was correctly identified out of all instances where the system declared i. Conversely, recall is the fraction of events where speaker i was correctly identified out of all of the cases where the true scenario is i.

4.5 Results

Device dependent case: Training and testing are done on audio data of same device. IITG-MV database has data for five different devices, DVR (D01), Headset (H01) Tablet PC (T01), Nokia 5130c (M01), and Sony EricssonW350i (M02) So five accuracy figures are there, one for each device (Table 1).

Device-independent case: In this case, data captured by all of these five devices are mixed. Same experiment is performed again on this combined dataset and accuracy is noted. Maximum accuracy reported earlier was 64. CNN accuracy is significantly high compared with earlier accuracy (Table 2 and Fig 5).

Table 1 SI accuracy for device dependent scenario for IITG-MV SR

Device	NumSpeakers	GMM accuracy	CNN accuracy	Resnet accuracy	CNN precision	CNN recall
D01	100	96	97	97	0.9813	0.9780
H01	100	93	98	97	0.9888	0.9860
T01	100	91	97	96	0.9792	0.9760
M01	100	95	95	95	0.9583	0.9555
M02	100	90	92	92	0.9484	0.9380

Table 2 SI accuracy for device-independent scenario for IITG-MV SR

Device	NumSpeakers	GMM accuracy	CNN accuracy	Resnet accuracy	CNN precision	CNN recall
All	100	64	90	90	0.9187	0.9135

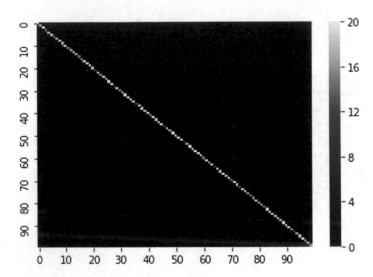

Fig. 5 Confusion matrix for device-independent case

5 Conclusion

CNN-based approach gave significantly high accuracy in device-independent speaker identification in noisy dataset. Experimental setup is close to real-life scenario, where speech data can be recorded in any environment or device or any language. So we can conclude that CNN-based approach can successfully be applied for real-world voice recognition. Though hyperparameter values can further be tuned to improve analysis on short frames. IITG-MV audio database has many real-life scenarios like environment mismatch, language mismatch, session mismatch, etc. In this paper, only device mismatch scenario was tested. Experiment needs to be done for other scenarios.

Acknowledgements This project is partially supported by the CMATER laboratory of the Computer Science and Engineering Department, Jadavpur University, India, TEQIP-II, PURSE-II, and UPE-II projects of Govt. of India. Subhadip Basu is partially supported by the Research Award (F.30-31/2016(SA-II)) from UGC, Government of India. This work is also supported by the project sponsored by SERB (Government of India, order no. SB/S3/EECE/054/2016) (dated 25/11/2016)

References

1. Barai, B., Das, D., Das, N., Basu, S., Nasipuri, M.: An ASR System Using MFCC and VQ/GMM with Emphasis on Environmental Dependency (2018)
2. Barai, B., Das, D., Das, N., Basu, S., Nasipuri, M.: Closed-set text-independent automatic speaker recognition system using VQ/GMM. In: Intelligent Engineering Informatics, pp. 337–346. Springer (2018)
3. Campbell, W.M., Sturim, D.E., Reynolds, D.A.: Support vector machines using GMM super-vectors for speaker verification. IEEE Signal Proc. Lett. **13**(5), 308–311 (2006)
4. Dieleman, S., Schrauwen, B.: End-to-end learning for music audio. In: 2014 IEEE International Conference on Acoustics, Speech and Signal Processing (ICASSP), pp. 6964–6968. IEEE (2014)
5. Ghahabi, O., Hernando, J.: Restricted boltzmann machines for vector representation of speech in speaker recognition. Comput. Speech Lang. **47**, 16–29 (2018)
6. Gu, J., Wang, Z., Kuen, J., Ma, L., Shahroudy, A., Shuai, B., Liu, T., Wang, X., Wang, G., Cai, J., et al.: Recent advances in convolutional neural networks. Pattern Recognit. **77**, 354–377 (2018)
7. Haris, B., Pradhan, G., Misra, A., Shukla, S., Sinha, R., Prasanna, S.: Multi-variability speech database for robust speaker recognition. In: 2011 National Conference on Communications (NCC), pp. 1–5. IEEE (2011)
8. Jumelle, M., Sakmeche, T.: Speaker clustering with neural networks and audio processing, arXiv preprint arXiv:1803.08276 (2018)
9. Madikeri, S., Bourlard, H.: KL-HMM based speaker diarization system for meetings. In: 2015 IEEE International Conference on Acoustics, Speech and Signal Processing (ICASSP), pp. 4435–4439. IEEE (2015)
10. McFee, B., Raffel, C., Liang, D., Ellis, D.P., McVicar, M., Battenberg, E., Nieto, O.: librosa: Audio and Music Signal Analysis in Python (2015)

A Novel and Efficient Spatiotemporal Oxygen Production Estimation Based on Land Vegetation Using PNN and Convolutional Autoencoder

Anish Saha, Rajdeep Debgupta and Bidyut Baran Chaudhuri

Abstract Oxygen is a sensitive indicator of atmospheric compositional changes and is also a primary requirement for human life. With deforestation on the rise, dwindling level of oxygen concentration in the air is a concern for mankind. Our paper aims to estimate the oxygen production levels of a particular area of land captured by satellite imagery. Our algorithm aims to identify forests and agricultural land patches and analyzes the oxygen production in a very efficient manner. Further, the algorithm takes in images of leaves from these areas and processes them to identify the species, chlorophyll content, and nitrogen levels in the plant using feature selection and probabilistic neural networks (PNN). We have computed the oxygen level of the patch of land and these calculations were performed on images spanning over few years allowing us to calculate the changes in oxygen production. This not only helps to map the carbon footprint, but also acts as a curative measure for global warming.

Keywords Convolution · Feature extraction · Image analysis · Autoencoders · Validation · Remote sensing · Precision agriculture · Oxygen production · Spatiotemporal variations

A. Saha (✉)
School of Computer Science and Engineering, Vellore Institute of Technology, Vellore, India
e-mail: anishsaha12@gmail.com

R. Debgupta
School of Electronics Engineering, Vellore Institute of Technology, Vellore, India
e-mail: raj_debgupta98@outlook.com

B. B. Chaudhuri
CVPR Unit, Indian Statistical Institute, Kolkata, India
e-mail: bbc@isical.ac.in

© Springer Nature Singapore Pte Ltd. 2020
V. Bhateja et al. (eds.), *Intelligent Computing and Communication*,
Advances in Intelligent Systems and Computing 1034,
https://doi.org/10.1007/978-981-15-1084-7_29

301

1 Introduction

Atmosphere warming-related changes in climatic carbon dioxide (CO_2) content have pulled in overall consideration. Nonetheless, another change in environmental substance synthesis, a slow decrease in oxygen (O_2), has been to a great extent disregarded, despite the fact that it straightforwardly undermines all life on earth. The atmosphere mainly consists of three gases: nitrogen (78%), oxygen (21%), and other gases (1%). Oxygen is present in two forms: atmospheric oxygen and dissolved oxygen in water. Atmospheric oxygen is one of the most important factors in sustaining life on earth. Forests (plants) convert the CO_2 to O_2 [1] and with dwindling forest cover, there is a rise in carbon dioxide concentration which is harmful for sustaining the life on earth and it is contributing to global warming. Our paper aims to use image processing and machine learning techniques to estimate the change of vegetation over an area of study, and hence calculate the amount of oxygen released by it.

To achieve this, we have collected the satellite images from remote sensing satellite coupled with GIS (geographic information system) over a particular period of time. We used imagery from the LANDSAT satellites (the longest running enterprise for acquisition of satellite imagery of earth launched on July 23, 1972) [2] accessible through the "Earth Explorer" website [11]. Each Landsat scene is about 115 miles long and 115 miles wide.

In this paper, we have used PNN classification technique to classify the leaves of the particular sample of plants. PNN is probabilistic neural network which is a feedforward neural network which is widely used in classification and pattern recognition problems. By this method, the probability of misclassification is minimized [1]. This type of ANN was derived from the Bayesian network and a statistical algorithm called Kernel Fisher discriminant analysis.

The aim of the investigation is to simulate the worldwide yearly oxygen generation of vegetation somewhere in the range of 1999–2017 by assessing land vegetation, and afterward changing over these qualities into oxygen creation dependent on photosynthesis and breath conditions.

The section is organized into five sections. Section 1 gives a general introduction of the problem of global warming and the overview of methodology. Section 2 gives a brief history of work done in this area. Section 3 describes the proposed system to estimate oxygen production and Sect. 4 discusses the results. Section 5 is the conclusion portion where we have concluded our remarks.

2 Literature Survey

The components utilized in this paper have been extensively studied individually over the years in the domain of image processing and remote sensing. Leaf recognition for plant classification here is done using probabilistic neural network. Previous work on classifications of plant is done basically from the descriptions of botanists. The paper implements a leaf recognition algorithm using easy-to-extract features and highly efficient recognition algorithm. All features are extracted from digital leaf image. 12 features were orthogonalized by principal component analysis (PCA). As to classifier, we used probabilistic neural networks (PNN) for its simple structure and easy execution [1]. Convolution neural network was also used for plant identification and classification. Features of images are extracted, segmented, analyzed, and classified to determine the type of crop to be grown. This is then fed to a universal model to generate the results [2].

Land cover classification and forest change studies, such as the case study of Area of Zagros Mountain in Iran, employed a methodology to map and monitor the forest cover change using multitemporal Landsat thematic mapper (TM) over the period of 1990, 1998, 2006 [3]. The images obtained were classified using the Anderson land-use/cover classification system. In conjunction with geographic information system (GIS), [4] classifications scheme classified the land into forest land, rangeland, water bodies, agricultural land, and residual land using unsupervised image classification. This paper calculates the area of forest cover from the images collected from GIS during a given period of time and calculates the area on the basis of changes happening during that period of time.

With the formation of ICUN, there was a compliance of Red List which contained all the endangered flora and fauna around the world [5]. With the help of the list, we are able to identify the species of plant which are in urgent need for growth, which has different beneficiaries which in turn help in the growth of fauna around the region, also taking crops into consideration [6]. The images obtained from GIS require classification and identification of vegetation cover. Using several image processing techniques, the remote sensing satellite imagery underwent enhancement, transformation, and classification [7]. Enhancement techniques are applied to preprocessed data in order to effectively display the image for visual interpretation.

There have been multiple techniques to estimate oxygen production but hardly any make use of the intrinsic properties of a particular area, such as the naturally flourishing fauna and other factors such as sunlight availability. We therefore propose an efficient system to estimate oxygen using these intrinsic properties.

3 Proposed System

The aim of the study is to simulate the global yearly oxygen generation by vegetation between the range of years 1999 and 2017, by assessing land vegetation, and thereafter converting these quantities into oxygen production in the particular area dependent on photosynthesis rate using chlorophyll content [8], breath conditions, and lighting.

The paper is divided into three modules, these being land cover detection and forest cover change analysis using satellite imagery (LANDSAT), leaf recognition for plant classification in a particular area using probabilistic neural network, spatiotemporal changes, and drivers of global land vegetation oxygen production over 20 years (Fig. 1).

3.1 Forest Cover Detection and Forest Change

To detect the vegetation and forest cover in an area, the preprocessing involved is creation of a mask to filter out the green and greenish tinge from the satellite images. Cloud cover free images must be chosen to effectively [9] estimate forest cover change [10].

The colored intensity image is first converted to double precision, rescaling the data if necessary. Then, a 3 × 3 mask is created by looking where the blue and red channels are dark and the green channel is high.

Next, the green section of image is threshold, making use of the mask created to convert the three-channel image into a binary image with green pixels marked ON and the rest as background. The number of ON pixels is counted and converted into land area in km^2 using appropriate scaling factor as obtained from the map.

The forest cover change is detected by taking the threshold image of a place over a period of time and the intersection provides the area of no change [11]. The original minus intersection region gives the area of forests cover lost.

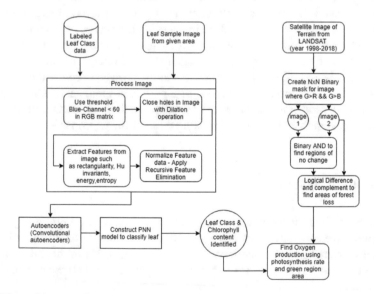

Fig. 1 The architecture of proposed system

3.2 Dimensionality Reduction

For the purpose of dimensionality reduction, we have used convolutional autoencoders on the dataset satellite images. The architecture of the convolutional autoencoder is as follows (Fig. 2).

In the result section, it has been shown that out of the autoencoder, deep autoencoder and convolutional autoencoder, it is the convolutional autoencoder which outperforms all others, for this dataset. The purpose for using convolutional autoencoder is that the generalization of performance of convolutional layers is usually higher than all fully connected layers of the network, and it appears that convolutional autoencoder can reconstruct accurately on all samples of images of the dataset.

3.3 Leaf Recognition for Plant Classification

It is an algorithm to identify leaves [12, 13] from a set of different leaves taken from the Flavia project dataset, and also to calculate its chlorophyll and nitrogen.

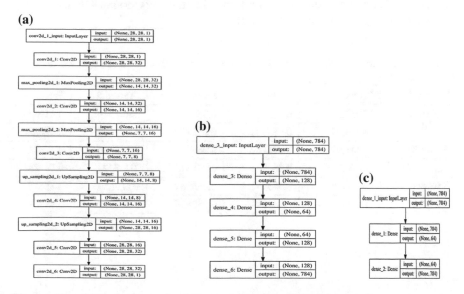

Fig. 2 a Convolutional autoencoder. **b** Deep autoencoder. **c** Normal autoencoder

---

Algorithm 1: PNN-Leaf-Classify

---

input: Captured 3-channel (RGB) leaf image: *original_image*
geometrical_shape = *original_image*.B < 60
leaf = complement(*geometrical_shape*)
leaf = closing_operation(*leaf*)
geometrical_features = regionprops(*leaf*) //a matlab function
textural features = graycoprops(*leaf*) //matlab function implementing GLCM
data = *geometrical_features*+ *textural_features*
data = max-min-normalization(*data*)
train_data, test_data = split(*data*, 0.7)
lamda = 0.011
foreach *class* **in** *train_data* **do**
    | theta[*class*] = onevsAll(*train_data, class*)
end
model = predictOneVsAll(theta, *test_data*)
accuracy = modelAccuracy(*model, test_data*)
return *accuracy*

The relevant features extracted [14] are aspect ratio, rectangularity, convex area ratio, eccentricity, diameter, zernike moment, Hu invariants, contrast, homogeneity, correlation, energy, and entropy (Table 1).

Table 1 This is used for the PNN classification of leaves. First 6 features displayed out of 12 features extracted

Feature 1	Feature 2	Feature 3	Feature 4	Feature 5	Feature 6
3.960147	13023.26	0.983861	0.967593	256.2539	0.419857
3.995767	10244.09	0.985981	0.968177	228.2926	0.434812
3.870087	12041.07	0.976162	0.96604	243.5836	0.424918
4.555877	8170.765	0.983721	0.975613	217.707	0.382781
3.988951	13409.04	0.985391	0.968067	260.9656	0.38615
3.297212	18434.36	0.981891	0.952899	278.1907	0.457599

Algorithm 2: Chlorophyll-Estimation

---

    input: Captured 3-channel (RGB) leaf image: *original_image*
    geometrical_shape = *original_image*.B > 90
    leaf_shape = complement(*geometrical_shape*)
    foreach *channel* **in** *original_image* **do**
        foreach *pixel* **in** *leaf_shape* **do**
            if *pixel*.color *!= black*
                sum[channel] += *original_image.pixel*.color
            end
        end
    end
    R = mean*(sum[*RED*])*
    G = mean*(sum[*GREEN*])*
    B = mean*(sum[*BLUE*])*
    Chlorophyll = G - (R/2) - (B/2)
    return *Chlorophyll*

Algorithm 3: Nitrogen-Estimation

---

input: Captured 3-channel (RGB) leaf image: *original_image*
geometrical_shape = *original_image*.B > 90
leaf_shape = complement(*geometrical_shape*)
foreach *channel* **in** *original_image* **do**
    foreach *pixel* **in** *leaf_shape* **do**
        if *pixel*.color *!= black*
            sum[channel] += *original_image.pixel*.color
        end
    end
end
R = mean*(sum[*RED*])*/255
G = mean*(sum[*GREEN*])* /255
B = mean*(sum[*BLUE*])* /255
max = max(R, G, B)
min = min(R, G, B)
if *max* = R **then**
    *Hue = ((G-B)/(max-min))*60*
end
else if *max* = G **then**
    *Hue = (((B-R)/(max–min)) + 2) *60*
end
else if *max* = B **then**
    *Hue = (((R-G)/(max-min)) + 4) *60*
end
saturation = (max - min)/ max
    Brightness = max
Nitrogen = ((H - 60)/60 + (1 - S) + (1 - B))/3
return *Nitrogen*

3.4 Spatiotemporal Changes in Global Oxygen Production

Photosynthesis is the procedure in which chlorophyll changes CO_2 and water (H_2O) into $C_6H_{12}O_6$ and O_2 within the sight of daylight. Green plants create OC by means of photosynthesis; the sum delivered per unit region and time is alluded to as essential efficiency (GPP). In the meantime, in any case, green plants additionally devour $C_6H_{12}O_6$ and O_2, and discharge CO_2 through respiration. The OC that remains, when conducting autotrophic respiration, is alluded to as NPP; while the OC that remains, while considering both autotrophic and heterotrophic respiration, is alluded to as NEP - which is figured by means of the oxygen production of plants. Along these lines, day-by-day oxygen creation was figured first with the goal that a yearly esteem could be assessed by increase.

Algorithm 4: Estimate-Oxygen-Level

---

    input: Captured 3-channel (RGB) leaf image: *leaf,*
           Captured 3-channel (RGB) LANDSAT image: *original_image*
   chloro = Chlorophyll-Estimation(*leaf*)
   area = sum(Green-Pixels(*original_image*))
   chlr = (chloro/1000)*10   //amt of ug of chlorophyll per 50 sq.cm
   Oxygen = chlr *9*365*15.652114/0.005*area   //mm³/year/m²
   // chlorophyll = multiply respiration rate with land area assuming 9 hours
   of sunlight in a day
   return *Oxygen*

---

4 Experimental Results and Discussions

In order to simulate the global yearly oxygen generation by vegetation between the range of years 1999 and 2017, by assessing land vegetation, the three sections are forest cover detection and change, leaf recognition for plant classification in an area, and spatiotemporal changes in oxygen production.

5 Forest Cover Detection and Change

1. The input image in this module is the change in the forest cover in four areas from 1999 to 2017 (LANDSAT images) (Figs. 3, 4, 5, 6).

   Output: As seen in the images, the forest cover is identified in green and the change is noted in the figures shown for Area 1 (Fig. 7).

5.1 Leaf Recognition for Plant Classification

Input images are the photos of various leaves of different plants taken from the Flavia leaf dataset (100 leaf taken—10 classes).

Input images are shown in Fig. 8.

The processed images used for feature extraction are shown in Fig. 9.

This output is obtained after the input leaf images from dataset and is converted into a monochromatic image.

Next, these images are used to extract relevant features: aspect ratio, rectangularity, convex area ratio, eccentricity, diameter, zernike moment, Hu invariants, contrast, homogeneity, correlation, energy, and entropy.

5.2 Autoencoder Performance Comparisons

The sample index relates to the losses computed on the validated image samples. The sample of epoch losses at stage 5 and 10 is as follows: (Figs. 10 and 11).

The performance of convolutional autoencoder with higher epochs stabilizes with respect to deep autoencoder and the use of convolutional autoencoder minimizes loss with less computational complexity.

Area 1: Semi urban area- Kansas, USA

| (a) | (b) | (c) | (d) |

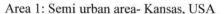
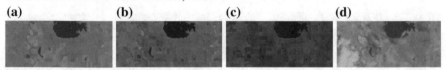

Fig. 3 The forest cover as observed on **a** September 2, 1999. **b** October 19, 2005. **c** November 07, 2010. **d** August 03, 2017

Area 2: Manhattan, USA

| (a) | (b) | (c) | (d) |

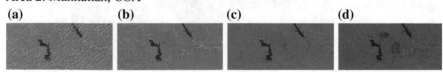

Fig. 4 The forest cover as observed on **a** August 08, 1999. **b** October 19, 2001. **c** December 13, 2005. **d** November 2, 2014

Area 3: Urban- Bangalore, India

(a) (b) (c) (d)

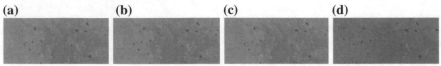

Fig. 5 The forest cover as observed on **a** November 09, 1999. **b** May 21, 2005. **c** May 21, 2010. **d** March 10, 2018

Area 4: Forest- Amazon rainforest

(a) (b) (c) (d)

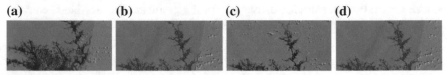

Fig. 6 The forest cover as observed on **a** September 16, 1999. **b** November 11, 2005. **c** Jan 11, 2010. **d** August 2, 2018

(a) (b) (c) (d)

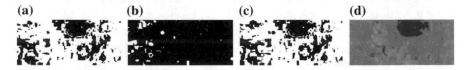

Fig. 7 Vegetation in **a** 1999—Thresholded image. **b** 2010—Thresholded image. **c** loss over 10 years. **d** Loss mask on original image

Fig. 8 Four sample images from the dataset used consisting of 100 leaf images

Fig. 9 Output of Algorithm 1 steps 1–4

Fig. 10 Loss at fifth epoch

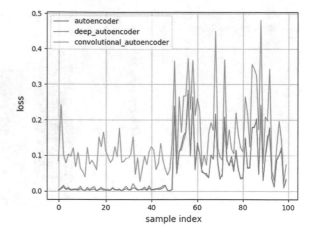

Fig. 11 Loss at tenth epoch

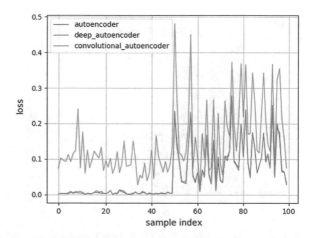

5.3 Spatiotemporal Changes in Global Oxygen Production

Using the outputs of chlorophyll and land area from the previous algorithms output, we use them as inputs for the third module and calculate the spatiotemporal changes in global oxygen production using the algorithm 4 (Figs. 12, 13, 14 and Table 2).

6 Conclusion

The results show that forest cover in urban areas has decreased drastically; however, in the crop fields and forest region, it has remained constant thus giving an overall slight decreasing rate of oxygen production. Advantages of PNN include PNNs are much faster than multilayer perceptron networks, PNNs can be more accurate

Fig. 12 Oxygen production in Manhattan area 1 over 20 years

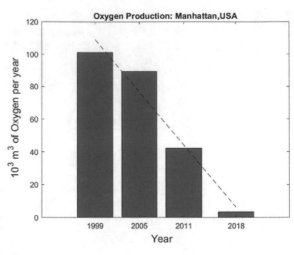

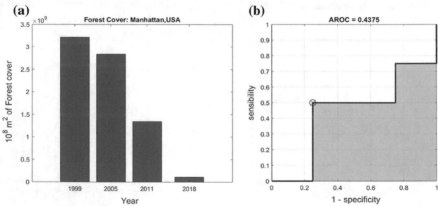

Fig. 13 **a** Forest cover in Manhattan Area 1 over a period of 20 years. **b** ROC for forest cover prediction in Area 1

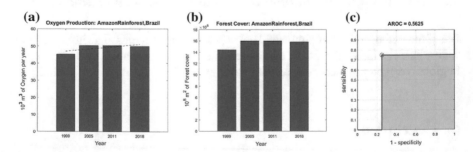

Fig. 14 **a** Oxygen production in Amazon Area 4 over 20 years. **b** Forest cover in Amazon Area 4 over 20 years period. **c** ROC for forest cover prediction in Area 4

Table 2 Efficiency of algorithm for two areas—prediction with linear regression

	Manhattan—Area 1	Amazon—Area 2
Distance	0.5590	0.3536
Threshold	82.9806	49.6602
Sensitivity	0.5000	0.7500
Specificity	0.7500	0.7500
AROC	0.4375	0.5625
Accuracy	62.5000%	75.0000%
PPV	66.6667%	75.0000%
NPV	60.0000%	75.0000%

than multilayer perceptron networks, and PNN networks are relatively insensitive to outliers, especially helpful as image thresholding gives multiple outliers.

In conclusion, we can say that this research has been beneficial in many sectors of interest such as agriculture and environment preservation. Through the land area of green trees estimation, it was possible to account for the deforestation. Knowing the oxygen output of a region enabled us to predict which kind of plant species to be grown in the area for maximizing the oxygen output. In future, this research has scope for application for town and urban planning while taking into account the environmental interests of the region.

References

1. Wu, S.G., Bao, F.S., Xu, E.Y., Wang, Y.X., Chang, Y.F., Xiang, Q.L.: A leaf recognition algorithm for plant classification using probabilistic neural network. In: 2007 IEEE International Symposium on Signal Processing and Information Technology, pp. 11–16. IEEE (2007, December)
2. Abdullahi, H., Sheriff, R., Mahieddine, F.: Convolution neural network in precision agriculture for plant image recognition and classification. In: 2017 Seventh International Conference on Innovative Computing Technology (Intech), pp. 1–3. IEEE, Londrés (2017, August)
3. Weng, Q.: Land use change analysis in the Zhujiang Delta of China using satellite remote sensing, GIS and stochastic modelling. J. Environ. Manag. **64**(3), 273–284 (2002)
4. Zhang, L., Zhang, B., Li, W., Li, X., Sun, L., Jiang, L., Liu, X.: Spatiotemporal changes and drivers of global land vegetation oxygen production between 2001 and 2010. Ecol. Ind. **90**, 426–437 (2018)
5. Tracewski, Ł., Butchart, S.H., Di Marco, M., Ficetola, G.F., Rondinini, C., Symes, A., Buchanan, G.M.: Toward quantification of the impact of 21st-century deforestation on the extinction risk of terrestrial vertebrates. Conserv. Biol. **30**(5), 1070–1079 (2016)
6. Karcher, D.E., Richardson, M.D.: Quantifying turfgrass color using digital image analysis. Crop Sci. **43**(3), 943–951 (2003)
7. Sowmya, D.R., Shenoy, P.D., Venugopal, K.R.: Remote sensing satellite image processing techniques for image classification: a comprehensive survey. Int. J. Comput. Appl. 161(11) (2017)
8. Ali, M.M., Al-Ani, A., Eamus, D., Tan, D.K.: An algorithm based on the RGB colour model to estimate plant chlorophyll and nitrogen contents. In: International Proceedings of Chemical, Biological and Environmental Engineering, vol. 57 (2013)

9. Congalton, R.G., Green, K.: Assessing the Accuracy of Remotely Sensed Data: Principles and Practices Boca Rotan. Lewis Publishers, Florida (1999)
10. Yang, X.: Satellite monitoring of urban spatial growth in the Atlanta Metropolitan area. Photogramm. Eng. Remote Sens. **68**(7), 725–734 (2002)
11. Jensen, J.R.: Digital Change Detection. Introductory Digital Image Processing: A Remote Sensing Perspective. Prentice-Hall, New Jersey (2004)
12. Chaki, J., Parekh, R.: Plant leaf recognition using shape based features and neural network classifiers. Int. J. Adv. Comput. Sci. Appl. (IJACSA) **2**(10) (2011)
13. Bama, B.S., Valli, S.M., Raju, S., Kumar, V.A.: Content based leaf image retrieval (CBLIR) using shape, color and texture features. Indian J. Comput. Sci. Eng. **2**(2), 202–211 (2011)
14. Ab Jabal, M.F., Hamid, S., Shuib, S., Ahmad, I.: Leaf features extraction and recognition approaches to classify plant. J. Comput. Sci. **9**(10), 1295 (2013)

A Faster Fuzzy Clustering Approach for Recommender Systems

Rajdeep Debgupta, Anish Saha and B. K. Tripathy

Abstract A recommender system has very important role nowadays, be it business, e-commerce, search engines, entertainment, etc. The need for faster, dynamic, and efficient recommender system arises with huge data on the Internet or website platform. A movie or anime recommender system has a major role in delivering improved entertainment. These recommender systems can provide much personalized recommendations, suggestion to a particular user based on one's watching habits, or other similar user's interests, ratings. Plentiful of recommendation techniques has been proposed but most are not able to provide useful recommendation within a very short span of time. In this paper, we aim to propose a fuzzy-clustering-based recommender system which is almost quite efficient and accurate as collaborative filtering (CF) technique but much faster than CF. We have achieved an improvement of approximately 4 s faster than CF techniques. Experimental data justifies the efficiency of the system.

Keywords Recommender systems · Collaborative filtering · Singular value decomposition · Fuzzy c-means · Ranking · User item · K-neighborhood clustering

1 Introduction

In today's world, we are flooded with data and its volume is increasing massively. Going back in the year of 1982, John Naisbitt said, "We are drowning in information but starved for knowledge" [1]. With growing data, we are enriched with information

R. Debgupta (✉)
School of Electronics Engineering, Vellore Institute of Technology, Vellore, India
e-mail: raj_debgupta98@outlook.com

A. Saha
School of Computer Science and Engineering, Vellore Institute of Technology, Vellore, India
e-mail: anishsaha12@gmail.com

B. K. Tripathy
School of Information Technology and Engineering, Vellore Institute of Technology, Vellore, India
e-mail: tripathybk@vit.ac.in

© Springer Nature Singapore Pte Ltd. 2020
V. Bhateja et al. (eds.), *Intelligent Computing and Communication*,
Advances in Intelligent Systems and Computing 1034,
https://doi.org/10.1007/978-981-15-1084-7_30

but we still lack to deduce valuable insights and knowledge out of it. In a website, e-commerce, or library, there are tons of products and books; however, it is still difficult for one to get the valuable items they actually want. A faster and efficient recommendation system can benefit both parties. In the research field for recommendation system, numerous approaches have been created [2]. Collaborative filtering [3] has proved one of the most accurate and fruitful approach both in research and practice.

Manos and Dimitris [4] had proposed random prediction algorithm, where items were chosen randomly and those corresponding items would be suggested to the user. As the algorithm works on random picking of items, it was quite not beneficial for any user or provider to take into consideration. Another approach was sequences in frequency, where customer or user would be recommended products on the basis of buying habits or number of items bought by that particular customer. The drawback was that if a user was new to the platform and had not bought anything before, then no efficient recommendation could be made for that user. Another approach is content-based recommendation [5] where items will be recommended on the basis of user's liking in the past. This algorithm works as a kind of search query for related items. Items history are stored and the one which has higher degree of correlation among items of user's preferences. It recognizes and analyzes the descriptions for items that are of particular interest to a user [6]. Chunhui et al. [7] proposed k-means approach and artificial bee colony method which showed better positive results and performance on experimental datasets. In the case of security issues in collaborative filtering methods, Lu et al. [8] had proposed a model for collaborative filtering which is dedicated to the user security threat minimizations and has shown positive recommendations and performances.

One of the major problems in recommendation systems is that an item cannot be added into recommendation metrics unless otherwise it has been rated by considerable number of users and it is disadvantageous for users of diverse tastes. This leads to a problem known as cold start. Yang et al. [9] had come up with a fused solution for this cold start problem in their hybrid recommender system for news recommendations. Yuan et al. [10] proposed deep learning models for collaborative filtering for massive datasets with a wide variance of cold start problem.

Sparsity problem in recommender system arises due to the result of rating that is very small in number of the total number of items available in the database in the presence of abundant highly active users. Huang et al. [11] had proposed a solution to overcome this sparsity problem through associative retrieval framework and related dispersal algorithms to discover transitive associations among users through their past habits, comments, and transactions.

Scalability is an important issue for recommendation systems especially when the user–item databases are massive. Although recommendation seems to be efficient and relevant, expensive nonlinear computational complexity and time result in overall degradation of the entire model. Cosley et al. [12] had proposed an open framework for experimental testing of recommender systems in an effort to deliver a standard for public to evaluate recommendation models for real-life circumstances.

This paper is categorized into four sections. Section 1 gives description of the dataset. Section 1.2 defines the data preprocessing steps involved. In Sect. 2, we put forward our methodology. In Sect. 3, we have given insights on our results and comparison, and finally Sect. 4 is the conclusion portion where we have concluded our remarks and discussed about the future scope for further improvement and scope of making the model faster.

1.1 Description of the Dataset

The data has been taken from myanimelist.net [13]. It is a website where users can rate anime and create profiles to write about the anime they have seen/watched or wish to watch in the future, and the anime they disliked or likely to drop. The number of users consisted is 9868 and for each user, the corresponding anime list is picked containing various sections and integer ratings ranging from 1 to 10 and 0 in case of non-rating values. The rating values of users for items have been taken into account for generating recommendations and optimizing the computations.

1.2 Data Preprocessing

The train–test anime has been split up to build five train and five test user–item files by taking the list of anime of each and every user. Scraping has been done to extract anime ID and status of user such as complete duration, currently watching, on hold, surfed-planned, or dropped. Subsequently, user–item files have been built for each user with additional field for average rate of the user. A user-cluster matrix has been generated which contains each user's probability value of belonging to each cluster. Aimed for each individual, these probabilities have been computed in the manner that for each anime in the list of user, the corresponding ranking (i.e., say status of anime "not planned") and sum to the result of the product of probability of the particular anime belonging to that cluster are valid. If an anime has not been dropped, ranking is equal to the mean of the ratings of user, else the rating is considered maximum between two and the mean rating divided by two.

2 Methodology

In this collaborative filtering [14], we have tried to recommend a user-based prediction. We have taken the rating or tastes of users who have liked the kinds of anime, movie, and forming a cluster that are among nearest neighbor and on this basis we make recommendation to a particular designated user. For any user A, we find a set of N number of different users whose ratings are similar with user A. Then, we estimate rating of user A, on the basis of N number of users. For similarity index measure, we have used centered-cosine similarity among users and items (anime/movies) so as to get rid of nonzero missing ratings which is a drawback of cosine similarity measure and values ignored in case of Jaccard similarity measure. The centered cosine gives better intuition for handling "tough raters"—those users who give usually less ratings and "easy raters"—those who are very generous in giving ratings.

The update of missing value for the user–item matrix is given by

$$A_i = \left(\sum_{i=1}^{n} A\right)/y - Ai,$$
(1)

where y is the number of existing values in the row matrix and n is the total number of columns.

We take k-nearest neighbor [15] on the user–item collaborative filtering matrix to avoid large computations for large number of users as described below. The recommendation equation used is

$$r_{u,i} = \overline{r_u} + k \sum_{u' \in U} \text{simil}(u, u')\left(r_{u',i} - r_{u'}^-\right),$$
(2)

where U is the set of all users in the matrix, $r_{u,i}$ is the "u" user's score of ith anime/movie, and r_u^- is the average score rating of user u. For the similarity function simil(A, B), A and B be two users, the way used for k-nearest neighbor for classification is

$$\text{simil}(A, B) = \begin{cases} 1 \text{ if } b \in N(A) \\ 0 \text{ if } b \notin N(B) \end{cases},$$
(3)

where N(A) belongs to the neighbors of user A and N(B) belongs to the neighbors of user B.

$$k = 1/\sum_{u \in U} \left|\text{simil}(u, u')\right|.$$
(4)

We retrieve the recommendation with highest values. But this method will eventually result in large computations and hence will become very slow when the number of users becomes very high in number. As a result, we add a neighbor-based searching that does not thoroughly compute the exact neighbors, but gets k number of users that are adequately closer to our targeted point. The distance of first those k number of users whose value is less than a specified threshold is considered.

Threshold = (average distance of k nearest neighbor) × relaxation factor.

To determine the threshold value, we take tests on approximately 1200 users and we have got average distance of nearest neighbors rounded off to 0.6. The threshold value is set as 0.6×1.1, i.e., 0.66.

Fuzzy clustering: When the number of users increases, traditional collaborative filtering algorithms will be inefficient due to scalability issues. One of the major hindrances in collaborative filtering is dimensionality reduction, as the CF is very sensitive to changes in the number of users. So, for fuzzy clustering algorithm in our model, the dimensionality reduction approach we have used is singular value decomposition (SVD) [16]. The SVD definition is given by

$$A_{max} = U_{[max]} \times \sum_{[rxx]} \times \left(V_{[nxr]}\right)^{T}, \tag{5}$$

where A: input data matrix of $m \times n$, where (for instance, m movies, n users)

U   Left singular vectors of size (m $\times$ r),
$\sum$ Singular values, r $\times$ r diagonal matrix (r is the rank of matrix A),
V   Right singular vectors (n $\times$ r matrix), and

U, $\sum$, V are unique, U and V are orthonormal, and $\sum$ is a diagonal matrix.

On this, we have the user–item matrix, say M of size (m $\times$ n), i.e., M $_{m \times n}$ where there are m number of items and n number of binary feature and corresponding every movie/anime has a binary array mapping to its features like genre, type, etc.

Fuzzy clustering C-means [17] is performed on the binary matrix for the purpose of soft clustering, i.e., assigning probabilistic membership value to the cluster points (items), which eventually gives an item–cluster matrix (M $\times$ C), where C is the number of clusters. The utility matrix (n $\times$ m), n be the number of users, i.e., user–item matrix and the item–cluster matrix are combined. The user–cluster matrix is computed by weighted mean, where each value represents the probability values of a user belonging to a particular cluster. Now, K-means is performed on the user–cluster matrix and prediction is done accordingly as mentioned above in the section of collaborative filtering (Fig. 1).

3 Results and Figures

For evaluation of our model, we have used root mean squared error (RMSE) and mean absolute error (MAE) measurement during the training period to observe which parameters perform better. The two error metrics are chosen for the fact that RMSE penalizes large errors and MAE is chosen for steady error measurements when the variance associated with the error distribution changes. The mathematical formulas are as follows:

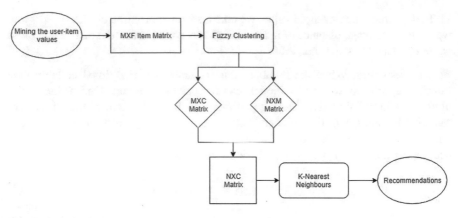

Fig. 1 The illustration of the entire architecture

$$\text{RSME} = \sqrt{\frac{\sum\limits_{i=1}^{n}\left(y_i' - y_i\right)}{n}}, \qquad (6)$$

n   number of recommendations,
y_i' the predicted rating of the movie/anime "i",
y_i the real rating a user provided to the movie/anime "i".

$$\text{MAE} = \frac{1}{n}\sum_{i=1}^{n}|f_i - y_i| = \frac{1}{n}\sum_{i=1}^{n}|e_i|, \qquad (7)$$

f_i the predicted rate of movie/anime "i", and
y_i the real rating a user provided to the movie/anime "i".

In the training period, the quality of recommendations is computed by number of neighbors for collaborative filtering and in case of fuzzy c-means, the number of clusters has been taken into consideration. The best parameter is chosen by looking into RSME values on test sets. The number of clusters for the recommender system has been fixed to 60 on the basis of fuzzy clustering. The comparisons of RSME and MAE with respect to number of neighbors of collaborative filtering (CF) in comparison with fuzzy c-means (FCM) clustering are plotted in Figs. 2 and 3.

The recommender system tests have been done by fixing the number of neighbors 12. The RMSE and MAE comparison with number of clusters on fuzzy clustering are as follows in Figs. 4 and 5.

Algorithm speed comparison: The model has been trained and tested on system of Core-i7-8th gen, Clock-2.2 GHz–4.1 GHz, Slot-6 core. GPU: NVIDIA Geforce GTX1050Ti. The comparison of time taken by CF and FCM with respect to number of neighbors is as follows in Fig. 6.

Fig. 2 RMSE of CF versus FCM

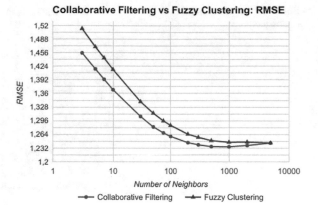

Fig. 3 MAE of CF versus FCM

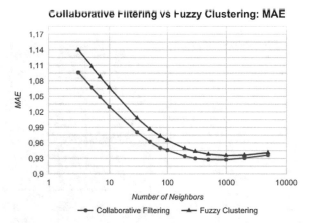

Fig. 4 RMSE versus number of clusters

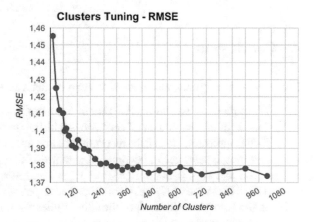

Fig. 5 MAE versus number of clusters

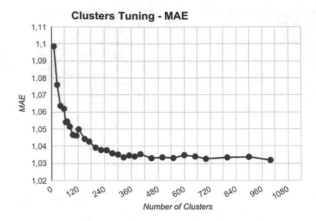

Fig. 6 Time (ms) versus number of neighbors

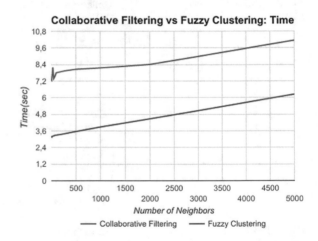

4 Conclusions

From the above figures, the conclusions can be derived as follows:

(i)   The collaborative filtering (CF) is a little bit more accurate than fuzzy clustering.

(ii) The recommendation quality enhances with the increase in number of neighbors K. It is quite notable that the recommendation enhances for values of K till 1000, and then it saturates or becomes constant till K = 5000 (at K = 50, RMSE = 1.282 and at K = 5000, RMSE = 1.238) and after it starts decreasing again. This high value of K considers cases for large number of users in the matrix.

(iii) In the fuzzy clustering algorithm, the higher the number of cluster we use, the better the performance of recommender system we can achieve.

(iv) The final and most significant observation is that the fuzzy clustering algorithm is faster than the collaborative filtering approach. In the graph, it is noticed that there is an improvement of 4 s and not depending on the number of clusters or neighbors. This difference in time is because of the clustering on K-neighborhood-based clustering on the basis of relatively smaller matrix, while, on the other hand, the collaborative approach is done on a relatively bigger matrix.

Even if the collaborative filtering-based recommender system is a bit better than fuzzy clustering one, the fuzzy clustering recommender system is much faster, and hence it can be considered as the better one than collaborative approach.

Future work: Further improvement can be made in similarity index measures between ratings of users and for matrix factorization of dimensionality measures, and loss convergence approaches can be implemented. Clustering on the basis of association rule mining can also be implemented.

References

1. Larose, D.: Discovering Knowledge in Data. Wiley Interscience (2008)
2. Deshpande, M., Karypis, G.: Item-based top-n recommendation algorithms. ACM Trans. Inf. Syst. **22**(1), 143–177 (2004)
3. Miyahara, K., Pazzani, M.J.: Collaborative filtering with the simple bayesian classifier. In: Pacific Rim International Conference on Artificial Intelligence (2000)
4. Papagelis, M., Plexousakis, D.: Qualitative analysis of user-based and item-based prediction algorithms for recommendation agents. Eng. Appl. Artif. Intell. **18**(7), 781–789 (2005)
5. Van Meteren, R., Van Someren, M.: Using Content-Based Filtering for Recommendation
6. Pazzani, M., Billsus, D.: Content-Based Recommendation Systems, pp. 325–341 (2007)
7. Ju C., Xu C.: A new collaborative recommendation approach based on users clustering using artificial bee colony algorithm. Sci. World J. (2013)
8. Lu Z., Shen H.: A security-assured accuracy maximised privacy preserving collaborative filtering recommendation algorithm. In: Proceedings of the 19th International Database Engineering & applications Symposium, pp. 72–80 (2015)
9. Yang W., Tang R., Lu L.: A fused method for news recommendation. In: International Conference on Big Data and Smart Computing, pp. 341–344 (2016)
10. Yuan J., Shalaby W., Korayem M., Lin D., AlJadda K., Luo J.: Solving cold-start problem in large-scale recommendation engines: a deep learning approach. In: IEEE International Conference on Big Data, pp. 1901–1910 (2016)
11. Huang, Z., Chen, H., Zeng, D. Applying associative retrieval techniques to alleviate the sparsity problem in collaborative filtering. ACM Trans. Inf. Syst. 22(1) (2004)
12. Cosley, D., Lawrence, S., Pennock, D.M.: REFEREE, an open framework for practical testing of recommender systems using Research Index. In: Proceedings of the 28th Very Large Data Bases Conference (2002)
13. Myanimelist, the World's Largest Anime and Manga Database and Community
14. Lee, J., Sun, M., Lebanon, G.: A Comparative Study of Collaborative Filtering Algorithms. arXiv:1205.3193v1 [cs.IR] (2012)

15. Guo, G., Wang, H., Bell, D.A., Bi, Y., Greer, K.: KNN Model-Based Approach in Classification. CoopIS/DOA/ODBASE (2003)
16. Sarwar, B., Karypis, G., Konstan, J., Riedl, J.: Application of Dimensionality Reduction in Recommender System—A Case Study. Minnesota Univ Minneapolis Dept of Computer Science (2000)
17. Bezdek, J.C., Ehrlich, R., Fill, W.: FCM: The Fuzzy c-Means Algorithm. Pergamom Press Ltd (1984)

Saliency Detection for Semantic Segmentation of Videos

H. Vasudev, Y. S. Supreeth, Zeba Patel, H. I. Srikar, Smita Yadavannavar, Yashaswini Jadhav and Uma Mudenagudi

Abstract There has been remarkable progress in the field of Semantic segmentation in recent years. Yet, it remains a challenging problem to apply segmentation to the video-based applications. Videos usually involve significantly larger volume of data compared to images. Particularly, a video contains around 30 frames per second. Segmentation of the similar frames unnecessarily adds to the time required for segmentation of complete video. In this paper, we propose a contour detection-based approach for detection of salient frames for faster semantic segmentation of videos. We propose to detect the salient frames of the video and pass only the salient frames through the segmentation block. Then, the segmented labels of the salient frames are mapped to the non-salient frames. The salient frame is defined by the variation in the pixel values of the background subtracted frames. The background subtraction is done using MOG2 background subtractor algorithm for background subtraction in various lighting conditions. We demonstrate the results using the Pytorch model for semantic segmentation of images. We propose to concatenate the semantic segmentation model to our proposed framework. We evaluate our result by comparing the time taken and the mean Intersection over Union (mIoU) for segmentation of the video with and without passing the video input through our proposed framework. We evaluate the results of Saliency Detection Block using Retention and Condensation ratio as the quality metrics.

H. Vasudev · Y. S. Supreeth · Z. Patel (✉) · H. I. Srikar · S. Yadavannavar · Y. Jadhav · U. Mudenagudi
KLE Technological University, Hubballi, India
e-mail: zeba99.patel@gmail.com

H. I. Srikar
e-mail: srikarindiresh@gmail.com

U. Mudenagudi
e-mail: uma@kletech.ac.in

© Springer Nature Singapore Pte Ltd. 2020   325
V. Bhateja et al. (eds.), *Intelligent Computing and Communication*,
Advances in Intelligent Systems and Computing 1034,
https://doi.org/10.1007/978-981-15-1084-7_31

1 Introduction

Semantic Segmentation is one of the important applications in the field of computer vision. Computer's version of visual perception is known as visual scene analysis. To analyze a visual scene, the computer must identify objects and various relationships between the objects, labeling each correctly. The process of associating each pixel of an image with a class label is described by semantic segmentation. Semantic segmentation is one of the key applications in Image Processing and Computer Vision. It finds its applications in Medical Imaging and Autonomous Driving systems. In recent years, there have been many methods proposed pertaining to Semantic Segmentation and efforts have been made in improving the accuracies.

There are many Deep Neural Network architectures, which are designed for semantic segmentation among which Enet, SegNet, ResNet, ContextNet, and BlitzNet have better accuracy. Enet architecture is specifically for tasks which requires low latency operation [6]. Segnet is a core trainable segmentation network. This consists of an encoder network and a corresponding decoder network with a pixel-wise classification layer [8]. There are many methods proposed to detect the salient frames of a given video. Padmavathi Mundur et al. proposed a method for Video Summarization using Delaunay Triangulation [2], which generates video summaries by capturing the visual content of the original videos in fewer frames. An Automatic Video Summarization and Quantitative Evaluation [5] proposed by Sandra et al. The K-means clustering algorithm is a simple method for estimating the mean (vectors) of a set of K-groups. These methods are mainly based on low level features such as color Histogram, edge Histogram, and frame correlation. Moratov et al. in [7] propose a method that operates with segments rather than with separate pixels. In case of saliency detection, there are two possible ways of applying segmentation: (i) by computing a saliency map and deriving an average saliency value over a segment, and (ii) by computing directly the saliency value of each segment. Yi et al. in [11] propose a deep learning architecture for saliency detection by fusing pixel-level and superpixel-level predictions. It investigates an elegant route to make two-level predictions based on a same simple fully convolutional network via seamless transformation. Bhateja et al. in [4] propose a robust algorithm based on median filtering for denoising where each and every frame in the video is extracted and robust decision is taken to selectively operate upon the corrupted pixels of the frame. The local statistical parameters are used to decide whether to restore the center pixel with median filter or with adaptively increment the kernel size of the filter for further operation. This process helps in restoration of structural content with minimal blurring at high noise densities present in the frame. Nguyen et al. in [3] propose a robust algorithm for feature selection using 1D discrete wavelet transform for video face recognition. An approach for optimal feature selection based on Min-max Ant system algorithm has been proposed based on 1D-DWT. The length of the culled feature vector is adopted as heuristic information. The optimal features are selected based on the shortest subset in terms of shortest feature-length and having best performance. Pablo et al. in [1] aim at providing solution for two major problems in computer vision domain, Contour detection, and Image segmentation. In Contour detection combination of multiple local cues into a globalization framework based on spectral clustering is used. Here Segmentation algorithm consists of generic machinery for transforming the output of detected contour into a hierarchical region tree which is a novel approach. In this manner, the paper aims at reducing the problem of image segmentation to that of contour detection in computer vision. In [9, 10], authors propose to use motion features and Gaussian mixture model to define saliency to arrive at the video summarization.

- We propose to consider the repeated frames in the video as non-salient frames. We propose a method to reduce the time taken for the segmentation of non-salient frames.
- We propose a saliency detection block to detect the salient frames present in the video based on changes in the position and values of pixels.
- We propose to map the segmented labels of the salient frames to the non-salient frames.

2 Proposed Approach

We propose a framework that builds upon the architecture "Resnet50" ResNet is one of the architectures for Semantic Segmentation of images, but adapting the same for processing of repeated frames limits its use in for its application in videos. The temporal change between the frames varies for various videos. The temporal consistency does not change immediately for every frame. We propose to use this concept to reduce the time for segmentation of frames in a video. Figure 1 represents our framework.

2.1 Saliency Detection Block

The foreground-background of the frame is extracted using MOG2—a robust background subtraction model for lighting conditions. We propose to extract the contours from the background subtracted frames in order to know the number objects present in the frame. We decide the number of objects by computing the area of the detected

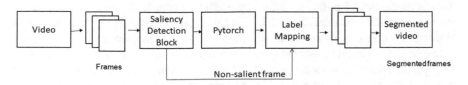

Fig. 1 Proposed framework

Fig. 2 Saliency detection
block

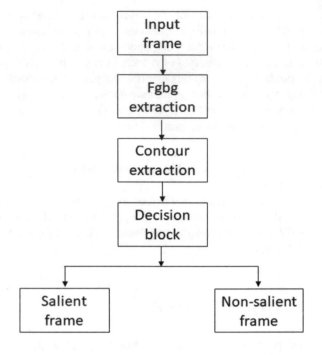

contours. If the area is greater than a heuristically set threshold, then we consider the object pertaining to the contour for the object count, else we neglect it. We propose to segment only the unique frames obtained from the decision block and map the label of the unique frames to the repeated frames.

Decision Block We compute the difference between the two frames by comparing the pixel values. If the object in the first frame has changed its position in the second frame, we propose to track the number of pixels of the second frame that is different from the first and then count the number of pixels. If this number of pixels is greater than a heuristically set threshold, the frame is considered to be unique (Fig. 2).

3 Pytorch Implementation Using Resnet50-dilated8-ppm-bilinear-deepsup

This implementation is mainly based on encoder-decoder architecture. The basic architecture used is ResNet50 which has 50 convolution layers. Along with this, it has eight dilated convolution layers. It uses pyramid pooling for sampling. The encoder network takes an input image and generates a high-dimensional feature vector. Later it aggregates features at multiple levels. For downsampling in the encoder network it uses methods such as Max-pooling, Average Pooling, and Strided Convolution. Then the images are sent to the decoder network where the decoder network takes the high-dimensional feature vector and generates a semantic segmentation mask. For upsampling in the decoder network methods such as Un-pooling and Deconvolution are used. Then it decodes the features aggregated by encoder at multiple levels. In the convolution blocks used, the filter weights are learned from the data.

4 Implementation Details

Hardware Used All the experiments were performed on a system with NVIDIA GEFORCE 840 M 2GB.

Softwares and Libraries OS-UBUNTU 16.04, CUDA 9.0 and supporting Cudnn 7.2.1, Keras 2.1.4, Tensorflow gpu 1.2, Pytorch.

Implementation ResNet50 architecture has been implemented using Pytorch on GPU with train accuracy and test accuracy. The implementation of the architecture is extended from images to video. The process of Semantic Segmentation took about 8 seconds per frame on GPU.

4.1 Results and Analysis

We demonstrate the results using the pretrained Pytorch model for semantic segmentation of images. We propose to concatenate the semantic segmentation model to our proposed framework. We evaluate our result by comparing the time taken for segmentation of the video with and without passing the video input through our proposed framework. We evaluate the results of Saliency Detection Block using Retention ratio as the quality metrics. Mean IoU of the Pytorch Implementation is 42.66. We test the same model by concatenating the same model with our proposed framework and obtain the same mIoU. The Retention ratio remains 100% for all the static videos. We compare the results of the state-of-the-art with and without the pipeline and observe that the mean IoU remains the same in both the cases, but the time taken for segmentation process reduces when we concatenate our framework.

Retention ratio is the ratio of number of objects in the summarized video to the number of objects in the original video.

$$RR = \frac{number\ of\ objects\ in\ summarized\ video}{number\ of\ objects\ in\ input\ video}$$

Condensation ratio is the ratio of length of summarized video to length of the input video.

$$CR = \left(1 - \frac{length\ of\ summarized\ video}{length\ of\ input\ video}\right) * 100$$

Table 1 Comparison of time taken for a video to get segmented with and without passing through the proposed framework

Resolution	State-of-art ResNet50		Proposed framework	
	Performance	Accuracy in mIoU	Performance	Accuracy in mIoU
	GPU	42.66%	GPU	42.66%
512 × 256	6 s/frame		6 s/frame	
1024 × 512	8 s/frame		8 s/frame	
Video (30 fps) (1024 × 512)	40.2 min (512 × 256)		13.6 min (512 × 256)	

Below table describes the results obtained for a video of duration 10 minutes of resolution 1024 × 512 and also segmentation results of individual frames of two different resolutions. The input video taken consists of 302 frames and by concatenating the saliency detection block, we observe that only 87 frames to be segmented. This reduces the segmentation time (Table 1).

4.2 Conclusion

In this paper, we propose a contour detection-based approach for detection of salient frames for faster semantic segmentation of videos. We propose to detect the salient frames of the video and pass only the salient frames through the segmentation block. Then, the segmented labels of the salient frames are mapped to the non-salient frames. The salient frame is defined by the variation in the pixel values of the background subtracted frames. The background subtraction is done using MOG2 background subtractor algorithm for background subtraction in various lighting conditions. We demonstrate the results using the ResNet50 model for semantic segmentation of images. We propose to concatenate the semantic segmentation model to our proposed framework. We evaluate our result by comparing the time taken and the mIoU for segmentation of the video with and without passing the video input through our proposed framework. We evaluate the results of Saliency Detection Block using Retention and Condensation ratio as the quality metrics. We observe that the retention ratio is always 100% for any input video. The condensation ratio depends on how dynamic the input video is, since the framework proposed mainly depends on the change in pixel values of the frames.

References

1. Arbelaez, P., Maire, M., Fowlkes, C., Malik, J.: Contour detection and hierarchical image segmentation. IEEE Trans. Pattern Anal. Mach. Intell. **33**(5), 898–916 (2011)
2. Badrinarayanan, V., Kendall, A., Cipolla, R.: Segnet: a deep convolutional encoder-decoder architecture for image segmentation. CoRR (2015). abs/1511.00561
3. Bao. L., Le, D.-N., Nhu, N., Bhateja, V., Satapathy, S.: Optimizing feature selection in video-based recognition using max-min ant system for the online video contextual advertisement user-oriented system. J. Comput. Sci. **21**, 361–370 (2017)
4. Bhateja, V., Malhotra, C., Rastogi, K., Verma, A.: Improved decision median filter for video sequences corrupted by impulse noise. In: 2014 International Conference on Signal Processing and Integrated Networks (SPIN), February (2014)
5. de Avila, S.E.F., da_Luz, A., Araújo. A.D., Cord, M.: Vsumm: an approach for automatic video summarization and quantitative evaluation. In: 2008 XXI Brazilian Symposium on Computer Graphics and Image Processing, pp. 103–110, October (2008)
6. Mundur, P., Rao, Y., Yesha, Y.: Keyframe-based video summarization using delaunay clustering. Int. J. Digit. Libr. 6(2), 219–232 (2006)
7. Muratov, O., Zontone, P., Boato, G., De Natale, F.G.B.: A segment-based image saliency detection. In: 2011 IEEE International Conference on Acoustics, Speech and Signal Processing (ICASSP), pp. 1217–1220, May (2011)
8. Paszke, A., Chaurasia, A., Kim, S., Culurciello, E.: Enet: a deep neural network architecture for real-time semantic segmentation. CoRR (2016). abs/1606.02147
9. Sujatha, C., Chivate, A.R., Ganihar, S.A., Mudenagudi, U.: Time driven video summarization using gmm. In: 2013 Fourth National Conference on Computer Vision. Pattern Recognition, Image Processing and Graphics (NCVPRIPG) (2013)
10. Sujatha, C., Mudenagudi, U.: Gaussian mixture model for summarization of surveillance videos. In: 2015 Fifth National Conference on Computer Vision, Pattern Recognition, Image Processing and Graphics (NCVPRIPG), pp. 1–4, December (2015)
11. Yi, Y. Su, L., Huang, Q., Wu, Z., Wang, C.: Saliency detection with two-level fully convolutional networks. In: 2017 IEEE International Conference on Multimedia and Expo (ICME), pp. 271–276, July (2017)

Real-World Anomaly Detection Using Deep Learning

Unnati Koppikar, C. Sujatha, Prakashgoud Patil and Uma Mudenagudi

Abstract In this paper, we have carried out a comparative study on two deep learning models for detecting real-world anomalies in surveillance videos. Anomalous event is the one which deviates from the normal behavior. The anomalies considered are related to thefts such as robbery, burglary, stealing, and shoplifting. A framework is set up using supervised learning approach to train the models using the weakly labeled videos. The deep learning models considered are VGG-16 and inception model which are trained with both anomalous and normal videos to detect any anomalous activity in the video frame. UCF-Crime dataset is used which comprises long, untrimmed surveillance videos. The deep learning models are evaluated using various metrics. The experimental results show that the Inception V3 model performs significantly better in detecting the anomalies as compared to the VGG-16 model with an accuracy of 94.54%.

Keywords Surveillance · Anomaly detection · Theft · Deep learning · Convolutional Neural Networks · VGG-16 model · Inception V3 model

U. Koppikar (✉) · C. Sujatha (✉)
Department of CSE, KLE Technological University, Hubballi, India
e-mail: unnatikoppikar@gmail.com

C. Sujatha
e-mail: sujata_c@kletech.ac.in

P. Patil
Department of MCA, KLE Technological University, Hubballi, India
e-mail: prpatilji@gmail.com

U. Mudenagudi
Department of ECE, KLE Technological University, Hubballi, India
e-mail: uma@kletech.ac.in

© Springer Nature Singapore Pte Ltd. 2020
V. Bhateja et al. (eds.), *Intelligent Computing and Communication*,
Advances in Intelligent Systems and Computing 1034,
https://doi.org/10.1007/978-981-15-1084-7_32

333

1   Introduction

In this paper, we focus on the problem of anomaly detection in surveillance videos. Surveillance videos capture different activities 24/7 and 365 days, based on which various applications are developed [1, 2]. It is a challenging task to detect rare activities in the videos. Anomalous event is the one which deviates from the normal behavior. It becomes one of the prime applications in the area of video surveillance. One of the critical tasks in video surveillance is to detect anomalous events such as criminal or illegal activities and traffic accidents. A survey [3] was carried out to learn how severe and rampant were the crimes such as theft, robbery, burglary, etc. that happened in India. The statistics showed that during the year 2015 (555 robbery cases, 12,592 theft cases, and 555 dacoity cases), the year 2016 (441 robbery cases, 14,619 theft cases, and 218 dacoity cases), in 2017 (415 robbery cases, 18,936 theft cases, and 27 dacoity cases) and in 2018 (89 robbery cases and 159 theft cases) were reported, respectively. A study [4] was also carried out on the number of thefts and robberies that happened in Indian Banks. The statistics showed that 337 cases, 334 cases, and 188 cases of robbery were reported by Bandhan Bank, State Bank of India, Bank of Baroda, respectively. This statistics motivates us to detect anomalies which are recorded by the surveillance cameras such that timely action can be taken. Therefore, to mitigate the loss of labor and time, there is a need for developing intelligent computer vision algorithms that can automatically detect an anomaly from the videos. Many researchers have carried out work in this field, Sultani et al. [5] presented a C3D network which is a fully connected layer (FC6) as the deep learning architecture and proposed multiple instance learning algorithm solution for anomaly detection and detected 13 real world. The work proposed by Hu et al. [6] shows how anomaly can be detected in industrial system. Several deep learning algorithms are applied to detect real-world anomalies [7]. Authors [8–10] have presented proposed ML algorithms in this direction. Background subtraction method [11] was used to detect violence in videos such as hitting with objects, kicking, fist fighting, etc. Karpathy et al. [12] presented a research work on large-scale video classification with convolutional neural networks. Detection of violence in crowded scenes is challenging [13–15] have addressed this issue. Survey on anomaly detection has provided insights to this issue [16].

The aim of a real-time anomaly detection system is to timely signal an activity which diverges from the normal behavior and identifies the frame where the anomaly has occurred. The single biggest challenge in anomaly detection is to find out which observations are truly anomalous observations, as the interpretation of the words "abnormal" or "normal" may frequently change. To find a solution to this challenge, we intend to develop a system by training the neural networks with normal and anomalous videos. So that the network will be able to learn the internal features of videos consisting of the events which will build the model that will accurately predict any kind of abnormality. We detect anomalies by the pipeline as shown in Fig. 1. We have carried out a study on two widely used deep learning models, namely, VGG-16 and Inception V3 models.

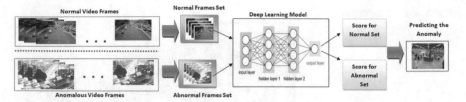

Fig. 1 Framework for the anomaly detection

We summarize our work as follows:

- We focus on detection of anomalies related to theft such as robbery, burglary, and shoplifting.
- We propose to build an anomaly detection framework using VGG-16 and Inception model and made a comparative study on their performance.
- We demonstrate the results on real-world anomaly detection using the UCF-Crime dataset and show the results achieved.

2 Framework for Anomaly Detection

In this section, we will discuss about the designed framework and implementation for the anomaly detection.

2.1 Approach

We present our framework for anomaly detection system as shown in Fig. 1. Initially, we create normal and abnormal sets by extracting 30 frames per second. These frames are then divided into training and testing set. These video frames are then subjected to one of the two models (VGG-16, Inception V3). The deep learning models have been trained and tested using Tensorflow framework. The process begins with the extraction of features from the input video frame. The visual features further get refined as they pass from the deeper layers.

As discussed in Sect. 2.3, for the VGG-16 model, visual features (4096D) are passed to the three-layer FC neural network. And as discussed in Sect. 2.4 Inception model bottleneck layers are used for feature extraction. The video frame is classified among one of the two classes (abnormal, normal) by providing the score against which the image is classified, as shown in Fig. 1.

Fig. 2 Feature extraction

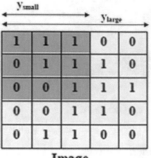

Image   Convolved Feature

2.2 Feature Extraction

The basic building block of the image is pixel. Assuming each pixel has 8-bit value, for a 640×480 image, it will have $640 \times 480 \times 8$ bits of information. To process such large information, we perform feature extraction as shown in Fig. 2, wherein we find out which parts of an image are peculiar, like lines, corners, and unique patches that can solely describe the image. These features are the unique property of the image. Features extracted are further used by the deep learning models to study the features based on which it can classify the image into the predictive class.

Given an image y_{large}, as shown in Fig. 2, the sparse autoencoder is trained on small patch y_{small} sampled from these images, which learns k features.

$$f = \sigma\big(W(1)y_{small} + b(1)\big), \tag{1}$$

where σ is the sigmoid function, $W(1)$ signifies weights, and $b(1)$ signifies biases of hidden units. For every patch y_{small} in the large image y_{large}, we compute Eq. (1) giving us $f_{convolved}$, $k \times (r - m + 1) \times (c - n + 1)$ array of convolved features.

2.3 VGG-16

In this section, we discuss in detail about the VGG-16 Model. VGG stands for Visual Geometry Group. VGG-16 helps in image localization and image classification.

The VGG-16 Model is built using convolution layers (3 * 3), Max pooling layers (used only 2 * 2 size), and fully connected layers at end. It comprises a total of 16 layers. At the end of convolutional layer, pooling layers are added which reduces the amount of data that will be sent to the next layer and increases the performance and reduces the training time, after which fully connected layers and Softmax layers are added as shown in Fig. 3. All hidden layers were equipped with the rectification nonlinearity. The rectified linear unit (ReLU) [17] is used as the activation function.

Fig. 3 VGG-16 model

Fig. 4 ReLU

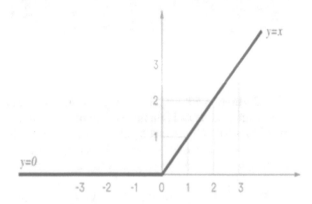

The function returns a value of 0 for any negative input and for any positive value x it returns the value.

Mathematically it can be defined as

$$f(x) = \max(0, x). \tag{2}$$

ReLU can be graphically indicated in Fig. 4. If ReLU function is nonlinear around 0, even then its slope is always either 0 (for negative values) or 1 (for positive values). In this way, it accounts for limited type of nonlinearity.

2.4 Inception V3

In this section, we discuss in detail about the Inception-V3 model. The Inception model is a deep convolutional architecture that was introduced as GoogLeNet. The model is constructed with a number of symmetric and asymmetric building blocks and it also includes convolutions, max pooling, average pooling, dropouts, concats, and fully connected layers. The basic architecture of Inception model is as shown in Fig. 5 from which it is evident that there are a variety of convolutions present specifically, and there are $1 \times 1, 3 \times 3,$ and 5×5 convolutions as well as with a 3×3

Fig. 5 Inception V3 model

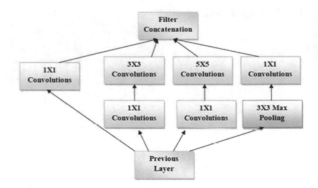

max pooling layers in this architecture. The larger convolutions are computationally expensive, and hence we consider to include 1×1 convolution for reducing the dimensionality of its feature map.

The Softmax output layer at the end employs the logits function. The Logistic Function also called as Sigmoid is an algorithm that uses linear equation with independent predictors to predict a value. This value can lie between $-\infty$ and $+\infty$. In order to restrict the value between 1 and 0, that is, 1 for anomalous event and 0 indicating normal event, the sigmoid function is used.

$$z = \theta_0 + \theta_1 \cdot x_1 + \theta_2 \cdot x_2 + \cdots. \tag{3}$$

$$g(x) = \frac{1}{1 + e^{-z}}. \tag{4}$$

The output z of Eq. 3 of the linear equation is given to the function $g(x)$ of Eq. 4 which returns a squashed value h, and the value h ranges from 0 to 1. This can be graphically visualized as shown in Fig. 6.

Fig. 6 Logistic function graph

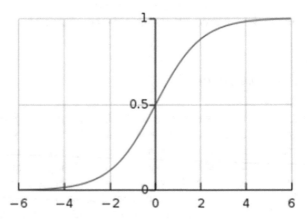

Fig. 7 Examples of anomalous frames in the dataset

Table 1 Number of videos in training dataset

Dataset	Category	No. of videos
Normal	Normal	20
Abnormal	Robbery	15
Abnormal	Shoplifting	5
Abnormal	Stealing	7
Abnormal	Burglary	3
	Total	50

2.5 Datasets

We have used UCF-Crime dataset for detecting the real-world anomalies. The dataset is constructed by the researchers which consists of lengthy surveillance videos which capture many real-world anomalies. The four real-world anomalies of our interest include burglary, robbery, stealing, and shoplifting [1]. These anomalies have been chosen as they have a compelling impact on social safety. For implementation, the number of anomalous videos considered is 30 anomalous video and 20 normal video sample which are shown in Fig. 7 and details of videos subjected for training are shown in Table 1.

3 Results and Discussions

In this section, we will discuss about the results obtained by our anomaly detection using different evaluation parameters as shown in Table 2. The system is built to recognize normal and anomalous events from the crime dataset using the two deep learning pretrained models, i.e., VGG-16 model and Inception V3 model.

We compare the performance of both the models as shown in Table 2. We infer that Inception V3 model has performed significantly better than VGG-16 model. Inception V3 model gives the error rate to be 0.054 which is close to the ideal benchmark of 0.0 compared to VGG-16 model. Similarly, Inception V3 model gives

Table 2 Comparison of results obtained from VGG-16 and Inception model

Parameter	VGG-16	Inception V3
Comparison based on evaluation metric		
Train accuracy	99.4%	99.7%
True positive	19	22
False positive	3	2
True negative	29	30
False negative	4	1
Testing accuracy	87.27	94.54
Error rate	0.1272	0.054
False positive rate	0.093	0.062
Specificity	0.906	0.9375
Recall	0.76	0.956
Comparison based on architecture		
Depth of the model	23	129
Parameter extracted	Comparatively less	More than VGG-16

recall and specificity to be 0.956 and 0.9375, respectively, which is close to the ideal benchmark of 1.0 compared to VGG-16 model.

We observe that the Inception model detected the abnormal events correctly as shown in Fig. 9i–k, whereas the VGG-16 model failed to detect as shown in Fig. 8a–c. For normal events, Inception model detected is as shown in Fig. 9l, m and VGG-16 failed to detect as shown in Fig. 8d, e. Further, in Fig. 8f, g and Fig. 9n, o, the events were detected correctly as abnormal by both the models and in Figs. 8h and 9p were detected correctly as normal events by both the models. However, the Inception V3 model has lower false alarm rate of 0.062 as compared to the VGG-16 model, and hence we infer that Inception V3 model performs better than VGG-16 model.

4 Conclusion

We have carried out a comparative study between the two deep learning models VGG-16 and Inception V3 for detecting real-world anomalies in surveillance videos. The system has been trained to detect anomalies related to thefts, such as robbery, burglary, stealing, and shoplifting by using a supervised learning approach. We proposed a framework for training the models to detect theft anomalous activity in the video frame and localize the anomaly. Through the evaluation metrics, it was evident that

Fig. 8 Results obtained from VGG-16 model

Fig. 9 Results obtained from Inception model

Inception V3 model performed significantly better in detecting the anomalies compared to VGG-16 model. Further, this work can be extended to detect different crime anomalies in real-time scenarios.

References

1. Sujatha, C., Akshay, C., Altaf, G., Uma, M.: Time driven video summarization using GMM. In: NCVPRIPG 2013, Jodhpur, ITT, pp. 1–4, December (2013)
2. Sujatha, C., Mudenagudi, U.: Gaussian mixture model for summarization of surveillance Video. In: NCVPRIG, December 2015, IIT Patna (2015)
3. https://economictimes.indiatimes.com/industry/transportation/railways/over-55000-thefts-reported-in-trains-in-last-3–5-years-railways-data/articleshow/65746357.cms
4. https://www.youthkiawaaz.com/2017/07/bank-robbery-india-least-safe-bank. Bank Robbery

5. Sultani, W., Chen, C., Shah, M.: Real-world anomaly detection in surveillance videos. In: CVPR, January (2018)
6. Hu, X., Subbu, R., Qiu, H., Iyer, N.: Multivariate anomaly detection in real-world industrial systems. In: IEEE Conference, June (2008)
7. Krizhevsky, A., Sutskever, I., Hinton, G.E.: Imagenet classification with deep convolutional neural networks. In: Advances in Neural Information Processing Systems, January (2012)
8. Wu, S., Moore, B.E., Shah, M.: Chaotic invariants of Lagrangian particle trajectories for anomaly detection in crowded scenes. In: 2010 IEEE Computer Society Conference on Computer Vision and Pattern Recognition, pp. 2054–2060 (2010)
9. Zhu, Y., Nayak, I.M., Roy-Chowdhury, A.K.: Context aware activity recognition and anomaly detection in video. IEEE J. Sel. Top. Signal Process. **7** (2013)
10. Zhicheng, J., Gao, X., Wang, Y., et al.: A deep feature based framework for breast masses classification. Neurocomputing **197**, 221–231 (2016)
11. Datta, A., Shah, M., Da Vitoria Lobo, N.: Person-on person violence detection in video data. In: ICPR, August (2002)
12. Karpathy, A., Toderici, G., Shetty, S., Leung, T., Sukthankar, R., Fei-Fei, L.: Large-scale video classification with convolutional neural networks. In: CVPR, April (2014)
13. Mohammadi, S., Kiani, H., Perina, A., Murino, V.: Violence detection in crowded scenes using substantial derivative. In: IEEE Conference, August (2015)
14. Kooij, J., Liem, M., Krijnders, J., Andringa, T., Gavrila, D.: Multi-modal human aggression detection. In: Computer Vision and Image Understanding (2016)
15. Mehran, R., Oyama, A., Shah, M.: Abnormal crowd behaviour detection using social force model. In: CVPR, June (2009)
16. Naik, A.J., Gopalakrishna, M.T.: Violence detection in surveillance video—a survey. In: IJLRET, pp. 11–17 (2017)
17. https://en.m.wikipedia.org/wiki/Rectifier_(neural_networks)

Automatic Facial Expression Recognition Using Histogram Oriented Gradients (HoG) of Shape Information Matrix

Avishek Nandi, Paramartha Dutta and Md Nasir

Abstract Automatic Recognition of Facial Expression (AFRE) has many potential applications in Human–Computer Interaction (HCI), security and surveillance, smart emotion-driven tutoring, etc. Human beings often display their emotions in various media of communication like language, voice, gesture, and facial expression. The contribution of facial expression is maximum among other means of communication. We have proposed a novel shape-oriented representation of facial structure using appearance information of the face. Active Appearance Model is used to select 25 salient landmark points on the face and a Shape Information Matrix (SIM) is formed by forming triangles on the face using landmark points. Histogram oriented Gradients feature of the Shape Information Matrix (SIM) is used to extract local orientation features of facial image. The proposed machine is trained using MultiLayer Perceptron (MLP) Neural Network and well tested over CK+, JAFFE, MMI, and MUG benchmark databases. Results obtained are impressive and encouraging.

Keywords Facial Expression Recognition · Histogram oriented Gradients · Active Appearance Model · Shape Information Matrix (SIM)

1 Introduction

Human emotion detection is a current trend of computing and has gained its popularity in recent days for real-world application by the research on primary emotions by Ekman and Friesen [13]. This paper conveys that the six basic facial expressions of

A. Nandi (✉) · P. Dutta (✉) · Md Nasir (✉)
Department of Computer and System Sciences, Visva-Bharati University,
Santiniketan 731235, India
e-mail: avisheknandi10@gmail.com

P. Dutta
e-mail: paramartha.dutta@gmail.com

Md Nasir
e-mail: nasir.vb@gmail.com

© Springer Nature Singapore Pte Ltd. 2020   343
V. Bhateja et al. (eds.), *Intelligent Computing and Communication*,
Advances in Intelligent Systems and Computing 1034,
https://doi.org/10.1007/978-981-15-1084-7_33

human being are having unique distinctive features in cross-culture study of human face. But the analysis of universality of facial expression on human and also on animals goes back to nineteenth-century Darwin's paper on "The Expression of the Emotions in Man and Animals" [11]. Distinguishing the rectangular area covering facial region in a given image plays a critical role in recognition of the facial expressions. The work in [15, 16] addresses this issue face recognition. In the pioneering work of Mase and Pentland [19] they have used optical flow-based skin movement extraction technique for automated facial expression recognition. Facial Action Unit Coding (FACS) is another delegate way of describing facial motions where facial expression is encoded with 44 Action Units (AUs) presented in the paper [14]. In contrast to FACS, Active Appearance-based Model (AAM) [8, 20] is a shape-based model modified to detect facial appearance structure. An AAM-based facial point model is used to detect facial expression by Datcu and Rothkrantz [12]. Histogram oriented Gradient (HoG) can also be used to extract facial appearance orientations represented by gradient magnitude by calculating grid separated local histogram on the face [9]. In the paper [15] authors have introduced a genetic algorithm-based approach for feature selection. Barman and Dutta has introduced a group of novel distance, shape and texture-based features for effective and efficient representation of facial shape/appearance relevant features in their innovative research on facial expression representation [2–7].

Here, we have suggested a novel Facial Expression Recognition System using Active Appearance Model (AAM) in combination with Histogram oriented Gradients (HOG). Active Appearance Model (AAM) is used to detect positions of eyes, eyebrows, nose, lips, and face in an image by plotting salient landmark points over the facial regions. The AAM model generates 68 landmark points represented in cartesian coordinate and we have selected only 25 salient landmark points out of them for our purpose. Triangles are formed on the face by selecting any three points from the 25 salient landmark points. Shape Information Matrix (SIM) is calculated by forming all possible single triangles over the face using salient landmark points. Dalal et al. popularized Histogram oriented Gradients (HoG) technique for feature extraction for detecting upright people [10]. We have used Histogram oriented Gradients (HoG) for extracting gradient features from Shape Information Matrix (SIM). Extracted Feature vectors are trained using Multilayer Perceptron with Scaled Conjugate Backpropagation learning. The proposed machine is tested over popular facial expression databases, like the JAFFE, CK+, MMI, and MUG databases. Efficient and effective training of the MLP has brought forth interesting and encouraging results.

2 Proposed Methodologies

The computation steps of the model are a *six* stage process. A flowchart of the proposed model is presented in Fig. 1.

Fig. 1 Flow chart
illustration of our method

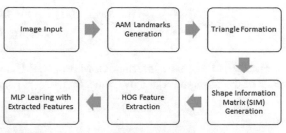

Fig. 2 Salient landmark
selection and triangle
formation demonstrated

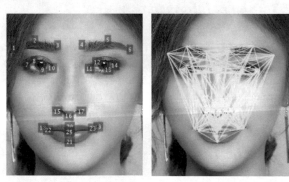

(a) Salient landmarks   (b) Traingle formation

2.1 Active Appearance Model

AAM uses *68* points to describe the appearance of the face *twelve* of them are used
to identify the eyes, having *six* for each, *ten* for the eyebrows with *five* for each of the
left and right one, *ten* for the nose, *twenty* points describe the mouth region and the
rest of them are on the circumference of the face. Now, *twenty five* salient landmarks
out of *68* are selected for our purpose. *Four* of them are on left eye and *four* on
right, *three* on the left eyebrow and *three* on right, *three* on the nose and *eight* on the
mouth. Selection of points are shown on Fig. 2a.

2.2 Triangle Formation

Those salient landmark points generated by AAM model are used to form triangles
over the face. Triangles are formed over the face by selecting all possible combinations of *three* points out of *twenty five* points in Fig. 2b. Three points from the 25
salient landmarks are selected in a topological order by fixing one point and varying
rest of the two points. Triangle shape information is calculated from the Eq. 1.

$$\kappa = \sum_{i}^{25} \sum_{j}^{25} \sum_{k}^{25} \frac{\max(\alpha_{ijk}, \beta_{ijk}, \gamma_{ijk}) - \min(\alpha_{ijk}, \beta_{ijk}, \gamma_{ijk})}{\alpha_{ijk} + \beta_{ijk} + \gamma_{ijk}} \tag{1}$$

Here κ is the Shape Information Matrix (SIM) and α_{ijk}, β_{ijk}, γ_{ijk} are the three sides of the triangle formed by connecting ith, jth and kth landmark points.

2.3 Shape Information Matrix

Shape information Matrix (SIM) is a 3 dimensional matrix of size $25 \times 25 \times 25$, which is flattened to a 2-dimensional matrix. Histogram oriented Gradients (HoG) are generated by forming grid having of 2×2 block size over SIM Matrix. Visualization of texture and gradients of SIM matrix are shown in Fig. 3a and 3b respectively.

2.4 MLP Learning

MLP for classification with one input layer, one hidden layer with *twenty* hidden neurons, and one output layer is used. SoftMax is used in output layer for class level prediction. Normalized HoG Feature vector of 1724 dimension is used as input. The perceptron is learned with Scaled Conjugate Gradient (SCD) backpropagation learning algorithm. Input data is divided in 70, 15, and 15% ratio for training, validation, and testing purposes, respectively. *Six*fold cross-validation is used for effective learning and quick convergence of error.

Fig. 3 Shape information matrix and gradients

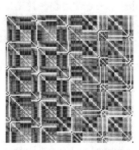 

(a) Flattened SIM Matrix.

(b) Gradient visualization of SIM.

3   Results and Discussion

We have tested our method of HoG feature extraction from Shape Information Matrix in four well-known databases, namely, JAFFE [18], CK+ [17], MMI [21], and MUG [1].

3.1   JAFFE Dataset

Japanese Female Facial Expression (JAFFE) dataset is a posed expression database of 213 images from 10 different subjects expressing 3 or 4 different variations of 6 basic expressions and 1 natural expression from Human Information Processing lab, Japan [18]. Efficient learning with Multilayer Perceptron(MLP) achieved an overall accuracy of 98.12% achieved in JAFFE dataset with 93.75% validation and testing accuracy displayed in Fig. 4.

3.2   CK+ Dataset

Extended Cohn-Kanade Dataset (CK+) is a mix dataset of posed and spontaneous expression from 182 subjects containing 2105 digitized image forming 327 sequence from CMU-Pittsburgh [17]. This dataset contains 6 basic expressions and 1 contempt expression. We have selected peak intensity image of each sequence for our analysis. Effective MLP Learning with CK+ database produced overall accuracy of 98.17% with validation and testing accuracy of 93.88% and 93.88%, respectively, displayed in Fig. 4.

3.3   MMI Dataset

MMI dataset is a combination of basic expression and single FAU expression of 75 subjects having 2900 videos and still images. MMI database is a multimodal dataset categorized in five different phases [21]. The Phase-I contains single Action Unit excitation videos of 20 adults. The Phase-II dataset contains 238 clips of basic

Fig. 4 Training, testing, validation, and total-dataset accuracy of CK+, JAFFE, MMI, and MUG databases

	CK+	JAFFE	MMI	MUG
Training	100	100	100	100
Validation	93.88	93.75	96.97	93.33
Testing	93.88	93.75	87.88	91.66
Total	98.17	98.12	95.95	97.76

expression of 28 adults. The Phase-III of MMI database contains 484 posed images of 5 subjects. Phase-IV and V contain video clips of spontaneous expressions. We have used a portion of Phase-III of the dataset for our purpose. Learning with MLP in MMI dataset, we have achieved an overall accuracy of 95.95% with testing and validation performance of 87.88% and 96.97%, respectively, as shown in Fig. 4. It is clearly evident from that the performance of MMI database is slightly low compared to results of CK+ in Sect. 3.2, JAFFE in Sect. 3.1, and MUG in Sect. 3.4, this is because of facial occultation condition occurring on the face in the form of facial hair, eyeglasses, hair falling, etc. at the time of capturing the expression images.

3.4 MUG Dataset

Multimedia Understanding Group (MUG) is a mixed collection of posed and non-posed collection of frontal images by 86 models displaying 6 basic expressions [1]. Participants are 51 men and 35 women of age varying from 20 to 35 years. The publicly available portion of the dataset contains 401 images of 26 subjects showing 6 atomic expression with 2 or 3 variations for each expression. Backpropagation learning with MLP achieved an overall classification rate of 97.76% with 91.66 and 93.3% testing and validation classification accuracy as shown in Fig. 4 (Fig. 5).

4 Conclusion

MLP learning in CK+, JAFFE shows overall good classification accuracy of almost 98.1%, followed by 97.7% in MUG dataset. MMI performs at 95.95% which is the lowest compared to others. 100% recognition accuracy is achieved in Anger, Happy,

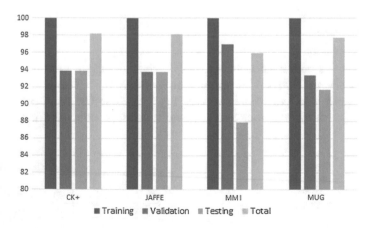

Fig. 5 Bar chart representation of training, testing, validation, and total-dataset accuracy of CK+, JAFFE, MMI, and MUG database

and Surprise expressions in JAFFE, Happy and Surprise expressions in CK+, Happy, Surprise, and Sad expressions in MMI and Surprise, Disgust and Fear in MUG database. Lowest accuracy as we see is 95.4% in anger expression of MMI database. Proposed method of extracting HoG features from Shape Information Matrix (SIM) and learning with MLP shows person independent learning of atomic facial expression. The system maintains an overall accuracy in between 95 and 98% in all four databases which implies generalization capability of our method. The significance of Shape Information Matrix is that it well captures the local and global shape changes in a single matrix representation. It is evident from the results in Table 4 that the texture of SIM matrix is a key feature describing the expression relevant shape changes in the face. Future scope of this work includes efficient generation of Shape Information Matrix (SIM) with more classification accuracy using other different types of shape features. Authors are presently engaged in different triangle side length features and incenter-circumcenter-centroid distance of a tringle which promise effective and efficient classification of expressions (Tables 1, 2, 3 and 4).

Table 1 Confusion JAFFE [18]

AN	**30**	0	0	0	0	0	0
FE	0	**29**	0	0	0	0	0
DI	0	0	**31**	0	0	0	1
HA	0	0	0	**30**	1	0	0
SA	0	1	1	0	**29**	0	0
SU	0	0	0	0	0	**30**	0
NU	0	0	0	0	0	0	**30**
	AN	FE	DI	HA	SA	SU	NU

Table 2 CK+ Confusion [17]

AN	**44**	0	0	0	0	1	0
CO	0	**18**	0	0	0	0	0
FE	1	0	**58**	0	0	0	0
DI	0	1	1	**23**	0	0	0
HA	0	0	0	0	**69**	0	0
SA	0	0	0	0	0	**28**	0
SU	1	0	0	1	0	0	**81**
	AN	CO	FE	DI	HA	SA	SU

Table 3 MMI Confusion [21]

AN	62	0	0	0	0	0
FE	3	21	0	0	0	0
DI	0	0	24	0	0	0
HA	0	0	0	45	0	0
SA	0	0	1	0	27	0
SU	0	1	0	0	0	38
	AN	FE	DI	HA	SA	SU

Table 4 MUG Confusion [1]

AN	24	0	0	0	1	0
FE	1	56	0	0	0	0
DI	1	0	70	0	0	0
HA	0	0	0	47	0	0
SA	2	0	1	1	81	0
SU	0	0	0	0	0	48
	AN	FE	DI	HA	SA	SU

Acknowledgements The authors like to avail this opportunity to express their gratitude to Dr. A. Delopoulos for providing MUG database and Prof. Maja Pantic for providing MMI database free for carrying out this work. The authors would also like to thank the Department of Computer and System Sciences, Visva-Bharati, Santiniketan for their infrastructure support. The authors gratefully acknowledge the support of UGC NET-JRF Fellowship (UGC-Ref. No.: 3437/(OBC)(NET-JAN 2017)) for pursuing Doctoral Research in Department of Science and Technology, Ministry of Science and Technology, Government of India.

References

1. Aifanti, N., Papachristou, C., Delopoulos, A.: The mug facial expression database. In: 11th International Workshop on Image Analysis for Multimedia Interactive Services WIAMIS 10, pp. 1–4. IEEE (2010)
2. Barman, A., Dutta, P.: Facial expression recognition using distance and shape signature features. In: Pattern Recognition Letters (2017)
3. Barman, A., Dutta, P.: Texture signature based facial expression recognition using narx. In: 2017 IEEE Calcutta Conference (CALCON), pp. 6–10. IEEE (2017)
4. Barman, A., Dutta, P.: Facial expression recognition using distance signature feature. In: Advanced Computational and Communication Paradigms, pp. 155–163. Springer (2018)
5. Barman, A., Dutta, P.: Facial expression recognition using distance and texture signature relevant features. Appl. Soft Comput. **77**, 88–105 (2019)
6. Barman, A., Dutta, P.: Influence of shape and texture features in facial expression recognition. IET Image Process. (2019). https://doi.org/10.1049/iet-ipr.2018.5481
7. Barman, A., Dutta, P.: Influence of shape and texture features in facial expression recognition. IET Image Process. (2019)

8. Cootes, T.F., Edwards, G.J., Taylor, C.J.: Active appearance models. IEEE Trans. Pattern Anal. Mach. Intell. **6**, 681–685 (2001)
9. Dahmane, M., Meunier, J.: Emotion recognition using dynamic grid-based hog features. In: Face and Gesture 2011, pp. 884–888. IEEE (2011)
10. Dalal, N., Triggs, B.: Histograms of oriented gradients for human detection. In: International Conference on computer vision & Pattern Recognition (CVPR'05), vol. 1, pp. 886–893. IEEE Computer Society (2005)
11. Darwin, C., Prodger, P.: The Expression of the Emotions in Man and Animals. Oxford University Press, USA (1998)
12. Datcu, D., Rothkrantz, L.: Facial expression recognition in still pictures and videos using active appearance models: a comparison approach. In: Proceedings of the 2007 International Conference on Computer Systems and Technologies, p. 112. ACM (2007)
13. Ekman, P., Keltner, D.: Universal facial expressions of emotion. In: Segerstrale, U., Molnar, P. (eds.) Nonverbal Communication: Where Nature Meets Culture, pp. 27–46 (1997)
14. Friesen, E., Ekman, P.: Facial action coding system: a technique for the measurement of facial movement. Palo Alto **3** (1978)
15. Guha, R., Ghosh, M., Kapri, S., Shaw, S., Mutsuddi, S., Bhateja, V., Sarkar, R.: Deluge based genetic algorithm for feature selection. In: Evolutionary Intelligence, pp. 1–11 (2019)
16. Kumar, N., Behal, S.: An improved lbp blockwise method for face recognition. Int. J. Nat. Comput. Res. (IJNCR) **7**(4), 45–55 (2018)
17. Lucey, P., Cohn, J.F., Kanade, T., Saragih, J., Ambadar, Z., Matthews, I.: The extended cohn-kanade dataset (ck+): a complete dataset for action unit and emotion-specified expression. In: 2010 IEEE Computer Society Conference on Computer Vision and Pattern Recognition-Workshops, pp. 94–101. IEEE (2010)
18. Lyons, M., Akamatsu, S., Kamachi, M., Gyoba, J.: Coding facial expressions with gabor wavelets. In: Proceedings Third IEEE International Conference on Automatic Face and Gesture Recognition, pp. 200–205. IEEE (1998)
19. Mase, K.: Recognition of facial expression from optical flow. IEICE Trans. Inf. Syst. **74**(10), 3474–3483 (1991)
20. Tzimiropoulos, G., Pantic, M.: Optimization problems for fast aam fitting in-the-wild. In: Proceedings of the IEEE International Conference on Computer Vision, pp. 593–600 (2013)
21. Valstar, M., Pantic, M.: Induced disgust, happiness and surprise: an addition to the mmi facial expression database. In: Proc. 3rd International Workshop on EMOTION (satellite of LREC): Corpora for Research on Emotion and Affect, p. 65 (2010)

Quality Enhancement of The Compressed Image Using Super Resolution

Mohammed Bilal, V. C. Naveen, D. Chetan, Tajuddin Shaikh,
Kavita Chachadi and Shilpa Kamath

Abstract Advancement in computing power and the accessibility to huge training data sets have gained interest in the implementation of deep learning CNN's (Convolution neural networks) to define image recognition and image processing operations. The enhancement of the visual quality of the compressed image and video compression has been considered a significant problem. The paper fundamentally focuses on the visual quality improvement of a compressed image using super resolution as the up-sampling technique. To address the problem of visual quality of the compressed image, the input image is bicubic down sampled. The downsampled image is encoded using image codec (JPEG). Finally, the decoded image is up-sampled using proposed up-sampling technique to reconstruct the original image with high quality. Experiments show our proposed up-sampling technique can improve the image both qualitatively and quantitatively and also the model achieves reconstruction performance with less computation resources and processing time compared to DCSCN [1].

Keywords Bicubic · Deep CNN · Super resolution · Image Codec

M. Bilal (✉) · V. C. Naveen · D. Chetan · T. Shaikh · K. Chachadi · S. Kamath
KLE Technological University, Hubballi, India
e-mail: mohammedbilal06011998@gmail.com

V. C. Naveen
e-mail: naveenchikka093@gmail.com

D. Chetan
e-mail: chetandhaduti16@gmail.com

T. Shaikh
e-mail: s.m.taj1000@gmail.com

K. Chachadi
e-mail: 17kavi17@gmail.com

S. Kamath
e-mail: shilpakul@gmail.com

© Springer Nature Singapore Pte Ltd. 2020
V. Bhateja et al. (eds.), *Intelligent Computing and Communication*,
Advances in Intelligent Systems and Computing 1034,
https://doi.org/10.1007/978-981-15-1084-7_34

1 Introduction

Image Super Resolution has very high importance in the areas such as security video surveillance and medical applications. It is the technique used to increase the resolution of the image. Now it is widely needed in areas where digital image processing is involved. The deep learning has led to multiple breakthroughs in image representation such as image compression, super-resolution, and image regeneration. We present a super-resolution model that can perform the enhancement of the quality of compressed image. The contributions of this study are as follows:

- Our custom dataset contains 1000 training and 100 testing images of different resolutions. It also includes artifacts (compressed images).
- The complexity of the model i.e. total number of training parameters has been reduced from 1,754,942 to 6,19,680 upon reducing the number of layers from 17 to 6 layers compared to the existing DCSCN [1] architecture.
- The model achieves similar reconstruction performance for the up-sampling of the compressed images with less computation resources and processing time compared to DCSCN [1].

2 Related Work

Single image super-resolution (SISR) tasks are showing remarkable performance based on Deep Learning techniques. C. Dong et al. proposed the first method of Super-Resolution Convolutional Neural Network (SRCNN) [2]. This method performed very well on single image super-resolution (SISR) tasks with the use of 2–4 convolutional neural networks. It also derives the conclusion that a larger filter size is recommended instead of deep CNN layers for appreciable performance. Another method, Deeply Recursive Convolution Network for Image Super Resolution (DRCN) [3] has large number of training parameters and operates on 20 deep CNN layers. To lower the number of parameters, each CNN weights are shared. Thus, they accomplished in training the deep CNN network and the method achieved great performance on image super-resolution tasks.

FSRCNN [4] was proposed by C. Dong et al. which accelerated the previous version SRCNN [2]. It uses transposed CNN for the direct processing of input image. Processing speeds of RAISR and FSRCNNs are much greater (10–100 times) than other Deep Learning-based super-resolution approaches. But, they do not perform like other deep CNN methods such as DRCN, VDSR [5] and RED [6]. Impact of feature selection algorithms on blind image quality assessment [7] examines the impact that a good feature selection algorithm has in the task of BIQA techniques. A noteworthy point compared to the proposed approach is the use of JPEG codec to introduce distorions/noise on the image. Multiple distortion pooling image quality assessment [8] presents a novel quality metric Q which is used to assess the quality of the image with different distorions. It also uses the structural similarity measure and mean square error for comparison with quality metric Q.

Human visual system-based unsharp masking for enhancement of mammographic images [9] investigates the use of nonlinear polynomial filters (NPF) in the quality improvement of digital mammographic images. Nonlinear filters are the combined model consisting of linear and quadratic filter components. An up-sampling technique [10] uses an efficient subpixel CNN that rearranges the elements of H × W × C*r*r tensor to form rH × rW × C tensor. The operation removes the hand-crafted bicubic filter from the pipeline with little increase of computation. The model DCSCN [1] is a fully connected convolutional neural network composed of feature extraction and reconstruction network. Feature extraction includes 12 CNN layers and reconstruction is five layers network.

3 Motivation

Convolutional neural networks (CNNs) have been used in low-level computer vision because of their significant performance in image processing tasks. Many attempts are being made to enhance the subjective quality of the decoded image using an up-sampling technique. Recent Deep Learning-based approaches have accomplished good performance in the area of single image super resolution,. The arrangement of convolutional neural networks and nonlinear activation function, i.e, ReLU units in a sequence can help to extract all important features of the image. However, deep and fully connected convolutional neural networks depend upon more number of computations and take more time for processing. So, we propose image super-resolution framework as the up-sampling network which uses compressed image as the input.

4 Proposed Method

As shown in the Fig. 1, The input RGB image is down sampled and passed to JPEG codec for encoding and decoding. Only Y-channel of the decoded image is used for up-sampling. The up-sampling architecture is made up of feature extraction and reconstruction network. Output of each hidden layer is connected as skip connection to the reconstruction network for the extraction of both the global and the local features. The concatenated features are up-sampled using pixel shuffler and then reconstructed using 3 × 3 CNN. The last layer outputs four channels. At the end, the up-sampled (scale = x2) output is generated by the summation of these four channels to the image up-sampled using bicubic.

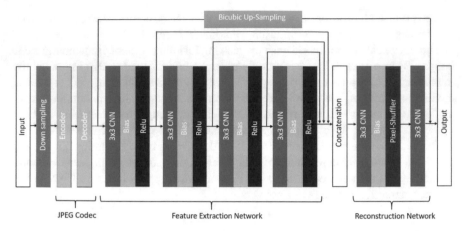

Fig. 1 Super resolution architecture

4.1 Feature Extraction

The feature extraction network consists of four sets of 3×3 CNN, bias and parametric ReLU units which are cascaded. Each output of the sets is passed to the next set and simultaneously passed to the reconstruction network. The other super-resolution techniques use deep CNN network, and hence computations are high. So, the number of features is reduced by reducing the number of layers. This reduces the complexity of the model. Dying ReLU problem [11] is handled by using the PReLU as activation units. This can guide weights for better performance.

4.2 Image Reconstruction Network

Compressed images are directly processed by the model for efficient feature extraction. The reconstruction of final high-resolution image is done at the end of the model. The reconstruction network outputs four channels. Each of these four channels represents each corner-pixel of the up-sampled pixel. Finally, it is combined with the bicubic up-sampled original input image. Pixel shuffler is used as the up-sampling layer.

5 Experiments

5.1 Dataset Analysis

The dataset contains 1000 training images and 100 testing images of different resolutions, which includes artifacts (compressed images). Data augmentation is performed on the training images. The image is rotated in four directions and each image is mirrored to produce two more images, totally 8 images are generated. Thus, the training dataset consists total of 8,000 images. Input RGB images are transformed to YCbCr to decrease the training process time. Y-channel images are used for processing. The batch size of each training image is 48 × 48 generated with stride 16. 64 patches are used as a mini-batch. SET 5 and SET 14 datasets are used for testing.

5.2   Training Setup

The weights for each convolutional neural network are loaded using the approach recommended by He et al. [11]. for all biases and PReLUs, Weights are loaded as 0. While training, the output of each PReLU layers is applied p = 0.8 dropout [12]. Adam [13] optimizer is used as optimization algorithm to minimize loss with an initial learning rate = 0.002. When there is no significant decrease in loss following 5 epochs of training steps, the learning rate is reduced by a factor of 2 and training is done up to the learning rate 0.00002.

6   Results

6.1   Quantitative and Qualitative Comparison

6.1.1   Accuracy Comparison

Quality metrics, Peak Signal-to-Noise Ratio (PSNR) and structural similarity measure (SSIM) are used to compare the objective quality of the up-sampled image. Output of the DCSCN and bicubic up-sampling for the input compressed images of the datasets SET5 and SET14 are used for the quantitative comparison (PSNR and SSIM values). Up-sampling is done on both compressed and decompressed images. But the quantitative comparison provided in the table is for the up-sampling of compressed images. Table 1 shows quantitative comparisons for 2× SISR (Fig. 2).

Table 1   Quantitative comparison (scale = x2)

Dataset	DCSCN (PSNR(dB)/SSIM)	Bicubic (PSNR(dB)/SSIM	Proposed (PSNR(dB)/SSIM)
set5	33.9284/0.8865	31.590/0.8680	33.9180/0.9100
set14	30.1937/0.82145	28.8476/0.7859	30.1613/0.8192

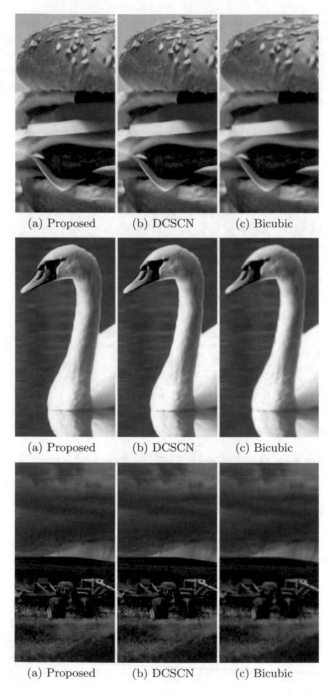

Fig. 2 Comparison of our results with results of DCSCN and Bicubic of scale = x2

Results provided in the Table 1 show that our proposed algorithm has similar performance for the datasets SET5 and SET14. Qualitative results of the images for DCSCN, bicubic and proposed up-sampling network are also shown above.

7 Conclusion

The proposed paper implements the Image Super-Resolution technique for the compressed image using CNN with skip connection. The results of the proposed up-sampling technique are similar to DCSCN [1], but compressed image is given as input to the up-sampling network. The total number of training parameters has been reduced from 1,754,942 to 6,19,680 by reducing number of layers from 17 to 6 compared to the existing DCSCN [1] architecture. Thus, the model can process original size images and achieves similar reconstruction performance of DCSCN with less computations.

References

1. Jin, Y., Kuwashima, S., Kurita, T.: Fast and accurate image super resolution by deep CNN with skip connection and network in network. In: In International Conference on Neural Information Processing, pp.217–225. Springer, Cham (2017)
2. Dong, C., Loy, C.C., He, K., Tang, X.: Learning a deep convolutional network for image super-resolution. In: European Conference on Computer Vision, pp. 184–199 (2014)
3. Kim, J., Lee, J.K., Lee, K.M.: Deeply-recursive convolutional network for image super resolution. In: Computer Vision and Pattern Recognition, pp. 1637–1645 (2016)
4. Dong, C., Loy, C.C., Tang, X.: Accelerating the super-resolution convolutional neural network. In: European Conference on Computer Vision (2016)
5. Kim, J., Lee, J.K., Lee, K.M.: Accurate image super-resolution using very deep convolutional networks. In: Computer Vision and Pattern Recognition, pp. 1646–1654 (2016)
6. Mao, X.J., Shen, C., Yang, Y.B.: Image restoration using very deep convolutional encoder-decoder networks with symmetric skip connections, In: Neural Information Processing Systems (2016)
7. Nizami, I.F., Majid, M., Afzal, H., Khurshid, K.: Impact of feature selection algorithms on blind image quality assessment. Arab. J. Sci. Eng. **43**(8), 4057–4070 (2018)
8. Gupta, P., et al.: Multiple distortion pooling image quality assessment. Int. J. Converg. Comput. **1**(1), 60–72 (2013)
9. Bhateja, V., Mishra, M., Urooj, S.: Human visual system based unsharp masking for enhancement of mammoographic images. J. Comput. Sci. **21**, 387–393 (2017)
10. Shi, W., Caballero, J., Huszr, F., Totz, J., Aitken, A.P., Bishop, R., Rueckert, D., Wang, Z.: Real-time single image and video super-resolution using an efficient sub-pixel convolutional neural network (2017)
11. He, K., Zhang, X., Ren, S., Sun, J.: Delving deep into rectifiers: surpassing human-level performance on imagenet classification. In: IEEE International Conference on Computer Vision, pp. 1026–1034 (2015)

12. Hinton, G.E., Srivastava, N., Krizhevsky, A., Sutskever, I., Salakhutdinov, R.R.: Improving Neural Networks by Preventing Coadaptation of Feature Detectors (2012). arXiv preprint arXiv:1207.0580
13. Kingma, D.P., Ba, J.L.: Adam: A method for stochastic optimization. In: International Conference on Learning Representations (2015)

Power Allocation Mechanism in Uplink NOMA Using Evolutionary Game Theory

Kewal Kumar Sohani, Anjana Jain and Anurag Shrivastava

Abstract The rapid advancement in communication systems and Internet of Things (IOT) has increased the demand for high data rates, reliability, better Quality of Service (QoS), and reduced interference. As the upcoming 5G aims to accommodate more number of users in the limited available spectrum, Non-Orthogonal Multiple Access (NOMA) outperforms the accommodation of users on the channels. In this paper, an algorithm to optimize the uplink power allocation of the user pairs in NOMA for the two-user system model is discussed. Typically, NOMA uses same frequency slot, same time slot, but different power levels to transmit user's data, resulting in higher spectral efficiency. Successive interference cancellation is the major challenge in this scheme, which is enhanced with proper selection of power levels at which the users transmit the data. The optimum power allocation in NOMA can be modeled as a game in which Evolutionary Game Theory (EGT) helps the Base Station (BS) to provide the best optimal power level for the user pair to transmit their message signals simultaneously on the same channel under the profound channel conditions and position of that user pair in the cell. It is shown that EGT is a successful tool in deciding power levels of participating mobile stations.

Keywords Non-orthogonal multiple access scheme · Power allocation · Evolutionary game theory · Payoff · Utility function · Replicator dynamics

K. K. Sohani (✉) · A. Jain · A. Shrivastava
Shri Govindram Seksaria Institute of Technology and Science, Indore, India
e-mail: sohanishanu@gmail.com

A. Jain
e-mail: jain.anjana@gmail.com

A. Shrivastava
e-mail: ashrivastava827@gmail.com

© Springer Nature Singapore Pte Ltd. 2020
V. Bhateja et al. (eds.), *Intelligent Computing and Communication*,
Advances in Intelligent Systems and Computing 1034,
https://doi.org/10.1007/978-981-15-1084-7_35

1 Introduction

The upcoming wireless communication scenario aims to achieve high data rate, ultra-high reliability, and efficient spectrum efficiency. However, as the number of users is increasing magnificently, there is now a great challenge to accommodate these users in limited available spectrum. Presently, Orthogonal Multiple Access (OMA) scheme is in use. A new Medium Access Control (MAC) technique, namely, NOMA is under development phase, is considered to be the future of 5G system as NOMA compared to OMA outperforms in terms of spectral efficiency, energy efficiency, and user data rate [1]. The principle working of NOMA is to transmit and receive multiple users' data simultaneously within the same frequency slot. So, basically NOMA captures the advantages of both FDMA and TDMA with difference in power levels for the users. Control of power allocation to various transmitters is the main parameter to reduce the interferences in this NOMA scheme [2]. Evolutionary Game Theory (EGT) helps to select the best pairs of power levels for transmitting the message signals simultaneously in the same channel. The remaining paper is organized as follows: Sect. 2 describes the system model which is taken into consideration for modeling the uplink NOMA scenario of two-user system. Section 3 gives the conceptual overview of NOMA. Section 4 covers the mathematical overview of EGT using replicator dynamics. The proposed methodology for power allocation mechanism using EGT in Uplink NOMA is discussed in Sect. 5. The simulation results and analysis are given in Sect. 6. Finally, the paper is concluded in Sect. 7.

2 System Model

We have considered two-user system model, where two mobile users or stations (MS) use the same channel at same time with different power levels to transmit their message signals to the base station (BS) and BS decodes using Successive Interference Cancellation (SIC) [1]. This scenario is shown in Fig. 1a. Figure 1b shows SIC decoding at BS [1]. In this, the key idea is to equip the BS with a mechanism

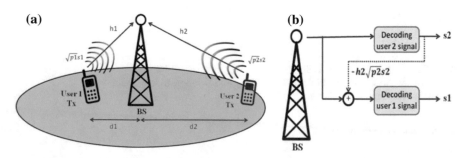

Fig. 1 **a** Two-user uplink NOMA model [1]. **b** SIC decoding at the BS [1]

to assign the power levels to these MS pairs such that both MSs can transmit their data to the BS reliably using NOMA scheme. This mechanism follows Evolutionary Game Theory (EGT) which considers the effect of noise, shadowing and attenuation, resulting in the best possible power level for the mobile stations.

3 Basics of NOMA Two-User Model

NOMA is suggested as a radio access scheme for future wireless mobile communication, which has the ability to reliably uplink or downlink the multiple users' message signals through single channel at the same time, saving a lot of spectrum to accommodate more users further. As it is non-orthogonal, which means all those non-orthogonal channels which were incapable to be used in OFDMA can also be serviced, resulting in optimum utilization of resources. The mathematical formulation for the downlink NOMA given by Xin et al. [1] proves the possibility of existence of NOMA. The basic equations for NOMA are given as follows:

The transmitted signal of MS_i ($i = 1, 2$) is

$$x_i = \sqrt{p_i} s_i, \tag{1}$$

where p_i is the transmitting power level of MS_1 and S_i is the message symbol of MS_1.

At the BS, the received signal y is a superposed signal of MS_1 and MS_2 with channel coefficient h and assuming AWGN n,

$$y = hx + n, \text{ where } x = x_1 + x_2. \tag{2}$$

Throughput, or Rate of Users, at the downlink end with noise power density of n_i as $N_{0,i}$, assuming ideal SIC, is given as

$$R_1 = \log_2\left(1 + \frac{p_1|h_1|^2}{N_{0,1}}\right), \quad R_2 = \log_2\left(1 + \frac{p_2|h_2|^2}{N_{0,2}}\right). \tag{3}$$

From Eqs. (1), (2), and (3), we can observe that the power levels p_1 and p_2 are very important parameters to control the interference in the NOMA system. Thus, a mechanism is needed to choose these powers such that the rate of each user gets enhanced. For this, we have developed a mechanism for the base station to sense the condition of user and assign the best possible available power level to transmit signal.

4 EGT with Replicator Dynamics

Game Theory is a tool which models the situation as a game with three parameters, namely, players, strategies, and payoff functions that give utility to each player [3, 4]. A simple game is an approach of Traditional Game Theory in which the game is played once. When the game is iterated several times and each time the player learns from the outcome of previous actions and updates the strategic behavior, it is known as evolution of the game and this approach is said to be Evolutionary Game Theory (EGT).

In physical layer of wireless network model, game theory is used in power control or resource allocation as the network nodes show strategic random behavior which is modeled as a game [5, 6]. EGT focuses on Evolutionary Stable Strategy (ESS), the concept better than Nash equilibrium, which was defined in [7], and is calculated against mutations [8], with following characteristics:

(1) The proportions of every population remain unchanged with time, when ESS is achieved.
(2) There is no effect on a population due to all other remaining populations, at ESS. The ESS point of a game model is an important parameter used in various engineering problems [9].

The basic mathematics of EGT required for the model development in paper is given below.

4.1 Strategy

The behavioral selection of move in a game is called as a strategy. In a game, there will be a set of strategies for each player denoted by S [10]:

$$S = \{s_1, s_2, s_3, \ldots\} \tag{4}$$

4.2 Payoff

It is a function $\pi: S \to R$ that associates a numerical value with every strategy $s \in S$ [10]. A utility function is a function u: $\Omega \to R$, where ω is the wealth for that strategy, such that [10]

$$u(\omega_1) > u(\omega_2) \Longleftrightarrow \omega_1 > \omega_2 \tag{5}$$

$$u(\omega_1) = u(\omega_2) \Longleftrightarrow \omega_1 \sim \omega_2. \tag{6}$$

The replicator dynamics assume individuals are programmed to use only pure strategies from a finite set $S = \{s_1, s_2, \ldots, s_k\}$. Let n_i be the number of individuals using s_i, then the total population size is

$$N = \sum_{i=1}^{k} n_i. \tag{7}$$

And the proportion of individuals using s_i is

$$x_i = \frac{n_i}{N}. \tag{8}$$

The selection probability of population is then described by a matrix X,

$$X = \begin{bmatrix} x_{11} & \cdots & x_{1N} \\ \vdots & \ddots & \vdots \\ x_{\rho 1} & \cdots & x_{\rho N} \end{bmatrix}, \tag{9}$$

where ρ is the number of users in single population, x_{11} represents the selection probability of user 1 to choose power level 1 [10].

The average payoff of the population is given by

$$\bar{\pi} = \sum_{i=1}^{N} x_i \mathcal{J}(s_i, X), \tag{10}$$

where $\mathcal{I}(s_i, X)$ denotes the user with strategy i, which encounters the strategies of rest of the players.

The individual payoff is given by

$$\pi = \sum_{i=1, j=1}^{N} x_i \mathcal{J}(s_i, x_j), \tag{11}$$

where $\mathcal{I}(s_i, x_j)$ denotes the expected payoff when user with strategy i encounters the user with strategy j.

The rate of change of number of individuals using s_i is

$$\dot{x}_i = x_i(\pi - \bar{\pi}). \tag{12}$$

This is also known as *replicator equation*. As can be seen from Eq. (12), the proportion of individuals $\dot{x}_i$ using strategy s_i increases (decreases) if its payoff $\pi(s_i, x)$ is bigger (smaller) than the average payoff $\bar{\pi}(x)$ in the population [10]. There are other algorithms also, based on Evolutionary Game Theory, which have different game models, as discussed by Patibandla et al. [11], Pacifico et al. [12], and Prado et al. [13].

5 Proposed Method

From the research work presented by various researchers [1, 2, 14], it is clear that
NOMA outperforms OMA scheme for all the power levels ideally. In practical scenario, NOMA needs to optimize the power levels to gain high spectral and energy
efficiencies. Each user needs optimum power level so as to transmit their message
signal efficiently under the present channel condition and attenuation effect. Here,
the channel estimation is a very important factor as the payoff matrix is dependent
on channel characteristics.

To select the best possible set of power levels for the given channel and attenuation
condition, there is a game between MSs to get the best possible power from the BS.
Therefore, we have simulated EGT for solving this game at the base station. For
this, we consider two-user NOMA system model as shown in Fig. 1, in which the
cell tower (BS) sends control signals to all the users in its coverage area to estimate
the channel coefficients [15] and attenuation factor for each user. It is assumed
that the channel is Rayleigh distributed and the user pair is selected by the base
station which has significant difference in the attenuation factor (i.e., for $d_1 \gg d_2$ or
$d_1 \ll d_2$, attenuation will be significantly different). It is assumed that total power
levels $= 6$.

The algorithm designed for the base station is divided into three phases, namely, (a)
initialization, (b) replicating process, and (c) distribution of power levels to MSs. The
algorithm can be simulated using PDX Toolbox in MATLAB [16] and NETLOGO
[17]. This mechanism is shown in Table 1.

6 Simulation Results and Analysis

The simulation parameters are mentioned in Table 2.

Here, the total power levels (i.e., $N = 6$) are initialized with equal probability
and are arranged in increasing order (i.e., six strategies corresponding to six power
levels are arranged in order S6 < S5 < S4 < S3 < S2 < S1). In the graphs shown in
Figs. 2, 3, 4, and 5, the selection probability and payoff variation for all the strategies
corresponding to each power level are obtained. As the game starts for channel
condition case-1 (i.e., worst channel is configured), the game iterates to evolve as
seen in Fig. 2, and updates the vector X till the difference reaches to value "ε". Here,
we see that, for worst channel, the evolution of this game shows maximum power
strategy gets the high probability of selection, which means the user is assigned with
the highest power level strategy-(S1). The payoff profile for all the strategies for
case-1 during evolution, which is the behavior of strategies, is shown in Fig. 3.

Similar for case-2, i.e., relatively better channel, evolution can be seen in Fig. 4.
As here the channel is quite better, theoretically, less power should be assigned
to the user, and practically also we get the power level with strategy-(S5) getting
more proportion for selection. The payoff profile for all the strategies for case-2 is

Table 1 Proposed power allocation mechanism using EGT

1.) Initialization :
1. Compute the position of the MSs from the BS using control signal in terms of attenuation factor. 2. Select the two MSs which have attenuation difference of greater than 1 (*i.e.*, $|\alpha 1 - \alpha 2| > 1$). 3. Obtain the channel coefficients h_i for both the users using channel estimation. 4. Map the N=6 power levels on the channel distribution.
2.) Algorithm for replicating process :

1. set population state vector $x = \{\frac{1}{N}, \frac{1}{N}, ... upto\ N\ terms\}$.
2. Form the selection probability matrix p *of size* N×p.
3. for i=1 to 6:
    Find individual payoff $\pi = \sum_{i=1,j=1}^{N} x_i \mathcal{J}(s_i, x_j)$ and average payoff $\bar{\pi} = \sum_{i=1}^{k} x_i \mathcal{J}(s_i, X)$.
  end
4. while $|x(t) - x(t-1)| > \epsilon$, where ϵ is 0.005
    Solve $\dot{x}_i = x_i(\pi(t) - \bar{\pi}(t))$
    Update population state vector x with new $\dot{x}$.
    End
5. when $|x(t) - x(t-1)| \leq \epsilon$, ESS is achieved, select strategy (i.e. the power level) with highest value in x.

3,) Distribution of power levels to MSs
1. The two selected power levels with higher selection probabilities are assigned to the two users, such that, the MS located far from the BS gets more power than the other user, to combat the attenuation and noise effect. 2. If the power level selected is not free at BS, assign the next higher power level. 3. if no power level is free at BS, the user is discarded from the service and there is a call drop behavior.

shown in Fig. 5. This shows that for different channel conditions, the payoff matrix gets updated and the replicator dynamics equation optimizes the strategies selection probability, which results in the selection of most favorable power levels to the users.

7   Conclusion

In this paper, we presented mechanism for power allocation in Uplink NOMA using Evolutionary Game Theory (EGT) in future 5G networks. For power selection in multiple access NOMA scheme, the distribution of transmit power among multiple users must be optimized to achieve minimum interference effect. The simulation results show that power selection using EGT improves the user signal response under the given channel condition. We have shown the advantages of using evolutionary game theory in power allocation algorithm, which takes into account the channel characteristics to select the most suitable power level(s) for the user(s). This paper shows that EGT is a successful tool to mitigate the interference between the user pairs by making effective power allocation mechanism.

Table 2 Simulation parameters for power allocation based on EGT in uplink NOMA

Number of cells in a network	2
Cell radius	389 m
No. of users in single population	2
Users locations	Uniformly distributed [18]
Minimum and maximum power levels: [p min, p max]	[20 dBm, 30 dBm]
No. of power levels	N = 6
Max. users per sub-carrier (simultaneously transmitting users)	2
Propagation model	COST231 Hata model [19]
Modulation scheme	BPSK
Shadowing distribution	Rayleigh shadowing [18], $\sigma^2 = 8$ dB
Noise power	−173 dBm/Hz
Stopping criterion (ε)	0.005
Payoff dependency	Power levels, channel characteristics, attenuation factor
Payoff matrix size	6 × 6
Channel condition cases	1. Worst-case channel 2. Relatively better channel

Fig. 2 The evolved selection probability distribution per millisecond as the game iterates under worst channel condition

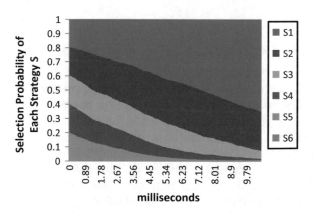

Fig. 3 The expected payoff values for each strategy as the game iterates in the process of evolution under worst channel condition

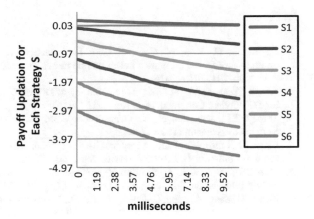

Fig. 4 The evolved selection probability distribution per millisecond as the game iterates under relatively better channel condition

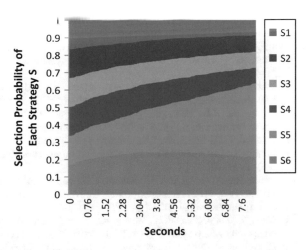

Fig. 5 The expected payoff values for each strategy as the game iterates in the process of evolution under relatively better channel condition

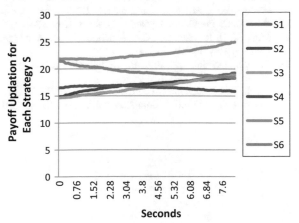

References

1. Su, X., Yu, H., Kim, W., Choi, C., Choi, D.: Interference cancellation for non-orthogonal multiple access used in future wireless mobile networks. EURASIP J. Wirel. Commun. Netw. Article no. 231, 2016 (2016)
2. Wei, Z., Yuan, J., Ng, D.W.K., et al.: A Survey of Downlink Non-orthogonal Multiple Access for 5G Wireless Communication Networks (2016)
3. Turocy, T.L., von Stengel, B.: Game Theory, pp. 403–420. Encyclopedia of Information Systems by Academic Press (2002)
4. Bacci, G., Luise, M.: Game theory in wireless communications with an application to signal synchronization. Adv. Electron. Telecommun. 1(1) (2010)
5. Mehta, S., Kwak, K.S.: Application of game theory to wireless networks. In: Convergence and Hybrid Information Technologies, Intechopen, Chap. 24 (2010)
6. Zhang, T., Zeng, Z., Feng, C., Zheng, J., Ma, D.: Utility fair resource allocation based on game theory in OFDM systems. In: ICCCN, pp. 414–418 (2007). https://doi.org/10.1109/icccn.2007.4317854
7. Smith, J.M., Price, G.R.: The logic of animal conflict. Nature 246(5427), 15–18 (1973)
8. Tembine, H., Altman, E., El-Azouzi, R., Hayel, Y.: Evolutionary games in wireless networks. IEEE Trans. Syst. MAN Cybern.-Part B Cybern. 40(3) (2010)
9. Vincent, T., Vincent, T.: Evolution and control system design. IEEE Control Syst. Mag. 20(5), 20–35 (2000)
10. Webb, J.N.: Game theory-decisions, interaction and evolution. Springer Undergraduate Mathematics Series. Springer, London Limited (2007). ISBN-10:1-84628-423-6
11. Patibandla, R.L., Veeranjaneyulu, N.: Performance analysis of partition and evolutionary clustering methods on various cluster validation criteria. Arab. J. Sci. Eng., 1–12 (2018)
12. Pacifico, L.D., Ludermir, T.B.: Improved evolutionary extreme learning machines based on particle swarm optimization and clustering approaches. Int. J. Nat. Comput. Res. (IJNCR) 3(3), 1–20 (2012)
13. Prado, R.S., Silva, R.C.P., Guimarães, F.G., Neto, O.M.: A new differential evolution based metaheuristic for discrete optimization. In: Nature-Inspired Computing Design, Development, and Applications, IGI Global, pp. 104–119 (2012)
14. Chen, Z., Ding, Z., Dai, X., Zhang, R.: A mathematical proof of the superiority of NOMA compared to conventional OMA. IEEE Trans. Signal Process. XX(X) (2016)
15. Chavan, M.S., Chile, R.H., Sawant, S.R.: Multipath fading channel modeling and performance comparison of wireless channel models. Int. J. Electron. Commun. Eng. 4(2), 189–203 (2011)
16. Barreto, C.: Population Dynamics Toolbox. Game Theory Society, October 24 (2015)
17. Izquierdo, L.R., Izquierdo, S.S., Sandholm, W.H.: Abed (agent-based evolutionary dynamics). Game Theory Society (2017)
18. Popa, S., Draghiciu, N, Reiz, R.: Fading types in wireless communications systems. J. Electr. Electron. Eng. 1 (2008)
19. Singh, Y.: Comparison of Okumura, Hata and COST-231 models on the basis of path loss and signal strength in Intl. J. Comput. Appl. 59(11), 37–41 (2012)

Performance Analysis of Convolutional Neural Networks for Exudate Detection in Fundus Images

Nandana Prabhu, Deepak Bhoir and Uma Rao

Abstract In the diagnosis of diabetic retinopathy and macula edema, the extraction of exudates is a crucial task, particularly in the presence of optic disc and cotton wools which have similar intensity. To add to this some of the reflectance around the vessels originating from the optic disc also needs to be eliminated. In this paper, supervised method with deep learning neural network is proposed for exudate classification. The network is trained with large number of sub-images, around 48,000 for each image, which are fed as inputs to the network. The original image is pre-processed before the patches are extracted. The database e-Ophtha is used. It has 47 images with exudates and ground truths. The performance of the system is good with an accuracy of 98.67%, area under the curve 97.29, sensitivity of 72.26% and specificity of 98.76%. Analysis of the system with respect to depth of network and patch size is also performed.

Keywords Diabetic retinopathy · Convolutional Neural Network · Image split · Patches

N. Prabhu (✉)
Research Scholar, Fr. Conceicao Rodrigues College of Engineering, Mumbai 400050, Maharashtra, India
e-mail: nandanaprabhu@somaiya.edu

D. Bhoir
Professor, Department of Electronics Engineering, Fr. Conceicao Rodrigues College of Engineering, Mumbai 400050, Maharashtra, India
e-mail: bhoir@fragnel.edu.in

U. Rao
Retired Professor, Department of Electronics Engineering, Shah and Anchor Kutchhi Engineering College, Mumbai 400088, Maharashtra, India
e-mail: uma.rao@sakec.ac.in

© Springer Nature Singapore Pte Ltd. 2020
V. Bhateja et al. (eds.), *Intelligent Computing and Communication*,
Advances in Intelligent Systems and Computing 1034,
https://doi.org/10.1007/978-981-15-1084-7_36

1 Introduction

Diabetes mellitus has been growing at an alarming rate all over the world. An estimation has been made that in 2000, the prevalence of diabetes for all age groups worldwide was 2.8%, and 4.4% in 2030 [1]. An estimation has been made that people with diabetes is expected to rise to 366 million by 2030 [1]. Factors responsible for the rise in diabetes are population growth, aging, obesity, urbanization and reduced physical activity combined with mental stress.

In the long run, diabetes affects heart, nervous system, eyes and kidneys, leading to various complications such as diabetic retinopathy, nephropathy causing damage of kidney and macrovascular problems. The complications in the eye can be glaucoma, cataracts and Diabetic Retinopathy (DR). In urban areas in the working age group many people have diabetes. Moreover, in many people, diabetes is not diagnosed. Therefore for early diagnosis, screening is essential. Particularly in developing countries, this will help to improve the health care systems. In the present scenario, the number of ophthalmologists is very less as compared to the number of diabetic patients, with an estimated count of 1 per million population [2]. The ratio is poorer especially in rural areas of our country. Currently, the diagnosis is done through manual examination of fundus images. In this scenario, an automated system for diagnosis of condition of retina is highly desirable as it helps the ophthalmologists in quick and easy diagnosis of DR or as an assisting tool for preliminary examination.

DR is one of the effects of diabetes on eyes. The images normally used for diagnosis are either fundus images or Optical Coherence Tomography (OCT) images. The fundus images are normally two-dimensional and OCT images are three-dimensional. Fundus photography provides accurate and precise information of the anatomical structure of the retina in two dimensions [3]. Figure 1 shows the fundus image with exudates [4]. The two broadly classified stages of DR are: NonProliferative DR (NPDR) and Proliferative DR (PDR). The symptoms of DR are: changes in diameter of blood vessel and symptoms such as dark lesions microaneurysms and hemorrhages, bright lesions such as exudates and cotton wools, and in later stage (PDR) formation of new brittle blood vessels. Different supervised and unsupervised algorithms are proposed by researchers for the detection of these symptoms. Over a decade, with the advancement in medical imaging and the field of computer vision and artificial intelligence, a lot of new dimensions have been added to the diagnosis.

1.1 Related Works

Automated segmentation has been performed using morphological operations, region-growing techniques, multiscale approach and using both supervised and unsupervised classification. The results of research are published using various evaluation parameters, such as accuracy, sensitivity, specificity, area under the curve (AUC), F1 score, and Positive Predictive Value (PPV). In the conventional methods, prior to

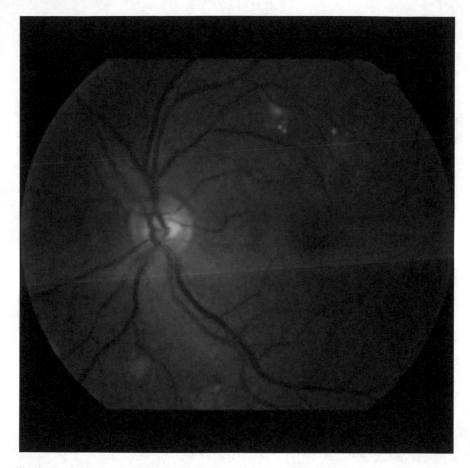

Fig. 1 Fundus images with exudates [4]

the process of extraction of exudates, it is necessary to pre-process the input image, extract the Optic Disc (OD) and blood vessels and eliminate them, thereby avoiding false positives; that is, they should not be mistaken for the lesions.

Sinthanayothin et al. used region-growing technique for exudate detection [5]. Sopharak et al. used morphological operations for extracting the exudates [6]. Franklin et al. have used multilayer perceptron for classification [7]. To extract the exudates, Jaya et al. first performed morphological operations and Hough transform for removal of OD. To represent the exudates, color and texture features are used and fuzzy support machine is used for classification [8]. Agurto et al. used optimal thresholding of instantaneous amplitudes for exudate extraction and partial least squares for classification [9].

Deep learning has been used for various applications including DR [10]. Convolutional Neural Network (CNN) is used for blood vessel extraction [11–13]. Tan et al. used CNN for simultaneous extraction of OD and fovea [14]. CNN is also used for dark lesions detection [15]. Chen et al. used two stacked CNNs for deep image salience computing. The first CNN performed coarse-level observations and second CNN for fine-level observations in the image [16]. CNN is also used for detection of exudates [17–19].

The proposed algorithm uses supervised learning technique with deep neural network. The original images are first pre-processed using resizing, histogram equalization, normalization, gamma correction and median filtering. Large number of patches are then extracted from an image and fed as input to the network for training. The classification is performed to determine patch with or without exudates and evaluation parameters are measured.

2 Materials and Methods

The algorithm is developed and tested using the publically available dataset e-Ophtha [1]. It contains 47 retinal images with exudates. These images are available with their ground truths. The size of the image is 1024×1024 pixels. Majority of the images are used for training. To be more precise, 40 are used for training and 7 for testing. From the images, 48,000 sub-images are extracted and given as input to the network. The proposed architecture is trained using the training images and tested using the testing images.

Hardware and software: The CNN is executed on Intel 8 core CPU, with 60 GB SDD and Tesla K80 GPU. The framework used for deep learning is Karas package with tensor flow at the backend. This helps in speeding up the network.

The proposed work used the CNN for segmenting and classifying the exudates. The pre-processing, CNN architecture and details of training given to the network are explained below.

2.1 Pre-processing

It is a very important stage in any image processing application. It is used to enhance the quality of the image. Some of the pre-processing techniques are resizing, color model conversion, contrast enhancement, correction of non-uniform illumination and noise. The neural network depends a lot on the raw image provided to it for learning. In the pre-processing stage, the images are cropped to retain the region corresponding to a bounding box containing only the fundus image, thereby eliminating the background. Resizing to 512×512 pixels is done to reduce the computational cost and yet maintain the features that the software needs to handle. The input image in our case is an RGB image. As green channel shows better reflectance to background, it is used in further processing. To remove the salt and pepper noise, median filtering of size 3×3 is applied. Contrast enhancement is performed using image intensity adjustment with gamma correction of 1.2 and Contrast Limited Adaptive Histogram

Equalization (CLAHE). Batch normalization is performed for eliminating the effect of non-uniform illumination.

2.2 Architecture

The architecture contains a number of convolutional layers. Large number of convolutional layers helps the network to learn the features better. The architecture follows an "M" shape. The implementation is adopted from the work done by Cortinovis [13]. It has a total of nine blocks, as shown in Fig. 2. Each block has two convolution layers with a dropout between the two. The dropout helps to increase the performance. Up-sampling is performed after first, fifth, sixth and seventh blocks. Max pooling is performed after each of the blocks 2, 3, 4 and 8. The purpose is to reduce the size of representation and speed up the computation. All the convolutional layers are followed by Rectifier Linear Unit (Re LU) as the activation function, except the last layer, which uses softmax function to predict the exudates.

In each of the iterations, the image patch is presented as input to the network; it propagates through all the layers to the output layer. The errors calculated in output layer are back propagated through the network and used for weight updating. However, the updating is done for each batch.

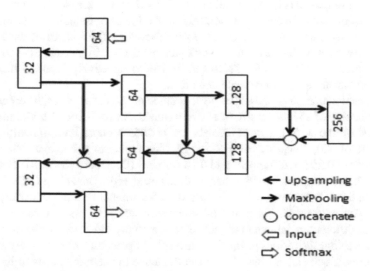

Fig. 2 Architecture for exudate extraction

2.3 Training

A total of 48,000 patches of 16 × 16 are extracted from the fundus image. In the retina image, the exudates are very small, less in number and do not have a fixed shape. The neural networks have a severe problem of overfitting as most of the time the images mostly have features belonging to a class, which are not exudates. To handle this problem, training is performed using image split of 0.25 for first 200 epochs. Then image split of 0.15 is used and the network is trained further for another 100 epochs. The batch size used in each case is 300.

The rectifying linear unit provides the effective training of network. The dropout helps to eliminate some units temporarily. With the emergence of GPUs, the time required for training is also much reduced because of parallel processing.

Training a deep neural network requires a large amount of labeled training data, which remains a challenge for medical images because of the expense of expert annotation and proportion of lesions when compared to entire image.

3 Results and Discussion

The extraction of exudates giving patches to the network is a challenging task as the number of exudates is very small and their total area is very less as compared to the area of the entire image. However, this method does not require the elimination of blood vessels and optic disc as we do it otherwise. The images of database e-Ophtha are used as inputs. Our network achieved an AUC, accuracy, sensitivity and specificity of 97.29, 98.67, 72.26 and 98.76%, respectively. Figure 3 shows the outputs at various stages of exudate detection.

The results in Table 1 shows the performance of the network for different patch size with depth of 256, batch size of 150 and image size of 512 × 512. The results in Table 2 show the performance of the network at different deep level of learning. The batch size of 300, image size of 512 × 512, the number of sub-images 48,000, the image split of 0.25 for 200 epochs and 0.15 for next 100 epochs are used for training. Table 3 shows comparison with other methods used earlier including CNN.

To finalize the above network we started with a smaller network with only seven blocks. The performance in terms of sensitivity was very poor. This was tested then with different variations in blocks. The sensitivity did not improve much. The network was run for various patch sizes and split of patches. The sensitivity still did not improve much and at times deteriorated. Hence the network needs to include a deeper structure, as shown in Fig. 2. The performance improved.

Selecting the size of the patches is also important. Large size of patch makes the data sparse and the neural network difficult to learn to distinguish between exudates and non-exudates. Before we finalized the patches to 16 × 16, we tested it on sizes 32 × 32, 64 × 64 as well. With 32 × 32, the computational time was less and the sensitivity was much less. The 64 × 64 patch size had a problem when batch size

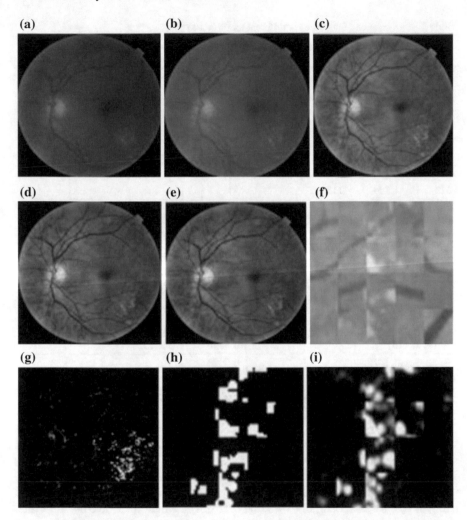

Fig. 3 Outputs at various stages of exudate detection. **a** Original image, **b** gray image, **c** adaptive histogram equalized image, **d** gamma correction applied image, **e** median-filtered image, **f** sample patches extracted image, **g** predicted image, **h** original mask image, **i** predicted mask image patch

Table 1 Performance of network for different patch size

Level	Epochs	Image split	No. of sub-images	AUC	Accuracy	Sensitivity	Specificity
16 × 16	200	0.25	120,000	0.9764	0.9926	**0.6183**	0.9940
32 × 32	200	0.25	120,000	0.9837	0.9957	0.5985	0.9972
64 × 64	200	0.25	120,000	0.9855	0.9957	0.6041	0.9972

Table 2 Performance of network for different levels of learning

Level	Epochs	Image split	No. of sub-images	AUC	Accuracy	Sensitivity	Specificity
16 × 16 (down to 128)	200 + 100	0.25 + 0.15	48,000	0.9402	0.9867	0.6096	0.9909
16 × 16 (down to 256)	200 + 100	0.25 + 0.15	48,000	**0.9729**	**0.9867**	**0.7226**	**0.9876**
16 × 16 (down to 512	200 + 100	0.25 + 0.15	48,000	0.9633	0.9873	0.6300	0.9886

Table 3 Comparison with other methods used earlier including CNN

Author	AUC	Accuracy	Sensitivity	Specificity	PPV	F1 score
Sopharak et al. [6]	–	–	0.8000	0.9946		–
Welfer et al. [20]	–	–	0.7048	0.9884		–
Harangi et al. [18]	–	–	0.7100	0.6800	0.6600	
Prentašic et al. [19]	–	–	0.7800	–	0.7800	0.7800
Tan et al. [15]			0.8758	0.9873		
Quallec et al. [17]			0.5570			
Mo et al. [21]			0.9255	–	0.9100	0.9053
Proposed method	0.9729	0.9867	0.7226	0.9876		

was increased, which resulted in memory shortage. When we used 16 × 16 patch size, it gave better results.

Then to decide on the optimum image split, keeping all other parameters such as patch size, epochs, batch size, number of patches to be extracted constant, we tried using different splits, 0.75, 0.5, 0.25 and 0.15 out of which 0.25 gave better results. The accuracy improved further when we used combination of 0.25 and 0.15.

Training of the networks takes 14–20 h. However, once the training is done, it takes a few seconds to several minutes to test an image. The results obtained are impressive. To further refine the results post-processing can be done.

4   Conclusion

Extraction of exudates is very important task in the detection of DR. In this paper, a CNN-based approach is used for extraction of exudates. The pre-processing steps are used for improving the quality of the image. The blood vessel extraction and OD elimination required in the conventional method are not needed. The number of pixels representing exudates is very less compared to the entire image. Image split feature is used to input to the network sufficient number of patches with exudates to alleviate the problem of overfitting. The database e-Ophtha is used as it has masks provided with exudates. The achieved AUC, accuracy, sensitivity and specificity are 97.29, 98.67, 72.26 and 98.76%, respectively. The proposed work can be extended to improve the sensitivity and for classification of severity of DR.

References

1. Wild, S., Roglic, G., Green, A., Sicree, R., King, H.: Global prevalence of diabetes: estimates for the year 2000 and projections for 2030. Diabetes Care **2004**(27), 1047–1053 (2004)
2. Lundquist, M.B., Sharma, N., Kewalramani, K., Lundquist, M.B., Sharma, N., Kewalramani, K.: Patient perceptions of eye disease and treatment in Bihar India. J. Clin. Exp. Ophthalmol. **3**, 213 (2012)
3. Goh, J.K.H., Sim, S.S., Tan, G.S.W.: Retinal imaging techniques for diabetic retinopathy screening. J. Diabetes Sci. Technol. **10**, 282–294 (2016)
4. Decencière, E., Cazuguel, G., Zhang, X., Thibault, G., Klein, J.C., Meyer, F., Marcotegui, B., Quellec, G., Lamard, M., Danno, R., Elie, D., Massin, P., Viktor, Z., Erginary, A., Laÿ, B., Chabouis, A.: TeleOphta: machine learning and image processing methods for teleophthalmology. Innov. Res. BioMed. Eng. **34**(2), 196–203 (2013)
5. Sinthanayothin, C., Kongbunkiat, V., Phoojaruenchanachai, S., Singalavanija A.: Automated screening system for diabetic retinopathy. In: Proceedings of the third International Symposium on Image and Signal Processing and Analysis, pp. 915–920 (2003)
6. Sopharak, A., Uyyanonvara, B., Barman, S., Williamson T.H.: Automatic detection of diabetic retinopathy exudates from non-dilated retinal images using mathematical morphology methods. J. Comput. Med. Imaging Graph. **32**, 720–727 (2008)
7. Franklin, S.W., Rajan. S.E.: Diagnosis of diabetic retinopathy by employing image processing technique to detect exudates in retinal images. IET Image Proc. **8**, 601–609 (2014)
8. Jaya, T., Dheeba, J., Albert Singh, N.: Detection of hard exudates in colour fundus images using fuzzy support vector machine-based expert system. J Digital Imaging **28**, 761–768 (2015). Springer
9. Agurto, C., Murray, V., Yu, H., Wigdahl, J., Pattichis, M., Nemeth, S., Barriga, E.S., Soliz. P.: A multiscale optimization approach to detect exudates in the macula. IEEE J. Biomed. Health. Inf. **18**(4),1328–1336 (2014)
10. Sandur, P., Naveena, C., Aradhya, V.N.M., Nagasundara, K.B.: Segmentation of brain tumor tissues in HGG and LGG MR images using 3D U-net convolutional neural network. Int. J. Nat. Comput. Res. **7**(2), April–June (2018)
11. Liskowski, P., Krawiec, K.: Segmenting retinal blood vessels with deep neural networks. IEEE Trans. Med. Imaging **35**(11), 2369–2380 (2016)
12. Ngo, L., Han, J.H.: Multi-level deep neural network for efficient segmentation of blood vessels in fundus images. Electron. Lett. **53**(16), 1096–1098 (2017)
13. Cortinovis, D.: Retinal blood vessel segmentation with a convolutional neural network (U-net). https://github.com/orobix/retina-unet. Accessed 15 Sept 2018
14. Tan, J.H., Acharya, U.R., Bhandary, S.V., Chua, K.C., Sivaprasad, S.: Segmentation of optic disc, fovea and retinal vasculature using a single convolutional neural network. J. Comput. Sci. **20**, 70–79 (2017)
15. Tan, J.H., Fujita, H., Sivaprasad, S., Bhandary, S.V., Rao, A.K., Chua, K.C., Acharya, U.R.: Automated segmentation of exudates, haemorrhages, microaneurysms using single convolutional neural network. J. Inform. Sci. **420**, 66–76 (2017)

16. Chen, T., Lin, L., Liu, L., Luo, X., Li, X.: DISC: deep image saliency computing via progressive representation learning. IEEE Trans. Neural Netw. Learn. Syst. **27**(6), 1135–1149 (2016)
17. Quellec, G., Charrière, K., Boudi, Y., Cochener, B., Lamard, M.: Deep image mining for diabetic retinopathy screening. J. Med. Image Anal. **39**, 178–193 (2017)
18. Harangi, B., Lazar, I., Hajdu, A.: Automatic exudate detection using active contour model and region wise classification. In: Engineering in Medicine and Biology Society (EMBC), Annual International Conference of the IEEE, pp. 5951–5954 (2012)
19. Prentašic, P., Lončăric, S.: Detection of exudates in fundus photographs using deep neural networks and anatomical landmark detection fusion. Comput. Methods Programs Biomed. **137**, 281–292 (2016)
20. Welfer, D., Scharcanski, J., Marinho. D.R.: A coarse-to-fine strategy for automatically detecting exudates in color eye fundus images. Comput. Med. Imaging Graph. **34**, 228–235 (2010)
21. Mo, J., Zhang, L., Feng, Y.: Exudate-based diabetic macular edema recognition in retinal images using cascaded deep residual networks. J. Neurocomputing **290**, 161–171 (2018)

The Role of Parallelism for Rust Defect Recognition Using Digital Image Processing

Mridu Sahu, Tushar Jani, Maski Saijahnavi, Amrit Kumar and Vikas Dilliwar

Abstract Traditionally, most of the computer software were programmed for serial computing. To solve a problem, the algorithm is implemented and run in a serial manner. Parallel computation, on the other hand, uses multiple processing elements simultaneously to solve a problem. It uses threads to run the different instructions in a parallel manner. As it involves the handling of multiple program instructions by dividing them among multiple processors with the goal of running a program in very less time, it is widely used in domains like Artificial Intelligence, Digital Image Processing, Astrophysics, etc. In computer science, digital image processing is the use of different algorithms to perform image processing on digital images. This paper deals with the parallelization of the digital image processing method which determines the rust present on the images. The differentiation of the rusted and non-rusted images is done with the help of the Eigenvalues calculated for the Co-variance matrices of multiple pair of images.

Keywords Parallel computation · Digital image processing · Rust defect recognition · Image classification · Eigenvalues

This work is supported by National Institute of Technology, Raipur.

M. Sahu · T. Jani (✉) · M. Saijahnavi · A. Kumar · V. Dilliwar
Information Technology, National Institute of Technology,
Raipur 492001, Chhattisgarh, India
e-mail: tusharjani@live.com

M. Sahu
e-mail: mrisahu.it@nitrr.ac.in

M. Saijahnavi
e-mail: saijahanvi1198@gmail.com

A. Kumar
e-mail: amritkumar963@gmail.com

1 Introduction

As existing infrastructures are very old, there is a need for further developed approaches to check the quality of structure [1]. Digital Image Processing is an application of one of the approaches [2]. In the domain of civil engineering several digital image processing methods have been developed to the various areas of pavement conditions, underground pipeline inspection, and steel bridge coating assessment [3]. We opted for this domain because of the efficiency, accuracy, and objectivity provided by the methods of image processing [4]. The application of these various methods is to minimize the shortcomings of existing inspection practices [4].

The states of steel bridge painting surfaces can be assessed precisely and rapidly by application of digital image processing [5]. Likewise, machine vision-subordinate assessments can give more reliable examination results than human visual reviews. Since regular examination intensely depends on individual capacities, assessment results are blunder inclined and may have wide varieties between controllers [6]. The outcomes can be diverse relying upon individual inclinations, work encounters, and the remaining task at hand of the controllers. It is quite critical to create solid foundation condition appraisal for better upkeep of the advantages [7].

The basic idea behind the solution is the calculation of eigenvalues of a pair of images and then comparing the various eigenvalues which can provide the different results for different comparisons between images [8]. The eigenvalues can be used as a medium to distinguish between images [9].

The solution for the problem revolves around the analysis of the set of rusted images and non-rusted images which can provide an approximate comparison between a healthy bridge and a bridge which is rusted. This can be further used to determine the impact of the rusting on the bridges [10]. An important observation is that the calculation of eigenvalue for different pairs of images is independent of each other. So, various eigenvalues can be calculated in parallel threads resulting in increased efficiency of the program [1].

1.1 Parallel Processing

Parallel processing is a type of operation which splits the program and execute parts of program tasks simultaneously on multiple microprocessors, thereby reducing processing time [11]. Parallel processing might be practiced through a PC with at least two processors or by means of a computer network [12]. Simply we can define parallel processing as a type of computational operation in which many calculations or the execution of processes are carried out simultaneously [13].

As per the work requirement, the approaches that are commonly used in parallel programming are by executing code with the help of threads and also by using multiple processes [14]. Those threads will have access to the same shared memory areas [14]. If the processes are being written to the same shared memory location and at the same time then this very situation can lead to conflicts in case of improper synchronization [15].

2 Materials and Methods

2.1 Input Data

Images of Bridge Coat are processed to generate eigenvalues. For the testing purpose, we have created our own database containing 8 rusty images and 8 non-rusty images. Some of these images were taken from [16] images and some of the rusted images were clicked by us. Input images are shown in Fig. 1a, b.

2.2 Digital Image Processing Method

Implementing digital image processing method involves three stages 1. Image acquisition 2. Image processing 3. Data analysis.

Image Acquisition It involves the gathering of two types of images namely rusted and non-rusted images. In our project, we have gathered 8 rusted and 8 non-rusted images for further processing.

Image Processing In this stage, the RGB image is resized and converted to Grayscale image. A pairwise comparison is made between digital images, i.e., a non-rusted image and another non-rusted image or a rusted image and a non-rusted image to calculate eigenvalues. Procedure to calculate eigenvalues: A digital image can be represented as a two-dimensional spatial coordinate, $g(x, y)$ with a size of $M \times N$. Then, the value of g at any point (x, y) indicates the brightness of the image at

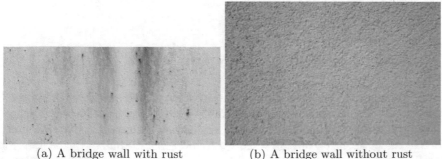

(a) A bridge wall with rust   (b) A bridge wall without rust

Fig. 1 Different images acquired

that point. Co-variance matrix of $Z(2 \times 2)$ can be calculated by adding one more reference image, i.e., $h(x, y)$, with the size same as $g(x, y)$.

$$Z_{11} = \left[\frac{1}{M \times N} \times \sum_{x=0}^{M-1} \sum_{y=0}^{N-1} g^2(x, y) \right] - (\overline{g})^2 \tag{1}$$

$$Z_{22} = \left[\frac{1}{M \times N} \times \sum_{x=0}^{M-1} \sum_{y=0}^{N-1} h^2(x, y) \right] - (\overline{h})^2 \tag{2}$$

$$Z_{12} = Z_{21} = \left[\frac{1}{M \times N} \times \sum_{x=0}^{M-1} \sum_{y=0}^{N-1} g(x, y) \cdot h(x, y) \right] - (\overline{g} \cdot \overline{h}) \tag{3}$$

There are two eigenvalues obtained from the symmetrical matrix, Z. The equations to get the values are as follows:

$$\lambda_L = \frac{1}{2} \times \left[Z_{11} + Z_{22} + \sqrt{(Z_{11} - Z_{22})^2 + 4Z_{12}^2} \right] \tag{4}$$

$$\lambda_S = \frac{1}{2} \times \left[Z_{11} + Z_{22} - \sqrt{(Z_{11} - Z_{22})^2 + 4Z_{12}^2} \right] \tag{5}$$

where $\lambda_L > \lambda_S$. The pairwise comparisons performed were: two non-rusted images and a rusted image and a non-rusted image. Since we have gathered 8 rusted and 8 non-rusted images. Total 28 and 64 eigenvalue pairs were obtained from the comparison of non-rusted images pairs and 1 non-rusted image and 1 rusted image, respectively. Eigenvalues were generated and distributed on a two-dimensional distribution map.

Data Analysis In data analysis, five values Maximum, Minimum, Mean, Variance, and Standard Deviation were calculated for the pair of eigenvalues in both the comparisons, i.e., comparison between two non-rusted images and a rusted image and a non-rusted image.

2.3 Parallelizing the Image Processing Method

There was a need to calculate 28 eigenvalues for the comparison of two non-rusted images and 64 values for the comparison of a non-rusted image and a rusted image. Calculating eigenvalue involves complex mathematical formula it will take more time to calculate one eigenvalue when we follow serial execution. Hence need for parallel programming arises here. There are various aspects where parallel processing can be applied. It can either be applied for parallelizing the tasks in processing of individual images, i.e., conversion into gray scale image, calculation of co-variance matrix, calculation of eigenvalues, etc. Other than this, the eigenvalue calculation for all the image pairs can be parallelized. In this, various image pairs are compared in parallel threads which makes the computation faster. For achieving parallelization multiprocessing library is used. This package of multiprocessing offers both local and remote concurrency along with effectively side-stepping the Global Interpreter Lock by using subprocesses instead of threads. It runs on both Unix as well as Windows [17]. In multiprocessing, first of all, the processes are spawned by the creation of a Process object and then by calling its start() method. The process follows the thread API [17]. Several other APIs are introduced within the multiprocessing module that does not have analogs in the threading module [17, 18]. We use the pool class from the multiprocessing module for our algorithm [19].

2.4   Algorithm

To the above algorithm, all pairs of non-rusted images are fed to generate the eigenvalues. Then all pairs of 1-rusted and 1-non-rusted images are fed to generate eigenvalues for the same.

Algorithm 1 Computation of Eigen Values

1: **procedure** CALCEIGENVALUE(image1, image2)

2:    $g \leftarrow$ grayScale($image1$)/255, $h \leftarrow$ grayScale($image2$)/255

3:    $Z_{11} \leftarrow [\dfrac{1}{M \times N} \times \displaystyle\sum_{x=0}^{M-1} \sum_{y=0}^{N-1} g^2(x, y)] - (\overline{g})^2$

4:    $Z_{21} \leftarrow Z_{12} \leftarrow [\dfrac{1}{M \times N} \times \displaystyle\sum_{x=0}^{M-1} \sum_{y=0}^{N-1} g(x, y).h(x, y)] - (\overline{g}.\overline{h})$

5:    $Z_{22} \leftarrow [\dfrac{1}{M \times N} \times \displaystyle\sum_{x=0}^{M-1} \sum_{y=0}^{N-1} h^2(x, y)] - (\overline{h})^2$

6:    $\lambda_L \leftarrow \dfrac{1}{2} \times [Z_{11} + Z_{22} + \sqrt{(Z_{11} - Z_{22})^2 + 4Z_{12}^2}]$

7:    $\lambda_S \leftarrow \dfrac{1}{2} \times [Z_{11} + Z_{22} - \sqrt{(Z_{11} - Z_{22})^2 + 4Z_{12}^2}]$

8:    $return(\lambda_S, \lambda_L)$

---

3   Results

In the Fig. 3, the output shows the plot of the eigenvalues for the 2 cases—all pairs of rusty images and all pairs of 1 rusty and 1 non-rusty image. In the same figure, λ_L is on the vertical axis while λ_S on the horizontal axis. The difference in the regions of the two figures can be observed (Fig. 2).

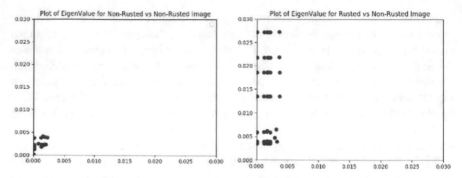

Fig. 2 Plot of eigenvalues

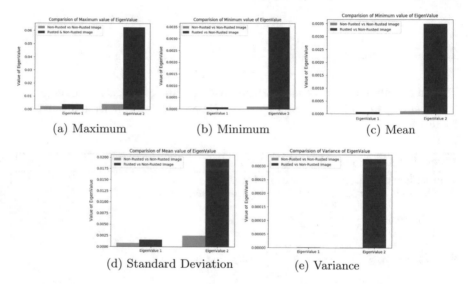

Fig. 3 Statistical parameters different eigenvalues

Table 1 Time comparison of the execution in serial and parallel environment

Processing time for	Serial execution (s)	Parallel execution (s)
2 non-rusty images	16.546	6.198
1 non-rusty image and 1 rusty image	31.013	17.285
Total processing time	47.559	23.483

Figure 3a–e show the maximum, minimum, mean, standard deviation, and variance of the different eigenvalues, respectively. Here it can be observed that λ_L changes drastically for the two cases (Table 1).

The time indicates the increased efficiency of the program while in the parallel environment. We run this program in an Ubuntu OS, i3 processor with 4 GB ram and quad core processor of clock speed 1.7 GHz.

4 Conclusion

The paper gives an insight on how can the eigenvalue calculated for a pair of images be useful for the distinction of the images with rust from the other non-rusted ones. The process consisted of three parts: image acquisition, image processing, and analysis and visualization of the parameters. In image acquisition, we generated two image sets namely defective and non-defective class. The defective images contained the images of bridge with rust while non-defective contained the images of bridges with no rust. In image processing stage, the generated images were given as input to the algorithm in a pairwise manner of 2 non-defective images and 1 defective and 1 non-defective image. A total of 28 eigenvalues were generated for the 1st comparison while 64 eigenvalues were generated for the second comparison. In the Analysis and visualization stage, we plotted the eigenvalues for both the cases and compared the results by the parameters Maximum, Minimum, Mean, Variance, and Standard Deviation. The results show that the two cases show a huge difference between both the cases and thus can be used as a parameter to separate images into the two categories of rusted and non-rusted images. To further improve the efficiency of the algorithm, we have implemented it in the multithreaded environment in python. The result indicates that when running in a parallel environment, the run-time of the algorithm is reduced to less than half of that of the serial execution. This result can further be improved if it is run on a system with better hardware and the number of cores is high.

References

1. Lee, S.: Automated defect recognition method by using digital image processing. In: Proceedings of the 46th Annual International Conference by Associated Schools of Construction (ASC), pp. 7–10 (2010)
2. Sinha, S.K., Fieguth, P.W., Polak, M.A.: Computer vision techniques for automatic structural assessment of underground pipes. Comput.-Aided Civ. Infrastruct. Eng. **18**(2), 95–112 (2003)
3. Lee, S., Chang, L.M., Chen, P.H.: Performance comparison of bridge coating defect recognition methods. Corrosion **61**(1), 12–20 (2005)
4. Bonnin-Pascual, F., Ortiz, A.: Corrosion detection for automated visual inspection. In: Developments in Corrosion Protection. In techOpen (2014)
5. Yeum, C.M., Dyke, S.J.: Vision-based automated crack detection for bridge inspection. In: Comput.-Aided Civ. Infrastruct. Eng. **30**(10), 759–770 (2015)
6. Koch, Christian, et al.: A review on computer vision based defect detection and condition assessment of concrete and asphalt civil infrastructure. Adv Eng Inf **29**(2), 196–210 (2015)
7. Jahanshahi, M.R., et al.: A survey and evaluation of promising approaches for automatic image-based defect detection of bridge structures. Struct. Infrastruct. Eng. **5**(6), 455–486 (2009)

8. Lee, S.: Color image-based defect detection method and steel bridge coating. In: Texas 47th ASC Annual International Conference Proceedings, Texas (2011)
9. Shen, Heng-Kuang, Chen, Po-Han, Chang, Luh-Maan: Automated steel bridge coating rust defect recognition method based on color and texture feature. Autom. Constr. **31**, 338–356 (2013)
10. Luiz, A.O.M., Flávio, L.C.P., Paulo, E.M.A.: Automatic detection of surface defects on rolled steel using computer vision and artificial neural networks. In: IECON 2010-36th Annual Conference on IEEE Industrial Electronics Society. IEEE, pp 1081–1086 (2010)
11. Stone, H.S.: Parallel processing with the perfect shuffle. IEEE Trans. Comput. **100**(2), 153–161 (1971)
12. Flatt, H.P., Kennedy, K.: Performance of parallel processors. Parallel Comput. **12**(1), 1–20 (1989)
13. Roosta, S.H.: Parallel Processing and Parallel Algorithms: Theory and Computation. Springer Science & Business Media, New York (2012)
14. Kirk, D.B., Wen-Mei, W.H.: Programming Massively Parallel Processors: a Hands-On Approach. Morgan kaufmann, Amsterdam (2016)
15. Yang, Chao-Tung, Huang, Chih-Lin, Lin, Cheng-Fang: Hybrid CUDA, OpenMP, and MPI parallel programming on multicore GPU clusters. Comput. Phys. Commun. **182**(1), 266–269 (2011)
16. Points. Rusted painted metal wall. Rusty metal background with streaks of rust. The metal surface rusted spots. https://depositphotos.com/203792762/stock-photo-rusted-painted-metal-wall-rusty.html
17. Multiprocessing—process-based parallelism. Python Software Foundation. https://docs.python.org/3.4/library/multiprocessing.html (2019)
18. Multiprocessing in python. Linux J. https://www.linuxjournal.com/content/multiprocessing-python (2019)
19. Raschka, S.: An introduction to parallel programming using Python's multiprocessing module - using Python's multiprocessing module. https://sebastianraschka.com/Articles/2014_multiprocessing.html (2019)

An Enhanced Bit-Map-Assisted Energy-Efficient MAC Protocol for Wireless Sensor Networks

Kumar Debasis and M. P. Singh

Abstract The proposed model, enhanced bit-map-assisted energy-efficient (EBEE) medium access control (MAC) protocol, adopts a centralized approach toward cluster head election. The base station elects the node with the highest residual energy as the cluster head in each grid. The elected cluster heads broadcast time-division multiple access (TDMA) schedules in their clusters. Each non-cluster-head node is allocated one control slot in the control period of a frame. If a node wants to claim one or more data slots within a frame, it has to send a control message in its control slot. However, the node should know how many data slots were claimed by source nodes owning the previous control slots, if any. So, the node keeps its radio ON from the beginning of the control period until the end of its control slot. The proposed model is compared with an existing TDMA-based MAC protocol, namely bit-map-assisted (BMA) MAC. The simulation results show that EBEE-MAC saves more energy compared to BMA-MAC.

Keywords Cluster head · Schedule · Control slot · Data slot · Energy

1 Introduction

A wireless sensor network (WSN) is a collection of numerous autonomous devices (sensor nodes) that are deployed in a region to monitor the physical or environmental conditions and report it to a base station (BS). Over the years, the popularity of WSNs has motivated researchers worldwide to work on the major issues that affect its design and performance. Some of the issues like energy consumption, time synchronization, localization, security, and so on, have attracted more interest than others. During the past 20 years, numerous energy-efficient routing protocols were proposed to set up low-cost paths between sensor nodes and the BS [1]. Similarly, researchers

K. Debasis (✉) · M. P. Singh
National Institute of Technology Patna, Patna 800005, Bihar, India
e-mail: debasis.cse14@nitp.ac.in

M. P. Singh
e-mail: mps@nitp.ac.in

© Springer Nature Singapore Pte Ltd. 2020
V. Bhateja et al. (eds.), *Intelligent Computing and Communication*,
Advances in Intelligent Systems and Computing 1034,
https://doi.org/10.1007/978-981-15-1084-7_38

also worked on clustering techniques for WSNs [2]. In Singh et al. [3], the authors proposed an energy-efficient cluster-based routing protocol. In Parwekar and Rodda [4], the authors used genetic algorithm to optimize the number of clusters in the network. The work in Mahapatra and Shet [5] focused on the position information of nodes in an indoor environment. To this end, a received signal strength (RSS) based algorithm was used. In Pegatoquet et al. [6], the authors proposed a wake-up radio-based medium access control (MAC) protocol for WSNs.

Low-energy adaptive clustering hierarchy (LEACH) [7] is a time division multiple access (TDMA) based MAC protocol. It follows a decentralized approach toward cluster head (CH) election. Elected CHs broadcast advertisement messages and receive join-request (JOIN-REQ) messages from interested non-cluster-head (non-CH) nodes. For any non-CH, the decision to join a CH depends on the signal strength of the advertisement message. After receiving JOIN-REQ messages, each CH broadcasts a transmission schedule for its cluster members. A source node (node that has data) transmits in the allocated slot and keeps its radio OFF in all other slots. After receiving data from all the source nodes, CH aggregates and sends it to the BS.

Bit-map-assisted (BMA) MAC [8] follows the same cluster formation process as LEACH. However, a source node has to send a control message in the allocated control slot. The CH broadcasts a transmission schedule based on the number of control messages received. Each source node transmits in the allotted data slot and keeps its radio OFF in the other slots.

Bit-map-assisted energy-efficient (BEE) MAC [9] also follows the same cluster formation process as LEACH. A source node transmits a control message in the allocated control slot. But, it has to keep its radio ON from the beginning of the control period (total duration of all control slots) until the end of its allocated control slot. This is necessary as the node needs to know how many data slots were booked by source nodes owning the preceding control slots, if any. Based on this information, the node can successfully claim one or more subsequent data slots.

The proposed model, enhanced bit-map-assisted energy-efficient (EBEE) MAC protocol, is an improvement over BEE-MAC. EBEE-MAC adopts a centralized approach toward CH election unlike LEACH and the other two protocols. This ensures that the CHs are evenly distributed in the network. If the residual energy of a node falls below a predefined threshold value, it is not considered in the CH election procedure henceforth. This ensures that a newly elected CH has sufficient energy to sustain until the end of the round.

2 Proposed Model

The proposed model assumes that the deployment area is composed of equal-sized grids. The BS is well aware of the dimensions and the boundaries of all the grids. All the sensor nodes that fall within a grid are by default considered to be the part of the same cluster. All the sensor nodes are homogenous, stationary, and fitted with global positioning system (GPS) devices that can be used to know their locations.

It is assumed that each node in the deployment area can transmit to the BS directly. This paper considers the energy consumption in the communication process only.

The proposed model divides a round into (i) set-up phase and (ii) steady-state phase. During the set-up phase, BS elects CHs for the round, and non-CH nodes join their respective CHs. During the steady-state phase, non-CH nodes send data to their CHs which are then forwarded to the BS.

Set-up Phase. After deployment, each sensor node transmits a HELLO message to the BS. This message basically consists of the identity (ID), location (LOC), and residual energy (RE) of the node. After receiving HELLO messages from all the sensor nodes, BS chooses the CH for each grid. In each grid, the node with the highest RE is elected as the CH. If two or more nodes fulfill this criterion, any node can be elected at random.

Next, BS broadcasts a duty allocation (DUTY) message that contains the CH identity (CH_ID) for each node. If the CH_ID for a node matches the ID of the node, then the node itself has been elected as a CH for the current round. If the CH_ID for the node does not match the node's ID, then the node has not been elected as a CH for the current round. In this case, the node has to join its CH by sending a JOIN-REQ message to it. After a timeout period, the CH broadcasts a schedule (SCH) message in its cluster. This message allocates control slots and provides timing and synchronization information to all the non-CH nodes in the cluster.

Steady-state Phase. The steady-state phase consists of k frames of equal size. Each frame is composed of a control period, a data transmission period, and an idle period. Each non-CH node is allotted one control slot in the control period. A source node turns its radio ON at the beginning of the control period, and keeps it ON until the end of its control slot. It is necessary for the node to learn how many data slots were claimed by source nodes owning the preceding control slots, if any. When this information is obtained, the node can safely claim one or more subsequent data slots for itself.

To claim data slot(s), a source node sends a control message in its control slot. After sending the control message, the node turns its radio OFF until the beginning of its data slot(s) in the data transmission period. As the CH was ON throughout the duration of the control period, it knows how many data slots have been claimed. The CH stays ON for that duration to receive data messages from all the source nodes. After collecting data from all the source nodes, the CH sends the aggregate value to the BS. During the idle period, all the nodes within the cluster turn their radios OFF to save energy.

There is a small difference between the set-up phase of the first round and the set-up phase of all the other rounds. In the beginning of the set-up phase of all the other rounds, each CH broadcasts a QUERY message in its cluster. All the non-CHs respond with their IDs and REs. The CH sends this information to the BS along with its own information (ID and RE). In each grid, the BS chooses the node with the highest RE as the CH for the current round. A node whose RE falls below a predefined threshold value is not considered in the CH election procedure henceforth. This condition ensures that the newly elected CH has sufficient energy to sustain in the current round. Figure 1 shows the workflow of the proposed model.

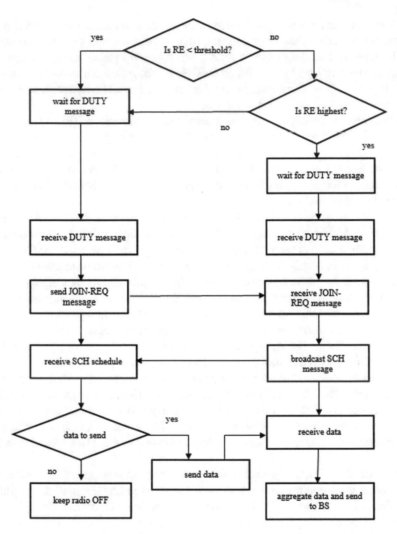

Fig. 1 Workflow of EBEE-MAC protocol

3 Results and Analysis

The python programming language has been used to carry out all the simulations. In all the simulations, the first-order radio model [7] is used to calculate the energy consumption in sensor nodes. Table 1 shows the parameters that are used in the simulations. The first simulation is done by setting the number of frames to 10 and the second simulation is done by setting the number of frames to 20. In both the cases, the desired percentage of cluster heads is 12% and the number of nodes is 100. In the proposed model, both the cases are implemented by dividing the deployment area

Table 1 Simulation parameters

Parameter	Value
Deployment area	$100 \times 100 \, \text{m}^2$
Initial energy of node	5 J
Eelec (energy spent in running the transmitter or receiver circuitry)	50 nJ/bit
Eidle (energy spent in idle mode)	40 nJ/bit
ϵamp (energy spent in running the transmit amplifier)	$100 \, \text{pJ/bit/m}^2$
Threshold energy	0.1 J
Data/schedule message size	100 Bytes
Control message size	20 Bytes
Desired percentage of CHs	12%
Number of frames	10/20
Number of nodes	200/300

into 12 equal-sized grids. The BS is assumed to be located outside the deployment area but very close to it.

Figure 2 shows the graph that is generated after the first simulation and Table 2 shows when the first node and the last node of each protocol dies in the first simulation.

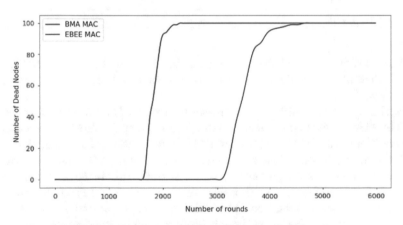

Fig. 2 Number of dead nodes versus number of rounds, when number of frames is 10, desired percentage of cluster heads is 12%, and number of nodes is 100

Table 2 Round number in which the first and the last node of each protocol dies, when number of frames is 10, desired percentage of cluster heads is 12%, and number of nodes is 100

Protocol	Round number in which the first node dies	Round number in which the last node dies
BMA-MAC	1619	2280
EBEE-MAC	3076	4540

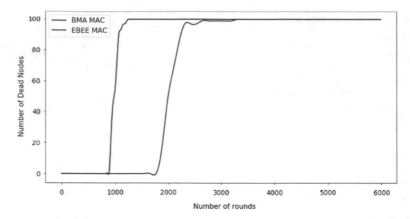

Fig. 3 Number of dead nodes versus number of rounds, when number of frames is 20, desired percentage of cluster heads is 12%, and number of nodes is 100

Table 3 Round number in which the first and the last node of each protocol dies, when number of frames is 20, desired percentage of cluster heads is 12%, and number of nodes is 100

Protocol	Round number in which the first node dies	Round number in which the last node dies
BMA-MAC	859	1220
EBEE-MAC	1766	3260

Figure 3 shows the graph that is generated after the second simulation and Table 3 shows when the first node and the last node of each protocol dies in the second simulation.

Table 2 shows that EBEE-MAC increases the network lifetime by 99% compared to BMA-MAC in the first simulation. Table 3 shows that EBEE-MAC increases the network lifetime by 167% compared to BMA-MAC in the second simulation.

The third simulation is done by setting the number of nodes to 200 and the fourth simulation is done by setting the number of nodes to 300. In both the cases, the number of frames is 10 and the desired percentage of cluster heads is 12%. In the proposed model, the first case is implemented by dividing the deployment area into 24 equal-sized grids. Similarly, the second case is implemented by dividing the deployment area into 36 equal-sized grids.

Figure 4 shows the graph that is generated after the third simulation and Table 4 shows when the first node and the last node of each protocol dies in the third simulation. Figure 5 shows the graph that is generated after the fourth simulation and Table 5 shows when the first node and the last node of each protocol dies in the fourth simulation.

Table 4 shows that EBEE-MAC increases the network lifetime by 57% compared to BMA-MAC in the third simulation. Table 5 shows that EBEE-MAC increases the network lifetime by 42% compared to BMA-MAC in the fourth simulation.

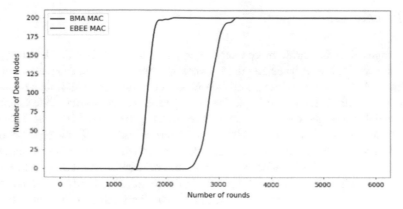

Fig. 4 Number of dead nodes versus number of rounds, when number of frames is 10, desired percentage of cluster heads is 12%, and number of nodes is 200

Table 4 Round number in which the first and the last node of each protocol dies, when number of frames is 10, desired percentage of cluster heads is 12%, and number of nodes is 200

Protocol	Round number in which the first node dies	Round number in which the last node dies
BMA-MAC	1432	2110
EBEE-MAC	2425	3312

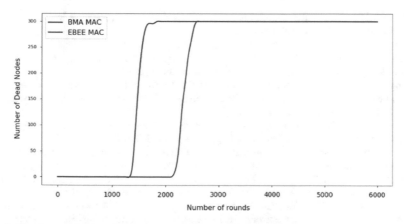

Fig. 5 Number of dead nodes versus number of rounds, when number of frames is 10, desired percentage of cluster heads is 12%, and number of nodes is 300

Table 5 Round number in which the first and the last node of each protocol dies, when number of frames is 10, desired percentage of cluster heads is 12%, and number of nodes is 300

Protocol	Round number in which the first node dies	Round number in which the last node dies
BMA-MAC	1335	1802
EBEE-MAC	2110	2561

4 Conclusion

The proposed model saves more energy compared to BMA-MAC. Therefore, the proposed model is able to extend the network lifetime compared to its counterpart. In BMA-MAC, source nodes as well as non-source nodes lose a significant amount of energy in idle listening during the control period. A non-source node listens to the schedule message even if it does not have anything to send. EBEE-MAC reduces idle listening by allowing a non-source node to turn its radio OFF during the control period. A source node too turns its radio OFF after sending the control message. In EBEE-MAC, a CH broadcasts the schedule message in the set-up phase of a round. This schedule is followed throughout the round. EBEE-MAC also adopts a centralized approach toward cluster head election. This ensures that the CHs are evenly distributed in the network.

Acknowledgements The authors would like to acknowledge the Ministry of Electronics and Information Technology (MeitY), Government of India, for supporting the research work through "Visvesvaraya Ph.D. Scheme for Electronics and IT".

References

1. Parwekar, P.: Comparative study of hierarchical based routing protocols: wireless sensor networks. In: ICT and Critical Infrastructure: Proceedings of the 48th Annual Convention of Computer Society of India, vol. I, pp. 277–285. Springer (2014)
2. Kalla, N., Parwekar, P.: A study of clustering techniques for wireless sensor networks. Smart Computing and Informatics, pp. 475–485. Springer, Singapore (2018)
3. Singh, D.P., Bhateja, V., Soni, S.K., Shukla, A.K.: A novel cluster head selection and routing scheme for wireless sensor networks. Advances in Signal Processing and Intelligent Recognition Systems, pp. 403–415. Springer, Cham (2014)
4. Parwekar, P., Rodda, S.: Optimization of clustering in wireless sensor networks using genetic algorithm. Int. J. Appl. Metaheuristic Comput. **8**(4), 84–98 (2017). IGI Global
5. Mahapatra, R.K., Shet, N.S.V.: Localization based on RSSI exploiting gaussian and averaging filter in wireless sensor network. Arab. J. Sci. Eng. **43**(8), 4145–4159 (2018). Springer
6. Pegatoquet, A., Le, T.N., Magno, M.: A wake-up radio-based MAC protocol for autonomous wireless sensor networks. IEEE/ACM Trans. Netw. **27**(1), 56–70 (2019). IEEE
7. Heinzelman, W.R., Chandrakasan, A., Balakrishnan, H.: Energy-efficient communication protocol for wireless microsensor networks. In: Proceedings of the 33rd Annual Hawaii International Conference on System Sciences. IEEE (2000)
8. Li, J., Lazarou, G.Y.: A bit-map-assisted energy-efficient MAC scheme for wireless sensor networks. In: Proceedings of the 3rd International Symposium on Information Processing in Sensor Networks, pp. 55–60. ACM, New York (2004)
9. Debasis, K., Singh, M.P.: Bit-map-assisted energy-efficient MAC protocol for wireless sensor networks. Int. J. Adv. Sci. Technol. **119**, 111–122 (2018). NADIA

Intuitionistic Fuzzy Logic Controller Over the Irrigation Field to Optimized Water Utilization

S. Rajaparakash, R. Jaichandran, S. Muthuselvan, K. Karthik and K. Somasundaram

Abstract The intuitionistic fuzzy set and intuitionistic fuzzy logic was introduced by Atanassov. It will give the very good output over the uncertainty because it contains membership and non-membership function. The intuitionistic fuzzy system provides a mechanism for communication between computer system and humans. In this work we describe that the development of an intuitionistic fuzzy logic controller control the motor switch on and off for irrigation. In irrigation field and well water level measured by the sensors. The intuitionistic fuzzy system and mamdami intuitionistic fuzzy rules and Takagi Sugni Formula for the defuzzification are used to get the crisp output from an intuitionistic fuzzy input. The well water level and water level in agriculture land are calculated by applied using intuitionistic fuzzy rules applied in an inference engine using defuzzification method.

Keywords Intuitionistic · Fuzzy logic · Irrigation · Utilization · Optimized

1 Introduction

The water system is a fundamental procedure in agriculture that impacts efficiency of production in crop. Well water is one of the primary hot spots for the water system.

S. Rajaparakash (✉) · R. Jaichandran · S. Muthuselvan · K. Karthik
Aarupadai Veedu Institute of Technology, Vinayaka Mission's Research Foundation, Salem, India
e-mail: srajaprakash_04@yahoo.com

R. Jaichandran
e-mail: rjaichandran@gmail.com

S. Muthuselvan
e-mail: csmuthuselvan@gmail.com

K. Karthik
e-mail: karthik@avit.ac.in

K. Somasundaram
Dept. of CSE, Sri Venkateswara College of Engineering and Technology, Chittoor, Andhra Pradesh, India
e-mail: soms72@yahoo.com

© Springer Nature Singapore Pte Ltd. 2020
V. Bhateja et al. (eds.), *Intelligent Computing and Communication*,
Advances in Intelligent Systems and Computing 1034,
https://doi.org/10.1007/978-981-15-1084-7_39

Ranchers for the most part visits their developed fields intermittently to check soil dampness level and in light of prerequisite, water system engine changed to flood particular fields. Agriculturist needs to sit tight for a certain period before turning off the engine with the goal that adequate water streams in the field. This manual water system technique takes a part of time and exertion, especially when rancher needs to inundate numerous farming fields circulated in various topographical regions.

Robotization in the water system framework influences ranchers to work substantially less demanding. Sensor based computerized water system framework gives promising answer for ranchers where the nearness of agriculturist in the field is not compulsory to perform a water system process. Robotized water system frameworks are electromechanical customized to control mechanical gadgets like a water system engine, water channels, open/close valves, and so on in light of the criticism of sensor hub set in water systems fields. PCs can be utilized as a part of observing and controlling water system movement. In any case, progressively, ranchers require shabby and straightforward UI for controlling sensor based computerized water system framework. These days a cell phone is the basic gadget utilized by farmers. Along this line cell phone is utilized as a part of sensor based robotized water system framework. This causes ranchers to control the water system process effectively from remote areas.

Already there is a built up model for programmed controlling and remote for getting water system engine. In the model, sensor hub sense and send soil dampness in the field of client cell phone. Here a client sends signs utilizing enrolled cell phone to switch on/off water system engine are necessary. Existing framework considers water in a well is constantly accessible. But that as it may not be valid continuously. Especially in amid summer season, water in the well is constrained and may dry out in the center of the water system process. Along these lines we propose a remote sensor organize model for programmed controlling programme and getting remote water system engine for successful well water system in water level unnerve conditions. The proposed model will turn off the water system engine when water in the well is low. Again the water system engine will be exchanged on when regular springs discharges adequate water in well. This procedure proceeds in a round-robin way till all fields get adequate water from well.

1.1 Intuitionistic Fuzzy Set

The generalization of fuzzy sets is Intuitionistic fuzzy sets (IFS). In 1983, Atanassov introduced the concept of IFS. He introduces the new component which determines the degree of nonmember ship function. Fuzzy sets give the degree of membership for an element in a given set; the nonmember ship of degree equals gives the degree of membership.

Intuitionistic fuzzy sets, which are higher order of five fuzzy sets, for both a degree of membership and a degree of nonmember-ship, which are more-or-less independent of each other; the only requirement is that the sum of these two degrees is not greater than 1. Application of higher order fuzzy sets makes the solution-procedure more complex, but if the complexity of computation-time, computation-volume and memory-space are not concern now a days we can be achieved better results. Agarwal et al. Presented the design of a probabilistic intuitionistic fuzzy rule based controller.

Let U be the universe and A be the subset of U then the intuitionistic fuzzy set (IFS) is defined by $A^* = \{\langle x, 0 \leq \mu_A(x) + \nu_A(x) \leq 1 \rangle | x \in U \}$

where $0 \leq \mu_A(x) + \nu_A(x) \leq 1$.

Thefunctions $\mu_A(x) : U \to [0, 1]$ and $\nu_A(x) : U \to [0, 1]$ represent the Membership and Non Membership functions. $\pi(x) = 1 - \mu_A(x) + \nu_A(x)$ represent the degree of indeterrminacy or hesitation degree [1]. Where $\pi(x) : U \to [0, 1]$. Obviously every Fuzzy set has the form $\{\langle x, \mu_A(x), 1 - \mu_A(x) \rangle | x \in U \}$.

- Triangular Intuitionistic Fuzzy Number

A triangular intuitionistic fuzzy number $A(x)$ is an intuitionistic fuzzy set in R with membership function and nonmembership function as given below.

$$\mu_A(x) = \begin{cases} \frac{x-a_1}{a_2-a_1} & \text{for } a_1 \leq x \leq a_2 \\ \frac{a_3-x}{a_3-a_2} & \text{for } a_2 \leq x \leq a_3 \\ 0 & \text{otherwise} \end{cases} \tag{1}$$

$$\nu_A(x) = \begin{cases} \frac{(a_2-x)}{a_2-a_1'} & \text{for } a_1' \leq x \leq a_2 \\ \frac{x-a_2}{\beta'} & \text{for } a_2 \leq x \leq a_3' \\ 1 & \text{otherwise} \end{cases}$$

where $a_1' \leq a_1 \leq a_2 \leq a_3 \leq a_3'$

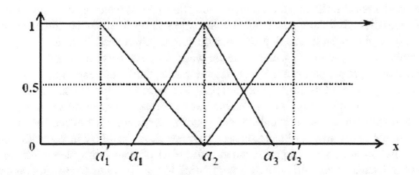

Definition 2.1 ([1]) For every common fuzzy subset An on X, intuitionistic fuzzy set of x in A is defined as. Is also called the degree of hesitancy or the degree of uncertainty of the element x in A.

2 Literature Survey

In this section a survey of the existing method of irrigation, intuitionistic set in uncertainty are discussed. In the Sect. 2.1 a detail study of the irrigation using various technologies is given and in the Sect. 2.2 detail studies about the intuitionistic fuzzy set in various fields is discussed. In the Sect. 2.3 contain gap identification in the existing system is given.

2.1 Irrigation Using Various Technologies

In 2016 Manimaran and Yasar Arfath [2] proposed new model, in that they discussed if whether the free power has been utilized barring electric engines for pumping water without power is being abused, it close the aggregate reserve for the agriculturists through a stumbling circuit. By utilizing remote systems the power board about mal tradition can be insinuated. The advancement of this task at exploratory level inside provincial territories is introduced and the execution needs to display that the programmed water system can be utilized to diminish water utilize.

Khriji et al. [3] introduced an entire water system answer for the ranchers in view of WSN. The robotized water system framework utilizing minimal effort sensor hubs having lessened power utilization can decrease the water squander and is financially savvy. A hub is conveyed utilizing Telos bit and sufficient sensors/actuators. Field hubs are utilized to distinguish the level of dampness and temperature in the dirt. Climate hubs screen the climatic changes, and the hubs associated with actuators are utilized to control the opening of the water system valve whenever required.

Mahir and Ozden [4] proposed a proficient water utilization framework by pump control decrease utilizing sun oriented controlled trickle water system framework in a plantation. Soil dampness content is investigated by Artificial Neural Networks (ANN) to give even circulation of water to the required area. This will keep the pointless water system and decrease the use water request. This framework decreases the use of water plantation every day water use and vitality utilization by 38 rates.

Goumopolulos et al. [5] depicted a versatile choice emotionally supportive network and its combination with a remote sensor/actuator system to execute independent shut circle zone-particular water system. Cosmology is utilized for characterizing the application rationale stresses framework adaptability in versatility and backings the utilization of programmed inferential of approval systems. Machine learning is connected with rules by investigating logged datasets for separating new information and expanding the framework.

Coates et al. [1] advanced a business remote detecting and control systems utilizing valve control equipment and programming. The valve incitation framework included advancement of custom hub firmware, actuator equipment firmware, a web portal with control, correspondence and web interface programming. The framework utilizes single jump radio range utilizing a work connects with 34 valve actuators for controlling the valves and water meters.

Nolz et al. [6] contemplated two kinds of soil water potential sensors. One is the Watermark sensor by Irrometer Co., and the other is MPS-1 by Decagon Devices, Inc. The framework was incorporated with sensors into a remote observing system to decide and assess adjustment capacities for the coordinated sensors. The framework thinks about the estimating range and the response time of both sensor writes in a dirt layer amid drying. Information is transmitted more than a few kilometers and made accessible by means of Internet get to.

Goumopoulos et al. [7] depicted the outline of a versatile choice emotionally supportive network and its combination with a remote sensor/actuator system to actualize independent shut circle zone-particular water system. Utilizing cosmology for characterizing the application rationale stresses framework adaptability and versatility and backings the utilization of programmed inferential and approval instruments. A machine learning process is connected for instigating new principles by breaking down logged datasets for extricating new information and expanding the framework cosmology with a specific end goal to adapt.

Chate and Rana [8] proposed framework has exhibited the new inventive water system framework. This framework contains the live spilling of products utilizing android telephones and programmed engine on/off framework, these two frameworks make the water system completely programmed.

2.2 Survey on Intuitionistic Fuzzy Set in Various Field

In the year 1996, Burillo and Bustince [9] has been learned around two hypothesis in intuitionistic fuzzy set, the hypothesis which permits to build an intuitionistic fuzzy set from two fuzzy sets. The creators additionally examined about the recoup the fuzzy sets utilized as a part of the development from the intuitionistic fuzzy sets. The new meaning of separation between two intuitionistic fuzzy separates from that Atanassov examined by Szmidt and Kacprzyk [10] in the year 2000. Intuitionistic fuzzy combine, intuitionistic fuzzy couple and intuitionistic fuzzy esteem have been contemplated by Atanassov et al. [11] in 2013. Intuitionistic Fuzzy Delphi Method used as evaluating instrument in light of ace's proposal. They used triangular cushy number and combination process in light of the sentiment of the ace proposed (Tapan Kumar Roy et al. [12] in 2012). The idea of intuitionistic fuzzy t-standard and t-co norm is as found in 2002 by Deschrijver et al. In 2007 Xu propose the intuitionistic fuzzy triangular standards. Using the intuitionistic fuzzy sets as a tool diagnosis of disease D based on the symptoms S was studied by Szmidt and Kacprzyk [13] in 2001. In that work the author used set of symptoms as a database. A non-probabilistic type entropy measure for intuitionistic fuzzy set was proposed by Szmidt and Kacprzyk [13] and in this work the measure is consistent with ordinary fuzzy sets. Utilizing the Intuitionistic fuzzy rationale built up an intuitionistic fuzzy framework is built up to control the radiator fans. In this work the spew Chris d of the radiator fan is computed utilizing intuitionistic fuzzy standards connected in a derivation motor and defuzzification strategy by Akram et al. [14] in 2013. Socio metric questionnaire has been developed to identify of the social status of a pupil in a school class using the intuitionistic fuzzy set. It has been studied by Magdalena Rencova in 2009. The investigation of capital pointer and positioning by utilizing IFAHP is utilized to assess the four fundamental markers of Human capital marker simultaneously by means of master judgment it has been finished by Chate and Rana [8] in 2013.

Anusha and Gouthami [15] have been study about the framework comprises of a remote system of dirt dampness sensor and temperature sensor put under the dirt where plants roots are achieved in conveyed organize. The framework has a water level sensor which will demonstrate the nearness of water level in tank. A portal unit deals with the data identified with sensors which triggers the actuators, and information is transmitted in information utilizing GSM module. The remotely detected picture where distinctive sorts of districts are removed utilizing intuitionistic fuzzy set hypothesis, consider by Chaira [16]. In his work the creator contrasted and fuzzy and non-fuzzy strategy, the new extricated picture utilizing IFS were better since of the dithering degree. Cornelis et al. [17] made an endeavor for execution of Intuitionistic fuzzy set hypothesis and second are interim esteemed fuzzy set hypothesis. Following these models one end stays aware of an imperative convention of mathematical structures for creating sensible and utilizing the grid that they characterize on and the other hand to uncover in a reasonable way the two models formal equality. Utilizing the established and fuzzy ramifications to start an arrangement system for the subsequent administrators in light of additional intelligent criteria. At least the creators giving the two methodologies as models of imprecision and apply them in a common sense setting. Utilizing the Intuitionistic fuzzy set decide the understudy expected execution was contemplated by Vasanthi and Viswanadham [18] in that of 2015 and in her work utilizing standardized Euclidean separation strategy by estimating the separation between each understudy and each subject execution is utilize separately. Jose and Kuriakose [19] proposed a strategy for taking care of multi-quality basic leadership issues in intuitionistic fuzzy condition. Likewise, an advancement demonstrate with appraise the relative level of significance of enrollment, non-participation and faltering and correcting the disadvantages in the exacting work are given by Zhi Pei and Li Zheng. Varghese and Kuriakose [20] have proposed different properties in the intuitionistic fuzzy set like symmetry, reflexivity, transitivity and so on.

2.3 Identify the Gap in Existing System

Many authors discussed about the irrigation using wireless sensor network. In that decision is taken based on the crisp value only not on the optimum solution.

Based on the survey it can be observed that many authors discussed fuzzy logic controller while decision making. It is not give the optimum solution because of several factors such as:

- If water level is less in the field automatically the water pumping motor has to run at the same time the water level of well less vice versa.
- In the fuzzy logic decision taken value belong to non-membership value, then we can't get the optimum solution.

3 Methodology

In this work it discussed design an intuitionistic fuzzy controller for irrigation system. The system contains three attributes such as well water level, field water level and motor running time. The intuitionistic fuzzy controller received the two input (well water level and field water level) and generates a single output (Motor running timing). There are three triangular membership functions for each input like low, Medium and High.

Mamdani inference method is used for membership function and non-membership function. The proposed system consists of the following modules.

- Intuitionistic fuzzifier
- Intuitionistic fuzzy inference system
- Intuitionistic defuzzifer (Fig. 1).

Intuitionistic Fuzzifier: its contain two input functions, well water level and Field water level. For each input three membership and non membership function is formed using Eq. (1).

Intuitionistic Fuzzy Inference System: it contains set of intuitionistic fuzzy rules with the combination of field water and well water level.

Intuitionistic Defuzzifer: Using the Takagi Suing Formula, intuitionistic fuzzy value converted into crisp value which is the running time of the mortar.

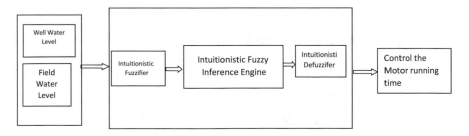

Fig. 1 Intuitionistic fuzzy system

Table 1 Comparison

S. No.	Existing system	Proposed system
1	Many authors discussed about the on and off the pumping motors with the help of crisp value	In this work on and off the pumping motors with the help of Intuitionistic fuzzy set. IFS give the optimum solution
2	Some authors used the fuzzy logic controller to control the motor. In fuzzy logic controller membership functions are used by the authors. It is discussed the gaps in the existing system in Chap. 3	In this proposed system, pumping motor is controlled by membership function, non membership function and discriminant value. So it gives the more optimum solution to utilize the water in irrigation field

3.1 Algorithm

1. Begin
2. Creation of intuitionistic fuzzy inference system
3. Identify the linguistic values of input and output variable (field water level, well water level, motor running timing).
4. Read the input values
5. Find the degree of membership function for field water level and well water level [Low, Medium, and High].
6. Find the degree of non-membership function for field water level and well water level [Low, Medium, and High].
7. Compute the intuitionistic fuzzy logic rules for degree of truth for output variable in Motor running timing.
8. Apply defuzzification to get the crisp output of motor running time.
9. Display the motor running time.
10. End.

4 Comparison with Existing Work

See Table 1.

5 Implementation Work

In the sample work based on the one field, one well and one motor running time. The field water level and well water level is fixed which is given below. In this work it is implemented in paddy field. The water levels are measured by sensor node and similarly the well water level is also measure and transmit to controller node for decision making. Based on the water level the triangular intuitionistic fuzzy membership function and non-membership are fixed. The controller node switch on or off the water pumping motor based on the intuitionistic fuzzy controller are described as follows: The dimension of the Field 50 m length, 40 m breadth and the dimension of the well is 10 m radius and 20 m depth from the surface. The water capacity of field and well water chart given below (Table 2).

References Water pumping capacity calculated by the following steps (Table 3).

- Water horsepower = minimum power required to run water pump
- TDH = Total Dynamic Head = Vertical distance liquid travels (in feet) + friction loss from pipe
- Q = flow rate of liquid in gallons per minute
- SG = specific gravity of liquid (this equals 1 if you are pumping water)
- Water horsepower = (TDH * Q * SG)/3960 (Fig. 2).

Based on the Table 1 three membership function and non membership function for each input low, medium, high is given. Based on the condition intuitionistic fuzzy rules are framed which is given below.

Intuitionistic Fuzzy Rule

1. If (F is low and W is low) then Motor running time S1
2. If (F is low and W is Medium) then Motor running time S2
3. If (F is low and W is Medium) then Motor running time S2
4. If (F is Medium and W is Medium) then Motor running time S3
5. If (F is Medium and W is High) then Motor running time S1
6. If (F is Medium and W is low) then Motor running time S1.

Table 2 Field water capacity chart in liter

S. No.	Length (m)	Breadth	Water height	Volume	Water (l)
1	50	40	0.01	20	20,000
2	50	40	0.02	40	40,000
3	50	40	0.03	60	60,000
4	50	40	0.04	80	80,000
5	50	40	0.05	100	100,000
6	50	40	0.06	120	120,000
7	50	40	0.07	140	140,000
8	50	40	0.08	160	160,000
9	50	40	0.09	180	180,000
10	50	40	0.1	200	200,000
11	50	40	0.11	220	220,000
12	50	40	0.12	240	240,000
13	50	40	0.13	260	260,000
14	50	40	0.14	280	280,000
15	50	40	0.15	300	300,000
16	50	40	0.16	320	320,000

Table 3 Well water capacity chart in liter

S. No.	Radius	Water height	Volume	Water (l)
1	10	1	31.4	31,400
2	10	2	62.8	62,800
3	10	3	94.2	94,200
4	10	4	125.6	125,600
5	10	5	157	157,000
6	10	6	188.4	188,400
7	10	7	219.8	219,800
8	10	8	251.2	251,200
9	10	9	282.6	282,600
10	10	10	314	314,000
11	10	11	345.4	345,400
12	10	12	376.8	376,800
13	10	13	408.2	408,200
14	10	14	439.6	439,600
15	10	15	471	471,000
16	10	16	502.4	502,400

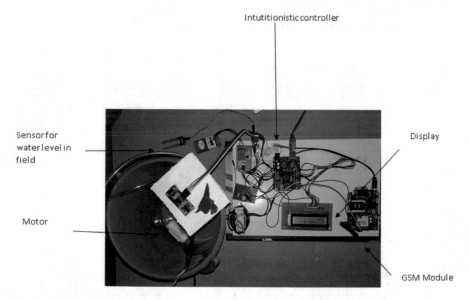

Fig. 2 Automated irrigation system using intuitionistic fuzzy logic controller in lab

Table 4 Linguistic variable

S. No.	Field water level (W)	Well water level (F)	Motor running time (min)	Stage
1.	Low	Low	15	S1
2.	Low	Medium	30	S2
3.	Low	High	60	S3
4.	Medium	Medium	30	S2
5.	Medium	High	15	S1
6.	High	High	0	S0
7.	High	Low	0	S0
8.	Medium	Low	15	S1

In this work the fuzzy sets defined by triangular membership function. Likewise, three non-membership functions are used. The membership functions are defined as follows:

Intuitionistic fuzzifiers: Based on the functional definitions of the fuzzy set three triangular membership functions are fixed. On the other hand three corresponding nonmember ship function is fixed by adding some value to the corresponding functions. The membership and nonmembership function defined by (Table 4),

Field

$$\mu_{LOW} = \begin{cases} \frac{15-x}{15-5} & \text{if } x \in (5, 15) \\ 0 & \text{else} \end{cases} \qquad \nu_{LOW} = \begin{cases} \frac{5-x}{20-5} & \text{if } x \in (5, 20) \\ 1 & \text{else} \end{cases}$$

$$\mu_{MEDIUM} = \begin{cases} \frac{x-10}{25-10} & \text{if } x \in (10, 25) \\ \frac{30-x}{30-25} & \text{if } x \in (25, 30) \\ 0 & \text{else} \end{cases} \qquad \nu_{MEDIUM} = \begin{cases} \frac{25-x}{25-10} & \text{if } x \in (10, 25) \\ \frac{x-25}{35-25} & \text{if } x \in (25, 35) \\ 1 & \text{else} \end{cases}$$

$$\mu_{HIGH} = \begin{cases} \frac{x-15}{35-15} & \text{if } x \in (15, 35) \\ \frac{35-x}{40-35} & \text{if } x \in (35, 40) \\ 0 & \text{else} \end{cases} \qquad \nu_{HIGH} = \begin{cases} \frac{35-x}{35-15} & \text{if } x \in (15, 35) \\ \frac{x-35}{45-35} & \text{if } x \in (35, 45) \\ 1 & \text{else} \end{cases}$$

Well

$$\mu_{LOW} = \begin{cases} \frac{x-600}{1200-600} & \text{if } x \in (600, 1200) \\ 0 & \text{else} \end{cases} \qquad \nu_{LOW} = \begin{cases} \frac{600-x}{1300-600} & \text{if } x \in (600, 1300) \\ 1 & \text{else} \end{cases}$$

$$\mu_{MEDIUM} = \begin{cases} \frac{x-600}{1500-600} & \text{if } x \in [600, 1500] \\ \frac{1500-x}{1600-1500} & \text{if } x \in [1500, 1600] \\ 0 & \text{else} \end{cases} \qquad \nu_{LOW} = \begin{cases} \frac{600-x}{1500-600} & \text{if } x \in [600, 1500] \\ \frac{x-1500}{1700-1500} & \text{if } x \in [1500, 1700] \\ 1 & \text{else} \end{cases}$$

$$\mu_{\text{HIGH}} = \begin{cases} \frac{x-1200}{1800-1200} & \text{if } x \in [1200, 1800] \\ \frac{1500-x}{1700-1800} & \text{if } x \in [1800, 1700] \\ 0 & \text{else} \end{cases} \quad \nu_{\text{LOW}} = \begin{cases} \frac{1200-x}{1800-1200} & \text{if } x \in [1200, 1800] \\ \frac{x-1800}{2000-1800} & \text{if } x \in [1800, 2000] \\ 1 & \text{else} \end{cases}$$

Output Motor Running Timing

$$\mu_{S1} = \begin{cases} \frac{15-x}{15-0} & \text{if } x \in (0, 15) \\ 0 & \text{else} \end{cases} \quad \nu_{S1} = \begin{cases} \frac{0-x}{20-0} & \text{if } x \subset (0, 20) \\ 1 & \text{else} \end{cases}$$

$$\mu_{S2} = \begin{cases} \frac{x-15}{45-15} & \text{if } x \in (15, 45) \\ \frac{45-x}{50-45} & \text{if } x \in (45, 50) \\ 0 & \text{else} \end{cases} \quad \nu_{S2} = \begin{cases} \frac{15-x}{45-15} & \text{if } x \in (15, 45) \\ \frac{x-45}{55-45} & \text{if } x \in (45, 55) \\ 1 & \text{else} \end{cases}$$

$$\mu_{S3} = \begin{cases} \frac{x-30}{60-30} & \text{if } x \in (30, 60) \\ \frac{60-x}{65-60} & \text{if } x \in (60, 65) \\ 0 & \text{else} \end{cases} \quad \nu_{S3} = \begin{cases} \frac{30-x}{60-30} & \text{if } x \in (30, 60) \\ \frac{x-60}{70-60} & \text{if } x \in (60, 70) \\ 1 & \text{else} \end{cases}$$

To generate the crisp out Takagi Sugni Formula is used in Intuitionistic fuzzy inference engine. The intuitionistic defuzzification formula given below.

6 Result and Discussion

In this work paddy field water level and well water level are measured and transmitted to controller node for decision making. Based on the water level the triangular intuitionistic fuzzy membership function and non membership are fixed. The controller node switch on or off the water pumping motor is based on the intuitionistic fuzzy controller. Here using the mamdami intuitionistic fuzzy rules and Takagi Sugni Formula for the defuzzification of the intuitionistic fuzzy set if can be observe from the Tables 5 and 6. The motor running time was calculated. From the above Table 3 motor running time was 5.19 min. Based on the water level in the well and in the field the running time is changed.

Table 5 Defuzzification field and well

Field			Well			And		Defuzzification of intuitionistic fuzzy set				
x1	μLOW	vLOW	x2	μLOW	vLOW	μLOW	vLOW	π	$A = (1-\pi)\mu$	$B = \pi\mu$	A+B	X*(A+B)
1	1.4	0.266	605	0.008	−0.007	1.4	−0.0071	−0.392857143	1.95	−0.55	1.4	2.8
2	1.3	0.2	606	0.01	−0.0085	1.3	−0.0085	−0.291428571	1.678857143	−0.37886	1.3	2.6
3	1.2	0.133	607	0.011	−0.01	1.2	−0.01	−0.19	1.428	−0.228	1.2	2.4
4	1.1	0.066	608	0.013	−0.0114	1.1	−0.0114	−0.088571429	1.197428571	−0.09743	1.1	2.2
5	1	0	609	0.015	−0.0128	1	−0.0128	0.012857143	0.987142857	0.012857	1	2
6	0.9	−0.06	610	0.016	−0.0142	0.9	−0.0666	0.166666667	0.75	0.15	0.9	1.8
7	0.8	−0.133	611	0.018	−0.0157	0.8	−0.1333	0.333333333	0.533333333	0.266667	0.8	1.6
8	0.7	−0.2	612	0.02	−0.0171	0.7	−0.2	0.5	0.35	0.35	0.7	1.4
9	0.6	−0.26667	613	0.021	−0.01857	0.6	−0.2666	0.666666667	0.2	0.4	0.6	1.2
10	0.5	−0.33333	614	0.023333	−0.02	0.5	−0.33333	0.833333333	0.083333333	0.416667	0.5	1
11	0.4	−0.4	615	0.025	−0.02143	0.4	−0.4	1	0	0.4	0.4	0.8
12	0.3	−0.46667	616	0.026667	−0.02286	0.3	−0.46667	1.166666667	−0.05	0.35	0.3	0.6
13	0.2	−0.53333	617	0.028333	−0.02429	0.2	−0.53333	1.333333333	−0.066666667	0.266667	0.2	0.4
14	0.1	−0.6	618	0.03	−0.02571	0.1	−0.6	1.5	−0.05	0.15	0.1	0.2
15	0	−0.66667	619	0.031667	−0.02714	0.031667	−0.66667	1.635	−0.020108333	0.051775	0.0316667	0.06333333
16	−0.1	−0.73333	620	0.033333	−0.02857	0.033333	−0.73333	1.7	−0.023333333	0.056667	0.0333333	0.06666667
17	−0.2	−0.8	621	0.035	−0.03	0.035	−0.8	1.765	−0.026775	0.061775	0.035	0.07
Result flow											10.6	21.2
												2

Table 6 Defuzzification motor

Motor			Defufuzzification of intuitionistic fuzzy set				
x	μ_{LOW}	ν_{LOW}	π	$A = (1-\pi)\mu$	$B = \pi\mu$	$A+B$	$X*(A+B)$
1	0.933	−0.05	0.116666667	0.824444444	0.108889	0.9333333	0.93333333
2	1.3	0.2	−0.5	1.95	−0.65	1.3	2.6
3	1.2	0.13333	−0.33333333	1.6	−0.4	1.2	3.6
4	1.1	0.06666	−0.16666666	1.283333333	−0.18333	1.1	4.4
5	1	0	0	1	0	1	5
6	0.9	−0.0666	0.166666667	0.75	0.15	0.9	5.4
7	0.8	−0.1333	0.333333333	0.533333333	0.266667	0.8	5.6
8	0.7	−0.2	0.5	0.35	0.35	0.7	5.6
9	0.6	−0.2666	0.666666667	0.2	0.4	0.6	5.4
10	0.5	−0.3333	0.833333333	0.083333333	0.416667	0.5	5
11	0.4	−0.4	1	0	0.4	0.4	4.4
12	0.3	−0.4666	1.166666667	−0.05	0.35	0.3	3.6
13	0.2	−0.5333	1.333333333	−0.06666666	0.266667	0.2	2.6
14	0.1	−0.6	1.5	−0.05	0.15	0.1	1.4
15	0	−0.6666	1.666666667	0	0	0	0
16	−0.1	−0.7333	1.833333333	0.083333333	−0.18333	−0.1	−1.6
17	−0.2	−0.8	2	0.2	−0.4	−0.2	−3.4
Result flow							5.1917808

References

1. Coates, R.W., Delwiche, M.J., Broad, A., Holler, M.: Wireless sensor network with irrigation valve control. J. Comput. Electron. Agric. **96**, 13–22 (2013)
2. Manimaran, P., Yasar Arfath, D.: An intelligent smart irrigation system using WSN and GPRS module. Int. J. Appl. Eng. Res. 11(6) (2016). ISSN 0973-4562
3. Khriji, S., Houssaini, D.E., Jmal, M.W., Viehweger, C., Abid, M., Kanoun, O.: Precision irrigation based on wireless sensor network. IET J. Sci. Meas. Technol. **8**(3), 98–106 (2014)
4. Dursun, M., Ozden, S.: An efficient improved photovoltaic irrigation system with artificial neural network based modeling of soil moisture distribution—a case study in Turkey. J. Comput. Electron. Agric. **102**, 120–126 (2014)
5. Goumopolulos, C., O'Flynn, B., Kameas, A.: Automated zone-specific irrigation with wireless sensor/actuator network and adaptable decision support. J. Comput Electron. Agric. **105**, 20–33 (2014)
6. Nolz, R., Kammerer, G., Cepuder, P.: Calibrating soil water potential sensors integrated into a wireless monitoring network. J. Agric. Water Manag. **116**, 12–20 (2013)
7. Goumopoulos, C., O'Flynn, B., Kameas, A.: Automated zone-specific irrigagion with wireless sensor/actuator network and adaptable decision support. Comput. Electron. Agric. Elsevier **105**, 20–33 (2014)
8. Chate, B.K., Rana, J.G.: Smart irrigation system using raspberry PI. Int. Res. J. Eng. Technol. **3**(5) (2016)
9. Burillo, P., Bustince, H.: Construction theorems for intuitionistic fuzzy sets. **84**(3), 271–281 (1996)
10. Szmidt, E., Kacprzyk, J.: Distances between intuitionistic fuzzy sets. Fuzzy Sets Syst. **114**(3), 505–518 (2000)

11. Atanassov, K., Szmidt, E., Kacprzyk, J.: On intuitionistic fuzzy pairs. Notes Intuit. Fuzzy Sets **19**(3), 1–13 (2013)
12. Tapan Kumar Roy, A.G.: Intuitionistic fuzzy delphi method: more realistic and interactive forecasting tool. Notes Intuit. Fuzzy Sets **18**(50), 37–50 (2012)
13. Szmidt, E., Kacprzyk, J.: Intuitionistic fuzzy sets in some medical applications. In: Fifth International Conference on IFSs, Soa. NIFS 7-4, 22–23 September 2001, pp. 58–64 (2001)
14. Akram, M., Shahzad, S., Butt, A., Khaliq, A.: Intuitionistic fuzzy logic control for Heater fans. Math. Comput. Sci. **7**(3), 367–378 (2013)
15. Anusha, A., Gouthami, D.: Wireless network based automatic irrigation system. Int. Adv. Res. J. Sci. Eng. Technol. **3**(7) (2016)
16. Chaira, T.: Intuitionistic fuzzy set approach for color region extraction. J. Sci. Ind. Res. **69**, 426–432 (2010)
17. Cornelis, C., Deschrijver, G., Kerre, E.E.: Implication in intuitionistic fuzzy and interval-valued fuzzy set theory: construction, classification, application. Int. J. Approx. Reason. **35**, 55–95 (2004)
18. Vasanti, G., Viswanadham, T.: Intuitionistic fuzzy set and its application in student performance ꞁꞁꞁꞁꞁꞁꞁꞁꞁꞁꞁꞁꞁꞁꞁꞁꞁ ꞁꞁ ꞁꞁꞁꞁꞁꞁꞁ ꞁꞁꞁ ꞁꞁꞁꞁꞁꞁꞁꞁꞁꞁ ꞁꞁꞁꞁꞁꞁꞁꞁꞁ ꞁꞁꞁꞁꞁꞁꞁꞁ ꞁꞁꞁꞁꞁꞁ ꞁꞁꞁ ꞁ ꞁꞁꞁꞁꞁꞁꞁꞁꞁꞁꞁ ꞁꞁꞁ Emerg. Res. **4** (2015)
19. Jose, S., Kuriakose, S.: Note on multiattribute decision making in intuitionistic fuzzy context. Notes Intuit. Fuzzy Sets **19**(1), 48–53 (2013)
20. Varghese, A., Kuriakose, S.: More on intuitionistic fuzzy relations. Notes Intuit. Fuzzy Sets **18**(2), 13–20 (2012)

Energy-Efficient Hybrid Firefly–Crow Optimization Algorithm for VM Consolidation

Nimmol P. John and V. R. Bindu

Abstract Virtual machine consolidation (VMC) is a successful approach to enhance resource utilization and reduce energy consumption by minimizing the number of active physical machines in a cloud data center. It can be implemented in a centralized or a distributed fashion. In this paper, an efficient multi-objective-based VM consolidation using hybrid firefly-crow optimization algorithm (HFCOA) is proposed. The proposed HFCOA is a novel approach developed by combining firefly optimization algorithm (FA) with crow search optimization algorithm (CSA). A new multi-objective fitness function is derived based on energy consumption, migration cost, and memory utilization. To analyze the performance of the algorithm, the simulation is carried out in the ClousdSim simulator. The proposed HFCOA is compared with FA and CSA in the same simulation environment. Experimental results show that the proposed hybrid algorithm significantly outperforms the original firefly and crow search algorithms.

Keywords Virtual machine consolidation · HFCOA · Crow search algorithm · Firefly algorithm · Energy consumption · Load balancing

1 Introduction

Cloud computing is a moderately latest computing technology. It uses a few standard ideas and advancements, for example, data centers and equipment virtualization, and results in another side of view. Cloud computing offers mainly three service models and four deployment models [1]. Infrastructure as a Service (IaaS), Platform as a Service (PaaS), and Software as a Service (SaaS) are the three service models, and private cloud, community cloud, public cloud, and hybrid cloud [2, 3] are the commonly available deployment models. Besides, because of the regularly expanding

N. P. John (✉) · V. R. Bindu
School of Computer Sciences, Mahatma Gandhi University, Kottayam, Kerala, India
e-mail: nimmolpjohn96@gmail.com

V. R. Bindu
e-mail: binduvr@mgu.ac.in

© Springer Nature Singapore Pte Ltd. 2020
V. Bhateja et al. (eds.), *Intelligent Computing and Communication*,
Advances in Intelligent Systems and Computing 1034,
https://doi.org/10.1007/978-981-15-1084-7_40

cloud framework requests, there exists a huge increment in the number and energy usage of cloud data centers [4]. As per Amazon it reveals that the expense related to energy at its data centers is 42% of the absolute working expense. Besides, high vitality utilization implies a high working cost, just as prompts higher carbon surges. The comprehensive office of relocating the running VMs with no perceivable personal time, from the over-stacked hubs to under-stacked hubs, manages the exceptional main job to restrain the imperativeness use. The reduction in the devoured vitality is because of the improved hub usage that outcomes from a balanced circulation and execution of remaining task at hand on the hubs [5].

Server consolidation using virtualization has turned into a vital innovative approach for enhancing the energy productivity of data centers [6]. The fundamental theory is to reduce the number of virtual machines (VMs) to as few in each physical machines (PMs) as could reasonably be expected, and afterward turn off the various lower-loaded PMs. Physical machines on data centers are ordered into three classifications depending on their utilization, in particular (i) overloaded, (ii) underloaded, and (iii) ordinary hosts. This characterization depends on host's use, for example, the usage in excess of a specific esteem (regularly named as maximum limit) might be considered as overloaded hosts and the usage not exactly of a specific esteem (normally known as lower edge) might be considered as underloaded hosts. Every single other hosts aside from these two classes are considered as ordinary hosts. As indicated in [7], under the typical situation, a data center works just at 10% of half of the peak limit and these underloaded hosts need no explanation on the misuse of power. Thus, it is required to diminish the vitality utilization by updating resource usage in cloud server through workload consolidation [8].

On the other hand, a disseminated or decentralized VM solidification approach uses a passed on figuring or a circled plan for PMs [9] or offers assistance for various, geographically appropriated server farms [10]. Coursed VM union is a tedious subject in continuous VM solidification approaches [9, 11]. Circulated approaches are picking up popularity since they have benefits over incorporated methodologies. Virtual machine consolidation (VMC) is a NP-hard issue [12–14]. Many existing investigation tries to have associated improvement in meta-heuristic approaches for taking care of VMC issue [15–17]. Their proposed calculations optimize the number and areas of the VMs to limit power consumption and normal CPU usage of running PMs. We concur that the absolute energy consumption of VMC calculation relies upon the energy consumption amid VMs migration, so decrease of the quantity of migrations alone cannot diminish the all-out energy consumption of VMC. In any case, the amount of movements may be less. Anyway, when the storage capacity of moving VMs is tremendous the vitality utilization during VM's relocation will increment.

The main objective of the proposed HFCOA algorithm is to migrate and place the VMs and switching of unused PMs to conserve energy. To improve the energy in data center, the consolidation approach is utilized. The migration of virtual machine is the key part of consolidation. The main contributions of the proposed methodology are given as follows:

- A multi-objective fitness function based on VM migration cost, memory utilization, and energy for selecting a suitable PM for consolidation.
- A hybrid firefly–crow algorithm by integrating FA with CSA for getting optimal PM for consolidation.

The remaining of the paper is arranged as follows: Sect. 2 is about literature review of VM migration and consolidation. Section 3 describes the motivation. The proposed methodology is given in Sect. 4 and simulation analysis results are given in Sect. 5. Lastly, conclusions are made in Sect. 6.

2 Related Works

Numerous researchers have used different VMC procedures to increase resource utilization and decrease energy consumption in data centers, particularly as of late. Among them some of the research works are analyzed here.

Xu et al. [18] have clarified an ideal VMs arrangement in cloud condition dependent on distributed parallel ant colony optimization (DPACO). This strategy expands execution of VMs, decreases vitality cost, and boosts asset usage; however, it needs a few enhancements with respect to its vitality cost. In [19], Azra et al. have proposed an ant colony system framework to tackle VM solidification issue expecting to spare the vitality utilization of cloud data center. This technique essentially lessens the quantity of relocations and the dynamic physical machines that result in the decrease of all-out vitality utilization of server farms, but this strategy is very mind boggling. Similarly, Li et al. [20] have introduced a multi-objective optimization for rebalancing virtual machine placement. This method mainly focuses on reducing cost on each VM in cloud provider, but requires high processing capability. Li et al. [21] presented a solution of double threshold with maximum resource usage to boost the transportation of VMs. The revised particle swarm optimization pattern was implemented in the reduction of VMs to escape from going into local optima that is considered as a common problem in normal heuristic algorithms.

Likewise, Moorthy and Fareentaj [22] have developed a method to solve VM arrangement in data center relied on ACO. This method is particularly aimed at minimizing the cost in each VM and the response time of each provider. ACO reduces the cost and response time but fails to reduce the number of migrations. Lin et al. [23] detailed the QoS-aware VM arrangement (QAVMP) issue as an integer linear programming (ILP) model by thinking about the three factors as the benefit of the cloud supplier and utilized a bipartite chart model to propose a polynomial-time heuristic calculation. Mosa and Paton [24] have acquainted a strategy which selects ideal VM arrangement in data center dependent on GA. This technique attempts to discover ideal VM position to lessen power consumption in servers, diminish reaction time, and boost assets use. In [25], Dashti et al. have presented a vitality productive VMs position utilizing changed PSO. This methodology finds the ideal VM situation by changing the PSO calculation to spare power utilization and by workload balancing. In [26], Praveen proposed a method for allocation of resources and scheduling of task. The proposed method is compared with GA and found it to be better.

3 Motivation

By giving an appropriate platform for data centers, cloud computing is now the best effective mechanism for data storage. Though there are many merits of cloud computing, still there exist some management problems that need to be investigated, such as energy consumption and cost. In order to serve the clients, there need to be thousands of server machines functioning in the cloud data centers. Even at 20% utilization of servers, it consumes 80% of the total power. Apart from increasing the operating cost, high energy consumption results in huge emission of carbon. To avoid this, the cloud providers are trying to find an effective mechanism through which IT infrastructure reconfiguration can be dynamically performed for lowering power depletion. One of the methods to consolidate virtual machines on one physical server for lowering the power consumption is known as virtualization consolidation [27].

A VM is selected for migration based on its load. The PMs are categorized into overloaded and underloaded depending on the threshold value. The load of each PM is an important factor while considering the problem of consolidation. The load balancer migrates the VM, based on its load, to an underloaded PM. An example of the consolidation of virtual machines on underloaded PMs for saving energy is shown in Fig. 1.

To minimize the energy consumption of VM migration, in this paper, a multi-objective fitness function is designed based on three important criteria, namely VM migration cost, memory utilization, and energy. Consider a cloud network containing N number of PMs PM = $\{PM_1, PM_2, \ldots, PM_N\}$ and each PM consists of M number of VMs VM = $\{VM_1, VM_2, \ldots, VM_M\}$. Each task T_i $T = \{T_1, T_2, \ldots T_i\}$ is randomly assigned to different VMs. The VM in the cloud consists of various components, namely number of CPUs, memory, and bandwidth. Similarly, each task has different size, energy, and memory. Based on the size the task is assigned to VM. To save the energy, we set a few number of PMs in sleep mode. To achieve minimum energy consumption, a multi-objective function is designed which is given in Eq. (1).

$$\text{Fitness Function} = \text{Min}[E + MC + M] \tag{1}$$

The fitness function relied on the three factors: energy (E), migration cost (MC), and memory utilization (M). The issue can be figured by utilizing condition (1). Energy consumption is the first parameter of optimal VM consolidation. It is calculated by the total Euclidean distance (ED) of all the active PMs at the same time. The PM is switched off, when no task is executing in it. For each active node, the energy-efficient factor ED is calculated as given in Eq. (2).

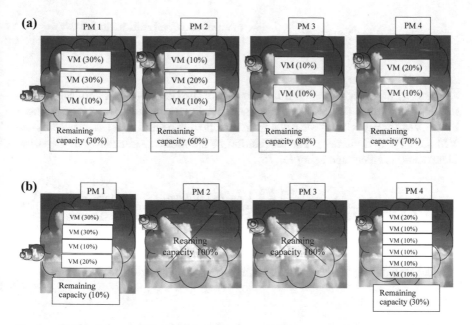

Fig. 1 **a** VMs before consolidation. **b** VMs after consolidation

$$ED = \sqrt{\sum_{j=1}^{d} \left(u_j - u\,\text{best}_j\right)^2} \tag{2}$$

Here, j represents the resources, namely processor and memory, U_j represents the current utilization, and ubestj represents the fine usage of resource j for energy controlling in each active node. The system energy efficiency (E^t) at slot t is represented as the summation of ED of all active PM at that slot. The E^t is calculated using Eq. (3).

$$E^t = \sum ED \tag{3}$$

Within slot T the energy efficiency is traced out by (4)

$$E = \sum_{t=0}^{T} E^t \tag{4}$$

Migration cost is the second parameter of VM consolidation. The migration cost of VM is increased when the number of migrations increases. A good consolidation system should maintain minimum migration. The calculation of the migration cost of the entire cloud setup is calculated using Eq. (5).

$$\text{Migration Cost}(MC) = \frac{1}{N} \sum_{i=1}^{N} \left(\frac{\text{Number of migration in } VMs}{\text{Total number of } VMs} \right) \quad (5)$$

Here N is the maximum PMs.

Memory utilization is the third parameter of the objective function. Memory is nothing but a load. The load of the framework directly depends on the assets used by the VM for completing the tasks from the client. The different assets used by the VM are CPU and memory. The calculation of the memory utilization of the entire cloud setup is calculated using Eq. (6).

$$\text{memory}(M) = \frac{1}{\text{PM} \times \text{VM}} \left[\sum_{i=1}^{\text{PM}} \sum_{j=1}^{\text{VM}} \frac{1}{2} \left(\frac{CPU \text{ utilized}_{ij}}{CPU_{ij}} + \frac{\text{memory utilized}_{ij}}{\text{memory}_{ij}} \right) \right]$$

$$(6)$$

Here PM and VM is maximum of active physical and virtual nods.

4 Proposed VM Consolidation Using Hybrid Firefly–Crow Search Algorithm

The core object of proposed method is to design an energy-efficient virtual machine migration and consolidation based on multi-objective hybrid firefly–crow optimization algorithm (HFCOA) [28]. In this consolidation process, we can save energy and cost. The good consolidation system has minimum number of migrations and minimum number of active PMs. To attain this, in this paper HFCOA is used. The proposed HFCOA migrates the VM in the underloaded PM to the overloaded PM and switch off the unwanted PM to save energy. The proposed VM consolidation using hybrid firefly–crow search algorithm is given in Fig. 2. And the flowchart is shown in Fig. 3.

Step 1: Initialization: Initialization is an important process for all optimization algorithms. Consider the solution consisting of PMs, VMs, and tasks. At first, the incoming tasks are randomly assigned to VMs. Then, we check the overloaded and underloaded PMs. The proposed HFCOA selects the optimal VM to migrate.

Step 2: Fitness calculation: After solution initialization, fitness is calculated for each solution, which is a combination of energy, migration cost, and memory utilization. It is given in Eq. (1).

Step 3: Updation using HFCOA: After fitness calculations, we update the solution using HFCOA. In this updation process, two cases are available.

Case 1: The proprietor crow n of food hiding place B_n^t does not realize the cheat crow m that follows it, so the cheat crow attains to the shroud location of proprietor crow. The cheat crow position updation is provided in Eq. (7).

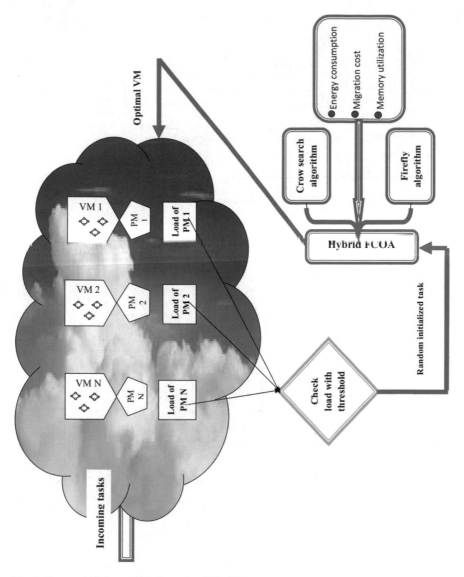

Fig. 2 Proposed VM consolidation using HFCOA

$$S_m^{t+1} = S_m^t + R_m \times FL_m^t \times \left(B_n^t - S_m^t\right) \qquad (7)$$

where R_m is an arbitrary number in the interval [0, 1]; FL_m^t is the flight length of crow m at cycle t.

Case 2: The proprietor crow n realizes that the cheat crow m follows it, then proprietor crow misleads crow m by moving toward to any other location of food hiding place. In this case the place of crow m is replaced using Eq. (8).

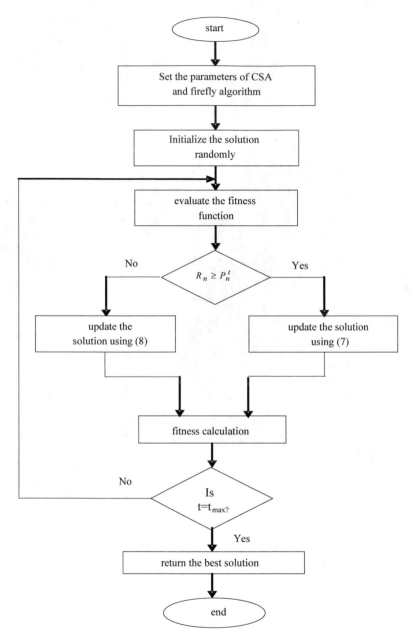

Fig. 3 Flowchart of proposed HFCOA

$$S_m^{t+1} = S_i^t - \beta_t \left(S_j^t - S_i^t \right) + \sigma_t \mu_i^k \tag{8}$$

In the Eq. (8), S_m^{t+1} gives the reframed solution, S_i^t and S_j^t are the ith and jth solutions, whereas σ_t shows the arbitrary factor and μ_i^k is a random number from Gaussian distribution at slot t and β_t is a constant value. The fitness updation condition is given in Eq. 9.

$$S_m^{t+1} = \begin{cases} \textit{if } R_n \geq P_n^t \text{ update position by using Eq. (7)} \\ \text{else} \qquad \text{update position by using Eq. (8)} \end{cases} \tag{9}$$

where P_n^t is the chance of understanding of crow m at cycle t, and R_n is a random number. The positions are updated using Eq. (9) and after that the latest objective function is evaluated.

Step 4: VM consolidation: After solution updation, check the load of each PM. A threshold value is calculated as the half of total capacity of corresponding PM. If the load of PM is more than the threshold means the corresponding PM is overloaded and VMs may be migrated to it. Similarly, if the PM is underloaded with minimum load means, we migrate the VM to another PM and switch off the PM.

Step 5: Termination criteria: The calculation ceases its execution just if a greatest number of cycles are accomplished and the arrangement which is scoring the good fitness function is chosen and it is indicated as the better VM consolidation.

5 Results and Discussion

In this area, we inspect the result got from the proposed HFCOA-based VM migration and consolidation. We have built up our proposed calculation utilizing Java (jdk1.6) with the CloudSim simulation toolbox. The tests were performed on a PC with Windows 7 Operating system at 2 GHz double center PC with 4 GB RAM running a 64-bit variant.

5.1 Evaluation of Results

The basic idea of HFCOA is to assign tasks to virtual machines, and if possible switch off few PMs to conserve energy. The objective function is based on energy consumption, migration cost, and memory utilization. To evaluate the performance we compared the work with FA and CSA. The experimental results are analyzed using three different configurations that is (i) PM = 5, VM = 15 and 50 tasks, (ii) PM = 10 and VM = 30 and 75 tasks, and (iii) PM = 20 and VM = 50 and 100 tasks.

- **PM = 5, VM = 15 and 50 tasks**

Figure 4 shows the performance analysis based on memory. The memory utilization is to be less for the method to be effective. In first iteration, the memory consumed by HFCOA is 14,475 bits, by FA is 17,357 bits, and by CSA is 16,862 bits. On fifth iteration, the memory consumed by HFCOA, FA, and CSA is 15,754, 19,474, and 17,324 bits, respectively. The order of the values is the same for all the remaining iterations in this analysis. HFCOA has the lowest memory utilization, which makes it more efficient than FA and CSA. The performance analysis of energy consumption for five iterations is shown in Fig. 5. The method with low energy consumption can be concluded as the efficient one. Here on iteration 20, the energy consumption by HFCOA is 6.4632, by FA is 9.8538, and by CSA is 8.7348. At iteration 40, the energy consumed by HFCOA, FA, and CSA is 6.5352, 9.6483, and 8.7483, respectively. It is clear that HFCOA consumed less energy. Figure 6 shows the performance analysis

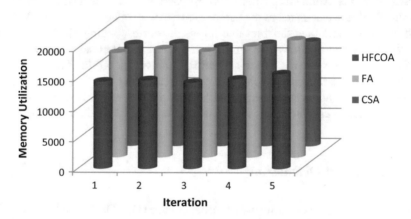

Fig. 4 Performance analysis based on memory

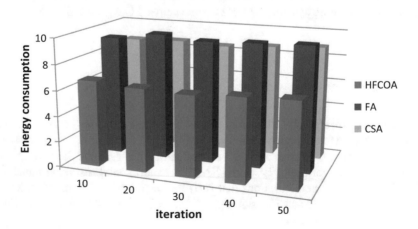

Fig. 5 Performance analysis based on energy consumption

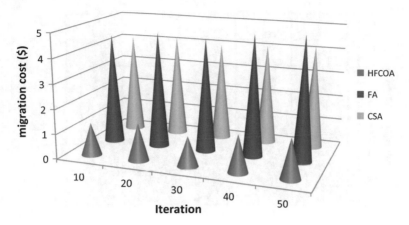

Fig. 6 Performance analysis based on migration cost

based on migration cost. The migration cost of HFCOA is 1.2, that of FA is 4.6, and for CSA it is 4. HFCOA has the lowest migration cost and this makes it superior. Therefore, from the results it is clear that HFCOA outperforms FA and CSA methods.

- **PM = 10 and VM = 30 and 75 tasks**

The performance analysis based on memory utilization is shown in Fig. 7. In iteration 40, the memory utilized by HFCOA is 19,865 bits, that by FA is 26,439 bits, and by CSA is 23,943 bits. In iteration 50, the memory utilization for HFCOA, FA, and CSA is 20,754, 27,474, and 24,324 bits, respectively. Figure 8 shows the performance analysis based on energy consumption. For HFCOA it is 7.7246, FA 11.3572 and that for CSA is 10.6347. From the results it is clear that HFCOA has the lowest

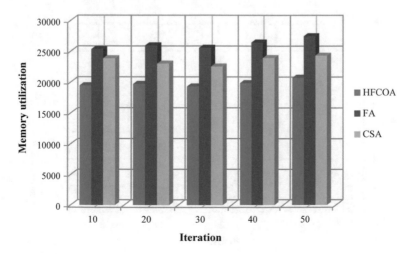

Fig. 7 Performance analysis based on memory utilization

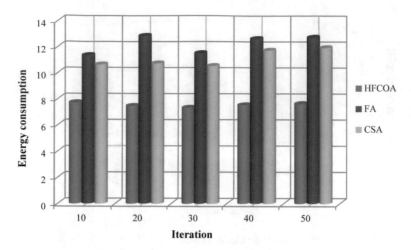

Fig. 8 Performance analysis based on energy consumption

energy consumption. Figure 9 shows the performance analysis based on migration cost. When the iteration is 50, the migration cost of HFCOA is 1.7, for FA it is 5.8, and for CSA it is 5.2. Therefore, from the results it can be concluded that HFCOA outperformed both FA and CSA approaches.

- **PM = 20 and VM = 50 and 100 tasks**

As shown in Fig. 10, the memory utilized by HFCOA is 22,475 bits, by FA is 27,357 bits, and by CSA is 26,862 bits, when the iteration is 10. When the iteration is 30, it is 22,312, 28,573, and 26,534 bits, respectively. Figure 11 shows the performance analysis based on energy consumption. When the iteration is 20, the energy consumed by

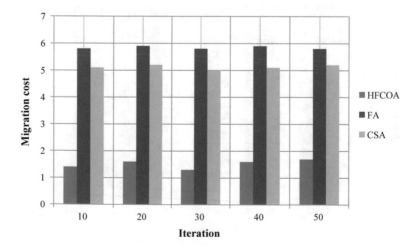

Fig. 9 Performance analysis based on migration cost

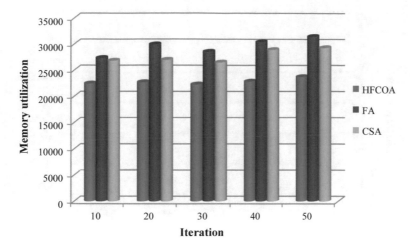

Fig. 10 Performance analysis based on memory utilization

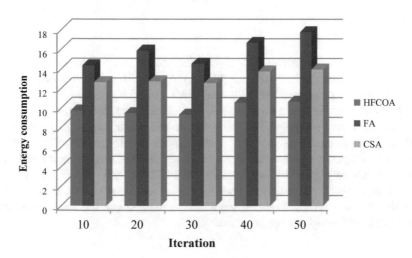

Fig. 11 Performance analysis based on energy consumption

HFCOA is 9.4632, by FA is 15.8538, and by CSA is 12.7348. The energy consumed by HFCOA, FA, and CSA is 10.5352, 16.6483, and 13.7483, respectively, when the iteration is 40. HFCOA consumed lesser energy than FA and CSA approaches. Therefore, HFCOA is superior to other two methods. The performance analysis based on migration cost is shown in Fig. 12. When iteration is 10, the migration cost of HFCOA, FA, and CSA is 1.8, 8.8, and 7.1, respectively. The migration cost has to be low for a method to be effective. Here, HFCOA has the lowest migration cost in all iterations. Therefore, from the result it is clear that HFCOA method is superior to FA and CSA methodologies.

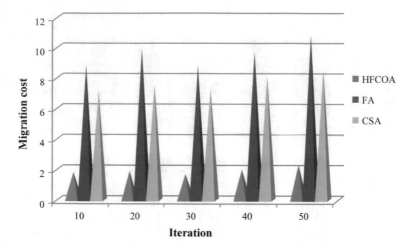

Fig. 12 Performance analysis based on migration cost

6 Conclusion

We proposed a vital, effective virtual machine solidification dependent on hybrid firefly-crow optimization algorithm (HFCOA) to diminish the vitality utilization of servers by keeping least number of dynamic PMs. A multi-objective fitness function was designed based on energy consumption, migration cost, and memory utilization to attain maximum sleeping PMs. To evaluate the performance of the proposed method, three problem instances are analyzed. The simulation results of the proposed model are compared against the existing techniques. HFCOA outperformed other models by achieving the minimum values of 14,312 bits, 6.3215%, and 1.2\$ for the memory, energy consumption, and migration cost, respectively. In future, this algorithm can be enhanced on real workload traces and can use more parameters such as bandwidth and quality of service (QoS).

References

1. Mell, P., Grance, T.: The NIST definition of cloud computing. In: Recommendations of the National Institute of Standards and Technology, pp. 800–145 (2011)
2. Motta, G., Sfondrini, N., Sacco, D.: Cloud computing: An architectural and technological overview. In: 2012 International Joint Conference on Service Sciences (IJCSS), pp. 23–27 (2012)
3. Sriram, I., Khajeh-Hosseini, A.: Research agenda in cloud technologies. Large Scale Complex IT Systems (LSCITS), Technical report (2010)
4. Beloglazov, A., Buyya, R., Lee, Y.C., Zomaya, A.: A taxonomy and survey of energy-efficient data centers and cloud computing systems. Adv. Comput. **82**, 47–111 (2011)
5. Kansal, N.J., Chana, I.: Energy-aware virtual machine migration for cloudcomputing—a firefly optimization approach. J. Grid Comput. (2016)

6. Meng, X., Pappas, V., Zhang, L.: Improving the scalability of data center networks with traffic-aware virtual machine placement. In: Proceeding of IEEE International Conference on Computer Communications, pp. 1–9 (2010)
7. Barroso, H.: The case for energy-proportional computing. J. Comput. **40**, 268–280 (2007)
8. Masoumzadeh, S.S., Hlavacs, H.: A cooperative multi agent learning approach to manage physical host nodes for dynamic consolidation of virtual machines. In: 2015 IEEE Fourth Symposium on Network Cloud Computing and Applications, NCCA, pp. 43–50 (2015)
9. Farahnakian, F., Ashraf, A., Pahikkala, T., Liljeberg, P., Plosila, J., Porres, I., Tenhunen, H.: Using ant colony system to consolidate VMs for green cloud computing. IEEE Trans. Serv. Comput. **8**(2), 187–198 (2015)
10. Lucanin, D., Brandic, I.: Pervasive cloud controller for geo-temporal inputs. IEEE Trans. Cloud Comput. **99**, 80–195 (2015)
11. Marzolla, M., Babaoglu, O., Panzieri, F.: Server consolidation in clouds through gossiping. In: 2011 IEEE International Symposium on a World of Wireless, Mobile and Multimedia Networks, pp. 1–6 (2011)
12. Ferdaus, H., Murshed, M., Calheiros, R.N., Buyya, R.: Virtual machine consolidation in cloud data centers using ACO metaheuristic. In: Euro-Par 2014 Parallel Processing, pp. 306–317 (2014)
13. Li, X., Qian, Z., Lu, S., Wu: Energy efficient virtual machine placement algorithm with balanced and improved resource utilization in a data center. Math. Comput. Model. **58**, 1222–1235 (2013)
14. Jayasinghe, D., Pu, C., Eilam, T., Steinder, M., Whalley, I., Snible, E.: Improving performance and availability of services hosted on IaaS clouds with structural constraint aware virtual machine placement. In: IEEE International Conference on Services Computing (2011)
15. Feller, E., Rilling, L., Morin: Energy-aware ant colony based workload placement in clouds. In: Proceedings-2011 12th IEEE/ACM International Conference on Grid Computing, pp. 26–33 (2011)
16. Esnault, A., Feller, E., Morin, C.: Energy-aware distributed ant colony based virtual machine consolidation in IAAS clouds bibliographic study. Inf. Math. (INRIA) 1–13 (2012)
17. Gao, Y., Guan, H., Qi, Z., Hou, Y., Liu, L.: A multi-objective ant colony system algorithm for virtual machine placement in cloud computing. J. Comput. Syst. Sci. 1230–1242 (2013)
18. Xu, G., Dong, Y., Fu, X.: VMs placement strategy based on distributed parallel ant colony optimization algorithm. Appl. Math. Comput. **9**(2), 873–881 (2015)
19. Aryania, A., Aghdasi, H.S., Khanli, L.M.: Energy-aware virtual machine consolidation algorithm based on ant colony system. J. Grid Comput. (2018)
20. Li, R., Zheng, Q., Li, X. and Yan, Z.: Multi-objective optimization for rebalancing virtual machine placemen. J. Future Gener. Comput. Syst. **614**, 1–19 (2017)
21. Li, H., Zhu, G., Cui, C., Tang, H., Dou, Y., He, C.: Energy-efficient migration and consolidation algorithm of virtual machines in data centers for cloud computing. J. Comput (2015)
22. Moorthy, R., Fareentaj, U.: An effective mechanism for virtual machine placement using ACO in IaaS cloud. In: IOP Conference Series: Materials Science and Engineering (2016)
23. Lin, J.-W., Chen, C.-H., Lin, C.-Y.: Integrating QoS awareness with virtualization in cloud computing systems for delay-sensitive applications. J. Future Gener. Comput. Syst. **37**, 478–487 (2014)
24. Mosa, A., Paton, N.: Optimizing virtual machine placement for energy and SLA in clouds using utility functions. J. Cloud Comput. **5**, 2–17 (2016)
25. Dashti, S., Rahmani, A.: Dynamic VMs placement for energy efficiency by PSO in cloud computing. J. Exp. Theor. Artif. Intell. **1**, 1–16 (2016)
26. Praveen, S.P., Rao, K.T., Janakiramaiah, B.: Effective allocation of resources and task scheduling in cloud environment using social group optimization. Arab. J. Sci. Eng. **43**(8), 4265–4272 (2018)
27. Zain, A.M., Udin, A., Mustaffa, N.: Firefly algorithm for optimization problem. J. Appl. Mech. Mater. **421**, 512–517 (2013)
28. Abdelaziz, A.Y., Fathy, A.: A novel approach based on crow search algorithm for optimal selection of conductor size in radial distribution networks. Eng. Sci. Technol. Int. J. (2017)

Keyless, Robust and Cost-Effective Privacy Protection Algorithm for Videos Using Steganography

Jayati Bhadra, M. K. Banga and M. Vinayaka Murthy

Abstract In this paper, a novel reversible keyless steganography algorithm for video piracy protection which uses randomized pixel for hiding authentic information is proposed. Randomization at two different levels is considered which is not in any of the existing algorithms. It increases the security with low cost. This proposed approach enhances the randomization of the specific pixels where authentication information will be stored in a frame and the location of such modified pixels is stored in an immediate next frame. Each pair is identified with an embedded random number. Modified least significant bit (mLSB) is used to insert the authentication information bits which is cost effective due to simplicity and can withstand statistical attacks. During extraction, frame with pixel locations is used. The extracted information will be compared. Different quality metrics values proved that it can withstand visual attacks, whereas StirMark test proved its high robustness.

Keywords Data hiding · Modified LSB · Watermark · Steganography

J. Bhadra (✉)
School of Computer Science, REVA University, Rukmini Knowledge Park, Kattigenahalli,
Yelahanka, Bengaluru 560064, Karnataka, India
e-mail: jayatibhadra@sjc.ac.in; jayatib@yahoo.com

M. K. Banga
Department of Computer Science Engineering, Dayananda Sagar University, Kudlu Gate, Hosur
Main Road, Bangalore 560068, Karnataka, India
e-mail: chairman-cse@dsu.edu.in

M. Vinayaka Murthy
School of Computer Science Applications, REVA University, Rukmini Knowledge Park,
Kattigenahalli, Yelahanka, Bengaluru 560064, Karnataka, India
e-mail: mvinayakamurthy@reva.edu.in; dr.m.vinayakamurthy@revainstitution.org

© Springer Nature Singapore Pte Ltd. 2020   429
V. Bhateja et al. (eds.), *Intelligent Computing and Communication*,
Advances in Intelligent Systems and Computing 1034,
https://doi.org/10.1007/978-981-15-1084-7_41

1 Introduction

In this era of deep penetration of internet connectivity and smartphones, images and videos have taken the center stage as a media for communication, entertainment, and so on. At every instance, a mind-boggling number of images/videos are being transmitted using internet. Naturally, such large volume of image/video exchanges has brought copyright protection the most important aspect resulting in such contents being embedded with copyright information in various forms using different mechanisms.

Any video is composed of a sequence of still images and audio tracks. It can be broken up into image frames and audios. Each image frame consists of a set of pixels. Each pixel is having a set of intensity information bits. As such, it is easier to hide sensitive privacy information in a video compared to other media. LSB method is the most common, simple but effective method with low computational complexity in spatial domain.

Identification of pixels where the authentication information has to be embedded and selection of embedding method are the most important criteria for this scheme. After embedding, the resultant frames should have minimized distortion so that utmost protection from visual and statistical attacks shall be observed. The existing lossless embedding processes generally use pseudo random number generator (PRNG) [1] for selecting the pixels simple but effective method with low computational complexity to embed secret message and LSB [2] of those pixels' intensity values will store the bits of authentication information, though they are vulnerable to statistical attacks. Gupta et al. [3] and Bhateja et al. [4] invented two different quality evaluation processes of distorted color images.

In this paper a modified LSB instead of only LSB and the pixels of red color channel of the frames are used. Two layers of influence circles instead of energetic pixel [5, 6] concept are used to randomize the pixel selection for embedding, which increases the computational complexity. In turn, it increases the security of the embedded information and imposes minimum distortions and maximum resistance to the removal of authentication information for effective control of the rampant piracy menace that is prevalent currently.

2 Review of Literature

Video watermark algorithms are generally based on relationships between frames and different types of embedding. LSB steganographic method is a very simple but effective method with low computational complexity to embed secret message. Ramalingam [7] used modified LSB algorithm for better efficiency. Li-Yi et al. [8] proposed a data encapsulation method based on motion vectors by using matrix encoding in video. Cao et al. [9] proposed a video watermark algorithm based on motion vector as carrier for data hiding method and H.264 video compression process.

The principle of "linear block codes" is used here to reduce transformation rates of motion vectors. Kelash et al. [10] proposed an algorithm to directly embed data into video frames using color histogram, where pixels of each video frame will be partitioned into two halves: right half will contain hidden bits and left half will contain the counts of right half. Deshmukh and Pattewar [11] proposed "the syndrome trellis code (STC) with flipping distortion measurement" which compromises capacity but solves the security issue. Feng et al. [12] proposed "modified duel watermark scheme" which is in transform domain and gives excellent recovery capacity. The scheme is having computational complexity. Paul et al. [5] uses one-bit per pixel for LSB-based insertion inside energetic pixel. Paul et al. [6] proposed dynamic optimal multi-bit image steganography using energetic pixels and ising concept and hiding bits of the secret message in the higher bit planes of image pixels which are highly energetic in nature. This scheme can be considered as an embedding process. In this scheme, high embedding capacities are considered over grayscale image. However, all the three planes of color image are not exploited. Also, it is implemented on uncompressed images.

3 Proposed Authentication Scheme

Let us define the terms distance_row, distance_column and influence_distance as follows:

Definition 1 We define the distance_row of a particular pixel as the difference of the index value of the row of a particular pixel and that of other pixel.

Definition 2 We define the distance_column of a particular pixel as the difference of the index value of the row of a particular pixel and that of other surrounding pixels in influence circle.

Definition 3 We define the influence_distance as the maximum distance_row or distance_column between a specific pixel and its surrounding pixels.

Let us consider that value of a specific pixel be $d_{i,j}$ and an arbitrary pixel from it be $Q_{k,l}$. Let us assume a, b are the distance_row and distance_column, respectively. Then

$$\text{Either } k - i \leq 0 \text{ or } k - i = a \text{ where } 0 \leq a \leq 2 \text{ and } 0 \leq i, k \leq 2$$
$$\text{Either } l - j \leq 0 \text{ or } l - j = b \text{ where } 0 \leq b \leq 2 \text{ and } 0 \leq j, l \leq 2$$

Since influence_distance cannot be negative or zero, we can consider the $k - i = a$ or $i - j = b$. As per the definition of influence_distance, if $a > b$ then a is the influence_distance, otherwise b is the influence_distance.

Here each video consists of $\{I_i, i = 1, 2, ..., n\}$ image frames, where n is the total number of image frames in the video. Therefore,

$$\text{Video} = \sum_{i=1}^{n} (\text{image frame})_i + \text{audio track} \tag{1}$$

Each image frame has height h and breadth w. Therefore,

$$\text{image frame } I = hw \tag{2}$$

Each image frame consists of pixels with pixel values $P_{i,j}$ where $i = 0, 1, \ldots, h$ and $j = 0, 1, \ldots, w$, that is,

$$\text{image frame } I = \sum_{i=0}^{h} \sum_{j=0}^{w} d_{i,j} \tag{3}$$

Each pixel is surrounded by other pixels. The neighbors of a particular pixel $P_{i,j}$ is defined as

$$N = \left\{ d_{s,t} \in I, s \neq i, t \neq j, 0 < |s - i| \leq h, 0 < |t - j| \leq w \right\} \tag{4}$$

Inner neighborhood N_{ij} of any pixel P_{ij} is defined as

$$N_{ij} = \left\{ d_{s,t} \in I, s \neq i, t \neq j, 0 \leq |s - i| \leq 1, 0 \leq |t - j| \leq 1 \right\} \tag{5}$$

Outer neighborhood N'_{ij} of any pixel P_{ij} is defined as

$$N'_{ij} = \left\{ d_{s,t} \in I, s \neq i, t \neq j, 1 \leq |s - i| \leq 2, 1 \leq |t - j| \leq 2 \right\} \tag{6}$$

Equation (7) is used to get all the pixel values of the frame with respect to values of its influence_distance of a particular pixel with respect to inner and outer neighborhood as defined by Eqs. (5) and (6). The highest pixel value of the row of the image frame after the transformation is possible pixel to hide the information. All such pixels are used to securely store authentication information.

To implement the proposed randomization concept, inner and outer influence circle is used. The equation for the pixel value of $d_{i,j}$ is

$$f_{i,j} = \text{abs} \left(\sum_{s} \sum_{t} |d_{i,j} - d_{s,t}| - \left(\sum_{u} \sum_{v} |d_{i,j} - d_{u,v}| \right) \right) \tag{7}$$

where $i = 0, 1, \ldots, h$ and $j = 0, 1, \ldots, w$

$$s \neq i, t \neq j, 0 \leq |s - i| \leq 1, 0 \leq |t - j| \leq 1 \text{ and}$$
$$u \neq i, v \neq j, 1 \leq |u - i| \leq 2, 1 \leq |v - j| \leq 2$$

Let us consider an arbitrary 5×5 matrix of image I as follows (Table 1).

Table 1 Pixel values of inner and outer neighborhood of $d_{i,j}$

$d_{i-2,j-2}$	$d_{i-2,j-1}$	$d_{i-2,j}$	$d_{i-2,j+1}$	$d_{i-2,j+2}$
$d_{i-1,j-2}$	$d_{i-1,j-1}$	$d_{i-1,j}$	$d_{i-1,j+1}$	$d_{i-1,j+2}$
$d_{i,j-2}$	$d_{i,j-1}$	$d_{i,j}$	$d_{i,j+1}$	$d_{i,j+2}$
$d_{i+1,j-2}$	$d_{i+1,j-1}$	$d_{i+1,j}$	$d_{i+1,j+1}$	$d_{i+1,j+2}$
$d_{i+2,j-2}$	$d_{i+2,j-1}$	$d_{i+2,j}$	$d_{i+2,j+1}$	$d_{i+2,j+2}$

Table 2 Sample pixel values of $d_{3,3}$

12	10	11	42	06
8	09	15	08	05
9	08	09	06	04
10	12	10	40	03
07	09	08	01	02

Table 3 Pixel values of inner and outer neighborhood of $d_{3,3}$

5	31	7	41	20
11	59	25	10	19
10	62	25	8	34
2	27	5	57	9
1	29	18	32	17

Let us assume a sample pixel matrix with pixel values (Table 2).

Using Eq. (7) we get the following pixel values (Table 3).

The proposed scheme does not consider the actual image content while selecting the pixels to embed authentication information. This technique is devised to hide short text data as secret message for authenticity in spatial domain. Using the cover frame image of red color, the sender calculates pixel value matrix using Eq. (7) and extracts the hidden bits.

In this scheme, each such position value is converted into a bit stream and those bits are stored sequentially using the seventh bit of the pixel value of red color channel of each of the pixels, starting from the first pixel of the applicable row in the even numbered image frame.

3.1 Algorithm

1. Read the video
2. Read authentication information
3. Convert authentication information into binary

4. Split the video into image and audio
5. Split each image into image frames
6. Repeat

   6.1. Convert the pixel values of each frame using proposed conversion Eq. (7)
   6.2. Find highest value of each row
   6.3. Embed authentication information bit in seventh bit of the highest value
   6.4. Embed authentication information bit location in the next frame
   6.5. Generate a random number and embed that number in the last row of both
        the frames

7. Until end of video.

The ratio of the data which is communicated and the distortion which is introduced is called embedding efficiency. We get the proper definition of *embedding efficiency* as per [13]. The maximum embedding efficiency of an arbitrary bit-stream insertion and uniform distribution of 0's and 1's using LSB interchanging method is 2. So the embedding efficiency of the proposed method is also 2.

4 Results

The proposed method is novel as the randomization technique used here is not used in any of the state of art work in this domain. Randomization at two different levels is considered here. Also, we have considered color frames instead of converting color frames to grayscale frames.

4.1 Complexity

We are using randomization concept with modified LSB embedding which will enhance the time complexity as well as computational complexity at a marginal limit. Time complexity for randomizing the pixel positions is $O(n\log n)$ and hiding the bits using modified LSB is $O(n)$. So time complexity of this proposed method is $O(n\log n) + O(n) = O(n\log n)$. As time complexity of the proposed system is increased, security of the system also increased. We used 50 different uncompressed videos from Archive.com to test this algorithm. The results clearly established that the proposed video watermark process is resistant to different attacks and gives the optimal imperceptibility. For interpretation and presentation of the results here, as samples, we have used two files and the same are detailed below. The proposed method of hiding the message bits using LSB is because of its simplicity. Also, the proposed method's time complexity is $O(n\log n)$, whereas DCT, DFT is having time complexity as $O(n^2)$. Above all, LSB is a lossless encryption method which gives high embedding efficiency.

4.2 Keyless

As the proposed process is keyless, no overhead for key exchange is required which ensures data security. The locations of pixels are selected randomly for embedding. Different statistical analysis proves that the keyless embedding process is secured compared to embedding process with key.

4.3 Cost-Effectiveness

The proposed system uses modified LSB method which is very simple to implement. So computation takes less time. Introducing the authentication information in every alternate frames takes less time. We can conclude that the method is cost effective.

4.4 Robustness

Robustness refers to maximum amount of data that may be hidden into the host video without fidelity losing. Using our proposed method, we will be storing the authentication information in every alternate frames and also random number for each frame as two levels of authentication. We have used StirMark [14–16], which is a standard benchmark tool for checking whether the steganographic and watermarking algorithms used on the image is robust or not. The attack simulates image distortions that generally take place when an image is photocopied, printed or rescanned. There will be some distortions after photocopy/reprint/rescan in the resultant image. StirMark simulates image distortions of that type and checks whether the distorted image is robust or not. We executed StirMark 4.0 on the image frame and the results are shown in Table 4 that shows none of the tests failed. So the proposed embedding process is robust.

4.5 Resistance to Statistical Attack

By implementing Westfeld and Pfitzmann's test [17], we can prove that the proposed algorithm is resistant to first-order *statistical attacks*. Also Provos [18] proved that the color frequency test is not effective if the information is hidden in randomly selected pixels though it will work well if the information is hidden sequentially. We are not starting the hiding process from the beginning of the image frame, and the proposed process of hiding is not chronological based on pixel positions, so it can withstand this test. The proposed algorithm can withstand *duel statistics test* as flipping of LSB in this test will not be effective. In our proposed method, we are

Table 4 StirMark result for robustness

Name of the test	Value of distortion	PSNR	Resultant noise (dB)
Test_MedianCut	3	173.46	40.1901
Test_MedianCut	9	173.26	34.1105
Test_SelfSimilarities	1	172.57	41.0251
Test_SelfSimilarities	3	172.69	39.2516
Test_RemoveLines	10	172.822	NA
Test_RemoveLines	50	172.722	NA
Test_RemoveLines	100	172.824	NA
Test_Cropping	1	205.37	NA
Test_Cropping	20	197.739	NA
Test_Cropping	75	179.611	NA
Test_Rescale	50	172.875	NA
Test_Rescale	200	172.793	NA
Test_Rotation	−2	159.874	NA
Test_Rotation	5	143.98	NA
Test_RotationCrop	−0.5	173.18	NA
Test_RotationCrop	1	173.739	NA
Test_RotationScale	−1	173.738	20.326
Test_RotationScale	0.75	173.467	23.002
Test_Affine	1	169.97	NA
Test_Affine	8	169.22	NA
Test_SmallRandom Distortions	0.95	166.19	16.45
Test_SmallRandom Distortions	1.1	165.48	15.70
Test_LatestSmall Random Distortions	0.95	167.96	20.31
Test_LatestSmall Random Distortions	1.1	167.53	19.61
Number of tests which failed		Nill	

using the last but one-th bit, that is, seventh-bit position to hide the message bit in pixels at random positions in the image.

We have given here as an example, the histogram of the original frame5 of out.avi in Fig. 1 and stego frame5 with the authentication information in Fig. 2. In terms of histogram analysis, the stego image frames are found visually indistinguishable from their cover counterpart image frames. Figures 1 and 2 show high similarity.

From Table 5, it is clear that the cover and stego frames are very similar as Euclidean distance between the histograms is very small, correlation is less than one, chi-square test shows strong similarity, the intersection of two histograms is small, which established that the proposed method is visual attack resistant.

The difference between means of cover frame and the frame with authentication varies from 0.0001 to 0.0010, which is very small. Similarly, the standard deviation

Fig. 1 Histogram of frame5
of out.avi

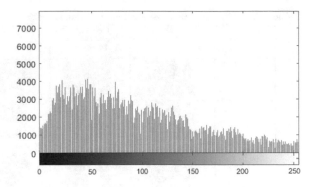

Fig. 2 Histogram of frame5
of out.avi with authentication
information

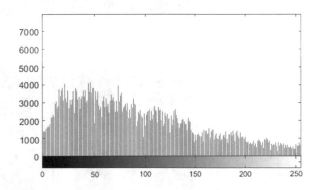

Table 5 Comparison of two histograms of original and with authentication information of frame 5

Euclidean distance	Correlation	Chi-square	Intersection
$2.5804e^{-10}$	0.302	3.523	0.403

of cover frame and the frame with authentication varies from 0.0001 to 0.0006, which is again very small. Both these difference value ranges tell us that the original frame and the frame with hidden authentication message are very similar to each other. So from Figs. 1, 2 and Table 5, the comparative values prove that the two frames are *very close to similarity*.

Bit planes of the cover and stego images can be analyzed as part of visual attack. We know that each pixel of a grayscale image is represented by 8 bits and 1-bit planes vary from zero-bit plane (LSB) to seven-bit plane (MSB). According to Wayner [18] and Westfeld [19], all bit planes are non-random. If each bit plane of cover and stego frame are non-random, then we can say that the proposed method can *withstand visual attack*. Let us consider $I(p)$ to be a frame with hidden information and p as the cover frame. Here, we can have two cases:

(i)   If $p = I(p)$ then the cover frame is unchanged
(ii) If $p \neq I(p)$ then the cover frame is changed with hidden information.

Table 6 Bit-plane analysis shows resistance to visual attack

Bit planes	Original Frame 10	Embedded Authentication in Frame 10	Original Frame 11	Embedded Location information in Frame 11
Plane 0				
Plane 1				
Plane 2				
Plane 3				
Plane 4				
Plane 5				
Plane 6				
Plane 7				

From Table 6, it is clear that the bit plane images are same for cover and stego frames. So we can say that our algorithm withstands bit plane test.

In Figs. 3 and 4, the normalized 2-D cross-correlation (NCC) of the image frame5 before and after the authentication information is embedded, and we find no visual difference which proves the resistance to the visual attack.

The hidden data is very difficult to extract without knowing the randomization logic which helps rise to the security of the proposed algorithm. Also, it is difficult to detect using HVS as the quality of the stego image is very high. So we can say that the proposed method can withstand visual attacks.

4.6 Analysis Through Quality Metrics

Image quality measurement – MSE, PSNR, SSIM

Mean Square Error (MSE)

$\text{MSE} = \frac{1}{M*N} \sum_{m=1}^{M} \sum_{n=1}^{N} [x(m,n) - y(m,n)]^2$, where M, N are rows and columns of image matrix, $x(m,n)$ is the original image, $y(m,n)$ is stego image.

Fig. 3 NCC of original
frame5 out.avi

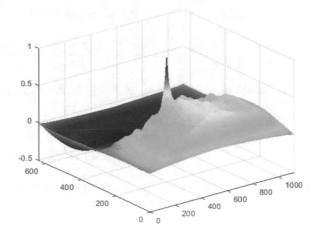

Fig. 4 NCC of stego frame5
of out.avi with authentication
information

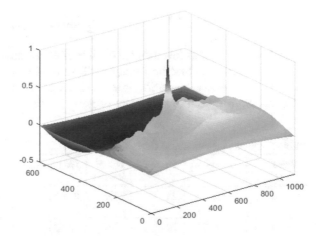

If the value of MSE is greater than 0, then the stego image will be of poor quality.

Peak Signal to Noise Ratio (PSNR)

PSNR $= 20\log_{10}$ [MAXPIX/RMSE], where RMSE $= \sqrt{\text{MSE}}$ is the root mean square error, MAXPIX is the maximum possible pixel value of the image. If the PSNR has large value then the stego image is of good quality.

Structural Similarity Index Metric (SSIM)

$$\text{SSIM} = \left((2a_x a_y + c_1)(2b_{xy} + c_2)\right/\left(a_x^2 + a_y^2 + c_1)(b_x^2 + b_y^2 + c_2)\right)$$

where "a", "b" and "b_{xy}" are mean, variance and covariance of the images, and "c_1, c_2" are the stabilizing constants. The value of SSIM will vary between 0 and 1. The value of SSIM for similar images will be near or equal to 1.

Average Difference (AD)

The formula for calculating the AD is as follows:

$$AD = \frac{abs\left(\sum\sum\left(f(u,v) - f'(u,v)\right)\right)}{m*n}$$

If the AD value is low then the quality of the resultant image will be good.
Maximum Difference (MD)
The formula for calculating the MD is as follows:

$$MD = max\left(max\left(f(u,v) - f'(u,v)\right)\right)$$

If the MD value is large then the quality of the resultant image will be poor.
Normalized Absolute error (NAE)
The formula for calculating the NAE is as follows:

$$NAE = \frac{\sum\sum abs\left(f(u,v) - f'(u,v)\right)}{\sum\sum abs(f(u,v))}$$

If the NAE value is low then the quality of the resultant image will be good.

In Table 7, we see that the average MSE value is less than 0, average PSNR value is more than 1 and average SSIM is less than equal to 1, which implies that the frames with authentication information are of good quality. Also in Table 8, we establish

Table 7 Quality metrics of test videos

Video name	Resolution	Frames/s	No. of frames	AverageMSE	AveragePSNR	AverageSSIM
Out.avi	720 × 1280 × 3	25	125	0.0020	77.5	1.0
Test_1.avi	320 × 870 × 3	20	120	0.0034	78.3	1.0
Test_4.avi	640 × 360 × 3	29.97	7691	0.00082	77.5	1.0
Test_5.avi	640 × 360 × 3	29.87	5597	0.00085	75.4	1.0

Table 8 Other quality metrics of out.avi

Frame type	Color channel	AD	MD	NAE
With authentication	Red	0.000050	2	2.2087
With authentication	Gray	0.000043	1	2.8083
With location of pixels	Red	0.000035	2	4.2843
With location of pixels	Gray	0.000018	1	5.0450

the facts that AD value is less than 1, MD and NAE values are ≥ 1, which show that the frame with embedded information is of good quality.

As we have seen in LSB implementation using similar type of concept called energy pixel by Paul et al. [5], PSNR value is 38.62, which is less than our average PNSR value of 78.73. This shows that the proposed randomized pixel-based watermark is better than the energetic pixel-based watermark.

5 Conclusion

Here, the time complexity of the algorithm is O($n$log$n$) indicating that it is non-computational-intensive. Experimental results show that the PSNR value is high, the MSE is very low and SSIM value is 1.0. Hence, we can infer that the stego image quality is close to cover image. Further, histograms show that it is difficult for the human visual to recognize any difference between a cover and Stego image frame, leading to the conclusion that optimal imperceptibility is achieved. Further, as the algorithm uses every odd-numbered image frame of the video to store the authentication information, robustness is implemented. We have considered authentication embedding in spatial domain only. In the sample out.avi that we used above, there were 3.69 MB of data for a very short video of duration 9 s. As such, for full-length feature films, data will be in big data scale. We can then consider the use of cloud for processing. We can also explore the same algorithm on transform domain.

References

1. Cem kasapbaşi, M., Elmasry, W.: New LSB-based colour image steganography method to enhance the efficiency in payload capacity. Security and integrity check. Sādhanā **43**, 68 (2018). https://doi.org/10.1007/s12046-018-0848-4
2. Singh, N., Bhardwaj, J.: Comparative analysis for steganographic LSB variants. In: Iyer B., Nalbalwar S., Pathak N. (eds.) Computing, Communication and Signal Processing. Advances in Intelligent Systems and Computing, vol. 810. Springer, Singapore (2019)
3. Gupta, P., Srivastava, P., Bhardwaj, S., Bhateja, V.: A modified PSNR metric based on HVS for quality assessment of color images. In: International Conference on Communication and Industrial Application, pp. 1–4. IEEE (2011)
4. Bateja, V., Srivastava, A., Kalsi, A.: Fast SSIM index for color images employing reduced-reference evaluation. In: Proceedings of the International Conference on Frontiers of Intelligent Computing: Theory and Applications (FICTA), pp. 451–458. Springer, Cham (2014)
5. Paul, G., Davidson, I., Mukherjee, I., Ravi, S.S.: Keyless steganography in spatial domain using energetic pixels. In: Venkatakrishnan V., et al. (eds.) Proceedings of the 8th International Conference on Information Systems Security (ICISS), vol. 7671, pp. 134–148. LNCS, Springer, Guwahati (2012). ISBN:978-3-642-35129-7
6. Paul, G., Davidson, I., Mukherjee, I., Ravi, S.S.: Keyless dynamic optimal multi-bit image steganography using energetic pixels. Multimed. Tools Appl. (2012). https://doi.org/10.1007/s11042-016-3319-0
7. Ramalingam, M.: Stego machine-video steganography using modified lsb algorithm. World Acad. Sci. Eng. Technol. **74**, 502–505 (2011)

8. Li-Yi, Z., Wei-Dong, Z., et al.: A novel steganography algorithm based on motion vector and matrix encoding. In: IEEE 3rd International Conference on Communication Software and Networks (ICCSN), pp. 406–409. IEEE (2011)
9. Cao, Y., Zhang, H., Zhao, X., Yu, H.: Video steganography based on optimized motion estimation perturbation. In: Proceedings of the 3rd ACM Workshop on Information Hiding and Multimedia Security, pp. 25–31. ACM, New York (2015)
10. Kelash, H.M., Wahab, O.F.A., Elshakankiry, O.A., El-sayed, H.S.: Hiding data in video sequences using steganography algorithms. In: International Conference on ICT Convergence (ICTC). IEEE, pp. 353–358 (2013)
11. Deshmukh, P.U., Pattewar, T.M.: A novel approach for edge adaptive steganography on LSB insertion technique. In: IEEE International Conference on Information, pp. 27–28 (2014)
12. Feng, B., Lu, W., Sun, W.: Secure binary image steganography based on minimizing the distortion on the texture. IEEE Trans. Inf. Forensics Secur. **10**(2), 243–255 (2015)
13. Fridrich, J., Lisonek, P., Soukal, D.: On steganographic embedding efficiency, information hiding. In: 8th International Workshop. Alexandria, vol. 4437, pp. 282–296 (2008)
14. Petitcolas, F.A.P., Anderson, R.J., Kuhn, M.G.: Attacks on copyright marking systems. In: Aucsmith D. (ed.) Information Hiding. Second International Workshop, IH'98, Portland, Oregon, U.S.A., 15–17 April 1998, Proceedings of LNCS, vol. 1525, pp. 219–239. Springer (1998). ISBN 3-540-65386-4
15. Petitcolas, F.A.P.: Watermarking schemes evaluation. IEEE Signal Process. 17(5), 58–64 (2000)
16. http://www.petitcolas.net/fabien/watermarking/stirmar\k
17. Westfeld, A., Pfitzmann, A.: Attacks on steganographic systems. Proceedings the 3rd International Workshop on Information Hiding. LNCS, vol. 1768, pp. 61–76. Springer, Berlin (1999)
18. Provos, N.: Defending against statistical steganalysis. In: 10th USENIX Security Symposium, pp. 325–335 (2001)
19. Bas, P., Filler, T., Pevny, T.: Break our steganographic system, the ins and outs of organizing BOSS. Lecture Notes in Computer Science, vol. 6958/2011. Information Hiding. Czech Republic, pp. 59–70 (2011). https://doi.org/10.1007/978-3-642-24178-915

Detection of Epileptic Seizures in EEG—Inspired by Machine Learning Techniques

K. Mahantesh and R. Chetana

Abstract The main objective of this paper is to diagnose epileptic seizures using diverse machine learning techniques. Since EEG signals are highly nonstationary, wavelet transformation is applied to raw EEG signals to extract spectral information in time–frequency domain. These spectral features are fed as input to machine learning techniques to train the model and to predict the epileptic seizures. Machine learning techniques such as Principal Component Analysis (PCA), Linear Regression (LR), Support Vector Machine (SVM), Linear SVC, k-Nearest Neighbor (kNN), and Naive Bayes are explored for comparison and to validate the results. Nave Bayes and SVM classifiers recorded high recognition accuracies of 95.86% and 98.30%, respectively, and has outperformed other ML techniques to detect epileptic seizures using wavelet features in EEG signals.

Keywords EEG · Epileptic seizures · Wavelet · Machine learning (ml) techniques

1 Introduction

Epilepsy is a nervous disease, which causes disturbance of the normal neuronal activity [1] resulting in extreme tremor and loss of control [2]. Seizures are random and whenever there is abnormal synchronous activity in the brain, a seizure attack occurs and the waveforms are temporary in nature. Electroencephalography (EEG) is a popular noninvasive test which checks the brain's electric activity by placing electrodes on the patient's scalp. An epileptic seizure is seen in the EEG as a changed waveform of very high amplitude and period, pulses of lesser time which indicates EEG wave variations during a seizure attack [3]. Assessing may be erratic by humans

K. Mahantesh (✉) · R. Chetana
Department of ECE, SJB Institute of Technology,
Visvesvaraya Technological University, Bangalore, India
e-mail: mahantesh.sjbit@gmail.com

R. Chetana
e-mail: chetana_73@yahoo.com

© Springer Nature Singapore Pte Ltd. 2020   443
V. Bhateja et al. (eds.), *Intelligent Computing and Communication*,
Advances in Intelligent Systems and Computing 1034,
https://doi.org/10.1007/978-981-15-1084-7_42

which is time-consuming and expensive procedure which does not give complete data [4, 5]. Sections here are segregated as follows: Sect. 2 gives literature review revealing the trends in seizure detection. In Sect. 3, current leading machine learning algorithms are described. Experiments and results are discussed in Sect. 4. Summary and gist are discussed at the end.

2 Related Works

In recent years, many techniques have been developed based on the stages which involve preprocessing, feature extraction, and classification [6–9]. Several times and T-F-based features are drawn out and given to LDA for classifying epilepsy in highly nonlinear EEG signals [10]. Wavelet db4 is used to decompose EEG into 1 approximation and 5 detailed coefficient with MATLAB toolboxes on EEG signals producing 12 statistically significant features [11]. Time–frequency components were found crucial in determining the epileptic features, hence wavelet, PCA combined with ANN classifier in compressed domain was proposed in [12]. Fourier transform is applied to individual windows and alpha, beta, gamma, delta, and theta obtained to measure the power spectrum density and classified seizures using deep neural nets [13].

The EEG signals are preprocessed using (i) an analog band-pass filter with [0.570] Hz and (ii) a 50 Hz notch filter to remove noise due to power line. A low pass-filter whose f_c is 16 Hz is given to the noise-free signal and further decimated to 32 Hz to bring down the load got by computations [14]. The unpreprocessed signal is sent through an infinite impulse response (IIR) Butterworth band-pass filter from 1–30 Hz which eliminates frequency components below 1 Hz and above 30 Hz [15]. The WT is better than Wiener filtering [16] where time-variant decomposition is got which is advantageous and SNR is bettered. Artifacts are major contaminations in EEG signals which are due to biological activities happening in human body, contact impedance from electrodes and electro magnetic field effects of electronic instruments. These EEG signals are continuous for 30 s which are converted into epochs of 2 seconds. Each epoch is sampled at 200 Hz and provides instantaneous amplitude. There are 400 values for each epoch which contributes to large dimension of data value [17]. Frequencies lie in the range of 0.50–29 Hz and each channel is divided into epochs each of 1 minute duration. First task is to preprocess, then extract distinguishing features or characteristics which aids in achieving EEG signals, later the features are fed into classifier which categorizes the EEG signals. Many methods have been designed and implemented in different domains like (a) time domain (T-D), (b) frequency domain(f-D), (c) time–frequency domain (T-D), and (d) nonlinear methods. Witnessing the trend, Time–frequency domain technique and Machine learning approaches are selected for quantitative evaluation and validation of epileptic seizure detection.

3 Proposed Methodology

Wavelet transforms are used for preprocessing in time–frequency domain to get significant spectral features. These spectral components are given as input to machine learning (ML) techniques for building training models. ML techniques such as PCA, LR, SVM, kNN, and Naive Bayes are explored for generating a training model and predicting epileptic seizures in EEG data.

3.1 Discrete Wavelet Transforms (DWT)

Spectral analysis is a primeval task in describing epileptic features. Since, energy concentration in time–frequency domain leads to more robust feature extraction of alpha, beta, gamma, and delta signals [18]. Wavelet transform decomposes a signal into different levels, and dominant frequencies are obtained on number of decomposition levels as shown in figure. Statistical time–frequency features of EEG signal sequences such as average of the absolute value, maximum absolute value, mean force coefficients, standard deviations in each sub-band are considered as features and given as input to machine learning techniques.

3.2 Machine Learning Techniques Theoretical Background

Machine learning algorithms are classified as parametric and nonparametric. Parametric algorithms are algorithms which learn using predefined mapped function Ex. LDA, Regression, Preceptor. The advantages of these algorithms are they are simple, easy to understand and can interpret results easily and fast. However assumptions and their parameters have poor distribution, it may be solved using nonarametric algorithms which do not make any assumptions and are free to learn from dataset provided such as, KNN, Naive Bayes, SVM, Decision Trees. Some of the contemporary machine learning algorithms are briefed in following sections.

Principal Component Analysis (PCA) is a dimension-reduction tool which reduces a large set to a small set containing most of the information by reducing the number of redundant features. A number of correlated variables are transformed into a number of uncorrelated variables called principal components using PCA which also compresses a lot of data. PCA is used to reduce overfitting problem. In feature elimination the feature space is reduced by eliminating features. The following are the steps to find PC:

1. Take the mean of the given attributes and make a covariance matrix C and solve for the elements of covariance to find Eigenvalue using $C - \lambda I$

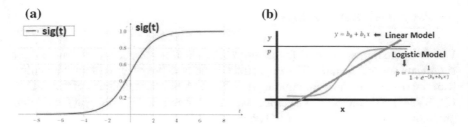

Fig. 1 Graph showing **a** sigmoid and **b** logistic regression curves

2. Using eigenvalue find eigenvectors and put in matrix and observe the eigenvalues. Pick the one which is greater
3. The eigenvector got by selecting the largest eigenvalue is the principal component.

Logistic Regression is a fast training linear model based on supervised learning and so the classes or clusters are expressed in probabilities as good or bad, true, or false. Logistic regression is used toward predicting relationships [5]. Regression analysis is predictive modeling technique, It estimates the relation between dependent (target) and independent variable (predictor) in a straight line equation $y = mx + c$, by selecting arbitrary value of x, y is predicted. This is used when the output is in binary. To get a categorical value, output should be discrete and hence solves classification problem. If the range is from $-\infty$ to $+\infty$ sigmoid curve is used to convert into discrete values, Fig. 1 shows sigmoid and logistic regression curves. A threshold value is fixed and the range is categorized based on the threshold. This can provide probability estimates, by scaling down the data, coefficients of data can provide a clue to select important features. However, this can be updated with new training data and is easy to parallelize. Equations 1 and 2 are linear and logistic models, respectively, of x attributes.

$$Y = b_0 + b_1 x \tag{1}$$

$$P = 1/1 + e^{-(b_0 + b_1 x)} \tag{2}$$

Support Vector Machine Machine has the capability to learn on its own and improve from past histories without being programmed. SVM classifies data more effectively than logistic regression and separates the data using hyperplane. It trains on a set of labeled data both for classification and regression problems which is one of the important features of SVM [19]. The data is classified by finding a line which separates training dataset and does margin maximization, i.e., it maximizes the distance between various classes involved. Math kernel trick is used and transformation from lower dimension to higher dimension is done using SVM kernel functions. Transformation may be by taking the square or by taking the logarithmic value of the data point. By doing so, new data points are got and a line can be drawn now. The margin between the decision boundary or hyperplane and closest point is taken and the mar-

(a) **(b)**

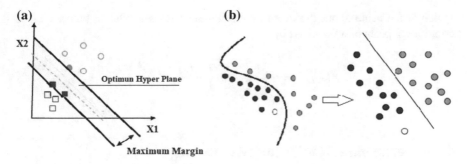

Fig. 2 Input data and transformed feature space

gin has to be maximum and these points are called as support vectors. Ultimately, SVM has to choose right kind of kernel as shown in Fig. 2.

k-Nearest Neighbor is a nonparametric method used for classification and regression, and hence pattern recognition is done. Categories are made by setting the K value to a numeric and more the value of K more clarity is got. KNN provides nonlinear decision boundary and can provide probability estimates. KNN can be updated with new dataset and can act parallel. It is fast to train, but slow to classify and hence need to define similarity in a good way [20].

Gaussian Naive Bayes is a machine learning algorithm used as a classifier. Text classification is the prime focus involving high-dimensional training datasets. Nave Bayes Algorithm models can be built fast and quick predictions can be made effectively. This algorithm gets to know the probability of object with distinguishing features belonging to a specific group or class. It is called as naive because it presumes the occurrence of a certain feature is self dependent of the other and works on the principle of conditional probability. This algorithm is based on Bayes theorem which is stated as given in equations below.

$$P(A/B) = \frac{P(B/A)P(A)}{P(B)} \tag{3}$$

where $P(A/B)$ is the possibility of occurrence of event A with given event B is true and also known as posterior probability. $P(A)$ is prior likelihood of the response variable, $P(B)$ is The possibility of training data or the evidence and $P(B/A)$ is known as the likelihood of the training data.

$$Posterior = \frac{Prior \, X \, Likelihood}{Evidence} \tag{4}$$

In Gaussian Naive Bayes, values which are continuous are related with each feature and supposed to be scattered according to a Gaussian distribution. A Gaussian distribution is also called Normal distribution. A bell shaped curve is got when plotted which is the same about the mean of the feature values as shown below and the conditional probability is given by

$$P(x_i/y) = \frac{1}{\sqrt{2\pi\sigma_y^2}}\exp\left(-\frac{(x_i - \mu_y)^2}{2\sigma_y^2}\right) \tag{5}$$

4 Experimental Results and Discussions

Experiment is carried out using Python 3 along with numpy, pandas, matplotlib, scikit-learn, and seaborn Packages on Intel(R) Core(TM) i3-6006U CPU at 2.00 GHz, 2000 MHz PC. In this section, dataset, experimental setup and performance analysis are explained.

4.1 Dataset

EEG data used consists of 400 EEG samples (of which 200 samples are epileptic and remaining 200 samples are normal) captured at a sampling rate of 173 Hz sampling frequency from international 1020 electrode placement for 23.6 s. Data is passed through band-pass filter [freq range 0.5340 Hz] to obtain clinically interested range. We selected EEG segments containing spike and wave complex with background artifacts and normal EEG. The original dataset consists of 4097 time-series data points for 23.6 s. These data points are mixed and divided into 23 bins, each contains 178 data points per second and hence obtained EEG recording at different points in time domain. All 178 data points are placed in columns as a row and labeled the data from in the range 1–5 as last column (column y).

4.2 Experimental Setup

All of the proposed Machine learning techniques are implemented using following steps:
Step 1: Defining a problem: Diagnosing epileptic seizures in EEG signals.
Step 2: Preparing Data and Feature Extraction: Frequency components are extracted using Discrete Wavelet Transform (DWT).
Step 3: Training Model: Features extracted in Step 2 is given as input to training model to learn with/without labels.
Step 4: Making some Predictions: Predictions are made on query data seizure detection and classification and finally improving and presenting results at the end.

Table 1 Performance analysis of machine learning techniques

Methods	Accuracy (%)
Principal component analysis [12]	89.00
LDA [10]	79.20
Logistic regression	82.83
ANN [13]	93.92
KNN [20]	93.92
Wavelet+PCA+ANN [12]	91
Wavelet+Gaussian Nave Bayes	**95.86**
Wavelet+support vector machines	**98.30**

4.3 Performance Analysis

The emphasis of this paper is to investigate the performances of machine learning techniques and choice of actual features to better the outcomes of seizure detection. Proposed techniques learn from the dataset without assumption and also found flexible in fitting a large number of features results in high performing prediction models. Earlier studies have suggested the use of various frequency domain techniques for feature extraction and distance measures as classifiers to confine the seizure onset areas. In this paper, an effort is made toward learning features automatically and generated training models and also to predict seizures on query EEG data. Performance Evaluation of proposed Machine Learning Techniques is recorded in Table. Nave Bayes and SVM are able to detect more irregularities in seizures with high variations with minimum amount of labeled data and outperforms when compared with PCA, LDA, and logistic regression techniques (Table 1).

With reference to the above table, its evident that Naive Bayes and SVM are very basic and not difficult to implement, needs less training data, handles both continuous and discrete data, highly scalable with number of predictor and data points and not sensitive to irrelevant features. Performance is good with categorical input variables rather than numerical variables.

5 Conclusion

EEG analysis is very critical in detecting Epileptic seizures and accuracy becomes a decisive metric to evaluate the performances of seizure detection systems. EEG signals are transformed using wavelets time–frequency domain to extract spectral information. Later, these spectral components are used specifically to build a training model with the labeled set of data using. Proposed parametric methods such as Nave Bayes and SVM techniques have resulted with highest accuracies and seem to be very promising in classifying epileptic and non-epileptic EEG data. Sometimes, parametric approaches may require large amount of learned data and affect the performance of system in terms of overfitting, processing time and flexibility, one can think about adopting deep learning techniques to overcome these as future research avenues.

References

1. Yaun, Y.: Detection of epileptic seizure based on EEG signals. In: 3rd International Congress on Image and Signal Processing, vol. 9, pp. 4209–4211 (2010)
2. National Center for Biotechnology Information. http://www.ncbi.nlm.nih.gov
3. Dingle, A.A., Jones, R.D., Carroll, G.J., Richard Fright, W.: A multistage system to detect epilepti from activity in the EEG. IEEE Trans. Biomed. Eng. **40**(12), 1260–1268 (1993)
4. Carey, H.J., III, Manic, M., Arsenovic, P.: Epileptic spike detection with EEG using artificial neural networks. In: IEEE (2016). 978-1-5090-1729-4/16
5. Bajaj, V., Pachori, R.B.: Classification of seizure and nonseizure EEG signals using empirical mode decomposition. IEEE Trans. Inf. Technol. Biomed. **16**(6), 1135–1142 (2010)
6. Kabir, E., Siuly, Zhang, Y.: Epileptic seizure detection from EEG signals using logistic model tress. Brain Inf. **3**, 93–100 (2016)
7. Sharma, R., Pachori, R.B.: Classification of epileptic seizures in EEG signals based phase space representation of intrinsic mode functions. Expert Syst. Appl. **42**, 1106–1117 (2015)
8. Kumar, Y., Dewal, R.L., Ananad, R.S.: Epileptic seizure detection using DWT-based fuzzy approximate entropy and support vector machine. Neurocomputing **133**, 271–279 (2014)
9. Xiang, J., Li, C., Li, H. et al.: The detection of epileptic seizure signals based on fuzzy entropy. J. Neurosci. Methods **243**, 18–25 (2015)
10. Khayrul Bashar, Md., Reza, F., Idris, Z., Yoshida, H.: Epileptic seizure classification from intracranial EEG signals: a comparative study. In: IEEE EMBS Conference on Biomedical Engineering and Sciences (IECBES), pp. 96–101 (2016)
11. Mamun, Md., Ahmad, M.: Epileptic seizure classification using statistical features of EEG signal. In: International Conference on Electrical, Computer and Communication Engineering (ECCE), pp. 308–312 (2017)
12. Mahantesh, K., Sindhu, R., Chetana, R., Rashmika, R.: An impact of time-frequency domain technique and neural networks for epileptic seizure detection. In: IEEE - 3rd International Conference on Electrical, Electronics, Communication Computer Technologies and Optimization Techniques (2019). [article in press]
13. Birjandtalab, J., Heydarzadeh, M., Nourani, M.: Automated EEG-based epileptic seizure detection using deep neural networks. In: IEEE International Conference on Healthcare Informatics, pp. 552–555 (2017)
14. Boashash, B., Ouelha, S.: Automatic signal abnormality detection using time-frequency features and machine learning: a newborn EEG seizure case study (2016). Elsevier
15. Carey, H.J., Manic, M., Arsenovic, P.: Epileptic spike detection with EEG using artificial neural networks. In: IEEE (2016)
16. Faust, O., Acharya, U.R., Adeli, H., Adeli, A.: Wavelet-based EEG processing for computer-aided seizure detection and epilepsy diagnosis 1059–1311 (2015). Epilepsy Association. Elsevier
17. Manjusha, M., Harikumar, R.: Performance analysis of KNN classifier and k-means clustering for robust classification of epilepsy from EEG signals. In: IEEE (2016). 978-1-4673-9338-6/16/
18. Mahantesh, K., Aradhya, M.: Coslets: a novel approach to explore object taxonomy in compressed DCT domain for large image datasets. In: The Proceedings of 3rd International Symposium on Intelligent Informatics, ISI14, vol. 320, pp. 39–48 (2015)
19. Guyton, A.C.: Text Book of Medical Physiology. Saunders, Philedelphia (1986)
20. Yol, S., Ozdemir, M.A., Akan, A., Chaparro, L.F.: Detection of epileptic seizures by the analysis of EEG signals using empirical mode decomposition. In: IEEE, pp. 258–262 (2018)

Automated Video Surveillance System Using Video Analytics

T. M. Rajesh and Kavyashree Dalawai

Abstract In this current trend, intellectual video scrutiny is playing a major role in research. It is best suited for various kinds of applications that include security and safety at public areas like airports, traffic control in metropolitan cities and detection of crime that include hit-and-run once and so on. In the hit-and-run case, detection and recognition of object in the video is one of the biggest challenging tasks. Even though several algorithms have been used to detect and recognize the object in the video, it is still considered as a challenging task because of change in illumination. In this work, we have proposed a method for detecting and recognizing a moving object which is involved in crime. We have used Kalman and Gaussian mixture analysis for detecting the object and SIFT and SURF and a classification algorithm like PCA to recognize the object, and continued by the analytics like false acceptance rate and false rejection rate for verification and validating the object during detection as well as recognition process. The investigational results show the efficiency of the algorithms used on various databases.

Keywords Object detection · Object recognition · SURF · SIFT · Connected component analysis (CCA) · Principle component analysis (PCA) · Hierarchical centroid

1 Introduction

Computer vision and video analytics are substantial fields in image processing. Computer vision is used to analyze and understand the image to extract the information out of the images. In other words, video analytics is used in various applications,

T. M. Rajesh (✉) · K. Dalawai
Department of Computer Science and Engineering, School of Engineering,
Dayananda Sagar University, Bengaluru, Karnataka, India
e-mail: rajesh-cse@dsu.edu.in

K. Dalawai
e-mail: kavyashree-cse@dsu.edu.in

© Springer Nature Singapore Pte Ltd. 2020
V. Bhateja et al. (eds.), *Intelligent Computing and Communication*,
Advances in Intelligent Systems and Computing 1034,
https://doi.org/10.1007/978-981-15-1084-7_43

which include broadcast network where we want to know how many viewers hooking up the show, to detect the crime in hit-and-run case and many. Video analytics is capable of inspecting the video and determining spatial and temporal events automatically. Object detection and recognition are two important steps to achieve success in various applications. Object detection is an initial stage that provides the information about background and foreground object from the scene. Based on the information we can just detect the object but this is not sufficient. We have to verify and validate the object detected by recognizing it based on the feature extracted during object detection. The main objective of this work is to study and analyze the novel method of detecting and recognizing the moving object. Detection of moving objects is one of the challenging task because of many reasons like moving object has to distinguished from the stationary background. In this paper, we have used the recorded videos as the database where each video is converted into image frame for our experiments. We have used some morphological techniques for enhancing the image frames. Along with morphological techniques, we have used OCR, PCA and SURF features. Optical character recognition (OCR) plays an important role in extracting the text from videos. It helps to convert text from video into digital text. Video OCR detects the text from the video file and generates text file that can be used in search engine. Search engine can easily index the file based on the metadata extracted by video or image. Principle component analysis (PCA) and other techniques are useful in extracting the good features of the objects. As we extract many features from the scenes, they help us to verify the object properly. The approach mentioned in this paper helps us to identify and verify the object involved in the crime with promising result.

2 Related Works and Literature Survey

Review of literature is an important stage for any kind of research. Survey of the previous works will give us proper, relevant and adequate knowledge in the specific domain. Here is the precise work in the field of moving object detection and recognition. Chen et al. [1] presented an online learning architecture to learn the templates and detect the moving objects in real time. The trained architecture was applied on highly cluttered background and this was difficult for 2D computer vision methods. Ahn et al. [2] used hardware architecture. The system was implemented on FPGA-based hardware architecture in which the CNN was used as the core function for recognizing the moving objects in the real-time video signals. In this system, a total of 150 standard multipliers were used with an average time of 6%. The proportion of hardware resources used other than multipliers was under 40%. Qiu et al. [3] proposed a novel obstruct object detection method using partial configuration object model (PCM) for HR-RS images. First, they defined a buffer layer of partial configurations based on semantic parts combination. Second, they compared semantic parts with DPM model to get a whole object. Third, using the partial configurations clustering, they used nonmaximum suppression (NMS) method to find exact locations of the objects. Yeh et al. [4] proposed a concept of hysteresis threshold and motion compensation that includes spatial and temporal compensation to detect the moving object. They have also measured other algorithms in their paper. Nimmagadda et al. [5] in their paper have used a real-time mobile robot surveillance system to recognize and track the moving object. They followed two steps for recognizing the objects: In the first step, features are extracted and in the next step features were searched from the database. They have shown base case with a set of fixed parameters, like distance of object, speed of object and so on, and also analyzed the decision when the parameters are overloaded. Nagendran et al. [6] in their main work tried to recognize various moving objects in the dynamic backgrounds. The proposed algorithm is resistant in changing the brightness. Further, it includes a module that decreases the impact of camera development. Finally, blob analysis is performed to detect a group of connected pixels that corresponds to the moving object. In this paper, the object detection and recognition with the complex backgrounds were successfully evaluated with four classes of objects and produced good recognition results

3 Working Mechanism and Proposed Methodology

3.1 Video Acquisition

Video acquisition is an essential step in recognizing the moving object. The raw video captured by camera should be converted into manageable entity.

3.2 Background Subtraction

Motion detection is one of the first steps of a multistage computer vision system (person recognition, car tracking, etc.). As a result, it is usually required to be very fast and as simple as possible. There are various methods used in background subtraction which are discussed below [7]:

3.2.1 Frame Differencing

The background of an image can be calculated by subtracting the present frame from the preceding frame or from an average image of the number of frames. According to this, pixels belong to foreground if

$$|I_i(r, k) - I_{i-1}(r, k)| > R \tag{1}$$

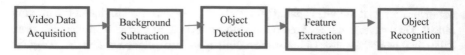

Fig. 1 Working mechanism

where I_i is the current frame, I_{i-1} is the previous frame and R is the selected threshold. The easiest way to model the background B is through a single grayscale or color image of moving objects. In order to cope with illumination changes and background modifications, that can be iteratively updated as follows:

$$B_i = \alpha \, I_{i-1}(r, k) + (1 - \alpha) B_{i-1}(r, k) \tag{2}$$

where α is a constant whose value ranges between 0 and 1.

The average can be expanded as follows:

$$B_i(x, y) = \begin{cases} I_{i-1}(r, k) + (1 - \alpha) B_{i-1}(r, k), & \text{if } I_{i-1}(r, k) \text{ is background} \\ B_{i-1}(r, k) & \text{if } I_{i-1}(r, k) \text{ is foreground} \end{cases} \tag{3}$$

At each new frame, each pixel is either classified as foreground or background. The selective background model ignores any pixels which are classified as foreground by the classifier.

3.3 Object Detection

It is a procedure on discovery illustrations of real-world objects such as buildings, cars, faces in video and images [8]. Objects are detected based on the features extracted in the previous step. In the classification step, certain classification algorithms are used which classify the real-world entities captured during acquisition process as mentioned in the Fig. 1.

3.4 Feature Extraction

It is a type of dimensionality reduction that competently represents a fascinating part of video (image) as a compact feature vector. Features are extracted during object detection and object recognition [9]. The techniques and methods used for object detection are discussed below:

Fig. 2 Original image

Fig. 3 Result of connected component

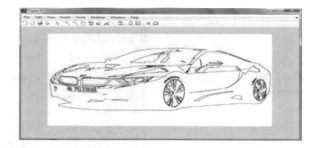

3.4.1 Connected Component

This method scans an image pixel-by-pixel in mandate to analyze combined pixel sections, that is, regions of adjoining pixels which share the set of strength values *V*. Once the image is scanned, equivalent label pair is arranged in similarity classes and exclusive label is assigned to each class [10]. Once the foreground regions are detected postprocessing techniques are applied to remove noise and illumination change (Figs. 2 and 3).

3.4.2 Hierarchical Centroid

This method computes the *x*-coordinate of the centre of mass for an image. The descriptor subdivides an image into two images by *x*-coordinate and calls itself (recursively) when each of the subimages is altered. The hierarchical method gives output in two forms: the first one, if the images are same and the distance between two images is considered as zero, and second one, when the image is slightly changed due to some variation. Then the distance between two images will have some value [11, 12] (Figs. 4 and 5).

Fig. 4 When the image has
the same value 0

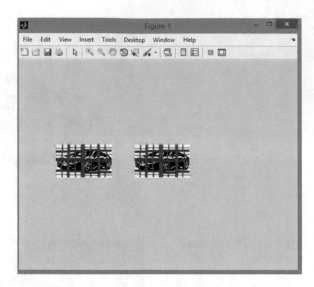

Fig. 5 When the image has
different value 0.933

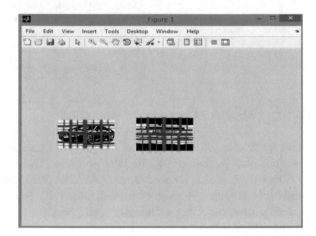

3.4.3 Principal Component Analysis

It is a demographic approach that practices an orthogonal renovation to change usual annotations of conceivably connected variables into a set of values of linearly uncorrelated variables called principal components. PCA can be performed by eigenvalue rottenness of the data covariance matrix or particular value decomposition of a data matrix, for the most part after mean focusing the information grid for each characteristic. This method is subtle to the scaling of the statistics and there is no accord in the matter of how to best scale the information to get ideal conclusions. PCA is one of the measurable systems that is, most of the time, used as a part of flag preparing to the information measurement decrease or to the information décor relation [13, 14].

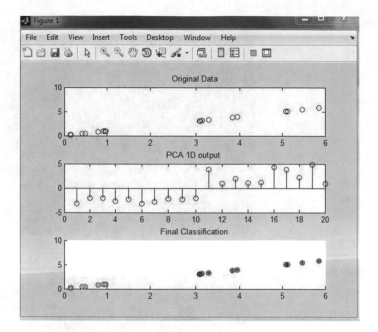

Fig. 6 Principal component analysis

3.4.4 Speed up Robust Features

The classic form of SURF in certain times is quicker than the SIFT, and it is more robust against different image transformations than SIFT. The objective of feature is to afford a distinctive and robust description of an image feature. When we run the code if the image 1 is matching with the image 2 and matching point is greater than 80%, then the result will be found matching and if the image 1 is not matching with the image 2, then the result will be found not matching [15, 16] (Fig. 6).

3.5 Object Recognition

Object recognition is also called as verification and validating the object that has been found in the image. Below are some of the terms which refer to verification and validation process.

Equal Error Rate: It is computed by taking the average values derived from both FRR and FAR. It is a good criterion for evaluating the accuracy of a method. The method with the lower ERR can be considered as the most accurate technique

$$EER = FAR + FRR/2$$

Fig. 7 Original image

Fig. 8 Dummy car image

Fig. 9 The output of both
Figs. 9 and 10

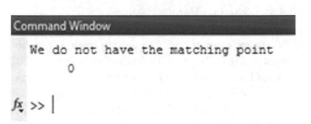

The recognition is done by comparing structures of an object to the features of the object of the database using Euclidean distance method and a fast nearest neighbor algorithm which is used to perform fast in bigger datasets [17, 18] (Figs. 7 and 8).

4 Results and Comparisons

Train frame of running is selected from the folder, converted into grayscale image for further processing. For moving object action, the identification system is performed efficiently when the threshold is 60:40 with good accuracy [19, 20].

After getting the linearity result by using EER technique to prove that accuracy of our proposed work is promising, we have evaluated the results with another metric

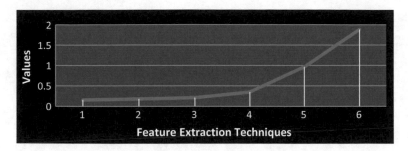

Fig. 10 EER chart

by comparing the FAR and FRR results with combination of different threshold point ratios as shown in Table 3. For stable object action, the identification system is performed efficiently when the threshold is 60:40 with good accuracy [21, 22]. This analysis result shows us that more number of features in the trained dataset against the training dataset gives more accuracy [23, 24].

5 Conclusion

Detection of moving objects in the area of forensic crime detection is still remaining as a challenging task. Although an enormous amount of work is done in this field, the accuracy in detecting the objects (in our case its car) which are involved in crime was substantial. This is purely because of extracting the less features from the scenarios. In our proposed work we have used as many as various techniques to extract the features like SURF, SIFT, CCA, PCA, hierarchical centroid for object detection and recognition which give more information on the scenarios. The more the number of feature extraction, the more will be the accuracy. Combination of different algorithms and techniques used in our work is promising and very much accurate. The analysis result of different techniques is as shown in Tables 1 and 2. To prove that combination of different techniques gives high accuracy, we have experimented with 10 different datasets and each dataset has got more than 1000 number of images with the combination of ratios like 20:80 till 60:40. The average percentage accuracy rates what we have got in our proposed system is 0.62% of EER. For different aspect ratios of threshold points, we have got an average rate of 0.24% of FRR and 0.45% of FAR. So the error rate in our case is linear and the accuracy rate is very high. In future, many more features will be considered for experiment to get 100% accuracy in identification and verification system (Table 3).

Table 1 Analysis of feature extraction

Database no.	SURF	Connected component	SIFT	Hierarchical centroid	PCA
1	0.0423	0.2291	0.340	4.7165	1.68
2	0.211	0.0565	0.258	2.6572	2.09
3	0.133	0.1449	0.4619	0.1814	1.8
4	0.181	0.2463	0.349	0.0383	4.565
5	0.043	0.1415	0.404	0.3123	1.94
6	0.387	0.3682	0.510	0.2517	0.56
7	0.252	0.5466	0.175	0.3985	0.98
8	0.050	0.0651	0.091	0.6645	1.04
9	0.166	0.1825	0.552	0.3664	0.42
10	0.07	0.1257	0.35	0.0641	2.75
Average	0.1535	0.2106	0.3491	0.9651	1.882

Table 2 EER table

SI no.	Feature extraction	EER
1	SURF	0.1535
2	Connected component	0.2106
3	SIFT	0.3491
4	Hierarchical centroid	0.9651
5	PCA	1.8825

Table 3 FRR and FAR

Threshold	FRR	FAR
20:80	1	0
30:70	0.13	0.47
40:60	0.07	0.66
50:50	0.04	0.67
60:40	0	0.89
20:80	1	0

References

1. Chen I.-K., Chi, C.-Y., Hsu, S.-L., Chen, L.-G.: An integrated system for object tracking,detection and online learning with real-time RGB-D video. In: IEEE International Conference on Acoustics, Speech and Signal Processing (ICASSP), pp. 6558–6562. https://doi.org/10.1109/icassp.2014.6854868. (2014)
2. Ahn, B.: Real-time video object recognition using convolutional neural network. In: IEEE International Conference on Neural Network, p. 17. Neurocoms, Seoul (2015). https://doi.org/10.1109/ijcnn.2015.7280718

3. Qiu, S., Wen, G., Fan, Y.: Occluded object detection in high resolution remote sensing images using partial configuration object model. In: IEEE Journal of Selected Topics in Applied Earth Observations and Remote Sensing, pp. 1–17 (2017). https://doi.org/10.1109/jstars.2017.2655098

4. Yeh, C.-H., Lin, C.-Y., Muchtar, K., Lai, H.-E., Sun, M.-T.: Threepronged compensation and hystersis thresholding for moving object detection in real-time video surveillance. In: IEEE Transaction on Industrial Electronics, pp. 1–1 (2017). https://doi.org/10.1109/tie

5. Nimmagadda, Y., Kumar, K., Lu, Y.-H., Lee, C.S.G.: Realtime moving object recognition and tracking using comp utation offloading. In: IEEE/RSJ International Conference on Intelligent Robot sand Systems, pp. 2449–2455 (2010). https://doi.org/10.1109/iros.2010.5650303

6. Nagendran, A., Ritika, V., Sharma, V.: Recognition and tracking moving objects using moving camera in complex scene. Int. J. Comput. Sci. Eng. Appl. (IJCSEA) **4**(2) (2014)

7. Shaikh, S.H.: Moving Object Detection Using Background Subtraction. Springer Briefs in Computer Science (2014). https://doi.org/10.1007/978-3-319-07386-6_2

8. Yokoyama, M., Poggio, T.: A contour based moving object detection and tracking. In: IEEE International Workshop on Visual Surveillance and Performance Evaluation of Tracking and Surveillance. pp. 271–276 (2005). https://doi.org/10.1109/vspets.2005.1570921

9. Cohen, I., Medioni, G.: Detecting and tracking moving objects for video surveillance. In: IEEE Computer Society Conference on Computer Vision and Pattern Recognition (Cat. No. PR00149), vol. 2, p. 325 (1999). https://doi.org/10.1109/cvpr.1999.784651

10. Deep, A., Goyal, M.: Moving object detection techniques. IJCSMC **4**(9), 345–349 (2015)

11. Kim, D.-S., Kwon, J.: Moving object detection on a vehicle mounted back-up camera. Academic Editor: Felipe Jimenez (2015). Accepted 22 Dec 2015, Published 25 Dec 2015

12. Lee, S.-H., Yang, C.-S.: A real time object recognition and counting system for smart industrial camera sensor. IEEE Sens. J. **99** (2017)

13. Rajesh, T.M., Manjunath Aradhya, V.N.: An application of GMM in signature skew detection. J. Pattern Recogn. (2015)

14. Rajesh, T.M., Manjunath Aradhya, V.N.: ICA and neural networks for Kannada signature identification. Int. J. Latest Trends Eng. Technol. (2016)

15. Oh, J., Kim, G.: A 320 mW 342GOPS real-time moving object recognition processor for HD 720p video streams. IEEE Int. 220–222 (2012). https://doi.org/10.1109/isscc.2012.6176983

16. Changhua, L., Ningning, C., Rui, F., Chun, L.: A novel algorithm for moving object recognition based on sparse bayesian classification. 135–139 (2006). https://doi.org/10.1109/mlsp.2006.275536

17. Zhou, W., Miura, Y.: Denoising method using a moving-average filter for the moving object recognition of low illuminance video image. In: IEEE International Conference on Consumer Electronics-Taiwan (ICCE-TW), pp. 1–2 (2016). https://doi.org/10.1109/icce-tw.2016.7521023

18. Kavyashree, D., Rajesh, T.M.: Analysis on text detection and extraction from complex background images. J. Pattern Recogn. 37–43 (2018)

19. Desai, G.G., Rajesh, T.M.: Analysis on moving object detection. Int. J. Sci. Adv. Res. Technol. (2017)

20. Gaurav, K.A., Viveka, Rajesh, T.M., Shaila, S.G.: Automated number plate verification system based on video analytics. Int. Res. J. Eng. Technol. (2018)

21. Kumar, A., Rajesh, T.M.: A survey on moving object recognition using video analytics. Int. J. Eng. Res. Technol. (2017)

22. Bhateja, V., Malhotra, C., Rastogi, K., Verma, A.: Improved decision median filter for video sequences corrupted by impulse noise. In: 2014 International Conference on Signal Processing and Integrated Networks (SPIN), IEEE (2014)

23. Srivastav, A., Bhateja, V., Moin, A.: Combination of PCA and contour lets for multispectral image fusion. In: Proceedings of the International Conference on Data Engineering and Communication Technology, Springer, Berlin, Singapore (2017)

24. Kumar, A., Rajesh, T.M.: A moving object recognition using video analytics. Int. J. Sci. Res. Dev. (2017)

On Decision-Making Supporting the Shift from Fossil to Renewable Energy Sources Within the Norwegian Maritime Transport Sector

Tom Skauge, Ole Andreas Brekke and Sylvia Encheva

Abstract Reduction of both CO_2 emissions and energy consumption is of high importance in Norway. One of the numerous areas where this can be achieved is maritime transport in general and ferry transport in particular. In this work, we present an approach for finding which of the existing solutions is closest to the one a new county would like to implement. Such a choice is effected by a number of factors among which are political, economic, social, to name a few. While some of them are nearly impossible to quantify, the effect of others is not known due to some confidentiality issues. In order to avoid the effect of bias assumptions and subjective guesses we base our work on publicly available data.

Keywords Maritime transport · Renewable energy · Decision-making

1 Introduction

One of the most important measures to reduce global CO_2 emissions is the shift from fossil to renewable energy sources within the transport sector. In later years, Norway has been in the forefront of introducing electric vehicles, with the highest density of electric vehicles per capita in the world. Almost all stationary energy production in Norway stems from hydropower, and electrifying the transport sector is a major means to lower the country's CO_2 emissions. This electric trend has also been introduced within maritime transport. Norway is a land of fjords, and approximately 200 vessels traffic 130 ferry routes throughout the country. Since 2015 when the first pilot electric ferry was put in operation, new contracts for electric or hybrid-electric vessels have been signed for so far 37 ferry routes with approximately 70

T. Skauge · O. A. Brekke · S. Encheva (✉)
Western Norway University of Applied Sciences, Bergen, Norway
e-mail: Sylvia.Encheva@hvl.no; sbe@hvl.no

T. Skauge
e-mail: Tom.Skauge@hvl.no

O. A. Brekke
e-mail: Ole.Andreas.Brekke@hvl.no

© Springer Nature Singapore Pte Ltd. 2020
V. Bhateja et al. (eds.), *Intelligent Computing and Communication*,
Advances in Intelligent Systems and Computing 1034,
https://doi.org/10.1007/978-981-15-1084-7_44

new ferries [1]. This amounts to a (silent) revolution of ferry transport, and could be the start of a major shift within maritime transport in general. The shift has been driven by political ambitions, with a parliamentary decision in 2016 stating that "The Government is requested by the Parliament, to ensure that zero-emission technology (and low-emission technology) are included in all future tenders for public ferries, when the technology allows for it." [2].

The county administrations are responsible for the ferry transport within their county except for the national roads. The first mover in the shift towards electric propulsion systems was Hordaland County, when they put a demand on minimum 55% reduction of CO_2 emissions on tenders for new ferries in 2015. Since then, several other counties have followed suit.

Decision-makers are expected to find optimal solutions for complicated problems. On one hand they have to take into consideration a number of various factors while on the other hand they are usually pressed for time. Numerous decision support systems both free and commercial have been developed and used in practice. The majority of them are ranking existing options based on the requirements of an expert committee.

In this article we aim at providing the decision-makers with an approach to select a small number of solutions that are as close as possible to the one they are interested to implement. This way they will spend less time and efforts looking at possible solutions that might be quite good in general but very difficult to apply locally.

2 Preliminaries

Since Hordaland County introduced emission reductions in their tender for new ferries in 2015, several other counties along the coast have followed suit. We have gathered data from the tender process for 4 counties, totaling 70 vessels with full-electric or hybrid-electric propulsion systems with variations of diesel/biodiesel back-up engines. Some of the ferries are already operative, others will be put in operation in the period 2019–2021. We have two sources of data. First, we use the Norwegian official platform for public tender—DOFFIN [3]. Our second source of data is the records from the counties' decision-making both from special committees and from the parliament and parliament's steering committee.

An interval number is often defined as

$$a = [a-, a+] = \{x | a- \leqslant x \leqslant a+, a- \leqslant a+,\}$$

and $a-$, $a+$ are real numbers, [4]. If $a+ = 0$, the interval reverts to a point, and thus we would return to the basic crisp model.

For further reading on interval numbers see [4–7].

Definition 1 Jaulin et al. [4] Let l be an extended order relation between the intervals $A = [a_L, a_R]$ and $B = [b_L, b_R]$ on the real line R, then for $m(A) \leq m(B)$, we construct a premise $(A \mid B)$ which implies that A is inferior to B (or B is superior to A) in terms of value. Here, the term 'inferior to' ('superior to') is analogous to 'less than' ('greater than').

Definition 2 Jaulin et al. [4] Let I be the set of all closed intervals on the real line R. An acceptability function is defined as follows $A: I \times I \rightarrow [0,\infty)$ such that $(A < B) = \frac{m(B)-m(A)}{w(B)+w(A)}, \quad w(B) + w(A) \neq 0.$

$A (A \mid B)$ may be interpreted as the grade of acceptability of 'interval A to be inferior to interval B'.

3 Selection of Closest Solutions

Suppose a county is interested to apply "green technologies" for the first time. It is quite natural for this county to benefit from similar experiences from counties that have already implemented similar technologies. One way to do it is to compare current data from the new county with data from each of the other counties. While this approach is seemingly doable and intuitive it is at the same time quite laborious. In addition it opens for inconsistencies since all available cases are different and it is difficult to decide which solution would help the most to avoid common investment pitfalls.

Below we propose an approach for first building sets of similar solutions based on available data and afterward applying an acceptability index for comparing an intended solution to the existing ones. When it is done it is easier to make a final recommendation.

Original data presented with natural numbers is converted into percentages to allow organizing it in sets with comparable properties.

Table 1 has rows corresponding to counties {H, S, M, T, N} and columns corresponding to evaluation criteria {Sb, Sa, Cb, Ca, Rn, Rb, Ra}. Notations in Table 1 are as follows

- Counties—Hordaland (H), Sogn and Fjordane (S), Møre and Romsdal (M), Nordland (N), referred also as attributes
- criterion 1—amount applied for green technology at the time of measurement where Sb stands for amount below 33% and Sa stands for amount above 33%

Table 1 Single ferry solutions

Regions	Amount of green technology		CO$_2$ reduction		Energy reduction	
	Sb	Sa	Cb	Ca	Rb	Ra
H	0.16			0.86–0.92		0.58–0.74
S	0.25		0.7			0.7
M		0.33		0.8	0.15	
N	0.07		0.65		0.2	

- criterion 2—expected reduction of CO_2 where Cb stands for amount below 86% and SC stands for amount above 86%
- criterion 3—expected energy reduction where Rn stands for unknown, Rb stands for amount below 58% and SC stands for amount above 58%.

Notations in Table 2—P1, ..., P5 are five package solutions and the criteria are like the ones already used in Table 1.

Based on the theory of interval arithmetics and data presented in Tables 1 and 2 we identify the following sets of objects and criteria they commonly possess in Table 3. This is later on summarized in Table 4 in terms of intervals.

In the presence of larger datasets, useful groups of regions and criteria they commonly possess (as in Table 3) can be identified by employing concept lattices, [8] or data mining techniques [9]. In our case lattices depicted from Tables 1 and 2 are presented in Figs. 1 and 2, respectively.

The question we try to answer is as follows: A new region has an amount of 0.2 green technology as a start and wants to achieve 0.83 of CO_2 reduction and 0.6 energy reduction. Which regions are closest to the region in question with respect to the initial conditions and goals?

Applying the acceptability index we conclude that the new region would benefit from the experience in counties (Hordaland, Sogn and Fjordane), (Nordland, Sogn and Fjordane), package solutions for (P1 and P3) in that particular order.

Table 2 Package ferry solutions

Packages	Amount of green technology		CO_2 reduction		Energy reduction	
	Sb	Sa	Cb	Ca	Rb	Ra
P1		0.14	0.87		0.6	
P2		0.2		0.9		0.65
P3	0		0.86		0.58	
P4	0		0.88			0.65
P5	0			0.92		0.74

Table 3 Regions and criteria

Sets of regions	Criteria
N, S	Sb, Cb
H, S	Ra, Sb
P1, P3	Cb, Rb
P3, P4	Cb, Sb
P2, P5	Ca, Ra

Table 4 Interval presentation of criteria

Sets of regions and packages	Amount of green technology	CO_2 reduction	Energy reduction
N, S	[0.16, 0.25]	[0.7, 0.92]	[0.58, 0.74]
H, S	[0.07, 0.25]	[0.65, 0.7]	[0.2, 0.7]
P1, P3	[0, 0.14]	[0.86, 0.87]	[0.58, 0.6]
P2, P5	[0, 0.2]	[0.9, 0.92]	[0.65, 0.74]
P3, P4	0	[0.86, 0.88]	[0.58, 0.65]

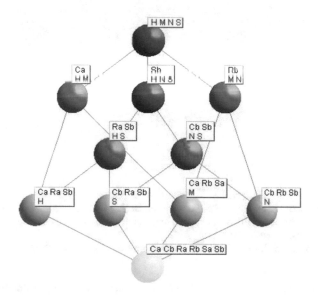

Fig. 1 Concept lattice derived from data in Table 1

4 Conclusion

The presented approach is easy to implement. It is flexible with respect to the number of options, criteria, and committee members. It allows inclusion or removal of alternatives, attributes, and experts without additional cost. As such, the approach could prove beneficial for ensuring a smooth transition towards defossilized maritime transport.

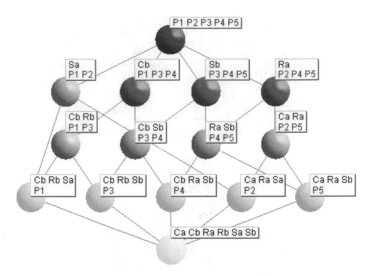

Fig. 2 Concept lattice derived from data in Table 2

References

1. Skauge, T., Anita Sjøseth Tordal A.S., Lummen, N., Brekke, O.A., Jensen, T.Ø.: Implementing carbon free ferry technology—electrical routes in the land of fjords. In: 12th International Scientific Conference on "Energy and Climate Change", Energy Policy and Development Centre (KEPA), Athens, Greece (2018)
2. Tillegg til Innst. 2S: til Stortinget fra Finanskomiteen. Parliament Printing Office, Oslo **37** (2014–2015)
3. https://www.doffin.no/
4. Jaulin, L., Kieffer, M., Didrit, O.: Applied Interval Analysis. Springer, Berlin (2001)
5. Mohiuddine, S.A., Raj, K., Alotaibi, A.: On some classes of double difference sequences of interval numbers. Abstract Appl. Anal. 1–8 (2014)
6. Walster, G.: William; Hansen, Eldon Robert, Global Optimization Using Interval Analysis, 2nd edn. Marcel Dekker, New York, USA (2004)
7. Zhang, J., Wu, D., Olson, D.L.: The method of grey related analysis to multiple attribute decision making problems with interval numbers. Math. Comput. Model. **42**(9–10), 991–998 (2005)
8. Davey, B.A., Priestley, H.A.: Introduction to Lattices and Order. Cambridge University Press, Cambridge (2005)
9. Tan, P.-N., Steinbach, M., Karpatne, A., Kumar, V.: Introduction to Data Mining, 2nd edn. (2018)

Context-Aware Personalized Mobile Learning

Radhakrishnan Madhubala and Akila

Abstract Increasing usage of mobile apps encourages the mobile application developers to add more services through specialized mobile applications. One of such service is m-leaning (mobile learning) application. Learning through mobile pro video unique experience to the learners to learn course/subject at anywhere and at any time. To increase the learning interest and success of learning is relying on the learners need and the learning environment. This requirement brings new research direction called context-aware personalized learning. This paper proposes the new m-learning system which organizes the course content, accommodating the dynamic nature of learner's preferences and state. Learners' preferences are identified through ILS (index of learning styles) test and subsequently updated based on learners activity log using apriori algorithm. The system finally will adopt the content based on the learner's environment context such as network and surrounding light.

Keywords Context · Personalized learning · Adapted learning · Mobile learning · Data mining · Association rule mining

1 Introduction

Mobile learning brings new opportunity to the learners to engage themselves in educational activities anytime and anywhere. It extends the teaching and learning spaces beyond our traditional classrooms by using mobile (portable) device as a mediator for the teaching and learning activities and disseminating quality learning materials through mobile devices. However, the success of any learning system is not only delivering educational contents to the learners but also understanding the

R. Madhubala (✉) · Akila
Department of Computer Science, Vels Institute of Science, Technology & Advanced Studies, Chennai, India
e-mail: balamadh@gmail.com

Akila
e-mail: akila.ganesh.a@gmail.com

© Springer Nature Singapore Pte Ltd. 2020
V. Bhateja et al. (eds.), *Intelligent Computing and Communication*,
Advances in Intelligent Systems and Computing 1034,
https://doi.org/10.1007/978-981-15-1084-7_45

learning nature of the learner and environment. This requirement brings new direction in mobile learning system known as context-aware personalized learning.

The term context refers to any information that can be used to characterize the situation of an entity. An entity is a person, place, or object [1]. Context awareness was introduced in 1994 by Schilit et al. [2], and they defined three important aspects of context such as computing context, user context, and physical context. Dey et al. [3] group the context in four categories: identity, location, status, and time. Wang [4] specifies six dimensions (identity, learner, activity, collaboration, temporal, and facility) of the M-learning context. Zervas et al. [5] categorize learning and mobile contextual elements in two dimensions such as learning context and mobile context. Rachid and Xiaoyun [6] divide the context into four groups such as learner, activity, device, and environment. Personalized learning considers the learning methods, and contents are tailored toward the learner's needs, skills, and interest and adapt the course content (customizing learning material/contents). Context-aware personalized learning means that personalize the m-learning application considers the learner's environments, learning style, situations, needs, and interest and delivers the customized learning activities/content to them.

The paper is organized as follows. Section 2 presents the overview of existing context-aware personalized mobile learning system. Section 3 describes functions of the major components of proposed system. Section 4 describes the proposed system architecture. Section 5 presents experiments and results and finally concludes the paper in Sect. 6.

2   Context-Aware Personalized M-Learning System

Researchers propose numerous personalized context-aware m-learning systems by considering different contextual parameters in order to increase the learner's interest and learning success [7].

Guabassi et al. [8] proposed an approach for providing personalized learning which includes learning style and context awareness. The system follows client-server architecture and has three components such as *learner context* (learning style, cognitive style, and cognitive state), *structure of the course material* (level course, section, concept, and content), and *device context* (device, activity, and environment). Adaptation engine follows the adaptation rules to select the course material based on learner's profile. In general, cognitive state of the learner (beginner/intermediate/expert) is dynamic but this system keeps it as static parameter throughout the learner's interaction and order of rules (adaptation rules) execution is not clearly defined.

Curum et al. [9] presented context-aware mobile learning system which divides the context into two: learner context and device context, for learning materials adaptation. Learning contents/materials are adapted based on the three age groups. But age alone is not being the correct parameter to decide learner's maturity. Moreover, there is less focus on content modeling.

Tortorella and Graf [10, 11] proposed an adaptive mobile learning system which includes learning style and other context parameters in order to select the appropriate formats for present learning content. The system consists of the following components: *ILS component* (administrating ILS questionnaire score of each user), context modeling (process the information gathered from device and sensors and the results are used for adaptation), *adaptive engine* (estimate the optimal delivery mode to deliver the content to the user), and course material. (Course is divided into section. The section is dived into small blocks. Each block is presented in different modes (video, audio, text, and ppt.) In this system, learning style is primary parameter (static) for selecting the course content in this system. But naturally, learning style is not static. Moreover, course assessments are also not considered.

3 Proposed System

Our proposed system includes three context attributes such as device, learner, and content as shown in Fig. 1.

3.1 Device Context

Device context considers the device and its surrounding condition of the learner's mobile device which includes data from device configurations, sensors, and speed of network connection. Data gathered from device and sensors requires preprocessing to obtain the meaningful information. Sensor data are varied depend on the manufacturer's specification [10]. Table 1 lists the sensors and its common uses in mobile learning system [11]

Fig. 1 Context attributes of m-learning system

Table 1 List of common sensors available in mobile devices

Sensor	Common use
Proximity	Detects how close the device is to the learner
Accelerometers	Detect the orientation of the phone
Ambient light	Detects the surrounding light of the device
GPS	Detects the location (geographical location) of the learner

Table 2 Network speed versus content type [9]

Speed range	Content
Low bandwidth (Up to 100 kbps)	Text only
Medium bandwidth (101–250 kbps)	Visual content (text and images)
High bandwidth (>1000 kbps)	Video, text, and images

Network connection speed decides the type of media files to be retrieved and present to the learners. Table 2 lists the relationship between network speed and content type.

Among all these device contexts, the proposed system considers only network and ambient light context information at the time of adaptation.

Adaptation Rule 1

- *If the course context is video, then network connection speed must be > 568 Kbps.*
- *If the course content is audio, the network connection speed must be > 350 Kbps.*
- *If the course content is textual, the network connection speed must be > 110 Kbps.*

Adaptation Rule 2

- *If outside brightness is high, the mobile light sensor adjusts the mobile display brightness to 40%.*
- *If outside brightness is medium, the mobile light sensor adjusts the mobile display brightness to 60%.*
- *If outside brightness is low, the mobile light sensor adjusts the mobile display brightness to 90–100%.*

3.2 Learner's Context

The learner's context is the important source of personalization in mobile learning. There are several context parameters to be considered for personalized learning such learner's situation, intention, nature of learning, preferences, state, age, etc. The proposed system considers most important context parameters that are learner state and learning style, which are having more research attention than the others. Most

Table 3 Learner's state assessment

Domain knowledge test score during registration	Learner's state	Assessment questions
Less or equal to 50% of marks	Beginner	True/false, yes/no, matching
50–75% of marks	Intermediate	Multiple choice, fill in the blanks, basic problems
Above 75% of marks	Advanced	Scenario-based questions and problems

of the m-learning systems consider these two parameters that are static but in nature, those are dynamic.

3.2.1 Learner's State

According to the Kazanidis and Satratzemi [12], the learner's state includes learner's knowledge level, goal, and personal characteristics. Premlatha et al. [13] divide the learner state as novice, beginner, intermediate, and experts. Schmidt and Winterhalter [14] classified the learner's state into two: beginner and intermediate or expert. So, in general, learner's state points to knowledge level of the learner. Our proposed system domain knowledge is assessed by short quiz (state assessment quiz) during registration. Based on the quiz, learners are classified as beginners or intermediate or advanced as shown in Table 3.

Subsequent assessment will be conducted for every course. At the end of each test, learner state will be reassessed. Minimum passing score is 60% in each state

Beginner: If course-wise test $\geq$ 90%, then allow to take intermediate state assessment. Otherwise scores will be stored in learner's assessment profile.

Intermediate: If course-wise test $\geq$ 75%, then allow to take advanced state assessment. Otherwise score will be stored in learner's assessment profile.

If the learner passes at least five course-wise tests at some learner state, then his/her personal profile learner's state information is updated accordingly if required.

3.2.2 Preferences Through Learning Style

Learning styles indicates the way the learner understands the course content and perception of the learner about the course. It also helps to find the learner's learning preferences and shows the positive results in terms of satisfaction, increasing score and fast learning [10]. There are two ways to find the learning styles [15]:

1. Questionnaire-Based Method;
2. Literature-Based Method.

Questionnaire-based method uses the questions to find the learner's learning style. Literature-based method helps to find the dynamic nature of the learners learning style by monitoring learner's activities and progress in the assessments.

Questionnaire-Based Method

There are several learning style models available in the literature some of them are Felder Silverman learning style model [FSLSM] [16], Myers–Briggs type indicator [17], Kolb [18], Pask [19], and Honey and Mumford [20]. But FSLSM combines several major learning style models that are most appropriate for hypermedia courseware [21]. FSLSM characterizes each learner according to four dimensions: active/reflective, sensing/intuitive, visual/verbal, and sequential/global.

Literature-Based Method

In general, learning style of the learner is always dynamic and must be assessed based on the learner's learning activities such as learning object usage, duration of learning, test performance and navigations, etc. [13]. Our proposed system considers this dynamic nature. Every learner's learning style based on FSLSM is assessed at the time of registration. Learner's learning style information is stored in the learner's profile, and it will be updated based on learner's activities.

3.3 Contents Context

Most important part of mobile learning is course content model. Figure 2 shows the course content model for the proposed system. Based on curriculum and objective,

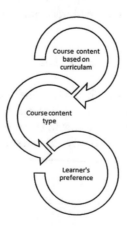

Fig. 2 Parts of m-learning content model

Table 4 Content types for different learning styles (FSLSM)

No	Course content type	A	R	S	I	V	Ve	Se	G
1	Text (concept) description		✓		✓		✓	✓	
2	Exercises	✓						✓	✓
3	Practical problems			✓					
4	PDF		✓	✓	✓		✓		
5	PPT		✓	✓			✓		✓
6	Audio		✓				✓	✓	✓
7	Video	✓	✓	✓	✓	✓	✓		✓
8	Assignments	✓						✓	✓
9	References		✓		✓	✓		✓	✓

the course contents are designed. Each course has a number of chapters, and each chapter includes different content types as shown in Table 4 and is organized in three levels:

Base Level (B): text/audio/video;

Preference Level (P): exercise, PDF, PPT, practical, assignment, and reference.

Every learner can have at least one of the base level content type in every chapter and allow to access the other preferences (P). If more than one, base level content types are selected, and then adaptation engine will select the suitable content type based on the network speed. If the network speed is below the minimum requirement to the selected content type, then system will notify to the learner either to wait for network speed or provide other suitable content type to the specific situation. The preference level (P) is further divided into S1 and S2.

- L (local preference specific to the learner): Content type are selected based on learner's ILS test and updated periodically based on learner interaction with each chapter in courses algorithm [22].
- G (global preference list): Content type is selected other than S1 which is most frequently used by the global learners to that course. It is computed by using apriori algorithm [22].

Let us consider I be the set of different content type as shown in Table 4.

$$I = \{ai : i = 1, 2 \ldots .n\} \tag{1}$$

Base level (B) has maximum of three content type, which is denoted as

$$B = \{aj : j = 1, 2, 3\} \text{ where } n(B) < 3 \tag{2}$$

Local preference (L) and global preference (G) for the learner are denoted as

$$L = \{ak : k \neq j\} \text{ and } G = \{ag : g \neq k\} \text{Or } L = I - B \text{ and } G = 0 \tag{3}$$

Every learner has a separate learner preference list in the server. Every interaction with L and G in each chapter is stored. Initially, learner's preference will follow the ILS test result. After that, at the beginning of every new course, preferences will be computed using apriori algorithm [22] based on previous interaction records. Table 4 shows the preferred content type for different learning.

4 System Architecture

Proposed system follows the client-server architecture in which server retains all the computing and storage components. Learner can interact with server through the mobile application. Figure 3 shows the details of the proposed system.

Course Content: Course content with different types.

User Profile: User personal details, current learners preference, and current state.

Assessment: Store the test result of each learner.

Question Bank: QB bank for course wise test, state assessment quiz, and ILS test.

Adaptation and Deliver Engine: Selected course contents or assessments questions will be adapted based on device network and ambient light and deliver to the learner.

Content Engine: Select the course content based on learner's preference. Also reassess the local and global preference.

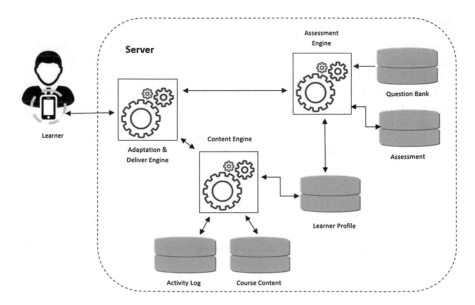

Fig. 3 Proposed system

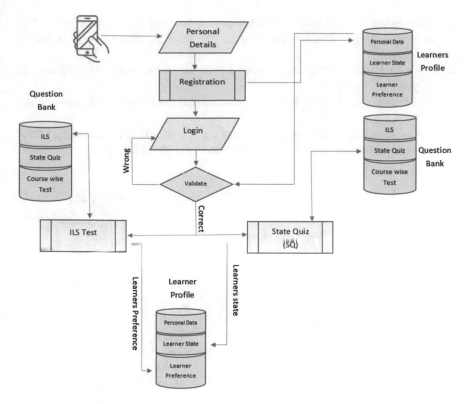

Fig. 4 Registration process flow

Activity Log: It stores the chapter-wise learner interaction on different content types.

Assessment Engine: Select the questions from question bank and compute score and store it. Also it updates the learner state in learner profile if required.

Stage 1: Registration

- Every learner must register before using the system.
- During registration, personal details are stored in the learners profile.
- After registration, learner can log in using his/her login credentials.
- Initially, every learner, must attend two test such as ILS test and state quiz. ILS test is used to find the learners learning preference (active, sensing, intuitive, visual, verbal, sequential, and global). The result of ILS test is stored in learners profile.
- State quiz is used to assess the domain knowledge level of the learner (beginner/intermediate/advanced). The result is also stored in the learners profile (Figs. 4, 5 and 6).

Stage 2: Course content engine

1. After successful login, learner can access the course.

2. Every course is organized as number of chapters and each chapter is in different type (audio, video, text description, ppt, etc.) are stored in course content repository.
3. Course content in every chapter is delivered based on learner's preference and adapted by the device.
4. Every interaction with different course content types in each chapter is recorded in activity log.
5. At the beginning of every course, learner's preference is computed using apriori algorithm based on his/her previous interaction with other courses.

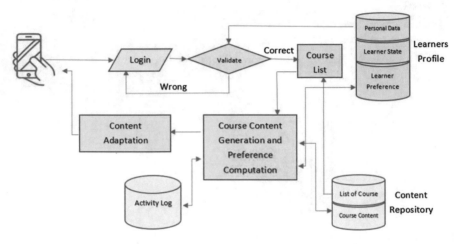

Fig. 5 Content access flow

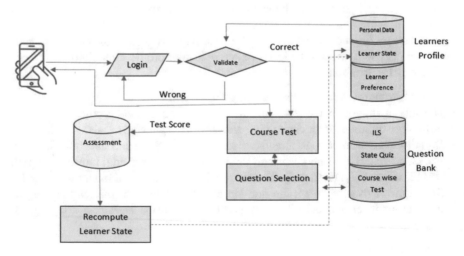

Fig. 6 Assessment flow

6. Global preference is also computed using apriori algorithm at the beginning of every course and will be the input to content generation.

Stage 3: Assessment engine

1. After successful login, learner can attend the course test.
2. At the end of the course test, result will be stored in the assessment repository.
3. Also recompute the learner state, if required, learner state will be updated in the learner profile.

5 Experiments

ILS test result of 30 students in a practical course (Unix) in IT department is shown in Fig. 7a. Based on the learning style, the preferred course content types are selected from content repository. If the learner is having more than one learning style, then preferred course content types are combined as listed in the Table 4. After completing 100% full practical course (Unix) through m-learning, slight changes in the learning style (learner preferences) are displayed in Fig. 7b.

Global preference of the learners for the 100% full practical course (Unix) in IT is displayed in Fig. 8. Most frequently (preferred) used content types among all the learners of the course are video, PPT, and practical problems.

No of records: 60
Number of attributes: 9
Minimum support count: 60%.

Similarly, for 100% theory course (Software Engineering) in IT, preferred content type for all the learners is PPT, PDF, and video. If the content types are already in local preference or base level, then global preference is null. Otherwise global preference is also available for the learner.

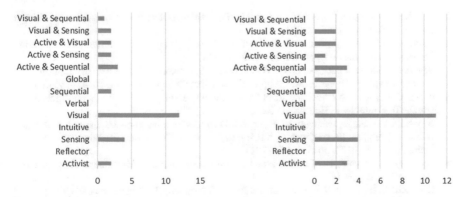

Fig. 7 **a** Learning style (ILS test) during registration and **b** learning style after completing course

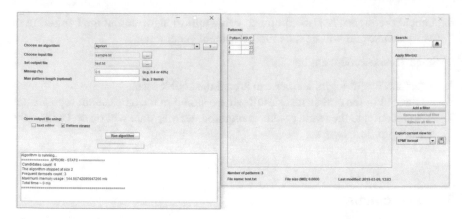

Fig. 8 Assessing the global preference

6 Conclusion

Existing context-aware personalized m-learning systems in the literatures focused on learning style and device context but lacking in content organizing and assessing the learner's preferences and state. Our proposed system modifies the existing design and pays attention to learner's preference and state and also considers dynamic nature of learner's preference and state. Contents are organized in three levels such as base, local preference, and global preference. The learner's preferences are expanded through apriori algorithm. So that learners can have freedom of choosing different content types for the courses. This will help the learners to understand the chapters well and improve the performance in the course.

References

1. Dey, A.K.: Understanding and Using Context, Personal and Ubiquitous Computing, vol. 5 issue 1, pp. 4–7. Springer, Berlin (2001)
2. Schilit, B., Adams, N., Want, R.: Context-aware computing applications. In: Mobile Computing Systems and Applications, pp. 85–90. WMCSA (1994)
3. Dey, A.K., Abowd, G.D., Salber, D.: A conceptual framework and a toolkit for supporting the rapid prototyping of context-aware applications. Hum. Comput. Interact. **16**, 97–166 (2001)
4. Wang, Y.: Context awareness and adaptation in mobile learning. In: Workshop on Wireless and Mobile Technologies in Education (2004)
5. Zervas, P., Gómez, S., Fabregat, R., Sampson, D.: Tools for context-aware learning design and mobile delivery. In: Proceedings 11th IEEE International Conference on Advanced Learning Technologies (ICALT), pp. 534–535 (2011)
6. Benlamri, R., Zhang, X.: Context-aware recommender for mobile learners. Hum. Centric Comput. Inf. Sci. (2014)
7. Triantafillou, E., Pomportsis, A., Demetriadis, S., Georgiadou, E.: The value of adaptivity based on cognitive style: an empirical study. Br. J. Edu. Technol. **35**(1), 95–106 (2004)

8. El Guabassi, I., Al Achhab, M., Jellouli, I., El Mohajir, B.E.: Personalized ubiquitous learning via an adaptive engine. J. Emerg. Technol. Learn. **13** (2018) (Elsevier B.V.)
9. Curum, B., Chellapermal, N., Khedo, K.K.: A Context-aware mobile learning system using dynamic content adaptation for personalized learning. Emerging Trends in Electrical, Electronic and Communications Engineering, Lecture Notes in Electrical Engineering, vol. 416, Springer, Berlin (2017)
10. Tortorella, R.A.W., Graf, S.: Considering learning styles and context-awareness for mobile adaptive learning. Educ. Inf. Technol. **22**(1), 297–315 (2017) (Springer)
11. Tortorella, R.A.W., Graf, S.: Personalized mobile learning via an adaptive engine. In: IEEE Conference on Advanced Learning Technologies (2012)
12. Kazanidis, I., Satratzemi, M.: Adaptivity in ProPer: an adaptive SCORM compliant LMS. J. Distance Educ. Technol. **7**(2), 44–62 (2009)
13. Premlatha, K.R., Dharani, B., Geetha, T.V.: Dynamic learner profiling and automatic learner classification for adaptive e-learning environment. Interact. Learn. Environ. **24**(6) (2016)
14. Schmidt, A., Winterhalter, C.: User context aware delivery of elearning material: approach and architecture. J. Univ. Comput. Sci. **10**(1), 38–46 (2004)
15. Kolekar, S.V., Pai, R.M., Manohara Pai, M.M.: Modified literature based approach to identify learning styles in adaptive E-learning. Adv. Comput. Netw. Informat. **1**, 555–564 (2014) (Springer)
16. Felder, R.M., Silverman, L.K.: Learning and teaching styles in engineering education. Eng. Educ. **78**, 674–681 (1988)
17. Briggs Myers, I.: Manual: The Myers-Briggs Type Indicator. Consulting Psychologists Press, Palo Alto, CA (1962)
18. Kolb, D.A.: Experiential Learning: Experience as the Source of Learning and Development. Prentice-Hall, Englewood Cliffs, NJ (1984)
19. Pask, G.: Styles and strategies of learning. Br. J. Educ. Psychol. **46**, 128–148 (1976)
20. Honey, P., Mumford, A.: The manual of learning styles. Peter Honey, Maidenhead (1982)
21. Carver, C.A., Howardl, R.A., Lane, W.D.: Addressing different learning styles through course hypermedia. IEEE Trans. Educ. **42**, 33–38 (1999)
22. Saleem Raja, A., George, E.: Compact bitTable based adaptive association rule mining using mobile agent framework. J. Comput. Sci. Softw. Eng. **4**(9), 224–229 (2015)

An Evolutionary Algorithmic Framework to Solve Multi-objective Optimization Problems with Variable Length Chromosome Population

K. Devika and G. Jeyakumar

Abstract In the last decade, Evolutionary Algorithms (*EAs*) have been widely used to solve optimization problems in the real world. *EAs* are population-based algorithms, starting the search with initial set of candidates or chromosomes, for the optimal solution of a given optimization problem. Traditional *EAs* use a population with Fixed Length Chromosomes (*FLCs*). In *FLCs,* all the chromosomes will have same length, whereas, in Variable Length Chromosomes (*VLCs*), a population can have chromosomes of different lengths. This paper proposes to use *VLCs* in the context of Multi-Objective Differential Evolution (*MODE*) algorithm. The *MODE* with *VLCs* is to solve *RFID* reader placement problem for the buildings with multiple rooms of different sizes. The type of coverage of *RFID* readers considered is elliptical. Based on the dimensions of each room, the number of *RFID* readers required is varied, which warrants the deployment of *VLCs*. This paper also presents the consequence of *VLCs,* in solving the *RFID* reader placement problem using different weight vectors.

Keywords Multi-objective optimization · Variable length chromosomes · Evolutionary algorithms · *RFID* reader placement · Weight-vector approaches

1 Introduction

An optimization problem is a problem in which a best possible solution has to be selected from a set of feasible solutions. A set of feasible solutions is referred to as population space. The Evolutionary Computation (*EC*) field of Computer Science includes a set of algorithms, termed Evolutionary Algorithms (*EAs*), for solving optimization problems. For *EAs,* a population with a set of individuals or solutions

K. Devika · G. Jeyakumar (✉)
Department of Computer Science and Engineering, Amrita School of Engineering, Amrita
Vishwa Vidyapeetham, Coimbatore, India
e-mail: g_jeyakumar@cb.amrita.edu

K. Devika
e-mail: k_devika@cb.students.amrita.edu

© Springer Nature Singapore Pte Ltd. 2020
V. Bhateja et al. (eds.), *Intelligent Computing and Communication*,
Advances in Intelligent Systems and Computing 1034,
https://doi.org/10.1007/978-981-15-1084-7_46

is supplied with limited resources. Individuals must compete with each other for the resources, which result in survival of the fittest individual. To appraise the suitability of candidate solutions, a quality function is applied. Based on the fitness values, candidates are chosen to seed the next generation, a concept derived from Darwin's theory of natural selection [1, 2].

In *EA*, the population creates the unit of evolution. Candidates or individuals within the population are objects that undergo change. It is imperative to define a population by specifying the number of individuals (population size) and its dimensions. The dimension of a candidate is problem dependent, indicative of the number of design variables it represents. It is called as the length of the chromosome. In most cases, the population size and the length of the chromosomes are kept constant such that all the chromosomes in the population would have the same length. Some optimization problems require solutions with varying number of design variables; hence, the usage of populations with *FLC* doesn't suite them. In those cases, *VLCs* are used to deduce expected solutions. Populations with *VLCs* are used in various applications, such as satellite constellations, topology optimizations, detection of network intrusions, and composition of melodies. Generally, problems in the real world demand optimal solutions for more than one objective. Based on the number of objectives, the optimization problem can be classified as Single Objective Optimization Problem (*SOOP*), Multi-objective Optimization Problems (*MOOP*), and Many-Objective Optimization Problem (*MaOOP*) [3, 4]. In this paper, a real-world *MOOP* with *VLCs* is solved using the Differential Evolution (*DE*) algorithm.

The remaining of the paper is organized as follows: Sect. 2 reviews related works, while Sect. 3 covers the design of the experiment, and Sect. 4 presents the results of the experiments, followed by closing remarks in Sect. 5, to conclude the paper.

2 Related Works

Researches related to *VLCs*, multi-objective optimization problems, Multi-objective Differential Evolution (*MODE*) and population initialization techniques are elaborated in this session.

The variable length genome, addressing the problems of *FLCs* in canonical *EAs*, was proposed in [5]. The authors compared their plan with other variable length representations. In the context of adaptive genetic algorithms, the concept of *VLCs* was propounded in [6] for the optimization of structural topology. Using *VLCs*, the authors resolved the short cantilever and bridge problem. Initially, their study was launched with smaller length chromosomes to find the optimal solution, prior to the pursuit of longer chromosomes. To enhance the quality of solutions, in terms of quantum cost, a Variable Length Chromosome Evolutionary Algorithm for Reversible Circuit Synthesis (*VLEA RC*) was proposed in [7]. Verifications performed on certain benchmarking functions, revealed that the proposed method was capable of producing results of high quality. For detection of network intrusions, a *VLC* system integrated with Genetic Algorithm (*GA*) was proposed in [8]. The proffered method was evaluated on Defense Advanced Research Project Agency's (*DARPA*) 1998 data. The system successfully detected network intrusions. With the intention of enhancing scalability and model sparsity, minor changes were introduced in the formulation of standard chromosomes [9]. For high dimensional datasets, the indexed chromosome formulation showed high computational efficiency and sparsity. Usage of *VLCs* for composition of melodies, in the context of *GA,* was assessed in [10]. As musical notes are of variable lengths, they are the representative candidates of diverse lengths, in the population. Hitomi and Selva [11] demonstrated the application of *VLCs* to optimize satellite constellations. Previously, *EAs* utilized *FLCs* explicitly to optimize the constellations [11].

If a problem is comprised of two or more conflicting objectives, then it is called a *MOOP*. Vargas et al. [12] displayed Generalized Differential Evolution (*GDE3*), along with Adaptive Penalty Method (*APM*), for the perception of advantages and limitations dealing with *MOOPs*. Results of their method were compared with *NSGA-II,* which verified the promising results of *GDE3* for *MOOPs*. Combining multi-objective problems with uncertainty is a challenging task. A new method, put forth in [13], combines various population-based algorithms into a single algorithmic structure. Results showed that the proposed scheme performed better than other algorithms. In [14], a *MOOP*-based framework was proposed, which yielded improvements in the search capability and speed of convergence. The advantages of the proposed framework were compared with the Bayesian updating and verified its superiority. A comprehensive survey of the existing multi-population-based methods, issues related to it, and its applications were presented in [15]. This paper also puts forward alternative solutions to many of the open issues in this area. Liang et al. [16] proposed two reference vector adaptation strategies, named as Scaling of Reference Vectors (*SRV*) and Transformation of Solutions Location (*TSL*). The *SRV* and *TSL* are used to solve the problems dealing with irregularities in the Pareto fronts (*PFs*). Effectiveness of the proposed strategies was compared with the state-of-the-art algorithms and found that it is performing better than the other algorithms.

Population-based algorithms are undertaken for deriving solutions to many of the discrete optimization problems in the real world. Mahdavi et al. [17] defined application of population-based algorithms to election-based systems. A new terminology called "*president candidate solutions*" is introduced and the candidates of it are created from the current generation of candidates. Majority voting-based logic was used for generation of president candidate solutions. Two algorithms: Majority Voting-based Discrete Differential Evolution (*MVDDE*) and Majority Voting-Based Discrete Particle Swarm Optimization (*MVDPSO*) were evaluated on chromosomes of different lengths: 10, 30, 50, 100, 200, and 500 candidates, respectively. A better performance was achieved by majority voting-based discrete optimization algorithms, on different benchmarking functions in the *EC* domain. Devika and Jeyakumar [18] submitted various Population Initialization (*PI*) techniques in *EAs*. Commonly used *PI* schemes, Random population Initialization (*RPI*), and Oppositional-Based Learning *PI* (*OBLPI*), were compared to identify the best *PI* technique. *RPI* and *OBLPI* practices are usable on different population size and chromosome lengths. Results showed that for larger population sizes, *OBLPI* performs better than *RPI*. *PI* techniques can also affect the performance of *EAs*. In [19], the authors evaluated the performance of *MOOP*, using two different *PI* techniques, with the *DE* algorithm. Weight-vector approach was used for comparing multi-objective benchmarking functions with different *PI* techniques, population sizes, and chromosome lengths; the performance of *DE* was found to vary with different *PI* techniques.

Opara and Arabas [20] surveyed theoretical analyses of existing *DE* variants. Current trends and promising research areas in *DE* algorithm were also discussed in this paper. *DE* has proven to be easily adaptable for real-world engineering problems, for accuracy, and faster solutions [21, 22]. For the solution of multi-objective Radio Frequency Identification (*RFID*) reader placement problem, Weight-Vector approach with two different *PI* techniques was presented in [23]. In this work the proposed approach is evaluated first with single objectives and then with combined objectives but for a single room.

On review of different aspects related to *VLCs*, reported in literature, *MOOP*, *PI* techniques, and *RFID* reader placement, this paper is set to demonstrate how a population with *VLCs* can be handled, by *DE* algorithm with *RPI*. The goal was to solve multi-objective *RFID* reader placement problem in a building with multiple rooms; the objectives analyzed in the study were Maximum coverage, Minimum cost, and Minimum Interference.

3 Design of Experiments

Initial step of the experiment was to construct a four-sided rectangular region in the computer screen using a program, with an area of 400×400 pixels. The rectangular region was assumed to be the whole area of a building in the simulation environment. Within the rectangular region, three different-sized rectangular regions were created; the three sizes denote large, medium, and small rooms in a building. The areas of these three rooms were (370×170) pixel, (300×100) pixel, and (200×130) pixel, respectively. The positions of the *RFID* tags, attached with the objects, were assumed to be static, and the tag positions were treated as user input.

The number of *RFID* tags in each room varies with the size of the room. 12, 8, and 4, respectively, were the number of tags placed in the large, medium, and small rooms. As most of the real-time *RFID* readers at present use elliptical coverage of the tag area, in our simulation environment also the readers are assumed to have elliptical coverage. The major and the minor axis of the ellipse were fixed as 50 and 30 pixels, respectively, in x- and y- directions, which signified the maximum coverage capacity of the *RFID* readers. The initial number of readers, required in each room, was empirically obtained, as 32, 21, and 15, respectively. *RFID* reader placement problem is defined to be an instance of a *MOOP*, with the following multiple objectives: maximum coverage (f_1), minimum cost (f_2), and minimum interference (f_3). Multi-objective optimization for each room is done separately.

Coverage for a room was calculated as the number of tags covered out of all tags placed. The cost of readers is calculated multiplying the optimal number of readers placed in each room and the unit cost of a reader. The interference, in each room, is calculated as the number of tags covered by more than one reader. Thus, this optimization problem includes one maximization and two minimization objectives which are to be combined. This was achieved in this study by negating the values of the minimization objectives, and converting them into maximization objectives. Hence it becomes an instance of a multi-objective maximization problem.

The *DE* variant used in the experiment was *DE/rand/1/bin*. Adopting heuristics, multiple objectives were combined, and the weight-vector approach was used to generate the weights. The weights were assigned at random, such that sum of the weights must be equal to one. Maximum weight can be assigned to the most important objective. The weight vectors (w_1, w_2, w_3) for the objective functions (f_1, f_2, f_3) are

- (0.5, 0.3, 0.2)—Highest weightage for coverage
- (0.3, 0.5, 0.2)—Highest weightage for cost
- (0.3, 0.2, 0.5)—Highest weightage for interference
- (0.33, 0.33, 0.34)—Equal weightage for all the 3 objectives.

The setup of the *DE* algorithm's parameters, for solving the multi-objective *RFID* problem with *VLCs* is listed in Table 1.

Henceforth in this paper, the *DE* framework with *VLCs* is renamed as *DE_MORFID_VLC*. The candidates in the population of *DE_MORFID_VLC* will show only the presence or absence of the readers in each room.

Table 1 Parameter setup for the experiment

S no	*DE* parameters	Value
1	Population size (NP)	150
2	Dimension (D)	32, 21 and 15
3	Crossover rate (Cr)	0.9
4	Mutation factor (F)	0.1 to 0.9
5	Maximum number of generation (MaxGen)	150
6	Number of runs (MaxRun)	1

4 Results and Discussions

The *DE_MORFID_VLC* used Differential Evolution with Random Population Initialization (*DE_{RPI}*). The *DE's* evolutionary process was launched with an initial population, containing candidates with components, randomly initialized between the numerical values of 0 and 1. Since *DE_MORFID_VLC* searched multiple solutions, for each room with different sizes, the length of candidates in the population varied with the size of the room. Each component in a candidate means the presence of a *RFID* reader. The components with random values greater than 0.5 were assumed to be 1 (indicative of the presence of the corresponding *RFID* reader), and all others were assumed to be 0 (which means those *RFID* readers were absent for the placement). The proposed *DE_MORFID_VLC* used different weight-vector approaches, where the weight values were generated heuristically, for 3 objectives with four schemes, as follows:

- Scheme 1 (S_1)—(0.5, 0.3, 0.2)
- Scheme 2 (S_2)—(0.3, 0.5, 0.2)
- Scheme 3 (S_3)—(0.3, 0.2, 0.5)
- Scheme 4 (S_4)—(0.33, 0.33, 0.34).

The performance of all the four schemes were compared, by recording the best objective function value, (*ObjValue*), attained by the *DE_MORFID_VLC* algorithm along with the number of *RFID* readers (*NoR#*) selected for the placement in each room.

Table 2 shows the results obtained by the proposed *DE_MORFID_VLC* frame work with highest weightage for coverage (S_1). The optimal solutions obtained at the end of the run are visualized in Fig. 1a. From the figure it is clear that the proposed framework provides 100% coverage of the *RFID* tags, with minimum number of *RFID* readers, for the 3 rooms. Scheme 2 sets more weightage to Cost whose computation is proportional to *NoR#*, the number of *RFID* readers. Thus, this scheme avoids the usage of unnecessary readers to reduce the number of *NoR#*. The results obtained using scheme 2 are also presented in Table 2 and the optimal solution is visualized in Fig. 1b. In Tables 2, 3 and 4, the *Room 0, Room 1,* and *Room 3* means the Large, Medium, and Small rooms in the simulation environment, respectively. In these tables the variable *G* means the generation number of the evolutionary process.

The results obtained from the implementation of Scheme 3, which give more weightage to interference, are presented in Table 3. The optimal result is shown in Fig. 2a. The optimized number of readers in large room has increased by 1. Table 3 also shows the results obtained for Scheme 4, with equal weightage for all the objectives. Optimized result for each room is shown in Fig. 2b.

Table 2 Results obtained by the *DE_MORFID_VLC* for S_1 and S_2

$S_1 - 0.5(f_1) - 0.3(f_2) - 0.2(f_3)$

G	Room 0		Room 1		Room 2	
	ObjValue (%)	NoR#	ObjValue (%)	NoR#	ObjValue (%)	NoR#
1	29.17	16	31.43	13	40	5
20	38.75	12	40.00	7	42	4
40	40.63	10	41.43	6	42	4
60	40.63	10	42.86	6	42	4
80	40.63	10	42.86	5	42	4
100	40.63	10	42.86	5	42	4
120	41.56	9	42.86	5	42	4
140	41.56	9	42.86	5	42	4

$S_2 - 0.3(f_1) - 0.5(f_2) - 0.2(f_3)$

G	Room 0		Room 1		Room 2	
	ObjValue (%)	NoR#	ObjValue (%)	NoR#	ObjValue (%)	NoR#
1	0.83	16	4.82	9	13.33	5
20	10.31	11	11.96	6	16.67	4
40	12.81	9	11.96	6	16.67	4
60	15.94	9	18.10	5	16.67	4
80	15.94	9	18.10	5	16.67	4
100	15.94	9	18.10	5	16.67	4
120	15.94	9	18.10	5	16.67	4
140	15.94	9	18.10	5	16.67	4

(a)   (b)

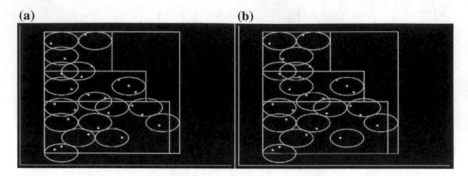

Fig. 1 *DE_MORFID_VLC's* optimal solution for multiple rooms: **a** S_1, **b** S_2

The comparison of optimal results obtained from different schemes at the end of each run is presented in Table 4. Results show that the S_1 attains the highest value, for the *ObjValue* parameter, with minimum number of readers, in all the three rooms (except the case, where S_3 attained good *ObjValue*, with more readers than S_1).

The proposed *DE_MORFID_VLC* was able to solve the *RFID* reader placement problem in a building with multiple rooms. The results were verified with different weight vectors for combing the multiple objectives. As per the weightage assigned for an objective, the optimal solution obtained is different at the end of each run.

5 Conclusions

This paper proffered an algorithmic framework (*DE_MORFID_VLC*) to solve multi-objective *RFID* reader placement problem, with different-sized rooms, using *DE* algorithm with *VLCs*. The design of the framework was verified following four different schemes by considering the objectives such as coverage, cost, and interference. The proposed algorithm was evaluated in the simulation environment of a building, with three different-sized rooms. The empirical results validated that the algorithmic framework is able to solve the problem of optimal placements of *RFID* readers.

The future work will involve verification of this investigation in the real-world *RFID* network in smart-building environment.

Table 3 Results obtained by the *DE_MORFID_VLC* for S_3 and S_4

$S_3 - 0.3(f_1) - 0.2(f_2) - 0.5(f_3)$

G	Room 0 ObjValue (%)	NoR#	Room 1 ObjValue (%)	NoR#	Room 2 ObjValue (%)	NoR#
1	15	16	17.68	9	23.33	5
20	23.13	11	25.24	5	24.67	4
40	23.75	10	25.24	5	24.67	4
60	23.75	10	25.24	5	24.67	4
80	23.75	10	25.24	5	24.67	4
100	23.75	10	25.24	5	24.67	4
120	23.75	10	25.24	5	24.67	4
140	23.75	10	25.24	5	24.67	4

$S_4 - 0.33(f_1) - 0.33(f_2) - 0.34(f_3)$

G	Room 0 ObjValue (%)	NoR#	Room 1 ObjValue (%)	NoR#	Room 2 ObjValue (%)	NoR#
1	11	16	14.73	9	22.00	5
20	18.91	11	23.57	6	22.00	5
40	22.69	10	23.57	6	22.00	5
60	23.72	9	23.57	6	22.00	5
80	23.72	9	25.14	5	22.00	5
100	23.72	9	25.14	5	24.20	4
120	23.72	9	25.14	5	24.20	4
140	23.72	9	25.14	5	24.20	4

Table 4 Comparison of schemes

Scheme No.	Room 0		Room 1		Room 2	
	ObjValue (%)	NoR#	ObjValue (%)	NoR#	ObjValue (%)	NoR#
S_1	41.56	9	42.86	5	42	4
S_2	15.94	9	18.1	5	16.67	4
S_3	43.75	10	25.24	5	24.67	4
S_4	23.72	9	25.14	5	24.2	4

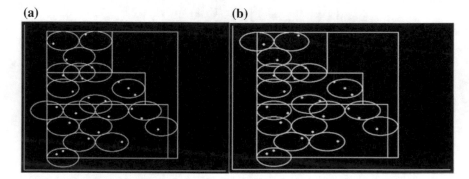

Fig. 2 *DE_MORFID_VLC's* optimal solution for multiple rooms: **a** S_3, **b** S_4

References

1. de Garis, H.: Introduction to evolutionary computing. Evol. Comput. **12**, 269–271 (2004)
2. Storn, R., Price, K.: Differential evolution-a simple and efficient adaptive scheme for global optimization over continuous spaces (1995)
3. Gong, W., Cai, Z.: A multi-objective differential evolution algorithm for constrained optimization. In: Proceedings of IEEE Congress on Evolutionary Computation, pp. 181–188. (2008)
4. Das, S., Suganthan, P.N.: Differential evolution: a survey of the state-of-the-art. IEEE Trans. Evol. Comput. **15**, 4–31 (2011)
5. Lee, C.-Y., Antonsson, E.K.: Variable length genomes for evolutionary algorithms. In: Proceedings of the GECCO (2000)
6. Kim, I.Y., de Weck, O.L.: Variable chromosome length genetic algorithm for structural topology design optimization. Struct. Multi. Optim. **29**(6), 445–456 (2004)
7. Wang, X., Jiao, L., Li, Y., Qi, Y., Wu, J.: A variable-length chromo-some evolutionary algorithm for reversible circuit synthesis. Mult. Value. Logic Soft Comput. **25**, 643–671 (2015)
8. Pawar, S.N., Bichkar, R.S.: Genetic algorithm with variable length chromosomes for network intrusion detection. Int. J. Autom. Comput. **12**(3), 337–342 (2015)
9. Gan, C.C., Learmonth, G.: An improved chromosome formulation for genetic algorithms applied to variable selection with the inclusion of interaction terms (2016). https://arxiv.org/abs/1604.06727
10. Nam, Y.-W., Kim, Y.-H.: A geometric evolutionary search for melody composition. In: Proceedings of the GECCO (2018)
11. Hitomi, N., Selva, D.: Constellation optimization using an evolutionary algorithm with a variable-length chromosome. In: IEEE Aerospace Conference, pp. 1–12. (2018)

12. Vargas, D.E., Lemonge, A.C., Barbosa, H.J., Bernardino, H.S.: Differential evolution with the adaptive penalty method for structural multi-objective optimization. Optim. Eng. 1–24 (2018)
13. Zaman, F., Elsayed, S.M., Sarker, R., Essam, D., Coello, C.A.C.: Multi-method based algorithm for multi-objective problems under uncertainty. Inf. Sci. **481**, 81–109 (2019)
14. Jin, Y.-F., Yin, Z.-Y., Zhou, W.-H., Huang, H.-W.: Multi-objective optimization-based updating of predictions during excavation. Eng. Appl. Artif. Intell. **78**, 102–123 (2019)
15. Ma, H., Shen, S., Yu, M., Yang, Z., Fei, M., Zhou, H.: Multi-population techniques in nature inspired optimization algorithms: a comprehensive survey. Swarm Evol. Comput. **44**, 365–387 (2019)
16. Liang, Z., Hou, W., Huang, X., Zhu, Z.: Two new reference vector adaptation strategies for many-objective evolutionary algorithms. Inf. Sci. (2019)
17. Mahdavi, S., Rahnamayan, S., Mahdavi, A.: Majority voting for discrete population-based optimization algorithms. Soft Comput. **23**(1), 1–18 (2019)
18. Devika, K., Jeyakumar, G.: Theoretical analysis and empirical comparison of different population initialization techniques for evolutionary algorithms. Indones. J. Elect. Eng. Comput. Sci. **12**(1), 87–94 (2018)
19. Devika, K., Jeyakumar, G.: Solving multi-objective optimization problems using differential evolution algorithm with different population initialization techniques. In: Proceedings o International Conference on Advances in Computing, Communications and Informatics, pp. 1–5 (2018)
20. Opara, K.R., Arabas, J.: Differential evolution: a survey of theoretical analyses. Swarm Evol. Comput. **44**, 546–558 (2019)
21. Abraham, K.T., Ashwin, M., Sundar, D., Ashoor, T., Jeyakumar, G.: Empirical comparison of different key frame extraction approaches with differential evolution based algorithms. In: Proceedings of 3rd International Symposium on Intelligent System Technologies and Applications (2017)
22. Rubini, N., Prashanthi, C.V., Subanidha, S., Jeyakumar, G.: An optimization framework for solving RFID reader placement problem using differential evolution algorithm. In: Proceedings of ICCSP-2017—International Conference on Communication and Signal Proceedings (2017)
23. Shinde, S.S., Devika, K., Jeyakumar, G.: Multi-objective evolutionary algorithm based approach for solving RFID reader placement problem using weight-vector approach with opposition-based learning method. Int. J. Recent Technol. Eng. **7**(5), 177–184 (2019)

Irrigation System Automation Using Finite State Machine Model and Machine Learning Techniques

H. K. Pradeep, Prabhudev Jagadeesh, M. S. Sheshshayee and Desai Sujeet

Abstract Irrigation practices can be upgraded by the aid of finite state machines and machine learning techniques. The low water use efficiency (WUE) is the universal problem encountered by the existing irrigation systems. The finite automata model provides an efficient irrigation system with input features such as soil properties, crop coefficient, and weather data. The K-Nearest Neighbor (KNN) algorithm predicts crop water requirement based on crop growth stage with accuracy of 97.35% and for soil texture classification with accuracy of 93.65%. The proposed irrigation automation model improves water productivity.

Keywords Finite state machine · Irrigation system · Machine learning · Soil texture · Water use efficiency (WUE)

1 Introduction

Fiscal improvement and increasing universal population extend the need for agriculture production. In accordance with the prediction of food and agriculture organization (FAO), the universal food demand will rise about 60% by the year 2050 [1]. The irrigated area contributes 40% to the universal food need [2]. World wide

H. K. Pradeep (✉)
JSS Academy of Technical Education, Visvesvaraya Technological University, Bangalore, Karnataka, India
e-mail: phk.contact@gmail.com

P. Jagadeesh
JSS Academy of Technical Education, Bangalore, Karnataka, India

M. S. Sheshshayee
University of Agricultural Sciences, GKVK, Bangalore, India
e-mail: msheshshayee@hotmail.com

D. Sujeet
Indian Council of Agricultural Research, Central Coastal Agricultural Research Institute, Goa
Velha, Goa, India
e-mail: desai408@gmail.com

© Springer Nature Singapore Pte Ltd. 2020   495
V. Bhateja et al. (eds.), *Intelligent Computing and Communication*,
Advances in Intelligent Systems and Computing 1034,
https://doi.org/10.1007/978-981-15-1084-7_47

Table 1 Evolution of irrigation technology

References	Technology	Utilization level	Implementation status
[12–14]	Multiclient electronic hydrants	Dispensation network	Mostly installed
	Variable frequency pumps	Pumping plant	Mostly installed
	Sprinkler and drip irrigation	Farm water control system/cropped plots	Marginally installed
	Subsurface drip irrigation (SDI)	Farm water control system/cropped plots	Very limited
	Deficit irrigation	Cropped plots	Very limited
	Machine learning irrigation	All the above levels	New era

it is analyzed that 70% of the water used for agriculture purpose, collated with 10% for civic usage and remaining put to use for industrial activity [3]. The universal irrigated area accounts for 302Mha and occupies only 16% of the cultivable area [1, 4]. Presently 36% of the universal land is concealed by arid and semi-arid regions [5] and predicted that aridity is expected to again rise because of global warming [6, 7]. In the design and management of irrigation systems, water use efficiency is the primary concern along with yield [8], since less than 65% of the supplied water is consumed by plants [9]. A strategy concerning right volume of water supply to plants upon right moment improves water use efficiency. Effective management of irrigation water is challenging due to the elements such as atmospheric conditions, soil properties, crop type, and irrigation method adopted [10, 11] (Table 1).

2 Irrigation Automation Model

The proposed irrigation framework has input features such as soil texture, crop coefficient, and weather data. The soil texture we have considered has four types such as loam, clay loam, silty loam, and sandy loam are represented in the model as s1, s2, s3, and s4, respectively. The growth stages are initial stage, crop development stage, mid-season, and late-season are represented in the model as g1, g2, g3, and g4, respectively. In weather data the parameters such as maximum temperature, minimum temperature, and relative humidity are considered to estimate crop water requirement. The model has three weather data variables w1, w2, and w3 the corresponding water requirement values are 0–3 mm, 3–6 mm, and >6 mm, respectively. For all the valid input patterns, system will halt at one of the final states and each final state describes the depth of water needs to be supplied based on the input features (Fig. 1) (Tables 2 and 3).

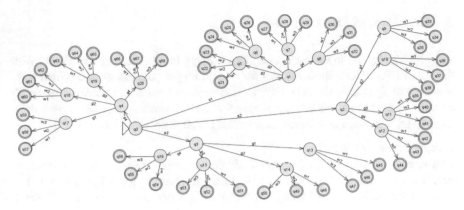

Fig. 1 Deterministic finite state machine (DFSM) model for irrigation automation

Table 2 Deterministic finite state machine (DFSM) model variables

DFSM attributes	Description
States	{ q_0 to q_{68} }
Input symbols	{Soil type—s1, s2, s3, s4; Crop growth stage—g1, g2, g3, g4; Weather data—w1, w2, w3}
Start state	q_0
Final states	{ q_{21} to q_{68} }

Table 3 Transition table instance for DFSM model

Current state	Input symbol	Next state
q_0	s1	q_1
q_0	s2	q_2
q_0	s3	q_3
q_0	s4	q_4
q_1	g1	q_5
q_1	g2	q_6
q_1	g3	q_7
q_1	g4	q_8
q_1	g1	q_5
q_5	w1	q_{21}
q_5	w2	q_{22}
q_5	w3	q_{23}

3   Results and Discussions

The efficient irrigation system is simulated using a finite state machine model with the aid of JFLAP tool. In this model for all valid input patterns we reach final state represented by double circle in DFSM model. Each final state describes the depth of water needs to be supplied considering soil type, growth stage, and weather data. The model validation is demonstrated here through state transitions (Fig. 2).

Sample Input - > s1g2w2
Soil texture and crop coefficient are vital factors to estimate crop water requirement, here we have classified the soil type based on silt, clay, and sand proportion. The crop coefficient is predicted using K-Nearest Neighbor (KNN) algorithm (Tables 4, 5 and 6).

Fig. 2 Validation of DFSM model for the input pattern s1, g2, w2

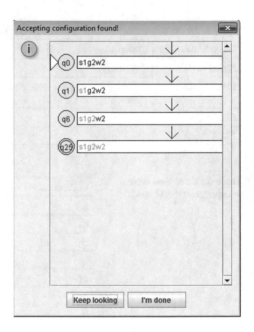

Table 4 Data set samples for Soil type identification

Sand	Silt	Clay	Type
85.5	11	9.5	Sandy loam
19.1	55	20.2	Silty loam
26.9	48.1	19.6	Loam
35.6	20.9	39.6	Clay loam

Table 5 Data set sample for corn crop water requirement based on growth stage

Crop Growth Stage	Crop coefficient
Initial stage	0.17
Development stage	1.06
Mid-season	0.96
Late season	0.15

Table 6 Soil texture and crop coefficient prediction accuracy using KNN algorithm

Model	Objective	Accuracy (%)
KNN	Soil texture classification	90.35
KNN	Crop coefficient prediction	97.43

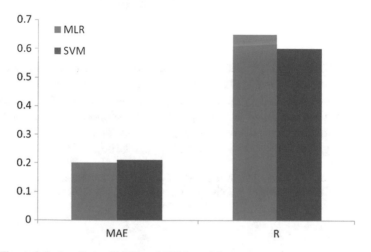

Fig. 3 The statistical analysis of MLR and SVM models

In this section, the crop water requirements based on weather data are estimated using multiple linear regression (MLR) and support vector machine (SVM) techniques. The models are also compared using statistical performance measures such as mean absolute error (MAE) and coefficient of correlation (R) (Fig. 3) (Tables 7 and 8).

4 Conclusion

The irrigation system can be automated using a finite state machine model and machine learning techniques. The K-Nearest Neighbor (KNN) algorithm is used for soil texture classification and crop coefficient prediction with an accuracy of 90.35% and 97.43%, respectively. The key input parameter for deterministic finite

Table 7 Data set samples for crop water requirement prediction based on temperature and humidity

Maximum temperature	Minimum temperature	Relative humidity	Crop water requirement
32.6	22.0	63	3.9
33.2	20.4	61	4.2
34.2	21.2	62	4.7
32.4	21.8	59	3.5
31.8	20.6	65	6.8
27.6	17.4	78	2.7

Table 8 The statistical performance analysis for MLR and SVM models

Method	MAE	R
MLR	0.20	0.65
SVM	0.21	0.60

state machine (DFSM) model is the weather data which plays crucial role in estimation of crop water requirement. The MLR and SVM techniques are used to estimate crop water requirement and the accuracy of prediction is statistically analyzed and proves that regression method is slightly better than support vector machine algorithm. The overall highlights of this novel approach exhibit that irrigation automation support for irrigation area expansion through efficient usage of available water.

Acknowledgements We would like to thank Dr. Darshan M B, Assistant Research Engineer, University of Agricultural Sciences, GKVK, Bengaluru for his valuable input and suggestions.

References

1. Alexandratos, N., Bruinsma, J.: World Agriculture Towards 2030/2050: the 2012 Revision, vol. 12. No. 3. FAO, Rome: ESA working paper (2012)
2. World Water Assessment Programme, The United Nations WorldWater Development Rep. 3: Water in a changing world, UNESCO and Earthscan, Paris and London (2009)
3. Provenzano, G., Sinobas, L.R.: Special Issue on Trends and Challenges of Sustainable Irrigated Agriculture (2014)
4. Playán, E. et al.: Solid-set sprinkler irrigation controllers driven by simulation models: Opportunities and bottlenecks. J. Irrigat. Drain. Eng. **140.1** (2013)
5. Safriel, U. et al.: Dryland systems. Ecosyst. Human Well-Being. Current State Trends **1**. Island Press (2006)
6. Alcamo, J., Martina, F., Michael, M.: Future long-term changes in global water resources driven by socio-economic and climatic changes. Hydrol. Sci. J. **52.2** (2007)
7. Arnell, N.W., van Vuuren, D.P., Isaac, M.: The implications of climate policy for the impacts of climate change on global water resources. Global Environ. Change **21.2** (2011)
8. Burt, C.M., et al.: Irrigation performance measures: efficiency and uniformity. J. Irrigat. Drain. Eng. **123.6** (1997)
9. Chartzoulakis, K., Maria, B.: Sustainable water management in agriculture under climate change. Agricult. Agricult. Sci. Procedia **4** (2015)

10. Dabach, S., et al.: Numerical investigation of irrigation scheduling based on soil water status. Irrigat. Sci. 31.1 (2013)
11. Soulis, K.X., Elmaloglou, S.: Optimum soil water content sensors placement for surface drip irrigation scheduling in layered soils. Comput. Electron. Agricult. **152** (2018)
12. Torres-Rua, A.F. et al.: Machine learning approaches for error correction of hydraulic simulation models for canal flow schemes. J. Irrigat. Drainage Eng. **138**(11) (2012)
13. Niu, C.-J. et al.: Real-time irrigation forecasting for ecological water in artificial wetlands in the Dianchi Basin. J. Informat. Optimizat. Sci. **38.7** (2017)
14. Levidow, L., et al.: Improving water-efficient irrigation: prospects and difficulties of innovative practices. Agricultural Water Manage. **146**

EVaClassifier Using Linear SVM Machine Learning Algorithm

V. Vinothina and G. Prathap

Abstract Evaluating descriptive answer scripts is one of the challenging tasks for academicians along with their routine works and increase in the number of students enrolling in educational institution. It involves various factors such as man power, time, cost, and mental health. These factors are directly proportional to students' strengths. Hence, evaluation scheme needs to be automated to ease the work of staff. Many research activities have been carried out to automate the evaluation process and easier the work of staff. In this paper, an attempt is made to propose two classes Eva classifier using Support Vector Machine Supervised Machine Learning algorithm for auto evaluating short answers and performance of the classifier is evaluated using accuracy of answer classification.

Keywords Support vector machine · Answer script evaluation · Supervised machine learning · Linear kernel · Java

1 Introduction

The publication of results on time after evaluation is a critical task for examination section staffs in any educational institution. A number of revaluation cases reveals the quality of evaluation done by the staffs. It has been discussed forcefully by media [1]. The reasons behind this are lack of qualified examiners, lack of time, and mental health due to overload of work assignment. The major significant reason for revaluation cases is overloading valuators with more answer scripts in limited time. This results in variation of marks in revaluation of answer scripts [2, 3].

Mostly, the descriptive answer scheme is followed at undergraduate and postgraduate level in arts and science educational institution. The descriptive answer may be

---

Please note that the LNCS Editorial assumes that all authors have used the western naming convention, with given names preceding surnames. This determines the structure of the names in the running heads and the author index.

V. Vinothina (✉) · G. Prathap
Department of Computer Science, Kristu Jayanti College Autonomous, Bengaluru, India
e-mail: Vinothina.v@kristujayanti.com

© Springer Nature Singapore Pte Ltd. 2020
V. Bhateja et al. (eds.), *Intelligent Computing and Communication*,
Advances in Intelligent Systems and Computing 1034,
https://doi.org/10.1007/978-981-15-1084-7_48

503

classified as short answer or long answer based on marks. As we know, Computer Science deals with fundamental theory of facts, execution, and application in computer systems. The descriptive answer for particular question may include text, diagrams, calculations, programs, input, output, and formal specification (syntax) of programming language statements. The paper is intended to propose classifier for evaluating computer science subjects answer scripts.

It has been proved that a holistic view of individuals learning can be acquired by examining them through descriptive answers. The objective type examination is not sufficient to obtain a holistic view of individuals [4]. Moreover, the softwares available for evaluating descriptive answers are non transparent. Therefore, there is a need for auto-evaluation software for evaluating descriptive answers. In the real scenario, the descriptive answers are evaluated by predefined answer key, scheme of valuation, and experience of evaluators. Marks will be awarded based on how much the answer is closest to the scheme. As it is a time- consuming process, there is a need for a system that does auto-evaluation of answer scripts.

The aforementioned evaluation process can be replaced by supervised machine learning classifiers. The classifier learns from the set of training data provided. After the training, the system classifies the new answer based on the training it got. Various researches have been carried out to automate the evaluation process. But the prime objective of this paper is to propose a two-class classifier for automating the evaluation of short answers using supervised machine learning algorithm Support Vector Machine (SVM). The accuracy obtained from classifier is calculated to evaluate the performance of the proposed classifier. Obtaining a dataset for training a model is a challenging task. Hence, short answers are collected from limited number of students to train the classifier.

Similar and related research works carried out by various researchers are described in Sect. 2. The proposed classifier is discussed in Sect. 3 along with experimental setup and the tools used. Section 4 describes the results obtained from the experiments and Sect. 5 describes conclusion and future directions.

2   Related Work

Though there were many researches done on text classification, it seems to be minimal research done on auto-evaluation of descriptive short answers. Madhumitha et al. used a discrete model to automate the discrete mathematics domain relations classification. It outperformed well as pellet reasoner is used as a classifier which uses predefined classification rules to classify the types of relations. By using the matching ratio for the keywords in instructor and student answers, Alla et al. [6] proposed an E-assessment system. But the approach proposed in [6] is based on semantics and document similarity.

Kirithika et al. [7] addressed short answers evaluation using two-class averaged perceptron, linear, and isotonic regressions in Microsoft Azure platform and used a combination of many semantic and graph alignment features. They provided first attempt to use graph alignment features at sentence level. The results have shown that technique gave the precise classification of right and wrong answers.

In paper [8], the clustering approach using Modified K-Means algorithm for training a similarity metric is proposed. Stanford dependency parser is used in paper [9] for feature extraction and training the graph alignment features of each node pair from all students' answers and the corresponding instructor answer by using averaged perceptron is found in [10]. Grading is done using isotonic regression in the scale of [0.5] after scoring done by SVM regression and ranking model. Mohler et al. [11] also used text similarity approach, where a grade is given based on a measure of relatedness between the student and the instructor answer using corpus-based measures and knowledge-based measures.

Syed et al. [12] tested the prediction accuracy of three recommended machine learning algorithms in LightSIDE, namely Naïve Bayes, Sequential Minimal Optimization (SMO), and J48. The conclusion from their research is that although differences between human and machine classification for transcription variables were generally not large, they are fair enough that they should not be ignored. Sunil Kumar et al. [13] experimented with various training sample sizes in order to determine the best training sample size required for automated evaluation of descriptive answers through sequential minimal optimization. It was determined that when the training sample size is 900, the best prediction accuracies were obtained [14].

3 Proposed Classifier

In this section, related processes such as Linear Kernel SVM and text processing are discussed to design and implement EVaClassifier. SVM is a supervised machine learning algorithm for classification and regression problems. It solves both linear as well as nonlinear problems. The algorithm creates a line or a hyperplane which separates the data into classes.

Text processing is applied for preprocessing the answers. Then, the answers had undergone various steps as per SVM algorithm for classification. EVaClassifier will be trained to classify whether a given answer, x, is correct ($y = 1$) or wrong ($y = 0$). The steps involved in classifier construction are depicted in Fig. 1. When there is large number of features and each letter is a new feature, linear Kernel SVM gives better result. The advantages of linear kernel are faster and use only one regularization parameter called c for optimization [15].

As given in diagram 1, the first step in building the classifier is collecting answers from students. This step also includes digitizing the answers in the form of text files. The following processes have been implemented as a part of preprocessing the answers for effective classification.

- Lowercasing: Students' answers may be in lower case or uppercase. But both should be treated as same. Hence, the entire answer is converted into lower case,

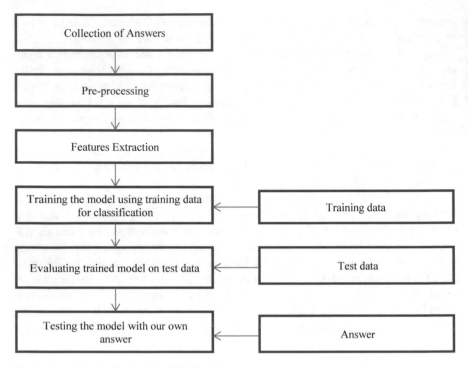

Fig. 1 Steps involved in building EVaClassifier

so that capitalization is ignored. For example, Java, Object, and PATH is treated as java, object, and path, respectively.

- Stemming: Words are transformed into their stemmed form. For example, "thread", "threads", "threaded", and "threading" are all replaced with "thread". This step identifies the relationships and similarities in the large dataset.

  Additional characters such as "e", "'dant", "ed", and "tions" are also removed from the end of word by Stemmer.

- Deduction and Deletion: Words which are disapproved and punctuation have been stripped off and white spaces have all been trimmed to a single space character.

For the result of the above processes, the process of feature extraction is much easier. After preprocessing, vocabulary list is created based on the question and answer chosen for automatic evaluation. Next, the words to be considered to use in classifier is determined. For sample, the question chosen for automatic evaluation is, "What are the significant characteristics of java programming language?" For this question, we have chosen only the most frequently used words for answering as the vocabulary list. Otherwise, words which occur rarely lead the model to overfitting.

The vocabulary list used to build a classifier is shown in Fig. 2. Using this, word indices are created by mapping each word in the preprocessed answer to the index of the word in the vocabulary list. Figure 3 shows sample answer for the aforementioned question and the corresponding word indices.

java	architectur	develop	becom
is	neutral	futurist	choic
a	secur	desktop	top
simpl	portabl	packag	rate
languag	emerg	to	build
ha	as	web	complex
an	on	enterpris	it
object	of	mobil	thread
orient	the	embed	unmatch
model	most	system	featur
platform	popular	there	
independ	and	ar	
robust	power	varieti	
support	program	us	
multi	for	that	
high	build	apprec|	
perform	rang	that	
offer	applic	appreci	
strong	busi	thi	
distribut	from		

Fig. 2 Vocabulary list

Fig. 3 Sample answer and
word indices

```
Java is a simple language
It has an object oriented model
It is platform independent
```

1 2 3 4 5 64 6 7 8 9 10 64 2 11 12

Feature extraction converts each answer into a vector in R^n. n is equal to the number of words in the vocabulary list. The feature $xi \in \{0, 1\}$,i.e., $x_i = 1$ if the i-th word is in the answer otherwise, $xi = 0$. After completing feature extraction, the preprocessed training data is given as input to train a classifier. After loading the dataset, SVM is used to train the classifier to classify between right $(y = 1)$ and wrong $(y = 0)$ answers. Then, the model is loaded with test data to predict the model performance on unseen data.

4 Experimental Setup and Discussion

MATLAB is used to build EVaClassifier. All the steps depicted in Fig. 1 are implemented as per the description given in Sect. 3. As mentioned above, answers are collected from students. Initially, the training dataset is loaded with 12 sample answers which include both right and wrong answers and test data is loaded with approximately 30% of training data. And also, the number of features used for training the model is 101. At the initial stage, our model did not yield good accuracy on both training and test datasets. To override it, as a first step, the number of model answers in training and test datasets is increased. The accuracy results are given in Table 1. From that, it is shown that till seventh attempt, model is overfitted. Overfit is a case in which the training datasets are classified very well, but fail to generalize to new examples.

Overfitting can be avoided by reducing the number of features and by regularization parameter c in liner kernel. Hence, the number of features is reduced from 101 to 68. After training the model with new features, the classifier yields 98.7% on training data and 96.5% on test data. Finally, the model is tested with our answer. The sample answer and classification results are shown in Fig. 4.

Table 1 Classification Results

Attempt	Number of training data (answers)	Number of testing data (answers)	Training accuracy (%)	Testing accuracy (%)
1	12	4	66.6	40.3
2	18	6	71.2	52.5
3	22	7	74	59.1
4	28	9	83.33	64.6
5	36	12	88.7	72.3
6	42	14	91.2	79
7	48	16	94.6	86
8	52	18	98.7	96.5

```
Java is a complex software which converts high
level language program
in to machine language program.suitable for web
application development.
```

Processed wrongans.txt
correct Answer Classification: 0
(1 indicates correct, 0 indicates wrong)

```
Java is a simple language
It has an object oriented model
It is platform independent
```

Processed ans1_1.txt
 correct Answer Classification: 1
(1 indicates correct, 0 indicates not spam)

Fig. 4 Sample answers and classification results

5   Conclusion and Future Direction

An attempt is made to classify short answers whether it is correct or not using Linear SVM supervised machine learning algorithm. All the mentioned steps in Sect. 3 are implemented in MATLAB and obtained optimal results. The limitation of EVaClassifier is that it is a two-label classifier and it classifies only texts. But descriptive answers in computer discipline may contain numbers, diagrams, programs, and syntax. So, it would be better to classify the answers based on how far the answer is relevant to the model answer used for evaluation. For example, in the case where ten points answer is to be written for a question, if only two points are given, marks can be distributed based on the number of points written. Hence, there is a need for multilabel classifier for automatic evaluation of descriptive answers. Hence, constructing multi-label classifier is the future direction of this proposed classifier.

References

1. Protest Over Delay in Evaluation Work. http://www.thehindu.com/news/cities/
2. 80 out of 83 Score More After Revaluation. http://articles.timesofindia.indiatimes.com
3. Revaluation Fails 100 'Passed' PU Students. http://www.bangaloremirror.com/index.aspx
4. Siddhartha, G.: e-Examiner: A System for Online Evaluation and Grading of Essay Questions. http://elearn.cdac.in/eSikshak/eleltechIndia05/PDF
5. Ramamurthy, M., Krishnamurthi, I., Ilango, S., Palaniappan, S.: Discrete Model Based Answer Script Evaluation Using Decision Tree Rule Classifier, pp. 1–12 (2019)
6. Alrehily, A.D., Siddiqui, M.A., Buhari, S.M.: Intelligent electronic assessment for subjective exam, ACSIT, ICITE, SIPM, pp. 47–63 (2018)
7. Kirithika, R., Jayashree, N.: Learning to Grade Short Answers using Machine Learning Techniques, WCI '15, August 10–13, 2015
8. Basu, S., Jacobs, C. and Vanderwende, L.: Powergrading: a clustering approach to amplify human effort for short answer grading. Trans. Associat. Computat. Linguistics (2013)
9. Nlp.stanford.edu. The stanford nlp (natural language processing) group (2015)
10. Mohler, M., Bunescu, R. and Mihalcea, R.: Learning to Grade Short Answer Questions using Semantic Similarity Measures and Dependency Graph Alignments. Association for Computational Linguistics (2011)
11. Mohler, M., Mihalcea, R.: Text-to-text Semantic Similarity for Automatic Short Answer Grading. Association of Computational Linguistics (2009)
12. Shermis, M.D., Hamner, B.: Contrasting State-of-the-Art Automated Scoring of Essays: Analysis. Contrasting Essay Scoring, pp. 1–54 (2012)
13. Latifi, S.M.F., Guo, Q., Gierl, M.J., Mousavi, A., Fung, K., Lacroix, D.: Towards Automated Scoring using Open-Source Technologies. Annual Meeting of the Canadian Society for the Study of Education, pp. 13–14 (2013)
14. Kumar, S. and Sree, R.R.: Experiments towards determining best training sample size for automated evaluation of descriptive answers through sequential minimal optimization. ICTACT J. Soft Comput. **4**(2), 710 –714 (2014)
15. Text Categorization with Support Vector Machines. http://www.cs.cornell.edu/people/tj/publications/joachims_98a.pdf

MFCC-Based Bangla Vowel Phoneme Recognition from Micro Clips

Bachchu Paul, Himadri Mukherjee, Santanu Phadikar and Kaushik Roy

Abstract Speech recognition has developed highly and different solutions are available, most of them in English and few other non-Indian languages. People face difficulty in handling them who are not proficient in such languages. Bangla is the sixth most spoken language in world [1] and speech-based solutions are not fully available due to complex nature of the language. Thus Bangla speech recognizer is important. Every language has atomic sound called phonemes. Vowel phonemes are one of the most important aspects of a language, as most words are constituted with them. In this paper, we have categorized Bangla vowel phonemes with MFCC features and knn-based classification. The phonemes were split into micro clips of 30 ms before categorization in order to uphold real-world scenario which often consists of incomplete or extremely short duration data. We had experimented with disparate classifiers on our dataset of 92,649 short-duration clips and obtained a highest accuracy of 98.87%. We have compared the accuracy with standard MFCC features for our data set and found better result.

Keywords Phoneme · Grapheme · ASR · Zero crossing · MFCC · KNN

B. Paul (✉)
Department of Computer Science, Vidyasagar University,
Midnapore 721102, West Bengal, India
e-mail: ableb.paul@gmail.com

H. Mukherjee · K. Roy
Department of Computer Science, West Bengal State University,
Kolkata 700126, West Bengal, India
e-mail: himadrim027@gmail.com

K. Roy
e-mail: kaushik.mrg@gmail.com

S. Phadikar
Department of Computer Science and Engineering, Maulana Abul Kalam
Azad University of Technology, Kolkata 700064, India
e-mail: sphadikar@yahoo.com

© Springer Nature Singapore Pte Ltd. 2020   511
V. Bhateja et al. (eds.), *Intelligent Computing and Communication*,
Advances in Intelligent Systems and Computing 1034,
https://doi.org/10.1007/978-981-15-1084-7_49

1 Introduction

Speech recognition is a very challenging task, since the quality of the speech production is dependent on the training and test data set. It is very difficult since the pronunciation style and accent is mostly depends on the region of the people. Even, the pronunciation far differs from people of the same state. There are so many popular algorithms to recognize the speech in different languages has been developed. For Bangla language, it is a very difficult task since; the number of vowels and consonants is more than the English and many other languages. In Bangla, there are 11 vowels and nearly 35 consonants in grapheme form but their corresponding phonemes have 7 for vowels and about 29 for consonants [2]. As far as our concern, there is not any standard Bangla Phoneme dictionary or speech corpus available. The written characters of Bangla are the same for both West Bengal and Bangladesh, but their pronunciation is different [2]. So, it is necessary to research in Bangla language and need to construct a Bangla ASR.

The Bengali speech recognition system can be applied in different applications like voiced command interpretation, hands free voiced-based typing, voiced-based biometric signature, gender identification, etc. Most of the peoples in Bangladesh and West Bengal have access to mobile phones [3]. However, they are uncomfortable in English language. Thus, Bangla ASR-based system will help them to use of the mobiles and computers more user friendly [2, 3].

2 Literature Review

Formants frequency, Mel Frequency Cepstral Coefficients (MFCC), and many other features-based system has been used previously for Bangla phoneme recognition. Sultan et al. [4] proposed rule base with the help of Microsoft speech API and got accuracy of 74.81%. Sumit et al. [5] has been proposed noise robust End-to-End Speech Recognition for Bangla Language using CTC-based deep neural network and achieved 12.31% and 9.15% Character Error Rate (CER) for clean and noisy speech, respectively. Sourav et al. [6] showed Bangla Speech Recognition for Voice Search using Gaussian Mixture Model–Hidden Markov Model (GMM-HMM) based and Deep Neural Network-based Model (DNN-HMM) and have obtained 3.96% and 5.30% Word Error Rate (WER) respectively. Das Mandal et al. [7] proposed F0 contours for Bangla readout speech of prosodic information and obtained a highest of 4.70 mean opinion score using their five point scale. Mandal et al. [8] proposed Bengali speech corpus for continuous speech recognition system and they used Hidden Markov Model Toolkit (HTK) for aligning the speech data. They obtained a 85.3% accuracy using their 39-dimensional feature extraction (LPC, MFCC and PLP). Mukherjee et al. [1] has proposed a MFCC-based Bangla phoneme recognizer and used four levels of classification and obtained a highest accuracy rate of 98.22%.

But, our research more effectively and efficiently can recognize a Bangla vowel phoneme, since we obtained the highest accuracy of 98.87%. Our research can handle arbitrary length of vowel phoneme. Our dimensions of features are very limited, thus not necessary to dimensionality reduction.

Our paper is structured as: Sect. 3 explains the proposed method, Sect. 4 explains the result of the method, and finally, Sect. 5 discusses the conclusion.

3 Proposed Method

The proposed method is given in the following schematic diagram in Fig. 1. The steps taken are illustrated given in subsection 3.

3.1 Data Collection

This was one of the challenging tasks, since the quality of the data affects the accuracy of the result and average error rate.

We have recorded 700 raw speech vowels through the audacity software [9] with the help of 10 volunteers from different regions of the state West Bengal, India; among them 6 are male and 4 female with different age groups from 10 to 60. We recorded the vowels in normal room environment. Each vowel has almost 100 pronunciations uttered 10 times by each speaker with .wav format in sampling frequency of 16 kHz with mono channel.

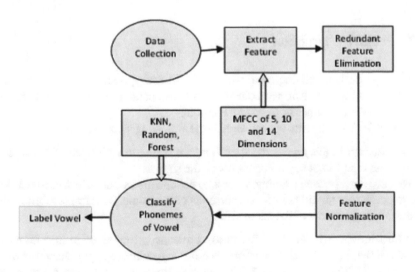

Fig. 1 Schematic diagram of the proposed method

Table 1 International phonetics alphabets for Bangla vowels

Bangla phoneme	IPA	Bengali pronunciation	Equivalent English pronunciation	Place and manner articulation
অ	/ɔ/	অসীম	audio	Low mid back rounded
আ	/aa/	আমাদের	umbrella	Low central unrounded
ই / ঈ	/i/	ইস্তেহার	individual	High front unrounded
উ / ঊ	/u/	উদয়	mooch	High back unrounded
এ	/e/	এসো	error	Mid high front unrounded
ও	/o/	ওল	throat	Mid high back rounded
অ্যা	/ae/	ব্যায়াম	apple	Low mid front unrounded

From the 700 raw speech data, we divided each of the uttered phonemes with a 30 ms short clips with 75% overlap. The number of engendered clips now become 92,649 samples or frames. Actually, the number of phonemes is very low. To enlarge out data sets we worked out with such micro clips. The monophthongs Bangla vowels and their International Phonetic Alphabet (IPA) are given in Table 1.

3.2 Feature Extraction

Since a speech is an analog data, we can't feed raw speech data into our model. Instead of giving raw data, we converted this into parametric form. We used here MFCC of different dimensions as our features.

To find MFCC from the speech signal, the following steps are used.

Preprocessing: In this phase, the speech signal is sent to a high pass filter to compensate the high-frequency components of the signal.

We have already done framing process to enlarge the data set. Next used windowing, to get the continuity at two extreme ends of the frame, each frame is multiplied by a hamming window of the same size.

Fast Fourier Transform: The FFT is used to convert time domain into frequency domain of the signal. Fourier transform is a fast algorithm, to apply Discrete Fourier Transform (DFT) [10]. Assume $S_i(n)$ is our ith frame after applying window, then the DFT is given by Eq. 1 [10, 11].

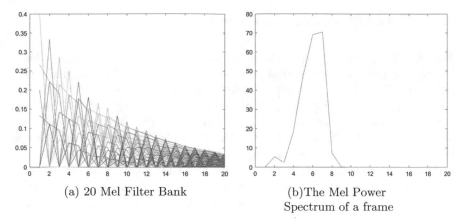

(a) 20 Mel Filter Bank

(b) The Mel Power Spectrum of a frame

Fig. 2 The 20 triangular filter bank and corresponding power spectrum of a frame

$$S_i(k) = \sum_{n=1}^{N} s_i(n)e^{-\frac{j2\pi kn}{N}} \quad 1 \le k \le K \quad (1)$$

where, N is the frame length and K is the DFT length.

Since the DFT is complex real number, we then find the power spectrum from each of the frame by calculating the absolute value of the DFT and took the mean of the squared the result is given by the Eq. 2 [11].

$$P_i(k) = \frac{1}{N}|S_i(k)|^2 \quad (2)$$

Mel Scale Filter Bank: In this step, the above-calculated spectrum is mapped on Mel Scale using 20 numbers of triangular overlapping filter banks to find the energy at each spot of the signal. The formula to convert the frequency in Hz to frequency in Mel is given by Eq. 3 [10] (Fig. 2).

$$m = 2595 \log_{10}\left(1 + \frac{f}{700}\right) \quad (3)$$

Discrete Cosine Transform: In this step the Mel frequency spectrum is back into time domain using Discrete Cosine Transform (DCT) of the logarithmic Mel power spectrum. Assume E_k is the Mel power spectrum of the frame i. Then, the equation of DCT is given in Eq. 4 [10].

$$C_m = \sum_{k=1}^{M} \cos\left[m\left(k - \frac{1}{2}\right)\frac{\pi}{M}\right]E_k \quad (4)$$

here M is the number of filter bank and L is the number of MFCC coefficients.

3.3 Elimination of Redundant Feature

Now for each of the frame, we obtained 5, 10 and 14 MFCCs as our feature. Thus, roughly for a one-second speech signal with 30 ms frame and 22.5 ms overlapping (75%) we got 130 frames or clips and for each phoneme has 5×130, 10×130 and 14×130 features respectively. We have 700 phonemes, each of duration from 0.5 to 2 s. After framing, we got 92,649 samples. Thus in our raw data set we obtained $5 \times 92,649$, $10 \times 92,649$ and $14 \times 92,649$ features respectively. Without elimination of any feature we got highest of 86.25% accuracy for 14 dimension MFCCs. Since vowel phonemes are atomic sound and within a vowel there are no pauses or silences. Thus we can eliminate the features for the unvoiced clips or noise like clips. The uncorrelated frames are calculated by either the frame has low power or an average zero-crossing rate is very high. The average power of a frame is calculated by the formula in Eq. 2. The average zero crossing of a frame at time instant 'i' is calculated by the formula given in Eq. 5.

$$ZCR = \frac{1}{2N} \sum_{j=i-N+1}^{i} |\mathrm{sgn}(x(j) - \mathrm{sgn}(x(j-1))| \, w(i-j) \qquad (5)$$

where, $\mathrm{sgn}(x)$ gives 1 for positive x and 0 for negative x; and $w(i)$ is the window parameter.

Then all such frames which have very small power and zero crossing are high which are discarded from the features data set. We got reduced features of order $5 \times 34,298$, $10 \times 34,298$ and $14 \times 34,298$ respectively.

3.4 Feature Normalization

The MFCCs features are highly varying in range. Most of the machine learning algorithms use Euclidean distance between two data points for computation. This creates a problem. To overcome this problem, feature normalization is needed to extract the features in a close interval instead of a wide range of intervals by mean and standard deviation.

4 Result and Discussion

We used five Fold Cross-alidation on the different dimensions of features set and applied different classifiers got a highest accuracy of 98.87% on reduced data set using 14-dimensional MFCCs feature. On our data set of 700 phonemes, we got 92,649 samples of feature with dimensions 5, 10, and 14, respectively. The highest accuracy obtained for non-reduced data set is 86.25% using KNN classifier. After

Table 2 Different classifier based accuracy

MFCC order	Feature dimension	Accuracy rate (%)				
		Random forest	Random tree	MLP	SVM	KNN
5 Dimension (Type 1)	92,649 × 5	73.02	65.04	50.38	64.35	72.88
5 Dimension (Type 2)	34,298 × 5	88.64	82.58	72.51	82.13	89.47
10 Dimension (Type 1)	92,649 × 10	83.05	71.93	60.96	75.75	84.35
10 Dimension (Type 2)	34,298 × 10	96.43	90.45	87.24	95.26	98.16
14 Dimension (Type 1)	92,649 × 14	84.64	71.91	65.45	78.29	86.25
14 Dimension (Type 2)	34,298 × 14	97.66	90.39	90.75	97.22	98.87

Table 3 Confusion matrix in non-reduced data set

a	86.22	3.191	2.303	1.336	1.483	2.532	2.253
aa	3.762	**86.43**	2.407	1.776	1.34	2.033	2.251
ae	1.774	1.931	**89.83**	1.484	1.162	2.033	1.782
e	1.563	2.236	1.741	**86.51**	2.284	2.623	2.847
i	1.734	1.82	1.758	2.731	**86.21**	2.362	3.39
o	2.548	2.258	1.776	2.522	1.736	**86.63**	2.528
u	2.233	2.083	2.011	2.158	2.763	2.938	**85.79**
	a	aa	ae	e	i	o	u

elimination of redundant features discussed in Sect. 3.3, number of samples become 34,298. Here, we used the notation type 1 for non-reduced data set of 92,649 samples and type 2 for reduced data set of 32,298 samples. For each of the samples we got 5, 10, and 14-dimensional MFCCs followed by a phoneme label. Then these samples are trained on different classifiers on WEKA tool [12] and got highest accuracy of 98.87% using KNN classifier with 14-dimensional MFCC feature. Table 2 shows the comparison between different dimensions and different classifier based accuracy levels on our feature set.

The Confusion Matrix for accuracy obtained in non-reduced data set is given in Table 3.

With the analysis of the confusion matrix given in Table 3, the miss classification label contains some arbitrary value in all cell. The main reason is that any unvoiced clip or a clip with low power can be matched by any other unvoiced clips. That's why the miss classification is equally distributed among all the phoneme labels. The higher values in the miss classification indicate close pronunciation with the actual phoneme label.

The Confusion Matrix for accuracy obtained in reduced data set is given in Table 4.

We got some of the wrong predictions for those phonemes whose pronunciations are closely related. For example, some mismatch occurs between /a/ and /aa/, simi-

Table 4 Confusion matrix in reduced data set

a	**98.14**	0.9027	0.1766	0.0196	0.0392	0.6083	0.1177
aa	1.104	**97.97**	0.7838	0.0356	0	0.1069	0
ae	0.2483	1.043	**97.82**	0	0	0.1986	0.6951
e	0.0972	0	0.0648	**98.64**	0.777	0.2916	0.1296
i	0.0530	0.0796	0.1327	1.062	**98.27**	0.1858	0.2124
o	0.3643	0.0700	0.1261	0.0840	0.0280	**98.81**	0.5185
u	0.0769	0.1385	0.3385	0.0461	0.0615	0.8924	**98.45**
	a	aa	ae	e	i	o	u

larly between /o/ and /u/ and between /aa/ and /ae/. We also extracted more than 14 dimension MFCCs features, such as 15 and 19 dimensions were extracted. But our accuracy was saturated at 14 dimensions.

5 Conclusion

Monophthongs Bangla Vowel recognition using MFCC features is able to recognize vowel phonemes almost accurately. The same steps will be followed for larger data set in future. We think that the method will able to recognize consonants also. In the future, we will emphasis on Bangla diphthong vowels. The MFCCs and ΔMFCCs will be incorporated for a large number of phonemes. The dimensions of the features can also be reduced using PCA for faster computation in future. This model is only to detect monosyllable vowel phonemes. To make the system more reliable we will apply for outdoor data that will able to handle noisy speech. Correctly vowel recognition indicates recognition of words and continuous speech will give a better result in future.

References

1. Mukherjee, H., Phadikar, S., Rakshit, P., Roy, K.: REARC-a Bangla phoneme recognizer. In: 2016 International Conference on Accessibility to Digital World (ICADW), pp. 177–180. IEEE (2016)
2. Hasan, M.M., Hassan, F., Islam, G.M.M., Banik, M., Kotwal, M.R.A., Rahman, S.M.M., Muhammad, G., Mohammad, N.H.: Bangla triphone HMM based word recognition. In: 2010 IEEE Asia Pacific Conference on Circuits and Systems, pp. 883–886. IEEE (2010)
3. Sumon, S.A., Chowdhury, J., Debnath, S., Mohammed, N., Momen, S.: Bangla short speech commands recognition using convolutional neural networks. In: 2018 International Conference on Bangla Speech and Language Processing (ICBSLP), pp. 1–6. IEEE (2018)
4. Sultana, S., Akhand, M.A.H., Das, P.K., Rahman, M.H.: Bangla speech-to-text conversion using SAPI. In: 2012 International Conference on Computer and Communication Engineering

(ICCCE), pp. 385–390. IEEE (2012)

5. Sumit, S.H., Al Muntasir, T., Zaman, M.A., Nandi, R.N., Sourov, T.: Noise robust end-to-end speech recognition for Bangla language. In: 2018 International Conference on Bangla Speech and Language Processing (ICBSLP), pp. 1–5. IEEE (2018)

6. Saurav, J.R., Amin, S., Kibria, S., Rahman, M.S.: Bangla speech recognition for voice search. In: 2018 International Conference on Bangla Speech and Language Processing (ICBSLP), pp. 1–4. IEEE (2018)

7. Mandal, S.D., Warsi, A.H., Basu, T., Hirose, K., Fujisaki, H.: Analysis and synthesis of F0 contours for Bangla readout speech. In: Proceedings of Oriental COCOSDA (2010)

8. Das, B., Mandal, S., Mitra, P.: Bengali speech corpus for continuous automatic speech recognition system. In: 2011 International Conference on Speech Database and Assessments (Oriental COCOSDA), pp. 51–55. IEEE (2011)

9. Sichivitsa, V.: Audacity in vocal improvisation: motivating elementary school students through technology. Teach. Music 14(4), 48 (2007)

10. Gupta, S., Jaafar, J., Ahmad, W.W., Bansal, A.: Feature extraction using MFCC. Signal Image Proc. Int. J. (SIPIJ) 4(4), 101–108 (2013)

11. Nahid, M.M.H., Purkaystha, B., Islam, M.S.: Bengali speech recognition: a double layered LSTM-RNN approach. In: 2017 20th International Conference of Computer and Information Technology (ICCIT), pp. 1–6. IEEE (2017)

12. Hall, M., Frank, E., Holmes, G., Pfahringer, B., Reutemann, P., Witten, I.H.: The WEKA data mining software: an update. ACM SIGKDD Explor. Newslett. 11(1), 10–18 (2009)

Supporting QoE-Aware Live Streaming Over Peer-to-Peer Network

Swati Kumari Tripathy and Manas Ranjan Kabat

Abstract QoE-aware (Quality of Experience) dynamic adaptive streaming over HTTP in Peer-to-Peer (P2P) networks has attracted researchers during the last decade due to the increasing multimedia applications over Internet. In this paper, we propose a super-peer architecture which can reduce the delay of live streaming as well as increase the delivery ratio. Furthermore, we also propose a rate migration control algorithm over our proposed system model which enhances QoE of clients by switching different video quality versions depending upon their current network status. This model can able to provide different quality versions of the same video as representation, where each representation streams to all super-peers over our proposed multi-overlay system model. The performance of proposed live streaming mechanism in super-peer network is studied through simulation and it is observed that it outperforms state-of-the-art live streaming in peer-to-peer networks.

Keywords Dynamic adaptation streaming over HTTP (DASH) · Quality of experience (QoE) · Peer-to-peer (P2P) · Live streaming

1 Introduction

In the present scenario, video is the constitutional transmission form over the Internet which has accomplished the astonishing growth. YouTube and Netflix are two popular websites for communicating videos, transmitting movies, and TV channels of different programs, respectively. Live Streaming these increases large multimedia file usage which impacts the amount of network traffic rotate around streaming servers [1]. One of the solutions for this increasing users video requests, differing network conditions, and conserving servers HTTP via a standardized approach is the deployment of Dynamic Adaptive Streaming over HTTP (DASH) [2, 3]. A

S. K. Tripathy · M. R. Kabat (✉)
Department of CSE, Veer Surendra Sai University of Technology, Burla, India
e-mail: manas_kabat@yahoo.com

S. K. Tripathy
e-mail: swatitripathy57@gmail.com

© Springer Nature Singapore Pte Ltd. 2020
V. Bhateja et al. (eds.), *Intelligent Computing and Communication*,
Advances in Intelligent Systems and Computing 1034,
https://doi.org/10.1007/978-981-15-1084-7_50

DASH streaming server conceals multimedia contents at different bit rates and gives unambiguous files for the structure of the video segments, and makes distinct video representations available [4]. Peer-to-Peer (P2P) network is a decentralized server where each node acts as server for storing and sharing data among them [5]. All peers are equally privileged over participation in various applications. They are said to design a pure peer-to-peer network of various nodes. A kind of research based on the uses of the Internet demonstrates [6, 7], here we are observing a meteoric development of video contents with different versions. Many dominion solutions are very first flourish, it is from smooth video stream of Microsoft to adaptive Live Streaming over HTTP by Apple [8] and dynamic adaptive Streaming by Adobe [6]. They bring out the intent of the Dynamic Adaptive Streaming over HTTP (DASH).

This paper proposes the design of P2P streaming architecture consists of three levels of FAT Tree model. Our proposed model follows some features: I Migrating neighbor peers from overlays can be able to know about the Streaming position of super-peers by sharing their buffer map cache. II Streaming algorithm for our proposed system model. III Rate adaptation control algorithm to receive requested video chunks from super-peers and servers by adopting the quality level according to their available bandwidth and network condition, where we can measure the QoE.

2 Proposed System Model

This section reviews various methods that have been proposed for data balancing and fraud detection in this domain. The peer-to-peer network is designed as a graph $(G) = (V, E)$ where $V = \{P_1, P_2, ..., P_N\}$ is a set of peer nodes and $E \subseteq V \times V$ is the set of links connecting between them. The P2P network is organized on application layer for sharing multimedia files. All users in this virtual network relay the multimedia contents available in their own local buffers to more peers using different virtual connection links. In this paper, it is considered that K number of DASH representations is available for streaming a distinct video channel for a population of N users. Furthermore, we also assume that each DASH representation is distinguished by its streaming rate (r) and also we set $r_j < r_{j+1}$ for $1 \leq i \leq K$ within each overlay j a super-peer network is built which is represented by a graph $G^{Si} = (V^i, E^{Si})$, where $V^{Si} \subseteq V_i$ and $E^{Si} \subseteq E_j$. In our proposed super-peer overlay network, the super-peer nodes are in V^{Si}, mapped to a number of peers in V_j. The super-peer nodes are responsible for overlay management. The super-peers are required to communicate with each other in each overlay. Therefore, it reduces the communication overhead in our model (Fig. 1).

In our architecture, each version of video is divided into a number of video chunks. Weak peers watching the same version with all similar interests will derive a connection among all neighbor super-peers form a P2P overlay. As a consequence, weak peers request the primary super-peers for same video chunks. In case if the primary super-peer fails to communicate with their weak peers, weak peers start sending requests to secondary super-peers. To overcome the same request problem,

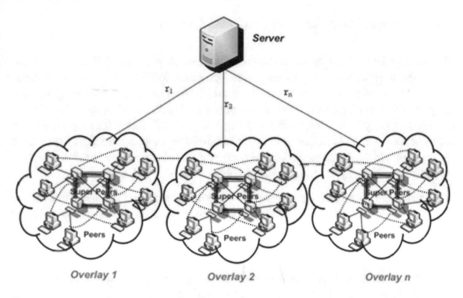

Fig. 1 Proposed system model

area recommends the acceptance of buffer map caches which stores the history of their recent searches. Super-peers cache the content's location information requested by their own weak peers. Each super-peer obtains video chunks from the connected super-peers, whereas peer stores the data item and their IP address or else collects from the server. Each peer is assigned to a fixed super-peer randomly and statically and all super-peers form bottlenecks to its fault tolerance [9]. The routing tables restore the system structure, after a super-peer crash. Then, it brings back to its consistent state and the peer requests secondary super-peer for a new video chunk. When a weak peer decides to migrate from overlay i to overlay i', it simply leaves the current overlay i to its desired overlay i'.

At first, a weak peer sends request to the super-peer and starts watching with the lowest video quality, r_1. It starts with a short period of time and influences by the user experiences for overall quality. In overlay 1, if the peer's requested quality representation is higher than the current quality representation, all DASH representations have the same chunk duration, notified by t_{chunk} with same n number of video chunks. In overlay i, each video chunk contains different sizes of media bits, notified by L_i, defined as

$$L_i = t_{chunk} \cdot r_i \tag{1}$$

It causes a smooth transmission between different quality representations, in case of switching.

2.1 Proposed Live Streaming Model Overview

The concept behind our proposed model is that weak peers get information statistically about the index of the super-peers. According to this procedure, weak peers can be able to take a decision where to connect. Here, we are considering streaming procedure on ith overlay; each super-peer can download different video chunks of similar quality level directly from server and stored in its buffer map cache. All super-peers share their buffer map to its neighbor super-peer and also stores their recent search history. Let us next consider that peer P_i requests for a video chunk to its super-peer in every Δt time as listed below.

(1) Peer node P_i requests super-peer Sp_i a video chunk for the first time; if the required representation is available, then assign it or else the super-peer has to check the availability in neighbor super-peers Sp_i'. If the requested video chunk is available in Sp_i', then Sp_i gets it and stores in their buffer map cache, further it assigns to its peer P_i.

(2) After searches complete in ith overlay, the super-peer downloads the requested video chunk directly from the streaming server.

Algorithm 1 Live Streaming over Proposed Super Peer Network
For super-peer in overlay j every Δt time
If ((*rep* (*req*)) is available Super-peer Sp_i) then,
; assign rep to peer in overlay i directly;
else
; verifies in other Super-peer i
If (((*rep*) *req*) is available Super-peer Sp'_i) then
; assign rep to peer in overlay i by super-peer overlay Sp'_i through Sp_i;
Else
; verifies for availability in server; exit;
End if
End if

3 Rate Adaptation Algorithm

This section defines the quality of video chunks, which strongly impacts the video streaming quality running in the entire overlay. It indicates both the status parameters and also the health of the current overlay and also periodically handled out streaming server to super-peers which rely on both to implement the rate migration control algorithm. Among parameters, we consider the following:

(1) The peer contribution index $\rho_j(t)$ of ith overlay, derived as

$$\rho_j(t) = \frac{U_j(t + \Delta t) - U_j(t)}{D_j(t + \Delta t) - D_j(t)} \quad (2)$$

where $U_j(t)$ is the number of video chunk uploaded by jth peer in ith overlay at time t, $D_j(t)$ is the number of video chunk downloaded by jth peer in ith overlay at time t.

(2) The overlay contribution index $\sigma_i(t)$ of ith overlay, derived as

$$\sigma_i(t) = \frac{U_{ij}(\Delta t) + \sum_{k \in j} U_{ik}(\Delta t)}{\sum_{k \in O_j} D_{ik}(\Delta t)} \quad (3)$$

where $U_{ij}(\Delta t)$ is the number of video chunks downloaded by jth peer in ith overlay at time Δt, $U_{ik}(\Delta t)$ is the number of video chunks uploaded by other peers (k) in ith overlay at time Δt, $D_{ik}(\Delta t)$ is the number of video chunks downloaded by other peers in ith overlay at time Δt.

(3) The peer request index $\gamma_i(t)$ of ith overlay is derived as

$$\gamma_i(t) = \frac{\chi_j + \partial_j}{X_i + Y_i} \quad (4)$$

where χ_i is the number of requests received from peer nodes in overlay i, ∂_i is the number of requests received from other super-peers, X is the total number of requests made by all peers to super-peers, and Y is the total number of requests made by all super-peers to super-peers.

(4) The average request γ_{avg} of n number of overlay is derived as

$$\gamma_{avg} = \frac{X_i + Y_i}{N} \quad (5)$$

where N is the total number of peers in ith overlay.

When the peer contribution index ρ_j is greater or equals to 1, it defines a good contributor or else it defines bad contributor. Let us next take peer node j satisfaction in i overlay.

(1) Peer node j checks for current streaming quality l_i running in ith overlay against $l_d(j)$, i.e., if $l_i < l_d(j)$, peer first verifies whether it will stay or leave the current overlay. This occurs, if $\sigma_i < 1$ and $\rho_j \geq 1$. This confines that peer does not migrate to higher quality level of overlay $i + 1$. If nothing affects the peer departure, peer node j further checks for positivity of future move to its destination overlay $i + 1$, i.e., if $\sigma_i > 1$, $\rho_j > \rho_{thres}$ (set threshold), and $\gamma_i > \gamma_{avg}$, peer is ready to migrate because it is beneficial to the next move $i + 1$ or else there will be no migration.

(2) For verifying the streaming quality of peer j, if $\gamma_i < \gamma_{avg}$ and $\sigma_i \leq 1$, peer viewing quality is not acceptable so it scales down to lower overlay $i - 1$.

Algorithm 2 Rate Migration Control Algorithm

A peer after each Δt second verifies the satisfaction

if ($l_i < l_d(j)$) then,

   ; verifies the status of current overlay

   if (($\varpi_i < 1$) and ($\upsilon_j A = 1$))

  no migration to overlay i+1; else

        ; verifies the status of target overlay

      if (($\varpi_i A1$) and ($\upsilon_j A\upsilon_{thres}$)) and ($\varphi_i A\varphi_{avg}$) then

        migration to overlay i+1; else

        no migration to overlay i+1;exit;

      end if; end if; end if

  ; verifies streaming quality

if (($\varphi_i < \varphi_{avg}$) and ($\varpi_i \leq 1$))

migration to overlay i-1; exit; else

stay in overlay i; exit;

end if

---

4 Simulation Results

The performance of our proposed system model is evaluated and implemented on event-driven simulator, which is coded in C++ available on [10]. Our model is a replica of multiple overlays with Dynamic Adaptive Streaming over HTTP (DASH). Here, each overlay streams a particular representation of DASH system. The current system population with $N = 2000$ peers (active in nature). These peers are very frequently entering and leaving their overlays. In every 20 s, peer nodes are joined into the system, whereas the arrival times of each peer should be balanced to exist in current overlay. All peers' session duration is fixed with 1500 s and simulation time takes 300 s. We have taken four different quality classes with four different bit rates of video representations (k) 700, 1500, 2500, and 3500 kb/s. The uploading and downloading capacity of all classes with average number of users are stemmed out from Akamai European average Internet connection. Initially, the server takes small number of upload capacity for all overlays, which is four times to the rate $C_{si} = 4 \cdot r_i$ focusing on a hybrid P2P system. We have taken the request window (X) size of 20 s and each video chunk durations is of $t_{chunk} = 200$ ms. In every 5 s of time t, the delivery ratio is computed here and defined three threshold values 0.5, 0.3, and 0.9 with weighted average of delivery ratio 1/3 and request window size 2/3. Here, we provide much priority to current time and situation of local delivery ratio and request window state, whereas the last periodicity of rate migration control algorithm takes Δt time of 4 s. We investigate the current system behavior whereas each peer is aiming to stream videos with best representation at higher bit rate, so that the results become lowest to its downloading capacity. All peer nodes are aiming to stream with higher at class 4, overlay 1 peers prefer representation 2. It is not possible to respond to all peer's requests, because it depends on the placement of peers in each overlay and their upload capacity, the peer, and overlay contribution index values $\rho_j = 0.48$ and $\sigma_i(t) = 0.91$ which are less than 1.

The upload and download capabilities of the peer nodes and the percentage of peers wish to remain in different classes as shown in Table 1. We consider three different scenarios that are conservative, uniform, and aggressive which are shown in Table 2. Figures 2, 3, 4 and 5 show the active nodes in four different overlays in three different scenarios. It is observed from the figures that there is more deviation in case of P2P DASH architecture in comparison to our proposed algorithm. In P2P DASH, there is a small deviation in both conservative and uniform scenarios. However in case of aggressive scenarios, there is a huge deviation, because of the unavailability of resources. In our proposed model, it is observed that there is less deviation compared to P2P DASH. Therefore, the satisfaction level of users in our model is better than the P2P DASH model. This is because in our model, the super-peers are involved in processing the request of the peers. Thus, bandwidth management is better and the communication overhead and the delay are less in our model.

Table 1 Peer capacities with percentages

	Class 1	Class 2	Class 3	Class 4
Upload capacity (kbit/s)	704	1024	1500	10,000
Download capacity (kbit/s)	2048	8192	10,000	50,000
% of peers	20	21	1042	17

Table 2 Peer number of nodes wishing to stream each representation

Scenario	Repr. 1	Repr. 2	Repr. 3	Repr. 4
Conservative	820	840	–	340
Uniform	600	600	400	400
Aggressive	–	400	–	1600

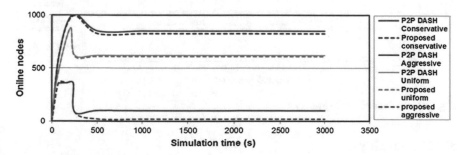

Fig. 2 Node distribution in overlay 1

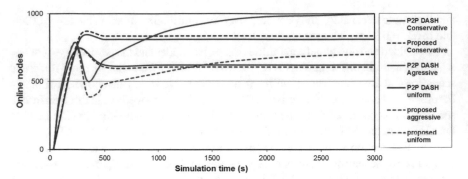

Fig. 3 Node distribution in overlay 2

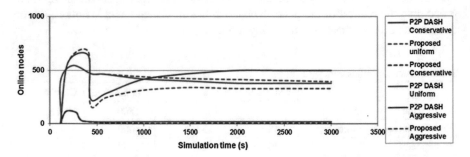

Fig. 4 Node distribution in overlay 3

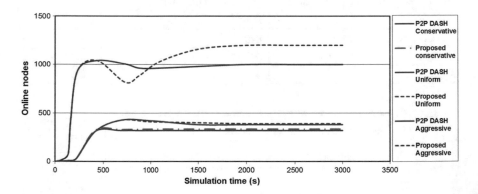

Fig. 5 Node distribution in overlay 4

5 Conclusions

This paper proposed a QoE-aware DASH over a P2P live streaming model of steaming server with additional to super-peers for monitoring all weak peers. This model can able to stream different quality of same video over a large number of users.

Each quality representation streams to all the super-peers of the multi-overlay system. Here, we proposed two algorithms of live streaming over proposed super-peer network and rate migration control algorithm. The second algorithm leads to the peer movements from current overlay to destination on the basis various parameters that reflect the current health of a peer and overlays. We demonstrated that our proposed algorithm gives best performance, achieves QoE, and also reduces the delay of peer migration and provides smooth transmission as compared to the rate switching control algorithm.

References

1. Ngai, Sandvine Inc., Waterloo, ON, Canada, Global Internet Phenomena, Latin America& North America. https://www.sandvine.com/downloads/general/global-internet-phenomena/2015/global-internet-phenomena-report-latin-america-and-northamerica.pdf (2015)
2. Mpeg, I.: Information technology-dynamic adaptive streaming over http (dash)-part 1: media presentation description and segment formats. ISO/IEC MPEG, Tech. Report (2012)
3. Stockhammer, T.: Dynamic adaptive streaming over HTTP: standards and design principles. In: Proceedings of the Second Annual ACM Conference on Multimedia Systems, pp. 133–144. ACM (2011)
4. Natali, L., Merani, M.L.: Successfully mapping DASH over a P2P live streaming architecture. IEEE Trans. Circuits Syst. Video Technol. 27(6), 1326–1339 (2017)
5. Lederer, S., Müller, C., Timmerer, C.: Towards peer-assisted dynamic adaptive streaming over HTTP. In 2012 19th International Packet Video Workshop (PV), pp. 161–166. IEEE (2012)
6. C. W. Paper: Cisco visual networking index: forecast and methodology, 2013–2018, June 2014
7. E. Commission: Digital agenda for Europe scoreboard 2012, June 2012, Directorate-General for Communication Networks, Content and Technology (CONNECT)
8. Apple: Http live streaming. https://developer.apple.com/streaming/
9. Garbacki, P., Epema, D.H., Van Steen, M.: Optimizing peer relationships in a super-peer network. In: 27th International Conference on Distributed Computing Systems (ICDCS'07), pp. 31–31. IEEE (2007)
10. Zhang, M., Zhang, Q., Sun, L., Yang, S.: Understanding the power of pull-based streaming protocol. IEEE J. Select. Areas Commun. 25(9) (2007)

Action Recognition Using 3D CNN and LSTM for Video Analytics

A. Umamakeswari, Jonah Angelus, Monicaa Kannan, Rashikha and S. A. Bragadeesh

Abstract With the advent of growing digital technology, large amount of video data is being generated, making video analytics a promising technology. Human activity recognition in videos is currently receiving increased attention and activity recognition systems are a large field of research and development with a focus on advanced machine learning algorithms, innovations in the field of hardware architecture, and on decreasing the costs of monitoring while increasing safety (Guo and Lai in Pattern Recognit 47:3343–3361, 2014, [1]). The existing system for action recognition involves using Convolutional Neural Networks (CNN). Videos are taken as a sequence of frames and frame-level CNN sequence features generated are fed to Long Short-Term Memory (LSTM) model for video recognition. However, the abovementioned methodology takes frame-level CNN sequence features as input for LSTM, which may fail to capture the rich motion information from adjacent frames or multiple clips. It is important to consider adjacent frames that allow for salient features, instead of mapping an entire frame into a static representation. Thereby, to mitigate this drawback, a new methodology is proposed wherein initially, saliency-aware methods are applied to generate saliency-aware videos. Then, an end-to-end pipeline is designed by integrating 3D CNN with LSTM, followed by a time series pooling layer and a softmax layer to predict the activities in video.

Keywords Saliency mask · Three-dimensional convolutional neural network · Action recognition · Long Short-Term Memory · Time series pooling · Softmax function

A. Umamakeswari (✉) · J. Angelus · M. Kannan · Rashikha · S. A. Bragadeesh
SASTRA University, Thanjavur, India
e-mail: a_umamakeswari@yahoo.com

J. Angelus
e-mail: jonahangelus@gmail.com

M. Kannan
e-mail: drizzlepopbb@gmail.com

S. A. Bragadeesh
e-mail: bragadeesh.prithvi@gmail.com; sa_bragadeesh@yahoo.co.in

© Springer Nature Singapore Pte Ltd. 2020
V. Bhateja et al. (eds.), *Intelligent Computing and Communication*,
Advances in Intelligent Systems and Computing 1034,
https://doi.org/10.1007/978-981-15-1084-7_51

531

1 Introduction

The human action and human activity recognition modules strive to detect not only the actions but also the end goals of the person given a series of observations in a particular context. This distinct field of research has received an abundant amount of interest and has been an active research topic in the recent past. Human action detection is one of the key elements in various applications such as visual surveillance, anomaly action detections, video retrieval, and many more. Human action recognition at its very base is little more than a pattern detection recognition problem with crafted features that have been used successfully in image processing. Hence, work done previously on image action detection can be extended and used on this. Spatial features are used to identify and characterize visual appearance while temporal features are used to characterize motion dynamics.

2 Literature Survey

To improve the performance of action recognition, recent methodologies have been proposed to add and apply direct deep learning models to learn and classify human actions [2] represented by videos. These methodologies delivered promising results. Based on previous works, it has been found that human action in sequences of the video is basically 3D signals containing visual appearance that changes continuously over time. Hence, attempts to change 2D Convolutional Neural Networks (2D CNN) to encode temporal information. The 3D CNN model proposed extracts both spatial as well as temporal features [3] not in only in single frames, but adjacent frames as well. This is due to 3D convolutions being performed. These frames of 3D CNN are sent to the Long Short Term Memory (LSTM) model. It has been found with recent studies that the LSTM model [4] has the capacity to learn when to forget and when to update the hidden states.

In previous studies, CNN sequence features are taken frame-by-frame at a frame level, which leaves out the important information about motion that can be inferred only by taking adjacent frames into consideration [5]. Also, the activity of the subject in the frame alone should be detected, not the activity in the entire frame.

These two problems are dealt with by using the abovementioned methodologies. A pipeline integrating 3D CNN and LSTM is proposed [6]. 3D CNN videos contain richer motion information while LSTM explores the temporal relationship. Saliency masks are introduced to extract only the necessary features in the given video frame [7].

From [8], we get to know that initial attempts for temporal and spatiotemporal action localization are based on a sliding-window scheme and focus on improving the search complexity. Soomro and Shah [9] was the first paper on unsupervised action detection problem. Discriminative clustering was used to discover which labels are presented in a dataset by the following techniques' usage of spectral clustering to get initial clusters and iteratively select videos from the nondominant set. To obtain spatiotemporal annotations, the following are done: by over-segmenting the video using supervoxel, constructing DAG, solving knapsack optimization with temporal constraints, and determining whether to include a supervoxel in the current "action" or not. It shows competitive performance (in terms of AUC) compared to supervised methods. It might be applied for weakly supervised action detection problem-solving. In [10], the detection of a moment of action completion is done. They want to separate pre-completion and post-completion of an action frame-by-frame. They define "completion" as the "goal" of an action is achieved. They use HMM and LSTM on top of ConvNet feature to detect the completion of an action. For HMM, they have two hidden states, pre and post. The parameters of HMM, initial and transition probs, covariance matrice,s and mean vectors are learned from training data. For LSTM, they feed fc7 feature and per-frame labels to LSTM as an input. Experimental results are quite trivial. Both models can detect the completion of an action with a reasonable accuracy, 75% within 10 frames, under strong assumptions: temporally trimmed sequences (no multiple actions per sequence), momentary completion (completion should be detected using only one frame even for human), and uniform prior for completion (50:50 chance of complete vs. incomplete). In [11], a attention weighted pooling method is proposed. With a rank 1 approximation of second-order pooling and manipulating the order of matrix multiplications, attention pooling can be viewed as a combination of class-agnostic bottom-up saliency and class-specific top-down attention. We can replace the average pooling operations in the ResNet architecture by the proposed attention pooling. With the attention pooling, we can get state-of-the-art performance on HMDB51 (video), HICO, and MPII (image) dataset. In [12], the "Temporal Activity Detection method incorporating temporal context" is developed. Temporal context means a temporal proposal with a temporal extent larger than the actual action extent. Propose temporal anchors with various scales for each temporal position. When encoding a feature, concatenate features from two scales to incorporate temporal context. Apply temporal convolution to further incorporate temporal context.

3 Model Architecture

The three-layered model which includes a saliency mask, 3D CNN, and LSTM is designed to perform the task to action recognition. Figure 1 shows the architecture of the model [3].

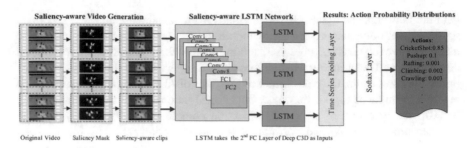

Fig. 1 Model architecture [3]

3.1 Algorithm

Analyze (video *V*):

1. For each video *V*, it has *n* frames $V = (v1, v2 \ldots vn)$ and it is divided into *T* splits
2. Each split is encoded with 3D CNN generating $X = (x1, x2, \ldots xT)$
3. Each encoded split is fed to the LSTM network to get vector *Z*
4. Mean pooling and max pooling are concatenated into a vector *Z* as the final video level descriptor as output vector *R*
5. Vector *R* is fed into a classifier loss layer
6. Output of classifier loss layer is sent to time series pooling layer to get Vector *Y*
7. Probability distribution of the action category is calculated using softmax layer over the output vector.

4 Model Specifications

4.1 Saliency Mask

The need to perform saliency check arises because, suppose that all the images obtained from video frames in the training set of the fish class contain a lake or sea with water, we will not be able to know if the CNN uses fish-related pixels or the pixels of the water in the sea. Apparently, this might happen very often in the case of small data sets. In the proposed methodology, we use a saliency-aware network. Saliency typically means computing the gradient of the result category with respect to the input. Through this, we get a rate of change in the output category with respect to the rate of change in input category. This gradient value can be used to find the regions in the input image that result in the highest modification in the output category. In the context of action recognition, the concept of saliency is introduced in order to extract the important features of video snippets which in turn enhances the performance of the 3D CNN. The recent methods have proved to be more efficient than the state-of-the-art methodologies due to the use of saliency-aware networks. We build saliency maps by using a saliency detection method that puts various parameters such as color, motion boundary, and edge. Only very few methods are available for saliency detection in videos. The method deployed here is one of the most efficient methods available for saliency detection in videos.

4.2 3D Convolutional Neural Network Layer

Temporal features are one of the most important aspects to be taken into account when analytics of video data comes into picture. Any analytics using 2-dimensional convolutional neural network take into account only the special features but it is important to capture motion information [13] in the contiguous frames of videos. It is to tackle this challenge, we use 3-dimensional convolutional neural network where even the important features of a video are also considered during analytics [14]. Each video in the KTH dataset is divided into frames and the frames are converted to grayscale to make the further processing of the video much easier. The kernel size of the 3D convolution is $5 \times 5 \times 5$. The 3D CNN gets help in incorporating the spatiotemporal features [15] and there also exists a 3D Max pooling layer in the model which performs the function of temporal features pooling. The hidden layers of the CNN perform a series of convulsions and pooling operations during which the features are detected. The fully connected layers in the CNN perform the function of class of the classifier on top of the features that are extracted. They assign a probability to each class and the class which is assigned with highest probability is taken as the predicted result. And since contiguous frame information is also taken into consideration in the constructed 3D CNN [16], it increases the accuracy of the developed model. The main motive to include spatiotemporal information is established using the 3D CNN layer after the saliency mask.

4.3 LSTM Layer

The classification of a particular task also depends on the recent information of the present task and it is this requirement that brings us to LSTM, a type of recurrent neural network. The problem of not able to remember long-term information during classification by any algorithm, commonly called the Long-Term Dependency Problem, is mitigated by LSTM's [17]. The architecture of LSTM network consists of an input gate, a memory cell, an output gate, and a forget gate. The activation layer of the gates is a logistic function. The input gate has the function of controlling the extent at which a new value flows into the cell whereas the forget gate controls the extent at which a new value remains in the cell and the output gate has control over the extent to which the value in the cell affects the output activation of the LSTM layer. The LSTM cell remembers the values over specific time intervals which helps

Fig. 2 KTH dataset
examples

our model to remember temporal features [18] while making the required analytics
on the video data.

5 Dataset

The KTH dataset has six action classes namely walking, jogging, running, boxing,
handwaving, and hand clapping. The videos are performed by 25 different people in
four different backgrounds to make the dataset more diverse and effective (Fig. 2).

6 Results

The model was developed in Python and run on a Nvidia GeForce 940 M GPU for
training. The accuracy obtained on the KTH dataset for the proposed model is 78%.
Figure 3 depicts the accuracy plot for the developed model. And Fig. 4 depicts the
loss function for the model during the learning phase of the model.

The results are obtained by tuning the CNN and LSTM parameters with the goal
of increasing the accuracy of the model (Table 1).

Fig. 3 Accuracy plot of the model

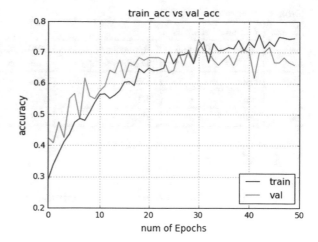

Fig. 4 Loss function of the model

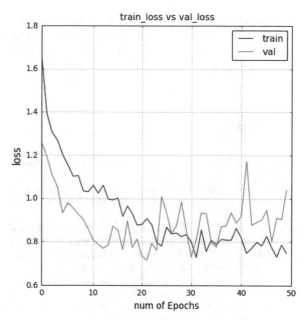

Table 1 Accuracy table

S. no.	Classified samples			
	Total no of samples	No of correctly classified samples	No of wrongly classified samples	Accuracy (in %)
1.	600	472	128	78.6
2.	450	284	166	63.1
3.	200	102	98	51

7 Conclusion

Here, a general architecture for action recognition from video snippets is proposed. We first use saliency-aware method [19] to generate a saliency-aware video that abstracts the foreground features and omits the background features so that the model performs optimally even for small datasets [20]. Next, this video is processed in two phases. During the first phase, the video is passed through a 3D CNN integrated with LSTM. In the second phase, the output is passed through an end-to-end pipeline with time series pooling and a softmax layer. The proposed model performs very well with the KTH dataset.

References

1. Guo, G., Lai, A.: A survey on still image based human action recognition. Pattern Recognit. **47**(10), 3343–3361 (2014)
2. Wang, H., Ullah, M.M., Kläser, A., Laptev, I., Schmid, C.: Evaluation of local spatio-temporal features for action recognition. In: Proceedings fo British Machine Vision Conference, vol. 127 (2009)
3. Wang, X., Gao, L., Song, J., Shen, H.: Beyond frame-level CNN: saliency-aware 3-D CNN with LSTM for video action recognition. IEEE Signal Process. Lett. **24**(4), 510–514 (2017)
4. Dollar, P., Rabaud, V., Cottrell, G., Belongie, S.: Behavior recognition via sparse spatio-temporal features. In: Proceedings of the 14th International Conference on Computer Communications and Networks, pp. 65–72 (2005)
5. Tran, D., Bourdev, L., Fergus, R., Torresani, L., Paluri, M.: Learning spatiotemporal features with 3d convolutional networks. In: Proceedings of 2015 IEEE International Conference on Computer Vision, pp. 4489–4497 (2015)
6. Donahue, J., et al.: Long-term recurrent convolutional networks for visual recognition and description. In: Proceedings of 2015 Computer Vision and Pattern Recognition (2015)
7. Ji, S., Xu, W., Yang, M., Yu, K.: 3D convolutional neural networks for human action recognition. IEEE Trans. Pattern Anal. Mach. Intell. **35**, 221–231 (2010)
8. Weinzaepfel, P., Martin, X., Schmid, C.: Human action localization with sparse spatial supervision. arXiv:1605.05197 (2016)
9. Soomro, K., Shah, M.: Unsupervised action discovery and localization in videos. In: Proceedings of the IEEE Conference on Computer Vision and Pattern Recognition (2017)
10. Heidarivincheh, F., Mirmehdi, M., Damen, D.: Detecting the moment of completion: temporal models for localising action completion. arXiv:1710.02310 (2017)
11. Girdhar, R., Ramanan, D.: Attentional pooling for action recognition. In: Advances in Neural Information Processing Systems (2017)

12. Dai, X., et al.: Temporal context network for activity localization in videos. In: 2017 IEEE International Conference on Computer Vision (ICCV). IEEE (2017)
13. Ji, S., Xu, W., Yang, M., Yu, K.: 3d convolutional neural networks for human action recognition. IEEE Trans. Pattern Anal. Mach. Intell. **35**(1), 221–231 (2013)
14. Zeiler, M.D., Fergus, R.: Visualizing and understanding convolutional networks. In: Proceedings of 13th European Conference on Computer Vision, pp. 818–833 (2014)
15. Wang, W., Shen, J., Porikli, F.: Saliency-aware geodesic video object segmentation. In: Proceedings of IEEE Conference on Computer Vision and Pattern Recognition, pp. 3395–3402 (2015)
16. Marszalek, M., Laptev, I., Schmid, C.: Actions in context In: Proceedings of 2015 IEEE Conference on Computer Vision and Pattern Recognition, pp. 2929–2936 (2009)
17. Karpathy, A., Toderici, G., Shetty, S., Leung, T., Sukthankar, R., Li, F.: Large-scale video classification with convolutional neural networks. In: Proceedings of 2014 Computer Vision and Pattern Recognition, pp. 1725–1732 (2014)
18. Ng, J.Y.-H., Hausknecht, M., Vijayanarasimhan, S., Vinyals, O., Monga, R., Toderici, G.: Beyond short snippets: deep networks for video classification. In: Proceedings of 2015 IEEE Conference on Computer Vision and Pattern Recognition, pp. 4694–4702 (2015)
19. Wang, W., Shen J., Shao, L.: Video salient object detection via fully convolutional networks. IEEE Trans. Image Process. **27**, 38–49 (2018)
20. Liu, Q., Cai, W., Shen, J., Fu, Z., Liu, X., Linge, N.: A speculative approach to spatial-temporal efficiency with multi-objective optimization in a heterogeneous cloud environment. Secur. Commun. Netw. **9**(17), 4002–4012 (2016)

A Large-Scale Implementation Using MapReduce-Based SVM for Tweets Sentiment Analysis

V. P. Lijo and Hari Seetha

Abstract Sentiment analysis is an interesting area of research due to the availability of sentiment data and opinion-oriented services. The efficiency and scalability of the sentiment analysis applications are important concerns as they expect accurate results in short period of time by processing a large amount of data. An efficient and scalable polarity detection method is proposed in this paper. The sequential minimal optimization with MapReduce (SMOMR) is used to achieve enhanced efficiency as well as scalability. The experiment results reveal that this method outperforms many existing methods.

Keywords Sentiment analysis · Polarity detection · Big data

1 Introduction

Nowadays, people use Twitter widely to get world news about political, environmental, recent technical advancements and so on. The data from social networks are growing extensively day-by-day, and data scientists are attracted towards analyzing the social network data such as tweets because of its richness of sentiment, economic and political aspects. Sentiment analysis on Twitter data is a challenge due to its big data characteristics, such as diversity, veracity and volume [22]. The literature offers a large number of methods to analyze the data, which include lexicon-based, machine-learning and a combination of both [12]. Although many techniques are developed and used in recent years, they all let down the stakeholders with their limitations.

Detecting the inclination of sentiment towards a tweet is essential in many applications which are looking for a feedback from its users. The user's opinion about

V. P. Lijo (✉)
Vellore Institute of Technology, Vellore, Tamil Nadu, India
e-mail: lijo.vp@vit.ac.in

H. Seetha
VIT-AP, Amaravati, Andhra Pradesh, India
e-mail: hariseetha@gmail.com

© Springer Nature Singapore Pte Ltd. 2020
V. Bhateja et al. (eds.), *Intelligent Computing and Communication*,
Advances in Intelligent Systems and Computing 1034,
https://doi.org/10.1007/978-981-15-1084-7_52

their service or product is very important to enhance the features, or rectify the existing flaws. The opinion may be positive or negative, which is termed as polarity of the opinion. The polarity detection [3] techniques are developed to detect the text's polarity in effective and efficient manner, but most of the techniques are not sufficient to deal with real-time scenarios.

The volume of data set is a challenge in terms of computation cost, but it is a great opportunity to improve the efficiency of the algorithms which utilize large number of instances to get high accuracy and precision. A distributed or parallel sentiment classification model with scalability for efficient computing is to be developed.

This research paper proposes a distributed/parallel model which is implemented in Hadoop MapReduce [13] that transforms our tasks into MapReduce jobs. In this model, the algorithms are used to extract features, select relevant features, prepare feature vector space for training and testing sets, and classify the samples based on polarity. Additionally, we modify the sequential minimal optimization (SMO) [17] for typical distributed sentiment classification. We investigate the effects of various parameters on the algorithm's performance and computation cost, such as number of instances and number of computing nodes, by performing substantial experimental evaluation.

The remaining sections of the paper are arranged as follows.

In Sect. 2, the related works in the area of sentiment analysis are discussed. Section 3 gives the details about SMO and MapReduce. Section 4 gives insights on the proposed method, while Sect. 5 discusses experiments and results. Section 6 gives the conclusion.

2 Related Works

In the sentiment classification, the main categories of techniques are: supervised, unsupervised and hybrid [14]. Above all, the supervised machine-learning approach, SVM, is very popular due to its generalization characteristics and since it works well in even unbalanced data [2]. Mamgain et al. [15] illustrates the cases where SVM outperforms other classification techniques, such as naïve Bayes and multilayer perceptron.

The SVM training [identifying set of support vectors (SVs) and other related parameters] is considered as computationally expensive. The literature includes many approaches to improve the efficiency of the SVM by improving its training time. Cauwenberghs and Poggio [5] proposed an incremental SVM training scheme which follows online recursive training as with one vector at a time. In this method they propose a decremental unlearning which helps to visualize the data geometry and generalization. But this method is computationally tedious because of its sequential nature. The communication overhead is very critical in the incremental method.

Forero et al. [8] propose a distributed SVM which relies on the communication networks topology and the SVs are updated based on some consensus among nodes. This method is reducing computational cost by eliminating exchanging SVs among nodes multiple times, but its accuracy highly depends on the communication strategy in the given topology. Chang's [6] parallel SVM is an efficient method for SVM training but it has a high communication overhead to update the SVs.

Sequential minimal optimization is a fast training method relied on some heuristics but sequential in nature [17]. This algorithm selects the working set with minimum (two) elements and this makes the decomposition at maximum.

The SVM with MapReduce is used to classify data based on its opinion [19], and map functions in this method find SVs locally and combine them with other training subset and retrain to get new set of SVs. This will progress in this manner and finally get global SVs. Retraining is time-consuming and does not contribute much on accuracy. Kanavos et al. [11] give some experimental results on using MapReduce and Spark for Twitter sentiment analysis. The combination of various methods, such as lexicon-based and machine-learning, is useful to improve the efficiency of the sentiment analysis [12]. The challenges of approaches and techniques for sentiment analysis are explored in [10].

3   The Sentiment Analysis

In this section, we describe in detail the method used in this research. In the proposed method, the sentiment classification is achieved by splitting the training data in chunks and processing them in parallel and combines the results to build a global classification model. The main components used in this method are one of the fastest SVM training methods, SMO and the MapReduce framework, which are explained in detail in the following subsections.

3.1   Data Pre-processing for Polarity Detection

Here we discuss some of the common pre-processing methods which are used to clean the data. For improving the accuracy of the sentiment extraction models, it is necessary to go through data pre-processing (DPP). DPP offers to get better quality on text classification and reduces the computational complexities.

Stop-words removal [21], stemming and lemmatization [18], part-of-speech tagging (POS) [16], handling negations [7] and tokenization into N-grams [4, 20] are some of the important DPP steps.

3.2   Sequential Minimal Optimization (SMO)

The SMO algorithm is introduced by Platt [17] and it achieves speed by reducing the operations in the iterations. This makes SMO as one of the fastest SVM algorithms available. This algorithm selects the working set with minimum (two) elements and this makes the decomposition at maximum. The working set warrants the following conditions:

$$\sum_{i=0}^{n} A_i Y_i = 0 \tag{1}$$

where A_i is Lagrange multiplier and Y_i is the class name. This helps to get analytical solutions for subproblems. In SMO algorithm, to build a linear SVM, only one weight vector needs to be stored instead of the training examples that correspond to non-zero Lagrange multiplier values [1].

3.3 MapReduce Model

MapReduce framework is efficient for processing big data in parallel manner. The map function splits input records and transformed (key/value pairs) in initial step and reduce function collects resultant records from map methods and summarized. Map function iterates over part of data and group all intermediate values and repeats the operation on resulting set of key/value pairs, whereas reduce function will present a reduced set as result. Appache Hadoop MapReduce is a popular implementation, in which, java is used for robust execution of jobs. The Hadoop File System (HDFS) is designed to read large chunks, process and store the outputs. Replicate storages in HDFS support reliability, and it provides by two processes as follows:

NameNode: ensures management and control services and keeps metadata.
DataNode: provides storage and retrieval of block of data.

4 Proposed Method

Further, we discuss the distributed approach by using SMO and MapReduce framework, the Twitter sentiment analysis and Twitter feature representation.

4.1 Distributed Model

The SMO in a single node takes more than 10 h to process one million instances of samples. To train SVM in a fast manner, an efficient SMO algorithm is needed or a provision to train them in a distributive manner. The normal SMO algorithm spends 90% of the time for updating the optimality vector [1]. This can be eliminated by updating this vector in parallel. But it takes large amount of communication cost due to large number of iterations. To keep the communication cost minimum train the chunks of data in parallel and perform a combination of the result. This approach is simple and provides better result in terms of accuracy and the abstract view is illustrated in Fig. 1.

Another solution is to partition the training data into chunks, and train them in parallel and combine the results with other chunks' results and retrain the combined result. This retraining is time-consuming and not improving accuracy significantly.

The SMO with MapReduce (SMOMR) is implemented using Hadoop MapReduce framework and the basic methods are similar to [1]. The map task processes and optimizes the subsets of the data in parallel. The linear SVM expects the output from each as partial weight vector (W_{local}) for the local partition and threshold value b.

$$W_{local} - \sum w_i \alpha_i x_i \qquad (1)$$

The reducer sums up the partial weight vector to calculate the global weight vector and average b as in Eqs. (3) and (4), respectively. Equation (5) shows the SVM output in general.

$$W_{global} = \sum W_{local} \qquad (3)$$

$$b_{avg} = \left(\sum b \right)/n \qquad (4)$$

$$SVM_{output} = W_{global} * x - b_{avg} \qquad (5)$$

In linear SVM, the weight vector and threshold value are the output of each map function. But the alpha array and threshold value b are the outputs for nonlinear SVM. The global alpha array is the result from the reducer, which joints all the maps' local alpha arrays. The reducer also takes the average of b as in case of linear SVM. The output is calculated as follows:

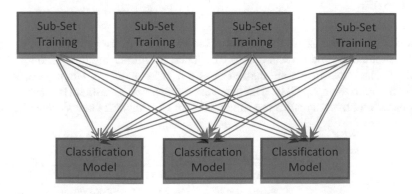

Fig. 1 The abstract model

$$u = \sum K(X_i, X) + b \qquad (6)$$

in which we need alpha array and the training data which corresponds to a > 0 and the threshold value b, with the kernel function K.

4.2 Twitter Sentiment Analysis

Twitter is very popular to convey opinions about service, products or even about people, and tweets' characteristics, such as short, straight to the point messages, made them a rich resource of sentiment analysis. A sentence-level sentiment analysis is very suitable to tweets as they are short in length and possess opinion about a single entity at a time. Twitter provides public API to retrieve tweets for real-time data. Go et al. [9] used emoticons in tweets to label tweets as positive or negative and created a Twitter data set with 1600 kilo tweets. Machine-learning algorithms support to achieve better accuracy and improve its performance by combining emoticon labels and manual labeled tweets.

4.3 Tweet Sentiment Representation

In this section, we describe the feature representation of tweets for sentiment classification purposes. The unigrams are used as features but the part-of-speech tags omitted due to less contribution in polarity detection. The presence of negation is noticed as binary data. The Natural Language Tool Kit (NLTK) is used for preprocessing the data. The collected features are ranked using information gain (IG) and considered only first 500 features.

5 Results

In this session, we discuss the results of the experiments carried out in this paper. The computation time (in seconds) is calculated for measuring the speed up of the method. The experiments are carried out on 4 Hp computers, Ubuntu 16, RAM-8.00 GB, Processor 2.3 GHz. The data sets Semval and Sentiment140 are used in this paper and their statistics are shown in Table 1. Figure 2 shows scalability test

Table 1 Statistics of data sets

Data sets	Positive	Negative	Neutral
Semval	1458	3639	4585
Sentiment140	248,576	799,999	0

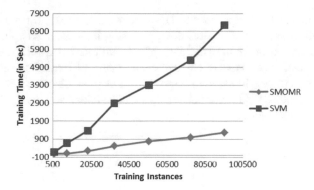

Fig. 2 Comparison of training times with SVM and SMOMR

Table 2 The computation time and accuracy for various data sets and methods

# of nodes	Data sets	Methods	Accuracy	Computational time (s)
1	Semval	NB	82.74	0.41778
		SMO	**85.21**	483.54
		Decision trees	84.78	3.7465
	Sentiment140	SMO	–	–
		NB	76.89	11.3785
		Decision trees	77.51	97.454
4	Semval	SMOMR	**84.49**	177.65
	Sentiment140	SMOMR	**78.75**	3328.76

results and comparison with SVM. Those data sets are publically available, which is ensuring the repeatability of experiments. The processing time is reduced and the scalability is achieved in this method.

Table 2 gives details about the performance of the algorithm. This is showing that the computation cost is reduced without affecting the accuracy. The proposed method processed the data sets with less computational cost and the reduction in accuracy is negligible. The proposed method is showing better accuracy than naïve Bayes (NB) and Decision Trees.

6 Conclusions

In this paper we proposed a distributed classification method for sentiment analysis. The extensive experiments show that this method outperforms the sequential approaches for polarity detection with respect to the computation cost without compromising accuracy. The publically available data sets are used in this method to facilitate repeatability of experiments. The distributed environment and the modified SMO algorithms support scalability for processing large data sets. Our method is mainly focused to improve the efficiency of the sentiment classification in terms of computation cost; speed ups are ensured even in case of large data sets. There are many challenges to be addressed to improve accuracy which are not in the scope of this work. As a future work, those challenges will be addressed to improve accuracy of sentiment classification.

References

1. Alham, N.K., Li, M., Liu, Y., Hammoud, S.: A MapReduce-based distributed SVM algorithm for automatic image annotation. Comput. Math Appl. **62**(7), 2801–2811 (2011)
2. Burges, C.J.C.: A tutorial on support vector machines for pattern recognition. Data Min. Knowl. Discov. **2**(2), 121–167 (1998)
3. Cambria, E., Das, D., Bandyopadhyay, S., Feraco, A. (eds.): A Practical Guide to Sentiment Analysis. Cham, Springer International Publishing, Switzerland (2017)
4. Caropreso, M.F., Matwin, S., Sebastiani, F.: A learner-independent evaluation of the usefulness of statistical phrases for automated text categorization. In: Chin, A.G. (ed.) Text Databases and Document Management, pp. 78–102. IGI Global, Hershey, PA, USA (2001)
5. Cauwenberghs, G., Poggio, T.: Incremental and decremental support vector machine learning. Adv. Neural Inf. Process. Syst. 409–415 (2001)
6. Chang, E.Y.: PSVM: parallelizing support vector machines on distributed computers. In: Foundations of Large-Scale Multimedia Information Management and Retrieval. Springer, Berlin, Heidelberg, pp. 213–230 (2011)
7. Councill, I.G., McDonald, R., Velikovich, L.: What's great and what's not: learning to classify the scope of negation for improved sentiment analysis. In: Proceedings of the Workshop on Negation and Speculation in Natural Language Processing, NeSp-NLP'10, Stroudsburg, PA, USA, pp. 51–59. Association for Computational Linguistics (2010)
8. Forero, P.A., Cano, A., Giannakis, G.B.: Consensus-based distributed support vector machines. J. Mach. Learn. Res. **11**(May), 1663–1707 (2010)
9. Go, A., Bhayani, R., Huang, L.: Twitter sentiment classification using distant supervision. Technical report, Stanford University (2010)
10. Hussein, D.M.E.D.M.: A survey on sentiment analysis challenges. J. King Saud Univ. Eng. Sci. **30**(4), 330–338 (2018)
11. Kanavos, A., Nodarakis, N., Sioutas, S., Tsakalidis, A., Tsolis, D., Tzimas, G.: Large scale implementations for twitter sentiment classification. Algorithms **10**(1), 33 (2017)
12. Kolchyna, O., Souza, T.T., Treleaven, P., Aste, T.: Twitter sentiment analysis: Lexicon method, machine learning method and their combination. arXiv preprint arXiv:1507.00955 (2015)
13. Lam, C.: Hadoop in Action. Manning Publications Co. (2010)
14. Lijo, V.P., Seetha, H.: Text-based sentiment analysis: review. Int. J. Knowl. Learn. **12**(1), 1–26 (2017)
15. Mamgain, N., Mehta, E., Mittal, A., Bhatt, G.: Sentiment analysis of top colleges in India using Twitter data. In: International Conference on Computational Techniques in Information and Communication Technologies (ICCTICT), New Delhi, India, March, pp. 525–530. IEEE (2016)
16. Manning, C.D., Schütze, H.: Foundations of Statistical Natural Language Processing. MIT Press, Cambridge, Massachusetts (1999)
17. Platt, J.: Sequential minimal optimization: a fast algorithm for training support vector machines, pp. 1–21 (1998)
18. Porter, M.F.: An algorithm for suffix stripping. Program **14**, 130–137 (1980)

19. Priyadarshini, A., Agarwal, S.: A map reduce based support vector machine for big data classification. Int. J. Database Theory Appl. **8**(5), 77–98 (2015)
20. Raskutti, B., Ferrá, H. L., Kowalczyk, A.: Second order features for maximizing text classification performance. In: Proceedings of the 12th European Conference on Machine Learning, EMCL'01, London, UK, pp. 419–430. Springer-Verlag (2001)
21. Salton, G., McGill, M. J.: In Introduction to Modern Information Retrieval. McGraw Hill Book Co. (1983)
22. Wu, X., Zhu, X., Wu, G.Q., Ding, W.: Data mining with big data. IEEE Trans. Knowl. Data Eng. **26**(1), 97–107 (2013)

Chatbot: Integration of Applications Within Slack Channel

Suresh Limkar, Sayli Baser, Yamini Jhamnani, Priyanka Shinde, R. Jithu and Prajwal Chinchmalatpure

Abstract Nowadays, huge amount of data are available which are used to get information. For extracting this information, according to user's requirements, there are many platforms available for searching incomprehensible or inconceivable data, for example, Quora, Google. But none of them provides optimized, accurate, precise, relevant answers. Even the problem of platform switching is observed. By developing our application and integrating it with Slack, this problem can be overcome and is beneficial for the Slack users. Slack software is cloud-based collaboration software and is designed to enable users to communicate easily and eliminate the app fatigue associated with using multiple communication applications. Slack software is a collaboration hub for work, no matter what work you do. It's a place where conversations happen, decisions are made, and information is always at your fingertips. It is a collaboration chat tool used both in and out of organizations to help teams communicate and coordinate in a more effective manner. Vizerto is a software application designed to get high-quality answers to our questions. In this, integration of these two platforms is done using APIs. The Slack conversation API provides your app with a unified interface to work with all the channel-like things encountered in Slack, public channels, private channels, direct messages, group direct messages, and our newest channel type, shared channels. Similarly, Vizerto API provides your

S. Limkar · S. Baser (✉) · Y. Jhamnani · P. Shinde · R. Jithu · P. Chinchmalatpure
AISSMS's Institute of Information Technology, Pune 411001, Maharashtra, India
e-mail: saylibaser1824@gmail.com

S. Limkar
e-mail: sureshlimkar@gmail.com

Y. Jhamnani
e-mail: jhamnaniyamini@gmail.com

P. Shinde
e-mail: priyankashinde247@gmail.com

R. Jithu
e-mail: jithraj97@gmail.com

P. Chinchmalatpure
e-mail: prajwalvvc@gmail.com

© Springer Nature Singapore Pte Ltd. 2020
V. Bhateja et al. (eds.), *Intelligent Computing and Communication*,
Advances in Intelligent Systems and Computing 1034,
https://doi.org/10.1007/978-981-15-1084-7_53

app with a unified interface to get the search results in no time. Our chatbot will be having all these functions that can be used and made knowledge available at your fingertips.

Keywords Vizerto · Slack · Chatbot · API

1 Introduction

In recent years, in IT industries revolution has been seen at great extent. We have started many start-ups and have established many new companies. In this, there are some companies which have got success but there are huge amount of start-ups that are lagging because of financial support, knowledge, team work, and so on. Many platforms are available for communicating and gaining knowledge from the internet but all these are not precise and not dictated with Slack. Slack software is a collaboration hub for work, no matter what work you do. It's a place where conversations happen, decisions are made, and information is always at your fingertips [1]. With Slack, your team is better connected and can share files among each other. In coming future, Slack will be having a great boom in the market. Slack has proved to be a paradise for the developers as it serves many functionalities, such as it provides interactive and collaborative platform to work as a team.

Application fatigue is the most common problem faced by all the users accessing the internet, which consumes a lot of efforts as well as time of the user. Our system mainly aims to eliminate the problem of switching platforms for scrutinizing their queries. Slack is an open source tool, and consequently, gathering of Slack APIs can be utilized to improve and redesign its functionalities. Imagine you are building an application which allows the end-user to view themselves on a map and then book a taxi nearby (like Uber [2]). Now you need access to certain information to make this application work (such as an accurate map). What do you do? Instead of making a map-based application (which requires massive data collection), you search for alternatives, and say suppose you stumble upon something called the **Google Maps** [3] **API**. This API allows you to access accurate map data at a particular price. You develop your application, use this API, and the application gets deployed. By taking this example, our application will function in a similar manner. It will provide many functionalities, such as Google [4] search, handling different types of app, setting alarm, setting reminder and notifications popup.

Have you ever visited any sites that shows an option for signing up through Facebook [5] or Google? How do you think you are able to login and proceed ahead with the application without even worrying how that code was written? It's because of the API that has simplified all your work. So all the information being provided for Google sign up are done through the API [6]. But good things require precautions and some measures to remain good. So do the APIs! Since most of the APIs are provided for free, they need some kind of security to keep them safe from various nonproductive purposes. Hence, a new concept called API key was introduced. This key can be considered as a permission that you take from the supplier, like from Google, to use their API like for signing up in your website or any other application. There are some APIs that are free to use, but some APIs need to be bought.

2 Proposed Model

In this model, we are planning to create a chatbot which will be exciting and also propose an interesting solution to solve the problem of "App fatigue" which most of the internet users suffer from. App fatigue can be better explained as overload of information or app which leads to user not using smartphones for the real use and also gets fatigued by many notifications that smartphones have these days. Chatbot is the solution which solves this problem by integrating two or three apps which each other. Therefore, our system will be helpful for multiple users and will be center of attraction in the market. We are developing this system in user's interest. System will reduce the user's effort to find the most accurate and precise answer of the queries in the very interactive manner. Our system will be marked as successful once our chatbot is ready to interact with the audience of the Slack platform. We are planning to add functionality of buttons as well as interactive messages in our system which will provide the experience of communicating with an intelligent, sensible, and helpful human being. We will be using different databases, that is, Google, Twitter [7] as well as Vizerto's database for accessing the queries of the user. Suppose a user has a doubt for some technical term, let's take encryption and this occurs while chatting with some friends in Slack. The user will most probably open a new tab, search it in Google, or go for Google chrome to get this job done. But, suppose we have a chatbot which constantly listens to the chats and responds when someone specifically calls it by using some trigger. In this case the user may call the chatbot, and chatbot will simply search it in the Google and provide the answer it gets from the Google to the user. It will solve the problem for user without him wasting too much time of his in opening new tabs or app. Our chatbot does the same, and the above-mentioned example is one of the case studies of our proposed bot (Fig. 1).

3 Working

Our system aims to reduce the complexity of the user by proving precise and accurate answers to user's query without changing the platform. We are assuming that the user is already registered to Slack [1], else first of all he will create an account in Slack platform to use it. Once the account is created, the user will login in Slack platform. Different applications are integrated with Slack. The user can independently install any application according to his/her needs. One of the applications that we are going to develop is Vizerto (Question and Answer Application) which we are going to integrate with Slack. The user can install this application independently as

Fig. 1 Architecture

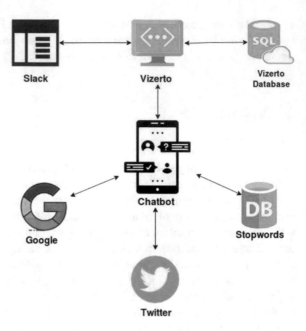

per the need. Once installed, the user has to register with this application to continue. Once registered successfully, the user will login into the application which will keep track of user's history and will permit to access the database/results searched in this application's database (Fig. 2).

Fig. 2 Working

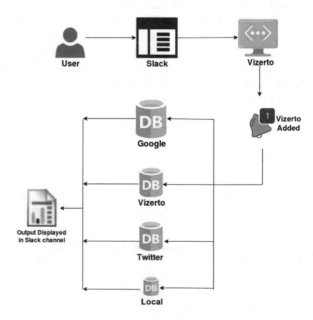

Once the required permissions are granted to application, chatbot will be automatically added to the user's workspace. This bot could only be accessed using the keyword Vizerto, that is, chatbot will only respond if the user has mentioned its name (e.g. Hey @Vizerto). Query will be taken as input and will be processed using NLP for optimized output. We will be removing stop words that are stored in our local database, from the query to generate optimized query.

Results of all the queries will be generated by Vizerto's (our application name) database by default. We have provided options to the user to search in Vizerto or Google [4] or Twitter [7]. On choosing the desired option, the processing will take place and outputs will be displayed.

The scope of our system is solely based on the idea to reduce the application fatigue and provide the user a friendly interface to reduce the complexity of incomprehensible or inconceivable search results of the query. It empowers the team work. It is mainly designed in the benefit of Enterprise or IT workers where they can easily share their work and bot will be an add-on feature which provides them the result of any query in a moment of time. Bot is designed in such a way that its frontend will look like interactive messages which will keep track of your day-to-day activities and will provide different notifications accordingly. You can provide any task to the bot and accordingly it will update or produce the result. Our system has used Vizerto's database which is upcoming application in the market which will surely make a great impact in IT Industry which will increase the demand of our system. This can be further used in multiple applications of different domains to enhance the learning as well as creativity in young minds.

4 Results and Analysis

The perceived output of the system is the successful integration of the different applications, bot creation, and precise and accurate output of the queries. That is, without switching to the different platforms or tabs, one can access multiple applications together. It is also called as removal of application fatigue from the platform. Through our system, the user can use Vizerto, Google as well as Twitter application for scrutinizing their queries by simply asking questions in Slack channel in a specific format, which includes the bot-name. Mentioning the name of the bot signifies that the particular query is for bot and not meant for any other user and hence bot will generate the precise, accurate, and optimized output of the query accordingly. For example: Hey Vizerto! How is the weather today? In this way, Vizerto will act like another member of the team.

This is how the working takes place. Step wise procedure is shown as follows:

(1) First, we get the dropdown menu, then we have got three options to select such as Google search, Twitter, Our App (we had built our own API in terms to know the current weather details). "Our App" can be any other application that can be integrated with slack channel. This is shown in Fig. 3.

Fig. 3 Home page

(2) In this, we have selected "Twitter" option, after selecting we got some dynamic buttons which consist of some functionalities of Twitter, such as twitting message, tweet analysis, and get friends list. If you tweet the message you will get a reply as "Your message has been tweeted". This is shown in Fig. 4.

(3) Now we can even perform "tweet analysis". Tweet analysis includes positive and negative tweets related to some particular topic. This functioning is shown in Fig. 5a, b.

(4) In this next option, that is "Our App" we have searched for the temperature details of Pune. Figure 6 shows about the longitude, altitude, and temperature of the searched content (Pune).

Fig. 4 Tweet message

(a)

(b)

Fig. 5 a Tweet analysis. **b** Tweet analysis

5 Conclusion

The motto of this system was to demonstrate how users in the Slack group are
engaging with third-party application that creates a seamless experience without a
user ever needing to install another application, or create an alternate user id and get
oriented on how to use. Most users are concerned about sign in credentials, security

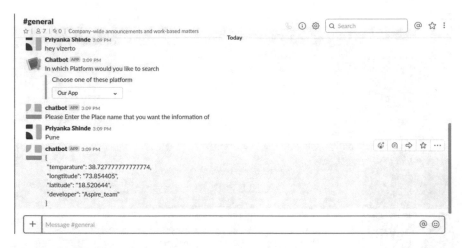

Fig. 6 Our own API to display temperature details

aspects of different applications, and time-consuming nature of switching from one application to another. These truly frictionless experiences can be achieved through engaging directly with Slack. Hence, we have successfully integrated our chatbot (application) with Slack as well as with Vizerto application.

References

1. Slack: https://slack.com/
2. https://www.uber.com/in/en/
3. https://www.google.com/maps/
4. https://www.google.com/
5. https://www.facebook.com/
6. https://en.wikipedia.org/wiki/Application_programming_interface
7. https://www.twitter.com/

A Cost-Effective Sampling Strategy for Monitoring Soil pH Value in Small-Scale Smart Farming

Ishita Bhakta, Santanu Phadikar, Himadri Mukherjee, Koushik Majumder, Kaushik Roy and Arkaprabha Sau

Abstract Field variability assessment is an important part of smart farming. The precise application of agricultural resources depends on the precise estimation of field variability. Traditionally, field variability is measured by grid-based soil sampling strategy. This method is not cost-effective for small-scale fragmented land structure. The cost can be reduced for small-scale farming by minimizing the sample size and using standardized sensors for field variability assessment. An effective sampling strategy can reduce the sample size. This study compares non-aligned systematic and simple random techniques with standard grid sampling method to measure the pH variability of a small-scale field of size 20 m × 10 m in West Bengal, India. The comparison result shows that non-aligned systematic sampling can assess the whole field pH variability with only 10 samples (p-value 0.650). Thus, the proposed method reduces the cost of soil testing by minimizing the sample size.

Keywords Precision agriculture · Cost-effective sampling strategy · Non-aligned systematic sampling

I. Bhakta (✉) · S. Phadikar · K. Majumder
Department of Computer Science, Maulana Abul Kalam Azad University of Technology, Kolkata, WB, India
e-mail: ishita.official@gmail.com

S. Phadikar
e-mail: sphadikar@yahoo.com

K. Majumder
e-mail: koushikwbutcse@gmail.com

H. Mukherjee · K. Roy
Department of Computer Science, West Bengal State University, Kolkata, India
e-mail: himadrim027@gmail.com

K. Roy
e-mail: kaushik.mrg@gmail.com

A. Sau
Factory Advice Service and Labour Institutes, Regional Labour Institute, Kanpur, India
e-mail: arka.doctor@gmail.com

© Springer Nature Singapore Pte Ltd. 2020
V. Bhateja et al. (eds.), *Intelligent Computing and Communication*,
Advances in Intelligent Systems and Computing 1034,
https://doi.org/10.1007/978-981-15-1084-7_54

1 Introduction

India is one of the fastest developing and densely populated countries in the world. According to 15th Indian Census, 2011, India is the second largest populated country with over 1.2 billion people [1]. Economy of this democratic developing country mostly depends on Agriculture, Industry, and Services [2]. More than 50% of population is engaged in agricultural activities here [3] and it contributes 13.9% of India's gross domestic product (GDP) [3]. Though population increases day by day, field area, water, and other natural resources remain the same. Providing quality food product to this large number of population with constraint resources is a challenging issue. Moreover, traditional farm management practices measure field variability by observation and store the information in memory or paper. In this case, farmers apply resources at constant rate in large amount to get better production. But this indecorous use of resources causes low fertility, soil erosion, and environmental pollution. Indian agriculture and crop production are affected by various factors like climatic changes, natural calamities, pest and disease infestation, and limited water resources due to global warming, and so on [4]. Precision agriculture (PA) is a modern concept of automated site-specific farming, which can overcome these challenges ensuring sustainable development [5]. PA uses modern information and communication technologies to fulfill its goal. Indian farmers have small-scale and fragmented field in which adoption of PA's standard technologies is challenging in terms of cost, time, scope, and so on [6]. Successful acceptance of PA in Indian agronomy needs large amount of experiments, analysis, and interventions [7]. PA concept depends on information collected from field for precise resource allocation. Data collection for large field is a tedious job. For this reason, sampling strategies are applied to estimate the characteristics of whole field with a small number of samples. So to make PA cost-effective and accurate for small-scale farmer, efficient soil sampling strategy is needed.

The objective of this research work is to propose an efficient, cost-effective, and user-friendly soil sampling method for small-scale fragmented field management by measuring field and environmental variables automatically with a low-cost system. The purpose of this work is to serve farmers in remote area of India with fragmented land structure in low-cost budget.

To fulfill this objective, two sampling strategies are compared with respect to the standard grid sampling method for a small field of Paschim Medinipure district in West Bengal. At first, a hand-held device is used to find geographical locations of the field area with GPS. Then sampling points are chosen with simple random sampling methods and non-aligned systematic sampling methods. As a case study, here the pH value of the considered field is selected for the experiment. The result shows that non-aligned systematic sampling gives better result with minimum sample points. Statistical significant differences among these methods are figured out with t-test result. The main advantage of this system is that it is free from disadvantages of traditional sampling strategies and cost-effective. Unavailability of good and cost-effective sampling strategy for small and fragmented land structure encourages us to propose such a method.

Rest of the paper is organized as follows: Sect. 2 describes background of the study reviewing literature in terms of existing sampling methodologies with their advantages and disadvantages. Section 3 defines the materials and methods used for the study. Section 4 discusses the results. Section 5 concludes the article with proper future directions in Sect. 6.

2 Research Background

Precision agriculture introduces a new era in agricultural field all over the world [8]. The first, and the most important, stage of PA is field variability assessment. It is a tedious job as soils from all over the field need to be collected to test the properties of the soil. But it is not cost-effective to test the soil from the whole field. Statistical method is required to estimate whole field variability from some sample location. Thus, a good sample strategy is needed to estimate the field variability in real time and cost-effective way.

In this section literatures are reviewed to find out the contribution of research in sampling methodology for PA application. All the methods are analyzed with respect to small and fragmented land structure. The goal of this section is to find out the disadvantage of existing soil sampling methods and proposed a solution for small and fragmented land structure. Table 1 presents the considered sampling strategy, parameters, field, and sample size of each reviewed papers.

Most of the sampling strategies in the reviewed literatures followed grid approach and laboratory testing of soil. This traditional method has some disadvantages like it averages the variability of soil properties within a cell, causes systematic error like uneven fertilizer distribution and biased error due to localized irregularity, no appropriate grid size, and so on. Except that, traditional way measures soil properties in laboratory which is a very costly process. For this reason, an unaligned, unbiased, and cost-effective sampling method is required for small-scale farmers in remote village. Statistically, this can be achieved through simple random sampling and non-aligned systematic sampling methodology.

3 Materials and Methods

A simple sampling technique has been proposed in this study for precise estimation of field variability for small-scale field with the goal of predicting whole field variability from a minimum number of samples with minimum cost and effort. Two statistical sampling techniques, simple random sampling and non-aligned systematic sampling are compared in this study with the standard grid sampling method with respect to

Table 1 Comparison of reviewed literatures with respect to sampling strategy, considered parameters, sample size, and field size

References	Sampling strategy	Parameters	Sample size	Field size
[9]	Grid	pH, P and K	413	40 acre
[10]	Grid	N, P, K, Mg, Ca and S in plants and soil	124	375 m × 250 m
[11]	Intensive grid and zone	K, P, pH and other organic component	2496	20.8 ha
[12]	Grid and zone	Electrical conductivity and nutrient content	224 and 396	15 ha
[13]	Grid and transect sampling	pH	334 for Grid. 72 along north–south transect and 73 along east–west transect	50 ha
[14]	Strip and simple random sampling	Nitrogen and phosphorus	Strip length 62 and 37 and 81 sample points in simple random sampling	4 ha

pH value of a field in a remote village of West Bengal, India. The experimental environment with used technologies is described in Table 2.

At first, the field of study (Fig. 1) needs to be delineated using a GPS device. An Arduino-based GPS-enabled hand-held device is used to find out the geo-positions of the field boundary. These positions are saved for further calculation. A field polygon

Table 2 Description of the experimental environment

Environmental parameters	Description
Field geo-location	22.530588, 87.756648
Field location	Vill.–Chak-sultan, P.O.–Panchberia, Dist.–Paschim Medinipur, Pin–721146, West Bengal, India
Field size	20 m × 10 m
Device and sensor used	Global positioning system (GPS) device (U-blox NEO-6 M GPS Module), Arduino Mega 2560, radio frequency (RF) transmitter and receiver module (FS1000A 433 MHz), pH sensor
Software used	R 3.3.2, Matlab 2014b, Arduino IDE 1.8.9
Standard database for pH value of the field	From http://www.bhoomigeoportal-nbsslup.in/ website [15]

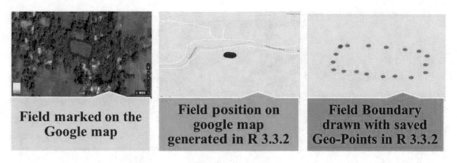

| Field marked on the Google map | Field position on google map generated in R 3.3.2 | Field Boundary drawn with saved Geo-Points in R 3.3.2 |

Fig. 1 Field position on Google map with defined boundary

has been generated with these points in R 3.3.2 software surrounded with red dot at its boundary (Fig. 1).

In statistics the magic number 30 is used as minimum number of sample size for simple random sampling method [16]. So, in this study, first 30 sample points are selected over the field in both non-aligned systematic sampling and random sampling methods. Then the sample size is reduced gradually to 25, 20, 15, 10, and 5 for non-aligned systematic sampling method. Figure 2 shows the field with selected sample points for non-aligned systematic sampling method. Figure 3 shows the field with sample points for simple random sampling. The soil pH value is measured with the sensor at each sample point and stored in a database using the Arduino-based data collection module for further application. This collected value is compared with the standard pH value of the selected field available in ICAR-NBSS&LUP, Nagpur, Govt. of India database [15]. Then the difference between the two sampling methods and standard grid method is shown with respect to mean pH value at sample points and number of samples. As the pH value is directly related to all other variabilities in a field and standard pH value of the field is already available of this field, so only this parameter is chosen for the case study.

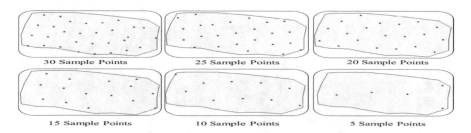

Fig. 2 Non-aligned systematic sampling method with 30, 20, 15, and 10 sample points generated in R 3.3.2

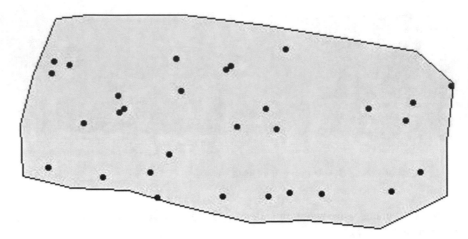

Fig. 3 Simple random sampling method with 30 sample points generated in R 3.3.2

4 Results and Discussion

At first, 30 samples were taken with simple random sampling method and non-aligned systematic sampling method. The mean pH value for all these sample points is compared with the standard one (8.6). Then the number of sample points is reduced with five intervals in non-aligned systematic sampling method until the number of sample points reaches 5. The mean pH value in each case is compared with the mean pH value of simple random sampling method for 30 sample points and standard one. Figure 4 shows this comparison of mean pH value. The result shows that if the number of sample points is smaller (5) than 10, then there is statistical significant difference between the standard value and the measured one. The statistical significance of the results is shown in Table 3 with the summary statistics. T-statistics is used to find out the statistical significance between the mean values. It shows that there is no statistical

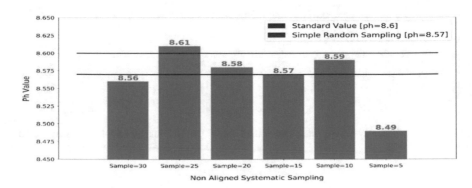

Fig. 4 Comparison of sampling strategies with respect to mean pH value at sample points

Table 3 Summary statistics of the comparison between non-aligned systematic sampling and standard grid sampling method

Number of samples	Non-aligned systematic sampling (mean ± SD)	Standard pH value (mean)	p-value at 95% confidence interval
30	8.56 ± 0.12	8.6	0.078
25	8.61 ± 0.14		0.698
20	8.58 ± 0.09		0.233
15	8.57 ± 0.11		0.146
10	8.59 ± 0.12		0.651
5	8.45 ± 0.11		0.038

significant difference between the mean pH value in non-aligned systematic sampling and standard pH value for 30 ($p = 0.078$), 25 ($p = 0.698$), 20 ($p = 0.233$), 15 (0.146), and 10 ($p = 0.651$) number of sample points as the p-value is >0.05 in each case with 95% confidence interval. But for five sample points the statistical significant difference is found between standard population mean and calculated mean with p-value (0.038) < 0.05 at 95% confidence interval. So, the minimum number of sample points chosen is 10 for the small-scale field with respect to pH value. Table 4 presents the summary statistics of the comparison between non-aligned systematic sampling and simple random sampling with respect to pH value. According to the statistics in this table, the minimum number of sample points is 10. There is also no statistical significant difference between standard value and mean pH value in simple random sampling method with p-value 0.1814 at 95% confidence interval for 30 samples. As according to statistics the sample size cannot be <30 for simple random sampling method [16], so the sample size is not reduced further for this method.

The non-aligned systematic sampling is more advantageous and cost-effective than simple random sampling method and grid sampling method as it spreads evenly all over the population and reduces sample size [17]. Thus it minimizes the error due to the un-sampled and over-sampled area of the field. It is cost-effective because it

Table 4 Summary statistics of the comparison between non-aligned systematic sampling and simple random sampling method

Number of samples	Non-aligned systematic sampling (mean ± SD)	Simple random sampling (mean ± SD)	p-value at 95% confidence interval
30	8.56 ± 0.12	8.57 ± 0.12	0.748
25	8.61 ± 0.14		0.259
20	8.58 ± 0.09		0.752
15	8.57 ± 0.11		0.999
10	8.59 ± 0.12		0.650
5	8.45 ± 0.11		0.041

reduces the huge cost of soil testing in laboratory by using low-cost standardized smart sensors at the selected sample points. It also reduces the number of sensors used.

5 Conclusion

This study helps to find out the minimum number of sample points for a small-scale fragmented land. A case study is done based on pH value of a field in a remote village, ChakSultan of Paschim Medinipure, West Bengal. The non-aligned systematic and simple random techniques and the standard grid sampling method are compared to measure the pH variability of this field with minimum sample points. The comparison result shows that minimum 10 samples (p-value 0.650) are required for this field to measure the field variability with respect to pH value in non-aligned systematic sampling method. This sample size is very low compared to the traditional grid sample method and simple random sampling method (30). The statistical significance of this result has also been carried out to show the reliability of the method. The proposed method shows that non-aligned systematic sampling is more efficient than grid sampling and simple random sampling for sample point calculation in a small-scale field. It is a cost-effective method as it reduces the number of sample point for the selected field and thus reduces the number of sensing module used for field variability assessment. More experiments are required for other field variabilities to prove the efficacy of the proposed model.

6 Future Work

The proposed method can be applied for calculating sample points for smart farming. Field variability can be measured using this method with minimum sample points in a small-scale field. The measured variability will help in taking decision about precise agricultural resource allocation in a cost-effective way. Moreover, pH is an important variable of soil which affects many chemical processes related to soil and plants. Plant growth, nutrient content and so on are affected by the pH variability. So, this study will be very helpful in designing smart technology for precision farming.

References

1. Census of India: Size, Growth Rate and distribution of Population. Provisional Population Totals (2011)
2. Misra, S., Puri, V.: Indian Economy (2011)
3. Wagh, R., Dongre, A.: Agricultural sector: status, challenges and it's role in Indian economy. J. Commer. Manag. (2016). www.indianjournals.com

4. Kumar, K.S.K., Parikh, J.: Indian agriculture and climate sensitivity. Glob. Environ. Chang. **11**(2), 147–154 (2001)
5. Bongiovanni, R., Lowenberg-Deboer, J.: Precision agriculture and sustainability. Precis. Agric. **5**(4), 359–387 (2004)
6. Mondal, P., Basu, M., Bhadoria, P., Emama, A., Salih, M., Adegbite, A.: Critical review of precision agriculture technologies and its scope of adoption in India. Am. J. Exp. Agric. **1**(3), 49–68 (2011)
7. Mandal, S., Maity, A.: Precision farming for small agricultural farm: Indian scenario. Am. J. Exp. Agric. **3**(1), 200 (2013)
8. Srinivasan, A.: Handbook of Precision Agriculture: Principles and Applications. CRC Press (2006)
9. Franzen, D.W., Peck, T.R.: Field soil sampling density for variable rate fertilization. JPA **8**(4), 568 (1995)
10. Jordan, C., Shi, Z., Bailey, J.S., Higgins, A.J.: Sampling strategies for mapping 'within-field' variability in the dry matter yield and mineral nutrient status of forage grass crops in cool temperate climes. Precis. Agric. **4**(1), 69–86 (2003)
11. Mallarino, A.P., Wittry, D.J.: Efficacy of grid and zone soil sampling approaches for site-specific assessment of phosphorus, potassium, pH, and organic matter. Precis. Agric. **5**(2), 131–144 (2004)
12. Li, Y., Shi, Z., Wu, C., Li, H., Li, F.: Determination of potential management zones from soil electrical conductivity, yield and crop data. J. Zhejiang Univ. Sci. B **9**(1), 68–76 (2008)
13. McCormick, S., Jordan, C., Bailey, J.S.: Within and between-field spatial variation in soil phosphorus in permanent grassland. Precis. Agric. **10**(3), 262–276 (2009)
14. Hu, W., Schoenau, J.J., Si, B.C.: Representative sampling size for strip sampling and number of required samples for random sampling for soil nutrients in direct seeded fields. Precis. Agric. **16**(4), 385–404 (2015)
15. Government of India and Nagpur: Bhumi Geoportal—A Gateway to Soil Geospatial Database. NBSS&LUP (2002). http://www.bhoomigeoportal-nbsslup.in/. Accessed 20 Mar 2019
16. Kar, S., Ramalingam, A.: Is 30 the magic number? Issues in sample size estimation. Natl. J. Community Med. **4**(1), 175–179 (2013)
17. Thompson, S.K.: Sampling. Wiley (2012)

Hybrid Music Recommendation System Based on Temporal Effects

Foram Shah, Madhavi Desai, Supriya Pati and Vipul Mistry

Abstract Use of music recommendation system is fully grown because of large number of online music websites. The most challenging gap we found is that there is an utmost need to consider more contextual information like weekdays, session of day, time, and frequency of listening songs. In this paper, we propose a hybrid music recommendation system which is a combination of two approaches. In the first approach we propose to use cosine similarity measure for ranking of music. The second approach considers the graph-based approach. In the graph-based approach we propose using particle swarm optimization with differential evolution to get optimized ranking of music. We recommend top-n songs by combination of these two approaches. Standard Last.fm dataset is considered for experimental purpose. Data pre-processing operation is performed on dataset to remove the noisy and inconsistent data. Comparison of our proposed model with the state-of-the-art model shows the effectiveness in the form of recall rate.

Keywords Hybrid · Music · Recommendation system · Collaborative · Context · Filtering · Temporal effects · Cosine similarity · Pearson similarity · Differential evolution · Particle swarm optimization

F. Shah (✉) · M. Desai · S. Pati
Department of Computer Engineering, Chhotubhai Gopalbhai Patel Institute of Technology,
Uka Tarsadiya University, Bardoli, India
e-mail: foramshah943@gmail.com

M. Desai
e-mail: desaimadhavi30@gmail.com

S. Pati
e-mail: supriya.pati@utu.ac.in

V. Mistry
Electronics and Communication Engineering Department, S.N.P.I.T & R.C, Umrakh, Gujarat,
India
e-mail: vipul.vpl@gmail.com

© Springer Nature Singapore Pte Ltd. 2020
V. Bhateja et al. (eds.), *Intelligent Computing and Communication*,
Advances in Intelligent Systems and Computing 1034,
https://doi.org/10.1007/978-981-15-1084-7_55

1 Introduction

There are number of choices available online. The recommendation system helps to find out the items related to the user's interest. There are many applications of recommendation system. One of them is music recommendation system. It tries to predict which songs users would like to listen [1]. Music has many attributes since everyone may have different choices. For example, someone likes listening to songs of particular artist or actor, or according to particular genre. Hence, it is not easy to find out which song will be listened next. There are number of songs available online and it is difficult to find out which song the listeners can listen according to their interest. Here, there is a need of a music recommendation system which can recommend songs to the listeners based on different parameters like artist, time, genre, and so on. A proposed hybrid music recommendation system model is discussed in this paper which removes the research gap based on our observations of existing literature.

Rest of the paper is organized as follows: Sect. 2 describes the related works in music recommendation system. Section 3 describes proposed hybrid music recommendation system. Experimental setup and result analysis of proposed method are described in Sects. 4 and 5, respectively. Conclusion and future work is described in Sect. 6.

2 Related Works

In the last decade, researchers worked on music recommendation system and found there is a requirement of more contextual information of songs. Ricardo et al. [2] proposed a general meta-heuristic differential evolution approach to get the recommended songs. Aristomenis et al. [3] proposed a cascade-hybrid music recommendation system in the context of mobile. Recommendations are done on the basis of genre and personality. They have used two quantitative parameters: one is mean absolute error (MAE) and the other ranked scoring (RS).

Negar et al. [4] presented a hybrid context-aware music recommendation system. It summarizes the sequence of listened songs to get better results. The evaluation parameter used here is hit ratio. They have used Last.fm dataset. Thomas et al. [5] presented a weighted hybrid recommendation method and added serendipity to the generated recommendations. By this, users can benefit from fact-finding diversification using Last.fm dataset. They have calculated track-similarity, tag-similarity, and time-similarity. The authors have suggested to analyze the full dataset of Last.fm.

Marcelo et al. [6] proposed an approach for concurrently tuning various types of controllers doing more than one operation at the same time. They have used parallel computing approach and mentioned that the use of parallel computing is more beneficial than non-parallel structure. Ben et al. [7] presented a hybrid music recommendation system (HMRS) that performed integration on the pseudo-tags for representation of a track. They have taken CAL500 dataset consisting of 500 songs, and annotations given on 174 tags (e.g. mood, instrument, genre, vocal). Manju et al. [8] presented an actual evaluation of the execution of various evolutionary computational techniques to anticipate the limitation of the software over multiple datasets.

Rahul et al. [9] proposed a hybrid music recommendation system by a graph-based approach using sessions and temporal features. They have used Pearson correlation coefficient and graph-based approach for recommendation of songs. For improvement of ranking of recommendation of songs, the author has used particle swarm optimization. Asela et al. mentioned in their work that Pearson correlation coefficient gives lower accuracy [10]. Keunho et al. mentioned that the use of cosine similarity measure gives good results [11]. Wang et al. mentioned that popular differential evolution algorithm improves the search capability and speed [12].

From the above literature survey, we observed that more contextual information with cosine similarity measures and differential evolution can give better recommendation system.

3 Proposed Hybrid Music Recommendation Model

Schematic diagram of the proposed recommendation system is shown in Fig. 1. There are mainly three layers: user-context layer, decision-context layer, and song-context layer. User-context layer works on the basis of which kind of songs users would like.

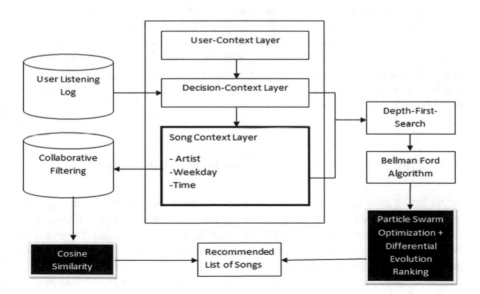

Fig. 1 Schematic diagram of hybrid music recommendation system

It includes feature of the song (e.g. note, tempo, pitch, harmony, rhythm). Decision-context layer takes user's listening logs, and on the basis of that, it will take decision that which songs can be recommended to the user. Song-context layer contains three attributes: artist, weekday, and time.

Here are major components of our proposed approach explained as follows.

3.1 Collaborative Filtering Using Cosine Similarity Measure

Collaborative filtering is a more effective technique in recommendation system which works on similarity score of other users. To find out track-similarity, we have used cosine similarity measure. In [9] implementation of music recommendation system is done using Pearson correlation coefficient, but cosine similarity measure is the most commonly used method in collaborative filtering in the recommended systems [13].

$$\text{Cos Sim}(x, y) = \frac{\sum_i x_i y_i}{\sqrt{\sum_i x_i^2}\sqrt{\sum_i y_i^2}} = \frac{\langle x, y \rangle}{\|x\|\|y\|} \tag{1}$$

Here, as shown in Eq. 1, x and y are two users, x_i and y_i are ratings of the songs given by the users. Cosine similarity finds how two vectors x and y are related to each other measuring cosine angle between these vectors.

Pearson correlation coefficient calculates similarity by the use of all items. It considers that rating has been given to each item. Thus, this system uses negative (did not use) scores, which considerably outnumber positive (used) scores. Therefore, it can be said that the Pearson correlation coefficient results in lower accuracy. Hence, cosine similarity gives better accurate results compared with Pearson correlation coefficient, as mentioned in [10, 11]. It also gives less error ratio than the Pearson correlation coefficient.

3.2 Depth-First Search and Bellman-Ford Algorithm

Depth-first search (DFS) technique is used to search different nodes in a graph. Initially, it finds out the newly designed vertex's edges. After traversing all the edges, if any vertex remains unvisited, depth-first search takes one as a latest source and begins to search from that node. This algorithm does not leave any vertex unvisited. Different edges and nodes are used to indicate the different parameters (users, artists, weekdays, and time). We have also taken different colors to differentiate those values. Bellman-Ford algorithm is used to find out the single-source shortest or longest path. Like depth-first search, it does not find the path from source vertex, but also considers neighbor vertex. This helps to decide whether the graph needs to be explored [9]. It searches for the artists who are less listened by the users, or at which weekday, or on which time, the user listened less. It eliminates those negated edges and does not take those edges (those artists, weekdays, and timing sessions). Remaining edges can be positively taken into consideration for recommendation. With the help of that, a list of songs will be recommended to the user, which is based on the artists, weekdays, and time according to the user preferences.

3.3 Optimization Using Particle Swarm Optimization and Differential Evolution

Particle swarm optimization calculates the fitness value based on birds and fishes. This algorithm calculates two values for each user: (i) *pbest* value which is personal best value. It is a group of all the neighbor users and that group has another (ii) *gbest* value, which is known as global best value. After completion of each iteration, the *pbest* and *gbest* values are changed. If the *pbest* value is better than the *gbest* value, then it replaces the *gbest* value with the *pbest* value. Hua Wang et al. [12] mentioned that particle swarm optimization with differential evaluation enhances the performance. Differential evolution counts the difference between *gbest* and *pbest* value. If the *pbest* value of particular user is more than the *gbest* value, then it replaces the *gbest* value with the *pbest* value. Rahul et al. have only used particle swarm optimization [9].

4 Experimental Setup

For the experiment setup of our proposed approach, we have used real-time Last.fm dataset [14]. It represents data of the 992 unique users' records till May 5, 2009. Total rows are 19,150,868. The main four things that have been taken into consideration are: user, timestamp, artist, and song. It consists of the attributes: userid, timestamp, artistid, artistname, trackid, and trackname. In timestamp the first part contains information about date and the second part contains information about time. The first part indicates the date, and from that it reveals the day of the week, the month, and also season. We have performed data pre-processing and data cleaning. This process removed the noisy and inconsistent data and also songs which are listened less than 10 times. The final dataset includes 992 users and 12,286 songs. April to May records are used as a training set and 1–4th May records are taken as a testing set. We have used python for implementation. Quantitative parameters used here are: MAE, root mean square error (RMSE) and recall rate.

(1) Mean absolute error (MAE): It calculates an average measure of errors in a set of predictions [15]. It is an average of an absolute difference between prediction and actual observation.

$$\text{MAE} = \frac{1}{n} \sum_{1}^{n} |e_i| \tag{1}$$

(2) Root mean square error (RMSE): It calculates an average magnitude of an error [15]. It is square root of an average of squared differences between prediction and actual observation.

$$\text{RMSE} = \sqrt{\frac{1}{n} \sum_{i=1}^{n} e_i^2} \tag{2}$$

In Eqs. (1) and (2), n is the number of songs, i is the number of users and e means number of errors. Less error ratio indicates the result is better.

(3) Recall: Recall is the percentage of total relevant results correctly classified by our algorithm.

$$\text{Recall} = \frac{Nrs}{Nr} \tag{3}$$

In Eq. (3), N_{rs} is the number of recommended songs that user prefers and N_r is the number of recommended songs. As the value of the recall is more, the results can be considered as more accurate and precise.

5 Experimental Results and Analysis

We have performed three different experiments to validate our proposed approach. The results are validated with MAE, RMSE, and recall parameters. Three different experiments are explained in the following:

Experiment 1: This experiment is the comparison of Pearson and cosine correlation coefficient on our proposed hybrid music recommendation approach. We observed from related works [10, 11] that cosine similarity measure can be used instead of Pearson correlation coefficient.

It can observed from Table 1 that both MAE and RMSE are using cosine similarity measure lesser than Pearson coefficient, Hence we can conclude that the use of cosine similarity measure will improve the recommendation system.

Table 1 Comparative analysis of Pearson and cosine similarity

Method	MAE	RMSE
Pearson correlation coefficient	1.7441	1.8392
Cosine similarity measure	**1.7335**	**1.8225**

Table 2 Comparative analysis of MAE and RMSE

Method	MAE	RMSE
Cosine similarity measure	1.7335	1.8225
Particle swarm optimization	2.4013	2.6131
Particle swarm optimization + Differential evolution	2.3190	2.5640
Rahul et al. hybrid music recommendation system [9] Pearson similarity measure + Particle swarm optimization	1.9413	2.0040
Proposed hybrid music recommendation system Cosine similarity measure + Particle swarm optimization + Differential evolution	**1.7158**	**1.8117**

Experiment 2: This experiment calculates the error rate using MAE and RMSE for our proposed hybrid music recommendation system. The results can be found in Table 2.

It can be observed from Table 2 that combining differential evolution approach with particle swarm optimization can reduce the error ratio. MAE and RMSE of the proposed approach are lesser than the existing system of Rahul et al. [9].

Experiment 3: This experiment validates that our proposed hybrid music recommendation approach gives better recall rate than existing HMRS of Rahul Katariya et al. [9].

We analyzed Table 3 and observed that the score of the recall ranking for top-1, top-5, top-10, and top-20 recommended songs of the proposed HMRS is comparatively good than the existing HMRS [9].

From the experiments mentioned above in Sect. 6, we found the ranked list of top songs as mentioned in Fig. 2. List of the recommended songs according to ranking is:

1. Karma police.
2. Somebody told me.
3. Paranoid android.

In the following section, we have given concluding remark and future direction.

Table 3 Comparative analysis of recall ranking in song recommendation

Method	Top-1	Top-5	Top-10	Top-20
Rahul et al. HMRS [9]	0.0355	0.1621	0.2158	0.3321
Proposed HMRS	**0.4782**	**0.5652**	**0.6956**	**0.7142**

```
For  Paranoid Android , top matches are:
0 Paranoid Android | Sim 0.9423076923076923 | Support 49
1 Karma Police | Sim 0.33860465116279065 | Support 40
2 Somebody Told Me | Sim 0.0975483870967742 | Support 28
For user user_000012  top choices are:
Paranoid Android | Sim 1.0 | Support 49
Karma Police | Sim 0.364 | Support 40
Somebody Told Me | Sim 0.108 | Support 28
Similar songs to Paranoid Android:
   Somebody Told Me  0.051535108551375663
   Karma Police  -0.06769925184477987
```

Fig. 2 Ranked list of first three recommended songs

6 Conclusion and Future Work

In this paper, we have introduced a hybrid music recommendation system that uses cosine similarity measure and graph-based approach with optimization technique. We have proposed to use two methods for ranking of music in recommendation system. The first method uses cosine similarity measure to find track similarity. In the second approach we proposed to use particle swarm optimization with differential evolution to get optimized ranking on ranked list by graph-based model of depth-first search and Bellman-Ford algorithm. From the comparative analysis with the state-of-the-art model, we can observe that the use of cosine similarity measure instead of Pearson correlation coefficient gives less MAE. Moreover, the use of differential evolution with particle swarm optimization gives better recall ranking. In future, one may consider the duration of time or social behavior as contextual information for better music recommendation system.

References

1. Deshmukh, P., Kale, G.: A survey of music recommendation system. Int. J. Sci. Res. Comput. Sci. Eng. Inform. Technol.© IJSRCSEIT, ISSN: 2456-3307 **3**(3), 1721–1729 (2018)
2. Sergio, R, Cesar, R., Gadelha, F., Magela, O.: A new differential evolution based metaheuristic fordiscrete optimization. Int. J. Nat. Comput. Res. 15–32 (2010)
3. Lampropoulos, A., Lampropoulou, A., Tsihrintzis, G.: A cascade-hybrid music recommender system for mobile services based on musical genre classification and personality diagnosis. Multimed. Tools Appl. **59**(1), 241–258 (2012)
4. Hariri, N., Mobasher, B., Burke, R.: Using social tags to infer context in hybrid music recommendation. In: ACM Twelfth International Workshop on Web information and data management, pp. 41–48 (2012)
5. Hornung, T., Ziegler, C.-N., Frenzy, S., Przyjaciel-Zablocki, M., Schatzle, A., Lausen, G.: Evaluating hybrid music recommender systems. In: 2013 IEEE/WIC/ACM International Conferences on Web Intelligence (WI) and Intelligent Agent Technology (IAT), pp. 57–64 (2013)
6. Favoretto, M., Carlos, S., Sipoli, D., Ribeiro, C., Andrade, R.: Automatic tuning of PSSs and PODs using a parallel differential evolution algorithm. Int. J. Nat. Comput. Res. 1–16, (2014)

7. Horsburgh, B., Craw, S., Massie, S.: Learning pseudo-tags to augment sparse tagging in hybrid music recommender systems. Artif. Intel. **219**, 25–39 (2015)
8. Khari, M., Kumar, P.: Evolutionary computation-based techniques over multiple data sets: an empirical assessment. Arab. J. Sci. Eng. **43**(8), 3875–3885 (2018)
9. Kataria, R., Verma, O.P.: Efficient music recommender system using context graph and particle swarm. Multimed. Tools Appl. **77**(2), 2673–2687 (2018)
10. Gunawardana, A., Shani, G.: A survey of accuracy evaluation metrics of recommendation tasks. J. Mach. Learn. Res. **10**, 2935–2962 (2009)
11. Choi, K., Suh, Y.: A new similarity function for selecting neighbors for each target item in collaborative filtering. Knowl. Based Syst. **37**, 146–153 (2013)
12. Wang, H., Zuo, L.-L., Liu, Z., Yi, W.-J., Niu, B.: Ensemble particle swarm optimization and differential evolution with alternative mutation method. Nat. Comput. 1–14 (2018)
13. Agarwal, A., Chauhan, M.: Similarity measures used in recommender systems: a study. Int. J. Eng. Technol. Sci. Res. (IJETSR) ISSN 2394–3386 **4**(6), 619–626 (2017)
14. Music Recommendation System Dataset. http://www.dtic.upf.edu/~ocelma/MusicRecommendationDataset/
15. Chai, T., Draxler, R: Root mean square error (RMSE) or mean absolute error (MAE)?—arguments against avoiding RMSE in the literature, pp. 1247–1250 (2014)

Differential Huffman Coding Approach for Lossless Compression of Medical Images

Arjan Singh, Baljit Singh Khehra and Gursheen Kaur Kohli

Abstract Medical images form a vital part of a patient's record in medical centers. Medical imaging devices generate data with huge memory requirements. Medical image compression is mandatory for storage and communication of medical data for the purpose of diagnosis. Compression removes the extraneous and redundant data in an image to reduce the storage cost as well as data transmission cost. Compression involves removing coding, interpixel or psychovisual redundancy in an image and, at the same time, retaining the integrity of the information required for the diagnosis in medical images. Lossless compression assures exact reconstruction of the original image after decompressing it and provides greater quality but lesser compression ratio. This paper presents two approaches for lossless compression of medical images. In the first approach, Huffman coding is implemented directly on medical images, whereas in the second approach, differential coding is applied on medical images before implementing Huffman coding. Experimental results show that differential Huffman coding improves the compression ratio.

Keywords Lossless compression · Huffman coding · Differential Huffman coding

1 Introduction

Medical imaging technique involves creation of visual representations of internal body structures used to assist the diagnosis or treatment of different medical conditions. Various types of medical imaging include X-ray, ultrasound, magnetic resonance imaging (MRI), computerized tomography (CT) and so on.

A. Singh
Punjabi University, Patiala, Punjab, India
e-mail: arjanpu@gmail.com

B. S. Khehra (✉) · G. K. Kohli
BBSB Engineering College, Fatehgarh Sahib, Punjab, India
e-mail: baljitkhehra@ieee.org

G. K. Kohli
e-mail: gursheenkohli@gmail.com

© Springer Nature Singapore Pte Ltd. 2020
V. Bhateja et al. (eds.), *Intelligent Computing and Communication*,
Advances in Intelligent Systems and Computing 1034,
https://doi.org/10.1007/978-981-15-1084-7_56

Data generated by medical imaging devices have huge memory requirement. The diagnostic images for radiologic analysis need efficient storage and transmission for future medical diagnosis or legal purposes [12]. Huge electronic files are produced by digital medical image processing that contains a large amount of data. For efficient electronic transmission and to reduce memory requirement, compression of medical images into files of smaller size is essential. Medical data must be compressed in a way that no medical information is lost and its diagnostic capabilities are not affected. Lossless compression assures that originality of the data should be maintained during the compression and decompression process. Lossy compression methods suffer from loss of quality to obtain more compression ratio. Lossless compression provides greater quality but lesser compression ratio [2].

Image compression removes the extraneous and redundant data in an image to reduce both storage space and transmission time required for archival and communication of data. It involves the removal of redundant data. Data redundancy means containing data that either provides irrelevant information or restate already known information. Compression involves removing coding, interpixel or psychovisual redundancy in an image.

Data compression process involves removal of redundant data that leads to low storage and transmission cost. There are two major categories of data compression methods: lossless and lossy methods [2]. Lossless methods assure exact restoration of data file from the compressed one. They remove the redundant data during the compression process and add it during decompression. These methods eliminate both coding and interpixel redundancy and are used in applications where no data loss is acceptable, such as legal and medical documents and computer programs. Lossy compression methods regenerate an approximation of the original data file. They provide higher compression ratio as compared to lossless methods. Lossless methods provide better quality as compared to lossy methods. Lossy methods are used in applications where a certain amount of loss is acceptable, such as broadcast television and videoconferencing. Compression techniques may be classified as substitution, dictionary and statistical-based compression techniques. Different data compression techniques can be combined to achieve better compression ratios. For example, RLE is combined with Huffman coding in fax machines to yield compression ratios of about 10:1.

1.1 Huffman Coding

Huffman coding is an effective lossless data compression method. It belongs to the class of statistical-based compression technique. It is a variable length encoding scheme that involves assignment of fewer bits to symbols that occur more frequently and more bits to symbols that appear less often. A Huffman code is built by rearranging the characters in increasing order of frequency of occurrence and then a binary tree structure is constructed. First of all two least-frequent characters are chosen, logically grouped as one, and their frequencies are added up. Then, the two elements with lowest frequencies are selected (considering the earlier grouping as a single element), grouped together and their frequencies are added. Then the same process is continued until only one element is left [9].

1.2 Differential Coding

Differential coding is simply a pre-processing method that is applied on original image. It converts an image into its difference image, that is, it stores the difference between adjacent pixels in an image. As a result, the neighboring pixels with same intensity are stored with pixel value zero. Differential coding proves successful in cases where there exist identical areas in an image-like background of an image. Here neighboring pixels have identical intensities resulting in more consequent zeroes in differentially coded image. So encoding schemes like run-length encoding (RLE) and Huffman coding can be applied on the differentially coded image to improve compression ratio [10].

2 Review of Literature

Al-Bahadili et al. [2] described the ACW algorithm as a complementary algorithm to Huffman coding and the effect of optimum character wordlength (b) on compression ratio was examined. ACW algorithm enhanced the compression ratio by a factor of b/8, and decreased its code rate by a factor of 8/b. Liang et al. [10] implemented four lossless encoding schemes: run-length encoding, LZ77 coding, LZW coding and Huffman coding using Hilbert space-filling curve ordering. Combined encoding schemes were implemented to compare the results of different compression techniques. Combination of LZW coding and Huffman coding gave the best compression result.

Al-Wahaib et al. [3] proposed a lossless image compression algorithm using duplication-free run-length coding and rule-based generative coding method. Duplication problem of the traditional RLC algorithms was removed. The number of occurrences that could be encoded by a single run was increased to infinity. DF-RLC resulted in better compression ratio as compared to TRLC1 and TRLC2.

Ouafi et al. [13] applied Mojette transform in image compression and implemented differential coding to improve compression results. Results showed that the proposed method achieved better compression ratio and entropy. Also, Aneja et al. [4] modified the weighted coefficients of bi-orthogonal filter "bior4.4" and developed an algorithm with the new weighted coefficients. Implementation of the proposed algorithm increased the retain energy and compression ratio at the cost of little degradation in PSNR. Ramteke et al. [14] proposed LOCO-R algorithm for 16-bit image that leads to reduction in the implementation complexity and compression ratio. Use of LOCO-R algorithm for 16-bit (gray-scale) images provided improved compression results for lossless image compression as compared with other current methods. LOCO-R algorithm provided up to 50% compression for 16-bit gray-scale images.

Nandi et al. [11] proposed windowed Huffman coding which used a fixed-size window buffer to store the recently encoded symbols. Hasan et al. [7] explained a spatial domain image compression technique which uses simple arithmetic operations to attain lossless compression for digital images. An improved algorithm was presented that reduced the overhead associated with the existing algorithm and improved the compression ratio. Hasan et al. [8] also presented a customized technique for lossless image compression in spatial domain along with the limitations of the existing methods. The proposed modification achieved better compression ratio as compared to the existing method for both the gray-scale images as well as color images. Hasan et al. [6] also proposed a lossless location-based image compression algorithm. This method also gave better results than that of existing lossless image compression schemes for both the gray-scale images as well as color images.

Rehna et al. [15] presented a method for compression of binary images. Compression code was designed by taking the first pixel value and calculating the run-length. Compressed data required less memory space and lead to fast transmission over communication channel. Abdmouleh et al. [1] developed a lossless image compression algorithm that combines arithmetic coding with RLE. Higher weight bit planes were encoded with RLE and the rest of the bit planes with AC. Results were compared with the static and adaptive model. The proposed method provided better mean compression ratio for gray-scale images as compared to other arithmetic coders. Pinto et al. [12] made comparison of JPEG, JPEG 2000 with SPIHT encoding for MRI and CT images based on compression ratio and compression quality. JPEG compression method provided greater PSNR and degree of compression for CT scan images than wavelet compression method. SPIHT provided the best compression ratio and PSNR value.

Thorat et al. [17] proposed an image compression algorithm based on anatomic symmetries present in medical images. This lowered energy of sub-bands and provided resolution and quality scalability by using run-length encoder and Huffman coder to encode the residual data generated after prediction. The proposed algorithm provided 3D scalable lossless compression for medical images and improved compression ratio along with decrease in complexity. Vrindavanam et al. [18] made comparison of Huffman coding and proposed position-based coding scheme (PBCS) implemented for JPEG. PBCS is based on the position of elements in the quantized transform coefficient matrix. PBCS achieved better PSNR and image quality as compared to Huffman coding.

Saranya et al. [16] proposed a method for compression and decompression of dual-color images by removing spatial redundancy using an algorithm that uses k-means clustering, bit-map generation and run-length encoding. The proposed algorithm sharpened the image and removed border color artifacts and can be used for different image file types. Results showed that the proposed algorithm improved compression ratio and PSNR.

3   Proposed Approach

Let $A(i, j)$ be the pixel corresponding to row i and column j of an image A_{m*n}. Difference image dA_{m*n} is defined as:

$$dA(i, j) = \begin{cases} A(i, j+1) - A(i, j) \ if \ j < n \\ A(i, j) \qquad\qquad if \ j = n \end{cases} \tag{1}$$

Let $a_1, a_2, ..., a_i$ are the existing intensity values of pixels in input image of size $m*n$. Let $P(a_1), P(a_2), ..., P(a_i)$ are their corresponding probabilities of occurrences.

Let A_{m*n} be the original image. A differential Huffman code is constructed using the following algorithm:

Step 1. Apply differential coding algorithm on the original image and let dA_{m*n} be the resulting image.

Step 2. Apply Huffman coding algorithm on dA_{m*n} to get Huffman code.

Step 3. Reconstruct image from Huffman code by applying Huffman decoding and let H_{m*n} be the resulting image.

Step 4. Reconstruct original image by applying differential decoding on H_{m*n}.

Block diagram of differential Huffman coding is shown in Fig. 1. Differential Huffman coding consists of four blocks: differential coder, Huffman coder, Huffman decoder and differential decoder. An image is input to the differential coder that converts it into its difference image. This difference image is then fed into the Huffman coder that generates Huffman code for the image. This Huffman code is transmitted to receiver. The receiver uses Huffman decoder to generate an image from the Huffman code. This image is then decoded by differential decoder to reconstruct the original image.

Pseudo code for differential Huffman coding is given as follows:

```
for i=1:m
{
```

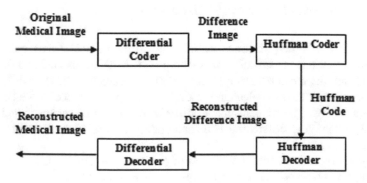

Fig. 1   Block diagram of differential Huffman coding

$dA(i,n)=A(i,n)$
$for\ j=1:n-1$
$\{$
$dA(i,j)=A(i,j+1)-A(i,j)$
$ifdA(i,j)<0$
$\{$
$dA(i,j)=dA(i,j)+256$
$\}\}\}$

$list=\{(dA_1,\ P(dA_1),\ null,\ null),\ (dA_2,P(dA_2),\ null,\ null),\ ....,\ (dA_i,\ P(dA_i),\ null,\ null)\}$

$while\ |list|>1$
$\{$
$a=min_p list$
$list=list-\{a\}$
$b=min_p list$
$list=list-\{b\}$
$c=(null,P(a)+P(b),b,a)$
$list=list+\{c\}$
$\}$
$hcode="\ "$
$for\ i=1:m$
$for\ j=1:n$
$\{\{$

$hcode=hcode+Huffman\ code\ for\ pixel\ corresponding\ to\ i^{th}\ row\ \&j^{th}\ column\ in\ image$

$\}\}$

4 Experimental Results and Discussion

Two encoding schemes, Huffman coding and differential Huffman coding, are applied on 15 different medical images of type CT, MRI, ultrasound and PET. The algorithms are implemented using MATLAB. The parameters PSNR, MSE, RMSE and compression ratio are calculated to evaluate the performance of the two compression schemes. The original image, difference image and reconstructed image obtained by applying differential Huffman coding on the 15 medical images are shown in Fig. 2.

Compression ratio defines the amount of compression which has been achieved by a compression algorithm. Compression ratio is given as [16]:

Fig. 2 Original image, difference image and reconstructed image obtained from differential Huffman coding

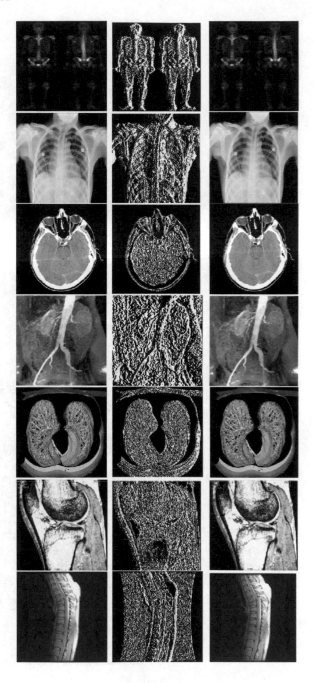

Fig. 2 (continued)

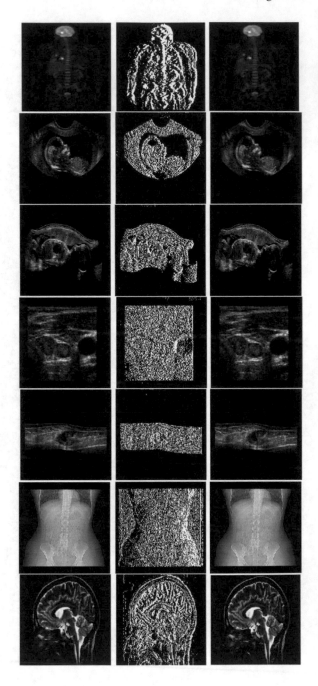

Fig. 2 (continued)

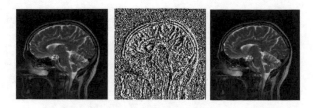

$$CR = \frac{S_{\text{orig}}}{S_{\text{comp}}} \tag{2}$$

Here, S_{orig} is the size of the original image and S_{comp} is the size of the compressed image.

Mean squared error (MSE) is the cumulative square of the difference between the compressed and the original image and can be calculated by using the formula given as [5]:

$$MSE = \sum_{i,j=0}^{n-i} \frac{\left[f'(i, j) - f(i, j)\right]^2}{n \times n} \tag{3}$$

Here, $f'(i, j)$ and $f(i, j)$ represent the compressed and the original images of size $n*n$, respectively.

Peak signal-to-noise ratio (PSNR) measures the peak error between the compressed image and the original image. It can be calculated using the formula given as [5]:

$$PSNR = 10 \log_{10}\left(\frac{R^2}{MSE}\right) \tag{4}$$

Here, R is the maximum possible pixel value of the image; and MSE is the mean squared error.

Compression ratio helps in analyzing the amount of compression achieved by an image compression technique. Peak signal-to-noise ratio (PSNR), mean squared error (MSE) and root mean squared error (RMSE) are used to analyze the quality of image obtained after compression.

Compression ratios obtained from the two schemes on medical images are shown in Table 1. Results are obtained by applying Huffman coding algorithm on original medical images and implementing differential coding on original images followed by Huffman coding algorithm. Medical images 1–12 are of type .TIF, image 13 of type .dcm (standard dicom image) and images 14 and 15 are of type .jpg.

Highest compression ratio is achieved for image 12 by both Huffman coding and differential Huffman coding. Lowest compression ratio is obtained for image 6 by differential Huffman coding and for image 4 by Huffman coding.

Table 1 Compression ratio of images

Image no.	Compression ratio	
	Huffman coding	Differential Huffman coding
1	1.7034:1	2.2468:1
2	1.2175:1	1.8084:1
3	1.9176:1	2.1428:1
4	1.0747:1	1.6215:1
5	1.4098:1	1.5301:1
6	1.1493:1	1.3299:1
7	1.4709:1	1.8981:1
8	1.629:1	2.9141:1
9	2.4356:1	2.6985:1
10	2.2325:1	2.5492:1
11	1.526:1	1.8348:1
12	2.7353:1	3.1447:1
13	1.2223:1	1.7183:1
14	1.4718:1	1.7599:1
15	1.2063:1	1.5083:1

Peak signal-to-noise ratio (PSNR) for all 15 images is infinity. Mean square error (MSE) and root mean square error (RMSE) for all images are zero. Infinite PSNR and zero MSE and RMSE values show that the compression achieved for the 15 medical images is lossless. Lossless compression assures exact reconstruction of the compressed image to the same as the original image.

The compression ratios achieved by applying differential Huffman coding are greater than those achieved by directly applying Huffman coding on original medical images. Pre-processing the original image with differential coding resulted in more consequent zeroes at neighboring pixels. As a result, application of Huffman coding on the difference image improved the compression ratio.

5 Conclusion and Future Scope

This paper presents two approaches for lossless compression of medical images. In the first approach, Huffman coding is implemented directly on medical images, whereas in the second approach, differential coding is applied on medical images before implementing Huffman coding. Results have shown that the second approach provides better compression ratio as compared to first. The use of differential coding before applying Huffman coding has lead to increase in compression ratio. So pre-processing the original medical images with differential coding has resulted in enhanced compression ratio.

In future, differential coding can be combined with encoding schemes other than Huffman coding to further improve the compression ratio. Also, pre-processing procedures other than differential coding could be proposed to achieve better compression ratio.

References

1. Abdmouleh, M.K., Masmoudi, A., Bouhlel, M.S.: A new method which combines arithmetic coding with RLE for lossless image compression. J. Softw. Eng. Appl. **5**, 41–44 (2012)
2. Al-Bahadili, H., Hussain, S.M.: An adaptive character wordlength algorithm for data compression. Comput. Math Appl. **55**, 1250–1256 (2008)
3. Al-Wahaib, M.S., Wong, K.: A lossless image compression algorithm using duplication free run-length coding. In: 2010 Second International Conference on Network Applications, Protocols and Services, pp. 245–250. IEEE Press, New York (2010)
4. Aneja, R.D., Gupta, S., Batra, V.: An enhanced bi-orthogonal wavelet filter for image compression. Int. J. Comput. Sci. Commun. **2**(2), 351–354 (2011)
5. Gupta, P., Srivastava, P., Bhardwaj, S., Bhateja, V.: A modified PSNR metric based on HVS for quality assessment of color images. In: 2011 International Conference on Communication and Industrial Application. IEEE Press, New York (2011)
6. Hasan, M., Nur, K.M.: A lossless image compression technique using location based approach. Int. J. Sci. Technol. Res. **1**(2), 101–105 (2012)
7. Hasan, M., Nur, K.M.: An improved approach for spatial domain lossless image data compression method by reducing overhead bits. Int. J. Sci. Eng. Res. **3**(4), 700–703 (2012)
8. Hasan, M., Nur, K.M., Noor, T.B., Shakur, H.B.: Spatial domain lossless image compression technique by reducing overhead bits and run length coding. Int. J. Comput. Sci. Inf. Technol. **3**(2), 3650–3654 (2012)
9. Huffman, D.A.: A method for the construction of minimum-redundancy codes. Proc. IRE **4**(9), 1098–1101 (1952)
10. Liang, J.Y., Chen, C.S., Huang, C.H., Liu, L.: Lossless compression of medical images using hilbert space-filling curves. Comput. Med. Imaging Graph. **32**(3), 174–182 (2008)
11. Nandi, U., Mandal, J.K.: Windowed Huffman coding with limited distinct symbols. Procedia Technol. **4**, 589–594 (2012)
12. Pinto, S.J., Gawande, J.P.: Performance analysis of medical image compression techniques. In: Third Asian Himalayas International Conference on Internet. IEEE Press, New York (2012)
13. Ouafi, A.E., Taleb-Ahmed, A.A., Zitouni, A., Baarir, Z.E.: Lossless image compression based on Mojette transform and differential coding. J. Appl. Comput. Sci. Math. **11**(5), 81–84 (2011)
14. Ramteke, K., Rawat, S.: Lossless image compression LOCO-R algorithm for 16 bit image. Int. J. Comput. Appl. **7**, 11–14 (2011)
15. Rehna, V.J., Jalall, K.S., Manjunath, T.C., Thomas, A.A.P.: Digital compression technique—a novel method of implementation. Special Issue of Int. J. Comput. Appl. 1–5 (2012)
16. Saranya, D., Balasubramani, G.: An efficient image compression algorithm to improve the compression-ratio of dual-color image. Int. J. Emerg. Technol. Adv. Eng. **4**(2), 419–422 (2014)
17. Thorat, M.Y., Bairagi, V.K.: Hybrid method to compress slices of 3D medical images. Int. J. Electron. Commun. Eng. Technol. **4**(2), 250–256 (2013)
18. Vrindavanam, J., Chandran, S., Mahanti, G.K., Vijayalakshmi K.: Huffman coding and position based coding scheme for image compression: an experimental analysis. Int. J. Appl. Inf. Syst. (2), 1–5 (2013)

Detecting Toxicity with Bidirectional Gated Recurrent Unit Networks

Vinayak Kumar and B. K. Tripathy

Abstract As large amounts of data keep being generated by users on social media platforms, some of the information can be considered as harmful. These kinds of textual information can be generated in forums, online discussions or any other communication exchange in an online medium. As such, it is sometimes difficult to filter out what information is actually meaningful. Detecting these harmful pieces of information can help in providing a means of online moderation so that a safe discussion can be maintained, which helps in preventing issues such as cyber bullying. Using the Kaggle Jigsaw dataset of comments that are classified as toxic labels, we can implement deep learning models to implicitly extract textual features from the comments and solve this supervised learning problem. This paper focuses on using the variations of recurrent neural networks, with the main focus on using bidirectional gated recurrent units, and evaluating their performances against each other.

Keywords Recurrent neural networks · Toxic · Multi-label text classification · Long short-term memory · Gated recurrent units

1 Introduction

Online moderation and content filtering have become more prevalent in recent years due to more intelligent and automated systems being used in social platforms. However, it is still quite difficult to identify and filter specific personal attacks on people in a public forum. The type of attack could also play a role in determining the severity of the attack. Based on an online public experiment, through Wikipedia, user-generated

V. Kumar (✉)
School of Computer Science and Engineering, Vellore Institute
of Technology, Vellore, Tamil Nadu, India
e-mail: sivakumarvinayak@gmail.com

B. K. Tripathy
School of Information Technology and Engineering, Vellore Institute
of Technology, Vellore, Tamil Nadu, India
e-mail: tripathybk@vit.ac.in

© Springer Nature Singapore Pte Ltd. 2020
V. Bhateja et al. (eds.), *Intelligent Computing and Communication*,
Advances in Intelligent Systems and Computing 1034,
https://doi.org/10.1007/978-981-15-1084-7_57

comments had been collected and labelled through machine learning classifiers [10]. There are six labels for each comment, each determining the type of attack. The main label is the toxic label, based on which it could fall under the other labels.

There has been quite a substantial increase of research in deep learning due to its applicability in various areas, such as medical, artificial intelligence, computer vision, signal processing and many more. Another particular area of its application is in natural language processing. Deep learning provides a means to solve natural language processing problems without worrying too much about additional processes to extract the information. Multi-label text classification is a natural language interpretation problem where text of varying length is required to be labelled under multiple labels, where a single piece of text can belong to more than one label.

Recurrent neural networks (RNN) fall under a special type of deep neural networks, where the networks have an internal memory state that processes sequences of inputs and reuses the output processes in the hidden layers. In other words, it remembers the previous state and adjusts the network based on it as well as the current state and the input state. This allows it to handle temporal data, as the previous information is compared with the current information. Based on this principle, there have been many variants of recurrent neural networks, such as long short-term memory (LSTM) and bidirectional long short-term memory, which have been used based on the sequential nature of the problem [4]. A newer variant which is similar to LSTM models called the gated recurrent unit (GRU) model has also started to be used in deep networks [12] and behaves very similar to LSTM. The bidirectional part of the recurrent neural network basically passes the input in both the forward and backward directions, hence making it able to predict the context of a long sequence of data. Using this we can apply it to the existing GRU, along with other hidden layers.

In Sect. 2, we will compare text classification approaches proposed in different studies. In Sect. 3, the proposed system for classifying the toxic data is explained and finally, in Sect. 4 we compare the classifier using various metrics.

2   Literature Survey

Yang [11] proposed a deep neural network model that focuses on the hierarchical nature of documents with attention mechanisms. This model is meant to classify documents by using the idea that not all documents are contextually relevant, hence finding certain parts that describe the context of the document can be achieved with attention-level mechanisms on both the word and sentence level. GRU cells are used to encode the words, which are then sent to a word attention mechanism to extract the important context words. Then the GRU cells are used to encode the sentences and are then sent to a similar sentence attention mechanism. Finally, the document is then classified using a softmax function.

In another study, an abusive language detection system was made by sample comments on several Yahoo! News and Finance pages [8]. Here more linear classifiers were used, and there was more emphasis on the different aspects of textual feature extraction. This included data preprocessing steps to handle linguistic, syntactic and semantic features by using n-gram models. Word embedding models were also used to create a larger comment embedding model which was based on the idea that surrounding word vectors, as well as the comment the word is currently in, play a large role in the occurrence of that word. The comments were represented by using low-dimension vectors, and the evaluation of the model was compared against other pretrained embedding models.

Joulin proposed a text classification model [5] that focuses on its efficient architecture and performance by using regular linear baseline models with a large corpus. Word representations were taken and then averaged into sentence representations, along with a hierarchical softmax function to calculate the probability of the occurrence of the class. A fast and efficient n-gram model with hashing was used for the textual feature extraction. This system was named fastText and was meant to scale large amounts of data, compared to deep learning models. Extensive performance testing was done on this system by comparing its usage in sentiment analysis and tag prediction.

The system [7] uses LSTM and convolutional layers to convert text from word vectors to a new "short-text" representation. In both instances, max pooling layers are done to obtain the final vector. This new short text vector is then fed to a feed-forward artificial neural network for that particular short text, which finally gives the probability distribution over a set of classes based on the input vector. This system was then used for dialog act prediction and was evaluated. A similar study [6] used a similar combination of deep neural network architectures, where recurrent cells were used in the convolution layer to capture the context of each word input, which was then passed to a max pooling layer to give an output. This model is meant to deal with the biasing problems of recurrent neural networks by being able to give more importance to certain word meanings, instead of just looking at the importance of a previous word input and the current word input. It was then compared with the traditional machine learning models with feature extraction methods such as bag of words bigrams, as well as other popular neural networks.

3 Proposed System

3.1 Exploratory Data Analysis and Preprocessing the Dataset

The dataset used to train and test the model is obtained from Kaggle. This was the Jigsaw Toxic Comment dataset which had several comments labelled as toxic, severe toxic, obscene, threat, insult or identity hate. The dataset had 159,571 comments, out of which 15,294 were labelled as toxic. Most of the comments had 50 words, while a few comments with the maximum number of words peaked at 200 words. This is important as this would help determine the feature vector of the input to the deep neural network.

The data was then tokenized based on a high average number of words per comment. This represents the input feature vector of the data. Unlike most linear classifiers, deep neural networks do not require more feature engineering, such as handling syntactic or semantic features. This is because we are able to use word embeddings based on the tokenized text.

3.2 Word Embedding Layer

Once the words are indexed with numbers, each comment becomes a vector with numeric features. This is fed into the deep neural network via the input layer. To be able to process the words for classification, we use an embedding layer that is able to project these words into a vector space, where similar words tend to occur next to each other, hence capturing the semantic meaning of that word. In other words, if a word is close to another word in this vector space, then it is highly likely that this word conveys a similar meaning. As such, the output of the word embedding are the coordinates for each word processed [9]. This is also a better option, in general, instead of using an entire vocabulary of words and using a large dimension vector to indicate whether that word is present or not.

There are several approaches to word embeddings. There is the baseline approach where our own embedding layer is created based on the tokenized text. Pretrained models such as GloVe and word2vec are also good alternatives that use neural network layers or matrix factorization to try to predict the word representation based on the given data.

3.3 Bidirectional Gated Recurrent Units Layer

Recurrent neural networks (RNN) consist of neural network layers comprising nodes in a directed graph, where the information of the input is fed recursively to the neuron. This is done through a transition function. These RNN units use dependent activations by generating the same weights and biases to all the layers, while giving the output as input to the next hidden layer [2]. The activation function of this current neural network neuron h_t is dependent on the value of the previous state h_{t-1} and the current input neuron x_t. Based on this activation function, the layer output can be calculated using the output layer weights. Normally, a tanh function is used as the activation function.

This was addressed in another study [3] by introducing a new memory-based learning unit in the recurrent layer called long short-term memory network (LSTM). This network was able to learn dependencies over a long period of time. This was achieved by having the LSTM unit consisting of a separate memory unit that updates the content of the LSTM cell only when it is required. This is done by using gates, represented by sigmoid functions "σ". There are three gates, each represented by a sigmoid function. The first gate is the forget gate that reads h_{t-1} and x_t. Based on this value it creates a probability distribution between 0 and 1 on whether to discard this input or retain it. Similarly, there is an input gate unit that decides which values to update, with the tanh function creating new candidate values. Finally, there is the output gate which checks what parts of the cell state need to be put in the output function. It is then finally sent to the activation function to create a new h_t.

Another system was used in [1] which also meant to remove the vanishing gradient problem, without the need of a memory unit though. This was called gated recurrent units (GRU). Unlike LSTM, GRU has only two gates, and there isn't a memory storage unit generated as well. The figure for a GRU unit is given in Fig. 1.

The first gate (sigmoid function σ) is called the update gate. It is calculated from the following formula:

$$z_t = \sigma\left(W^{(z)}x_t + U^{(z)}h_{t-1}\right) \tag{1}$$

x_t is the incoming input at instance t, and h_{t-1} is the state of the unit at instance $t-1$. $W^{(z)}$ and $U^{(z)}$ represent the weights of the input neuron and previous state at the update gate. They are added and then a sigmoid activation function is used. The update gate z_t is used to identify how much of the previous information needs to be passed and processed.

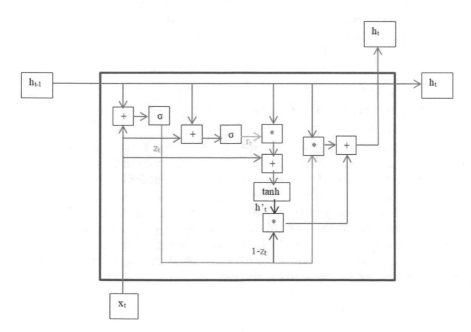

Fig. 1 Single GRU unit

The reset gate is the second gate and is calculated as follows:

$$r_t = \sigma\left(W^{(r)}x_t + U^{(r)}h_{t-1}\right) \tag{2}$$

$W^{(r)}$ and $U^{(r)}$ represent the weights of the input neuron and previous state at the reset gate. A temporary memory content (h'_t) is created which uses the reset gate to store relevant past information. It is calculated as:

$$h'_t = \tanh(Wx_t + r_t * Uh_{t-1}) \tag{3}$$

Note that "*" denotes the Hardmond matrix product operation. The element h'_t determines what to remove from the preceding time steps. Using this, we can find the current h_t state as:

$$h_t = z_t * h_{t-1} + (1 - z_t) * h'_t \tag{4}$$

If the update gate z_t is a value close to 1, which represents that the new input is important, then $1 - z_t$ will determine as how much of the current information to lose when calculating h_t.

Using these GRUs, we can then apply a mechanism called bidirectional RNNs [13] to improve how the text information is going to be processed and classified. Here there are two layers of recurrent units, one processing the information in the forward direction and the other layer of units processing the information in the backward direction. When a prediction on what words to use is required, the past information as well as the future information are used. Using this mechanism we will use a bidirectional layer of GRU units.

3.4   Pooling and Dense Layers

We then use a max pooling layer to reduce the dimension for the next set of layers. In this layer, each batch of incoming input data is processed and the maximum value of each batch is taken. The output would then be a smaller batch of data which makes it simpler to process in the succeeding dense layers. These dense layers use the rectified linear unit (ReLU) activation function. We also use dropout layers to try to constrain the network (avoiding overfitting of the neural network) and help in removing non-essential neurons from the network during that pass.

The final dense layer has a sigmoid activation function that generates a six-dimensional vector, representing the labels.

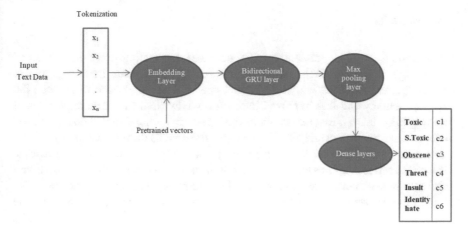

Fig. 2 System architecture

3.5 Final System Design

For training the model, the Jigsaw dataset was used and split into a pair of training and testing sets. It was then tokenized to be given as input to the neural network for training. After training the data, the classifier model should be able to label any text into the six labels by generating the probabilities of that label occurring on that specific text, as shown in Fig. 2.

4 Results and Discussion

4.1 Hardware and Software Setup

The system used was a quad core Intel i5 processor running at 2.2 GHz with 8 GB RAM, with a 2 GB Nvidia 920M graphics card. Python 3.6 was used along with libraries Pandas, Numpy, TensorFlow, Keras and Sklearn.

4.2 Evaluation

We use the proposed design and tested the model against other variants of RNNs. We also compared them against traditional machine learning models. The loss and accuracy was based on the binary cross-entropy function to handle the multiple labels. F1, precision and recall measures were also used to evaluate the model.

4.3 Results

When training the deep models, in all four cases, the testing loss results seemed to converge after two epochs. After the two epochs, the result became prone to overfitting. As seen from Fig. 3, the model was still prone to overfitting since the training accuracy and loss improved over time while there was little improvement for the testing dataset. Regularization or more hyperparameter tuning could be used to get a better testing accuracy and loss across the epochs during evaluation.

As seen from Fig. 4 all the deep learning models produced favourable results. Most of the differences in scores between the variations are minor. The bidirectional GRU was better than the remaining models in terms of its accuracy and loss. The LSTM models seemed to be better in terms of precision and recall scores. All the RNN variants seem to perform better than the other machine learning models. The remaining results are compiled in Table 1.

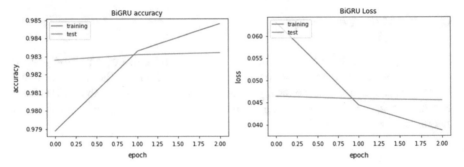

Fig. 3 Bidirectional GRU accuracy results (left), bidirectional GRU loss results (right)

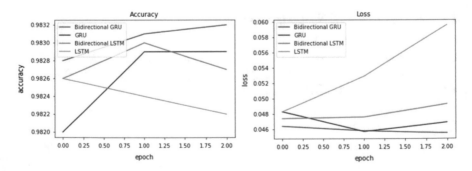

Fig. 4 Testing accuracy (left), testing loss (right)

Table 1 Evaluation of testing data

Model	Evaluation metrics				
	Accuracy	Loss	F1 score	Precision	Recall
Bidirectional GRU	0.9830	0.0459	0.6651	0.7585	0.6293
GRU	0.9826	0.0470	0.6613	0.7541	0.6259
Bidirectional LSTM	0.9828	0.0481	0.6583	0.7654	0.6131
LSTM	0.9824	0.0536	0.6589	0.7414	0.6293
Logistic Regression	0.9711	0.0808	0.2549	0.6998	0.1682
Random Forest Classifier	0.8789	0.1665	0.4093	0.6575	0.3411
Naïve Bayes	0.9135	0.0769	0.2748	0.5969	0.1954

5 Conclusion

As we have seen, the deep learning models have provided favourable results despite not focusing on the data preprocessing part and feature engineering that traditional classifiers need in order to perform decently. A lot of improvements can be made to the model, such as adding an attention-based mechanism to better handle the toxicity.

Using this kind of learning system, we can identify specific aspects of toxicity in a system and then probe into dealing with those issues. Toxicity isn't just limited to abusive behavioural comments as well. Anything unrequired in an information exchange could also be considered as toxic, and by understanding how to use deep learning in this system, we can help improve public discussions and online moderation.

References

1. Chung, J., Gulcehrel, C., Cho, K., Bengio, Y.: Empirical Evaluation of Gated Recurrent Neural Networks on Sequence Modeling, Cornell University. arXiv:1412.3555 (2014)
2. Elman, J.: Finding structure in time. Cogn. Sci. Multidiscip. J. **14**(2), 179–211 (1990)
3. Hochreiter, S., Schmidhuber, J.: Long short-term memory. Neural Comput. **9**(8), 1735–1780 (1997)
4. Huang, Z., Xu, W., Yu, K.: Bidirectional LSTM-CRF models for sequence tagging, Cornell University. arXiv:1508.01991 (2015)
5. Joulin, A., Grave, E., Bojanowski, P., Mikolov, T.: Bag of tricks for efficient text classification, Cornell University. arXiv:1607.01759 (2016)
6. Lai, S., Xu, L., Liu, K., Zhao, J.: Recurrent convolutional neural networks for text classification. In: AAAI'15 Proceedings of the Twenty-Ninth AAAI Conference on Artificial Intelligence, pp. 2267–2273 (2015)
7. Lee, J., Dernoncourt, F.: Sequential short-text classification with recurrent and convolutional neural networks. In: Proceedings of the 2016 Conference of the North American Chapter of the Association for Computational Linguistics: Human Language Technologies, pp. 515–520. Association for Computational Linguistics (2016). https://doi.org/10.18653/v1/n16-1062
8. Nobata, C., Tetreault, J., Thomas, A., Mehdad, Y., Chang, Y.: Abusive language detection in online user content. In: Proceedings of the 25th International Conference on World Wide Web,

WWW '16, Republic and Canton of Geneva, Switzerland, pp. 145–153. International World Wide Web Conferences Steering Committee (2016)

9. Shen, D., Wang, G., Wang, W., Min, M., Su, Q., Zhang, Y., Li, C., Henao, R., Carin, L.: Baseline needs more love: on simple word-embedding-based models and associated pooling mechanisms. In: Proceedings of the 56th Annual Meeting of the Association for Computational Linguistics, P18-1041, pp. 440–450. Association for Computational Linguistics (2018)

10. Wulczyn, E., Thain, N., Dixon, L.: Ex machina: personal attacks seen at scale. In: 26th International Conference on World Wide Web (WWW '17), pp. 1391–1399. International World Wide Web Conferences Steering Committee, Republic and Canton of Geneva, Switzerland (2017)

11. Yang, Z., Yang, D., Dyer, C., He, X., Smola, A., Hovy, E.: Hierarchical attention networks for document classification. In: Proceedings of the 2016 Conference of the North American Chapter of the Association for Computational Linguistics: Human Language Technologies, N16-1174, pp. 1480–148. Association for Computational Linguistics (2016)

12. Zhao, R., Wang, D., Yan, R., Mao, K., Shen, F., Wang, J.: Machine health monitoring using local feature-based gated recurrent unit networks. IEEE Trans. Ind. Electron. **65**(2), 1539–1548 (2018)

13. Zhou, P., Shi, W., Tian, J., Qi, Z., Li, B., Hao, H., Xu, B.: Attention-based bidirectional long short-term memory networks for relation classification. In: Proceedings of the 54th Annual Meeting of the Association for Computational Linguistics, P16-2034, pp. 207–212. Association for Computational Linguistics (2016)

A Design of 12-Bit Low-Power Pipelined ADC Using TIQ Technique

B. K. Vinay, S. Pushpa Mala, S. Deekshitha and M. P. Sunil

Abstract A CMOS 12-bit pipeline analog to digital converter (ADC) is designed for improved speed, resolution and low power consumption. The design incorporates 12 stages of 1-bit ADC cascaded to form pipelined architecture, with each stage containing a sub-ADC with a new approach of threshold inverter quantizer (TIQ) which substitutes the resistor array implementation and a multiplying digital to analog converter (MDAC) for quantized approximation of input voltage. The residue voltage is amplified in gain stage by closed-loop differential amplifier to have appropriate quantized output at the next stage. The sampling frequency is 200 MHz in 180 nm technology, and speed is 100 MSps and power is 50 mW, 0.5 pJ/step with 12-bit resolution.

Keywords Residue voltage · Threshold inverter quantizer · Differential amplifier · Flash ADC · Pipelined stage

1 Introduction

Analog to digital converter is very essential to convert continuously varying signal in time and amplitude into discrete time, discrete amplitude signal by sampling, quantization and encoding. In wireless communication, ADC is an integral part, and

B. K. Vinay · S. Deekshitha (✉)
Department of Electronics and Communication, CMRIT, Bengaluru, India
e-mail: sdeekshitha.1999@gmail.com

B. K. Vinay
e-mail: vinay.bk@cmrit.ac.in

S. Pushpa Mala
Department of Electronics and Communication, Dayananda Sagar University, Bengaluru, India
e-mail: pushpasiddaraju@gmail.com

M. P. Sunil
Department of Electronics and Communication, School of Engineering
and Technology, Jain University, Bengaluru, India
e-mail: mp.sunil@jainuniversity.ac.in

© Springer Nature Singapore Pte Ltd. 2020
V. Bhateja et al. (eds.), *Intelligent Computing and Communication*,
Advances in Intelligent Systems and Computing 1034,
https://doi.org/10.1007/978-981-15-1084-7_58

hence it is required to have high speed and high resolution for portable data acquisition devices [1]. The higher resolution ensures better conversion, and an input signal is encoded into 2N code with an ADC of N-bit resolution. Improved speed can be achieved by inculcating pipelined ADC which has 12 stages of 1-bit ADC, since this reduces the time delay by simultaneously processing the requests and hence reduces the variation in digital output. Flash ADCs are used in sub-ADC module to reduce the processing time [1].

The design shows pipelined ADC which offers speed, resolution, low power consumption and small die size. The digital quantization of analog input is realized by each pipelined stage which computes and amplifies the residue voltage. Each stage has sub-ADC and multiplying DAC. Multiplying DAC is a low noise precision DAC which produces current as output signal which is the product of given reference voltage and the code that is string of 0's and 1's flowing through it. The resistor ladder approach is replaced by TIQ approach in sub-ADC which eliminates static power consumption making the design more power and space efficient and hence it can be used in battery-powered applications [2, 3]. The power dissipation can also be reduced by using optimized MDAC architecture.

In the paper the sections are categorized as follows: Sect. 2 describes the existing design architecture, and Sect. 3 describes the proposed design architecture simulated in cadence virtuoso for power and speed analysis. Section 4 contains the results and discussion of the proposed design. Section 5 presents the conclusion of the paper.

2 Existing Design

The various types of ADCs are flash, pipelined, dual-slope and successive approximation register (SAR). A synchronous pipelined ADC is implemented by clocking N-stages at a lower frequency to obtain higher sampling frequency which is increased by a factor of N, but with proper combination of time-interleaving, for 12-bit resolution, signal is sampled at 3.2 GHz, and high sampling rate is obtained with a compromise on power consumption, that is, 900 mW at 180 nm process [4]. The proposed design has 50 mW power consumption for 12-bit resolution and the sampling rate is 200 MHz. Flash ADCs use TIQ technique instead of traditional series resistor ladder at comparators for providing reference voltage to reduce power consumption and area [2]. In a 12-bit SAR ADC architecture, hybrid DAC circuit is implemented in which 10-bit CDAC and 2-bit RDAC are incorporated to get 12-bit resolution and the speed is boosted by asynchronous control logic. This methodology has less sampling rate, that is, 1 MHz and power supply of 1.8 V. The maximum deviation of ADC transfer function from best fine line is defined by integral nonlinearity (INL), which is 0.92 LSB, hence this design has a trade-off between low power consumption and integral nonlinearity [5, 6] and the proposed pipelined ADC architecture incorporating MDAC circuit has lesser INL of 0.5 LSB at power supply 1 V. In dual-slope ADC, the sampling frequency is increased by sampling and integrating the input signal at two non-overlapping clock phases by double sampling technique at the sampling rate of 640 MHz. Here, the thermal noise is suppressed by using differential structure, but this design has more signal-to-noise ratio, that is, 72 dB for 12-bit resolution [3, 7] as compared to pipelined ADC presented in this paper with SNR of 50 dB for 12-bit resolution. Pipelined ADC serves as an optimum architecture for obtaining balance between power dissipation, speed and resolution.

3 Proposed Methodology

The operating principle of the pipelined ADC is multi-stage conversion in order to achieve high speed and resolution. Each stage realizes the digital quantization of analog input, which computes and amplifies the residue voltage [1]. Each stage has B-bit resolution, where B represents effective stage resolution. First k − 1 stages use similar architecture and the last stage does not have any redundancy, hence it has flash ADC with B-bit resolution.

The block diagram of pipelined ADC is shown in Fig. 1. Figure 2 shows the single stage of the pipelined ADC, which consists of a sample-and-hold, sub-ADC, a DAC, subtractor and an amplifier.

Analog input is converted to digital form by sub-ADC. This digital form is passed to MDAC for conversion of analog signal in order to get appropriate output at next stage. The analog output of DAC is subtracted with output of sample-and-hold circuit and the resulting residue voltage is amplified to pass it to the next stage.

Fig. 1 A block diagram of pipelined ADC

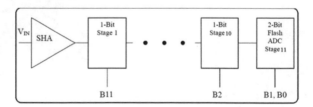

Fig. 2 Residue stage architecture for the 1-bit per stage pipelined ADC

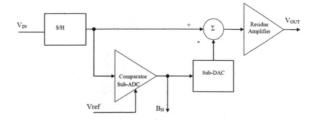

3.1 Sample-and-Hold Circuit

A sample-and-hold circuit samples the varying analog input voltage and holds these samples for specific period of time. During the positive half cycle of the clock input signal is tracked and during the negative half cycle tracked input is held stable, hence it is called track-and-hold circuit. The clock signal turns the switch on and off. When switch is a closed capacitor, C_H is charged to the level of input signal, and when switch is opened, the tracked input is present at V_{out}, and the sampling speed is determined by the on-resistance of the switch. In order to achieve high speed, a switch in sample-and-hold circuit is replaced with complementary MOS transistor, as shown in Fig. 3, to have capacitor charged to final value in finite time. Schematic of sample-and-hold circuit is shown in Fig. 4.

Fig. 3 Complementary switch

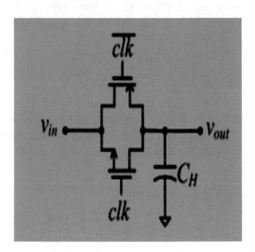

Fig. 4 Schematic for sample-and-hold circuit

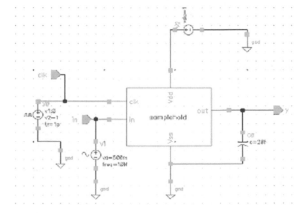

3.2 One-Bit Sub-ADC

The output of sub-ADC is one bit, namely 0 or 1. The resistor array implementation is replaced by the threshold inverter quantizer (TIQ) technique which eliminates static power consumption in analog signal quantization, since this technique relies on systematic transistor sizing. A differential mode ADC is realized with comparator. The safe analog input range is: Analog range $= V_{dd} - (VTN + |VTP|)$, where VTN and VTP are the threshold voltages for large NMOS and PMOS devices, respectively. The VTHO value from the model parameter data set is used during the entire design process. The signal is compared with reference voltage using comparators. In 12-bit pipelined ADC, the last stage is 2-bit differential mode single-ended flash ADC without any MDAC stage, as shown in Fig. 6. The other stages contain 1-bit differential mode flash ADC, as shown in Fig. 5.

In flash ADC, the resistive network with 2N resistors, which are connected in series ladder, is used to provide reference voltage for comparator by dividing V_{ref} into four levels. We use four resistors from analog library in cadence tool. The output of this network is connected to comparator block. Each comparator produces an output "1" when its input voltage is greater than the reference voltage. For N-bit

Fig. 5 Schematic for sub-ADC comparator

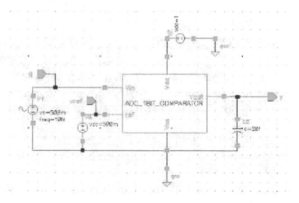

Fig. 6 Schematic for 2-bit flash ADC

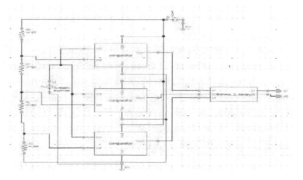

Fig. 7 Dashed box
represents MDAC

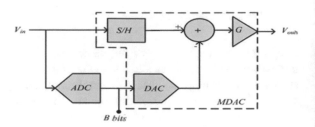

resolution, $2N - 1$ comparators are used in flash ADC, and the reference voltage
of each comparator is greater than the reference voltage of comparator immediately
below it by one least significant bit (LSB).

The thermometric output is converted to decimal equivalent in the File-writer
module, which is implemented as high-level model in Verilog-A. The behavior of
analog circuit is described by using modules in this high-level language. The ther-
mometer code is converted to equivalent binary code with the help of truth table and
relevant equations for direct conversion by using encoder circuit.

3.3 Multiplying Digital to Analog Converter (MDAC)

In a pipelined stage, multiplying digital to analog converter (MDAC) converts digital
signal to analog signal from sub-ADC circuit and subtraction of sampled analog
signal by sample-and-hold circuit results in residue voltage and this residue voltage
is amplified by gain stage. MDAC consists of sample-and-hold (S/H), sub-digital-
to-analog converter (sub-DAC), a subtractor and a residue amplifier, as shown in
Fig. 7.

Traditionally, an MDAC is implemented using the switched capacitor technique.
Since the S/H circuit is not integrated in MDAC implementation, in this thesis study,
we refer the term MDAC to a unit performing D/A conversion, subtraction of D/A
output from signal held by S/H and residue amplification. The output MDAC is given
by Eq. 1.

$$V_{out} = -(D/2^n) \cdot V_{in} \tag{1}$$

The differential amplifier provides gain for the residue voltage which is ($V_{in} - V_{dac}$), in order to have sufficient input range at the next stage of quantizer. The closed-
loop differential amplifier is used at gain stage to overcome nonlinearity effect of
open-loop differential amplifier and hence precise linear gain of the residue voltage
is obtained, and the schematic for this functionality is shown in Fig. 8.

Figure 9 shows the schematic for a single-stage 1-bit ADC and 12 such stages are
cascaded to form 12-bit pipelined ADC, which is shown in Fig. 10.

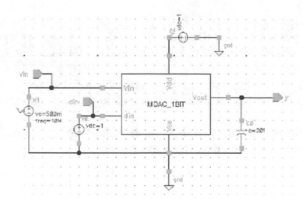

Fig. 8 Schematic for 1-bit MDAC

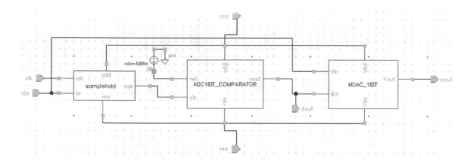

Fig. 9 Single-stage 1-bit ADC

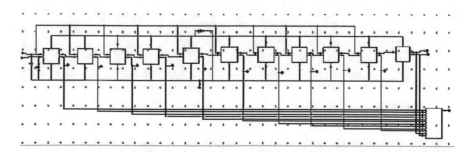

Fig. 10 12-Bit pipelined ADC

4 Results and Discussion

The purpose is to investigate the design of 12-bit pipelined ADC with sampling frequency of 200 MHz and 180 nm CMOS technology. The proposed design is simulated in virtuoso suite of cadence tool. The resolution is 12-bit and speed is 100

MSps power approximated to 50 mW, 0.5 pJ/step. Signal-to-noise ratio (SNR) is +70 dB and integral nonlinearity (INL) is ±0.5 LSB, with power supply of 1 V.

Figure 11 shows the output of sample-and-hold circuit when the input signal is 1 V.

Figure 12 shows the output of the comparator, which gives logic 1 at output when input voltage is higher than reference voltage and logic 0 when the input voltage is lower than the reference voltage.

Figure 13 shows the output of 2-bit flash ADC. The thermometric output is converted to binary output by direct conversion.

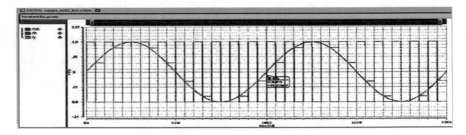

Fig. 11 Simulation for sample-and-hold circuit

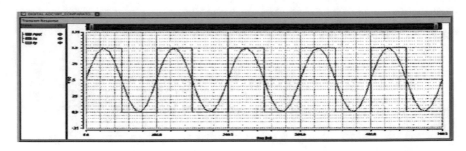

Fig. 12 Simulation for comparator

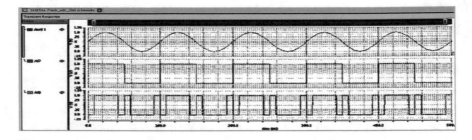

Fig. 13 Simulation for 2-bit flash ADC

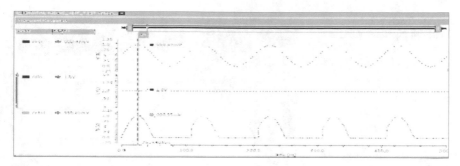

Fig. 14 Simulation for 1-bit MDAC

Fig. 15 Simulation for 12-bit ADC

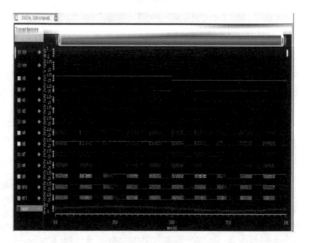

As shown in Fig. 14, the output of MDAC is digital to analog conversion of sub-ADC signal which is implemented using the switched capacitor technique, forming a SC integrator around an operational amplifier.

Simulation of each stage in 12-bit pipelined ADC is shown in Fig. 15 and its transient response is shown in Fig. 16.

5 Conclusion

A low-power CMOS 12-bit pipelined for 180 nm technology ADC was simulated using cadence virtuoso with power supply of 1 V and sampling frequency of 200 MHz. The time delay is reduced by using complementary MOS switch in sample-and-hold circuit. The static power consumption was reduced by including threshold inverter quantizer (TIQ) in sub-ADC circuit. An increased resolution is

Fig. 16 Transient response
for 12-bit pipelined ADC

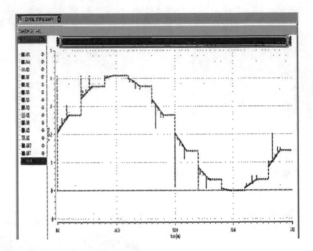

achieved through flash ADC architecture and a multiplying DAC produces low noise
at the output. The future work can be to reduce the area requirement and latency.

References

1. Yang, L., Wang, Z., Zhou, L., Feng, W.: A Low Power Pipelined ADC with Improved MDAC,
   pp. 624–627. Beijing Microelectronics Technology Institute, Apr 2016
2. Shylu, D.S., Jasmine, S., Jackuline Moni, D.: A low power dynamic comparator for a 12-bit
   pipelined successive approximation register (SAR) ADC. In: Fourth International Conference
   on Devices, Circuits and Systems, pp. 339–342, Mar 2018
3. Sedighi, B., Huynh, A.T., Skafidas, E., Micusik, D.: Design of hybrid resistive-capacitive DAC
   for SAR A/D converters electronics. In: IEEE International Conference on Circuits and Systems
   (ICECS), pp. 508–511, Dec 2012
4. Ren, S., Siferd, R.: CMOS 1.6 GHz bandwidth 12 bit time interleaved pipelined ADC. In:
   Eighth International Conference on Information Technology: New Generations, Dayton, OH,
   USA, pp. 785–790, Apr 2011
5. Talukder, A.-A., Sarker, Md.S.: A Three-Bit Threshold Inverter Quantization Based CMOS
   Flash ADC. Michigan State University, USA, Sept 2017
6. Wen, J.-Y., Chang, P.-H., Huang, J.-F., Lai, W.-C.: Chip Design of a 12-bit 5MS/s Fully Differential SAR ADC with Resistor-Capacitor Array DAC Technique for Wireless Application.
   National Communications Commission, Taipei, Sept 2015
7. Zhu, Y., Chan, C.-H., Chio, U.-F., Sin, S.-W., Martins, S.-P.U.R.P., Maloberti, F.: A 10-bit 100-
   MS/s reference-free SAR ADC in 90 nm CMOS. IEEE Solid-State Circuits **45**, 1111–1121
   (2010)
8. Ramalatha, M.: A High Speed 12-bit Pipelined ADC Using Switched Capacitor and Fat Tree
   Encoder, Zouk Mosbeh, Lebanon, pp. 391–395, July 2009
9. Torkzadeh, P., Javidan, J., Atarodi, M., Sharif, U.: A 640 MHz CMOS Switched-Capacitor
   $\Sigma\Delta$ ADC with 12 Bits of Resolution by Using Double Sampling Technique. EE Department,
   SICAS Group, May 2008

10. Yavari, M., Shoaei, O., Rodríguez-Vázquez, A.: Double-sampling single-loop modulator topologies for broad-band applications. IEEE Trans. Circuits Syst. II **53**(4) (2006)
11. Tu, W., Kang, T.: A 1.2 V 30 mW 8b 800 MS/s time-interleaved ADC in 65 nm CMOS. In: 2008 Symposium on VLSI Circuit Digest of Technical Papers, pp. 72–73 (2008)

Multiscale Template Matching to Denoise Epigraphical Estampages

P. Preethi, Anish Kasi, M. Manish Shetty and H. R. Mamatha

Abstract An epigraph is an engraved inscription from the past. Deciphering, recognizing and understanding the text in such epigraphical documents is mundane work for archaeologists. One of the major tasks in this is preprocessing the document and denoising it into segmentable form in an automated and fast manner. In this paper, we propose a pipelined solution which includes denoising techniques such as nested median filtering followed by a method of multiscale template matching to denoise noisy epigraphical documents. All epigraphs are usually in grayscale and they are converted to binary black and white images using otsu thresholding. Our template matching algorithm has been enhanced to create a mask of all the matched noise. A deep learning technique called Image Inpainting is then used in order to remove the noise and replace it with the respective background. A comparative study is performed with the output of the proposed method with use of novel denoising methods such as Gaussian blur and wiener filters. Human perception can determine that the proposed algorithm provides more satisfactory results. Standard quality measures such as the Structural Similarity Index (SSIM) and Peak Signal-to-Noise Ratio (PSNR) are used to show that the integrity of the image with respect to the text remains intact.

Keywords Template matching · Denoising · Epigraphical scripts

1 Introduction

The analysis of inscriptions on stones, rock pillars, temple walls, copper plates, and other writing material is called Epigraphy. The processing and preservation of these inscriptions are performed by the Archaeological Survey of India. Indian epigraphy in its raw form is widespread and belonged to the age when manuscript was the mode

P. Preethi (✉) · A. Kasi · M. Manish Shetty · H. R. Mamatha
PES University, Bengaluru, India
e-mail: preethip@pes.edu

H. R. Mamatha
e-mail: mamathahr@pes.edu

© Springer Nature Singapore Pte Ltd. 2020
V. Bhateja et al. (eds.), *Intelligent Computing and Communication*,
Advances in Intelligent Systems and Computing 1034,
https://doi.org/10.1007/978-981-15-1084-7_59

of communication. The engravings can be seen on the surfaces of rock beds of hills, on tall pillars, on tablets, and sketches on the walls in Ajanta and Ellora. Inscriptions are also written over palm leaves, coins, and copper plates. Temples even today show us the diverse history and culture of the country via these inscriptions.

Most inscriptions are written in a showy language. The information obtained from these inscriptions can be endorsed and validated with information from sources such as existing monuments or ruins, providing insight into India's dynastic history that lacks really old historical records. Thus, epigraphs present us with a lot of useful information. However, when creating such epigraphs the presence of noise is inevitable due to a variety of factors such as inconsistent texture of the rock and ink spread. In the mechanism of preserving such artifacts, initially, Estampages are developed, which are rubbed copies of the stone on a thick sheet of paper applying black ink and is then passed onto Epigraphers for deciphering [1]. Deciphering is a mundane task which takes more time and epigraphers who decipher are seldom. Another major issue with preserving such old inscriptions is their liability to get damaged over time due to various environmental conditions. Thus, digitization plays a major role. The digitized images act as inputs to the character recognition and automation pipeline.

At this very stage, the bridge to cross would be making sure that these images can actually be understood by the character recognition system and that the machine does not pick up mere noise during any learning steps. Reference [2] talks about the use of slant and slope correction methods for handwritten characters is used as a preprocessing technique. Till date, the character recognition is an open challenge because of the need to achieve high precision and accuracy. The image quality determines the accuracy of OCR performance. Ideally, OCR expects the image free from all types of noises.

Thus eliminating and discerning the evident noise from the background rock structure of the script is the scope of this research paper, as it diminishes the effects of noise which supports the correct working of systems in the larger scope of things like segmentation, feature extraction, and recognition. The input considered are the digitized Estampage images.

2   Literature Survey

The images of estampages do not contain any fixed or defined noise since it can arise due to texture, ink spitting or errors while writing. A few experiments have been conducted on documents containing similar noise and their techniques have been discussed below.

In [3], the authors talk about using a spatial filtering technique called bilateral filtering combined with grayscale mathematical morphology. The input images mainly consist of palm scripts and epigraphical script images. Reference [4] discusses how to remove noise created from ink spitting via PCA and K-means clustering by classifying pixels into three classes—background, original text or interfering text.

The authors of [5] perform anisotropic morphological dilation through a technique called implicit smoothing in order to restore the degraded character shapes of binarized images. The input images worked on in this case are Japanese historical documents. Horizontal and vertical RLC method is used to remove noise in the pipeline of script analysis [1, 6]. The authors of [7] propose a restoration model that looks for the optimal infra-red channel in which text is not present, and then extracts these degraded areas by a binarization transform, to generate a mask used to locate the target areas which have to be inpainted. References [8, 9] talks about using different binarization techniques to extract text and clearly distinguish between foreground and background.

Reference [10] dives into the use of template matching as a step after processing for pattern recognition so as to optimally localize more precisely the items in historic documents. These algorithms depend on a large scale on majorly the measures used to define similarity. Many times we use measures like cosine similarity or simply and link an distance in research areas like natural language processing. In the field of image recognition the similarity measure must identify the amount of overlap between template moving in a window and the source. The most popular is the Sum of either Squared Differences (SSD), Absolute Differences (SAD), and Normalized Cross-Correlation (NCC), primarily due to their computational efficacy and speed [11]. Multiple variants of aforementioned measures have been brought forward to deal with radiance/brightness changes and noise [12, 13].

In our research characterizing noise, approximating the extremity of noise plays a key role. The authors in [14] describe the existing noise approximations with the difference in pixel density-based, correlation-based, context-based, spectral distance-based, and human vision-based measures. But it is known that simple human observations are one of the best metrics to compare images and can be mathematically represented as structural similarity index. It simply tells us how different or same two visuals are. The authors in [15] compare and contrast between PSNR and SSIM indexing, and find that the two measures are to an extent equivalent or can at least be used to derive one another. The authors of [16] talk about an improved version of PSNR which takes into account Human Visual System (HVS) by considering the error sensitivity, structural distortion and edge distortion in the image.

The survey reveals that a myriad of different concepts and techniques have been applied on a set of inputs such as palm scripts, handwritten documents, manuscripts, and estampages.. Each document has different noise and the techniques mentioned above combat them using morphological operations, spatial and block filters, Machine learning techniques, etc. Estimators such as PSNR, MSE, and SSIM are used in order to determine the efficacy of noise removal. Existence of random noise, ink spots, stone texture impression in the images of estampages is the major hurdles to overcome in this topic.

3 Background Concepts

Standard denoising techniques involve the use of some type of filtering mechanism like frequency domain filters [17] and spatial domain filters [18] which are applied to remove general noise like Salt and Pepper, Dusut speckles, and Gaussian noise, depending on the type and the level of noise.

Template Matching [18] originates from being used in a plethora of learning-based applications. It is a machine vision technique that helps isolate positions or portions of images that match a reference, sometimes regardless of the direction, slope, noise, and other external parameters. These methods are flexible and easy to use when it comes to localizing or pinpointing items within other items. It can be used in systems to train recognition of number plates in real-time videos and many more applications. The only factor that limits the use of these methods is the increased requirement of CPU and RAM, as the complexity for the problem increases.

Template Matching methods are designed based on the intuition that, given a matching image aka the template image and a source image which needs to be analyzed for the local positions of the template, the algorithm identifies all the locations of the object referenced in the template image in the input image. Depending on the scenario of the question at hand, we can introduce inflections like orientation change, into the template.

This paper explains the novel use of such a technique in eliminating noise from images by providing a sample noise template to be matched over the input image. The matched areas are then patched up using either simple binary black and white patches or an ensemble technique by using Image Inpainting. The paper also displays the comparisons of results and standard estimators with respect to a few other methods.

4 Template Matching

4.1 Naive Template Matching

The operation of the template matching algorithm can be shown by looking at the sample pictures given below. Here, the reference image being selected is a template of the coin.

The algorithm now proceeds into a sequential forward search of matching pixels. That is, it positions the template at one location at a time and measures the similarity. This measurement is some numeric constant that tells how structurally similar two images are when placed on top of each other. Finally, it pinpoints the locations that provide the best similarity measures as the probable template occurrences. There can be multiple occurrences of the template as shown above (Fig. 1).

Fig. 1 Left: template. Right: input

4.2 Image Correlation Measures

One of the subproblems that occur in the design given above is finding the similarity index of the template image that is devoid of all external factors after alignment and the intersection portion of the two images are taken, which is analogous to computing a similarity measure of two images of identical dimensions in the same attribute space.

Normalized Cross-Correlation The basic method of computing the relation between two images is known as cross-correlation, which is basically the simple sum of the products of corresponding pixel values of the two images taken one at a time. The main drawback is its bias toward inflections in global illumination of the images, i.e., lustrousness or radiance of an image could simply magnify its cross-correlation with its partner image, even if the second image does not match with the first image at all. So the simple solution to that is to normalize and then interpret the autocorrelation. This has a good effect on the statistical properties of the estimated autocorrelations.

$$CrossCorrelation(ImageA, ImageB) = \Sigma_{xy} ImageA(x, y) * ImageB(x, y)$$
$$(1)$$

To tackle the disadvantage of using cross-correlation, we use an enhanced normalized formula that produces results in the range $[-1, 1]$, where one represents two images that are highly identical.

$$NCC(ImageA, ImageB) = \Sigma(A(x, y) - \mu(A)) * (B(x, y) - \mu(B))/N\sigma_1\sigma_2$$
$$(2)$$

Identification of matches Moving further, now the task is to decide which points of the reference image are viable in order to be classified as actual matches. A point is classified as a match if one of the three criteria given below is satisfied

Fig. 2 Marked locations after template matching

1. Correlation is greater than some predefined user threshold value (i.e, greater than 0.6)
2. Local maxima correlation (i.e., stronger correlation than the neighboring pixels)
3. Combination of both (Fig. 2).

5 Proposed Method

5.1 Data Collection

A set of 144 estampages was provided to us by the Archaeological Survey of India (ASI). Novel denoising techniques such as Gaussian, Median, and Wiener filters are not very effective. This is due to the fact that the noise is usually prominent in a certain area and this cannot be detected by such filters. A sample estampage is shown below. In most of these estampages the noise is distinct and looks like dots surrounding the characters. Some of the estampages being provided to us are in binary black and white while most are not.

5.2 Methodology

- A template is created which consists of the noise in the epigraph. The size of this template is about 7x7 pixels. A nested median filter is applied to the image as it provides clearer demarcation between text and background.
- The epigraph is binarized using a threshold value. The threshold value is determined by using Otsu's method. A binary image provides better results since it is clearer to identify noise and remove it by merging it with the background.
- The template matching algorithm is run and a mask is created of all matches. All images above a certain threshold value are matched. This threshold value has been found out to be approximately near 0.6. The algorithm is made to run for five iterations with an increase in the template size by 10% every iteration. Any further increase in size of the template results in information loss which is undesirable.
- In some cases, it may be undesirable to binarize the image. To handle such cases, we create a mask of all the matched noise and then apply a technique called image inpainting which merges the focus areas of the mask with the background of the image (Figs. 3, 4 and 5).

Algorithm 1 Multiscale Template Matching

1: **procedure** MSTM($input, template$)
2:    $img \leftarrow copy(input)$
3:
4:    **while** $ssim\_img\_prev > ssim(img)$ **do**
5:      $ssim\_img\_prev \leftarrow ssim(img)$
6:      $img \leftarrow median\_filter(img)$
7:    **end while**
8:    **while** $iterations < max\_iterations$ **do**
9:      $result \leftarrow matchTemplate(img, template)$
10:      $iterations \leftarrow iterations + 1$
11:      $threshold \leftarrow val$
12:      $locs \leftarrow getLocs(result, threshold)$ //final list of locations
13:      **if** $binary\_image$ **then**
14:        **for** point in locs **do**
15:          patch point
16:        **end for**
17:      **end if**
18:
19:      **if** non_binary_images **then**
20:        $mask \leftarrow copy(img)$
21:        create mask @points
22:        $img \leftarrow inpaint(mask, img)$
23:      **end if**
24:
25:    **end while**
26:
27:    **return** img
28: **end procedure**

(a) Source/Input

(b) Median Filter

(c) Otsu Thresholding

(d) Invert & Template Matching

(e) Result

(a) 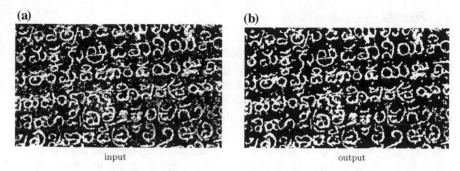 **(b)**

input output

Fig. 3 Sample 1 input and output

(a) 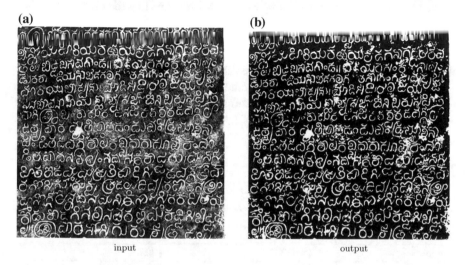 **(b)**

input output

Fig. 4 Sample 2 input and output

(a) 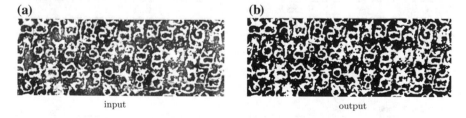 **(b)**

input output

Fig. 5 Sample 3 input and output

6 Results

The proposed algorithm is run on a set of 144 images and estimators such as Structural Similarity Index (SSIM) and Peak Signal-to-Noise Ration (PSNR) are computed to determine the loss in valuable data. The below table depicts the input and output image along with the value of the estimators. The average SSIM value obtained was 87%. Since SSIM estimator is being applied between the noise image and output image a value equivalent to about 85% is an indication that noise removal is being done whilst preserving the integrity of the characters. The proposed method is also time efficient since it did not take more than 1 sec to run on any input image. The estimators require that both the test images be of the same dimensions as well as have the same aspect ratio. Human vision can easily determine that the proposed method has provided better results than novel denoising techniques. Any further scaling up of the template will result in loss of character integrity. The below table depicts the value of the estimators for a sample of 10 images. The first three entries are those of the images depicted in the previous page (Tables 1 and 2).

These results were compared with other methods as well and have proved to remove more noise than other methods while relatively still maintaining the structural similarity of the image, i.e., As the average PSNR value for the proposed method

Table 1 Template matching output

Image name	SSIM	PSNR
Sample 1	0.84547	16.13389
Sample 2	0.85694	16.83282
Sample 3	0.89943	18.48997
Sample 4	0.94228	20.65224
Sample 5	0.88590	17.45861
Sample 6	0.95963	22.23564
Sample 7	0.85784	16.76800
Sample 8	0.97690	24.76808
Sample 9	0.81081	15.59777
Sample 10	0.93108	19.86717

Table 2 Method comparison

Method name	Average SSIM	Average PSNR
Weiner filter	0.9547	28.336
Median filter	0.9410	26.033
Nested RLC	0.9283	23.754
Multi-template matching	0.8966	18.8804

is the lowest it removes more noise than other standard methods, yet maintaining a relatively high SSIM by not removing the text.

6.1 Conclusion

The aim of this research paper is to complete one section of a larger pipeline that starts with simple script images and ends at character recognition. A good OCR is characterized by its precision. This can be achieved with better quality images which are denoised and intact in terms of their structure. This paper thus focuses on automating that denoising component using a novel algorithmic approach using template matching. This uses convolution of a template over the source and removes the matched template locations. The model was tested on the given dataset. Over and beyond that it was used to clear up synthetically created noisy images. To vindicate the efficiency of the model it has been compared to other existing approaches in terms of the realistic outputs produced. It is observed that our approach does a lot better than usual approached in removing the noise present in the image and yet maintaining the structural integrity of the foreground.

Acknowledgements The authors would like to thank the Director and staff members of Archaeological Survey of India, Mysore for providing dataset for this research and also PES university staff for their constant support.

References

1. Preethi, P., et al.: Denoising epigraphical estampages using nested run length count. In: International Conference on Ubiquitous Communications and Network Computing. Springer, Cham (2019)
2. Kar, R., et al.: Novel approaches towards slope and slant correction for tri-script handwritten word images. Imaging Sci. J. 1–12 (2019)
3. Gangamma, B., Srikanta Murthy, K., Singh, A.V.: Restoration of degraded historical document image. J. Emerg. Trends Comput. Inf. Sci. 3(5), 792–798 (2012)
4. Drira, F., LeBourgeois, F., Emptoz, H.: Restoring ink bleed-through degraded document images using a recursive unsupervised classification technique. In: Document Analysis Systems VII, pp. 38–49 (2006)
5. Shirai, K., Endo, Y., Kitadai, A., Inoue, S., Kurushima, N., Baba, H., et al.: Character shape restoration of binarized historical documents by smoothing via geodesic morphology. In: 2013 12th International Conference on Document Analysis and Recognition (ICDAR), pp. 1285–1289 (2013)
6. Karthik, S., Mamatha, H.R., Srikanta Murthy, K.: An approach based on run length count for denoising the Kannada characters. Int. J. Comput. Appl. 50(18), 42–46 (2012)
7. Hedjam, R., Cheriet, M.: Historical document image restoration using multispectral imaging system. PR 46, 2297–2312 (2013)
8. Bannigidad, P., Gudada, C.: Restoration of degraded historical Kannada handwritten document images using image enhancement techniques. In: International Conference on Soft Computing and Pattern Recognition (SoCPaR 2016), pp. 498–508 (2016)

9. Xiong, W., Xu, J., Xiong, Z., et al.: Degraded historical document image binarization using local features and support vector machine (SVM). Optik **164**, 218–223 (2018)
10. En, S., Petitjean, C., Nicolas, S., Heutte, L., Jurie, F.: Pattern localization in historical document images via template matching. In: 2016 23rd International Conference on Pattern Recognition (ICPR) (2016)
11. Ouyang, W., Tombari, F., Mattoccia, S., Di Stefano, L., Cham, W.-K.: Performance evaluation of full search equivalent pattern matching algorithms. In: PAMI (2012)
12. Hel-Or, Y., Hel-Or, H., David, E.: Matching by tone mapping: photometric invariant template matching. IEEE Trans. Pattern Anal. Mach. Intell. **36**(2), 317–330 (2014) (Online)
13. Elboher, E., Werman, M.: Asymmetric correlation: a noise robust similarity measure for template matching. IEEE Trans. Image Process. (2013)
14. Otsu, N.: A threshold selection method from gray-level histograms. IEEE Trans. SMAC **9**, 62–66 (1979)
15. Keiran, G.: Structural similarity index simplified, Occasional Texts in the Pursuit of Clarity and Simplicity in Research. Series 1, Number 1 (2015)
16. Gupta, P., et al.: A modified PSNR metric based on HVS for quality assessment of color images. In: 2011 International Conference on Communication and Industrial Application. IEEE (2011)
17. Mastin, G.A.: Adaptive filters for digital image noise smoothing: an evaluation. Comput. Vis. Graph. Image Process. **31**(1), 103–121 (1985)
18. Wang, D.C.C., Vagnucci, A.H., Li, C.C.: Digital image enhancement: a survey. Comput. Vis. Graph. Image Process. **24**(3), 363–381 (1983)

Security Issues Due to Vulnerabilities in the Virtual Machine of Cloud Computing

Swapnil P. Bhagat⬤, Vikram S. Patil⬤ and Bandu B. Meshram⬤

Abstract The virtual machine is a medium for provisioning cloud resources to customers. Cloud customers are accountable for configuration and security of the applications and the operating system running on virtual machines. The responsibilities of cloud customers include inbound and outbound traffic flow, access control, security configuration and practices. This work analyzes the vulnerabilities inside the virtual machine and security threats engendered due to it. Some of the security threats on a virtual machine are demonstrated with practical implementation on Amazon Web Service (AWS) as a cloud service provider. This paper also discusses general defense mechanisms for these security threats in a virtual machine.

Keywords Virtual machine · Vulnerabilities · Security threats · Defense mechanism

1 Introduction

Cloud computing assists in providing cost-effective, convenient, flexible and efficient services. Reasonable ownership cost and maintenance cost of resources in cloud computing entice organization towards it. According to Gartner study [1] public cloud services market is projected to grow $278.3 billion till 2021, up from $175.8 billion in 2018.

Accountability of cloud customers and cloud service provider changes with the service model and deployment model of cloud computing [2]. Both cloud customers and cloud users are accountable for security. The extent of security responsibilities

S. P. Bhagat (✉) · B. B. Meshram
Veermata Jijabai Technological Institute (VJTI), Mumbai, India
e-mail: spbhagat_p17@ce.vjti.ac.in

B. B. Meshram
e-mail: bbmeshram@ce.vjti.ac.in

V. S. Patil
SIES Graduate School of Technology, Navi Mumbai, India
e-mail: vikrams.patil@gmail.com

© Springer Nature Singapore Pte Ltd. 2020
V. Bhateja et al. (eds.), *Intelligent Computing and Communication*,
Advances in Intelligent Systems and Computing 1034,
https://doi.org/10.1007/978-981-15-1084-7_60

depends on the level of control in a cloud environment. More power brings more accountability for security.

Virtual machines are the combination of virtualized infrastructures like virtual CPU (vCPU), virtual RAM and virtual NIC. In the Infrastructure as a Service (IaaS) model, these virtualized infrastructures are allotted according to customer's requirement. The virtual infrastructures in cloud computing are engendered and managed by a cloud service provider. So, the customers do not need to bother about infrastructure and its maintenance (like power supply, cooling system, fault tolerance, natural hazards). Cloud customer needs an operating system to utilize these virtualized resources. However, cloud customer is responsible for the configuration and maintenance of the operating system. It includes security policies, regular updates, monitoring utilization of resources, and so on. Cloud customers are also entirely accountable for the applications installed on this operating system. Position of virtual machine in cloud computing stack and its adjacent components are depicted in Fig. 1.

Physical infrastructures are situated at the bottom of the cloud computing stack. All these physical infrastructures are converted into virtual infrastructure by the virtual machine monitor (VMM). It is also called as hypervisor. It provides synchronization between physical infrastructure and virtual infrastructure. According to the customer's demand these virtual infrastructures are combined to form a virtual machine. But it is nothing but a set of virtual hardware. The operating system needs to be installed to utilize this virtual hardware. Now the customer can install any software or application (APP) on it and provision services to his users.

Section 2 of this paper discusses the vulnerabilities and attacks in a virtual machine. Practical implementations of some the attacks on a virtual machine are also described in this section. Section 3 elaborates on general defense mechanisms for the attacks discussed in Sect. 2, whereas Sect. 4 concludes the work.

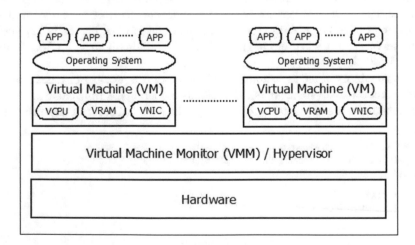

Fig. 1 Virtual machine and its position in the cloud computing stack

2 Vulnerabilities and Attacks on Virtual Machine (VM)

Vulnerabilities in the virtual machine can impact not only the services provided by the virtual machine but also the operations of the cloud environment. Such vulnerabilities and its attacks are discussed in this section.

2.1 Vulnerabilities in VM

Virtual machines behave like any standard standalone physical systems. Thus, virtual machines bear the same security threats as a physical system [3]. So, it can be estimated that the primary sources of vulnerabilities in virtual machines are operating systems, misconfiguration and applications installed on it.

Misconfiguration. System administrator is accountable for system configuration including the implementation of security policies. Disregarding the security configuration leads to loopholes in the system; for instance, unused open ports, use of default account and password, weak authentication and authorization (including missing or ineffective multi-factor authentication, weak access control of resources, weak or ineffective credential recovery and forget password process), enable or install unnecessary features [4]. Misconfiguration is one of the prime factors behind the attacks on IaaS [5]. Poorly configured permission leads to most of the data breach attacks on cloud web application [6]. A default installation of the MongoDB database could be accessed without any authentication or access control when browsing the open MongoDB 27017 port. Cybersecurity expert Chris Vickery found data stored in an Amazon Web Services (AWS) MongoDB database, including personally identifiable information (PII) and voting records of 93 million Mexican voters to be at-risk [7].

Bug in Operating System. Operating systems are made up of thousands of lines of code. Developers add debug code to test such vast code. After testing, such codes have to be eliminated, because it can result in a backdoor which can aid an attacker for unauthorized access [8]. Another potential consequence of large code is a bug. It is part of the code which gives an unexpected result. Bugs may assist an attacker in penetrating the operating system of a virtual machine [9]. Dirty copy-on-write (dirty COW) is an example of such a bug in an operating system. Dirty COW or CVE-2016-5195 is a privilege escalation vulnerability in the Linux Kernel which allows an unprivileged user to gain write access of read-only memory [7].

Software and Application Vulnerabilities. The customer may deploy applications like DBMS, web server, browser, phpMyAdmin. There may be vulnerabilities in such deployed applications. It may include unlimited authentication try, use of plain text, weakly encrypted password, forwarding data to the application without validating, filtering or sanitizing, security misconfiguration, insecure direct object references and missing function-level access control [10]. Potential security threats due to vulnerable applications are revealed in OWASP [4].

2.2 Attacks on VM

Vulnerable operating system and application make virtual machine vulnerable. Since there are multiple versions of operating systems and applications, the attack surface of a virtual machine is very large. Some of such attacks are given below.

Port Scanning. This attack is an example of exploitation of misconfiguration of security policies. Ports act as an entry point of the system. Port scanning attack does not harm the victim rather it gives information about victim machines like open, filtered and closed ports, firewall rules, router filtering, IP address and MAC address [11]. It is a prerequisite for many attacks.

Brute Force Attack. This attack exploits the weak authentication system. It tries to find the credentials of the victim by inserting all possible credentials. The attacker tries all possible combinations of words, digits and symbols. This attack can be hindered by restricting the number of attempts or using two-way authentication.

Injection Attack. It bypasses the web application and executes the malicious query. This attack causes due to forwarding untrusted input data to a database without verifying it. The consequence of such attacks is the leakage of sensitive data. It includes data like username and password which further utilized to perform malicious activities in victim's account. Injection attack includes NoSQL injection, lightweight directory access protocol (LDAP) injection attack and OS injection [4, 10]. Injection vulnerabilities are a consequence of flaws in the design and architecture of application [12].

DoS Attack. Denial of service (DoS) or distributed DoS attack is any malicious activity which prevents the availability of service. In the case of a virtual machine, it prevents the availability of data or service provisioned by it. DoS attack devastates virtual machine resources by producing an enormous amount of workload [13]. SYN flood attack, IP address attack (packet contains the same source and destination IP), IP packet option attack (set all quality bit to increase processing), reflection-based DoS attack (insert victim's IP in source IP address field of packet), amplification attack are some examples of DoS attack possible on virtual machine. According to CISCO report, globally, the total number of DDoS attacks will double to 14.5 million by 2022 from 7.5 million in 2017 [14]. In February 2018, GitHub experienced DDoS attack generating record-breaking traffic at a rate of 1.35 terabits per second [15]. This data is sufficient to understand that DDoS attacks are not only getting bigger but devastating than ever before. A detail classification of DoS attacks in a cloud computing environment is elaborated in [3]. Since protocols like HTTP, XML are widely employed in cloud computing, DDoS attacks based on these protocols are more dangerous than traditional DDoS attack [16].

Malware. Malware, short form for malicious software, bypass operating system security to perform some sort of task (like leak sensitive data, create loopholes or backdoors, perform an attack on another system) for an attacker [16]. Malware is one of the most favorite methods to transform a normal system into bot or zombie. These bots are further used to perform a large-scale attack. One of such malwares is identified in May 2018 known as VPNFilter. It affected more than 500,000 routers worldwide. VPNFilter not only created massive botnet but also able to spy on and manipulate web activity on the compromised routers [17]. According to 2018 Cyberthreat Defense Report, malware is one of the greatest cyber threat concerns [18].

Backdoor. There are many causes of backdoor, including forgetting to remove debug code added during code testing, bugs and poor configuration [8]. Backdoors may result in illegal remote access without detecting [11]. Since it is added for testing, the developer builds it in such a way that the operating system will be unaware about it, which hides the attacker from detection. There are various techniques to gain control of a system such as port binding, connect-back technique, connection availability abuse, legitimate platform abuse, common service protocol/file header abuse, protocol/port listening, custom DNS lookup use and port reuse [9].

2.3 Experiments

This section provides practical knowledge of some of these security threats. For this experiment, Amazon Web Service (AWS) is used as a cloud service provider. An instance with Ubuntu Server 18.04 LTS operating system is created and used for the experiment. Instance type selected is t2.micro which contains one virtual CPU and 1 GB RAM. Its public IPv4 address was 13.126.99.125. WebGoat web application is deployed on a port 8080 of the instance. WebGoat is a deliberately insecure web application maintained by OWASP. In this practical implementation, it is used as a target web application. Port scanning, brute force attack and shopping cart concurrency attack are performed on the virtual machine.

Port Scanning. In this attack, the NMAP tool is used to perform port scanning. The instance created to launch the WebGoat application is scanned by the Nmap tool. The following command is used for scanning the ports.

$ nmap <Victim IP address>

It gives the list of open ports of the victim machine. The result of the scanning is shown in Fig. 2. Since the instance providing only the WebGoat web application runs on 8080 port, remaining ports need to be closed to avoid any intrusion.

Brute Force Attack. In the brute force attack, the attacker tries all possible combination of credentials to find the correct one. In this case, Burp tool is used to perform a brute force attack on WebGoat web service running on virtual machine. Burp tool tries all the credential provided by the attacker against WebGoat authentication system. By verifying the length of response, the attacker is able to find the matched credential which is further used to perform a malicious activity in the user's account. Different length of responses is shown in Fig. 3.

Shopping Cart Concurrency Flaw. Web applications have to handle many HTTP requests concurrently. There may be variables which are not secure from the perspective of concurrent threats. Web services especially e-commerce websites need to maintain valid states of concurrent activities. It is a consequence of design flaws.

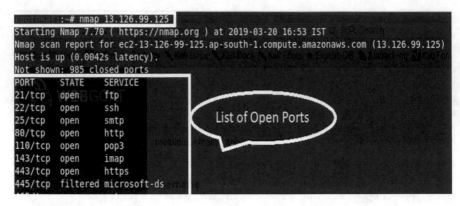

Fig. 2 Port scanning using Nmap tool

Request ▲	Payload	Status	Error	Timeout	Length
0		302	☐	☐	268
1	admin	302	☐	☐	268
2	password	302	☐	☐	268
3	local	302	☐	☐	268
5	guest	302	☐	☐	264

Fig. 3 A result of brute force attack

WebGoat web application elaborated this attack and provided the environment to perform it.

Attacker selects a low-cost item and clicks the purchase button. So, the low-cost item is added into the cart. But before proceeding for payment, the attacker opens another tab and select the high-cost item. When he clicks on the update, the cart is updated with the new high-cost item. Now the user goes back to the former tab and continues payment procedure. Since web developer uses the same variable in concurrent threads, the attacker gets the high-cost item at low-cost item price as depicted in Fig. 4.

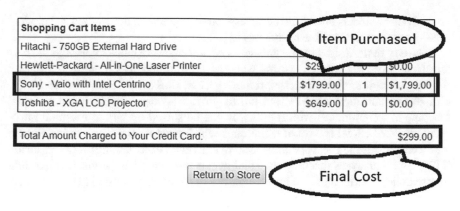

Fig. 4 Shopping cart confirmation page

3 Defense Mechanism for VM

Attacks on virtual machine influence the functionalities of it. It mostly impacts on confidentiality, integrity and availability of services provided by virtual machines. Such attacks can be prevented by the following mechanisms.

Secure Configuration. Effective and secure configuration can avoid most of the attacks possible on the virtual machine. It incorporates closing unused open ports, changing the default password, access restriction based on user role, minimal platform without any unnecessary features, sending security directives to the client [4, 6].

Regularly Update and Patch Software. Software update is an effective and easy way to remove vulnerabilities of software [10]. Operating system developers also supply patches to eliminate bugs. It assists in making the operating system more secure. Such updates can remove backdoor, bugs and replace insecure code [4, 8]. According to the 2014 Cyberthreat Defense Report, 75% of attacks use publicly known vulnerabilities in commercial software that could be prevented by regular patching [19].

Secure Software Development. Software developers concentrate more on the functional requirement like efficiency, speed and convenience. So, the nonfunctional requirement (like security, privacy and fault tolerance) is ignored. Eventually, it leads to security threats like software bugs, insecure code, backdoors and exposure of sensitive data. It can be prevented during software development. It is accomplished by examining security parameters in all possible phases of software development [2, 4, 8].

Access Control. Mechanisms, like multi-factor authentication, limit or increasingly delay failed login attempts, log all failed attempts and alert administrator, can improve the authentication system. It can avoid attacks like automated, credential stuffing, brute force attack and stolen credential reuse attack [4]. An effective access control policy hinders unauthorized access. Strong authentication system can prevent the system from attacks like backdoor [12].

Security tools. Software like anti-virus or anti-malware helps to prevent the system from viruses, malware, worms whereas firewall and IDS monitor the packet for the prevention of unauthorized remote access, DoS attack, malicious traffic [9, 13]. But monitoring tools like IDS fail to provide accurate results when a virtual machine is compromised with rootkits. It can be solved by virtual machine introspection (VMI). VMI is a set of techniques which empowers the user to inspect the VM from outside the guest operating system and analyze the running programs inside it [20, 21]. It analyzes machine state and events through the hypervisor. According to researchers, VMI can be an influential tool in the cloud forensic [22].

3.1 Tabular Representation of Vulnerabilities, Attacks and Defense Mechanism

Vulnerabilities of the virtual machine and its possible attacks are discussed in the previous section. Defense mechanism which can help to hinder these attacks is also elaborated. This section summarizes this information in tabular form as shown in Table 1.

Table 1 Summary of vulnerability, attack and defense mechanism

Vulnerability	Example	Attacks	Defense mechanism
Misconfiguration	Unused open ports, default credentials, weak authentication and authorization system, enable unnecessary features	Port scanning, brute force attack, DDoS attack	Secure configuration practice, security tools (IDS, IPS, firewall)
Bugs in operating system	Backdoor, dirty COW (copy-on-write), insecure code	Remote access control, malware, backdoor attack	Regular update and patch, robust access control
Software and application vulnerabilities	Unlimited authentication try, use of plain text, weakly encrypted password, lack of validation, filtering and sanitization of input	Brute force attack, injection attack, man in middle attack, DDoS attack	Robust access control, secure software development, security tools, update software regularly

4 Conclusion

In the Infrastructure as a Service (IaaS) model, cloud customers have more control over virtual machines and are accountable for operating system and application running over it. Negligence in security policies and practices can lead to security threats. These security threats can affect the functions of virtual machines or in the worst case can be remotely controlled by malicious users. Implementing robust security policies and strictly following security practices can prevent such attacks to a large extent. Along with it, security tools like firewall, IDS, IPS can be used for monitoring as well as detecting any malicious activities.

Many researches are concentrating on improving the security of cloud computing. Cloud service providers also concentrate on the security of customers but show less interest in digital forensic of cloud attacks. Digital forensics of security incidents occurred on a virtual machine is not much explored. So, there is a scope for future research in digital forensic of virtual machine attacks.

References

1. Gartner Forecasts Worldwide Public Cloud Revenue to Grow 17.3 Percent in 2019. https://www.gartner.com/en/newsroom/press-releases/2018-09-12-gartner-forecasts-worldwide-public-cloud-revenue-to-grow-17-percent-in-2019
2. Hashizume, K., Rosado, D.G., Fernández-Medina, E., Fernandez, E.B.: An analysis of security issues for cloud computing. J. Intern. Serv. Appl. (2013)
3. Masdari, M., Jalali, M.: A survey and taxonomy of DoS attacks in cloud computing. Secur. Commun. Netw. J. **9**(16), 3724–3751 (2016)
4. OWASP top 10, 2017: The OWASP Foundation (2017)
5. Xiao, Z., Xiao, Y.: Security and privacy in cloud computing. IEEE Commun. Surv. Tutor. **15**(2), 843–859 (2013)
6. Wueest, C., Barcena, M.B., O'Brien, L.: Mistakes in the IaaS cloud could put your data at risk. Symantec Report (2015)
7. Top Threats to Cloud Computing: Deep Dive. https://downloads.cloudsecurityalliance.org/assets/research/top-threats/top-threats-to-cloud-computing-deep-dive.pdf
8. Backdoor Attacks. http://www.omnisecu.com/security/backdoor-attacks.php
9. Chiu, D., Weng, S.H., Chiu, J.: Backdoor use in targeted attacks. A Trend Micro Research Paper (2014)
10. Most Common Web Security Vulnerabilities. https://www.toptal.com/security/10-most-common-web-security-vulnerabilities
11. Modi, C.N., Acha, K.J.: Virtualization layer security challenges and intrusion detection/prevention systems in cloud computing. Supercomputing **73**, 1192–1234 (2017)
12. Modi, C., Patel, D., Borisaniya, B., Patel, A., Rajarajan, M.: A survey on security issues and solutions at different layers of Cloud computing. J. Supercomput. **63**(2), 561–592 (2013). Publisher Springer, US
13. AWS best practices for DDoS resiliency. https://d1.awsstatic.com/whitepapers/Security/DDoS_White_Paper.pdf (2018)
14. Cisco Visual Networking Index: Forecast and Trends, 2017–2022. https://www.cisco.com/c/en/us/solutions/collateral/service-provider/visual-networking-index-vni/white-paper-c11-741490.html

15. Most Famous DDoS Attacks. https://www.a10networks.com/resources/articles/5-most-famous-ddos-attacks
16. Khalil, I.M., Khreishah, A., Azeem, M.: Cloud computing security: a survey. Computers **3**, 1–35 (2014)
17. The Worst Cybersecurity Breaches of 2018 so Far. https://www.wired.com/story/2018-worst-hacks-so-far/
18. Cyberthreat Defense Report. https://cyber-edge.com/wp-content/uploads/2018/03/CyberEdge-2018-CDR.pdf
19. The Treacherous 12—Top Threats to Cloud Computing + Industry Insights, Cloud Security Alliance Report (2017)
20. Rakotondravony, N., Taubmann, B., Mandarawi, W., Weishäup, E., Xu, P., Kolosnjaji, B., Protsenko, M., Meer, H.D., Reiser, H.P.: Classifying malware attacks in IaaS cloud environments. J. Cloud Comput. (2017)
21. Dykstra, J., Sherman, A.T.: Acquiring forensic evidence from infrastructure-as-a-service cloud computing: exploring and evaluating tools, trust, and techniques. Digit. Investig. J. **9** (2012)
22. Orr, D.A., White, P.: Current state of forensic acquisition for IaaS cloud services. J. Forensic Sci. Crim. Investig. **10**(1) (2018)

A TFD Approach to Stock Price Prediction

Bhabesh Chanduka, Swati S. Bhat, Neha Rajput and Biju R. Mohan

Abstract Accurate stock price predictions can help investors take correct decisions about the selling/purchase of stocks. With improvements in data analysis and deep learning algorithms, a variety of approaches has been tried for predicting stock prices. In this paper, we deal with the prediction of stock prices for automobile companies using a novel TFD—Time Series, Financial Ratios, and Deep Learning approach. We then study the results over multiple activation functions for multiple companies and reinforce the viability of the proposed algorithm.

Keywords Stock price prediction · Time series · Financial ratios · Deep learning

1 Introduction

Stocks are good metrics to evaluate the impending and current success of a company. Well-informed decisions regarding the purchasing and selling of stocks can help in preventing large losses. In the past, numerous approaches have been tried to predict the stock prices.

Skabar et al. [1] proposed a methodology by which neural networks can be trained indirectly, using a genetic algorithm-based weight optimization procedure, to determine buy and sell points for financial commodities traded on a stock exchange. Nayak [2] investigated the predictive power of the clustering technique on stock market data and its ability to provide stock predictions that can be utilized in strategies that

B. Chanduka (✉) · S. S. Bhat · N. Rajput · B. R. Mohan
Department of Information Technology, National Institute
of Technology Karnataka, Surathkal 575 025, India
e-mail: bhabesh.chanduka@gmail.com

S. S. Bhat
e-mail: swatibhat09@gmail.com

N. Rajput
e-mail: rneha7761@gmail.com

B. R. Mohan
e-mail: bijurmohan@gmail.com

© Springer Nature Singapore Pte Ltd. 2020
V. Bhateja et al. (eds.), *Intelligent Computing and Communication*,
Advances in Intelligent Systems and Computing 1034,
https://doi.org/10.1007/978-981-15-1084-7_61

outperform the underlying market using a brute force approach to the prediction of stock prices. A time series prediction approach which performs an evolutionary search of the minimum necessary number of dimensions embedded in the problem was put forth elegantly by Ferreira et al. [3]. This approach also worked on developing a novel clustering strategy to produce promising results. A neural model approach combining financial and economic theory was built in [4] and attempts to identify variables that drive stock prices were tried. In [5], Luo et al. tried PLR to decompose historical data into different segments. Temporary turning points (trough or peak) of the historical stock data were detected and inputted to the back propagation neural network (BPN)for supervised training of the model. Chang et al. [6] made further significant contributions to this field by trying an integrated system, CBDWNN by combining dynamic time windows, Case-Based Reasoning (CBR), and neural network for stock trading prediction. They successfully improved the forecasting results from BPN and presaged integrated systems for the literature. A Hybrid technique integrating a wide array of machine learning approaches was tried by Armstrong et al. [7] using technical analysis and economic analysis, leveraging various technical and economic indicators to identify and optimize the buy and sell triggers to maximize trading profits. Armstrong and Keevil [7] proposed a transit from the methodology of determining the decision from buying/selling to quantifying the tangible results in the form of regression results. Long-term profits were maximized based on S&P500 stock2 market index. However, this model had issues dealing with volatility in the stock prices. To handle this, Khaidem et al. [8] proposed a novel way to reduce stock market investment risk by predicting the returns using ensemble learning with indicators such as Relative Strength Index and stochastic oscillator.

In [9], Schoneburg showed that a deep learning model could produce promising results with regard to absolute value prediction even when a short time window spanning 10 days was used. We have extended this idea for multiple time windows—7, 13, 26, and 50 days. In [10], Lin et al. proposed a recurrent neural network approach to predict next day closing prices of the stocks. Their use of PCA helped further improve the results. In this paper, we have chosen feature selection over dimensionality reduction to achieve similar results. In [11], Huseyin et al. proposed the use of various technical indicators in conjunction with kernal principal component analysis and factor analysis to find the most relevant inputs for a multilayer perceptron and support vector regression. In [12], Lee applied TD(0), a reinforcement learning algorithm to Korean Stock price prediction. In [13], a nonlinear logistic function was applied to a FF-ANN for making predictions. Finally, in [14], a rough set theory and genetic algorithm approach was tried for making predictions.

The results of [15] show that artificial neural networks outperform traditional time series models when it comes to predicting stock prices. In this paper, we use a TFD—Time Series, Financial Ratios, and Deep Learning approach for predicting stock prices. Our approach is unique in that time series is not used to make the predictions but instead the parameters obtained by the analysis are used as features to make predictions, by passing as input to a deep neural network. The rest of the paper is organized as follows: Sect. 2 deals with the Methodology, Sect. 3 deals with the Results, followed by Sect. 4 with Conclusion and Future Work.

Fig. 1 Complete
methodology pipeline

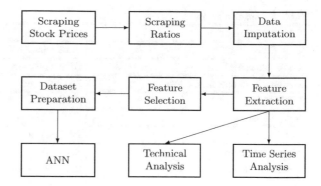

2 Methodology

Figure 1 elucidates the overall methodology proposed in this paper for the prediction of stock prices. The upcoming subsections throw more light on these individual steps in greater detail.

2.1 Scraping Stock Prices

As shown in the figure, the first step in the pipeline is scraping of the stock prices. For this purpose, a web scraper was used to collect the end of day opening, closing, high, and low prices of various automobile companies (source: Yahoo), spanning a time window of 5 years.

2.2 Scraping Ratios

The second step was scraping of the financial ratios for the corresponding stock prices, collected over a span of 5 years (source: money control) (Table 1).

Thirty-five different financial ratios were extracted (1). These ratios were later used for making predictions, after the required feature engineering.

2.3 Data Imputation

Missing values can adversely affect the accuracy of predictions. Missing data was particularly observed for the opening, closing, high, and low prices scraped for the past 5 years. The third step in the pipeline deals with the imputation of these values.

Table 1 Financial ratios

Basic EPS	Diluted EPS
Cash EPS	Book value/share
Book value/share	Dividend/share
Revenue from operations/share	PBDIT/share
PBIT/share	PBT/share
Net profit/share	PBDIT margin
PBIT margin	PBT margin
Net profit margin	Return on networth/equity
Return on capital employed	Return on assets
Total debt/equity	Asset turnover ratio
Current ratio	Quick ratio
Inventory turnover ratio	Dividend payout ratio
Dividend payout ratio	Earnings retention ratio
Cash earnings retention ratio	Enterprise value
EV/Net operating revenue	EV/EBITDA
MarketCap/Net operating revenue	Retention ratios
Price/BV	Price/net operating revenue
Earnings yield	

Since these values were Missing Completely at Random (MCAR), and the data was time series, a multiple imputation approach followed by averaging was taken. Markov Chain Monte Carlo technique was used for multiple imputations of these missing values. Fifty iterations were used for all rows where discontinuity in time series data was observed.

2.4 Feature Extraction

After data imputation, two different types of features were extracted: Time series features and technical indicators. Seven time series features were extracted for each company, by fitting a Seasonal Autoregressive Integrated Moving-Average with Exogenous regressors for the 5 years time series data. The seven features were the parameters (p, d, q)—the order of the model for the number of AR parameters, differences, and MA parameters, and (P, D, Q, s)—the order of the seasonal component of the model for the AR parameters, differences, MA parameters, and periodicity. These values were found using a brute force approach across all possible values. Those set of values were taken that maximized the Akaike Information Criterion for the corresponding time series data. After extracting time series features, technical analysis was performed on the closing prices using the following technical indicators:

Simple Moving Average (SMA), Exponential Moving Average (EMA), Regression Slope Moving Average Convergence Divergence (MACD), Bollinger Bands Rate of Change (ROC), Relative Strength Index (RSI).

Four features were extracted using each of the technical indicators (except Bollinger Bands, in which case $3 \times 4 = 12$ features were extracted), for a particular day, each feature being associated with one-time window. Time windows used were 7, 13, 26, and 50. Therefore, total number of technical analysis features extracted were 36 for a particular day.

2.5 Feature Selection

After the generation of features, there was a need to select only the most relevant features. In order to perform feature selection, the following were used:

1. Filter Methods

   (a) Pearson Correlation: selects k best, based on empirical relevance calculation.
   (b) F-Regression: selects k best, based on linear model to test individual effect of each of many regression coefficients.

2. Wrapper Methods

   (a) Recursive Feature Elimination: Recursive elimination of one feature every iteration based on the performance of the regressor until a desired number of k features remain.

3. Embedded Methods

   (a) Linear Regression: LASSO regression is done by augmenting an additional penalty term to the objective (cost) function of linear regression. This results in the coefficients of less important features approaching 0.
   (b) Random Forest: Using entropy measure, the information gain value for a subset of features can be ascertained. As it is performed by multiple random decision trees, the random forest model is known to produce good results.
   (c) LightGBM: a distributed and gradient boosted tree-based algorithm that selects the best features based on the information gain criteria.

These were applied independently of each other. If the algorithm took as input the number of features to be selected, 50 was used as the input. In all other cases, such as, for example regression, no such number was specified. Top 50 features, or the entire set of features selected by the feature selection techniques (whichever was minimum) were considered. As shown in Fig. 2, not all feature selection algorithms selected 50 features.

Those features that were selected by most number of feature selection techniques were given importance. To understand the importance of each feature, a voting type of system was established. This can be visualized with the aid of (Fig. 3).

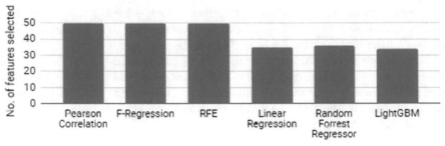

Fig. 2 Features versus number of feature selection algorithms that selected them

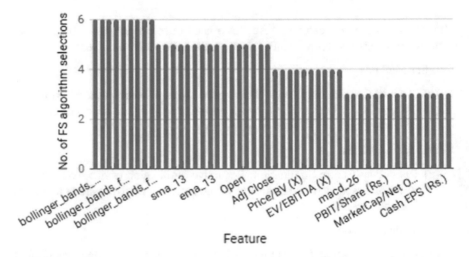

Fig. 3 Features versus the number of feature selection algorithms that selected them

2.6 Dataset Preparation

After feature selection, we obtain the most relevant subset of features obtained by performing time series analysis and technical analysis. In order to build an effective artificial neural network to make predictions, it is essential to have a meaningful and relevant dataset. For our purpose, this equates to having a set of features (independent variables) and having a column representing the predicted stock price (dependent variable). In order to make such a dataset, we build a dataset using the scraped prices. Say we are looking at the stock price of a particular company at Day_i (this constitutes our dependent variable or the predicted class). We use the features remaining after feature selection as a subset of features. We also use stock prices from Day_{i-50}, Day_{i-49}, ..., Day_{i-2}, Day_{i-1} as features in order to fully gauge the effect the past prices on the predictions. In total, we had 81 features (31 from feature selection and 50 previous day prices).

2.7 ANN

After building the dataset, the most important step is to build the artificial neural network. The specifics of the network are 81 input nodes, first hidden layer having 5 nodes, second hidden layer having 6 nodes, third hidden layer having 6 nodes, fourth hidden layer having 7 nodes, 1 output node.

The training dataset was split into training and testing sets, having ratio 4:1. Following this, the weights of the ANN were initialized using the Xavier initialization method [16]. The best activation functions for the input layer till the last hidden layer was found to be ReLU (among ReLU [17], Sigmoid and Tanh, as shown by Tables 2, 3 and 4). Since our output is predicted stock price, linear activation is used for the output layer. Adam Optimizer [18] was used as the optimizing hyperparameter. The model was trained using the backpropagation algorithm for 50 epochs and then the model was tested on the training set.

Table 2 Prediction results when Relu is used as the activation function

Company	Metric		
	RMSE	RMSLE	AR^2
Maruti Suzuki India Ltd.	28.67	0.43	0.27
Mahindra & Mahindra Ltd.	36.24	0.41	0.31
Bajaj Auto Ltd.	46.33	0.35	0.32
Hyundai Motor India Ltd.	31.26	0.39	0.26
Tata Motors Ltd.	44.18	0.33	0.25

Table 3 Prediction results when Sigmoid is used as the activation function

Company	Metric		
	RMSE	RMSLE	AR^2
Maruti Suzuki India Ltd.	33.20	0.45	0.31
Mahindra & Mahindra Ltd.	35.12	0.38	0.30
Bajaj Auto Ltd.	49.86	0.38	0.33
Hyundai Motor India Ltd.	32.29	0.39	0.29
Tata Motors Ltd.	44.23	0.34	0.25

Table 4 Prediction results when Tanh is used as the activation function

Company	Metric		
	RMSE	RMSLE	AR^2
Maruti Suzuki India Ltd.	31.25	0.43	0.28
Mahindra & Mahindra Ltd.	37.37	0.42	0.32
Bajaj Auto Ltd.	44.13	0.37	0.33
Hyundai Motor India Ltd.	28.02	0.37	0.25
Tata Motors Ltd.	43.75	0.32	0.24

3   Results

The model was tried on the training dataset for different companies, using different activation functions for the hidden layer nodes.

The metrics used were Root Mean Squared Error (RMSE), Root Mean Squared Logarithmic Error (RMSLE), and Adjusted R^2 (AR^2).

The model was tested for the following automobile companies: Maruti Suzuki India Ltd. (MSIL), Mahindra & Mahindra Ltd. (MNML), Bajaj Auto Ltd. (BAL), Hyundai Motor India Ltd.(HML), and Tata Motors Ltd. (TML).

To find the best activation function, the model was run using ReLU, Sigmoid, and Tanh as activation functions for the hidden layers. The results of our research are summarized in Tables 2, 3 and 4.

The results shown in Tables 2, 3, and 4 are those that are noted down after 50 epochs. Predictions are accurate to a minor margin of error as the permissible range of values for the training data is huge. It can also be inferred that ReLU activation function yields the best overall results when compared to the other activation functions.

Bar plots comparing the results of using each of the activation functions—ReLU, Sigmoid, and Tanh, across all metrics—RMSE, RMSLE, and AR^2 are shown in Figs. 4, 5, and 6.

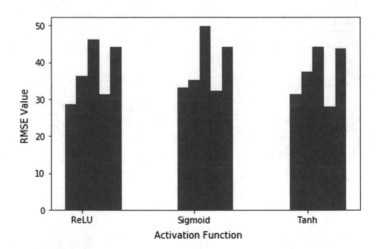

Fig. 4 Bar plot for varying RMSE values for different activation functions

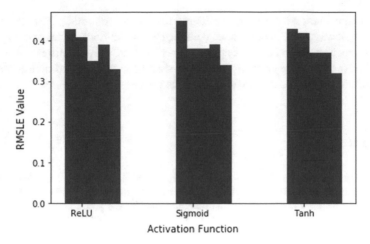

Fig. 5 Bar plot for varying RMSLE values for different activation functions

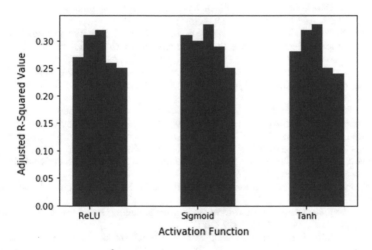

Fig. 6 Bar plot for varying AR^2 values for different activation functions

4 Conclusion and Future Work

A novel model based on time series analysis, technical analysis, and deep learning has been proposed for the prediction of stock prices. The algorithm performs well when there is enough past data available to make predictions, along with essential financial indicators. Multiple feature selection algorithms have been incorporated to select the most relevant features for predictions. The performance of the proposed model has been studied for different activation functions.

In our algorithm, we have not considered the macroeconomic parameters such as purchasing power parity of currencies with the Dollar and stock prices of competitors. In the future, a model incorporating these and other essential features will be tried. Besides this, other deep learning networks such as convolutional neural networks and recurrent neural networks will be tried as well.

References

1. Skabar, A., Cloete., I.: (2002) Neural networks, financial trading and the efficient markets hypothesis. In: ACSC 02: Proceedings of the Twenty-fifth Australasian Conference on Computer Science, pp. 241–249. Australian Computer Society, Inc., Darlinghurst, Australia
2. Nayak, R., Braak, P.: Temporal pattern matching for the prediction of stock prices. In: Proceedings of the Second International Workshop on Integrating Artificial Intelligence and Data Mining, pp. 95–103. Australian Computer Society, Inc, Darlinghurst, Australia
3. Ferreira, T., Vasconcelos, G.C., Adeodato, P.: A new evolutionary method for time series forecasting. In: ACM Proceedings of Genetic Evolutionary Computation Conference-GECCO, pp. 2221–2222. ACM, Washington, DC (2005)
4. Oliveira, F., Nobre, C., Zarate, L.: Applying artificial neural networks to prediction of stock price and improvement of the directional prediction index Case study of PETR4, Petrobras, Brazil. Expert Syst. Appl. 40(18), 7596–7606 (2013). ISSN 0957-4174
5. Luo, L., Chen, X.: Integrating piecewise linear representation and weighted support vector machine for stock trading signal prediction. Appl. Soft Comput. 13(2), 806–816 (2013). ISSN 1568-4946
6. Chang, P., et al.: A neural network with a case based dynamic window for stock trading prediction. Expert Syst. Appl. 36(3), 6889–6898 (2009). ISSN 0957-4174
7. Armstrong, P., Keevil, S.: A hybrid stock trading framework integrating technical analysis with machine learning techniques. J. Financ. Data Sci. 4(1), 42–57 (2016)
8. Khaidem, L., Saha, S., Keevil, S.: Predicting the direction of stock market prices using random forest, arXiv:1605.00003v1 (2016)
9. Schneburg, E.: Stock price prediction using neural networks: a project report. Neurocomputing 2(1), 17–27 (1990)
10. Lin, X., Yang, Z., Song, Y.: Short-term stock price prediction based on echo state networks. Expert Syst. Appl. 36(3), 7313–7317 (2009)
11. Ince, H., Trafalis, T.B.: Kernel principal component analysis and support vector machines for stock price prediction. Lie Trans. 39(6), 629–637 (2007)
12. Lee, J.W.: Stock price prediction using reinforcement learning. In: ISIE 2001. 2001 IEEE International Symposium on Industrial Electronics Proceedings (Cat. No. 01TH8570), vol. 1. IEEE (2001)
13. Tiwari, A., et al.: ANN-based classification of mammograms using nonlinear preprocessing. In: Proceedings of 2nd International Conference on Micro-Electronics, Electromagnetics and Telecommunications. Springer, Singapore (2018)
14. Rathi, R., Acharjya, D.P.: A framework for prediction using rough set and real coded genetic algorithm. Arab. J. Sci. Eng. 43(8), 4215–4227 (2018)
15. Adebiyi, A.A., Adewumi, A.O., Ayo, C.K.: Comparison of ARIMA and artificial neural networks models for stock price prediction. J. Appl. Math. (2014)
16. Glorot, X., Bengio, Y.: Understanding the difficulty of training deep feedforward neural networks. In: Proceedings of the Thirteenth International Conference on Artificial Intelligence and Statistics (2010)
17. Agarap, A.F.: Deep learning using rectified linear units (relu), arXiv:1803.08375 (2018)
18. Kingma, D.P., Ba, J.: Adam: a method for stochastic optimization, arXiv:1412.6980 (2014)

Recognizing and Refinement of Contorted Fingerprint Based on Various Unique Fingerprint Images

G. Padmapriya, M. Aruna, B. Arthi and V. Vennila

Abstract Nowadays there is a chance for false mismatch of fingerprints due to elastic distorted. This is the main problem which wholly affects the fingerprint recognition techniques and applications mainly in false recognition techniques in duplication recognition. In such applications the intruders modify their original fingerprints and they will try to mismatch their identification in all biometric security systems and applications. The proposed work mainly concentrates on developing a novel algorithm to identify and recognize the original fingerprint and to minimize the skin distorted. The proposed work contorted recognizing and refinement is proposed as classification problem for registered ridge orientation map class and period map of a fingerprint class which is considered as regression problem. Here the input image is contorted fingerprint image and the output obtained is distorted fields. To evaluate the crisis a database is created with large amount of distorted reference fingerprints with its distortion fields. If any nearest neighbour of the input fingerprint is identified then their respective distortion field is utilized to identify the normal fingerprint image. The evaluation results have show efficient output.

Keywords Elastic · Ridge orientation map · Rectification · Regression problem

G. Padmapriya (✉) · M. Aruna · B. Arthi
Department of Computer Science and Engineering, Saveetha School of Engineering, Saveetha
Institute of Medical and Technical Sciences, Chennai, Tamil Nadu, India
e-mail: gpadmapriyame@gmail.com

M. Aruna
e-mail: arunaraadhi@gmail.com

B. Arthi
e-mail: arthi1981@gmail.com

V. Vennila
K.S.R College of Engineering, Tiruchengode, Tamil Nadu, India
e-mail: vennilview@gmail.com

© Springer Nature Singapore Pte Ltd. 2020   645
V. Bhateja et al. (eds.), *Intelligent Computing and Communication*,
Advances in Intelligent Systems and Computing 1034,
https://doi.org/10.1007/978-981-15-1084-7_62

1 Introduction

Nowadays the digital image portrays a significant role in every aspects of the society. Many applications like administration, media, legal works are more depend on the digital images. In legal act digital photography proof plays a vital role mainly to take judgment result for any particular act. The main disadvantage of the digital image is, it can be matched with other image using the image editing software. Being aware of those images editing software we should provide the good authenticated digital images based on the application in which we work. Example of applications which requires the digital imaging are visualization, remote location sensing, investigation of bio medical image, self-directed navigation and energetic scene analysis. Delivering or transformation of the normal image to the enhanced version of digital image is a constraint for image processing techniques and its strategies. To identify a person, fingerprint detection technique is used in many applications. These techniques will have a huge database of fingerprints. Here if any of the images available in the database gets matched with the input fingerprint image then the person will be identified else the person will not be identified. The entire process is done by the computer system with help of the authenticated bio-metrics techniques. Each and every human being has unique and matchless fingerprints. These fingerprints will never change based on time and time factors. Finger print identification is utilized in many applications nowadays especially in legal community due to its popularity and its benefits. The experimental results shows better performance in varying the identifying and retorting strength, and detecting the behavior has been improved in proposed work.

2 Related Works

Strong knowledge about the fingerprints should be given in prior so as to improve the performance measures. In object detection method first the feature is extracted and later extracted feature is matched with object which is to be detected from the input image. The survey aim to detect the distorted object from the input image, based on the object comparison the final decision is made. The custom image is transformed onto common image plane by image rectification [1].

Based on the prior information that the fingerprints available in ridge orientations at remote locations will have unique characteristics, this paper proposed local dictionaries depending on the orientation field in estimation algorithm. The noise data are replaced by the original orientation based on the data available in the local dictionaries in the current location [2].

In-order to utilize the local dictionaries preconditions should be satisfied. The appearance of the latent fingerprint should be anticipated for the precondition. In this paper Hough transform based fingerprint pose estimation algorithm is projected.

In this approach, the fingerprint pose can be detected in all the orientations. The experimental results of the proposed work is efficient than the previous cases [3, 4].

The proposed work of this paper specifies a fingerprint orientation field calculation algorithm which is relayed on the former acquired information about the structure of fingerprints and the compatibility conditions of the nearby objects. Orientation field estimation is calculated based on the minimization of energy for the given latent fingerprint and it is solved by loopy belief propagation. The experimental result provides an excellent result for the NIST SD27 database for latent fingerprint while comparing with other conventional algorithms. This paper mainly focuses on addressing the negative fingerprint recognition system which is detected for low quality fingerprint so as to improve their quality and avoid the non matches or false match of the image which is an important problem to be discussed. The other problem addressed here is the photometric degradation which is used for evaluation and enhancement purpose [5, 6].

In this paper, a new algorithm is proposed which concentrates on statistical features of palm print, segment-based matching and fusion algorithm, here the input fingerprint image is given to the proposed algorithm were it is selectively applied for contextual filtering. Contextual filtering starts from high-quality regions which is automatically detected and then repeatedly done for the low quality region also. Here the proposed work does not expect the prior knowledge about the orientation or frequency. The output results obtained for both real and synthetic fingerprint images were effective while using the proposed algorithm. The fingerprint images are stored in binary values then it is allowed for thinning and identification process. Those works mainly concentrated on the reliable of the orientation estimation of the fingerprint images and also concentrates more parameters for achieving the effective matching results. They mainly concentrated on following conditions. The fingerprints are large image but they contains many number if minutiae. Palm prints are changeable while considering with fingerprints in fingertips. The quality of image varies from region to region in the palm prints. Based on these criteria the proper match is unable to provide by the existing system and the results may achieved with more number of noisy values.

3 Proposed System

The proposed system is examined in two levels namely, finger level and subject level. In finger level, natural and distorted image is compared and evaluation is done. In subject level, the natural fingerprints and unique objects which alter the fingerprints are taken and comparison is done. The current proposed work focuses on the distorted fingerprint recognizing and its refinement algorithm. During distortion recognizing, the first step is feature vector value is extracted from the registered ridge orientation map along with period map of a fingerprint. Second step is classification, here classifier is used for identify the given input fingerprint image as normal fingerprint image or distorted image.

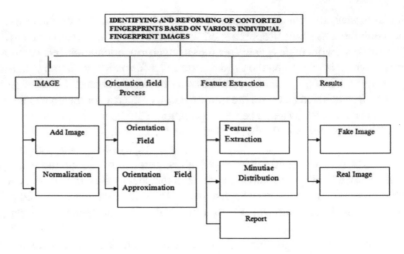

Fig. 1 System interface diagram

To recognize the distorted field from the given input fingerprint image, nearest neighbour regression approach is utilized. Finally the inverse method is utilized to refine the contorted fingerprint image into a normal fingerprint image (Fig. 1).

3.1 Methodology for Detection of Altered Fingerprint

(1) *Normalization*: With the help of NIST biometric Image Software (NBIS), the fingerprint images are normalized by slicing the fingerprint images into a regular rectangular images from center based on the longitudinal direction. Normalization is taken place in two steps namely, conversion and rotation of finger (Figs. 2, 3).

(2) *Orientation Field Estimation*: In order to analyze the fingerprint orientation, gradient based method is implemented. In this algorithm, initial orientation filed is smoothed moderating filter and gradually to modest the orientations in pixel blocks. Using the dynamic range of the gray value in the local blocks of the fingerprint images the foreground mask is created. For filling holes morphological methods are used and the isolated blocks are deleted here.

(3) *Orientation Field Approximation*: To obtain the polynomial model is near by the orientation field.

(4) *Feature Extraction*: Based on absolute difference between and used to create feature vector the error map is evaluated.

(5) *Analysis of Minutiae Distribution*: Every fingerprint recognition systems utilizes the minutiae for matching process. The proposed work, ridge behaviour like ridge ending or ridge junction is utilized in a minutia of a fingerprint image. In case if the minutiae distribution of the distributed fingerprints also leads to

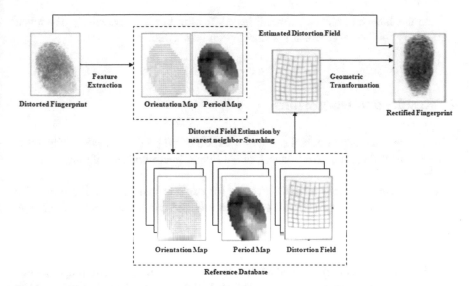

Fig. 2 Distorted fingerprint rectification

Fig. 3 The center (indicated by red circle) and direction (indicated by red arrows) of two fingerprints

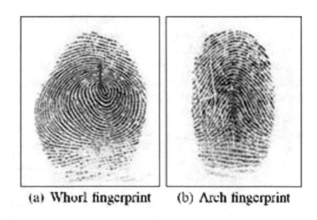

(a) Whorl fingerprint   (b) Arch fingerprint

the change from the normal original fingerprints, which is observed based on the orientation field. Using open source minutiae extractor in NBIS, the starting point of the minutiae can be taken out from the fingerprint.

The overall conditions applied for edge recognition involves, recognizing the edge with low error rate, suggests that the recognition ought to precisely hold several edges exposed within the image as potential. The edge purpose recognition from the operator ought to precisely localize on the middle of the sting. The proposed approach the fingerprint image ought to solely be noticed once, wherever potential, fingerprint distorted must not produce false edge.

(6) *Dermatoscopic Features*: If it is larger than a pencil eraser, which is about one-fourth inch or 6 mm, it needs to be examined by a doctor. This includes areas that

do not have any other abnormalities like color, border, asymmetry. However, size is not the only imperative since it can be smaller.

4 Results and Discussions

Table 1 shows the experimental results for existing and proposed system. The table contains image datasets, SAR image, Optical Finger Image and size of images details are shown.

Table 2 shows the experimental results for existing and proposed system. The table contains image datasets, size of images and EM and SVM number of change detection image details are shown.

Figure 4 shows the experimental results for EM and SVM algorithm system. The chart contains image datasets in y-axis, size of Finger images in x-axis and shows the pictorial representation of EM and SVM number of change detection image details.

Table 3 shows the experimental results for EM and SVM algorithm system. The table contains image datasets, similarity values and SAR image change detection duration (ms) details are shown.

Figure 5 shows the experimental results for EM and SVM algorithm system. The graph contains similarity values and SAR image change detection duration (ms) details are shown in detail.

Table 4 shows the experimental results for EM and SVM algorithm system. The table contains image datasets, similarity values and Optical finger image change detection duration (ms) details are shown.

Table 1 Dataset collections

Image dataset (n)	SAR image	Optical images	Size of images (MB)
25	Simage_1	Opimage_1	200
50	Simage_2	Opimage_2	250
75	Simage_3	Opimage_3	300
100	Simage_4	Opimage_4	350
125	Simage_5	Opimage_5	400

Table 2 Number of change detection images

Image dataset (n)	Size of images (MB)	EM (n)	SVMM (n)
25	200	11	13
50	250	23	30
75	300	35	41
100	350	47	50
125	400	58	62

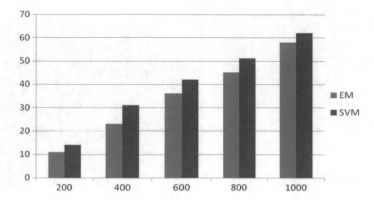

Fig. 4 shows the experimental results for EM and SVM algorithm system

Table 3 Experimental results for EM and SVM algorithm—change detection duration

Image datasets	Similarity (p)	EM (ms)	SVM (ms)
Simage_1	0.045	0.00.03	0.00.02
Simage_2	0.066	0.00.42	0.00.34
Simage_3	0.079	0.00.58	0.00.46
Simage_4	0.089	0.00.63	0.00.54
Simage_5	0.097	0.00.78	0.00.65

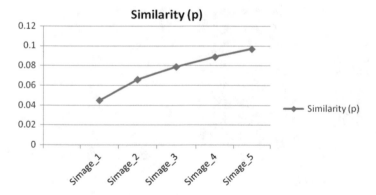

Fig. 5 Experimental results for EM and SVM algorithm—change detection duration

Table 4 Experimental results for EM and SVM algorithm system for optical finger image

Image datasets	Similarity (p)	EM (ms)	SVM (ms)
Opimage_1	0.056	0.00.05	0.00.03
Opimage_2	0.068	0.00.46	0.00.36
Opimage_3	0.082	0.00.62	0.00.55
Opimage_4	0.092	0.00.72	0.00.62
Opimage_5	0.096	0.00.81	0.00.73

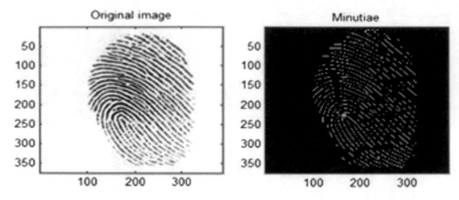

Fig. 6 Minutiae and bifurcation points

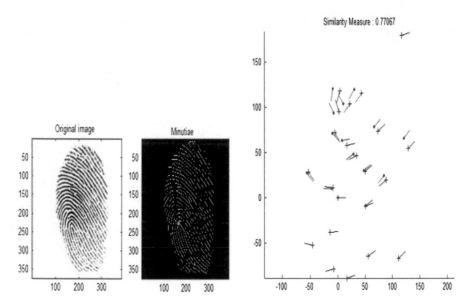

Fig. 7 Minutiae and bifurcation points

Figures 6, 7 shows the experimental results for EM and SVM algorithm. The figure shows the minutiae and bifurcation points and the red point denotes the changes.

5 Conclusion

Every fingerprint recognition and refinement system makes use of minutiae for matching for fingerprints. In the proposed work minutia in the fingerprint identifies the criteria like ridge ending or ridge bifurcation. Minutiae distribution of the modified fingerprints may create the difference between the natural fingerprints which is an additional abnormal problem identified based on the orientation field. The minutiae extractor in NBIS which is open source is used to extract the fingerprint then the density map is created based on the Parzen window technique based on uniform kernel function. False matches are identified based on the severely distorted fingerprints which are main disadvantage in automatic fingerprint identification and recognition system used by criminals and terrorist. This is the main aim to develop the fingerprint distortion recognition and refinement algorithms to fill the outlet. This paper successfully identifies the distortion field and the distorted points and displays a matching score list.

References

1. Meng, G., Wang, Y., Qu, S., Xiang, S., Pan, C.: Active flattening of curved document images via two structured beams (2014)
2. Si, X., Feng, J., Zhou, J., Luo, Y.: Detecting fingerprint distortion from a single image. IEEE (2012)
3. Kasaei, S., Deriche, M., Boashash, B.: Fingerprint feature extraction using block-direction on reconstructed images. In: IEEE TENCON—speech and image technologies for computing and telecommunication
4. Le, H., Bui, T.D.: Online fingerprint identification with a fast and distortion tolerant hashing. J. Inf. Assur Secur 4, 117–123 (2009)
5. Vatsa, M., Singh, R., Noore, A., Singh, S.K.: Combining pores and ridges with minutiae for improved fingerprint verification. Elsevier, Signal Process. **89**, 2676–2685 (2009)
6. Zhao, Q., Zhang, L., Zhang, D., Luo, N.: Adaptive pore model for fingerprint pore extraction. In: Proceedings of IEEE, 978-1-4244-2175-6/08 (2008)

Assessment of Groundwater Potential Using Neural Network: A Case Study

Sandeep Samantaray, Abinash Sahoo and Dillip K. Ghose

Abstract To properly utilize and manage exploitation of groundwater, application of recent methods is very much crucial to prepare an organized arrangement for protecting this valuable but dwindling natural resource. Groundwater potential was scrutinized by applying neural network models namely back-propagation neural network (BPNN) and Radial basis function network (RBFN) in Balangir watershed, India. The groundwater fluctuation is powerfully associated to hydrological elements, and therefore this relation allows groundwater potential mapping from hydrological elements by help of BPNN and RBFN model to simulate groundwater fluctuations. Also, multiple linear regressions (MLR) are used for observing the relationship amid reliant variable, i.e., depth of water table and free variable, i.e., the hydrogeological factors considered in this study. Three different types of transfer function tan-sig, log-sig are applied for comparing outcomes of the models for predicting fluctuation of groundwater. Results of present study show the supremacy of BPNN model than RBFN model for predicting groundwater potential.

Keywords BPNN · RBFN · Water table depth · Arid region

1 Introduction

Groundwater is essential for household, agriculture, and commercial utilisation. Mostly groundwater fulfills the necessity of water required in all sectors in India hence it has a vital significance. Presently, source of groundwater is fronting high danger because of unpredictability in climate and its variations. In this perspective, a viable water resources management strategy for predicting variations of

S. Samantaray (✉) · A. Sahoo · D. K. Ghose
Department of Civil Engineering, NIT Silchar, Assam, India
e-mail: samantaraysandeep963@gmail.com

A. Sahoo
e-mail: bablusahoo1992@gmail.com

D. K. Ghose
e-mail: dillipghose_2002@yahoo.co.in

© Springer Nature Singapore Pte Ltd. 2020
V. Bhateja et al. (eds.), *Intelligent Computing and Communication*,
Advances in Intelligent Systems and Computing 1034,
https://doi.org/10.1007/978-981-15-1084-7_63

groundwater depth is necessary that will help in ensuring justifiable utilization of a watershed's groundwater to supply water to city and country side areas. In West Orissa, water required for household, irrigation and commercial purpose add to extensive periodic variation in depth of water table. Pradhan et al. [1] studied conduct of groundwater under varying situations in subordinate portion of Ganga–Ramganga inter basin and investigated performance of co-active neuro-fuzzy inference system (CANFIS) and RBFN that were applied to predict height of water table. Aizebeokhai et al. [2] integrated geo-electrical resistivity soundings using 2D geo-electrical resistivity and time-domain IP imaging for characterization of water beneath the surface and define level of aquifer for assessing source of groundwater potential. Wang et al. [3] developed a BPNN for simulating three-dimensional dissemination of NO_3–N absorptions in groundwater with land cultivation evidence and site-definite hydro geological assets in Huantai County, China. Lee et al. [4] analyzed Groundwater productivity-potential (GPP) utilizing ANN models and support vector machine in Boryeong city, Korea. Li et al. [5] used RBFNN and GM (1, 1) model to predict groundwater depth variations in Longyan city, Fujian Province (South China). Al-Mahallawi [6] applied Multilayer Perceptrons (MLP), RBF, Generalized Regression Neural Network (GRNN), and a novel model called Linear Networks for estimating groundwater adulteration by nitrate. Arora et al. surveyed on RBFN in details based on its development and uses. Agrawal et al. [7] used BPNN as a classifier to perceive the fault and to treasure faulty pole. A new active queue management (AQM) controller is to control network congestion efficiently by stabilizing the queue length [8]. Nejad et al. [9] produced groundwater potential map by means of weights-of-evidence and evidential belief function model on basis of geographic information system (GIS) in Azna Plain, Iran. Chen et al. made a widespread study of most recent progresses in RBFN association methods which concentrated on simple thoughts, mathematical algorithms, and engineering uses. Lee et al. [10] applied an ANN and a GIS to map local GPP for a zone nearby City of Pohang, Korea. Shamsuddin et al. illustrated growth and use of ANNs for predicting level of groundwater in two upright boreholes positioned in confined aquifer neighboring to Langat River. Affandi and Watanabe [11] predicted variation of daily groundwater table to monitor the variation shape using ANFIS and data mining models known as Levenberg Marquardt (LM) and RBF. Ghose and Samantaray [12] predicted sediment concentration using regression and BPNN models. Ghose and Samantaray [12] found the opening linked to sensor networks and assimilated neural network (NN) set of rules to monitor circumstances for ground water sustainability. The objective of the study is to predict water table depth using ANN algorithm in rid watershed. Ground water potential is estimated and predicted by various ANN techniques [12, 13]. Machine learning approaches is a successive method to predict various climatic indices in a catchment [14–18]

2 Study Area

Balangir District is situated in western region of Odisha, in India as shown in Fig. 1. The district covers 5,165 km^2 area, and 1,335,760 inhabitants. Coordinates 20.6723° N, 83.1649° E. Mahanadi is the solitary persistent river in this district. Climate of the proposed study area is categorized to be tropical. In summer this area experiences more rainfall in comparison to winter. The mean temperature in summer is 34.8 °C and the warmest month is May whereas in winter mean temperature is 20.7 °C and the coolest month is December.

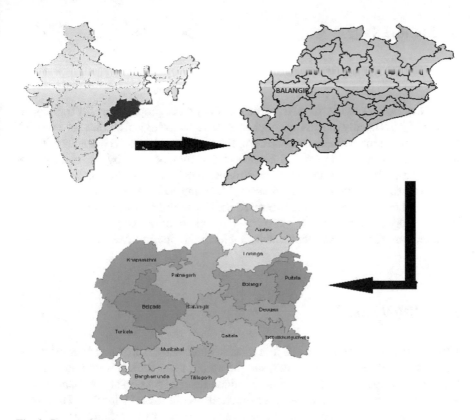

Fig. 1 Proposed study area

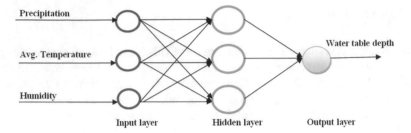

Fig. 2 Schematic diagram of BPNN

3 Methodology

3.1 BPNN Model

It is a distinctive multilayer ANN which utilizes back-propagation for training its network [19]. Usually the arrangement applied to hydrological studies for mapping the entire nonstop nonlinear function comprises of three layers: the input, the hidden, and the output layer (Fig. 2). NN consist of enormous nodes, by means of beginning values, stimulation functions and linking weights for characterizing structure of BPNN. BP algorithm is an administered learning technique on basis of steepest descent method for minimizing overall fault. For modifying beginning values and linking weights output faults are nurtured back via nodes. Lastly, optimum value is achieved through iterative modification. Objective function takes root mean square error.

3.2 RBFN Model

RBFN is comprises of huge number of modest and vastly interrelated neurons that is structured into the input, the hidden and the output layer [20]. It is nonlinear cross-breed network which contain a solitary hidden layer of processing elements (PEs). Hidden layer utilizes Gaussian transfer functions as nonlinearity for PEs. The output PEs is usually undeviating. Gaussian function reacts simply to a minor area of input space where the Gaussian is centered. Henceforth, accomplishment in implementation of RBFNs is dependent on appropriate centers for Gaussian functions that can be completed using administered learning. On the other hand, an unsubstantiated method frequently yields improved outcomes [21] (Fig. 3).

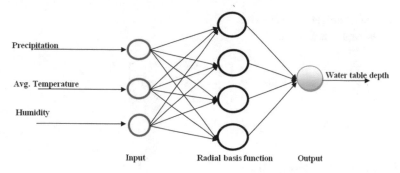

Fig. 3 Schematic diagram of RBFN

3.3 Processing and Preparation of Data

Precipitation once a month, average temperature, humidity is gathered from IMD and department of soil conservation Balangir provides data for water table depth for monsoon months (June to October), between 1978 and 2017. Data between 1978 and 2007 is utilized to train and data between 2008 and 2017 are utilized to test the network. Everyday data is transformed to once-a-month data, that is lastly utilized to train and test model. Input and output data are extent in such a way that individual data fall inside the quantified series before training. This procedure is known as normalization so as to normalize the values and bound them ranging from 0 to 1. Normalization equation utilized to scale the data is

$$X_t = 0.1 + \frac{X - X_{\min}}{X_{\max} - X_{\min}} \tag{1}$$

where X_t = distorted data, X = original data, $X_{\min}$ = minimum original data, $X_{\max}$ = maximum original data.

3.4 Evaluating Parameter

The evaluating criteria for ascertaining finest model are root mean square error (RMSE), mean square error (MSE) and coefficient of determination (R^2).

$$R^2 = \left[\frac{n\sum xy - (\sum x)(\sum y)}{\sqrt{[n\sum x^2 - (\sum x)^2][n\sum y^2 - (\sum y)^2]}} \right]^2 \tag{2}$$

$$\text{MSE} = \frac{1}{N} \sum_{i=1}^{N} (\widehat{y_i} - y_i)^2 \tag{3}$$

where, $\widehat{y_i}$ = Predicted runoff value
   y_i = Actual runoff value

$$\text{RMSE} = \left[\frac{1}{N} \sum_{i=1}^{N} (\widehat{x_i} - x_i)^2 \right]^{1/2} \tag{4}$$

4   Results and Discussions

Now for evaluating effectiveness of model for several designs, dissimilar transfer functions are utilized for establishing model potential. In case of every design, evaluating criteria is training, testing for MSE, RMSE and R^2.

For Tan-sig function in BPNN, 3-2-1, 3-3-1, 3-5-1, and 3-9-1 architectures are considered to compute performance and it is found that model architecture 3-2-1 that have MSE testing value 0.000419, MSE training value 0.000808, RMSE testing value 0.004165, training value 0.008392, and coefficient of determination for testing 0.9255, training 0.9914 is best. For Log-sig 3-2-1, 3-3-1, 3-5-1, and 3-9-1 architectures are considered to compute performance and it is found that model architecture 3-5-1 that have MSE testing value 0.000429 training value 0.000591, RMSE testing value 0.003460 training value 0.004271 and coefficient of determination for testing value 0.9417 training 0.9915 is best. The complete outcome is shown in Table 1.

Similarly, for RBFN in Tan-sig 3-2-1, 3-3-1, 3-5-1, and 3-9-1 architectures are considered to compute performance and it is found that model architecture 3-5-1 that have MSE testing value 0.000125, training value 0.000483, RMSE testing value 0.003085, training value 0.002316 and coefficient of determination value testing value 0.9611, training 0.9925 is best. For Log-sig 3-2-1, 3-3-1, 3-5-1, and 3-9-1 architectures are considered to compute performance it is found that model architecture 3-2-1 that have MSE training value 0.000236 testing value 0.000808, RMSE training value 0.003514 testing value 0.005299 and coefficient of determination value training 0.9861, testing value 0.9591 is best and the entire outcome is presented in Table 2.

4.1   Simulation

Graphs having finest value for water table depth from precipitation, average temperature, humidity using BPNN and RBFN having Tan-sig and Log-sig transfer function in Balangir. Actual versus predicted water table depths in concern to the best value of coefficient determination for testing phase are presented in Fig. 4.

Table 1 Results of BPNN

Model input	Sigmoid function	Architecture (L-M-N)	MSE		RMSE		R^2	
			Training	Testing	Training	Testing	Training	Testing
Precipitation average temperature humidity	Tan-sig	**3-2-1**	**0.000808**	**0.000419**	**0.008392**	**0.004165**	**0.9914**	**0.9255**
		3-3-1	0.000797	0.000559	0.008197	0.005876	0.9857	0.9075
		3-5-1	0.000449	0.000765	0.005282	0.001734	0.8691	0.8001
		3-9-1	0.000681	0.000309	0.002607	0.005542	0.9889	0.9103
	Log-sig	3-2-1	0.000465	0.000167	0.002153	0.004134	0.9847	0.9390
		3-3-1	0.000195	0.000408	0.003455	0.006394	0.9793	0.8114
		3-5-1	**0.000591**	**0.000429**	**0.004271**	**0.003460**	**0.9915**	**0.9417**
		3-9-1	0.000192	0.000739	0.003437	0.001133	0.9783	0.9167

Table 2 Results of RBFN

Model input	Sigmoid function	Architecture (L-M-N)	MSE		RMSE		R²	
			Training	Testing	Training	Testing	Training	Testing
Precipitation average temperature humidity	Tan-sig	3-2-1	0.000289	0.000298	0.001389	0.003601	0.8188	0.9270
		3-3-1	0.000751	0.000339	0.002737	0.005799	0.9875	0.9724
		3-5-1	**0.000483**	**0.000125**	**0.002316**	**0.003085**	**0.9925**	**0.9611**
		3-9-1	0.000394	0.000958	0.001987	0.003093	0.9761	0.9596
	Log-sig	**3-2-1**	**0.000236**	**0.000808**	**0.003514**	**0.005299**	**0.9861**	**0.9591**
		3-3-1	0.000940	0.000980	0.003064	0.007057	0.9678	0.9112
		3-5-1	0.000516	0.000255	0.001248	0.005021	0.7127	0.9536
		3-9-1	0.000437	0.000656	0.003789	0.008083	0.9792	0.9019

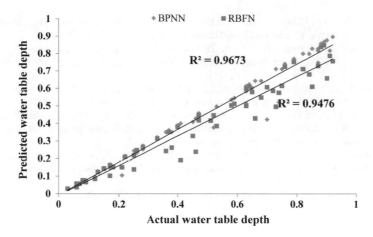

Fig. 4 Actual versus predicted water table depth in training phase

5 Conclusions

Present study depicts that BPNN is useful for assessing groundwater fluctuation for longer period in the arid region Balangir. This model provides suitable estimates for records of both small and big groundwater table in a zone which is a very important benefit while using this model. To improve development of water resource management the proposed BPNN model is ideal as this shows capable outcomes. This is likewise seen that prediction structures on basis of RBFN attain quicker conjunction comparing to structure on basis of BPNN but error predicting level is higher. In projected catchment, RBFN gives best results having architecture 3-5-1 succeeding Tan-sig transfer function while considering testing and training performance conditions amid both NNs taking into consideration. In similar manner, BPNN gives best outcomes with architecture 3-5-1 taking all evaluating condition into consideration.

References

1. Pradhan, S., Kumar, S., Kumar, Y., Sharma, H.C.: Assessment of groundwater utilization status and prediction of water table depth using different heuristic models in an Indian interbasin. Soft Comput. **23**(20), 10261–10285 (2019)
2. Aizebeokhai, A.P., Oyeyemi, K.D., Joel, E.S.: Groundwater potential assessment in a sedimentary terrain, southwestern. Nigeria. Arab. J. Geosci. **9**(7), 496 (2016)
3. Wang, M.X., Liu, G.D., Wu, W.L., Bao, Y.H., Liu, W.N.: Prediction of agriculture derived groundwater nitrate distribution in North China Plain with GIS-based BPNN. Environ. Geol. **50**(5), 637–644 (2006)
4. Lee, S., Hong, S.M., Jung, H.S.: GIS-based groundwater potential mapping using artificial neural network and support vector machine models: the case of Boryeong city in Korea. Geocarto Int. **33**(8), 847–861 (2018)

5. Li, Z., Yang, Q., Wang, L., Martín, J.D.: Application of RBFN network and GM (1, 1) for groundwater level simulation. Appl. Water Sci. **7**(6), 3345–3353 (2017)
6. Al-Mahallawi, K., Mania, J., Hani, A., Shahrour, I.: Using of neural networks for the prediction of nitrate groundwater contamination in rural and agricultural areas. Environ. Earth Sci. **65**(3), 917–928 (2012)
7. Agarwal, S., Swetapadma, A., Panigrahi, C.: An improved method using artificial neural network for fault detection and fault pole identification in voltage source converter-based high-voltage direct current transmission lines. Arab. J. Sci. Eng. **43**(8), 4005–4018 (2018)
8. Bisoy, S.K., Pattnaik, P.K.: An AQM controller based on feed-forward neural networks for stable internet. Arab. J. Sci. Eng. **43**(8), 3993–4004 (2018)
9. Ghorbani Nejad, S., Falah, F., Daneshfar, M., Haghizadeh, A., Rahmati, O.: Delineation of groundwater potential zones using remote sensing and GIS-based data-driven models. Geocarto Int. **32**(2), 167–187 (2017)
10. Lee, S., Song, K.Y., Kim, Y., Park, I.: Regional groundwater productivity potential mapping using a geographic information system (GIS) based artificial neural network model. Hydrol. J. **20**(8), 1511–1527 (2012)
11. Affandi, A.K., Watanabe, K.: Daily groundwater level fluctuation forecasting using soft computing technique. Nat. Sci. **5**(2), 1–10 (2007)
12. Ghose, D.K., Samantaray, S.: Integrated sensor networking for estimating groundwater potential in scanty rainfall region: challenges and evaluation. Comput. Intell. Sens. Netw. **776**, 335–352 (2019)
13. Das, U.K., Samantaray, S., Ghose, D.K., Roy, P.: Estimation of aquifer potential using BPNN, RBFN, RNN, and ANFIS. Smart Intell. Comput. Appl. **105**, 569–576 (2019)
14. Samantaray, S., Sahoo, A.: Appraisal of runoff through bpnn, rnn, and rbfn in tentulikhunti watershed: a case study. In: Satapathy S., Bhateja V., Nguyen B., Nguyen N., Le DN. (eds.) Frontiers in Intelligent Computing: Theory and Applications. Advances in Intelligent Systems and Computing, vol. 1014. Springer, Singapore (2020a)
15. Samantaray, S., Sahoo, A.: Estimation of runoff through bpnn and svm in agalpur watershed. In: Satapathy S., Bhateja V., Nguyen B., Nguyen N., Le DN. (eds.) Frontiers in Intelligent Computing: Theory and Applications. Advances in Intelligent Systems and Computing, vol. 1014. Springer, Singapore (2020b)
16. Samantaray, S., Sahoo, A.: Assessment of sediment concentration through rbnn and svm-ffa in arid watershed, India. In: Satapathy S., Bhateja V., Mohanty J., Udgata S. (eds.) Smart Intelligent Computing and Applications. Smart Innovation, Systems and Technologies, vol. 159. Springer, Singapore (2020c)
17. Samantaray, S., Ghose, DK.: Sediment assessment for a watershed in arid region via neural networks. Sādhanā 44(10), 219 (2019)
18. Samantaray, S., Sahoo, A., Ghose, DK.: Assessment of runoff via precipitation using neural networks: Watershed modelling for developing environment in arid region. Pertanika J Sci Technol 27(4), 2245–2263 (2019)
19. Fausett, L.V.: Fundamentals of neural networks: architectures, algorithms, and applications, vol. 3. Prentice-Hall, Englewood Cliffs (1994)
20. Haykin, S.: Kalman Filtering and Neural Networks, vol. 47. Wiley (2004)
21. Huang, G.B., Saratchandran, P., Sundararajan, N.: An efficient sequential learning algorithm for growing and pruning RBF (GAP-RBF) networks. IEEE Trans. Syst. Man Cybern. Part B (Cybernetics) **34**(6), 2284–2292 (2004)

Data Leakage Prevention and Detection Techniques Using Internet Protocol Address

A. Jaya Mabel Rani, G. Vishnu Priya, L. Velmurugan and V. Vamsi Krishnan

Abstract A data breach is the intentional or inadvertent exposure of confidential information to unauthorized users. In the digital world, data has become most important thing in enterprise. Data leaks provide serious threat to organizations, financial, and reputation of the company. As the data growing exponentially and data breaches are also increasing, detecting, and preventing data leaks become most one of the security concerns for company. Here we use the IP address to identify the users and hackers and administrator looks for the users IP address and block Hackers IP address so that data leakage is not possible from a third party. Hackers try to acquire information through SQL injection and our paper suggests a way to prevent it. This paper helps to detect and prevent the data breach while sharing the data with the consumer and improve the probability of preventing the unauthorized users while sharing the data.

Keywords Data breach · Confidential information · IP address · SQL injection · Detecting and preventing data leaks

1 Introduction

In the present business situation, information spillage is a major test as basic hierarchical information ought to be shielded from unapproved get to Information spillage might be defined as the inadvertent or deliberate conveyance of private hierarchical

A. Jaya Mabel Rani (✉) · G. Vishnu Priya · L. Velmurugan · V. Vamsi Krishnan
Department of Computer Science and Engineering, Jeppiaar Maamallan Engineering
College, Sriperumpudur, India
e-mail: ajayamabelrani@gmail.com

G. Vishnu Priya
e-mail: vishpriya28@gmail.com

L. Velmurugan
e-mail: velmurugan.vm261@gmail.com

V. Vamsi Krishnan
e-mail: vamsivemula11@gmail.com

© Springer Nature Singapore Pte Ltd. 2020
V. Bhateja et al. (eds.), *Intelligent Computing and Communication*,
Advances in Intelligent Systems and Computing 1034,
https://doi.org/10.1007/978-981-15-1084-7_64

information to the unapproved substances. It is imperative to shield the basic information from being abused by any unapproved use. Basic information incorporate scholarly duplicate right data, patent data, useful data and so forth. In numerous associations, this basic hierarchical information has been shared to numerous partners outside the authoritative premises [1]. Thusly, it is difficult to distinguish the guilty party, who has released the information. So in this proposed system, our goal is to identify the guilty user when the organizational data have been leaked by some agent. AES algorithm and RSA algorithm shows good performance among different symmetric and asymmetric encryption technique based on different performance factors of key data, computational time and speed and security. In RSA encryption time is more and also memory usage is very high [2]. In our paper we use the hybrid of RSA AES algorithm for encryption techniques (honey encryption can also be used). Our inspiration is to anticipate unapproved access to the information with the assistance of the IP following component. In future we can also upgrade the algorithm such that it provides more security.

2 Literature Review

Customarily, spillage recognition is managed by means of watermarking, e.g., an interesting code is implanted in each conveyed duplicate. If that duplicate is later found in the palms of an unapproved party, the leaker can be perceived. Watermarks can be extremely valuable at times, however once more, contain some revision of the real information [3]. Moreover, watermarks can occasionally be demolished if the information beneficiary is malignant. Be that as it may, at times it is significant now not to adjust the valid wholesaler's information. Generally, spillage location is dealt with by utilizing watermarking, e.g., an uncommon code is implanted in each apportioned duplicate. If that copy is found in the later picture of an unauthorized person, the leaker can be recognized. Watermarks can be exceptionally valuable sometimes, anyway once more, contain some difference in the bonafide data.

2.1 Issues in Existing System

Substances and Agents Let the merchant database claims a set $S = \{t1, t2, tm\}$ which comprises of insights objects. Let the no of operators be A1, A2, ..., A [2, 4, 5]. The wholesaler appropriates a lot of documents S to any operators put together absolutely two with respect to two their solicitation, for example, two example two or express two solicitation. Two of the Sample demand Ri = SAMPLE (T, mi): Any subset of mi data from T can be given to Ui [2].Explicit demand Ri = EXPLICIT (T;condi):Agent Ui gets all T protests that satisfies the Condition [2].The questions in T might need to be of a sort and size, for example they ought to be tuples in a connection, or two individuals from the family in two a database. In the way of

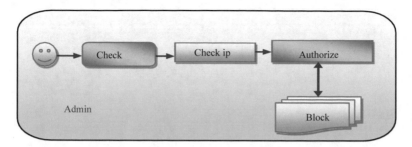

Fig. 1 Second design step of the system

offering article to specialists, the wholesaler finds that a set S of T has spilled. This is the information, it is genuine hoping to presume them releasing the information.

3 Proposed System

In this paper propose that the data leakage detection and prevention can be done with the help of the hybrid of RSA and AES encryption algorithm along with the IP tracking. The encryption algorithm hence encrypts the file which is send from the user to the receiver. When the user uploads the file in the database, we use the mechanism of IP tracking to avoid fake access and avoiding hackers from hacking the database with the help of SQL injection [6]. The file gets uploaded to the database and admin takes the full control of the data present in database. The admin can also able to block the hackers with the help of the IP. The file will be sent to the receivers with the key [7–9]. Hence data leakage in any manner is detected and prevented. This proposed system consists of the following steps.

1. **Interface Design**

This is the first step of the design model. The essential position for the consumer is to pass login window to user window. This design step mainly created for the protection purpose. In this login web page, we must enter login user identity and password. It will test username and password is in shape or no longer (valid person id and valid password). If we enter any invalid username or password, we cannot enter login window to person window it will shows error message [10].

So, this design step can stop unauthorized user coming into the login window to user window. Server comprises user identity and password server also tests the authentication of the user. It properly improves the protection and stopping from unauthorized user enters the network. In our mission we are the usage of JSP for creating design. Here we validate the login user and server authentication.

2. **File Upload**

Figure 1 shows the second design step of the proposed system. Here the person can upload the file that and get saved in the database.

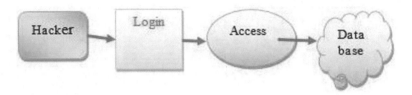

Fig. 2 Hacker login

3. Admin Login and Monitoring

This is the next step of the design process, right here symbolizes a unit of work performed within a database administration machine (or comparable system) against a database. After the login, admin will be monitoring the file user uploads.

4. User Authorization

In this module, the consumer will send the file to the admin. Here, the admin will be comparing the IP address that has been saved in the database and the person sends. If the IP address is matched by comparing both from the person and database, the admin will authorize them to access.

5. Hacker Login

In this step the hacker will login and enter the person port number, then get entry to the consumer files (Fig. 2).

6. Admin Block Unauthorizeduser

In this admin block the admin will be monitoring the files, if the hacker strives to access the person documents potential and it will be notified by means of the admin.

4 Hybrid RSA-AES Algorithm

Usually hybrid algorithms used to give efficient and proper result than individual algorithms [11]. In this proposed system used hybrid RSA-AES algorithm for encryption and decryption, specifically this hybrid RSA-AES algorithm provide more efficient in terms of security while the time of processing is little high compared to AES and RSA algorithms [12–14] (Fig. 3).

Algorithm:

Step 1: Import the required packages.
Step 2: Creating the file by using the concept of encryption and decryption.
Step 3: Generating the bit key by using encryption process with a key of size 128 bits.
Step 4: Then generate private key using RSA algorithm.
Step 5: Generate and load initialization vector.

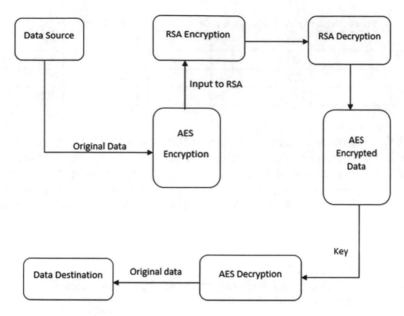

Fig. 3 Hybrid RSA-AES architecture

Step 6: Encrypt the file using AES key and RSA private key.
Step 7: Decrypt the file using RSA public key and AES secret key.
Step 8. Read the initialization vector for decryption process using AES secret key.
Step 9: Finally the output file content information is written into the file.
Step 10: Stop the process.

Now comes the actual decryption of the file contents using the AES secret key. The output is written into a file with the extension .ver for verification purposes.

5 Results and Discussions

The following results show the output results for encryption and decryption using RSA and AES and hybrid RSA AES algorithm and its performance decryption method. Tables 1 and 2 shows the performance efficient analysis of AES, RSA and hybrid AES and RSA algorithm.

Table 1 Performance analysis of encryption time of hybrid RSA-AES algorithm

Size (bytes)	140	1510	3330	6400
AES	09	12	12	13
RSA	658	1950	1661	2050
Hybrid	1683	1750	2080	2982

Table 2 Performance analysis of decryption time of hybrid RSA-AES algorithm

Size (bytes)	140	1510	3330	6400
AES	08	12	12	13
RSA	178	632	1300	2400
Hybrid	280	720	1400	2500

Table 1 represents the tabular and graphical representation of the encryption time in milliseconds. It is observed that compared to the AES algorithm and RSA algorithm the time of encryption is high in hybrid RSA-AES algorithm. Table 2 represents the tabular and graphical representation of the decryption time in milliseconds. It is observed that compared to the AES algorithm and RSA algorithm the time of decryption is high in hybrid RSA-AES algorithm. However, the security mechanism in hybrid RSA-AES is high compared to the AES algorithm and RSA algorithm.

The graphical representation of encryption time for AES, RSA and hybrid of AES RSA is given in the Fig. 4.

The graphical representation of encryption time for AES, RSA and hybrid of AES RSA is given the Fig. 5.

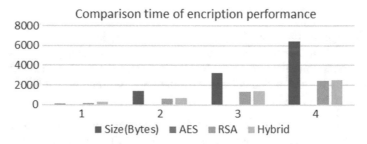

Fig. 4 Comparison time of encryption time for AES, RSA and hybrid of AES RSA

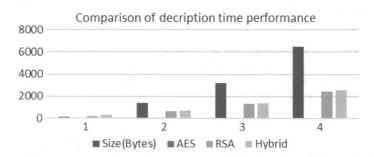

Fig. 5 Comparison time of decryption AES, RSA and hybrid AES-RSA

6 Conclusion

Data leakage is a soundless method of threat. In our organization hand as an insider can purposefully or unintentionally seepage sensitive information. This sensitive statistic can be electronically spread by means of e-mail, web sites, FTP, instant messaging, spread sheets, databases, and any other available skill methods. In an ideal world, there would be no need to hand over touchy facts to dealers that might also unknowingly or maliciously leak it [15–17]. In order to be greater effective and tightly closed we advocate the use of the encryption (hybrid RSA-AES) and additionally IP tracking mechanisms to observe and forestall the data leaking and it is one of the greatest ways to prevent the facts leakage mechanism. So here used the IP address to identify the users and hackers and administrator looks for the users IP address and block Hackers IP address so that data leakage is not possible from a third party. In order to hack the data SQL Injection is used. This paper provides the solution for protecting the database and preventing the hacker to enter it.

References

1. Desmet, L., Peissen, F., Joosen, W., Verbaeten, P.: Bridging the gap between web application firewalls and web applications. In: Proceedings of the fourth ACM workshop on formal methods in Security, pp. 67–77, Nov 2006
2. Choudhary, P.B., Waghmare, S., Patil, N.: Data transmission using AES-RSA based hybrid security algorithms. Int. J. Recent Innov. Trends Comput. Commun. (2015)
3. Xiang, S., He, J.: Database authentication watermarking scheme in encrypted domain (2017)
4. http://investors.imperva.com/phoenix.zhtml?c=247116&p=irol-newsArticle&ID=1596618
5. Ramamani, S., Balaguru Shankar, G., Sakthi Vignesh, K., Santhosh Kumar, G., Siva Surya, D.: Data leakage detection (2015)
6. Huang, X., Lu, Y., Li, D., Ma, M.: A novel mechanism for fast detection of transformed data leakage (2018)
7. https://searchfinancialsecurity.techtarget.com/tip/Data-leakage-detection-and-prevention
8. https://www.mimecast.com/de/content/dataleakage-prevention-solutions/
9. https://www.cyberoam.com/dataleakageprevention.html
10. Gendron, M.: Imperva introduces first network adaptive web application firewall (2006)
11. Jaya Mabel Rani, A., Pravin, A.: Multi-objective hybrid fuzzified PSO and fuzzy C-means algorithm for clustering CDR data. In: International Conference on Communication and Signal Processing, April 4–6, 2019, pp. 94–98. IEEE
12. Heiderich, M., Nava, E., Heyes, G., Lindsay, D.: Webapplication_obfuscation. Elsevier (2011). Accessed from https://doc.lagout.org/security/Web%20Application%20Obfuscation/Web%20Application%20Obfuscation.pdf
13. https://www.academia.edu/28193212/Data_Transmission_using_AESRSA_Based_Hybrid_Security_Algorithms
14. https://www.researchgate.net/publication/281165846_Data_Transmission_using_AESRSA_Based_Hybrid_Security_Algorithms
15. https://digitalguardian.com/dskb/data-loss-prevention
16. https://www.google.com/search?q=data+leakage+prevention+solution&source=lnms&tbm=isch&sa=X&ved=0ahUKEwiBpMfXgbzhAhWC8XMBHTQSDUAQ_AUIDygC&biw=1536&bih=754
17. https://www.surveilstar.com/prevent-data-leakage.html

Recognition of Handwritten Kannada Characters Using Unsupervised Learning Method

Manjunath Bhadrannavar, Omkar Metri, Chandravva Hebbi
and H. R. Mamatha

Abstract In the digital era, recognition of handwritten Kannada characters and words and converting them to machine editable form is a challenging research. It has applications like reading aid for blind, digitization of the handwritten documents, postal mail segregation and author, and handedness identification in forensics. The aim of this paper is to build the dataset of handwritten Kannada vowels using unsupervised learning technique. To build the dataset by moving the characters to their respective folders which is done by feature extraction methods like Histogram of Oriented Gradients (HOG), Run Length Count (RLC), Chain Code (CC), Local Binary Pattern (LBP). Classification of characters is achieved with clustering, an unsupervised learning method. To achieve a higher recognition rate, combination of feature extraction methods have been applied. The combined features of HOG and RLC yielded the best recognition rate of 86.92%. Experiment is carried out with the data collected from 100 people.

Keywords Character recognition · Kannada · Feature extraction · Unsupervised learning · Data set generation

1 Introduction

In the world of digitization, recognition of the characters from handwritten scanned documents is an inviting competition. Devanagari, Bangla, Kannada and Telugu languages are the most important literatures open to the area of handwritten character recognition. Kannada is a Dravidian and one of the official languages with 38 million native speakers. Consequence of which, it is the 32nd widely used and spoken language in the world. Modern Kannada (Hosagannada) has 13 vowels (swaragalu), 2 anuswaragalu, 34 consonants (vyanjanagalu) and 10 numerals.

M. Bhadrannavar · O. Metri (✉) · C. Hebbi · H. R. Mamatha
PES University, Bengaluru 560085, Karnataka, India

H. R. Mamatha
e-mail: mamathahr@pes.edu

© Springer Nature Singapore Pte Ltd. 2020
V. Bhateja et al. (eds.), *Intelligent Computing and Communication*,
Advances in Intelligent Systems and Computing 1034,
https://doi.org/10.1007/978-981-15-1084-7_65

Kannada language has been comprehensively and extensively used in many real-world applications, i.e., public sector banks, railways, transport systems, income tax department. Form filling at these places is mandatory and the database is kept in the digital format by scanning the forms and storing them as documents. Absence of recognition system will result in no option for editing of the scanned document and thereby becomes tedious task for correction of the document. Hence, automatic reading system can help save the time. For this reason, Handwriting Recognition (HWR) is the ability to present the computer with intelligible handwritten scanned input from database. The variability in the human handwriting, oblique angle and relative orientation of the style are hard situations for the recognition of offline and online handwritten characters by the computer.

The most important technique used to change the form of the text from the scanned documents into digital format is often referred as optical character recognition which in broad terms is referred to as Handwritten Text Recognition (HTR) systems. It involves scanning the image containing the text character-by-character, analysis and rendition of text into digital format. In other words, it is image recognition for handwritten text instead of landmarks or faces.

Good feature extraction methods play an important role in achieving the high recognition rate for the character recognition. These methods have been tested on handwritten Kannada vowels collected from various people. The rest of the paper is split into 4 sections with literature survey for recognition of the handwritten characters, proposed methodology, experiment results and conclusion.

2 Literature Survey

Some of the major works have been reported in the recognition of offline handwritten Kannada vowels and consonants. They are as follows.

The technique to divide image into zones and extract features using distance metric and pixel density is presented in [1]. K-Nearest Neighbor and Linear Discriminant Analysis (LDA) methods are used for classification. An accuracy of 94.6 and 84.7% is obtained for vowels and consonants respectively. Authors have not specified the size of the dataset used for testing.

Recognition of offline handwritten Kannada and Tamil scripts using Hidden Markov Model (HMM) is presented in [2]. HMM uses shape features and singular value decomposition coefficients. The rate of recognition is 76% for Kannada and 70% for Tamil. An accuracy of 40% for Kannada words and 30% for Tamil words is obtained.

Handwritten Kannada character recognition is presented in [3]. Features are extracted using moments, Zonal features, Zernike moments, Hu's Invariant and Fourier-Wavelet coefficients. Recognition of characters was done with Back Propagation Neural Network. Accuracy and the size of the dataset are not presented.

Genetic algorithm to recognize handwritten Kannada characters is presented in [4]. Features are selected using pixel-based method. Genetic algorithm is used for selecting the subset of features. Neural Networks is used for the classification of the characters. The recognition rate of system is 71.63% with size of dataset being 490 characters.

In [5] the method to recognize handwritten Kannada characters using structural features and SVM is presented. An accuracy of 89.84% is obtained for vowels and 85.14% for consonants. The size of dataset is 2490 samples of 49 characters of Kannada.

Recognition of Kannada characters with crack feature extraction and SVM classifier is presented in [6]. An accuracy of 80% was obtained on their dataset. Weighted Zone method was used to enhance the accuracy of recognition.

A technique to recognize basic Kannada characters with a two-dimensional discrete wavelet transforms and normalized chain code is presented in [7]. SVM with twofold cross validation is used for the classification. The accuracy of recognition is 90.09% for vowels and consonants. Experimentation is done with 9600 images.

In [8], experiment was conducted on Odia numerals with six feature extraction methods and the classification accuracy obtained was more than 93%.

Literature survey reveals that a lot of research has be done using supervised learning methods. Consequence of which is the dataset was labelled before feeding to the learning models. Hence, this process is time consuming and involves manual work of labeling. Adding to it, there is a need to increase the recognition rate for handwritten Kannada characters and convert them to machine editable form in less amount of time with the automation by machine. Hence in this paper, unsupervised learning method has been incorporated to recognize the offline handwritten Kannada characters, thereby saving the time of manual labelling of data along with automation and generation of dataset.

3 Proposed Methodology

The various phases involved in recognizing Kannada vowels are presented in the Fig. 1. The proposed system architecture consists of following steps:

A. Pre-processing
B. Feature extraction
C. Classification

Fig. 1 Architecture of the system

Fig. 2 Handwritten Kannada document

3.1 Preprocessing

Dataset Collection: Data is the collection of facts representing real world entities but in the field of computing, it refers to machine readable information. During literature survey, we have collected handwritten Kannada documents from different groups of people. Figure 2 represents single document in the dataset and experiment was conducted on 100 such documents. Documents scanned from the cameras or scanner will have noise and it is a raw image as shown in Fig. 2.

For the features to be extracted, images have to be preprocessed. As a part of pre-processing the horizontal and vertical lines are removed. The bounding box method is used on the images to extract the characters. Contours are drawn as part of bounding box method in which rectangle is drawn to each character in an input image. Character images are extracted after drawing bounding box. After that, image resizing and binarization based on the requirements of feature extraction method is carried out.

3.2 Feature Extraction

Histogram of Oriented Gradients (HOG)
HOG method is one of the famous techniques for object recognition and edge detection. HOG descriptor uses edge detection by gradient calculation and histograms of the gradients with magnitudes as weights [9].
Input: 13 * 100 Kannada vowel images.
Output: Vector of 329 features.

Algorithm

1. Begin.
2. The code uses [−1 0 1] kernel for gradient magnitude and orientation calculation.
3. Gradients are calculated in the range [0, 180].
4. Histograms of 9 bins are calculated with magnitudes as weights.
5. Each image is checked if its size is 32 * 32 pixels, else it is resized. The code reads images in greyscale.
6. The images are normalized for gamma, and then, for normal contrast.
7. Each 32 × 32 image pixel matrix is organized into 8 × 8 cells and histograms are calculated for each cell (16 * 9 matrix). Then, a 4 × 4 matrix with 9 bins for each cell is obtained (4 * 4 * 9 matrix) by reshaping.
8. This matrix is organized as 2 × 2 blocks (with 50 percent overlap) and normalized by dividing with the magnitude of histogram bins' vector. Hence every block will have 4 cells with each cell having 9 bins, i.e., 36 features per block.
9. Total of 9 blocks × 4 cells × 9 bins = 329 features.
10. End.

Run Length Count (RLC)

Run Length encoding is one of the well-known feature extraction and data compression techniques. Many variations of this algorithm are available [10]. In this paper, the following algorithm is proposed keeping in mind the structure and skewness of Kannada characters, i.e., mainly concentrating on the number of edge shifts in the image. Figure 3 illustrates the Run Length Count.

Input: 1300 Kannada vowel images.
Output: Vector of 100 features.

Algorithm

Fig. 3 Run length count for (5, 5) size block

1	0	1	1	0	1, 2
0	1	0	1	0	2, 2
1	1	1	1	0	0, 1
1	0	0	0	1	1, 1
0	0	1	0	1	2, 1
1, 2	1, 1	2, 2	0, 1	1, 0	[5,6,6,7]

1. Begin.
2. Image is resized to 50 * 50 pixels and binarized using Otsu's method.
3. Divide the image into block of 10 * 10 pixels, i.e., total of 25 blocks.
4. For each block, the vertical and horizontal run length count is being calculated, thereby resulting in 25 * 4 = 100 features.
5. The numbers are normalized for each block in the image.
6. Normalized run length counts are taken as features to the clustering algorithms.
7. End.

Chain Code (CC)

Chain code feature extraction method is used to extract the structural features of the characters. Hence, it is used for high speed feature extraction but demerit being large number of features. The algorithm concentrates on the structure and distribution of the contour points. Variation of algorithm presented in [11] is proposed below.

Input: 13 * 100 Kannada vowel images.

Output: Vector of 800 features.

Algorithm

1. Begin.
2. Normalize the binary image and resize the image to 100 * 100 pixels and binarize using Otsu's method.
3. Divide the image into blocks of size 10 * 10 pixels.
4. For each pixel in the image, it is considered as a contour point, if any of the eight of its neighboring pixels is a background point.
5. If it is a contour point, the direction of each of its background points is looked for and is added to the corresponding block's feature vector.
6. For each block, step 4 is repeated for all the contour points. E.g.: [3, 2, 5, 6, 2, 9, 15, 0] implies the block has 3 background points in total, in the direction 0 for the particular block as shown in Fig. 4, 2 for direction 1 and so on.
7. This results in an array of 8 features for each block. Thus, 800 (10 * 10 * 8) features are obtained for each of the characters.
8. End.

Local Binary Pattern (LBP)

Local binary pattern is a visual descriptor and feature extraction method in the field of image processing [12]. LBP is a simple feature extraction method giving labels to

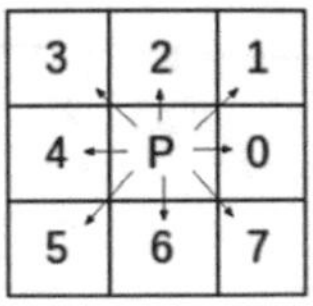

Fig. 4 Direction codes for the point P

3	2	1
4	P	0
5	6	7

the pixels of an image by looking into the neighborhood of each pixel by considering the result as a string of binary numbers which are then converted to single decimal number.

Input: 13 * 100 Kannada vowel images.

Output: Vector of 784 features.

Algorithm

1. Begin.
2. Resize the grayscale image to 30*30 pixels, and binarize using Otsu's method.
3. For each pixel in the image, compare each of its 8 neighbors (one pixel on each side is ignored due to white background).
4. Follow the pixels along a circle from top left corner in clockwise direction. This gives an 8-digit binary number (Fig. 5).
5. Convert the 8-digit binary number into decimal representation to represent that pixel.
6. This gives a feature vector for the entire image which contains 28 * 28 = 784 features per image.
7. End.

Hybrid Implementations

Hybrid algorithm is the method of combining features extracted by two or more algorithms to generate a new list of features. In this paper, higher recognition rate is obtained compared to pure implementations by giving weights, to features extracted from the above algorithms and then combining them.

There are quite a number of possible combinations of hybrid implementations. Since HOG gave the highest recognition rate among pure implementations, HOG features are always considered while clustering. Below-mentioned hybrid implementations are experimented on data and it which resulted in more than 3% increase in the recognition rate for the combination of HOG/RLC. The increase in the recognition rate pertains to the different types of features extracted by the algorithms. HOG uses edge detection, RLC handles the skewness, CC extracts structural features, LBP considers the change in the neighboring pixels.

Fig. 5 Local binary pattern

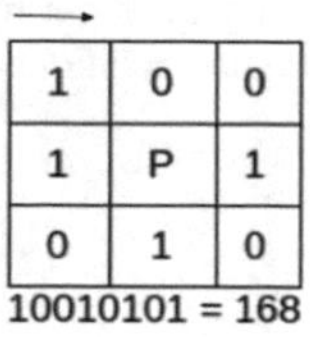

1	0	0
1	P	1
0	1	0

10010101 = 168

3.3 Classification

In this paper, the main aim was handling the data with unsupervised machine learning algorithms. K-Means clustering is used as the clustering method for classification of the characters. Clustering is the task of grouping the similar entities into single clusters based on standard measure of distance calculation (Euclidean distance) between the entities.

It is the most popular unsupervised classification algorithm because of its simplicity and speed of processing. Here, the mean of each cluster is calculated and then, each observation is classified into the cluster with nearest distance. The key limitation of K-Means is choosing the number of clusters and the initial centroid selection. Hence, the 13 (one for each vowel) initial centroid points were specified manually by choosing the features from the dataset. Clusters are of identical size because of the dataset being used, thereby giving no weights to any of the clusters.

4 Experiment Results

The proposed method is experimented on data collected from various types of scanned documents collected as a part of literature survey. We have experimented on 100 documents. Figure 6 shows the scanned image sample before and after pre-processing. Figure 7 and 8 shows the accuracies of pure and hybrid implementations respectively (Fig. 7).

We have obtained an accuracy of 81.26% for HOG and 83.87% for hybrid HOG/RLC. Table1 shows the best results obtained by the proposed methodology

Fig. 6 Image sample before and after preprocessing

Fig. 7 Accuracy comparison for hybrid implementations

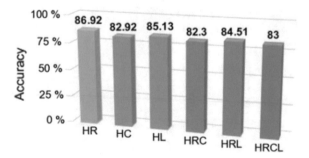

Fig. 8 Accuracy comparison for pure implementations

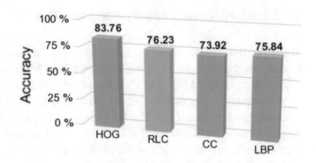

Fig. 9 Characters with 90% similarity

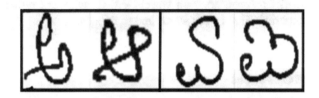

Table 1 Recognition rate of hybrid HOG/RLC

No.	1	2	3	4	5	6	7	8	9	10	11	12	13
Char	ಅ	ಆ	ಇ	ಈ	ಉ	ಊ	ಋ	ಎ	ಏ	ಐ	ಒ	ಓ	ಔ
Accuracy	85	59	100	93	100	87	96	97	89	58	93	96	77

on the features collected using hybrid HOG/RLC and there after using K-Means clustering as the classification scheme. But all the hybrid combinations didn't yield good results because of large number of feature vectors on combination.

The prime reason for the loss of 20% accuracy (approximate) is mainly because of the presence of similar characters, and the fact that each person's handwriting is unique and different. Figure 9 depicts the characters that approximately have 90% similar features.

The main aim of this paper is Kannada character dataset generation by automation and hence saving the time of manual labelling. After achieving the recognition rate of 86.92%, we have manually moved the incorrectly classified images to their respective folders and renamed the images, the data is converted into standard MNIST data format. Hence, the data can be directly fed to supervised models for training, validating and testing.

5 Conclusion and Future Enhancement

The proposed algorithm recognizes handwritten Kannada characters from the document images. Feature extraction method like Histogram of Oriented Gradients (HOG), Run Length Count (RLC), Chain Code (CC) and Local Binary Pattern (LBP) with clustering methods are used for the recognition. Good recognition is obtained with proposed methodology. The hybrid approach gave good results when compared with individual feature extraction methods. The proposed recognition system can be used to increase the accuracy of the existing systems. It can be extended to the other complex and similar handwritten Kannada character recognition as well. Currently, work has been done on vowels and future work remains to consonants and then experimenting on the words and sentences.

References

1. Kavya, T.N., Pratibha, V., Priyadarshini, B.A., Vijaya Bharathi, M., Vijayalakshmi, G.V.: Kannada characters and numerical recognition system using hybrid zone-wise feature extraction and fused classifier. Int. J. Eng. Res. Technol.(IJERT). **5**, 506–510 (2016)
2. Sandhya, N., Rahul, B., Krishnan, R., Ramesh Babu, D.R.: HMM based—character recognition for offline handwritten Indic scripts. Int. J. Eng. Res. Technol. (IJERT) **4**, 614–617 (2015)
3. Aravinda, C.V., Prakash, H.N., Lavanya, S.: Kannada handwritten character recognition using multi feature extraction techniques. Int. J. Sci. Res. (IJSR) **3**, 911–916 (2014)
4. Vishwaraj, Karthik, S.P., Sreedharamurthy, S.K.: Kannada handwritten character recognition using genetic algorithm. J. Mod. Trends Eng. Res (IJMTER) **02**, 369–374 (2015)
5. Angadi, S.A., Sharanabasavaraj, H., Angadi: Structural features for recognition of handwritten Kannada character based on SVM. Int. J. Comput. Sci., Eng. Inf. Technol. (IJCSEIT) **5**, 25–32 (2015)
6. Roopananda, M.K.: Weighted zone based Kannada character recognition using crack feature extraction technique and support vector machine classification. Int. J. Sci. Technol. Eng. (IJSTE) **1**, 180 (2015)
7. Shashikala, Parameshwarappa, B.V., Dhandra: Basic Kannada handwritten character recognition system using shape based and transform domain features. J. Adv. Res. Comput. Commun. Eng. J. Adv. Res. Comput. Commun. Eng. **4**, 495–499 (2015)
8. Majhi, B., Pujari, P.: On development and performance evaluation of novel Odia handwritten digit recognition methods. Arab. J. Sci. Engineering. **43**, 3887–3901 (2018)
9. Karthik, S., Srikanta, M.K.: Deep belief network based approach to recognize handwritten Kannada characters using histogram of oriented gradients and raw pixel features. Int. J. Appl. Eng. Res. **11**, 3553–3557 (2016). ISSN 0973–4562
10. Karthik, S., Mamatha, H.R., Srikanta, Murthy, K.: An approach based on the run length count for denoising the Kannada characters. Int. J. Comput. Appl. **50**, 43–44 (2012)
11. Mamatha, H.R., Srikanta, Murthy, K., Veeksha, A.V., Priyanka, S., Vokuda, Lakshmi, M.: Recognition of handwritten Kannada numerals using directional features and K-means. In: International Conference on Computational Intelligence and Communication Systems, pp. 645–646 (2011)
12. Gowthami, A.T., Mamatha, H.R.: Fingerprint recognition using zone based linear binary patterns. In: Second International Symposium on Computer Vision and the Internet (Vision-Net'15), Elsevier, vol. 58, pp. 552–557 (2015)

Despeckling of Synthetic Aperture Radar (SAR) Images Using Local Statistics-Based Adaptive Filtering

Vikrant Bhateja, Ankita Bishnu and Ankita Rai

Abstract Synthetic Aperture Radar (SAR) images are speckle corrupted right at the initial stage of their acquisition due to repeated backscattering signals from the uniformly distributed target sites. Addition of speckle noise in the images complicates visualization, image analysis and image translation. For this reason, despeckling of the image data is important to not only remove the speckle but also preserve the finer details without losing them. In this paper, a method of despeckling using local statistics-based adaptive filtering has been proposed which can be used in all kinds of speckle environment especially in higher speckle levels. The filter is adaptive in nature as the size of the filtering window which is being used in this method can be varied to despeckle those regions with fully developed speckle and protect minute details. The filtering competency of the proposed filter is then scrutinized and compared using various metrics like Peak Signal-to-Noise Ratio (PSNR) and Speckle Suppression Index (SSI) on different speckle variances.

Keywords SAR images · Speckle · Despeckling · Adaptive filtering · Local statistics

1 Introduction

Synthetic Aperture Radar is a radar type which is capable of two-dimensional- and three-dimensional imaging of Earth's landscapes. While acquiring images of landscapes, there is an interference phenomenon of the reflecting signals which occurs at

V. Bhateja (✉) · A. Bishnu · A. Rai
Department of Electronics & Communication Engineering, Shri Ramswaroop Memorial Group of Professional Colleges (SRMGPC), Faizabad Road, Lucknow 226028, UP, India
e-mail: bhateja.vikrant@gmail.com

A. Bishnu
e-mail: ankita.bishnu5@gmail.com

A. Rai
e-mail: ankita.rai3131@gmail.com

© Springer Nature Singapore Pte Ltd. 2020
V. Bhateja et al. (eds.), *Intelligent Computing and Communication*,
Advances in Intelligent Systems and Computing 1034,
https://doi.org/10.1007/978-981-15-1084-7_66

the synthetic transducer aperture. These backscattered signals add coherently result-ing in the formation of black and white dots in the images [1]. Therefore, speckle noise is also called granular noise. Speckle noise predominantly reduces the image resolution, image data extraction and classification. A significant step which is needed to be undertaken immediately after the SAR imaging is the despeckling process. The despeckling process helps in better study of SAR images. In the past few decades, various despeckling techniques had been devised for smoothing of images as well as for retaining the finer details of the edges. One of the many despeckling techniques, Anisotropic Diffusion plays a vital role in enhancing boundary information and reducing noise in the images considerably [2]. The Speckle Reduction Anisotropic Diffusion Filter (SRAD) and Detail Preserving Anisotropic Diffusion Filter (DPAD) were proposed which used local gradient magnitude and Kuan approximation to despeckle the images [3, 4]. The extensions of SRAD and DPAD filters are later modified and named as OSRAD filter [5]. Another set of speckle filters which belong to Local statistics filters are MMSE-based filters. The conventional filters of them are Lee, Frost and Kuan Filters [6]. The speckle filters quite recently evolved in recent years and yet another set of filters have come up in wavelet domain [7, 8]. Continu-ous Wavelet Transform (CWT) and Discrete Wavelet Transform (DWT) have been employed to successively denoise SAR images by analyzing the signals at different levels [9]. Over the years, a large magnitude of work and researches has been done on despeckling using adaptive filters [10]. In this paper, a filtering method has been established for denoising of SAR images using adaptive filtering where a larger win-dow is further divided into smaller sub-windows. The filter deploys such filtering window where local statistical neighborhood information like mean, variance and (Quadrature Corner Difference) QCDs of the sub-windows are used in the operation of the filter. The following paper is categorized in the following sections: Sect. 2 contains a brief description of adaptive filtering. Section 3 states the algorithm of the proposed filter. Section 4 contains the analysis of results and discussions which is followed by conclusion of the overall work stated in Sect. 5.

2 Despeckling Using Adaptive Filter

In general, for any filtering application there is a requirement of selection of mask or window size wherein all the pixels are weighted within the window of specified dimension around the central pixel [11]. In adaptive filtering technique, rather than the deployment of fixed window, varying window size is taken which adjusts to the spatial characteristics of the image. Though Lee filter protects the subtle edge details of the images [12], there still presents a certain degree of trade-off between the extent of speckle reduction and the filtering and edge-saving capability based on the window size [13]. As a result, the adaptive windowing technique is put forth such that the size of the window can be changed as and when required according to the amount of speckle present in the images and nature of the SAR images like single or multi-band SAR image data [11].

2.1 Local Statistics-Based Adaptive Filter

The local statistics-based adaptive filtering uses local statistics such as mean, variance and standard deviation of the homogeneous regions bounded by the adaptive window. In the novel approach of local statistics-based adaptive filtering technique which is proposed in this paper, there is a windowing and sub-windowing operation which runs over the entire image matrix and processes over each pixel and subsequently despeckles it. The despeckling function is based on various indices like Quadrature Corner Difference (QCD) and variance of the respective windows [14]. The usability of this filter is mainly in despeckling the SAR images corrupted with high speckle content where the images are destroyed by wholly developed speckle. The proposed filter is therefore evaluated and assessed in terms of its despeckling capability and efficiency using various image quality assessment parameters like Peak Signal to Noise Ratio (PSNR) [15, 16] and Speckle Suppression Index (SSI) [17, 18] and these factors are further compared in various speckle contents.

3 Proposed Design Methodology

In the proposed adaptive filter, sub-window approach within a larger window has been used to avoid unnecessary denoising which causes regions and edges to blur out. The filter uses neighborhood data to preserve edge details and of homogeneous blocks and despeckle the image at the same time. The filtering window (mask) is divided into five sub-windows such that each sub-window has the central pixel of the entire mask which is to be despeckled [14]. The size of the sub-window is determined by the following formula:

$$N = 2n - 1 \tag{1}$$

where, N is the window or mask size and n is the sub-window size. The Quadratic Corner Difference (QCD) of each of the five sub-windows is determined by the following equation:

$$QCD_i = \sqrt{\frac{1}{N^2} \sum_{k=1}^{N^2} \left(x_k^i - I_c\right)^2} \tag{2}$$

where, x_k^i is the kth pixel value in the ith sub-window, Ic is the central pixel value which is common to entire mask as well as the sub-windows.

An important aspect of this filter is the filtering ability factor which is denoted by η (eta). This is a significant feature of this filter as it adaptively changes with the homogeneous areas, heterogeneous areas and edgy regions of the SAR images which are taken under consideration. Multiple iterations using various values of η

have been performed and the final despeckled or restored image is obtained by the following formula:

$$R_{(i,j)} = \frac{\sum_{i=1}^{5} \mu_i \left(\frac{1}{QCD_i}\right)^{\eta}}{\sum_{i=1}^{5} \left(\frac{1}{QCD_i}\right)^{\eta}} \tag{3}$$

where $R(i,j)$ is the final despeckled pixel, μ_i is the ith sub-window mean. The reciprocal of QCD is responsible for the preservation of image attributes while mean of the sub-windows is responsible for the smoothing of the image textures. A window of a given dimension $N \times N$ is taken which is subdivided into five sub-windows in five different directions viz., North-West (NW), North-East (NE), South-West (SW), South-East (SE) and Central such that all sub-windows contain the central pixel of the larger window. This central pixel is also the pixel which is to be despeckled by the filter. Thus, all the sub-windows and the larger window or mask are convolved so that a single windowing operation is performed serially taking single pixel at a time and the neighborhood data are drawn like mean and $QCDs$ to obtain restored pixels, over the entire image matrix till it reaches the last pixel. The experimental value of η is taken as 5 which is sufficient enough to retain boundary information and despeckle the homogeneous areas. This value of η can vary according to the type of regions or landscapes which are being captured by SAR. It can also vary according to the single-polarized or multi-polarized SAR imagery or real and simulated SAR images. Qualitative analysis of the despeckled SAR images in various noise levels from lower to higher noise variance is depicted in subsequent section.

4   Results and Discussions

For testing the performance of the proposed filter algorithm, images are acquired from the dataset of RADARSAT-2 image gallery of Geospatial-International resources of MDA Corporation, Canada. The image that has been used for the experiment is a RADARSAT-2 image of Lake Claire, Alberta. In this image, the Northern tip of Lake Claire which is situated in Wood Buffalo National Park in the north of Alberta, Canada can be seen. This image is extracted from the wide ultra-fine RADARSAT-2 imager. Wide Ultra-Fine scenes from RADARSAT-2 provide high intensity resolution imagery (3 m) over a landscape size of 50 km $\times$ 50 km. The area shows the Peace River cutting across the top of the image [19]. The parameters which have been fixed for the proposed despeckling experiment is the window size (N) which is fixed to 5×5, sub-window (n) size which is fixed to 3×3 and value of filtering capability factor (η) which is fixed to 5 for efficient denoising as well as finer details preservation along the contours. The proposed local statistics based adaptive filter is applied on three different noise levels which are (i) Low Noise Variance, (ii) Medium Noise Variance and (iii) High Noise Variance. The low noise variance is taken to be

0.02. The medium noise variance which is slightly greater than low noise is taken as 0.04 while high noise variance which is comparatively higher than medium and low noise is taken as 0.06. The speckled SAR images in various noise levels are shown in Fig. 1a–c and their respective filtered images are shown in Fig. 1d–f.

The various performance indices like PSNR and SSI of the despeckled and noisy images are tabulated in Table 1 and it has been seen that the filter works reasonably

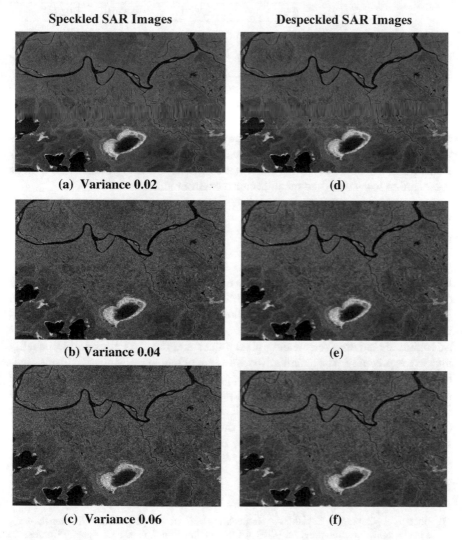

Speckled SAR Images   **Despeckled SAR Images**

(a) **Variance 0.02**   (d)

(b) **Variance 0.04**   (e)

(c) **Variance 0.06**   (f)

Fig. 1 Simulation results of the proposed adaptive filter. **a** Image with low speckle variance 0.02. **b** Images with medium speckle variance 0.04. **c** Images with high speckle variance 0.06. **d** Filtered image of low speckle variance. **e** Filtered image of medium speckle variance. **f** Filtered image of high speckle variance

Table 1 Simulation results of PSNR and SSI for despeckled SAR image

Noise variance	PSNR (in dB)	SSI (noisy)	SSI (filtered)
0.01	27.8873	0.2877	0.2415
0.02	25.7886	0.3066	0.2487
0.03	23.1358	0.3212	0.2564
0.04	21.6676	0.3389	0.2592
0.05	20.9203	0.3522	0.2640
0.06	19.9271	0.3701	0.2663

well in high speckled images. Higher the PSNR, greater is the quality of the image and lower the SSI, higher is the amount of speckle removed. It can be seen that in Fig. 1a, i.e., in already noiseless image, speckle is almost completely removed from the image as shown in Fig. 1d, whereas in medium and high noise variance the filter eliminates residual speckle fairly well from the image as can be seen in the Fig. 1e, f respectively. The SSI values in Table 1 show higher speckle content in noisy images which is suppressed to much lower speckle content in filtered images. Thus, it is evident that the benefit of using this filter is extended for high noise like 0.06 as compared to low-noise- and medium-noise environment.

5 Conclusion

In this paper, simulation of local statistics adaptive filter is achieved by incorporating different levels of noise in the original SAR image. The advantage of the proposed filter is that it prevents in smearing of the edge details and homogeneous regions information as it employs adaptive mask and sub-window size which takes in consideration the suitable surrounding pixel values and data. As a result, high quality images are procured. The quantitative results are obtained by estimating image quality parameters like PSNR and SSI for the speckled images at various noise levels and their respective filtered images by the proposed filter. The contribution of this paper is that the filter performs appreciably well even in the higher noise variance by aptly denoising the SAR images and restoring the image features.

References

1. Bhateja, V., Gupta, A., Tripathi, A.: Despeckling of SAR images in contourlet domain using a new adaptive thresholding. In: IEEE 3rd International on Advance Computing Conference (IACC), pp. 1257–1261 (2013)
2. Bhateja, V., Tripathi, A., Gupta, A.: Recent Advances in Intelligent Informatics, p. 23. Springer International Publishing, Switzerland (2014)
3. Aja-Fernández, S., Alberola-López, C.: On the estimation of the coefficient of variation for anisotropic diffusion speckle filtering. IEEE Trans. Image Process. **15**(9), 2694–2701 (2006)

4. Yu, Y., Action, S.T.: Speckle reducing anisotropic filtering. IEEE Trans. Image Process. 11, 1260–1270 (2002)
5. Krissian, K., Westin, C.F., Kikinis, R., Vosburgh, K.: Oriented speckle reducing anisotropic diffusion. IEEE Trans. Image Process. **16**(5), 1412–1424 (2007)
6. Lee, J.S.: Speckle analysis and smoothing of synthetic aperture radar images. J. Comput. Gr. Image Process. **17**, 24–32 (1980)
7. Kuan, D., Sawchuck, A., Strand, T., Chavel, P.: Adaptive noise smoothing filter for images with signal dependent noise. In: IEEE Transactions on Pattern Analysis and Machine Intelligence, vol. 7, pp. 165–177, February 1985
8. Bhuiyan, M.I.H., Omair Ahmad, M., Swamy, M.N.S.: New spatially adaptive wavelet-based method for the despeckling of medical ultrasound images. In: IEEE International Symposium on Circuits Systems, New Orleans, LA, USA, pp. 2347–2350, May 2007
9. Taquee, A., Bhateja, V., Shankar, A., Srivastava, A.: Pre-processing of cough signals using discrete wavelet transform. In: Second international conference on computing, communication and control technology, pp. 42 (2018)
10. Sivakumar, R., Nedumaran, D.: Implementation of wavelet filters for speckle noise reduction in ultrasound medical images: a comparative study. In: 16th International Conference on Signals, Systems and Communication, pp. 239–242 (2009)
11. Cozzolino, D., Parrilli, S., Scarpa, G., Poggi, G., Verdoliva, L.: Fast adaptive nonlocal SAR despeckling. IEEE Geosci. Remote Sens. Lett. **11**(2), 524–528 (2014)
12. Mahdavi, S., Saleni, B., Moloney, C., Huang, W., Brisco, B.: Speckle filtering of synthetic aperture radar image using filters with object sized adaptive windows. Int. J. Digit. Earth **11**(7), 703–729 (2018)
13. Lopes, A., Touzi, R., Nezry, E.: Adaptive speckle filters and scene heterogeneity. In: IEEE Transaction on Geoscience Remote Sensing, vol. 28, pp. 992–1000, November 1990
14. Park, J.M., Song, W.J., Pearlman, W.A.: Speckle filtering of SAR images based on adaptive windowing. Proc. IEEE Proc. Vision Image Sig. Process. 146(4), 191–197 (1999)
15. Shitole, S., Sharma, M., De, S., Bhattacharya, A., Rao, Y.S., Krishnamohan, B.: Local contrast based adaptive SAR speckle filter. J. Indian Soc. Remote Sens. **45**(3), 451–462 (2017)
16. Singh, S., Jain, A., Bhateja, V.: A comparative evaluation of various de-speckling algorithms for medical images. In: Proceedings of the CUBE International Information Technology Conference, Issue 3, pp. 32–37 (2012)
17. Wang, Z., Bovik, A.C., Sheikh, H.R., Simoncelli, E.P.: Image quality assessment: from error visibility to structural similarity. IEEE Trans. Image Process. **13**(4), 600–612 (2004)
18. Bhateja, V., Srivastava, A., Singh, G. and Singh, J.: A modified speckle suppression algorithm for breast ultrasound images using directional filters. In: Proceedings of the 48th Annual Convention of Computer Society of India, vol. 2, pp. 216–219 (2014)
19. RADARSAT-2 Images Database. https://mdacorporation.com/geospatial/international/resources/image-gallery

Performance Analysis of Classification Algorithm in Machine Learning over Air Pollution Database

N. Aditya Sundar, P. Samuel, A. Durga Praveen and D. RaghaVendra

Abstract Air quality is a vital necessity on the earth, it helps us to live on this planet. Every day, air is polluted heavily because of industries releasing the gases in the atmosphere, burning of e-waste, polluted exhaust from vehicles. It impact on public health causes diseases like cancer, asthma, lung diseases, etc. A study analysis has been done, by observing the increase of day-to-day air pollution, considering the data from the pollution board pertaining to two different years. Analysis was done on this data using Random Forest Algorithm implemented in R programming language, to observe the error rate between the 2 years of data so as to identify the component which impacts the environment and therefore public health.

1 Introduction

Air pollution is mixtures of various gases, particles and other activities. In this paper, we discuss about the breakpoints and the error rate among the data sets of the two different years. We apply the Random Forest algorithm [1] to calculate the error rate by applying the different formulas mentioned below using R programming language [2]. We calculate the breakpoint concentrations with the help of breakpoint table and different levels of AQI [3]. Air Quality Index is used to measure the air pollutants within the region. There are many methods in calculating the concentrations and different methods that have been used by different countries in calculating the breakpoint concentration levels of each individual pollutants. We have taken the data observed around a 24 hrs range. By considering this, we get to know not only concentrations, but also the pollutants that cause health effects, such as toxic pollutants, dangerous gases etc.

N. Aditya Sundar (✉) · P. Samuel · A. Durga Praveen · D. RaghaVendra
Department of Information Technology, Anil Neerukonda Institute of Technology & Sciences
– (ANITS), Sangivalasa, Bheemunipatnam, Vishakapatnam 531162, Andhra Pradesh, India
e-mail: adityasundar.it@anits.edu.in

© Springer Nature Singapore Pte Ltd. 2020
V. Bhateja et al. (eds.), *Intelligent Computing and Communication*,
Advances in Intelligent Systems and Computing 1034,
https://doi.org/10.1007/978-981-15-1084-7_67

2 Methodology

In order to understand the rise of air pollution in a particular region, a real time air quality monitoring was taken at the control board site for every region. In the present study we are going to discussed about radiations [4] that are causes due to the pollutants released in the atmosphere such as PM10, PM2.5, NO2, SO2, O3 and NH3. For this we have used the technique known as Beta Attenuation Monitoring (BAM). It is a widely used air monitoring technique for the absorption of beta radiation by solid particles extracted from the air. This technique [2] allows for the detection of PM10 and PM2.5. The main principle is based on *Bouguer* (*Lambert–Beer*) law: the amount by which the flow of beta radiation (*electrons*) is attenuated by a *solid matter* is exponentially dependent on its *mass* and not on any other feature (such as *density*, chemical composition or some *optical* or *electrical* properties) of this matter, and the component O3 is monitored based on the concentration that are released in the air. We take the index based on hourly or 8 hrs accurately [4]. For SO2 known as UV fluorescence, based on the emission of light by SO2 molecule observed by UV radiations is taken. While for NO2 the amount of nitrogen oxide in the air is measured by "Chemiluminescence Analyzer" [3].

Analysis of each component per day

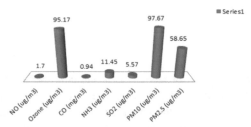

Methods I:
In this method the Air Quality Index (AQI) calculated through the concept of arithmetic mean along with the concentration [5] of air pollutants to the standard value of those concerned pollutants such as PM2.5, PM10, SO2, NO2, CO, O3 and NH3. From the obtained value, say the average is multiplied by 100 to get the index value. Thus, Air Quality Index is compared with the rating scale [4]. We can calculate AQI for the pollutants individually by the following formula:

$$\mathbf{AQI = (C/Cs) * 100}$$

where

AQI = Air Quality Index
C   = the observed value of the air quality parameters pollutant (PM10, PM2.5, NO2 and SO2)
Cs = CPCB standard for residential Area.

Method II:
In this method the Air Quality Index is calculated by observing the geometric mean to the ratio of the concentration of each pollutant to the standard value of that pollutants [2] such as PM10, PM2.5, SO2, NO2, CO and NH3. And the Air Quality Index is compared with the rating scale.

Method III:
Air Quality Index was done for combining qualitative measures with qualitative concept of the environment [6, 7]. The individual air quality index here is calculated as follows:

$$AQi = (W * C)/Cs$$

where

AQI = Air Quality Index
W   = Weighted of Pollutant
C   = the observed value of the air quality parameters pollutant (PM10, PM2.5, NO2 and SO2)
Cs   = CPCB standard for residential Area.

Method IV:
In this method, to calculate the AQI from the concentration of each individual pollutants were based on the breakpoint concentration table [8]. The individual air quality index for a given pollutant concentration (Cs) as based on linear segmented principle is calculated as

$$Ip = \big[\{(Ihl - ILo)/(BhI - BLO) * (Cp - BLo)\big] + ILo$$

where

BhI = Breakpoint concentration greater or equal to given concentration
BLo = Breakpoint concentration smaller or equal to given concentration
Cp   = The pollutant concentration
Ihl = AQI value corresponding to BhI
ILo = AQI value corresponding to BLo.

Note: In order to calculate Index of the multiple pollutants, we have to calculate the index of each pollutant, hence that pollutant with the highest value, or the "responsible pollutant/s" will determine the Air Quality Index and the category.

3 Result

The continuous monitoring of the air pollutants is undertaken from the GVMC corporation, Visakhapatnam city by the observation of Central Pollution Control Board during March to November 2017 with the reference to PM10, PM2.5, SO2, CO, NO2, O3 and NH3. Data obtained from the pollution board is used to calculate the air quality index (air pollution index) for critical parameter.

4 Air Quality Index (AQI)

In the present days, everyone in the society must have the awareness on the current air pollution levels and they must compare the past and present levels of air pollution. Air Quality Index is a tool, used to know the status of daily air quality along with the basic standards proposed by the pollution board in order to control the pollution levels. It gives an idea, how the index rate is increasing daily and tells the public to understand how clean the air to breathe daily. In our country we use Central Pollution Control Board standards in calculating the Air Quality Index from the data present by them. The AQI of specific pollutant is derived mainly from the physical measurement of pollutant like PM10, PM2.5, NO2, CO, NH3, O3 and SO2 etc. In the present study, different methods were used to calculate ambient air quality index.

Different AQI were estimated for various months and varying results were observed ranging from good to unacceptable for the same set of data. This may be due to eclipsing effect of the values used in the formulas. The statistical theory behind these AQI makes it more prone to variations, i.e., the use of means from simple arithmetic to logarithmic and weighted averages [6] to use of breakpoint concentration as basis of estimation. The breakpoint concentration based AQI is more robust and can be used for decision making. Accordingly, the AQI values are calculated based on Break point concentration [2] for 24 hourly averages for PM10, PM2.5, SO2, CO, O3 and NO2 concentrations and are categorized as satisfactory to moderate during the study.

The below figures represent the analysis combining data of 2 months, in 2018 (Fig. 1).

Below figures represent the analysis done through "Random Forest Algorithm" in R programming language (Fig. 2).

The above picture shows the data that is taken from the pollution board represent the different components such as PM10, PM2.5, NO, CO, VSO2, NH3, OZONE.

In order to use the random Forest model, we need two types of data sets. One is training data and the other one is testing data. We represent these data by applying Random Forest Algorithm (Fig. 3).

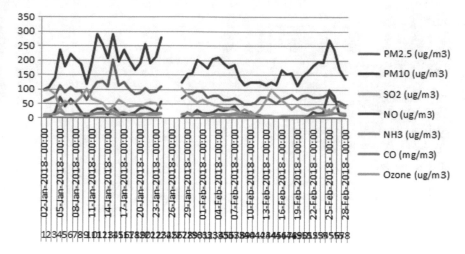

Fig. 1 Represent the Jan–Feb, 2018

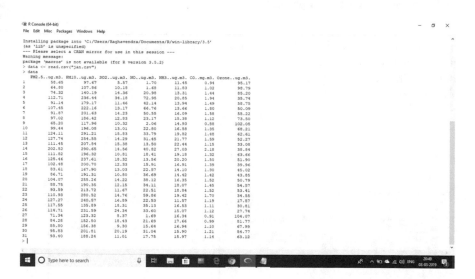

Fig. 2 Shows the data set, Jan–Dec, 2018

By considering the above-mentioned data, testing and training data we calculate the error rate among the data sets by applying Random Forest Algorithm through R programming language.

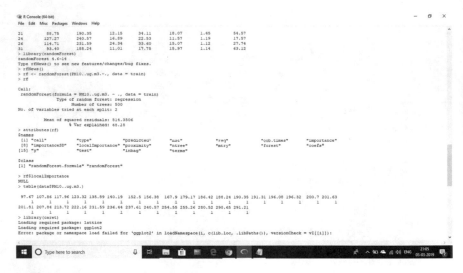

Fig. 3 Applying random forest algorithm

Above shown graphs Figs. 4 and 5 are represented through R programming language (following command "plot (filename)" for both training and testing data sets, indicates the levels of each components for the two different years 2018 and 2019 (Figs. 6 and 7).

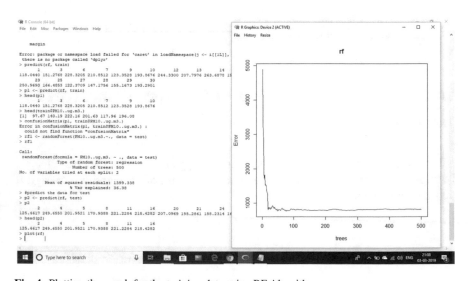

Fig. 4 Plotting the graph for the training data using RF Algorithm

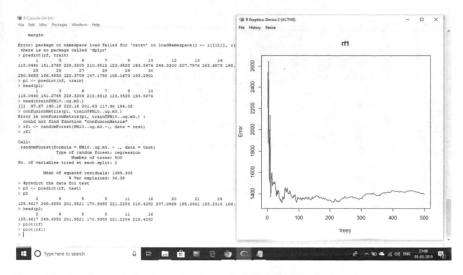

Fig. 5 Plotting the graph for the testing data using RF Algorithm

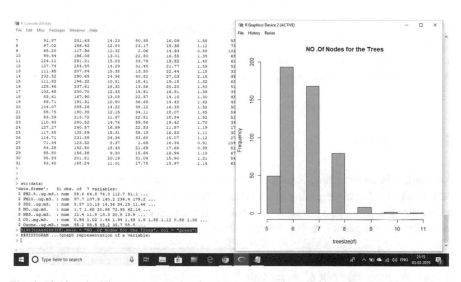

Fig. 6 Plotting the Histogram graph for the training data using RF Algorithm, Jan, 2018

The overall result will be showing the pollutants in a graph with month intervals and identifying the causes (Table 1).

While choosing the breakpoint table there are three cases to be followed.

1. We consider areas large in number and the AQI values based on the 8 h ozone values. However, for smaller number of areas AQI based on 1 hrs ozone value, that would be more precautionary. In these cases, both 8 and 1 h index value may be calculated and maximum two values are reported.

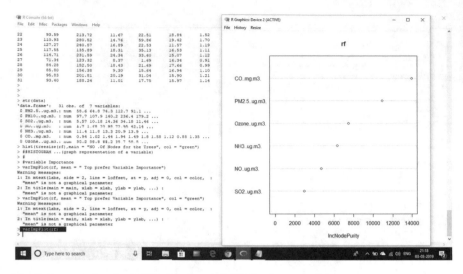

Fig. 7 Representing the Variable Importance, 2018 Jan data

2. NO2 has no short-term NAAQS (National Ambient Air Quality Standards) and can generate an AQI only above an AQI value of 200.
3. When 8-h O3 concentrations exceed 0.374 ppm, AQI values of 301 or higher must be calculated with 1-h O3 concentrations.

5 Conclusion

Air quality is an important aspect on the earth, because it is linked to our healthy survival. Everyday air is getting polluted heavily and it leads to health issues hence in an attempt to control it, a proper monitoring and analysis can be done by considering every 2 different years data set obtained from the Central Pollution Control Board. Making use of such data and by applying Random Forest Algorithm in R language, error rate has been identified and the result has been observed as provided in the above figures. As the quality of air is measurable its study and analysis can significantly impact public health.

Table 1 Breakpoints table for the AQI

Table 2-breakpoints for the AQI

These breakpoints							Equal these AQI's	
O_3(ppm) 8-h	O_3(ppm) 1-h[1]	$PM_{2.5}$(μg/m³) 24-h	PM_{10}(μg/m³) 24-h	CO(ppm) 8-h	SO_2(ppb) 1-h	NO_2(ppb) 1-h	AQI	Category
0.000–0.059		0.0–12.0	0–54	0.0–4.4	0–35	0–53	0–53	Good
0.060–0.075		12.1–35.4	55–154	4.5–9.4	36–75	54–100	51–100	Moderate
0.076–0.095	0.125–0.164	36.5–55.4	155–254	9.5–12.4	79–185	101–360	101–150	Unhealthy for sensitive groups
0.096–0.115	0.165–0.204	3.55.5–150.4	255–354	12.6–15.4	[4]186–304	361–64.	151–200	Unhealthy
0.116–0.374	0.205–0.4044	[3]150.5–250.4	355–424	15.5–30.4	[4]305–604	650–12–9	201–300	Very unhealthy
(2)	0.405–0.504	[3]250.5–350.4	425–504	30.5–40.4	[4]605–804	1250–1–49	301–400	Hazardous
(2)	0.505–0.604	[3]350.5–500.4	505–604	40.5–50.4	[4]805–1004	1650–2–49	401–500	

References

1. Building a random forest from scratch & understanding real-world data products (ML for programmers) **26**(2), 467–478 (2016)
2. Random Forest Algorithm for the Relationship between Negative Air Ions and Environmental Factors in an Urban Park **9**(12), 463 (2018)
3. Zhang, Y.-L., Cao, F.: Environ. Pollut. **202**, 217–219 (2015)
4. Cogliani, E.: Air pollution forecast in cities by an air pollution index highly correlated with the meteorological variables. Atmos. Environ. **35**, 2871–2877 (2001)
5. Gore, R.W., Deshpande, D.S.: An approach for classification of health risks based on air quality levels. In: International Conference on Intelligent Systems and Information Management (ICISIM), IEEE (2017)
6. (PDF) Using factor analysis to attribute health impacts to particulate pollution sources. Available from https://www.researchgate.net/publication/8502837_Using_Factor_Analysis_to_Attribute_Health_Impacts_to_Particulate_Pollution_Sources. Accessed 3 Jan 2019
7. World Health Organization: Air quality guidelines: global update 2005: particulate matter, ozone, nitrogen dioxide, and sulphur dioxide (2006)
8. Attri, A.K., Kumar, U., Jain, V.K.: Formation of ozone by fireworks. Nature (2001)

Face Recognition with Voice Assistance for the Visually Challenged

G. Sridhar Chakravarthy, K. Anupam, P. N. S. Harish Varma,
G. Harsha Teja and Sireesha Rodda

Abstract Visually impaired people face a lot of challenges in day-to-day life. Having seen the difficulties faced by them, our primary objective is to facilitate confidence and to empower them to lead a life free from threats related to their safety and well-being. The lack of ability to identify known individuals in the absence of auditory or physical interaction cues drastically limits the visually challenged in their social interactions and poses a threat to their security. Over the past few years many prototype models have been developed to aid this population with the task of face recognition. This application will reduce the inherent difficulty for recognition of a person. It will present a facial recognition application with an intuitive user interface that enables the blind to recognise people and interact socially. The carefully designed interface lets the visually challenged to be able to access and use it without any requirement for visual cues as the users are acquainted by a voice assistant to navigate through the application. The entire build is designed to run efficiently on a Raspberry Pi 3 model B module using the Android Things platform. The Open CV library has been used for the detection and recognition of people in this project. This enables the scope for the software to be run on a multitude of devices such as camera embedded glasses to warn users of their surroundings and identify people to interact safely. Since everything in the application is done in real time with no requirement for prior datasets to be hardcoded it drastically improves the versatility of the software. We hope to make the visually impaired feel closer, comfortable and more secure with the world surrounding them through our application.

Keywords Face recognition · Speech driven · Voice-based assistance · Pyttsx · Open CV

G. Sridhar Chakravarthy (✉) · K. Anupam · P. N. S. Harish Varma · G. H. Teja · S. Rodda
Department of Computer Science and Engineering, GITAM Institute of Technology, GITAM
(Deemed to Be University), Visakhapatnam, India
e-mail: sridharchakravarthyg@gmail.com

© Springer Nature Singapore Pte Ltd. 2020

701

V. Bhateja et al. (eds.), *Intelligent Computing and Communication*,
Advances in Intelligent Systems and Computing 1034,
https://doi.org/10.1007/978-981-15-1084-7_68

1 Introduction

Face Recognition has become an important utility in our day-to-day routine activities. It is really easy for an individual to develop applications on face recognition. Some of the well-known applications of facial recognition are used in the domains of Security, Authentication, Access, identity, Data Statistics. With the high availability of libraries like OpenCV, the dream to use face recognition and detection has become really easy. One can directly train, develop, implement, and use the libraries at ease to make their own working model of facial recognition and detection.

Since it has become really easy, one can think of making the best use of the technology in hand help mankind to grow gradually. This application focuses on the practical difficulty faced by visually impaired people in real time. Their disability to have a vision can be elevated by this as this application helps them to understand who has entered the room or the area. However, the application must be initially trained or must be trained individually after every necessity for a new user to be added in the database to be identified and detected. The main reason for the accurate working of facial recognition is that it is capable of capture and store all the unique set of data that are different for every individual.

The whole idea is help visually impaired people to get and utilize the Face Recognition and Detection to the best extent. This can be achieved by using and integrating OpenCV with text to speech libraries. This serves the purpose in the background, however there is an additional necessity to train the people who should be detected. This part of the implementation is done in different methods.

2 Related Work

Different authors have written regarding different technologies, in this study two different contributions have been combined together regarding face recognition and text to speech conversion which would help the visually disabled people.

Kar et al. [1] have proposed a well-developed system on verification of people based on their biometrics through a continuous video monitoring using Personal Component Analysis (PCA) algorithm. The result of this process has improved the overall accuracy and this project is carried using old black and white frames.

Zhang et al. [2] have developed a real-time face detection using the AdaBoost algorithm which has been modified and results are given to two methods timer and dual-thread among which dual-thread method demonstrates results which are much precise and simple.

Wilson and Fernandez [3] they have used a method for the detection of facial features developed by Viola and Jones. The main idea is to restrict the area of being focused for feature detection and concentrate on area where most probable features could be obtained which reduces the area to be considered and increases the speed to detect faces.

Kalas [4] have presented a difference between different methods and algorithms which are mainly used for the detection of face. The main algorithms that are used are Haar cascade and Adaboost on the OpenCV library.

Emami and Petruț [5] have created an application using the open source open cv library which gives access only to a certain category of people by examining their facial features.

Hudelist and Schoeffmann [6] most of the open cv applications are made on computer and there is a bridge between developing application on mobile and laptop and this has been overcame in this project. This project shows that the best performance on mobile is almost 80 up to the pc or rather laptop performance.

Degtyarev and Seredin [7] they worked on seven face detection algorithms. The aim of their work is to propose parameters of FD algorithms quality evaluation and methodology of their objective comparison, and to show the current state of the art in face detection.

3 Methodology

Recognition of fellow people is a major hurdle being faced by visually impaired. This caught our attention and we felt that we have to contribute to this cause for being a part of a society. This application has been developed from root, keeping in mind the inabilities and limited perception cues of the blind. It has a voice assistant that guides its user from the first step to intuitively navigate through the application.

This has been created with previous experience and interaction with the blind. The explained methodology is presented as a flowchart in Fig. 1.

The application starts with a voice prompt asking the user to either touch the right or left portion of the screen to recognise and train respectively as asking the visually impaired to select specific buttons will be too much of a hurdle. The entire user interface is designed to provide a user experience that is free from obstacles as navigation through the application is done minimal and basic interaction with the touch screen. As the left portion of the screen is selected the voice assistant takes over with prompts and asks the user to say out the name of the person who the model is going to be trained for. Once the name is stored, the assistant asks to proceed forward and guides the visually challenged carefully to train the model of given person's face.

This may take a few moments as images have to be scanned from various angles while Multilayer Perceptron (MLP), Naive Bayes and Decision tree models are trained. Any of the classes can be called based upon the requirement by CameraBridgeView.class and OpenCVLoader.class. Once the model has been trained the voice prompts the users to go back to the default screen by pressing the back button. The user's application will hence be trained with faces of multiple people that belong to the user's social circle in a similar fashion. On opening the default screen the voice assistant asks the user to touch the right portion of the screen to recognise people. Once the recognition mode has been turned on, the user can strap the phone to the chest pocket with a field of view similar to that of the eyes. The user

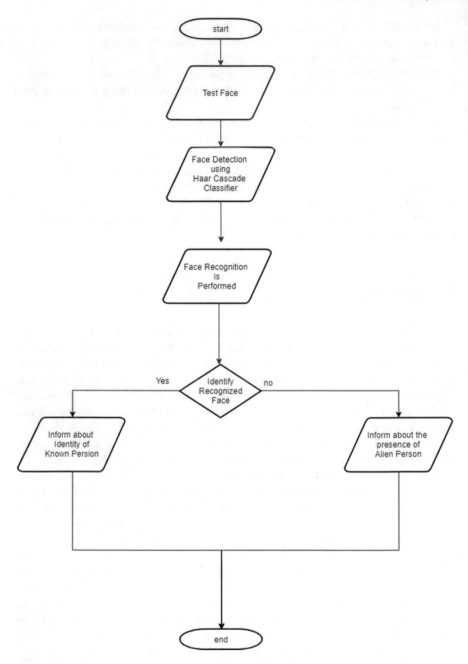

Fig. 1 Flow chart

now can now venture out into the public freely with having the confidence that they will be able to recognise people with the help of the application that has been trained to provide a voice prompt whenever a person's face has been recognised in the field of view of the camera and also whenever it detects an obstacle the user should avoid. In case of an emergency it can also send an alert to the local authorities. Hence this application allows the users to be comfortable and confident with social and outdoor interactions.

4   Experimental Setup

Android Studio (IDE) has been used to debug and run the software on a virtual emulator that is inbuilt into the IDE. Samsung Galaxy S8 and Xioomi Redmi Note 5 Pro have been used as prototype devices for testing and debugging. The support for this application has been set to android version 4.1 which means that every android version past that will be able to support the application and its libraries. The detection and recognition of the face is done with the help of the OpenCV library that is available as an open source repository. Various versions of this library have been called upon for the implementation of this application.

OPENCV_VERSION_2.4.2 to OPENCV_VERSION_2.4.10 are the libraries that been called by the OpenCVLoader.class to implement the features of object and facial detection and recognition. This is done for the real time training of the proposed model.

5   Results

The outcomes of each and every process involved in this development are shown here. Figure 2 depicts the desktop application version of the dataset that has been created which has 21 frames. Initially it takes the ID and saves the frames as the ID followed by the frame number. Figure 3 describes the recognition of the face searching from the dataset by displaying the name along with an audio feedback saying out the name for the identification of person to the disabled person. Figure 4 represents the default screen for the mobile application which has navigation options: train and recognise. Figure 5a describes the prompt to the user to say out the name of the person that the user is going to train the model for. Figure 5b shows how images are stored under the directory named by the person. Figure 6 represents the real time training of the proposed model. Various features will be extracted by the FeatureExtractor.class and concurrently train the application. Once the Recognize mode is selected Fig. 7 represents the application's voice assistant prompt the user whenever it recognises a person from the stored database. It is also trained to warn the users against walking into obstacles as it detects extreme proximity of objects and hurdles to the user.

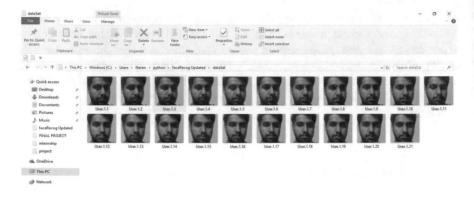

Fig. 2 Date set creation

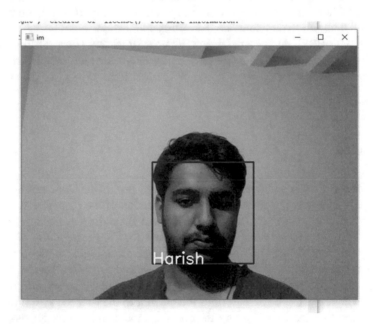

Fig. 3 Face recognition

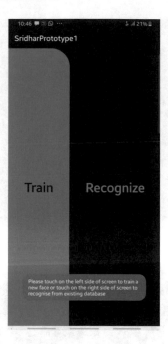

Fig. 4 Default screen

Fig. 5 **a** and **b**: Use for training of faces

Fig. 6 Real time training

Fig. 7 Recognition and speech in mobile application

6   Conclusions and Future Work

Visually impaired people face a lot of challenges in day-to-day life. Having seen the difficulties faced by them, our primary objective is to facilitate confidence and to empower them to lead a life free from threats related to their safety and well being. The lack of ability to identify known individuals in the absence of auditory or physical interaction cues drastically limits the visually challenged in their social interactions and poses a threat to their security. Over the past few years many prototype models have been developed to aid this population with the task of face recognition. This application will reduce the inherent difficulty for recognition of a person. It will present a facial recognition application with an intuitive user interface that enables the blind to recognise people and interact socially. The carefully designed interface lets the visually challenged to be able to access and use it without any requirement for visual cues as the users are acquainted by a voice assistant to navigate through the application.

The entire build is designed to run efficiently on a Raspberry Pi 3 model B module using the Android Things platform. The OpenCV library has been used for the detection and recognition of people in this project. This enables the scope for the software to be run on a multitude of devices such as camera embedded glasses to warn users of their surroundings and identify people to interact safely. Since everything in the application is done in real time with no requirement for prior datasets to be hardcoded it drastically improves the versatility of the software. We hope to make the visually impaired feel closer, comfortable and more secure with the world surrounding them through our application.

Declaration We have taken permission from competent authorities to use the images/data as given in the paper. In case of any dispute in the future, we shall be wholly responsible.

References

1. Kar, N., Debbarma, M.K., Saha, A., Pal, D.R.: Study of implementing automated attendance system using face recognition technique. Int. J. Comput. Commun. Eng. **1**(2) (2012)
2. Fan, X., Zhang, F., Wang, H., Lu, X.: The system of face detection based on OpenCV. In: IEEE Chinese Control and Decision Conference (CCDC), Taiyuan, China (2012)
3. Wilson, P.I., Fernandez, J.: Facial Feature Detection using Haar Classifiers. In: CCSC: South Central Conference, JCSC 21, April 2006
4. Kalas, M.S.: Real time face detection and tracking using OpenCV. Int. J. Soft Comput. Artif. Intell. **2**(1) (2014). ISSN: 2321-404X
5. Emami, S., Petruţ, V.: Facial recognition using OpenCV. J. Mob. Embed. Distrib. Syst. **4**(1) (2012)
6. Hudelist, M.A., Cobârzan, C., Schoeffmann, K.: OpenCV performance measurements on mobile devices. In: Proceedings of International Conference on Multimedia Retrieval, April 2014
7. Degtyarev, N., Seredin, O.: Comparative testing of face detection algorithms. In: International Conference on Image and Signal Processing, pp. 200–209 (2010)

An Biometric-Based Embedded System for E-Verification of Vehicle Users for RTO

D. Sailaja, M. Navya Sri, Nanda and P. Samuel

Abstract In this fast-growing population and generation use of motor vehicles (Four wheelers, three wheelers, two wheelers, etc.) have rapidly increased. In this circumstances unauthorized use of vehicles without license, speeding of vehicle over the limits and theft is major concern for authorities like RTO. The available manual verification is misused. For this cause we propose a biometric embedded system approach which identifies and verifies an individual without any hard copy documents which also integrates all the legal proceedings with automated fine generation and actions necessary on the case at any checkpoints with the help of internet without any delayed manual interventions.

Keywords RTO database · Aadhaar database · Embedded system · Central server · Biometrics · Finger print · Feature extraction

1 Introduction

The transport department, government frames a no of rules and regulations to be followed by the vehicle users on roads. Some of the important rules and regulations are do not drive without this documents (1) Valid driving license (2) Vehicle registration certificate (Form 23) (3) Valid vehicle's insurance certificate (4) Permit and vehicle's certificate of fitness (applicable only to transport vehicles) (5) Valid Pollution under

D. Sailaja (✉) · M. Navya Sri · Nanda · P. Samuel
Department of Information Technology, Anil Neerukonda Institute of Technology & Sciences
– (ANITS), Sangivalasa, Bheemunipatnam (Mandal), Vishakapatnam (District) 531162, Andhra
Pradesh, India
e-mail: dsailaja.it@anits.edu.in

M. Navya Sri
e-mail: navyasri.it@anits.edu.in

Nanda
e-mail: nrgkprasad@gmail.com

P. Samuel
e-mail: samuel.it@anits.edu.in

© Springer Nature Singapore Pte Ltd. 2020   711
V. Bhateja et al. (eds.), *Intelligent Computing and Communication*,
Advances in Intelligent Systems and Computing 1034,
https://doi.org/10.1007/978-981-15-1084-7_69

Fig. 1 Manual checking of hard copy

Control Certificate. These documents have to be produced for inspection on demand by the police officer. There are some other regulations to be fallowed such as wearing helmet while driving a two wheeler, seat belt while driving a four wheeler, following the speed limits on the road, following certain signaling at stop lines, not consuming alcohol while driving, etc., where in violation are subjected to prosecution or fining an individual. The verification of this all cases is done manually by the officers stopping the vehicles at any point checking the hard documents as shown in Fig. 1.

This would result in raising several misconceptions of the system where an individual can produce a fake document or tampered document which can leave the concerned non-subject to prosecution. This can also bring a difference if the driver forgets to carry his documents or in case of lost documents. All this problems can be addressed by the proposed approach where an individual can be authorized for his vehicles, license and any other documents by just using the biometric finger impression which would retrieve the entire details of the individual at once into an embedded system.

2 Literature Survey

The literature survey which is carried out regarding the technology developments in the "Electronic-verification Of Driving License through Aadhaar Database", as follows.

Gopi et al. [1] introduces an Automation of Road Transport Department through Cellular Network where in this process verification of the License and Vehicle documents electronically which reduces paper work and manual efforts. Sharma et al. [2] proposed a novel method called QR code in Smartphone. With this system, the driver goes through the verification process in a reliable and efficient manner. Bakale et al. [3] developed "Cross Verification of Driver and License for RTO", a system that facilitates for RTO officers to perform verification of license and vehicle documents through an android application. Nikam et al. [4] provide the facility that by having the image of the number plate and finger prints we can track the owner and vehicle information. Shelar et al. [5] presents an application which will facilitate the digitization of all documents which are required for the vehicle verification. From the above survey, it is clear, that work based on driving license verification is very less. Hence we are proposing a novel method called "E-verification Of Driving License through Aadhaar Database", and demonstrated its effectiveness for some test data. Experimental evidence shows that this technique is easier and faster than the other methods used in the survey.

In case of the electronic verification of driving license [1] the driver has to produce a hard copy of the document in order to verify using the Aadhaar database this would reduce the paper work but still can be improved in relation to complete automation of the system by using biometric approach as specified in the proposed model. The papers specified in survey [2] specifies a certain methodology called QR code but it does not specify and biometric approach. [3–5] specifies a related automated approach but does not specify the use of Aadhaar information using central server in processing the retrieving information. This proposed specifies the short comes of the present by introducing an embedded system with biometric equipment for automatic E-verification of an individual.

3 Proposed System

In this paper we propose a biometric based verification of an individual by an RTO officer for all its purposes which reduce false authentications and misuse of the manual system.

Our country has been with a unique identity card for every individual citizen called Aadhaar card which contains all information of the individual including the finger print images of the individual for verification. The finger print database of the Aadhaar can retrieve us all the information of the individual. If this information is linked with the information of the RTO database where the license, vehicle information, all the RTO related fining and traffic rule violations can be integrated on to a single application which can retrieve the present and history of an individual on the mark of his finger impression. This system would be liable for its use. This system can be explained by its individual models as follows:

i. *Biometric system*:
Biometrics measure individuals' unique physical or behavioral characteristics to recognize or authenticate their identity. Common physical biometrics is fingerprints. It can't be borrowed, stolen, or forgotten, and forging can be reduced. Biometrics can be integrated into any application that requires security, access control, and identification or verification of people. A fingerprint looks at the patterns found on a

fingertip. There are a variety of approaches to fingerprint verification. Some emulate the traditional police method of matching pattern; others use straight minutiae matching devices; and still others are a bit more unique, including things like more fringe patterns and ultrasonic characteristics. The features extracted from the fingerprint images are send to map with the available databases to retrieve corresponding information.

ii. *Matching*:

In this process features extracted from the finger print image are send for matching with the two databases such as RTO database and Aadhaar database verifying for a match of the fingerprints in the database. The enrollment of the fingerprint images and the data corresponding to it are stored in the database at the enrollment of the individual both at RTO office and Aadhaar centers. The features extracted from live fingerprint images are split to be matched with both the databases with proper authentications.

iii. *RTO database*:

RTO database is a generalized description of an information system where all the enrollment of the authorized vehicle users and license holders are stored with proper authentications and security for the authorities to consider in the case of requirement or verification. Here the fingerprint images of the individual vehicle users are recorded with their complete details like Name, DOB, Sex, gender, number of vehicles under his name, license details etc. including insurance and life time of them which can be accessed by the authorities. When the fingerprint feature extracted image is mapped with the database the details of the individual are retrieved at once into the server for further processing.

iv. *Aadhaar database*:

Aadhaar is a 12-digit unique identity number that can be obtained by residents of India, based on their biometric and demographic data. The data is collected by the Unique Identification Authority of India (UIDAI), a statutory authority established in January 2009 by the government of India, under the jurisdiction of the Ministry of Electronics and Information Technology. This available data on the Aadhaar database can be used in verification of the user to verify the identity of an individual. This database also store the individual biometric information which can be retrieved with a fingerprint image for further processing

v. *Central Server*:

This is central application server where the designed application is deployed at a global prospective with proper authentication and security. The application is designed to organize the data retrieved from the databases which has to be integrated with dynamic decisions to be taken automatically without the intervention of an individual. Some of the features the application has to provide are

(a) Integrating the data: the data retrieved from the two databases has to be integrated into a single page under certain columns for usability of the application users.
(b) Buffering: The integrated data has to be buffered in the server in order to produce it to the remote system requesting.

(c) Interfacing: The application developed for the usability of the authenticated user has to be interfaced with the system to access the server for client—server request response methodology using internet from any part of the end application user.

vi. *Embedded system*:

This is a user level embedded end system designed to handle every type of verification mechanism from the officials. This hand held system is embedded with finger print scanner to enable the functionality of the system.

Interfacing all the systems as shown in the figure would give us perfect results in producing a well-managed integrated system

The Block Diagram of the proposed methodology is as shown in Fig. 2. It explains the overall working process of the system. The traffic police read thumb impression of the driver through biometric enabled embedded device which is connected to the central server using internet with the help of an application. The central server integrates the information retrieved from the database to produce it to the user through a display module for the corresponding thumb impression. Here the two databases work as important bases for information retrieval.

A. The flow of the proposed system can be explained using the Fig. 3. In this approach we use already available databases (RTO and Aadhaar) for information regarding an individual. For a verification process to happen police logins into the system using a well defined APP available in the embedded system. He has to take the thumb impression of the driver, and it matches to the thumb impression present in the RTO database and the Aadhaar database. If it matches then the required information is integrated in the central server to be displayed in the end system. Information driving license is verified. If the verification is successful no action taken, otherwise based on the status of the driving license, fine amount is generated with necessary actions automated by the system. This application avoids the corruption in RTO department. Keeps the license documents safely and offer the drivers to be driving license.

B. Algorithm for driving license verification

Step 1: Start.

Step 2: Login page will be displayed.

Step 3: Authentication is done using central server.

Step 4: Read the thumb impression of the person.

Step 5: Match the thumb impression with template present in the RTO Data Base and Aadhaar Database.

Step 6a: If match is unsuccessful, necessary action would be taken
Goto step 7.

Step 6b: If match is successful, central server integrates the details and produces the required information through display module

Step 7: Based on the information official can make decision and can update the server accordingly.

Step 8: Logout.

Step 9: Stop.

Fig. 2 Architecture diagram

Biometric phase

Biometric system

Live Finger

| Fingerprint scanned | Feature extraction |

Finger print features extracted are matched with both the databases available over internet over proper authentication

Matching

Retrieving data from data Bases corresponding to finger mapping

RTO Database

Aadhaar Database

Processing at centralized server for the data received from the databases

Central server

Internet

Eg: Tab with fingerprint scanner

Fig: 2-Architecture Diagram

Application users can connect to the server through an APP in the system through available internet

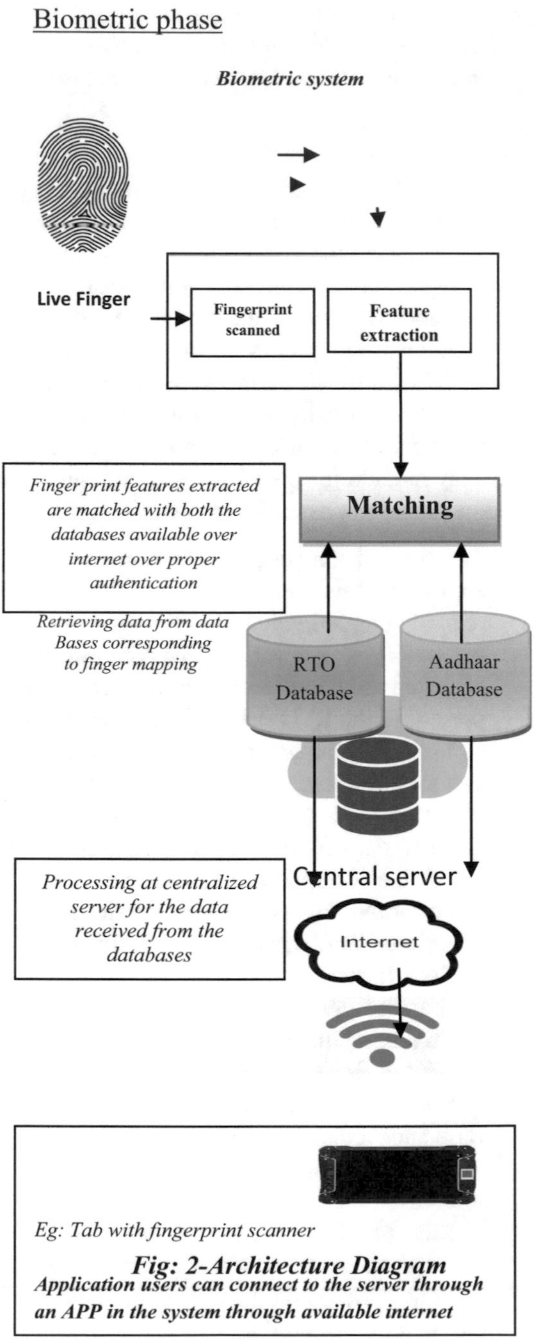

Process Flow

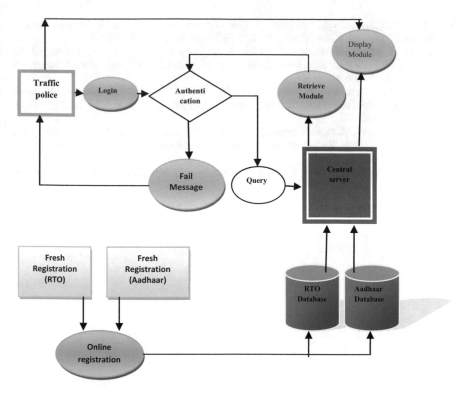

Fig. 3 Flow diagram

4 Conclusion

An biometric based embedded system for E-verification of vehicle users for RTO is an system which gives us an well developed biometric integrated embedded system where an official can verify an vehicle user using his biometric fingerprint images which retrieves the authenticated individuals at any point of time at any place just by using internet without the requirement of hard copy verification. This method can eradicates corruption and it can provide an automated billing system when required. This system can maintain history of an individual for proper decision-making.

References

1. Gopi, A., Rajan, L., Divya, P.R., Rajan, S.: E-RTO management system and vehicle authentication using RFID. Int. J. Eng. Res. Technol. (IJERT) (2013). ISSN: 2278-0181
2. Sharma, G., Sarade, A., Gupta, S., Janbhare, S., Mukhopadhyay, N.: Edriving license and RC book verification system using QR code. In: Proceedings of 65th IRF International Conference, ISBN:978 93, Nov 2016
3. Bakale, A.G., Awate, S.S., Megha, G.K., Pratibha, S.H., Hadapad, P.N.: Cross verification of vehicle and driver for RTO. Int. J. Emerg. Technol. Comput. Sci. Electron. (IJETCSE) (2015). ISSN: 0976-1353
4. Nikam, Y., Walunj, M.N., Paranjape, P.M., Kumbhar, R.A., Pawar, S.R.: Image processing and biometric approach for license and vehicle documents verification. In: Int. J. Eng. Comput. Sci. ISSN:2319-7242
5. Shelar, S., Sheikh, W., Shinde, P.: Vehicle Information System (IJCSIT). Int. J. Comput. Sci. and Inf. Technol. **6**(2), 1393–1395 (2015)

Comparative Analysis of Apache Spark and Hadoop MapReduce Using Various Parameters and Execution Time

Bhagavathula Meena, I. S. L. Sarwani, M. Archana and P. Supriya

Abstract Due to rapid growth in Information technology there is a lot of advancement in Electronics and communication. Every hour lot of data with various medium is getting generated which is referred as big data. Big Data and Hadoop are the trending terms nowadays. Storage and analysis of such a large data is becoming one of the challenges for computer science and Information Technology devotee throughout the world in the most recent couple of the years. As Apache Spark and Hadoop are the frameworks used for analyzing big data, our paper discusses a comparison of both the frame works by choosing different sizes of datasets and in terms of time comparison also. This comparison is made using word count algorithm. Although both the resources are relayed on an idea of significantly varying Big Data performance. This paper shows an analysis on both frameworks for word count algorithm over Hadoop MapReduce and Apache spark environment

Keywords Hadoop · Apache Spark · Big Data · HDFS · MapReduce

B. Meena (✉)
Associate Professor, CSE Department, Raghu Engineering College, Dakamarri, Visakhapatnam, Andhra Pradesh, India
e-mail: meenacserec@gmail.com

I. S. L. Sarwani
ANITS, Visakhapatnam, Andhra Pradesh, India
e-mail: sarwani.anits@gmail.com

M. Archana
CVR College of Engineering, Hyderabad, Telangana, India
e-mail: mogullaarchana23@gmail.com

P. Supriya
Raghu Engineering College, Dakamarri, Visakhapatnam, Andhra Pradesh, India
e-mail: supriya.puvvala@gmail.com

© Springer Nature Singapore Pte Ltd. 2020
V. Bhateja et al. (eds.), *Intelligent Computing and Communication*,
Advances in Intelligent Systems and Computing 1034,
https://doi.org/10.1007/978-981-15-1084-7_70

1   Introduction

Social networking and various industries in today's internet world are generating high voluminous data and data is immensely growing along with the increase in users. Big Data is classified into structured, unstructured and semi-structured form. Conventional data processing methods are not for storing and such a huge data Due to this reason traditional data processing methods face various challenges to process big data.

Nowadays, there are various sources of data generation such as web traffic, social networking content, system or machine generated data, social networking websites. Managing the data is one of the main challenges faced by various companies which are using these data for making Analysis. The data stored in data warehouse is a raw format and cannot be used directly for analysis; before it is processed it should be stored. It is very difficult to store, process and manage high voluminous data. There are many resources and tools that are used to handle structure and unstructured data.

Hadoop, [1] a distributed processing framework addresses these demands. It is built across more scalable clusters of commodity servers for processing, storing and managing data used in advanced applications. Hadoop has two main components-MapReduce and HDFS (Hadoop Distributed File System). HDFS is a file system used by Hadoop. MapReduce is a programming model of Hadoop.

2   Literature Review

There are various technologies for handling big data and their architectures. This paper discusses the challenges of Big Data (volume, variety, velocity, value, veracity) and various advantages and disadvantages of the technologies used for big data along with the architecture using Hadoop HDFS distributed data storage, real-time NoSQL databases, and MapReduce distributed data processing over a cluster of commodity servers. This paper presents the architecture and working of Hadoop and Spark and also brings out the differences between Hadoop and Spark along with challenges faced by MapReduce during processing of large datasets and working of YARN. It also gives a brief overview of the Apache Spark programming model, which includes RDDs, parallel computing etc.

Machine learning algorithm K-means is used for the comparative analysis of spark and Hadoop. The datasets are taken with a single node and two notes which are monitored in terms of time consumed for K-means clustering algorithm the result shows that the performance of Spark is considerably high in terms of processing time. For this a word count algorithm is used. It is found that Apache Spark is a very strong contender due its ability for in-memory computations, interactive querying and stream processing. Even though the report of the Spark says that it is 100 times faster and much better than Hadoop, in various situations it has no own storage system which is distributed in nature.

Hence it requires third-party. As spark has no own storage system, The process of spark installation above Hadoop will support spark to use distributed file system and many big data projects use the same pattern. For performance comparison word count and logistic regression is used. The performance is compared based on execution time.

3   Hadoop

Apache Hadoop [2] has a file system Hadoop Distributed File System and a MapReduce Engine. Each bunch has one main node and many slave nodes. The major node instructs the slave nodes and also ensures fault tolerances.

Architecture of Hadoop Distributed File system [3]
One of the open source software that processes big data very quickly is Apache Hadoop. It gets executed in different operating systems and handles various types of data like text, database, images, audio and video. It is scalable and flexible and fault tolerant. It has Hadoop Distributed File System. Figure 1 clearly shows that the name node has information of meta data and data node contains data. Each rack contains data node and also used for data replication.

MapReduce:
Google has given out Map to handle parallel processing jobs. Figure 2 explains the procedure. The client program submits the job and it is the responsibility of job tracker to assign the job. The Distributed file system [4] has input file and the mapper program splits the job and assigns to task tracker. The reducer combines the output and send to the output file.

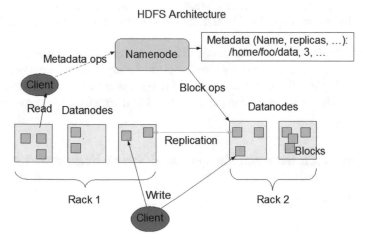

Fig. 1 HDFS architecture

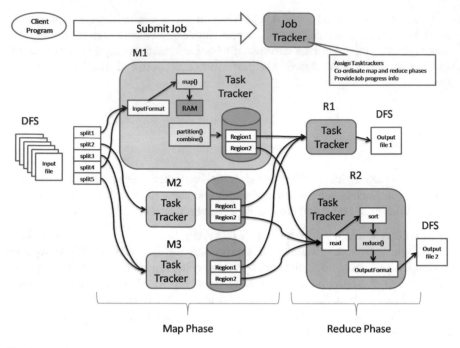

Fig. 2 MapReduce architecture

4 Spark

Spark is one of the open-source project given by apache. It is used to analyze huge datasets. When related to other big data technologies It has several benefits. Apache spark need a manager and storage system which is distributed. In order to manage cluster Spark has its standalone system that can interface with HDFS, Amazon and Cassandra.

Figure 3 explain Spark Architecture. This architecture has cluster [5] manager that interacts with master node and worker node through and driver program with the context of spark. The driver program and spark context are in master node where are worker node has corresponding task assigned to it with cache for processing.

5 Comparison Between Hadoop and Spark Frameworks

See Table 1.

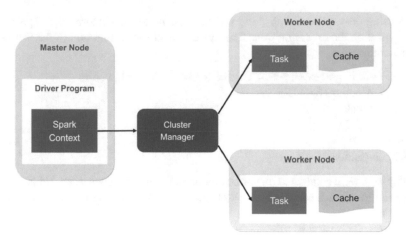

Fig. 3 Spark architecture

Table 1 Contrast between Hadoop and Spark frameworks

S. No.	Parameters	Hadoop	Spark
1	Distributed file system	Own file system	Depends on HDFS
2	Scalability	Highly horizontally Scalable	Horizontal
3	Message delivery guarantee	Exactly-once	Exactly-once
4	Streaming system	Do not support streaming	Micro batching
5	Data processing engine	At the core map reduce is batch processing	At the core spark is batch processing engine
6	Cost	Less expensive	More requirement of RAM increases cost
7	Data computation	Disk-based	In memory
8	Hardware requirement	Commodity hardware	Mid to high-level hardware
9	Auto-scaling	Yes	Yes
10	Languages supported	Mainly java and also supports Ruby, Groovy, Python and Perl	Java, Scala, python and R

6 Performance Evaluation

In this paper performance evaluation of Hadoop MapReduce and Apache Spark done and compared. For Comparison [6] Word count algorithm is used. In order to come to a conclusion about the practical comparison of Apache Spark and MapReduce, we performed a comparative analysis using these frameworks on a datasets of different sizes using the word count algorithm with single node and performance [7] is evaluated based on the execution time. The system used for evaluation has the following configurations:

- Intel-core i5 processor with 4 GBs of RAM 64-bit architecture
- Linux (Ubuntu 16.04 LTS).
- 500 GB disk
- Hadoop- 3.2.0
- Spark-2.4.9
- Eclipse-photon

The tests were conducted for various datasets having different sizes. Table 2 shows the execution time of both Hadoop MapReduce and Apache Spark on various datasets of different sizes.

The results related to time taken for execution has been noted for the different datasets for Apache Spark [8–10] and Hadoop. Figure 3 shows the visual representation of the results for the time taken for execution by both Hadoop and Apache Spark on the word count algorithm on different datasets (Fig. 4).

Table 2 Execution time comparison of Apache Spark and Hadoop MapReduce

S. No.	Dataset size	Execution time (in seconds)	
		Spark	Hadoop
1	30 MB	8	30
2	120 MB	25	55
3	382 MB	43	103
4	7591 MB	115	200
5	1.3 GB	200	250

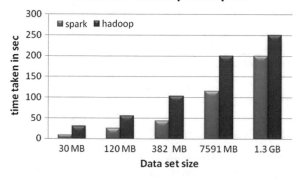

Fig. 4 Time taken for execution by Hadoop and Spark on different datasets

It has been observed that Apache Spark gives better performance in terms of execution time than Apache Hadoop when compared by using word count algorithm on datasets of different sizes on single node.

7 Conclusion

In this paper two programming model MapReduce and Apache Spark are compared for analyzing their performance using word count algorithm on datasets of different sizes on single node. By implementing both frameworks on various datasets of different sizes using the word count algorithm, performance of MapReduce and Apache Spark has been compared. In this paper, we have found that Apache Spark gives better performance in terms of execution time as compared to Hadoop MapReduce.

References

1. Jacob, J.P., Basu, A.: Performance analysis of Hadoop Mapreduce on eucalyptus private cloud. Int. J. Comput. Appl. **17** (2013)
2. Guanghui, X., Feng, X., Hongxu, M.: Deploying and researching Hadoop in virtual machines. In: Proceeding of the IEEE, International Conference on Automation and Logistics, Zhengzhou, China (2012)
3. Lakshmi, C., Nagendra Kumar, V.V.: Survey paper on big data. Int. J. Adv. Res. Comput. Sci. Softw. Eng. **6**(8) (2016)
4. Dahiya, P., Chaitra, B., Kumari, U.: Survey on big data using Apache Hadoop and Spark. Int. J. Comput. Eng. Res. Trends **4**(6) (2017)
5. Zaharia, M., Chowdhury, M., Franklin, M.J., Shenker, S., Stoica, I.: Spark: cluster computing with working sets. HotCloud **10**(10–10), 95 (2010)
6. Gopalani, S., Arora, R.: Comparing Apache Spark and Map Reduce with performance analysis using K-means. Int. J. Comput. Appl. (0975–8887) **113**(1) (2015)
7. Kaur, M., Dhaliwal, G.K.: Performance comparison of Map Reduce and Apache Spark on Hadoop for big data analysis. Int. J. Comput. Sci. Eng. **3**(11), 2015
8. Priya Ranjani, A.C., Sridhar, M.: Spark—an efficient framework for large scale data analytics. Int. J. Sci. Eng. Res. **7**(2) (2016)
9. Hazarika, A.V., Ram, G.J.S.R., Jain, E.: Performance comparison of Hadoop and Spark engine. In: International Conference on I-SMAC (IoT in Social, Mobile, Analytics and Cloud) (I-SMAC 2017)
10. Polato, I., Ré, R., Goldman, A., Kon, F.: A comprehensive view of Hadoop research—a systematic literature review. J. Netw. Comput. Appl. (2014)

A Fuzzy Approach to Spatio-temporal Analysis for Pedestrian Surveillance

Karan Jain and Rahul Raman

Abstract This article proposes a type-2 fuzzy system that works on the different spatio-temporal features (obtained by analyzing stills captured from surveillance videos) and uses it to differentiate a pedestrian from a human, a human from non-human etc. Existing systems propose a solution based on factors that can be influenced by several other environmental factors, for example data captured at infrared spectrum are influenced by weather conditions. Moreover, existing systems often neglect the differences between a human and a pedestrian. In this proposed work we propose a method specifying differences between non-human, human, and pedestrian as a case of human showing locomotion features in temporal domain. To achieve this, we have considered a type-2 fuzzy system because rather than calculating the probability of a subject being a human(human being pedestrian) we assess the subset of features that determine the subject's chances of being a human (human's chances of being a pedestrian), thereby emphasizing the need for a type-2 fuzzy system. Here the crisp output set is assumed to be {Pedestrian, Human, and Non-Human}. The achieved results are motivating towards these precise classifications and the use of fuzzy systems for the same.

Keywords Type-2 fuzzy system · Fuzzy logic · Pedestrian

1 Introduction

1.1 Theoretical Background

Fuzzy Logic is variation of multiple-valued logic in which the truth-values of variables may be a real number lying between 0 and 1. It has evolved to handle partial

K. Jain (✉) · R. Raman
Vellore Institute of Technology, Vellore, India
e-mail: kjain1496@gmail.com

R. Raman
e-mail: rahul.raman@vit.ac.in

© Springer Nature Singapore Pte Ltd. 2020   727
V. Bhateja et al. (eds.), *Intelligent Computing and Communication*,
Advances in Intelligent Systems and Computing 1034,
https://doi.org/10.1007/978-981-15-1084-7_71

truth values (in contrast to Boolean Logic). It can handle linguistic variables too. The way to handle while coming from crisp data:

- Fuzzify into membership functions (MF)
- Apply rules in database to compute fuzzy functions
- And finally De-Fuzzify to get outputs.

1.2 Type-1 Fuzzy Sets and Logic

Fuzzy Sets: Like Classical Sets, however the elements have degrees of membership. $\mu_A: X \varepsilon [0,1]$

Fuzzy Logic: Set of propositional variables that map into set of membership degrees.

Fuzzy Relations: Development based on Crisp Mathematical model and fuzzifying some quantities.

Fuzzification of quantities: Discretization and normalization, fuzzy partition of spaces, MF of primary fuzzy set.

Composition of fuzzy sets:

$$R = \{((x, y), \mu_R(x, y)) | \mu_R(x, y) = \min[\mu_A(x), \mu_B(y)] \tag{1}$$

or

$$\mu_R(x, y) = \mu_A(x) \bullet \mu_B(y)\} \tag{2}$$

Composition of fuzzy relations:

$$SR = S \, o \, R = \{((x, y), \mu_{SR}(x, z))\} \tag{3}$$

where

$$\mu_{SR}(x, z) = \max \min[\mu_R(x, y), \mu_S(y, z)] \tag{4}$$

De-fuzzification of quantities:

In daily practical applications of fuzzy logic, a control command is given as a crisp value

- a process to convert a fuzzy control action into a non-fuzzy control action that best represents its' possibility distribution.
- a good de-fuzzification strategy has no systematic method for choosing it,
- select any one in view of properties of applying to a particular case
  Example: Fuzzy Mathematical Model - Fuzzy K-means clustering

- Fuzzy Image Processing System which consists of Fuzzification i.e. Coding of the Image; Processing i.e. Modification, Aggregation, Classification, Modification by IF-Then Rules; Defuzzification, i.e., Decoding and lastly Image Fuzzification is a nonlinear transformation. The number, form and location of each MF should be optimized.

1.3 Type-2 Fuzzy Sets and Logic

Type-2 fuzzy logic was introduced because there are restrictions in the ability of type-1 FSs to model and/or minimize the effect of uncertainties as research shows. This is because a FS believes that is certain in the sense that its membership grades belong to a crisp value set. Due to the shortcomings of type 1 fuzzy logic namely its inability to deal with uncertainty as the MF returned crisp values, type-2 fuzzy sets were introduced by Prof. Lotfi A. Zadeh to tackle this problem, the degree of membership was represented itself as a type-1 fuzzy set instead of the regular crisp values between 0 and 1.

A type-2 fuzzy set can be defined as an ordered pair of the element x and the type-1 fuzzy set, which itself describes the degree of membership of x called as u_x and degree of membership of u_x, if the degree of membership of u_x (grade) is 1 then the entire type-2 fuzzy set reduces to a type-1 fuzzy set, it also represents uncertainty of the assigned value of u_x.

This uncertainty of the memberships of elements of X (primary membership), can be represented with the Footprint of Uncertainty (FOU), which is the union of all primary memberships of a type-2 fuzzy set. This allows us to visualize the uncertainties of a type-2 MF in 2D. The shaded portion of FOU represents all possible values of membership for a given element, i.e., the uncertainty. If all the distributions are equally spaced then the fuzzy set is called as interval type-2 fuzzy set.

Union and Intersection of type-2 fuzzy sets is not easy as their type-1 fuzzy sets counterparts, the same is derived via the extension principle. The union of two type-2 fuzzy sets say $\bar{A}$ and $\bar{E}$, for given elements with their degrees of memberships and grade, a new degree of membership is defined say w_x and is assigned the maximum value between the primary memberships of sets $\bar{A}$ and $\bar{E}$ for each element belonging to them, whereas the grade or the secondary membership value is decided on the supremum of the minimum of each corresponding possible values of grade for the above calculated w_x. Similarly, for the intersection the only difference is in the calculation of w_x, as the minimum value is calculated instead of the maximum value, the rest of the procedure remains the same.

Interval type-2 Fuzzy System (IT2 FS): Interval type-2 (IT2FS's), a special case of type-2 fuzzy sets, are the widely used for their reduced computing cost at the present. The membership value of an IT2 FS is an interval unlike a type-1 FS, whose membership value is a number. One can also perceive that an IT2 FS is bounded from the above and below by two type-1 fuzzy set, $\underline{X}$ and $\bar{X}$, which are called upper

MF and lower MF, respectively. The area between $\underline{X}$ and $\overline{X}$ is called the footprint of uncertainty (FOU). IT2 FSs are mainly needed when it becomes difficult to determine the exact MF, or in modeling the diverse opinions from different individuals. The MFs can be made from surveys and/or using optimization algorithms.

It is similar to its type-1 counterpart, the major difference being that at least one of the FSs in the rule base is an IT2 FS. Hence, the outputs of the inference engine are IT2 FSs, and a type-reducer is needed to convert them into a type-1 FS before defuzzification can be carried out. In run-through the number of computations in an IT2 FLS can be simplified. Considering that the rule-base of an IT2 FLS consisting of n rules assumes the following form:

IF X_n then Y_n, i.e., each rule consequent is a crisp number.

Spatio-Temporal Features: Spatial Features are those whose changes can be measured in space. Temporal features are those who accrue a change over time. Spatio-Temporal features are being exploited here to define MFs. These include *spatial* features extraction processes such as bounding-box, edge detection, gradients and contours and *temporal* features such as centroid movement detection. These have been already exploited before in a non-fuzzy system.

Motivation: An increasing need of surveillance in the world that can help predict behavior of pedestrians towards fellow pedestrians or the vehicles surrounding them, basically how they affect the environment around them.

2 Literature Review

2.1 Survey of Existing Models and Related Work

In recent times multiple researchers have stressed on pedestrian gait detection and monitoring system [1–8]. Most of those articles have stressed on spatio-temporal features (as described above) such as Raman et al. [4] whose approach has exploited spatial property for frame based estimation and further consolidated temporal features to generate overall pedestrian direction. Further few researchers such as Raman et al. [5], Mojdeh et al. [6], Akasaka et al. [7], have employed fuzzy approach to estimate direction of walk of a pedestrian however the authors in [5], have implemented a fuzzy type-1 system successfully which motivates us to go beyond and implement a fuzzy type-2 system to detect pedestrians. Apart from this, various other implementations such as by Mojdeh et al. [6], have presented person on foot guiding approach, which was displayed utilizing a fuzzy rationale approach foreseeing the effect of apparent attractive and unpleasant boosts, inside the person on foot field of view, facing abrupt environmental changes which are evaluated utilizing the social force methodology. These are executed in an office passageway comprising of a printer and leave entryway. Stochastic recreation utilizing the proposed fuzzy calculation produced sensible walking directions, shape guide of dynamic difference in natural impacts in each progression of development and high flow zones in the passageway.

Another paper by Akasaka et al. [7] proposes the walker route framework choosing a course in light of users own inclination for courses. The present framework comprises of a course determination part and a course direction part. Course choice part chooses the course with the best subjective fulfilment degree which is assessed by a road evaluation model (REM). The REM applies fuzzy measures and integrals to figure the subjective fulfilment degrees of a street. Path direction part gives users guidelines with phonetic articulations fitting to clients' own feeling of separation (SFD). Experimental outcomes demonstrate that the course chose by the present framework is desirable over different courses and the REM suitably mirrors users' own inclination. Completely shifting to soft environment outcomes, Stauffer et al. [8] builds up a visual checking framework that can inactively watch moving items in a site and takes in examples of movement from those perceptions. The key components of the framework were movement following, camera coordination, action order, and occasion recognition. Movement division yields a steady, constant outside tracker that dependably manages lighting changes, tedious movements from mess, and long haul scene changes. While a following framework is uninformed of the personality of any protest it tracks, the character continues as before for the whole following grouping. The framework uses this data by gathering joint co-events of the portrayals inside an arrangement. These joint co-event insights are then used to make a progressive parallel tree grouping of the portrayals. Employing several changes in electronic and algorithm, the implementation by Toth et al. [9] developed theoretical foundations and implementation algorithms, which integrate GPS, micro-electro-mechanical inertial measurement unit, etc., to provide navigation and tracking of military etc. This paper discusses a personal navigator prototype (special emphasis on dead-reckoning navigation supported by the human locomotion model). Based on ANN and Fuzzy Logic is trained while the GPS signal reception is stable and is used to maintain navigation under restrained GPS conditions. Step frequency and step length, the different human locomotion parameters are estimated during calibration period of the system then the predicted step length together with the heading information from the compass and gyro support the implementation.

2.2 Summary/Gaps Identified in the Survey

Almost all of the above papers simulate a fuzzy system nonetheless. Assessment of environmental stimuli, road evaluation and GPS tracking are novel solutions in themselves. But almost all of them stop at their solution. There was a need of a system that could add more factors and let them influence the system. For example many of the factors taken into consideration could be influenced by the weather because individual factors do not determine the decision. We decided to go for an IT2 FS where all factors could affect each other and still new factors could be used irrespectively of one affecting other.

3   Overview of the Proposed Solution

3.1   Introduction

Implementing a type-1 Fuzzy System to detect a pedestrian Example:
    Crisp Set: {Pedestrian, Human, Non-Human}
    MFs' Examples: Basically blob is to be measured w.r.t attributes
    Locomotive(x): If x is a object is moving or not
    Living(x): If x is a living being or not (heat detection, etc.)
    Human(x): If x is a human or not (feature detection such as average cross sectional area.)
    Pedestrian(x): If x is in/on a vehicle or not.
    IF-THEN rules:
    IF x is non-locomotive blob THEN reject y else | Centroid Detection etc.
    IF x is non-living THEN reject y else| Heat Detection etc.
    IF x is non-human THEN reject y else | Average Cross Sectional Area etc.
    IF x is on-road THEN accept.
    De-Fuzzification to the Crisp set: Humans can be directly fuzzified by the High Human(x) and High Living(x). Moving blob is fuzzified by high Locomotive(x), low Living(x) and low Human(x). Blobs can be fuzzified using low Locomotive(x), low Living(x). Pedestrian is to be fuzzified by high Pedestrian(x), High Locomotive(x), High Human(x) (which already includes high Living(x).
    Also Animals can be fuzzified by high Living(x) and low Human(x).

3.2   Attributes/Rules for the Proposed System

As depending on the features that are taken into consideration

- Centroid Motion Variance (Centroid Detection)

  Coefficient of Variation (c.o.v.) in the Centroid's Coordinates

- Bounding Box Ratio (Average Cross Sectional Area) Ratio of Bounding Box length to width

  Inputs: 2
      Locomotive(x): If x is an object is moving or not | if c.o.v. is greater than one ore not
      BoxRatio(x): If the box-ratio of x is good enough or not| if the box-ratio lies in the range
      IF-THEN rules:

- IF x is non-human THEN reject else | Average Cross Sectional Area
- IF x is non-locomotive THEN reject y else | Centroid Detection etc.

- Else accept.| The Blob is a Pedestrian.

De-Fuzzification to the Crisp set: Training data is taken into consideration and intervals of type-2 fuzzy system are recorded which then are used on the testing data to determine the outcome.

3.3 Architecture of the Proposed System

A IT2 FIS is proposed having two inputs *bbox* and *cvar,* i.e., Bounding Box Ratio and the Centroids Coefficient Of Variation (c.o.v.). The proposed FIS has 1 output function that has 3 MFs Human, Non-Human and Pedestrian

The input *bbox* has 2 MFs for each of the three orientations (0°, 45° and 90°).

One of the membership function is a UMF (highest average bounding box ratio) and the other is LMF (lowest average bounding box ratio).

Similarly the input *cvar* has 2 MFs for each of the three orientations (0°, 45° and 90°).

One of the membership function is a UMF (highest c.o.v. of the centroids position) and the other is LMF (lowest c.o.v. of the centroids position).

3.4 Objectives and Assumptions

This research article at last aims to develop a MATLAB program that identifies the features and fuzzifies them into a crisp set of outputs, i.e., Pedestrian, Non-Human and Human. It is robust to various inputs of datasets. Preprocessed (Background Subtracted and Resized 240×352) images are used here. Video has to be converted into frames and used as an input.

3.5 Reliability and Portability

The outcome of the research article, i.e., the MATLAB implementation here has a high reliability and high efficiency which is as usual for a IT2 FS FIS. It is portable and can be implemented on low computing dependent resources such as Raspberry Pi as MATLAB code can be executed on it.

3.6   Usage and Areas of Interests

The implementation can detect pedestrians and help in avoiding accidents or surveillance on them.

- The driver could be warned; for example: a notification on their mobile phones.
- The driver (autonomous or not) could be warned; for example: a notification on their entertainment screens.
- Vehicle could stop on its own detecting a pedestrian.
- This also could help in assisted living as people that are walking in their home could be monitored and could help them in various chores like illuminating parts of a room, detecting unwanted motions (break-ins) etc.

3.7   System Requirements

Hardware Requirements

- Data was collected using static cameras. Other cameras that can be used are Cyclops, Pan-tilt-zoom cameras.
- Proposed implementation can run on any existing architecture having camera
- Raspberry Pi can be also used in with a camera module

Software Requirements

- MATLAB is used here to work on silhouettes and the fuzzy inference system (defined above in Sect. 3.3)
- Dataset for training and testing, here we are using CASIA Dataset A [10] (background subtracted; if any dataset provides images please ensure a background subtraction before processing)
- IT2-FLS MATLAB/Simulink Toolbox provided by Taskin and Kumbasar [11].

3.8   Results and Discussion

Training

- For training silhouettes of each of 20 subjects were taken with 3 orientations of 0°, 45° and 90° each.
- Bounding boxes and centroids were located for each subject.
- Statistics of bounding box ratios and c.o.v. of centroids were calculated.
- These statistics were used to define an IT2 FS.

Fig. 1 CASIA Dataset A silhouettes of a moving subject

Fig. 2 Silhouettes of the same subject above; highlighting the bounding box and centroid of bounding box

Table 1 Coefficient of variation of x and y coordinates of centroid and bounding box ratios of above silhouettes

coordx	14.7500	34.7500	61.7500
coordy	165.0000	154.0000	150.5000
boxratio	0.0033	0.0042	0.0049

Testing

- For testing 10 samples of 4 subjects including a pedestrian, a moving car, a parked car and a human standing on the pavement were taken.
- These were inputs to evaluate the IT2 FS defined earlier.

Sample Test Cases
Silhouettes were taken from the above-mentioned databases (Figs. 1 and 2) (Table 1).

4 Conclusion, Limitations and Scope for Future Work

Out of 4 subjects × 10 samples each tested 37 were reported correctly, thereby indicating an accuracy of 37/40 ~ 93% which is quite high given the limited number of training samples. In future we could also add a feature for thermal detection and orientation detection. Detection could be made robust by using more training samples. With a type-2 fuzzy inter-dependency of the inputs increases, so any number of features might be added.

References

1. Goto, K., Kidono, K., Kimura, Y., Naito, T.: Pedestrian detection and direction estimation by cascade detector with multi-classifiers utilizing feature interaction descriptor. In: IEEE Intelligent Vehicle Symposium (IV), pp. 224–229 (2011)
2. Tao, J., Klette, R.: Integrated pedestrian direction classification using random decision forest. In: IEEE International Conference on Computer Vision (ICCV) Workshops, pp. 230–237 (2013)
3. Raman, R., Sa, P., Majhi, B., Bakshi, S.: Direction estimation for pedestrian monitoring system in smart cities: an HMM based approach. IEEE Access (2016)
4. Raman, R., Sa, P.K., Bakshi, S., Majhi, B.: Kinesiology-inspired estimation of pedestrian walk direction for smart surveillance. In: Future Generation Computer Systems. North-Holland (2017)
5. Raman, R., Boubchir, L., Sa, P.K., Majhi, B., Bakshi, S.: Beyond estimating discrete directions of walk: a fuzzy approach. In: Machine Vision and Applications. Springer, Berlin (2018)
6. Nasi, M., Nahavandi, S., Creighton, D.: Fuzzy simulation of pedestrian walking path considering local environmental stimuli. In: 2012 IEEE World Congress on Computational Intelligence, June 2012
7. Akasaka, Y., Onisawa, T.: Individualized pedestrian navigation using fuzzy measures and integrals. In: 2005 IEEE International Conference on Systems, Man and Cybernetics (2005)
8. Stauffer, C., Grimson, W.E.L.: Learning patterns of activity using real-time tracking. IEEE Trans Pattern Anal. Mach. Intell. **22**(8), 747–757 (2000)
9. Toth, C., Grejner-Brzezinska, D.A., Moafipoor, S.: Pedestrian tracking and navigation using neural networks and fuzzy logic. In: 2007 IEEE International Symposium on Intelligent Signal Processing, Alcala de Henares, pp. 1–6 (2007)
10. CASIA Gait Recognition Dataset; subset Dataset A silhouettes. Link
11. Taskin, A., Kumbasar, T.: An open source Matlab/Simulink toolbox for interval type-2 fuzzy logic systems. In: IEEE Symposium Series on Computational Intelligence – SSCI 2015, Cape Town, South Africa, Link (2015)

Human Action Recognition in Unconstrained Videos Using Deep Learning Techniques

G. G. Lakshmi Priya, Mrinal Jain, R. Srinivasa Perumal and P. V. S. S. R. Chandra Mouli

Abstract Human activity recognition is an active and interesting field in computer vision from past decades. The objective of the system is to identify human activities using different sensors such as cameras, wearable devices, motion and location sensors, and smartphones. The human actions are automatically identified through their physical activities in human–computer interaction. Determining the human action in an uncontrolled environment is a challenging task in human activity recognition system. In this paper, a novel approach is proposed to recognize human actions effectively in an uncontrolled environment. A frame for the video segment is selected by temporal superpixel, which acts as the input image for the model. Convolutional neural network techniques are applied to extract the features and recognize the human activities from the image. The proposed method has experimented on KTH database and it shows the performance of the method in terms of accuracy. However, the proposed method has attained better accuracy when compared to the existing methods.

Keywords Human activity recognition · Convolutional neural network · Classification

G. G. Lakshmi Priya · M. Jain · R. Srinivasa Perumal
School of Information Technology and Engineering, Vellore
Institute of Technology, Vellore, Tamil Nadu, India
e-mail: lakshmipriya.gg@vit.ac.in

R. Srinivasa Perumal
e-mail: r.srinivasaperumal@vit.ac.in

P. V. S. S. R. Chandra Mouli (✉)
Department of Computer Applications, National Institute
of Technology, Jamshedpur, Jharkhand, India
e-mail: chandramouli.ca@nitjsr.ac.in

© Springer Nature Singapore Pte Ltd. 2020   737
V. Bhateja et al. (eds.), *Intelligent Computing and Communication*,
Advances in Intelligent Systems and Computing 1034,
https://doi.org/10.1007/978-981-15-1084-7_72

1 Introduction

Human activity recognition is the most interesting research topic in the current era. Human activities are classified based on the function and the motion of the body. Under the motion category, sitting, running, and walking activities are identified. In the function category, some complex activities are identified such as boxing, hand clapping, and cooking [7]. Extensive research has been executed to recognize human activities but still many challenges have to be addressed in a complex human activity recognition system. Action recognition is based on the sequence of movements by patterns and their action type. In complex human action recognition, the activities are classified into four levels. They are actions, gestures, group activities, and interactions [3].

Action recognition system consists of feature extraction, classification, and recognition methods. The process of human activity recognition is shown in Fig. 1. The features are extracted from the input frames. However, the process requires deep knowledge to classify the human activities and hence the convolutional neural network (CNN) is applied to extract the discriminant features for each activity. Next, the activities are classified according to their actions and will be mapped into their respective class [1].

In an automated system, deep learning obtained better results in the activity recognition method by extracting more valued information from the input video. CNN has the capability to identify the spatial and temporal information among the sequence of frames [15]. The challenges in the unconstrained video are (i) removal of background and (ii) identification of the motion patterns for recognizing human activities. Generally, collecting enough training data in an unconstrained environment is not an easy task. In such scenarios, various approaches are used to collect the trained data but that may not be suitable for identifying the human activities [5].

Complex human activity recognition emerges due to the difficulty in understanding the variety of problems, ranging from gesture recognition to the identification of physical activities like running or climbing up the stairs. Recent breakthroughs state that deep learning is a substitute for hand-crafted feature extraction and classification techniques. CNN produces state-of-the-art results in both speech and image domains [6]. However, this end-to-end deep learning concept not only benefits image and speech but also can be used for video domains.

In this paper, the convolutional neural network was used to extract the discriminant features of human activities. This approach provides the relationship between the moving object and the background to identify the human activities effectively in an unconstrained environment. The rest of the paper is structured as follows: Sect. 2 summarize the existing methods related to human activity recognition. Section 3 explain the proposed method and demonstrate the approach. The experimental studies are described in Sect. 4 and the paper concludes with Sect. 5.

Fig. 1 Block diagram of human activity recognition system

2 Related Work

The research work related to human activity recognition is described in this section. An independent subspace is used to identify the features and actions in spatial and temporal domain. The cost of computation is too high [8]. Principal component analysis (PCA), Histogram of Oriented Gradient (HOG), Space Time Interest Points (STIP) are used for activity recognition but computation cost is high and it is suitable for 2D techniques. Hence linear SVM classifier is applied to classify various actions [10].

Chang et al. [2] proposed a Temporal Superpixels method for video representation by extracting local space-time features for recognition. This method performs better than supervoxel methods. Mostafa et al. [4] developed a hierarchical deep model to capture temporal dynamics using long short term memory model for group activity recognition. The two stage model identifies the temporal representation of individual and combines each representation to recognize the group activity. Wang et al. [13] proposed a robust and efficient video representation for action recognition which improves dense trajectory video features by explicit camera motion estimation and illustrated the efficiency of the proposed action recognition on a number of benchmark videos. A multi-level video represents the local motion, atomic actions, and composites to understand the human activity for recognition [14]. Zhao et al. [16] described latent variable models that learn mid-level representations of activities defined by multiple group members.

Bevilacqua et al. [1] used CNN to classify human activities. The raw data is collected from a set of sensors. Various types of actions are captured and CNN was used for action recognition. The results are very promising. Thakkar et al. [12] presented a 3-D Convolutional neural network-based hierarchical approach for human action classification. In general, human actions refer to positioning and movement of hands and legs and hence it can be classified based on those actions performed by hands or by legs or, other parts of the body. Lin et al. [9] introduced a spatiotemporal CNN to recognize the activities of humans and optimized using a learning algorithm in RGBD videos.

3 Proposed Method

In this work, a temporal super-pixel based convolutional neural network (TS-CNN) is introduced to extract the discriminant features and classify the actions by Softmax classification method. The videos are converted into a number of image frames. In the image frames, the redundant images will not be used for a training model to improve the speed of training. The frames are selected using temporal superpixels. A group of meaningful regions within the frame is referred to as superpixels and the temporal relationship between the superpixels of the same object across the frames is extracted as temporal superpixels. Based on the temporal coherence superpixels, a frame for the video segment is selected which acts as the input image for the model.

The input frame is divided into regions, the features are extracted from each region. Kernel determined the number of regions for the input frame and it convolved over both x-axis and y-axis. To extract discriminant features, the multiple filters are used for the image. Each filter produced one feature map for the region. These feature maps can be stacked into a 3-d array, which can then be used as the input to the layers followed by pooling layers, that reduce the spatial dimensions of the output.

A novel convolutional model is constructed to improve the accuracy rate in an unconstrained environment. The advantage of the model is that it can be used to encode the content of the image into a vector (with more depth, but less height and width). Sometimes average pooling layer is also used where the only difference is to take the average value within the window instead of the maximum value, but the maximum pooling layer model extracts the deep features from the region. Therefore, the convolutional layers increase the depth of the input image, whereas the pooling layers decrease the spatial dimensions (height and width). The combination of alternating convolutional and pooling layers is followed by a global Pooling layer. When there is no more spatial information left in the input to extract (the spatial dimensions of the input have cannot be further decreased by the pooling layers), the global pooling layer converts its input to a 1-d vector (having the same depth). The structure of the convolutional model is shown in Fig. 2.

The convolutional layer incorporates the max-pooling layers, that will extract the deep and discriminant features map for every frame based on the kernel. For each layer, the input of layers is padded properly then the loss of information may reduce in CNN operation. The normalization process is involved in each input layer after the convolutional layer. The pooling layers produce different features for every layer to recognize the activities. The activation function is used to calculate the loss of information with the cross-entropy function. The output layer is based on the number of activities in each layer. The softmax function will return the most likely class of the input windows in the multi-class classification task.

4   Experimental Results and Discussions

The proposed method was experimented on KTH dataset [11]. It consists of 600 video files, 100 videos for each of the 6 categories. Table 1 depicts the class labels of human activity in the dataset.

After the input layer, the convolutional layer extracts the deep features for the input. The depth of the vector obtained by the last convolutional layer is 1024. A Global Average Pooling layer determines the average value from the max-pooling layer and reduces the dimension from 3D to 1D to represent the entire video. The Global Average Pooling is followed by a fully connected layer containing 32 neurons. It has a dropout of 0.5, 50% of the neurons of this layer will be deactivated. The Global Average Pooling layer prevents the model from over fitting of the data. Finally, the output layer with 6 neurons is used to determine the six class output. Sample jogging class action frames are shown in Fig. 3.

Fig. 2 Structure of the
convolutional model

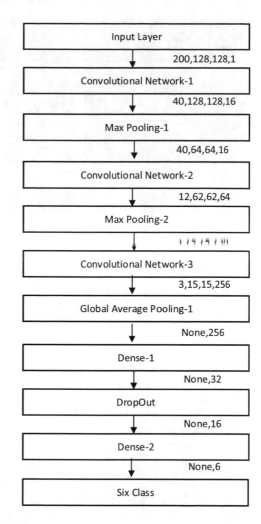

Table 1 Label of classes on
KTH dataset

Table 1 Label of classes on
KTH dataset

Class label	Mapped integer
Boxing	0
Handclapping	1
Hand waving	2
Jogging	3
Running	4
Walking	5

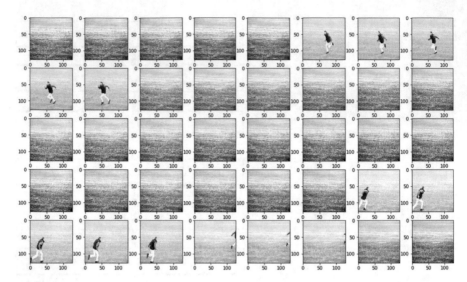

Fig. 3 Frames of jogging action

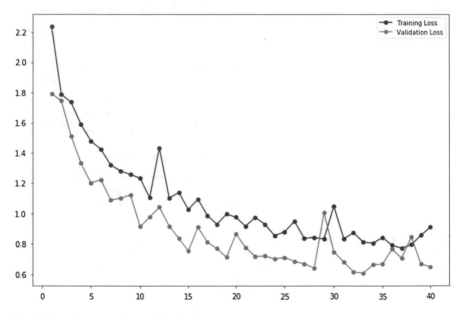

Fig. 4 Learning curve for model evaluation

This network gives a probability for the input video to belong to each of the 6 categories. The proposed model results in 64.5% accuracy. Figure 4, depicts the learning curve of the model for 40 epochs. The method is compared with existing methods that use the local features for recognizing human activities.

Table 2 Comparison of accuracy results of the proposed model with the benchmark model [16]

Action	Benchmark model [16]	Proposed model
Walking	0.98	0.94
Jogging	0.60	0.67
Hand waving	0.74	0.73
Hand clapping	0.60	0.98
Running	0.55	0.64
Boxing	0.98	0.58

A comparison of the results of the proposed method with the benchmark model [16] are shown in Table 2. From the results, the proposed method attained promising results than the existing method. The softmax classification classifies the class at par with the support vector machine. The results show that SVM and Softmax are equally contributing to activity recognition. For Example, in the Jogging class, the SVM and Softmax model results are 60% and 67%, respectively. Similarly, hand waving results are 74% and 73%, respectively and for hand clapping, the results are 60% and 98%.

In walking action, the movement of the person is slow so the empty frame is less if the motion is slow. Similarly, the movement is fast the number of empty frames is high. Describing the action through an empty frame is more difficult. Due to that accuracy rate for running, jogging, hand waving, hand clapping, and boxing is less compared to walking.

5 Conclusion

In this paper, a new temporal CNN was proposed to recognize human activity in unconstrained videos. The system observed a set of human activities extracted from unconstrained videos. The experimental result shows that the CNN model can address the challenging actions in complex human recognition. The proposed method has experimented with alternative convolutional and pooling layers such that it encodes the content of the image into depth vector. The method attained 64.5% accuracy on test data. The method was compared and the results show that the new temporal CNN with Softmax classification obtains better performance with benchmark methods. The method can be extended to address the real-world problems and a number of activities can be tested in the future with this model.

Acknowledgements The authors from Vellore Institute of Technology, Vellore thank the management for providing VIT SEED GRANT for carrying out this research work. Further, all the authors thank the NVIDIA for providing the Titan Xp GPU card under the NVIDIA GPU Grant scheme.

References

1. Bevilacqua, A., MacDonald, K., Rangarej, A., Widjaya, V., Caulfield, B., Kechadi, T.: Human activity recognition with convolutional neural networks. In: Joint European Conference on Machine Learning and Knowledge Discovery in Databases, pp. 541–552. Springer (2018)
2. Chang, J., Wei, D., Fisher, J.W.: A video representation using temporal superpixels. In: Proceedings of the IEEE Conference on Computer Vision and Pattern Recognition, pp. 2051–2058 (2013)
3. Cheng, G., Wan, Y., Saudagar, A.N., Namuduri, K., Buckles, B.P.: Advances in human action recognition: a survey. arXiv:1501.05964 (2015)
4. Ibrahim, M.S., Muralidharan, S., Deng, Z., Vahdat, A., Mori, G.: A hierarchical deep temporal model for group activity recognition. In: Proceedings of the IEEE Conference on Computer Vision and Pattern Recognition, pp. 1971–1980 (2016)
5. Jiang, Y.G., Dai, Q., Liu, W., Xue, X., Ngo, C.W.: Human action recognition in unconstrained videos by explicit motion modeling. IEEE Trans. Image Process. 24(11), 3781–3795 (2015)
6. Krizhevsky, A., Sutskever, I., Hinton, G.E.: Imagenet classification with deep convolutional neural networks. In: Advances in Neural Information Processing Systems, pp. 1097–1105 (2012)
7. Lara, O.D., Labrador, M.A.: A survey on human activity recognition using wearable sensors. IEEE Commun. Surv. Tutor. 15(3), 1192–1209 (2013)
8. Le, Q.V., Zou, W.Y., Yeung, S.Y., Ng, A.Y.: Learning hierarchical invariant spatio-temporal features for action recognition with independent subspace analysis (2011)
9. Lin, L., Wang, K., Zuo, W., Wang, M., Luo, J., Zhang, L.: A deep structured model with radius-margin bound for 3D human activity recognition. Int. J. Comput. Vis. 118(2), 256–273 (2016)
10. Patel, C.I., Garg, S., Zaveri, T., Banerjee, A., Patel, R.: Human action recognition using fusion of features for unconstrained video sequences. Comput. Electr. Eng. 70, 284–301 (2018)
11. Schuldt, C., Laptev, I., Caputo, B.: Recognizing human actions: a local SVM approach. In: 2004 Proceedings of the 17th International Conference on Pattern Recognition ICPR 2004, vol. 3, pp. 32–36. IEEE (2004)
12. Thakkar, S., Joshi, M.: Classification of human actions using 3-d convolutional neural networks: a hierarchical approach. In: Computer Vision, Pattern Recognition, Image Processing, and Graphics: 6th National Conference, NCVPRIPG 2017, Mandi, India, 16–19 December 2017, Revised Selected Papers 6, pp. 14–23. Springer (2018)
13. Wang, H., Oneata, D., Verbeek, J., Schmid, C.: A robust and efficient video representation for action recognition. Int. J. Comput. Vis. 119(3), 219–238 (2016)
14. Wang, L., Qiao, Y., Tang, X.: MoFAP: a multi-level representation for action recognition. Int. J. Comput. Vis. 119(3), 254–271 (2016)
15. Zeng, M., Nguyen, L.T., Yu, B., Mengshoel, O.J., Zhu, J., Wu, P., Zhang, J.: Convolutional neural networks for human activity recognition using mobile sensors. In: 6th International Conference on Mobile Computing, Applications and Services, pp. 197–205. IEEE (2014)
16. Zhao, F., Huang, Y., Wang, L., Xiang, T., Tan, T.: Learning relevance restricted Boltzmann machine for unstructured group activity and event understanding. Int. J. Comput. Vis. 119(3), 329–345 (2016)

A Multi-resolution Face Verification System for Crowd Images Across Varying Illuminations

S. Devi Mahalakshmi and B. Chandra Mohan

Abstract Human face detection and verification in crowd images is an active area of interest for researchers due to their roles in real-world applications. Most of the real-world photo images are taken under varying illumination and the scale of the face is not the same always. In this paper, a system to verify a specific identity of an individual from crowd images is presented. This system contains two main phases, face detection and verification. The process of face detection is performed by utilizing the Viola and Jones face detection algorithm combined with a skin color based face detection scheme. Face verification is carried out by using an illumination insensitive measure over image pyramids. The hierarchical information obtained at different levels of the pyramid naturally provides better representation at different scales. Experimental results show that the proposed method performs considerably well across varying illumination conditions and scales.

Keywords Contrast adjustment · Face verification · Illumination insensitive measure · Image pyramid · Noise filtering

1 Introduction

Since the advent of photography, faces have become the primary means of identification. The current growth in the computer vision community has led to a rapid development in automatic facial image analysis. Face as a biometric feature has significant advantage over other traditional authentication methods, as these biometric characteristics of an individual are nontransferable, unique to every person and cannot be lost, stolen or broken.

Many recent events, such as identity theft, exposed the serious need for preventing identity-related attacks. With the increase in need to prevent such attacks, face is used

S. D. Mahalakshmi
Mepco Schlenk Engineering College, Sivakasi, India

B. C. Mohan (✉)
VIT, Vellore, Tamil Nadu, India
e-mail: Dr.abc@outlook.com

© Springer Nature Singapore Pte Ltd. 2020
V. Bhateja et al. (eds.), *Intelligent Computing and Communication*,
Advances in Intelligent Systems and Computing 1034,
https://doi.org/10.1007/978-981-15-1084-7_73

as a biometric feature for verification. Therefore, face image can be used to secure access to restricted areas of airports or border areas, verify passengers' identities, scan terminals for suspected criminals and credit card authentication, etc.

Face detection is the basic step in any automatic facial image analysis. A general definition of the face detection problem is stated as follows: Given a still image scene, determine if any faces are present and locate an unknown number of faces (if any).

The solution, therefore, separates the detected faces from the scene's background. Over the past decade face detection has attracted researchers, as any facial image analysis requires automatic face detection as the first step. Much of the earlier work dealt with frontal faces, but now systems have been developed to detect faces with in-plane or out-of-plane rotations in real time. The reliability and accuracy of face detection had a direct impact on the usability of detected faces in face verification system. The number of false detection will reduce the system performance factor namely, speed.

Face verification is a biometric approach to verify the identity of a person based on his/her facial characteristics. It is the most important means used by human beings to verify individuals. Face verification is fundamentally different from face recognition. They differ in the number of subjects involved. In recognition, there are multiple subjects to compare with, and the output of recognition should be the identity corresponding to the input face (1: N matching). In verification, in contrast, there is only one subject involved and the output of verification should be a true or false decision (1:1 matching). Thus, in general, the recognition is a multi-class classification problem, whereas the verification is a two-class classification problem.

The need for reliable identification by law enforcement agencies and other security applications has led to the tremendous growth of face verification technology in the past few years. But face verification is still a research topic in order to design robust and reliable systems. In this work, we focus specifically on the problem of face verification in still images of a group of people. Obviously the most straightforward variety of this problem is the verification of a single face at a known scale and orientation. However, in real time, a more robust system to scale and illumination variations are required. The growing need for robust identity verification in many real-life security applications motivated this work.

One common application is a system, which verifies a known personality's presence in the still image of a group of individuals. Such images may be gathered on many public occasions. Examples include photographs of public gatherings in political rallies, election polling areas, still images obtained from video cameras of highly secure places like airports, border areas, etc. So when a criminal suspect's photograph is available, it may be often desirable for the authorities to verify if that individual is present in any such group images acquired. Furthermore, interesting applications for face detection and verification from crowd images include,

- **Identity Discovery**: Applications that can reliably verify one's appearance in a group by matching a probe image with the individuals detected.
- **Border Security**: Applications that verify the identity of border crossers from still images of border areas.

- **Security Alert Systems**: Access control systems that alert security authorities through face verification.

Therefore, the main goal of our research is deduced to, detect the human faces from a digital image and verify if a given identity is present among the detected individuals. In this work also, effort is dedicated to the problem of detecting and verifying the faces of individuals in group images with roughly frontal pose and is applicable only for still images. Also the system deals with scale and illumination variations during face verification while other challenges like occlusion, age are not taken into account.

2 Research Challenges

Although face verification tasks have received considerable attention from researchers, it still remains challenging in real-life applications. The research challenges primarily come from variations that affect the appearance of face. Generally, the challenges associated with face verification can be broadly categorized into interclass similarity and intrapersonal variations. Inter class similarity refers to the similarities in the appearance of different persons. The likeness of a mother and a daughter, resemblance of twins are all examples of interclass similarity. Intrapersonal variations are the variations between multiple images of the same person. These intrapersonal variations come from variations in,

- Illumination,
- Pose/rotation,
- Biometric changes (aging effects),
- Facial expressions,
- Occlusion.

The illumination variations may be due to the image acquisition conditions such as the direction of light source. Pose variations or rotations are attributed by the changes in the angle at which image is acquired. Aging effects are manifested in the form of facial hair, wrinkles, etc. Facial expression and occlusion due to the use of spectacles, scarves or turbans also impose considerable challenges to the face verification problem. Most of the face recognition researches has been developed to cope up with the illumination and pose variations, as they can cause serious performance degradation. In general, there is no complete system that is able to solve the face verification problem covering all the challenges imposed. However, all the obstacles for facial analysis should be taken into consideration when designing a more robust face detection and verification system.

The rest of the paper is organized as follows: The following section discusses the related work on face detection and verification tasks. The face verification architecture is explained in Sect. 2.1. The methodology is briefly discussed in Sect. 3. In Sect. 4, the results from a detailed experimental analysis of the proposed framework are presented and in Sect. 5 the paper concludes with a brief discussion and directions for future enhancements.

2.1 Related Work

Face verification methods can be broadly classified as "generative" and "non-generative". Generative methods apply transformations on the test image to reduce the differences and verify the claim whereas, non-generative approaches derive features from faces and use the same to perform face verification. Many generative and non-generative methods have been proposed for face verification that applies mostly to frontal faces.

Ramanathan et al. [1] in their work used a craniofacial growth model to characterize the shape variations undergone by human faces. However, the model was demonstrated for face recognition across age progression only on individuals under the age of eighteen. This limits the application of these methods since ages are often not available. It also lacks a textural model and does not consider the textural variations in the face recognition tasks. Lanitis et al. [2], presented a framework for simulating the aging effects on new face images.

Bruneli and Poggio [3] proposed a template matching technique for face recognition. In their work, face recognition was performed by matching templates of three facial regions i.e., the eyes, nose, and mouth. They compared the performance with geometrical matching algorithms and template matching algorithms on the same database of faces. The template matching was found to be superior in recognition of geometrical matching.

Anil Kumar et al. [4] proposed a template matching using edge-based representation of faces, for face verification tasks. In this technique template matching was performed using a correlation of images represented by the edge gradient. The approach was tested for illumination and pose variations.

Georgios Goudelis et al. [5] proposed a novel kernel discriminant approach for face verification. They developed a novel kernel criterion based on the fact that the distribution of facial images, under different viewpoints, illumination variations, and facial expression is highly complex and nonlinear. Verification results conducted on XM2VTS database showed promising results for their proposed method.

Biswas et al. [6] used coherency in facial feature drifts across ages as a measure to perform face verification across ages. Scale Invariant Feature Transform (SIFT) was used to obtain the facial drift maps as the features. If these computed feature drifts are coherent for a pair of face images, the verification result is true (face images belong to the same subject). Thus, it is a pure matching technique that does not require any facial synthesis.

Mian Zhao and Hong Wei [7] developed an algorithm based on Gabor wavelets and AdaBoost. Gabor wavelets were used to extract features from original face images. The face verification performance was improved by employing AdaBoost (Adaptive Boosting) for feature selection, in order to select the top 20 significant features which distinguish a specified client from other subjects in the face database. This approach was found to be robust to illumination and expression variations.

Haibin Ling et al. [8] proposed a discriminative approach to derive the Gradient Orientation pyramid (GOP) features of age separated faces for face verification.

The GOP features were illumination insensitive and the pyramid technique captured hierarchical information, which provided a better means for feature representation. SVM framework was used for the classification of facial images. This technique was found to verify age separated face images without any prior knowledge on age. Taipin Zhang et al. [9] proposed a novel method to extract illumination insensitive features from the face images. These features called the gradient faces were extracted in the gradient domain. Results indicated high recognition rates in the Yale Database B and PIE database. Few of our existing works are available in [10].

3 Proposed Work

Face detection and verification system is a computer application for automatically detecting human faces from crowd images and verifying a person's identity. One of the ways to do this is by comparing facial features of a query face image against the faces detected from crowd image. Thus. the overall system may be considered as two phases,

1. Face detection from crowd image
2. Face verification of identified faces against a single identity.

A face detection module determines the locations of human faces in input images. The input images may be acquired either as scanned photograph images through a desktop scanner or as a digital image. A typical face detection algorithm uses a sliding window technique to scan for faces in the input image. The detected faces from the face detection stage constitute the biometric data for face verification. A reference database contains the face images of claimed identities to be used for verification. The face verification module automatically verifies a person's identity from a digital still image or a video frame from a video source.

The primary steps in a face verification module are preprocessing, feature extraction, and face verification stages. The objective of the preprocessing step is to obtain a transformed face image of the input samples to increase the quality of the face image, retaining the important characteristics. Typical preprocessing includes alignment and cropping of the face images to the desired size. The common technique for alignment involves locating the eyes and then rotating the selected image so that their inclination is null. By cropping, all the test samples are transformed into a reduced uniform size for computational reasons. Some advanced preprocessing methods reduce the impact of lighting conditions and color variance.

After the preprocessing of images is completed, feature extraction methods are applied to them, resulting in a face feature vector, which is then used for face verification. In the feature extraction stage, the input data is transformed into a reduced representation set of features called feature vector. The features are extracted in such a way that the feature vector contains the relevant information from the input data, to perform the desired task using this reduced representation instead of the full-size input. The basic architecture of a typical face detection and verification system is shown in Fig. 1.

Face detection is the first module where we generally deal with the detection of frontal faces. In this work, a hybrid methodology that was proposed by Zahra Sadri Tabatabaie et al. [11] is utilized for the detection of frontal faces. This hybrid approach combines image-based and feature-based face detection systems. For face detection, initially the skin regions in an input image are identified using a skin detection algorithm and then the detected skin regions are classified as "face" or "non-face" using Viola and Jones face detection algorithm. These techniques are combined as an attempt to reduce the false rate furthermore. The algorithm for face detection is given in Fig. 2.

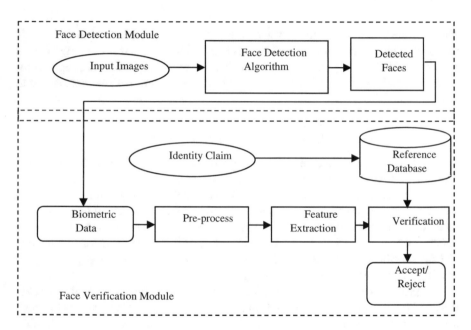

Fig. 1 Architecture of simple face verification system

Fig. 2 Algorithm for hybrid face detection

> **Algorithm: Hybrid Face Detection**
> **Input** : Group Image I
>   **Output :** Faces present in crowd image.
>   1. Detect the skin region in RGB colour space by using heuristics and black out the non-skin regions.
>   2. Apply Viola-Jones face detection algorithm to the detected skin regions.
>   3. Crop the detected face regions to generate the required faces for verification.

In many identity-based applications, color images have gained importance in recent years. Skin color information in color images is a valuable feature of human faces because it provides a more powerful means of classifying faces from non-faces. Therefore, the skin color pixels in the image are identified as the first step in face detection. Much recent work on skin color based face detection has been carried out in the YCbCr, RGB, and HSI color space. RGB color model is utilized for detecting skin color pixels as it is one of the most widely used color spaces for processing and storing digital images. Skin detection is carried out by explicitly defining boundaries for skin color through a number of heuristic rules in RGB color space. Zahra Sadri Tabatabaie et al. [11] used the heuristic rules as defined by Kovac et al. [12]. These rules are given below:

(R, G, B) is classified as skin if,

$R > 95$ and $G > 40$ and $B > 20$ and
$\text{Max}[R, G, B] \quad \text{Min}[R, G, B] > 15$ and
$|R - G| > 15$ and $R > G$ and $R > B$

The basic skin color information for the face regions is obtained from the above-mentioned rules. After applying the skin color classification algorithm, other insignificant non-skin regions are replaced with black color. This reduces the misclassification rate since most of the face-like background textures will be blacked out. The faces detected in hybrid approach are in rectangular areas comprising the face and some non-face regions too. Therefore, all the faces detected in the hybrid face detection step are subject to automatic cropping to single out the faces from other parts like shoulders, hair, etc. The face image cropping step separates the exact face region. This is done by enclosing the human faces within a vertical ellipse with a fixed ratio for the minor and major axes. Therefore, this processing consists of two steps,

- Creation of an elliptical mask to extract the face image.
- Reduction of images to a uniform size.

The elliptical mask is created by constructing a solid elliptical mask automatically around the face region and the other insignificant face regions are masked off. The ellipse is centered at the center of the rectangular face region. The ratios for the major and minor axes are set based on the column and row width of the rectangular face region which is defined as,

$$\text{Major\_axis} = \text{column\_width}/1.5$$
$$\text{Minor\_axis} = \text{row\_width}/1.5$$

The elliptical face is subsequently cropped as a square surrounding the face pixels to remove the excessive masked background. These cropped elliptical face sizes are reduced to 96×84 for processing efficiency. All these steps are automated and proceed without any human intervention. Figure 3 represents these steps carried on a Bao Face Database image.

Table 1 Hybrid face detection analysis under different factors

Criteria	Without preprocessing (%)	With contrast adjustment (%)	With noise filtering (%)
FPR	9.92	8.04	8.04
FNR	6.97	8.85	8.04
Accuracy	83.10	83.10	83.91

4   Result and Analysis

Most of the images that yield low face detection results are poor in contrast or noise corrupted, therefore, a preprocessing stage of contrast adjustment and median filtering are introduced for the misclassifications. This method, although produced no significant improvements in the face verification results, yielded better face detection by reducing the false positives. Table 1 indicates the results of hybrid face detection approach for various preprocessing factors.

From the results, it is evident that there exists a trade-off between the true detection rates and false detection rates. The decrease in the false positive rates leads to a slight increase in false negative rates. This is contributed by the good quality images, which when unnecessarily preprocessed produced a negative impact on the performance. However, the overall accuracy of the noise filtering is improved by 0.8%. The contrast adjustment produced no improvement over the hybrid approach without applying it. The experimental results of the face verification system have been discussed. The results have indicated that the hybrid approach for face detection efficiently decreases the false positive rate.

5   Conclusion

A novel face verification approach based on the gradient faces [9] and image pyramids called hierarchical gradient face technique was proposed. The results of the proposed technique showed a very good performance in Yale Database B images. As an attempt to improve the face verification, a series of contrast adjustment and noise filtering have been tried. The face detection can be further improved by applying the contrast adjustment and noise filtering selectively based on the image quality. Because, when these techniques are applied to all input image, sometimes the important features like edges will be diminished. Also, other color spaces and methods for skin color classification may be further studied to achieve improvements.

References

1. Ramanathan, N., Chellappa, R.: Modeling age progression in young faces. In: Proceedings of the IEEE Conference on Computer Vision and Pattern Recognition (CVPR), vol. 1, pp. 387–394 (2006)
2. Lanitis, A., Taylor, C.J., Cootes, T.F.: Toward automatic simulation of aging effects on face images. IEEE Trans. Pattern Anal. Mach. Intell. **24**(4), 442–455 (2002)
3. Brunelli, R., Poggio, T.: Face recognition: features versus templates. IEEE Trans. Pattern Anal. Mach. Intell. **15**(10), 1042–1052 (1993)
4. Sao, A.K., Yegnanarayana, B.: Face verification using template matching. IEEE Trans. Inf. Forensics Secur. **2**(3), 636–241 (2007)
5. Goudelis, G., Zafeiriou, S., Tefas, A., Pitas, I.: A novel kernel discriminant analysis for face verification. In: IEEE International Conference on Image Processing, ICIP 2007, pp. 493–496 (2007)
6. Biswas, S., Aggarwal, G., Ramanathan, N., Chellappa, R.: A nongenerative approach for face recognition across aging. In: Proceedings of the 2nd IEEE International Conference on Biometrics: Theory, Applications and Systems, 2008 (BTAS 2008), pp. 1–6, 29 Oct. 2008
7. Zhou, M., Wei, H.: Face verification using Gabor wavelets and AdaBoost. In: Proceedings of the 18th International Conference on Pattern Recognition, ICPR 2006, pp. 404–407, Sep 2006
8. Ling, H., Soatto, S., Ramanathan, N., Jacobs, D.W.: Face verification across age progression using discriminative methods. IEEE Trans. Inf. Forensics Secur. **5**(1), 82–91 (2010)
9. Zhang, T., Tang, Y.Y., Fang, B., Shang, Z., Liu, X.: Face recognition under varying illumination using gradient faces. IEEE Trans. Image Process. **18**(11), 2599–2606 (2009)
10. Chandra Mohan, B.: Restructured ant colony optimization routing protocol for next generation network. Int. J. Comput. Commun. Control **10**(4), 493–500, Agora University Press, 2015
11. Tabatabaie, Z.S., Rahmat, R.W., Udzir, N.I.B., Kheirkhah, E.: A hybrid face detection system using combination of appearance-based and feature-based methods. IJCSNS Int. J. Comput. Sci. Netw. Secur. **9**(5) (2009)
12. Kovac, J., Peer, P., Solina, F.: Human skin color clustering for face detection. The IEEE Region 8 Computer as a tool EUROCON 2003, vol. 2, pp. 144–148, Sept. 22–24 (2003)

A Wind Energy Supported DSTATCOM for Real and Reactive Power Support for a Linear Three-Phase Industrial Load

Rani Fathima Kamal Basha and M. L. Bharathi

Abstract Industrial loads like textile mills predominantly use three-phase induction motors. The load system is, therefore, reactive in nature. Since three phase squirrel cage induction motor is used, the load is considered as a balanced load. Therefore, in such a situation, the only requirement is a reactive power compensation. In this work a novel DSTATCOM is proposed that addresses the reactive power demand of the load. In addition, a PMSG based wind energy system is also included so that supplementary energy is also made available to the load. The functionality of the DSTATCOM is controlled by a set of two PI controllers. The MPPT for the PMSG is carried out using a Sliding Mode Controller. The objective of the work is to integrate renewable energy into the system as a supplementary source besides compensating the reactive power as well, using the single common Graetz bridge converter. The proposed idea has been validated with simulations carried out in the MATLAB SIMULINK environment.

Keywords DSTATCOM · WECS · PMSG · Sliding mode controller

1 Introduction

Reactive power compensation for standalone loads has been carried out by discrete power capacitors. The power capacitors are included or excluded by, electromagnetic relay initiated, circuit breakers. The process could be done manually as well as automatically. The addition of the capacitors is to be carried out in a discrete manner with those compensating capacitors of nearest values. In the industrial environment the motors may be turned on or off in a discretely switched manner. The mechanical

R. F. K. Basha (✉)
Al Musanna College of Technology, Al Mulladah Sultanate of Oman, Mulladah, Oman
e-mail: rani@act.edu.om

M. L. Bharathi
Sathyabama Institute of Science and Technology, Chennai, India
e-mail: bharathiml15@gmail.com

© Springer Nature Singapore Pte Ltd. 2020
V. Bhateja et al. (eds.), *Intelligent Computing and Communication*,
Advances in Intelligent Systems and Computing 1034,
https://doi.org/10.1007/978-981-15-1084-7_74

755

load on the motors could be a variable that changes over a continuous scale. This puts forward the demand for a novel two-step hybrid compensation.

In this work, therefore, a two-tier compensation is adopted. In the first step a discrete capacitor is automatically selected depending upon the reactive power demand. Then based on the fractional part of reactive power the DSTATCOM is activated so that an accurate compensation is provided with the harmonic disturbances minimized, that could otherwise be predominant if the DSTATCOM alone is used for reactive power compensation.

By using the hybrid compensation scheme, consisting of both capacitors and the DSTATCOM the electrical capacity of the DSTATCOM required, its size, the duty, the harmonic injection, and the losses can be minimized. The proposed control scheme is like a Vernier system in which the bulk requirement is provided by the capacitor and the fractional part is supplied by the DSTATCOM.

The DSTATCOM uses a three-phase converter [1]. It has the AC and the DC sides. The Graetz bridge converter used in the DSTATCOM can transact power in both directions and both real and reactive components as well. While the DSTATCOM can supply reactive power to the point of common coupling, it can simultaneously draw real power from the point of common coupling to top up the dc link capacitor voltage [2]. As a matter of possibility, the DSTATCOM can also act as a reactive power consumer while it supplies real power into the PCC in which case the dc link capacitor is supported from external energy sources like, batteries, solar power, fuel cell-based power or from a WECS or the combination of more than one of any of these sources [3–5]. The PMSG is a multiphase alternator and, mostly the three-phase configurations is widely used. The PMSG can be driven by the wind turbine typically of the three-bladed horizontal axis type [6–8]. The PMSG based wind energy harvesting system supplies real power to the point of common coupling and thus to the load. It is assumed that there is always a demand on the load side that the PMSG based WECS alone cannot meet out the demand on the load side. As such the grid integration of the harvested wind energy is ruled out. The three-phase Graetz bridge converter pumps real power as well as reactive power when wind velocity is sufficiently available and only reactive power when wind velocity is not sufficiently available [9, 10]. The WECS requires an MPPT scheme. The P and O or the hill-climbing method is the most popular one. While other methods such as the Fuzzy logic based MPPT for the WECS, the ANN based MPPT for the WECS are available in this work the SMC based MPPT is proposed. When the wind velocity is known from an external anemometer then from the lookup table the optimal turbine speed can be found. If the turbine is running at a speed more than the optimal speed, then the electrical load on the PMSG should be increased and vice versa.

When the harvested wind power is not continuously routed to the point of common coupling and thus to the load, then the DC link voltage will rise. When the wind power is not available the DC link voltage may fall [11, 12]. Thus, maintaining the DC link voltage at the pre-defined level ensures the proper transaction of the real power as well. The cases of reactive power support, the MPPT on the WECS and the real power support are demonstrated in this paper. The paper is arranged as follows. After the introduction given in Chap. 1, a review of the structure of the

PMSG based WECS and the various MPPT techniques used, and applicability of the sliding mode controller for the PMSG based WECS are discussed in Chap. 2. The various modes of operation of the DSTATCOM with a detailed description of the controllers used for the management of the DSTATCOM are discussed in Chap. 3. The MATLAB SIMULINK based simulations along with the results are discussed in Chap. 4 followed by the Conclusion and Reference sections.

2 The PMSG Based WECS and the Sliding Mode Controller Based MPPT

The PMSG based wind energy harvesting system is a variable speed system comes under the full conversion category. Under this category the entire power being harvested is processed through a power electronic converter or a set of power electronic converters and finally reach the load. Normally, since the PMSG based system is a variable speed system the frequency of the three-phase output of the PMSG is of variable nature. The variable frequency, as such, cannot be supplied to the load. Therefore, at the front end itself, the output of the PMSG is first rectified and converted into a DC potential. The DC potential thus obtained may not be of sufficient voltage level and hence it is boost converted before reaching the load. In this work the PMSG is connected to the DC side of a STATCOM through which the PMSG based WECS injects real power to the point of common coupling and thus to the load. The DSTATCOM supplies the required reactive power also to the point of common coupling so that even if a reactive load is connected to the AC load bus bar the source side power factor may be maintained high, near to unity. There are two power electronic converters that are directly associated with the PMSG. They are the Diode Bridge Rectifier (DBR) and the boost converter that follows the DBR.

With reference to the block diagram of the proposed system as shown in Fig. 1 the DBR is an uncontrolled converter. It just converts AC into DC. The DC to DC boost converter that follows the DBR is a controllable one. The output of this boost converter is connected across the DC link capacitor of the DSTATCOM. For the

Fig. 1 Block diagram of the proposed DSTATCOM with wind energy support

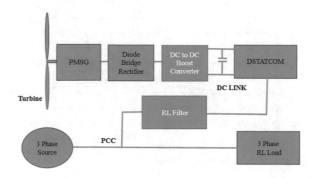

boost converter, the DC link is a sink, The DC link voltage is maintained by the DSTATCOM. The boost converter just pumps the power harvested by the wind turbine—PMSG system into this sink that is maintained at a fixed DC potential. The boost converter undertakes the job of maximum power point tracking of the PMSG.

2.1 MPPT for the PMSG

The PMSG based WECS gets the input power from the wind turbine. The nominal rating of the wind turbine is 3 kW. When the wind velocity is at a rated level of 9 m/s the wind turbine delivers the maximum power of 1 PU. For the rated wind velocity of 9 m/s, that is 1 PU, if the speed of the turbine is not maintained at 1 PU the power harvested is less than 1 PU and is as dictated by the speed at which the turbine is run.

The wind turbine converts the kinetic energy, contained in the wind, to rotational energy. The power output of the wind turbine is proportional to the density of wind, the total circular area swept by the blades, the performance coefficient and the cube of the wind velocity, the performance coefficient is a function of the tip speed ratio. This relationship is nonlinear. Even though the wind velocity is high, if it is greater than the critical value, when the tip speed ratio offers the peak coefficient of performance, the power output falls down. Thus, the wind turbine needs maximum power point tracking. With reference to Fig. 2 that gives a family of curves, relating the wind velocity, output power, and the turbine speed it is clear that for all wind velocities there is a unique speed at which the turbine should be run so that the power harvested is maximum. For a given wind velocity, for all speeds other than the unique MPP speed the power harvested will be less than the possible maximum value. Thus, the objective of the MPPT system used with the Wind Turbine—PMSG system, is, for the given wind velocity. to run the turbine_PMSG shaft at the specific speed that corresponds to the maximum power output harvestable for the prevailing wind velocity. The harvest of maximum power can be ascertained by two methods. In the

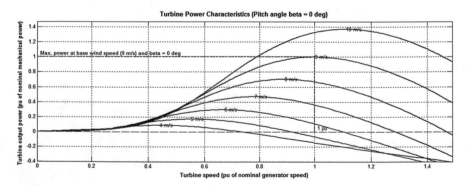

Fig. 2 Wind turbine characteristics

first method it should be done by measuring the wind velocity and the turbine or PMSG speed. This method needs two sensors. One to measure the wind velocity and the other to measure the shaft speed. However, the manipulated parameter is the duty cycle of the boost converter.

The second method is implemented on the basis of electrical parameters. The power being harvested by the PMSG is routed to the DC link of the DTSTACOM through the boost converter. Therefore, the power that reaches the DC link is a measure of the power being harvested by the WECS. The product of the DC link voltage and the boost converter output current to the DC link is the power that is harvested by the wind turbine, assuming zero losses at the PMSG, the DBR, and the boost converter. A simple perturb and observe type of MPPT can be used at the boost converter by manipulating the duty cycle of the boost converter.

2.2 Implementation of P and O method of MPPT

In this method, the power delivered to the DC link is the Observed parameter. The duty cycle of the Boost converter in between the DBR and the DC link is the disturbed or the manipulated parameter.

Step 1. Initialize duty cycle $D = 0.5$.; Step 2. Measure DC link voltage Vdc.; Step 3. Measure DC current from Wind subsystem to DC link Idc.; Step 4. Calculate Power flow to DC link Pdc1 = Vdc * Idc.; Step 5. Change $D(\text{new}) = D(\text{old}) + \text{del}(D)$; Step 6. Measure Pdc2; Step 7. If Pdc2 > Pdc1 then $D(\text{old}) = D(\text{new})$ and then $D(\text{new}) = D(\text{old}) + \text{del}(D)$; Else if Pdc2 < Pdc1 the $D(\text{old}) = D(\text{new})$ and then $D(\text{new}) = D(\text{old}) - \text{del}(D)$; Else if Pdc2 == Pdc1 then $D(\text{old}) = D(\text{new})$ and then $D(\text{old})$ $D(\text{old}) + 0$; Step 8 Pdc1 = Pdc2; Step 9 Go to step 6.

2.3 Steps is Sliding Mode Controller

In the SMC the Monitored variables are wind velocity and the turbine speed. The manipulated parameter is the state of the power electronic switch S used in the boost converter. If switch is turned On then $S = 1$; else $S = 0$. The duty cycle is initially fixed arbitrarily at 0.9, nearly the maximum level.

Steps: Step 1. Measure Wind velocity Vw.; Step 2. Find turbine speed $N(\text{Sp})$ from the lookup table.; Step 3. Measure actual speed of the turbine. $N(\text{act})$; Step 4. If $N(\text{act}) < N(\text{sp})$ turn Off the Switch S.; Step 5. If $N(\text{act}) > N(\text{sp})$ turn On the Switch S. This process is repeated and the turbine speed is pushed to stay close to the required speed that delivers the maximum power output for the given wind velocity.

3 The DSTATCOM

The DSTATCOM is a shunt element that comes across the feeder that feeds power from a three-phase source to a three-phase load. The point at which the DSTATCOM is connected to the feeder in the form of a three-phase T connection is called the Point of common Coupling (PCC). The three important elements that constitute the DSTATCOM are:

a. The three-phase, three leg Graetz bridge converter; b. The three-phase RL filter that lies between the AC side of the converter and the point of common coupling; c. The DC link capacitor on the DC side of the Graetz bridge converter. The three-phase power electronic converter may use either Sinusoidal PWM or Space Vector PWM. Besides, the DSTATCOM requires a comprehensive control system that guarantees the proper flow of bidirectional real and reactive powers so that the following objectives are achieved. 1. Flow of reactive power from the DSTACOM to the load through the point of common coupling; 2. Flow of real power from the PCC to the DC link capacitor, for maintaining the DC link voltage at the required level, when the wind energy harvesting system does not generate real power. 3. Flow of real power that is being harvested by the WECS, to the load through the DSTATCOM and the PCC. As for the DSTATCOM, by proper control strategies, both real and reactive powers can be transacted through the DTSTCOM in either direction simultaneously. For the maintenance of the DC link voltage a PI controller is used. For the reactive power support another PI controller is used. These two controllers, respectively, control the Phase angle and the Modulation Index of the three-phase reference signal used for the PWM operation of the core Graetz bridge converter of the DSTATCOM. For the maximum power point tracking, a sliding mode controller is used at the boost converter. The various mathematical aspects of the controls scheme, the subsystems used in the implementation of the proposed system in the MATLAB SIMULINK platform are presented in the following chapter.

4 The Scheme of Control and MATLAB SIMULINK Implementation; the PARK's Transformation

The signals associated with the three-phase DSTATCOM are usually three-phase sinusoidal. The PARKs transformation is used to convert the time-variant three-phase sinusoidal system of voltages or currents into a system of two orthogonal components of DC nature. In the PARK transformation the three-phase quantities are projected on to a rotating frame of $x\,y$ coordinates and the projections are called the d component and the q component. The d and the q components are time-invariant as long as the three-phase system is a balanced sinusoidal one. If the three-phase system is balanced, then the d and q quantities are constants and act at $90°$ in between. The PARK transformation is a transformation with reference to the synchronously rotating frame. The PARKs transformation leads to two advantages.

Table 1 MATLAB SIMULINK based system development, simulation, and results

Parameter	Value
System voltage	380 v line to line
Maximum active load	5 KW
Maximum reactive load	5 KVAR
Maximum active power support from wind	3 KW
DSTATCOM capacity	3 KW + 5 KVAR
DC link voltage	1200 V

They decouple the real and reactive components of the three-phase system under study. The controllers are designed to operate with less bandwidth and compared with the efficiency of controllers that could operate using sine waves of utility frequency. The PARKs transformation is carried out by the matrix expression as shown in equation x. With reference to the expression given as equation x the output of the PARK transformation is Vdq0 or Idq0 when the input is Vabc or Iabc, respectively When a three-phase load is connected to a three-phase voltage system with finite source impedance, as the load side reactive power demand gets increased the d component of the three-phase voltage at the load bus gets reduced. This aspect is shown in figures x and y. There are two controllers associated with the DSTATCOM and are as shown figures x and y. The controller shown in figure x is a PI controller and it takes care of the d component of the voltage at the point of common coupling. With purely resistive loads the d component of the PARK transformed three-phase voltage is at unity.

With the increase in reactive power demand by the load, the d component of the voltage at the point of common coupling falls down from unity. This fall is to be compensated by the DTSTCOM by the injection of reactive power by a quantity that will bring the d component of the voltage at the point of common coupling back to the unity level. This causes the source side power factor to unity as well (Table 1).

The block schematic of the proposed system in MATLAB SIMULINK is shown in Fig. 3. The DSTATCOM uses sinusoidal PWM and the switching pulses are generated after the three-phase reference signal is generated. The system includes the wind energy conversion system shown in a different color (Figs. 4 and 5).

An SMC is used for the Maximum Power Point Tracking. The DC link voltage, wind velocity versus power output of the Wind Turbine, the changes in the turbine rotational speed, the power output of the PMSG are shown in figure x. The source-side current and voltages with different reactive power consumption are shown in figure x. The power balance for various wind velocities and for various real and reactive power demands are tabulated and are as shown in figure x. The control system strategy. The main strategy of control or compensation of reactive power is that there are five capacitors of values 1 KVAR each. When all the capacitors are turned on the total three-phase reactive power connected will be 5 KVAR. If a demand of 1.45 KVAR reactive power is required at the load side, then only one capacitor Unit is connected

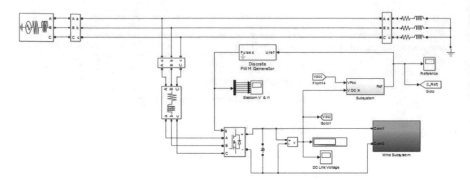

Fig. 3 MATLAB SIMULINK diagram of the proposed DSTATCOM with the WECS sub system

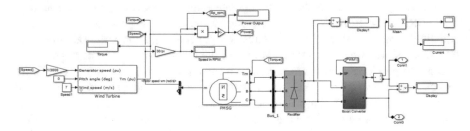

Fig. 4 The wind turbine, PMSG, DBR, and the boost converter sub systems

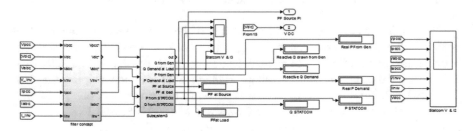

Fig. 5 Sub systems for the calculation and display of various parameters

across the PCC. The remaining 0.45 KVAR required by the load is supplied by the DSTATCOM.

Similarly, if the reactive power demand is 3.12 KVAR then the 3 KVAR demand is supplied by three capacitors connected across the PCC. The remaining 0.12 KVAR is supplied by the DSTATCOM. Thus, the main KVAR compensation is provided by the linear components and only a fractional part is supplied by the nonlinear DSTATCOM. This strategy improves the power quality and the source side PF is brought near unity as well the THD of the source current is much reduced. This improves the efficiency of the overall compensation system (Figs. 6, 7, 8, 9, 10, 11 and 12).

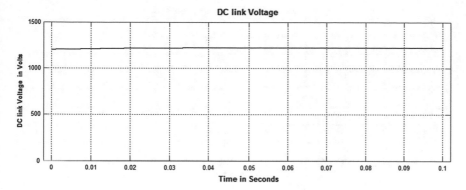

Fig. 6 The DC link voltage

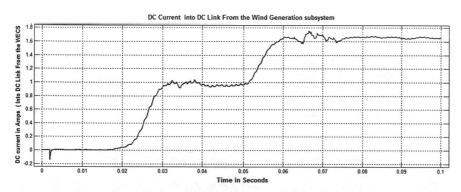

Fig. 7 The WECS output current to the DC link with a step change in wind velocity happening at time 0.05 s

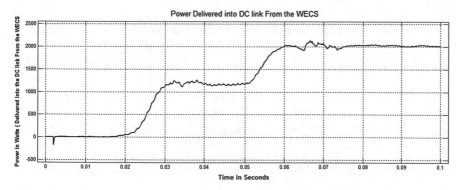

Fig. 8 The power output from the WECS delivered to the DC link indicating the change in the power delivered with change in wind velocity at 0.05 s

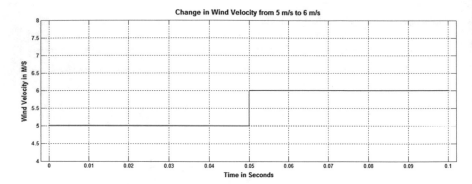

Fig. 9 Wind velocity with a step change from 5 to 6 m/s happening at time instant 0.05 s

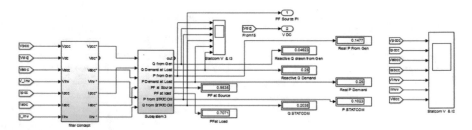

Fig. 10 The display system showing the real and reactive power demand, the magnitudes of *P* and *Q* drawn from the source and the magnitudes of *P* and *Q* supplied by the DSTATCOM

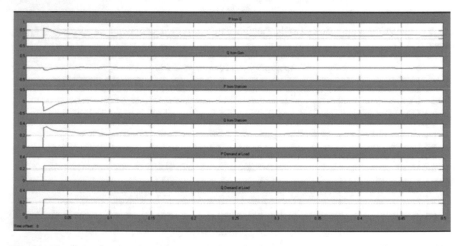

Fig. 11 The flow of real and reactive powers from the main source, the DTSTACOM, and the load side demand—shown from top to bottom respectively

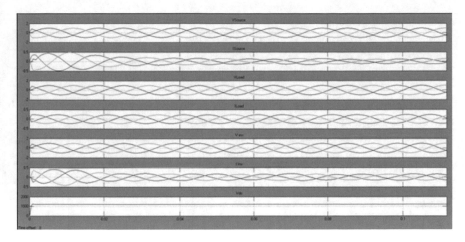

Fig. 12 The three-phase voltage and current waveforms pertaining to main source, load side, and the DSTATCOM. The DC link voltage is shown as the last one at the bottom. Demand at the load: $P = 0.25$ PU $Q = 0.25$ PU and PF $= 0.707$

Table 2 gives the real and reactive power balance with and without the wind source. With wind source a real power of 0.05 PU is supplied from the DSTATCOM and the real power drawn from the main source is reduced by 0.05 PU.

Table 3 gives the Power factor on the source and the load sides with and without wind source in action. It is clear that irrespective of the presence or absence of the wind source the load side power factor and the source side power factor and source side THD remain the same and are not affected by the injection of real power to the PCC from the DSTATCOM. Thanks to the decoupled control scheme.

Table 4 shows that when the reactive power demand is 4.5 KVAR and the whole reactive power demand is met by the DSTSTCOM then the source side current THD is 2.45. On the other hand with the inclusion of fixed KVAR capacitors and fractional with the DSTATCOM only meant for the fractional part the source current THD is improved to 1.25.

Table 2 Real and reactive power balance with and without the wind source

Wind velocity (m/s)	P main source	Q main source	P DSTATCOM	Q DSTATCOM
0	0.25	0	0.00	0.25
9	0.2	0	0.05	0.25

Table 3 Power factor on the source and load side

Wind velocity	PF source	PF load	THD I source
0	0.99	0.707	1.25
9	0.99	0.707	1.25

Table 4 Comparing the THD

Q demand	Fixed KVAR	Q DSTATCOM	THD of source current
4.5 KVAR	4 KVAR	0.5	1.25
4.5	0	4.5	2.45

Table 5 Comparing the efficiency of different compensators

Type of comp.	THD is	Demand P (KW)	Demand Q (KVAR)	PF load	Capacitive KVAR	PF source
No compen	0.00	5	5	0.707	0	0.707
Only capacitors	0.00	5	5	0.707	0	1.00
Only DSTATCOM	2.45	5	5	0.707	0	0.9969
Hybrid (proposed)	1.25	5	4.5	0.7433	4 KVAR	0.95

Table 6 With wind support wind velocity = 5 m/s

Type of comp.	Main source	DSTATCOM	Wind source
P	0.25	−0.02	0
Q	0.25	0.25	0

Table 7 Power balance with and without wind support

Type of comp.	Main source	DSTATCOM	Wind source
P	0.25	0.05	0.05
Q	0.25	0.25	0

Table 5 gives the details of various cases of operation. The case is considered with a reactive power demand of 5 KVAR and 4 KVAR met by the fixed capacitors and 1 KVAR only supported by the DSTATCOM. Without wind support Wind velocity = 0 m/s

Tables 6 and 7 give the power balance with and without the wind support, respectively. when the wind source can deliver a real power component of 0.05 PU the main source is relieved off this burden and the main source now delivers only 0.2 KW.

5 Conclusion

In this work, the functionality of the DSTATCOM for reactive power compensation is presented. Two PI controllers are used for this application. Besides the novelty of the work is that a wind power source is also added to the system and the wind generation system pumps in real power to the system through the DC link at the DC side of the three-phase Graetz bridge converter. Besides a simple reactive power compensator, the proposed system includes real power support from a renewable energy source.

References

1. Yuan, X., Wang, F., Boroyevich, D., Li, Y., Burgos, R.: DC-link voltage control of a full power converter for wind generator operating in weak-grid systems. IEEE Trans. Power Electron. **24**(9), 2178–2192 (2009)
2. Li, S., Haskew, T.A., Swatloski, R.P., Gathings, W.: Optimal and direct-current vector control of direct-driven PMSG wind turbines. IEEE Trans. Power Electron. **27**(5), 2325–2337 (2012)
3. Singh, M., Chandra, A.: Application of adaptive network-based fuzzy inference system for sensorless control of PMSG-based wind turbine with nonlinear-load-compensation capabilities, IEEE Trans. Power Electron. 26(1), 165–175 (2011)
4. Rajaei, A., Mohamadian, M., YazdianVarjani, A.: Vienna-rectifier based direct torque control of PMSG for wind energy application. IEEE Trans. Ind. Electron. **60**(7), 2919–2929 (2013)
5. Crescimbini, F., Lidozzi, A., Solero, L.: Highspeed generator and multilevel converter for energy recovery in automotive systems. IEEE Trans. Ind. Electron. **59**(6), 2678–2688 (2012)
6. Uehara, A., Pratap, A., Goya, T., Senjyu, T., Yona, A., Urasaki, N., Funabashi, T.: A coordinated control method to smooth wind power fluctuations of a PMSG-based WECS. IEEE Trans. Energy Convers. **26**(2), 550–558 (2011)
7. Chong, H., Huang, A.Q., Baran, M.E., Bhattacharya, S., Litzenberger, W., Anderson, L., Johnson, A.L., Edris, A.A.: STATCOM impact study on the integration of a large wind farm into a weak loop power system. IEEE Trans. Energy Convers. **23**(1), 226–233 (2008)
8. Bharathi, M.L.: FOPID controlled three stage interleaved boost converter fed DC motor drive. Int. J. Power Electron. Drive Syst. **8**(4),1771 (2017)
9. Bharathi, M.L., Kirubakaran, D.: Solar powered closed-loop controlled fuzzy logic-based three-stage interleaved DC-DC boost converter with an inverter. Int. J. Adv. Intell. Parad. **8**(2), 140–155 (2016)
10. Jayaprakash, S.: Comparison of solar based closed loop DC-DC converter using PID and Fuzzy Logic Control for Shunt motor drive. In: 2014 IEEE National Conference on Emerging Trends in New & Renewable Energy Sources and Energy Management (NCET NRES EM), IEEE, pp. 115–117 (2014)
11. Jayaprakash, S., Ramakrishnan, V.: Analysis of solar based closed loop DC-DC converter using PID and fuzzy logic control for separately excited motor drive. In: 2014 IEEE National Conference on Emerging Trends in New & Renewable Energy Sources and Energy Management (NCET NRES EM), IEEE, pp. 118–122 (2014)
12. Sindhuja, S., Abirami, P., Pushpavalli, M., Sivagami, P.: Implementation of multilevel inverter for controlling induction motor. Int. J. Recent. Technol. Eng. (IJRTE) **8**(1) (2019). ISSN: 2277-3878

Solving Combined Economic and Emission Dispatch Problem Using Hybrid RGA-DE Algorithm

C. N. Ravi, K. Vasanth, R. Harikrishnan, D. Godwin Immanuel, D. Lakshmi and G. Madhusudhana Rao

Abstract Combined Economic Emission Dispatch (CEED) is a bi-objective problem, needs to find a generating pattern to minimize generating cost and emission. Generating cost and emission are a function of real power generation. In CEED minimum generating cost along with minimum emission is calculated and subjected to real and reactive power balance equation, upper and lower bound of real and reactive power, voltage, and MVA power transfer in the transmission lines. In this paper, CEED is solved to find its bi-objective, real, and reactive power losses are calculated using Newton Raphson (NR). This bi-objective is converted into single objective function and subjected to equality and inequality constraints as same as OPF. The objective functions are quadratic function of real power generation. Real coded Genetic Algorithm (RGA) is superior in cross over operation and Differential Evolution (DE) is superior in mutation operation hence the hybrid RGA-DE algorithm will provide better minimum objective value and used to solve CEED.

Keywords CEED · Hybrid algorithm · Bi-objective · RGA · DE

C. N. Ravi (✉) · K. Vasanth · G. Madhusudhana Rao
Vidya Jyothi Institute of Technology, Hyderabad, India
e-mail: dr.ravicn@gmail.com

K. Vasanth
e-mail: vasanthecek@gmail.com

G. Madhusudhana Rao
e-mail: gurralamadhu@gmail.com

R. Harikrishnan
Symbiosis Institute of Technology Symbiosis International Deemed University, Pune, India
e-mail: rhareish@gmail.com

D. Godwin Immanuel
Sathyabama Institute of Science and Technology, Chennai, India
e-mail: dgodwinimmanuel@gmail.com

D. Lakshmi
Department of EEE, AMET Deemed to be University, Chennai, India
e-mail: lakshmiee@gmail.com

© Springer Nature Singapore Pte Ltd. 2020
V. Bhateja et al. (eds.), *Intelligent Computing and Communication*,
Advances in Intelligent Systems and Computing 1034,
https://doi.org/10.1007/978-981-15-1084-7_75

1 Introduction

Emission of power plants need to be minimized and included in the objective of cost minimization of Optimal Power Flow (OPF) [10] becomes a Combined Economic Emission Dispatch (CEED). CEED is a bi-objective problem optimize generating pattern in such a way to yield optimal generating cost at optimal emission level [7].

Many conventional mathematical optimization techniques have been used to solve cost minimization problem. They are Linear Programming (LP), Nonlinear Programming (NLP), Quadratic Programming (QP), Newton-based solution, and Interior Point (IP) methods. In recent days, intelligent techniques are commonly used to solve power system optimization problems [11, 12]. GA developed by John Holland and his collaborators in 1960s mimics biological-based evolution based on Charles Darwin's theory [1]. This artificial system uses crossover, mutation, and selection. The two main advantages are ability to solve complex optimization and parallelism. Storn R and Kenneth Price [2] introduced new heuristic approach to solve non-differential functions called Differential Evolution (DE). The nature of DE is minimization of optimization problem which helps to optimize OPF. Alsac and Stott [3] improved the Dommenl and Tinney work by incorporating new constraints for steady-state security. He considered IEEE 30 bus test system to demonstrate his work, this test case becomes standard system for many research work and even for this reported research work. They analyzed OPF with and without outage condition and considering security in both cases is presented in a neat way to understand in detail.

Venkatesh et al. [4] proposed evolutionary programming (EP) technique to solve bi-objective Combined Economic Emission Dispatch (CEED) problem. The work proposed novel price penalty factor to combine minimum emission objective cost minimization objective and the problem becomes a bi-objective problem. The price penalty factor is calculated using actual load demand using maximum fuel cost, maximum emission of each generator, and maximum capacity of individual generators [8]. Ajay Vimal Raj et al. [3] used PSO to solve OPF including minimizing emission level of nitrogen oxides NOx pollution. Hemamalini and Simon [5] used PSO to solve CEED problem. In this work valve point loading effect and NOX emission are considered. They used quadratic cost function with valve point loading and emission is a function real power generation in second-order is considered to solve the CEED problem.

Power System firm has to provide reliable, quality electric power supply to its consumer at the same time it has to minimize generating cost and emission. Optimal Power Flow solutions [6] help power system operation to yield optimal real power generation pattern of committed generators to minimize generating cost. Combined economic emission dispatch is a bi-objective problem that provides an optimal generation schedule for minimum generating cost and minimum emission. CEED ensures optimal generating cost, emission, power loss, and security of the power system.

2 Problem Formulation

Emission of air pollution such as sulfur oxides SO_X, carbon oxide CO_X, and nitrogen oxides NO_X are calculated in tons/hr. To convert this bi-objective problem into single objective the emission is multiplied by price penalty factor. CEED is an optimization problem has to minimize generating cost and air pollution and subjected to equality and inequality constraint.

2.1 *Objective Function*

$$\text{Minimize } F = F_1 + h * F_2 \quad \$/hr \tag{1}$$

$$F_1 = \sum_{i=1}^{NG} \alpha_i + \beta_i P_{Gi} + \gamma_i P_{Gi}^2 + \left|\zeta_i \, \sin\left[\lambda_i \left(P_{Gi}^{\min} - P_{Gi}\right)\right]\right| \tag{2}$$

$$F_2 = \sum_{i=1}^{NG} 10^{-2}\left(a_i + b_i P_{Gi} + c_i P_{Gi}^2\right) + d_i \, \exp(e_i P_{Gi}) \tag{3}$$

where

NG is number of generators
F_1 is total generating cost in United States Dollar ($)/hr
F_2 is total emission in tons/hr
h is price penalty factor in $/tons

Emission is computed in tons/hr is converted into $/hr by multiplying with price penalty factor. This final single objective function is calculated in terms of $/hr.

2.2 **Subject to**

Equality constraints
Power balance condition of power system

$$\sum_{i=1}^{NG} P_{gi} = P_D + P_L \tag{4}$$

$$\sum_{i=1}^{NG} Q_{gi} = Q_D + Q_L \tag{5}$$

Inequality constraints
Limits on control and dependant variables

$$P_{gi}^{min} \leq P_{gi} \leq P_{gi}^{max} \quad \text{for } i = 1 \text{ to NG} \tag{6}$$

$$Q_{gi}^{min} \leq Q_{gi} \leq Q_{gi}^{max} \quad \text{for } i = 1 \text{ to NG} \tag{7}$$

$$V_i^{min} \leq V_i \leq V_i^{max} \quad \text{for } i = 1 \text{ to NB} \tag{8}$$

$$T_i^{min} \leq T_i \leq T_i^{max} \quad \text{for } i = 1 \text{ to NT} \tag{9}$$

$$MVA_i \leq MVA_i^{max} \quad \text{for } i = 1 \text{ to Nbr} \tag{10}$$

where

P_{gi}, Q_{gi}	ith generator real and reactive power generation
P_D, Q_D	total real and reactive power demand
P_L, Q_L	total real and reactive power loss
V_i	ith bus voltage magnitude
T_i	ith transformer tap position
MVA_i	ith transmission line MVA flow
NG	number of generators
NB	number of bus
NT	number of transformer
Nbr	number of branch/transmission line

In the objective functions constants used α, β, γ are fuel cost coefficients and ζ, λ are valve point effect coefficients of the generator. Units of the coefficients α, β, γ, ζ and λ are \$/hr, \$/Mwhr, \$/Mw^{2}hr, \$/hr and \$/Mwhr, respectively. Constants a, b, c, d, and e are emission coefficients of the generator; units of these constants are ton/hr, ton/Mwhr, ton/Mw^{2}hr, ton/hr and ton/Mwhr, respectively.

3 Hybrid RGA-DE Algorithm to Solve CEED

Anumber of control variables are considered as genes and set of these genes forms a chromosome. The values of genes are taken as real value as used in the actual problem. Population is the group of these chromosomes. From this population some chromosomes are selected to mating pool based on their fitness value. In the implemented algorithm roulette wheel selection method is used to select mating pool chromosomes. Cross over is the operation of swapping better genes among the chromosomes in the mating pool. Single-point cross over method is used in this algorithm for this *cross over constant value 0.7* is considered. These selection and cross over process are taken from RGA. Since DE has better mutation operation in this hybrid RGA-DE algorithm [9], DE mutation operation is used. For the mutation *scaling factor 0.7* is considered. Mutation keeps the search robust and explores new areas in the search domain. Hybrid RGA-DE improves the solution iteration by iteration and the iteration has to be stopped by stopping criteria. In this work maximum number of *200 iterations* is considered as stopping criteria.

3.1 Algorithm

The procedure for Hybrid RGA-DE to solve CEED is given below

Step 1: 15 Control variables are selected and coded as genes
Step 2: Create initial population of 80 chromosomes
Step 3: Find fitness of chromosomes in the population
Step 4: Select chromosome using roulette wheel selection for mating
Step 5: Perform single-point crossover operation as like RGA
Step 6: Perform mutation operation as like DE
Step 7: Generate a new population for next generation
Step 8: Repeat Step 4 to Step 7 till stopping criterion is satisfied
Step 9: Print the optimal result after stopping criterion is satisfied

4 Case Study

To evaluate the performance of developed algorithms, benchmark test case IEEE 30 bus system is considered [7]. The system has 6 generators include slack bus, hence, 5-real power generation, 6 generator bus voltage magnitude, and 4 transformer tap position are considered as control variables, and hence these *15 control variables are genes* in a chromosome. *80 chromosomes* are considered as the population size. This system has 41 transmission lines whose MVA limit provides line flow limit constraint. Base MVA of the system is 100MVA. For CEED result comparison with other literature the following cost and emission coefficients given in Tables 1 and 2 are used.

5 Numerical Results

This section provides the numerical result of CEED using hybrid intelligent algorithms RGA-DE. Table 3 shows developed hybrid RGA-DE gives minimum generating cost and emission as compared to other algorithms in the literature.

Table 1 Generator limits and cost coefficients for CEED

Gen. no	P limit (MW)		Q limit (Mvar)		Cost coefficients				
	Min	Max	Min	Max	α $/hr)	β ($/Mwhr)	γ ($/Mw2hr)	ζ ($/hr)	λ ($/Mwhr)
1	5	50	−40	50	10	200	100	15	6.283
2	5	60	−40	50	10	150	120	10	8.976
3	5	100	−40	40	20	180	40	10	14.784
4	5	120	−10	40	10	100	60	5	20.944
5	5	100	−6	24	20	180	40	5	25.133
6	5	60	−6	24	10	150	100	5	18.480

Table 2 Generator emission coefficients for CEED

Gen. no	Emission coefficients				
	a (ton/hr)	b (ton/Mwhr)	c (ton/Mw2hr)	d (ton/hr)	E (ton/Mwhr)
1	4.091	−5.554	6.490	2e-4	2.857
2	2.543	−6.047	5.638	5e-4	3.333
3	4.258	−5.094	4.586	1e-6	8.000
4	5.426	−3.550	3.380	2e-3	2.000
5	4.258	−5.094	4.586	1e-6	8.000
6	6.131	−5.555	5.151	1e-5	6.667

Table 3 Comparison of hybrid RGA-DE CEED solution

Generation (MW)	Without valve point loading				With valve point loading	
	SPEA [2]	DE [8]	PSO [15]	RGA-DE	PSO [15]	RGA-DE
PG1	29.96	25.2758	17.613	21.8777	14.089	5.7696
PG2	44.74	40.6968	28.188	35.5781	34.415	40.5630
PG3	73.27	56.1153	54.079	58.0203	67.558	47.9630
PG4	72.84	66.9946	76.963	74.0125	83.971	79.9877
PG5	11.97	53.6240	65.019	54.4699	49.043	55.1418
PG6	53.64	43.6732	44.569	41.5252	39.797	56.7346
Fuel Cost $/hr	629.394	617.9962	612.35	**611.325**	639.6507	**617.862**
Emission ton/hr	0.21043	0.1999	0.20742	**0.20459**	0.21205	**0.21173**

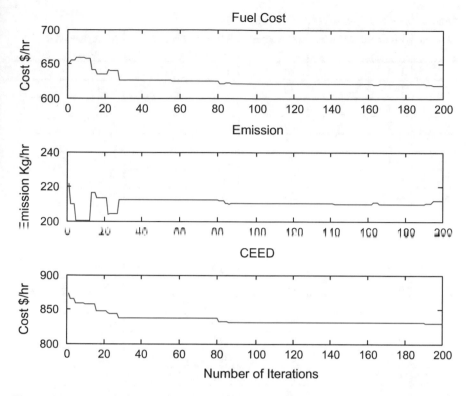

Fig. 1 Convergence curve with valve point loading

Convergence curve of CEED without valve point and without checking line flow limits are given in Table 3. Other limits of real and reactive power generation, voltages in all buses and transformer tap positions are within the limits.

Convergence curve for the same system with the same constraints but valve point loading is considered as shown in Fig. 1. In this approach, the line flow limits are not forced only the generating cost, emission and CEED are minimized. This optimization minimizes generating cost and emission but violates the line flow limit in the line number 11 which is connected between bus 6 and 10 as given in Table 4. MVA rating of the line is 32 MVA but actual MVA flow is 40.4 MVA violates the limits, and hence the generation pattern has to be rescheduled to avoid the violation. Reschedule of generation pattern removes the violation and the system will operate in secured condition. After rescheduling the MVA flow in violated line is reduced to 27.3 MVA which is less than the line rating 32 MVA. The generation reschedule is given in Table 5. In this case, the generating fuel cost is increased as compared to violated generation schedule but it is practical implementable generation schedule.

Table 4 MVA flow violation with valve point loading

line no	Line between buses	MVA rating	MVA flow before constraint	MVA flow after constraint
1	1–2	130	17.3	5.31
2	1–3	130	5.21	4.29
3	2–4	65	1.96	3.7
4	2–5	130	29.5	20
5	2–6	65	2.84	2.85
6	3–4	130	2.61	1.54
7	4–6	90	10.1	20.6
8	4–12	65	11.2	19.6
9	5–7	70	23.9	12.1
10	6–7	130	42.3	30
11	6–8	32	40.4	27.3
12	6–9	65	25.4	24.7
13	6–10	32	3.26	4.96
14	6–28	32	3.4	13
15	8–28	32	9.83	8.46
16	9–11	65	57.1	62.1
17	9–10	65	41	45.5
18	10–20	32	9.95	12.3
19	10–17	32	9.2	13.5
20	10–21	32	18.4	16.7
21	10–22	32	8.93	8.18
22	12–13	65	59.4	46.9
23	12–14	32	8.38	7.04
24	12–15	32	21	17
25	12–16	32	9.26	6.39
26	14–15	16	2.28	2.08
27	15–18	16	6.7	4.7
28	15–23	16	8.39	8.74
29	16–17	16	6.23	5.66
30	18–19	16	3.6	2.4
31	19–20	32	7.51	9.76
32	21–22	32	6.02	9.75
33	22–24	16	9.84	12.8
34	23–24	16	6.62	9.39
35	24–25	16	6.4	6.27

(continued)

Table 4 (continued)

line no	Line between buses	MVA rating	MVA flow before constraint	MVA flow after constraint
36	25–26	16	4.27	4.26
37	25–27	16	2.28	9.77
38	27–29	16	6.43	6.4
39	27–30	16	7.3	7.27
40	28–27	65	12.7	22.5
41	29–30	16	3.76	3.75

Table 5 Real power generations reschedule

Generation (MW)	Before rescheduling	After rescheduling
PG1	5.7696	0.0023
PG2	40.5630	40.4567
PG3	47.9630	67.8302
PG4	79.9877	65.4647
PG5	55.1418	60.8947
PG6	56.7346	42.1097
Fuel cost $/hr	617.862	629.839
Emission ton/hr	0.21173	0.20699

6   Conclusion

Hybrid intelligent algorithms RGA-DE is developed to provide better optimal solution to CEED problem. IEEE 30 bus system is used to evaluate the performance of the hybrid RGA-DE algorithm; the results are compared with other algorithms in the literature. In addition to CEED problem, valve point loading and constraints on MVA line flow are imposed and their results are presented in this paper. This developed algorithm provides better optimal solution as compared to other algorithms. It is evident that whenever more constraints are imposed the minimum value objective function is increased based on the complexity.

In the research work, all generators are considered as thermal power generator; it may extend to cogeneration thermal power plant and wind power generators for further research.

References

1. Attia, A.-F., Al-Turki, Y.A., Abusorrah, A.M.: Optimal power flow using adapted genetic algorithm with adjusting population size. Electr. Power Compon. Syst. **40**(11), 1285–1299 (2012)
2. Storn, R., Price, K.: Differential evolution—a simple and efficient heuristic for global optimization over continuous spaces. J. Glob. Optim. **11**, 341–359 (1997)
3. Alsac, O., Stott, B.: Optimal load flow with steady state security. IEEE Trans. Power Apparat. Syst. **PAS-93**, 745–751 (1973)
4. Venkatesh, P., Gnanadass, R., Padhy, N.P.: Comparison and application of evolutionary programming techniques to combined economic emission dispatch with line flow constraints. IEEE Trans. Power Syst. **18**(2), 688–697 (2003)
5. Hemamalini, S., Simon, S.P.: Emission constrained economic dispatch with valve-point effect using particle swarm optimization. In: IEEE Conference—TENCON 2008, pp. 1–6 (2008)
6. Abido, M.A.: Multi-objective optimal VAR dispatch using strength pareto evolutionary algorithm. In: IEEE Conference Proceeding, July 16–21, pp. 730–736 (2006)
7. Abido, M.A.: Multiobjective particle swarm for environmental economic dispatch problem. In: 8th International Power Engineering Conference, pp. 1385–1390 (2007)
8. Ajay-D-Vimal Raj, et al.: Optimal power flow solution for combined economic emission dispatch problem using particle swarm optimization technique. J. Electr. Syst. 3–1, pp. 13–25 (2007)
9. Bhattacharya, A., Chattopadhyay, P.K.: Hybrid differential evolution with biogeography-based optimization for solution of economic load dispatch. IEEE Trans. Power Syst. **25**(4), pp. 1955–1964 (2010)
10. Dommel, H.W., Tinney, W.F.: Optimal power flow solutions. IEEE Trans. Power Appar. Syst. **PAS-87**(10), pp. 1866–1876 (1968)
11. Gopalakrishnan, R., Krishnan, A.: Intelligence technique to solve combined economic and emission dispatch. In: Proceedings of the International Conference on Pattern Recognition, Informatics and Medical Engineering, pp. 1–6 (2012)
12. Liang, H., Liu, Y., Li, F., Shen, Y.: A multiobjective hybrid bat algorithm for combined economic/emission dispatch. Int. J. Electr. Power Energy Syst. **101**, 103–115 (2018)

Performance Analysis of ST-Type Module for Nearest Level Control Scheme Based Asymmetric 17-Level Multilevel Inverter Fed Single-Phase Induction Motor

P. Ramesh, M. Hari, M. Mohamed Iqbal, K. Chitra and D. Kodandapani

Abstract A novel design of ST (Square type)-type module for asymmetric multilevel inverter (MLI) fed induction motor is approached in this research. The two back-to-back ST-type inverters and few different switching modules produce 17-level output by 12 switches with four unequal DC sources and filter. The advantages of this proposed technique are first, output voltage of the inverter can be is increased by cascading the inverter module. The main aspect of proposed inverter is natural conception of both +ve and −ve voltages without any auxiliary circuit. With the Nearest Level Control (NLC) switching strategy, good quality of output voltage with lower total harmonic contents can be accomplished. The response of the proposed module is analyzed by using MATLAB/SIMULINK. The simulation results are validated with prototype experimental results.

Keywords ST-Type back-to-back inverters · Asymmetric multilevel inverter · Nearest level control

P. Ramesh (✉) · M. Hari · K. Chitra · D. Kodandapani
Department of Electrical and Electronics Engineering, CMR Institute of Technology, Bengaluru, India
e-mail: ramesh8889@gmail.com

M. Hari
e-mail: harimaaperuman@gmail.com

K. Chitra
e-mail: chitrapeee@gmail.com

D. Kodandapani
e-mail: kodandapani.depa@gmail.com

M. Mohamed Iqbal
Department of Electrical and Electronics Engineering, PSG Institute of Technology and Applied Research, Coimbatore, Tamilnadu, India
e-mail: mohdiq.m@gmail.com

© Springer Nature Singapore Pte Ltd. 2020
V. Bhateja et al. (eds.), *Intelligent Computing and Communication*,
Advances in Intelligent Systems and Computing 1034,
https://doi.org/10.1007/978-981-15-1084-7_76

779

1 Introduction

Various industrial applications have started to require higher power rating machines [1] and many utility applications required medium voltage and high voltage levels. It is difficult to associate just single power semiconductor switch legitimately for the medium voltage drives. As a result, an alternative solution for large and medium voltage situations, a structure of multilevel inverters is preferred [2]. The multilevel inverters are highly suitable for the usage of renewable energy sources [3]. The vast majority of renewable power sources like photovoltaic, wind, and fuel cells are effectively interfaced with MLI topology for high-power applications [4, 5]. With the multiple dc sources and proper switching combinations of power switches, the high-power output voltage is achieved. Power semiconductor devices voltage ratings depend on only the total peak value of voltage source that is connected to the device. Multilevel inverters are categorized as diode clamped, flying capacitor, and H-bridge cascaded types. In [6], T-source with most extreme steady boosted PWM control was conferred. In [7], an Envelop Type (E-Type) asymmetric multilevel inverter is proposed and PWM (Pulse Width Modulation) scheme is used. These are used in high voltages and high-power applications and have reduced Total Harmonic Distortion (THD). In [8] a scheme for nearest level control (NLC), sub-module is proposed. The value of THD is reduced due to large number of steps present in the output voltage. In [9], the system used for 1-phase 5-level cascaded H–bridge multilevel inverter, constant K- filter used at a different cutoff frequency, output voltage, and current harmonic distortion is analyzed at different conduction angles. In [10], various topology comparisons of asymmetric CHB, SHE, NLC, PWM, multi-carrier method are presented and it produces excellent output voltage quality and lower THD using NLC scheme.

The branch current replication based energy balancing technique for the modular multilevel matrix converter (M3C) is proposed [11]. A new topology has presented a Simplified NLC based voltage balancing method [12].

Modular MLI converter is used under normal and emergency conditions. LSPWM balancing algorithm is used in this module [13]. NLC scheme is used for closed-loop and industrial applications. In [14], proposed section is a new module multilevel inverter using NLC technique by reduced power switches. In [15], a Level Doubling Network (LDN) topology produces a double number of output voltage levels and it produces self adjusting during positive and negative cycles with no closed loop control/algorithm. In [16], a submodule hybrid cascaded multilevel inverter (HCMLI) with gate drive requirement switches are reduced and efficiency will be increased. It is used for lower voltage applications. In [17], the harmonic reduction techniques for 2-level inverter fed induction motor were presented. The PWM technique is used for the performance of multilevel inverter are presented [18, 19].

The objective of this paper is to generate 17-level output voltage with 12 switches and also to get good quality of output voltage of the MLI with reduced number of switches. This paper also describes a novel design of square T-type multilevel inverter fed induction motor using NLC scheme to achieve maximum voltage levels and improve economic implementation and high quality of power [20, 21].

2 Proposed System

2.1 Block Diagram of NLC Controller

Figure 1 shows that asymmetrical multilevel inverter can be modulated with nearest level control (NLC). In this method, the voltage levels are chosen nearest to the reference voltage [12]. The nearest level controller gives an excellent output voltage quality, and it produces an inverse relationship between frequency and distributed power from each unit.

NLC based inverter is used to produce higher number of levels and to simplify the calculation steps.

Figure 2 shows a point of reference voltage (V_{ref}) and then round off to the nearest voltage level (V_{aN}). The sampling is rehashed for each sample time (T_s) [20].

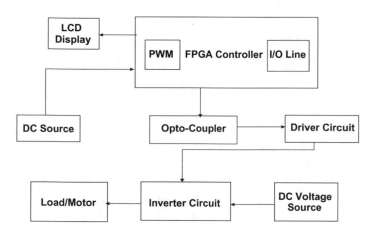

Fig. 1 Block diagram of NLC controller

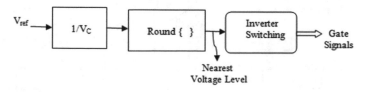

Fig. 2 Control diagram of NLC

The proposed module produces 17-levels such as eight positive, eight negative and one zero level without any need of additional circuit for creation of negative voltage levels. The NLC uses an excellent output voltage quality, and it produces an inverse relationship between frequency and distributed power from each unit.

2.2 Circuit Configuration

Figure 3 shows the proposed circuit diagram for 17-level ST–type inverter. The circuit configuration of proposed 17-level inverter consists of four unequal DC sources (two 10 V and two 30 V) and less number of power electronic switches (12 switches), which is parallel to filter and single-phase induction motor. Surrounding switches (S_1–S_6) are single-directional switches and middle switches (S_7–S_9)

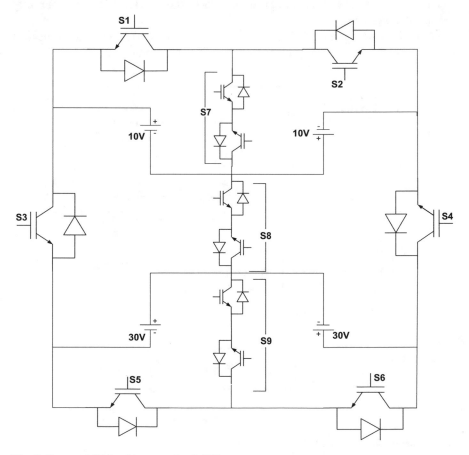

Fig. 3 Proposed 17-level inverter circuit [20]

are two-directional switches. The different modes of operations are used to define the current flow in switching components to obtain the various output voltage levels across "*A*" and "*B*". It consists of two levels of output voltages, positive and negative level, each level consist of eight modes of operation [20]. The output voltage of the inverter is fed to single-phase induction motor and analyzed the response with and without filter.

3 Simulation Results of NLC Based MLI

The proposed 17-level inverter is simulated with filter and without filter. In the existing system, the THD value of voltage is 4.87% and the output voltage is 81.27 V at the fundamental frequency of 50 Hz. In the existing system, the THD value of current is 2.09% and the output current is 8.11 A at the fundamental frequency of 50 Hz [20].

Figures 4 and 5 show the simulation output of proposed 17-level voltage and current. An imperative issue in MLI configuration is to generate sinusoidal output voltage waveform and to take out lower order harmonics as did in [4, 5].

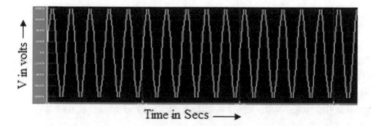

Fig. 4 Output voltage waveform of 17-level inverter

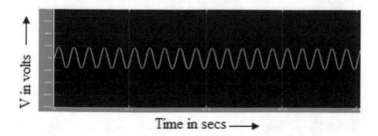

Fig. 5 Output current waveform of 17-level inverter

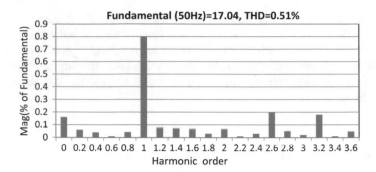

Fig. 6 FFT analysis of output current without filter

3.1 NLC Based MLI Without Filter

Figure 6 shows that the FFT harmonics spectrum is sampled with the harmonics up to 20th order. The measured output current THD is 0.51%. From the results, it's clear that proposed topology has slightly higher THD compared to topology with filter.

3.2 NLC Based MLI with Filter

Figure 7 shows the output of FFT analysis of the proposed system current with filter. From FFT analysis, the proposed system current THD without filter is 0.51% and with filter is 0.3% for the fundamental frequency of 50 Hz. From the result, it's clear that proposed topology with filter has better THD compared to topology without filter.

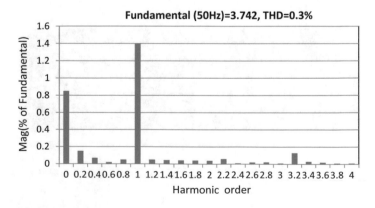

Fig. 7 FFT analysis of output current with filter

4 Experimental Setup and Analysis

The 17-level ST-type multilevel inverter fed 300 W single-phase induction motor is shown in Fig. 8. The experimental setup includes inverter circuit with 12 numbers of MOSFET (IRF540), FPGA, power supply, filter, and single-phase induction motor. FPGA controller generates sinusoidal PWM pulses. Switching pulse to the multilevel, it is generated from FPGA Controller [22, 23]. DC source is applied to inverter circuit from single-phase AC supply. The opto-coupler output signal is inverted from original PWM input signal. Load is connected across the power circuit and the output voltage and current are measured using DSO. Figures 9 and 10 show the hardware output waveform of proposed system current and voltage.

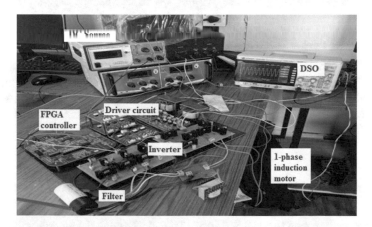

Fig. 8 Experimental setup of NLC based MLI

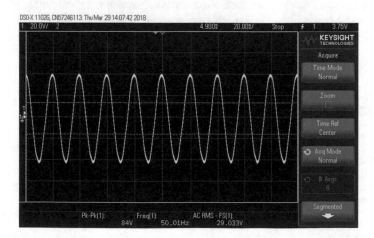

Fig. 9 Output current waveform of NLC based MLI

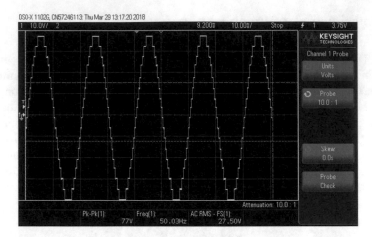

Fig. 10 Output voltage waveform of NLC based MLI

The ST-type module for NLC scheme based 17-level MLI proposed in this paper with a reduced number of switches has resulted in reduction of THD and switching losses. Based on the analysis, it is witnessed that the circuit is extremely simple compared to that of classical topologies.

5 Conclusion

This paper has presented a novel design of 17-level ST-type MLI module with reduced components using NLC Scheme. It is clear from the analysis that the proposed module can be utilized in high-voltage controlled drive applications by cascading principle. The proposed system has low switching stress on the switches which will improve the reliability of the inverter. The THD in current is very low as 0.3% with a simple LC-filter.

References

1. Rodriguez, J., Lai, S., Peng, F.Z.: Multilevel inverters: a survey of topologies, controls, and applications. J. IEEE Trans. Ind. Electron. **49**(4), 724–738 (2002)
2. Abu-Rub, H., Holtz, J., Rodriguez, J., Baoming, G.: Medium voltage multilevel converters—state of the art, challenges, and requirements in industrial applications. IEEE Trans. Ind. Electron. **57**(8), 2581–2596 (2010)
3. Deepika, G., Elakkiya, M., Mohamed Iqbal, M.: Harmonics reduction using multilevel based shunt active filter with SMES. Int. J. Eng. Tech. **9**(2), 1001–1011 (2017)
4. Ramesh, P., Sharmeela, C.: FLC based closed loop control of MLI fed induction motor drive. Int. J. Comput. Theor. Nanosci. **14**(3), 1259–1264 (2017)

5. Hari, M., Verma, A., Halakurki, R.R., Ravela, R.R., Kumar, P.: A dynamic analysis of SVM based three-level NPC for a 3-phase induction motor. In: IEEE Conference on Power Energy, Environment and Intelligent Control (PEEIC), pp. 162–167 (2018)
6. Chitra, K., Jeevanandham, A.: Three-phase on-line UPS by implementing T-source inverter with maximum constant boost PWM control. Iran. J. Sci. Technol. **39**(E2), 209–215; Trans. Electr. Eng. Springer, 2364–1827 (2015)
7. Samadaei, E., Gholamian, S.A., Sheikholeslami, A., Adabi, J.: An envelope type (E-Type) module: asymmetric multilevel inverters with reduced components. IEEE Trans. Ind. Electron. **63**(11), 7148–7156 (2016)
8. Park, Y.H., Kim, D.H., Kim, J.H., Han, B.M.: A new scheme for nearest level control with average switching frequency reduction for modular multilevel converters. J. Power Electron. **16**(2), 522–531 (2016)
9. Kumar, R.J., Choubey, A.: Harmonic elimination in multilevel inverter. Int. J. Rec. Res. in Electr. Electron. Eng. (IJRREEE) **4**(3), 11–15 (2017)
10. Soltani, M., Hoseinkhanzade, M.: Comparison of switching methods for asymmetric cascaded H-bridge multilevel inverter. IOSR J. Eng. (IOSRJEN) **05**(08), 52–60 (2015)
11. Fan, B., Wang, K., Wheeler, P., Gu, C., Li, Y.: A branch current reallocation based energy balancing strategy for the modular multilevel matrix converter operating around equal frequency. IEEE Trans. Power Electron. **33**(2), 1105–1117 (2018)
12. Meshram, P.M., Borghate, V.B.: A simplified nearest level control (NLC) voltage balancing method for modular multilevel converter (MMC). IEEE Trans. Power Electron. **30**(1), 450–462 (2015)
13. Thirumurugan, P., Vinothini, D., Arockia Edwin Xavier, S.: New model multilevel inverter using nearest level control technique by reduced power switches. Int. J. Inn. Res. Sci. Eng. Technol. **5**(4) (2016)
14. Chitra, K., Jeevanandham, A.: Design and implementation of maximum constant boost switched inductor z-source inverter for three-phase on-line uninterrupted power supply. J. COMPEL: Int. J. Comp. Math. Electr. Electron. Eng. **34**(4), 1101–1121 (2015)
15. Lee, S.S., Sidorov, M., Lim, C.S., Idris, N.R.N., Heng, Y.E.: Hybrid cascaded multilevel inverter with improved symmetrical 4-level submodule. IEEE Trans. Power Electron. **33**(2), 932–935 (2018)
16. Pramanick, S., Karthik, R.S., Azeez, N.A., Gopakumar, K., Williamson, S.S., Rajashekara, K.S.: A harmonic suppression scheme for full speed range of a two-level inverter fed induction motor drive using switched capacitive filter. IEEE Trans. Power Electron. **32**(3), 2064–2071 (2017)
17. Ramesh, P., Sharmeela, C., Bhavanibai, T.: Reduction of harmonics in an induction motor using THPWM technique. Int. J. App. Eng. Res. **10**(30), 22768–22773 (2015)
18. Kumar, C.G.A., Babu, V.A., Rao, K.S., Ali, M.H., Suresh, K.S., Hussain, M.M.: Performance and analysis of hybrid multilevel inverter fed induction motor drive. Int. J. Mod. Eng. Res. **2**(2) (2012)
19. Wang, L., Wu, Q.H., Tang, W.: Novel cascaded switched-diode multilevel inverter for renewable energy integration. IEEE Trans. Energy Convers. **32**(4), 1574–1582 (2017)
20. Samadaei, E., Sheikholeslami, A., Gholamian, S.A., Adabi, J.: A square T-type (ST-Type) module for asymmetrical multilevel inverters. IEEE Trans. Power Electron. **33**(2), 987–996 (2018)
21. Khan, A.A., Cha, H., Ahmed, H.F., Kim, J., Cho, J.: A highly reliable and high-efficiency quasi single-stage buck–boost inverter. IEEE Trans. Power Electron. **32**(6), 4185–4198 (2017)
22. Tzou, Y., Hsu, H.: FPGA realization of space-vector PWM control IC for three-phase PWM inverters. IEEE Trans. Power Electron. **12**(6), 953–963 (1997)
23. Velrajkumar, P., Senthilpari, C., Ramanamurthy, G., Wong, E.K.: Proposed adder and modified LUT bit parallel unrolled CORDIC circuit for an application in mobile robots. Asian J. Sci. Res. **6**(4), 666–678 (2013)

Facial Landmark Localization Using an Ensemble of Regression Trees on PC Game Controlling

Naren Sai Krishna Kandimalla, Soumya Ranjan Mishra, Goutam sanyal and Anirban Sarkar

Abstract Traditional PC gaming system has remained unmodified over the years. While GPU's and high graphics machines have taken the gaming industry to the next level, there is a need to physically improve the gamers' experience. Facial landmarks are the localization of certain points on the face. This paper proposes a novel model to control the computer games. It identifies the face and facial landmarks of the player sitting in front of the PC. To do this an ensemble regression trees are used with cascaded regression model. It performs various computations on the extracted landmarks of the input face to estimate the action to be performed. Then, it gives input to the game accordingly. Likewise, many of the keyboard and mouse controls can be mapped with facial movements. This paper gives an overview of head, eyes, and mouth mapping with the game controls. It eliminates the use of a keyboard or any such hardware device, thus creating a realistic gaming experience. Results show that with a continuous frame by frame input of a single face, the proposed model takes 209 milli-sec in poor lighting conditions and 12 milli-sec in good lighting conditions to identify and perform computations on the facial movements. Moreover, the observed gameplay of the final model is very smooth and robust.

Keywords Facial landmark detection · PC game controlling · Regression trees

1   Introduction

This paper is the detailed description of face-based computer game controlling model. The main objective of this paper is to present how a game can be controlled with a face using a webcam. Computer vision enables computers to identify and process images

S. R. Mishra (✉) · G. sanyal · A. Sarkar
Department of Computer Science and Engineering, NIT Durgapur, Durgapur, India
e-mail: soumyaranjanmishra.in@gmail.com

N. S. K. Kandimalla
Department of Computer Science and Engineering, ANITS Vishakhapatnam, Bheemunipatnam, India

© Springer Nature Singapore Pte Ltd. 2020   789
V. Bhateja et al. (eds.), *Intelligent Computing and Communication*,
Advances in Intelligent Systems and Computing 1034,
https://doi.org/10.1007/978-981-15-1084-7_77

in the same way that human vision does. Open Source Computer Vision Library is a library which deals with real-time computer vision to process digital data. We have implemented our model in OpenCV with python programming language. The facial landmark detection algorithms are grouped into three categories:Holistic methods are models that are built explicitly to represent the global facial shape and are prone to local corruption. The Constrained Local Models (CLM) are those which uses global shape model but built on local appearance model.Their inefficiency lies in the fact that the global appearance cannot be easily implemented into a local search framework and if implemented, it becomes expensive. The third is regression-based methods which capture facial shape implicitly. It is more efficient and robust framework because global appearance is implemented at the start itself.

2   Related Work

2.1   Active Appearance Model

This (AAM) [1, 2] uses principal component analysis (PCA) to build the global facial shape model and holistic facial appearance model. It fits less number of coefficients to the facial images thereby, controlling facial appearance and shape variations as well. The limitation with single AAM model which is linearity, i.e., it assumes the face to be linear. To overcome this, we go to ensemble models to improve the performance.

2.2   Local Appearance Model

This is a CLM model which gives a probability $p(x_d \mid I; \theta_d)$ that indicates that the specific pixel location x_d of the landmark d by considering the local appearance information around x_d of image I. It can be divided into two categories: classifier-based LAM and the regression-based LAM [3].

2.3   Classifier-Based Local Appearance Model

The classifier-based LAM binary classifier trained to distinguish the positive and the negative patches at the ground truth locations and far away from the ground truth locations, respectively. The classifier applied to different pixel locations during detection to generate the probabilities $p(x_d \mid I; \theta_d)$ through voting.

2.4 Regression-Based Local Appearance Model

This model predicts the displacement vector $\triangle x_d^* = x_d^* - x$, which is the difference between location of a pixel x of image I and the location of the ground truth landmark x_d^* from the local appearance information around x using the regression models. The regression model is used to patches at different locations x in a region of interest to predict $\triangle x_d$, which can be added to the current location to obtained x_d. Multiple patches prediction is combined to calculate the final prediction of the probability through voting. However, the accuracy-robustness trade-off is a major issue in local appearance model.

2.5 Cascaded Regression Methods

The cascaded regression method starts its prediction with an initial assumption of facial landmark locations (e.g., mean face), and then, in an iterative process (stages), different regression functions are learned by the regression model at each stage and are reflected to the next stage to update the shapes accordingly. During testing phase, the trained regression models are sequentially applied to update the shapes through iterations. For example, in [4], the shape-indexed pixel intensity features are used. It is the pixel intensity difference between pairs of pixels whose locations are defined by their relative positions to the current shape. In [5], the discriminative binary features are learned independently by the regression forests for each landmark. Then, concatenation of the binary features from all landmarks followed by a linear regression function is used to learn the joint mapping from appearance to the global shape. In [6], an ensemble of regression trees is used as regression models for face alignment in one-millisecond which is extremely fast. However, for cascaded regression methods, it is difficult to generate the initial landmark locations. One choice is to use the mean face, which is suboptimal in the case of images with large head poses. There are some hybrid methods that can generate the initial landmarks for cascaded regression methods using direct regression methods and it can solve the problem to some extent. These algorithms are efficient to some extent but there still exist some problems in facial landmarks localization like extreme variations in head poses, illumination, occlusion, etc.After localizing the facial landmarks, the task is to identify the appropriate geometric positions of the facial parts obtained. Then, we identify the action to be performed and map it with the keyboard controls of the game in hexadecimal form by applying a transitive closure and finally, send the input (command) to the game.

Algorithm 1: Cascaded regression detection

Initialize mean facial landmarks x_0;
for $t=1,2, ..., T$ **do**
  Update landmark locations, given image and its current landmark location.;
  $f_t : I, x_{t-1 \to \Delta x_t}$;
  $x_t = x_{t-1} + \Delta x_t$;
  Output the estimated landmark locations x_T.
end

---

3 Proposed Model

The proposed model works on facial landmarks localization. The person sits in front of a PC, we continuously capture his face through a webcam. Each frame thus obtained is then processed as per the algorithm suggested in Algorithm-1. Here, we must ensure that only a single person is present within the frame to avoid ambiguity.

3.1 Facial Landmarks Localization

First, the image is preprocessed, and Viola-Jones Algorithm is applied to extract the region of interest (Face Region). Then we deploy an ensemble of regression trees to predict the face's landmark positions from a sparse subset of intensity values [6]. Based on the trained model (iBUG 300-W) [7–9], 68-facial landmarks are obtained on the individual face parts that can be located. We then check for the appropriate face gesture from the methods provided below and decide the action to be performed. Finally, we instruct the game to perform the action accordingly.The proposed model works in the following steps:

Algorithm 2: Proposed Algorithm

Take frame by frame input from the webcam.
if *Each frame has only one face* **then**
  Resize the image to 400x400 pixels.
  Convert the image to grayscale.
  Apply Viola-Jones Algorithm (Face Detection).
  Extract Region of Interest (Face Region).
  Plot the facial landmarks onto the ROI.
  Obtain the coordinates of eye and mouth from the image;
  Identify the appropriate positions.
  Decide the command to be given to the game.
  Give the input to the game.
end
else
  | Do nothing
end

3.2 Identifying Head Tilt Position

The direction left, or right is determined by the direction of tilting of head with respect to coordinate axis. To do this, Locate the region of the eyes from the indexes Then, compute the mean of those points to get the centroid for each eye.

Now we get the coordinates of centroid's of each eye say (x_1, y_1) and (x_2, y_2).

Draw a line between $(x_1, y_1))$ and (x_2, y_2).

Find the slope of the line $\frac{\partial y}{\partial x} = \frac{y2-y1}{x2-x1}$ (say m)

Then, find $\theta = \tan^{-1}(m)$

Based on the angle calculated above, we decide a threshold angle (θ_1) on each side of the central vertical line to determine the direction: (Fig. 1).

Note: In Fig. 1, Clockwise Rotation (−ve) is Right direction and Counter-Clockwise Rotation (+ve) is Left Direction. So, the 2D space on either side is positive and negative.

3.3 Identifying Mouth Open/Close

To identify the mouth closing and opening gestures, we first obtain the coordinates of the upper lip and the lower lip Fig. 2a, b. Then we compute the mean of those coordinates to get the centroid for each lip. We then calculate the Euclidean distance between the two centroids say d_{actual} Fig. 2c. A threshold value $d_{threshold}$ to decide whether the mouth open or close.

(a)　　　　　　　　　**(b)**　　　　　　　　　**(c)**

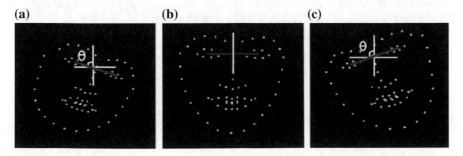

Fig. 1 Comparison of various head tilting movements **a** left tilt **b** straight **c** right tilt. Right → $\theta < -\theta_1$, Straight → $\theta_1 < \theta < -\theta_1$, Left → $\theta > \theta_1$

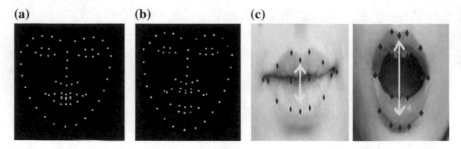

Fig. 2 **a** Mouth close **b** mouth open **c** mouth Euclidean distance (close vs. open)

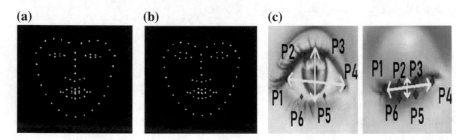

Fig. 3 **a** Eye close **b** eye open **c** eye Euclidean distance (close vs. open)

Mouth Open: $d_{actual} > d_{threshold}$

Mouth Close: $d_{actual} < d_{threshold}$

3.4 Identifying Eye Open/Close

We identify the eye close and open gestures Fig. 3a, b as mentioned in [10]. Here, we label the eye landmarks in sequence (p_1, p_2, p_3, p_4, p_5, p_6) as shown in Fig. 3c. Then, compute the eye aspect ratio (EAR). It is given by:

$$EAR = \frac{|P_2 - P_6| + |P_3 - P_5|}{2|P_1 - P_4|}$$

3.5 Giving Input to the Game

We identify the most appropriate facial movement by the above proposed methods and decide on the action to be performed shown in Table 1. Then we map the action to

Table 1 Facial movement and action to be performed

Facial movement	Action to be performed
Head tiltright	Move right
Head tiltleft	Move left
Eyeclose	Stop
Mouth close	Move front
Mouth open	Move backwards

Table 2 Mapping the action to be performed with the keys of the keyboard in hexadecimal form

Action to be performed	Hexadecimal key (scan code)
Move right	0 × 20
Move left	0 × 1E
Stop	Release all keys
Move front	0 × 11
Move backwards	0 × 1F

be performed with the keys of the keyboard in hexadecimal form. Thus, a transitive closure can be applied between these two tables and the hexadecimal key (scan code) for each key is given as inputs to the game accordingly shown in Table 2. These controls are subjective to the user's interest. The game takes in these inputs and performs the desired operation on the game.

4 Results and Analysis

We have implemented the proposed model on 'Grand Theft Auto 5' game and obtained the following results. With a PC having the configuration of 16GB LP-DDR3 Ram, i7-8550U processor, and 8 GB Intel UHD 620 graphics under 1280x1024 resolution, we have observed an average frame rate of the game as 29 fps and that of camera input is 28 fps. The graph indicating the frames/second observation for 20 s is shown in Fig. 4. Under good lighting conditions, the model is taking 12 milliseconds to perform the computations involved in the proposed model and under poor lighting conditions, the model took 209 milliseconds.

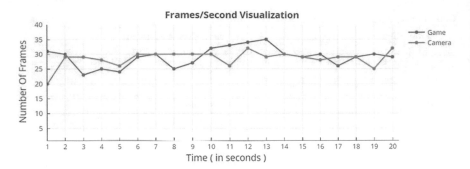

Fig. 4 Frames/second observation for 20 s

5   Conclusion

The proposed model removes the barrier between the game and the gamer. As all the game controls can be mapped with facial movements, we can develop a new robust model which eliminates the use of a keyboard or a mouse. Thus, our proposed model is a reliable and accurate mechanism that can replace all the old techniques of controlling the game in PC. The lag observed while giving inputs is very less 1.017 % which can be further minimized when run on a GPU. This model can further be extended to other gaming consoles and devices with a higher refresh rates than observed which rejoices the user's experience.

References

1. Edwards, G.J., Taylor, C.J., Cootes, T.F.: Interpreting face images using active appearance models. In: IEEE International Conference on Face and Gesture Recognition, pp. 300–305. IEEE Computer Society (1998)
2. Cootes, T.F., Edwards, G.J., Taylor, C.J.: Active appearance models. IEEE Trans. Pattern Anal. Mach. Intell. **23**(6), 681–685 (2001)
3. Wu, Y., Ji, Q.: Facial landmark detection: a literature survey. Int. J. Comput. Vis. 1–28 (2018). Springer, New York LLC
4. Cao, X., Wei, Y., Wen, F., Sun, J.: Face alignment by explicit shape regression. Int. J. Comput. Vis. **107**, 177–190 (2014)
5. Ren, S., Cao, X., Wei, Y., Sun, J.: Face alignment at 3000 FPS via regressing local binary features. In: IEEE Conference on Computer Vision and Pattern Recognition (CVPR), pp. 1685–1692 (2014)
6. Kazemi, V., Sullivan, J.: One millisecond face alignment with an ensemble of regression trees. In: Proceedings of the IEEE Computer Society Conference on Computer Vision and Pattern Recognition, pp. 1867–1874 (2014)
7. Sagonas, C., Antonakos, E., Tzimiropoulos, G., Zafeiriou, S., Pantic, M.: 300 faces In-the-wild challenge: database and results. Image Vis. Comput. (IMAVIS) (Special Issue on Facial Landmark Localisation "In-The-Wild") (2016)

8. Sagonas, C., Tzimiropoulos, G., Zafeiriou, S., Pantic, M.: 300 faces in-the-Wild Challenge: the first facial landmark localization Challenge. In: Proceedings of IEEE International Conference on Computer Vision (ICCV-W), 300 Faces in-the-Wild Challenge (300-W). Sydney, Australia, December 2013

9. Sagonas, C., Tzimiropoulos, G., Zafeiriou, S., Pantic, M.: A semi-automatic methodology for facial landmark annotation. In: Proceedings of IEEE International Conference Computer Vision and Pattern Recognition (CVPR-W), 5th Workshop on Analysis and Modeling of Faces and Gestures (AMFG 2013). Oregon, USA, June 2013

10. Soukupová, T., Cech, J.: Real-time eye blink detection using facial landmarks. In: 21st Computer Vision Winter Workshop, pp. 1–8, February 3 2016

Continuous Gait Authentication Against Unauthorized Smartphone Access Through Naïve Bayes Classifier

Praveen Kumar Rayani and Suvamoy Changder

Abstract Continuous authentication is one of the emerging feature to identify the genuine user of the smartphone passively and prevent information theft. Earlier works on entry-point authentication failed to identify the genuine user of the smartphone and were resilient to information theft. So, we observed that theft of personal information from a smartphone might not be the loss of confidential information. However, we identified that loss of sensitive information through an internal attack causes the user vulnerable to attack and blackmail from an anonymous user. In this paper, we proposed an algorithm on the continuous gait authentication to work against unauthorized smartphone access. The continuous gait authentication uses accelerometer and gyroscope data to classify user patterns via naïve Bayes. We systematically evaluated proposed work with fifteen users and three behavioral postures of the smartphone. Finally, we addressed open research challenges existed in continuous gait authentication of the smartphone.

Keywords Continuous authentication · Gait recognition · Smartphone sensors · Internal attacks · Naïve bayes classifier

1 Introduction

The progress in hardware technology, the smartphones are rapidly growing to reach various end users to make their modern life run smoother by applications of the smartphone. Also, applications of the smartphone are helping to store, access, or share personal information of the user. However, the personal information of the user "including but not by way of limitation to contacts, messages, and photos" [1, 9] might include e-banking credentials, emails, personal documents, social networking,

P. K. Rayani (✉) · S. Changder
National Institute of Technology, Durgapur, India
e-mail: rayanipraveenkumar@gmail.com

S. Changder
e-mail: suvamoy.changder@cse.nitdgp.ac.in

© Springer Nature Singapore Pte Ltd. 2020
V. Bhateja et al. (eds.), *Intelligent Computing and Communication*,
Advances in Intelligent Systems and Computing 1034,
https://doi.org/10.1007/978-981-15-1084-7_78

mobile payments, and mobile commerce. For accessing user's personal information in a smartphone, entry-point authentication plays an important role [5] to unlock the device. The existing techniques of entry-point authentication in smartphone include knowledge-based authentication[1] [1] and physical biometrics[2] [1, 8, 11]. The credentials of knowledge-based authentication are easy to steal [17, 20] because the maximum number of smartphone users use weak and expected passwords [6, 13]. However, the existing techniques of knowledge-based authentication are vulnerable to the internal attacks [2, 3, 9, 12]. The existing user's physical biometrics could be captured by an anonymous user in public places or conferences. An anonymous user can achieve his/her desires by redeveloping the similar biometrics. So, the entry-point authentication may not prevent unauthorized physical access of a smartphone.

To overcome existing issues with entry-point authentication, the user of the smartphone must authenticate continuously [12]. For continuous authentication, methods available in a smartphone are limited. So, behavioral biometric traits [1, 12] are preferable to verify continuous authentication of a smartphone without the user's active participation. Continuous gait authentication (CGA) has widely covered throughout the related work [4, 7, 10, 14, 18], and it is still a growing field of research. ul Haq et al. [7] identified that an entry-point authentication of a smartphone is vulnerable to different security and privacy threats. To counter those vulnerabilities, they used a support vector machine to recognize the physical activity patterns of a user. ul Haq et al. achieved an accuracy of 97.95%. Lin et al. [10] identified that an entry-point authentication of the smartphone is intrusive. And as a substitute, they implemented a non-intrusive authentication based on the motion sensors of a smartphone with K-Nearest Neighbor and majority voting scheme. Lin et al. achieved an equal error rate of 6.85%. Primo et al. [14] identified smartphone theft as a significant issue. To solve smartphone theft, they implemented continuous user authentication for a smartphone. To train the user authentication of the smartphone, the solution used logistic regression. Primo et al. achieved accuracy between 61.76 and 72.58%. Conti et al. [4] recognized movements of a human as a measure for biometric authentication of the smartphone. To validate a dataset, they used dynamic time warping (DTW). Conti et al. achieved impostor pass rate between 4.44 and 32.00%. Shen et al. [18] identified leakage of private information from the smartphone. To prevent information leaks, they implemented a user authentication of the smartphone by motion sensors. They used ten one-class classifiers to validate the dataset. Shen et al. achieved an equal error rate between 2.21 and 28.22%.

The novelty introduced in this work is designing of CGA for accessing information by verification of the user. The work done in this paper is continuously recognized by the smartphone to verify his/her gait behavior before the end of the session from subsequent entry-point authentication. Here, we have included a solution for unauthorized smartphone access. Also, we have solved stealing of the

---

[1] Knowledge-based authentication is a way to verify the identity of claimed sensitive information (such as PIN, password or draw-a-pattern) of the user before allowing access of smartphone.

[2] Physical biometric is a way to recognize the identity of claimed biological characteristic (such as fingerprint, face, or iris) of the user before allowing access of smartphone.

smartphone's sensitive information. Our future work will be considering to solve practical usability issues [14, 19] existing in CGA. In particular, this paper makes the following contributions:

- We proposed the architecture of smartphone's CGA. The proposed architecture helps to secure personal information of smartphone. The proposed architecture of CGA verifies user behavior through naïve Bayes (NB) classifier.
- We explained proposed CGA algorithm steps for both training and prediction of user gait behavioral patterns through continuous authentication. Also, the time complexity and space complexity of NB classifier is explained.
- We show a summary of a systematic evaluation of CGA with average-case and worst-case experimental results. Also, we explained our observations on experimental results. Finally, we justified our problem statement.

The following section describes the problem statement, Section 3 describes the system design. It includes the architecture of CGA, design of classifier, and proposed algorithm. Section 4 describes experimental results and analysis. Finally, Sect. 5 describes the conclusion and future work.

2 Problem Statement

This section describes the problem that existed in the smartphone. The smartphone contains a variety of personal information. To security and privacy of user's personal information, the smartphone is under possession of entry-point authentication. While unlocking the smartphone, knowledge-based passwords in entry-point authentication are visible to his/her surrounding people. Surrounding users (friends/relatives) can steal knowledge-based passwords through internal attacks.[3] Stealing of user passwords may lead to a potential physical access of a smartphone. Potential physical accessing of user's personal information from a smartphone might not be the stealing of confidential information. However, theft of user's confidential information through smartphone physical access may cause monetary or non-monetary gains and mischiefs.

3 System Design

In this section, we explain the proposed architecture of CGA [12]. Further, we contribute proposed algorithm for the continuous authentication of a smartphone, which uses behavioral biometric traits of gait recognition.

---

[3] Internal attacks–shoulder surfing attack, smudge attack, and video-based attack.

3.1 Architecture of Continuous Gait Authentication

The architecture of CGA is subdivided into two phases. They are the enrollment phase and verification phase as shown in Fig. 1 [12].

The enrollment phase [12] includes three stages such as data acquisition, data preprocessing and feature extraction, and database. Verification phase [12] includes four stages such as data acquisition, data preprocessing and feature extraction, matching and decision-making, and the database. Each stage in enrollment and verification phases are described below.

Data Acquisition In the data acquisition, the tiny sensors in the smartphone acquires the user's behavioral activity. In this stage, the accelerometer sensor (A_x, A_y, A_z) and gyroscope sensor (G_x, G_y, G_z) of the smartphone are chosen to acquire a user's behavioral biometric patterns. Assume,

$$A = [A_x, A_y, A_z]^T \tag{1}$$

$$G = [G_x, G_y, G_z]^T \tag{2}$$

where, $A_x, A_y, A_z, G_x, G_y, G_z$ are vectors. A_x, A_y, A_z are stored in two-dimensional array such as A. G_x, G_y, G_z are stored in two-dimensional array such as G.

$$D = [A, G] \tag{3}$$

where D is a two-dimensional array. Equations (1) and (2) are horizontally concatenated and stored in D [shown in Eq. (3)]. The resultant of data acquisition is forwarded to data preprocessing stage.

Preprocessing and Feature Extraction In the preprocessing stage, the missing values are removed from raw behavioral data for improving the quality of the behavioral pattern. The resultant of data preprocessing stage is forwarded to the feature extraction stage. The feature extraction stage extracts statistical features from preprocessed raw data. The list of extracted features is shown in Table 1.

Database In this case, the database stores behavioral templates, which are labeled with the user ID. In the continuous authentication, the enrollment and verification are logically connected with the database (as shown in Fig. 1). The enrollment phase

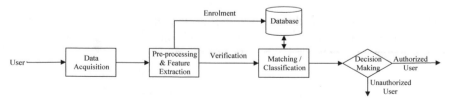

Fig. 1 Architecture of continuous gait authentication

Table 1 List of extracted features

S. no.	Name of the feature	S. no.	Name of the feature
1.	Mean of A_x	10.	Variance of G_x
2.	Mean of A_y	11.	Variance of G_y
3.	Mean of A_z	12.	Variance of G_z
4.	Mean of G_x	13.	Standard deviation of A_x
5.	Mean of G_y	14.	Standard deviation of A_y
6.	Mean of G_z	15.	Standard deviation of A_z
7.	Variance of A_x	16.	Standard deviation of G_x
8.	Variance of A_y	17.	Standard deviation of G_y
9.	Variance of A_z	18.	Standard deviation of G_z

uses the database to store user templates for verification. The verification phase gets a matching template from the database for classification.

Matching and Decision-Making The matching and decision-making decides whether the claimed identity belongs to the authorized user or unauthorized user. By comparing templates, matching algorithm calculates the matching score of claimed user. In this paper, we use a two-class NB classifier (refer Sect. 3.2) for producing a match score. The produced match score decides whether the claimed identity belongs to an authorized or unauthorized user. If a user is authorized, then the user can access resources available in a smartphone. Otherwise, deny access.

3.2 Design of Classifier

The classifier is a significant component in behavioral continuous authentication. Each classifier has two subphases which are the training phase and prediction phase. In the training phase, the classifier learns from the extracted features of gait behavioral patterns. In the prediction phase, extracted features from input gait behavioral patterns are compared with a learned gait behavioral pattern to predict the authorization of the user.

The proposed architecture uses the two-class NB classifier [15] to classify the behavioral patterns of user. At first, x_i, where $i = 1, 2, 3, \ldots, n$ represents the n features and c_i represents the class label $c_i \in (0, 1)$. The probability of observing x given the class label c can be computed by Eq. (4) [15]. To predict the label of instance x, the probability of instance x in each class label is computed. Here, the class with maximum probability is recognizing as class label of the instance x. Label estimation process of instance x defined in Eq. (5) [15].

$$p(x_1, \ldots, x_n|c) = \prod_{i=1}^{n} p(x_i|c) \tag{4}$$

$$C(x)_{NB} = \underbrace{\text{argmax}}_{c}\ p(c) \prod_{i=1}^{n} p(x_i|c) \tag{5}$$

Gaussian NB classification [15] works with an assumption of Gaussian distribution on feature values given the class label. The class label c is computed by Eq. (6).

$$p(x_i|c) = \frac{1}{\sqrt{2\pi\sigma_{c,i}^2}} \exp^{\left(-\frac{(x_i - \mu_{c,i})^2}{2\sigma_{c,i}^2}\right)} \tag{6}$$

3.3   Proposed Algorithm

In this subsection, we propose a CGA algorithm based on the motion behavioral sensors of the smartphone. The CGA procedure takes input as a Boolean flag. The detailed description of the CGA algorithm steps is shown in Algorithm 1.

First (lines 2–5), CGA procedure captures user behavioral patterns through motion sensors of a smartphone. A detailed description of data capture is explained in Sect. 3.1. Second (line 6), CGA procedure performs horizontal concatenation of two-dimensional array A and G. The resultant of horizontal concatenation is stored

---

Algorithm 1 Continuous gait authentication (CGA)

1: **procedure** CGA($flag$)
2:    **fork**
3:        $A \leftarrow$ readAccelerometer( )
4:        $G \leftarrow$ readGyroscope( )
5:    **join**
6:    $D \leftarrow [A, G]$
7:    $D_p \leftarrow$ dataPreprocessing(D)
8:    $D_f \leftarrow$ featureExtraction(D_p)
9:    $classifier \leftarrow$ NaïveBayes( )
10:   **if** $flag \neq 1$ **then**
11:       $score \leftarrow classifier$.fit($D_f, C_e$)
12:       **return** true
13:   **else**
14:       $C_p \leftarrow classifier$.predict($D_f$)
15:       $acc \leftarrow$ accuracy(C_a, C_p)
16:       **if** $acc \geq threshold$ **then**
17:           **return** true
18:       **else**
19:           **return** false

---

in two-dimensional array D. Where D indicates raw data captured from motion sensors of a smartphone. Third (line 7), CGA procedure performs data preprocessing operation on raw data (D). The resultant of data preprocessing is stored in two-dimensional array D_p. Fourth (line 8), CGA procedure performs feature extraction on preprocessed raw data (D_p). The resultant of feature extraction is stored in two-dimensional array D_f. The detailed list of features is shown in Table 1. Fifth (line 9), CGA procedure builds the naïve bayes classifier. The generated model is stored in *classifier*. The detailed description of naïve bayes is discussed in Sect. 3.2.

Subsequently, line 10 is used to switch between enrollment and recognition phase. Switching can be done through a Boolean $flag$. Where $flag$ "0" indicates enrollment and $flag$ "1" indicates recognition of CGA of a smartphone. By default, the $flag$ is set to "0". Consider $flag$ is "0", CGA procedure (line 11) performs training operation through fit function. The resultant scores can be stored in a two-dimensional array named as *score*. If the training process is completed, then line 12 in CGA procedure returns true.

Consider $flag$ is set to "1", CGA procedure (line 14) performs recognition of user motion behavioral patterns through predict function. The resultant predicted labels can be stored in C_p. Line 15 is used to calculate the predicted accuracy [16] of naïve bayes classifier. The resultant is stored in a variable acc. Lines 16–19, CAG procedure checks if $acc \geq threshold$ is satisfied then it returns true otherwise returns false. Where true indicates authorized user and false indicates unauthorized user. This procedure is implicitly repeated to verify the current user transparently.

Analysis of CGA Procedure CGA procedure uses NB classifier to classify the behavioral gait patterns of the user. The time required to classify behavioral pattern via NB classifier takes $O(ndc)$. The space complexity for NB classifier takes $O(dc)$. Where n indicates the number of training samples, d indicates dimensionality of features and c indicates the number of classes.

4 Experimental Results and Analysis

In this work, we systematically examined the reliability and usability of our approach. To conduct this experiment, we selected 15 users for data acquisition. We used three operations for collecting behavioral data, which would roughly cover a user's routine of the operating environment of the smartphone. The selected operations are hand-hold pattern,[4] table-hold pattern[5] and walking pattern.[6]

We set up a free experimental environment on an android smartphone for collecting behavioral data of the user. We developed a data-capture application on an android

---

[4] *Hand−hold pattern*: User holds smartphone and accesses resources of the smartphone while sitting on a chair.

[5] *Table−hold pattern*: User holds smartphone and accesses resources of the smartphone while sitting on a chair with table support.

[6] *Walking pattern*: User holds a smartphone and accesses resources while walking.

Table 2 Experimental results for selected operations

	Accuracy (%)		FAR (%)		FRR (%)		EER (%)	
	Average	Worst	Average	Worst	Average	Worst	Average	Worst
Pattern 1	98.67	97.4	0.34	1.16	2.34	4.06	1.34	2.61
Pattern 2	94.34	75.81	8.2	44.06	3.02	3.67	5.61	23.86
Pattern 3	98.55	98.15	0	0	2.92	3.74	1.46	1.87

Note Where, Pattern 1 indicates hand-hold pattern, Pattern 2 indicates a table-hold pattern, Pattern 3 indicates walking pattern, FAR indicates false acceptance rate, FRR indicates false rejection rate, and EER indicates equal error rate

smartphone which runs in the background. When the user unlocks his smartphone, the data-capture application starts recording user's accelerometer and gyroscope sensory data. Otherwise, it stops recording user activity. In this experiment, we collected 7000 * 3 samples of the accelerometer and 7000 * 3 samples of the gyroscope from each user. We extracted 18 features from time-series domain. To perform continuous gait authentication, we implemented NB classifier [15]. NB classifier is best suitable for small datasets. It takes less time for training the model. The experimental results for selected operations are shown in Table 2.

Table 2 shows both average-case and worst-case experimental results for three operations. In worst-case, we achieved accuracy between 75.81 and 98.15% and EER between 1.87 and 23.86%. In average-case, we achieved accuracy between 94.34 and 98.67% and EER between 1.34 and 5.61%. However, we observed that the hand-hold pattern achieved best average-case accuracy and EER. We observed that table-hold pattern achieved 44.06% for worst-case FAR. It means that some cases user's table-hold patterns may be identical. We achieved the best worst-case accuracy for recognition of the user's walking pattern. Also, we can say that the user's walking pattern is not identical. Because FAR of the walking pattern is 0%.

The proposed continuous gait authentication successfully works against unauthorized physical access of the smartphone. Stealing user gait behavioral pattern through the human eye is too difficult due to small movements. Also, the proposed method works against information theft after entry-point authentication of a smartphone. However, continuous gait authentication method might be compromised during table-hold scenario due to max FAR (shown in Table 2).

5   Conclusion and Future Work

In this paper, continuous gait authentication is proposed for verifying the user who is physically accessing a smartphone. The proposed architecture applies to any operating system of a smartphone where continuous authentication is required. Our research on the design of continuous gait authentication achieved to prevent unauthorized physical access of the smartphone and improved the security of personal information stored in the smartphone.

It has been observed that most of the earlier works [4, 7, 10, 14, 18] on continuous authentication captured data acquisition process in a conditional environment. Due to the conditional environment, practical usability issues [14, 19] would raise during authentication. These usability issues may lead to misclassifying the genuine user. At some instance, the genuine user cannot access the smartphone. In our future work, this exciting work will be addressed.

References

1. Alzubaidi, A., Kalita, J.: Authentication of smartphone users using behavioral biometrics. IEEE Commun. Surv. Tutor. 18(3), 1998–2026 (2016)
2. Aviv, A.J., Gibson, K., Mossop, E., Blaze, M., Smith, J.M.: Smudge attacks on smartphone touch screens. In: Proceedings of the 4th USENIX Conference on Offensive Technologies, pp. 1–10. USENIX Association, Berkeley, CA, USA (2010)
3. Biddle, R., Chiasson, S., Oorschot, P.V.: Graphical passwords. ACM Comput. Surv. 44(4), 1–41 (2012)
4. Conti, M., Zachia-Zlatea, I., Crispo, B.: Mind how you answer me! In: Proceedings of the 6th ACM Symposium on Information, Computer and Communications Security ASIACCS'11, pp. 249–259. ACM Press (2011)
5. Corcoran, P., Costache, C.: Biometric technology and smartphones: a consideration of the practicalities of a broad adoption of biometrics and the likely impacts. IEEE Consum. Electron. Mag. 5(2), 70–78 (2016)
6. Frank, M., Biedert, R., Ma, E., Martinovic, I., Song, D.: Touchalytics: on the applicability of touchscreen input as a behavioral biometric for continuous authentication. IEEE Trans. Inf. Forensics Secur. 8(1), 136–148 (2013)
7. ul Haq, M.E., Azam, M.A., Naeem, U., Amin, Y., Loo, J.: Continuous authentication of smartphone users based on activity pattern recognition using passive mobile sensing. J. Netw. Comput. Appl. 109, 24–35 (2018)
8. Jain, A., Ross, A., Prabhakar, S.: An introduction to biometric recognition. IEEE Trans. Circuits Syst. Video Technol. 14(1), 4–20 (2004)
9. Lee, W.H., Lee, R.B.: Multi-sensor authentication to improve smartphone security. In: Proceedings of the 1st International Conference on Information Systems Security and Privacy, pp. 1–11. SCITEPRESS (2015)
10. Lin, C.C., Chang, C.C., Liang, D., Yang, C.H.: A new non-intrusive authentication method based on the orientation sensor for smartphone users. In: 6th International Conference on Software Security and Reliability, pp. 245–252. IEEE (2012)
11. Liu, S., Silverman, M.: A practical guide to biometric security technology. IT Prof. 3(1), 27–32 (2001)
12. Mahfouz, A., Mahmoud, T.M., Eldin, A.S.: A survey on behavioral biometric authentication on smartphones. J. Inf. Secur. Appl. 37, 28–37 (2017)
13. Mazurek, M.L., Komanduri, S., Vidas, T., Bauer, L., Christin, N., Cranor, L.F., Kelley, P.G., Shay, R., Ur, B.: Measuring password guessability for an entire university. In: Proceedings of the ACM SIGSAC Conference on Computer and Communications Security CCS'13, pp. 173–186. ACM Press (2013)
14. Primo, A., Phoha, V.V., Kumar, R., Serwadda, A.: Context-aware active authentication using smartphone accelerometer measurements. In: IEEE Conference on Computer Vision and Pattern Recognition Workshops, pp. 98–105. IEEE (2014)

15. Rogers, S., Girolami, M.: A First Course in Machine Learning. CRC Press, UK (2011)
16. Saeed, K.: New Directions in Behavioural Biometrics. CRC Press, USA (2016)
17. Shen, C., Cai, Z., Guan, X., Maxion, R.: Performance evaluation of anomaly-detection algorithms for mouse dynamics. Comput. Secur. **45**, 156–171 (2014)
18. Shen, C., Chen, Y., Guan, X.: Performance evaluation of implicit smartphones authentication via sensor-behavior analysis. Inf. Sci. **430–431**, 538–553 (2018)
19. Theofanos, M.F., Stanton, B.C., Wolfson, C.A.: Usability and biometrics: ensuring successful biometric systems. NIST (2008)
20. Ye, G., Tang, Z., Fang, D., Chen, X., Kim, K.I., Taylor, B., Wang, Z.: Cracking android pattern lock in five attempts. In: Proceedings 2017 Network and Distributed System Security Symposium, pp. 1–15. Internet Society (2017)

DNA Sequence Alignment Using Dynamic Programming

Niharika Singh, Gaurav Rajput, Yash Dixit and Aastha Sehgal

Abstract In today's time, Molecular Science is increasingly dependent on Software Engineering Calculations in the department of Research and Development. Aligning sequences of DNA, RNA is becoming a major part of present-day natural sciences. A hereditary database holds a large amount of unprocessed data that resides very crucial information. A single pair of chromosomes contains approximately 3 billion DNA base pairs. To look through this information, retrieve the connections, and sub-connections in it is a very slow procedure. Therefore, scientists are looking forward toward computer science algorithms for faster retrieval of information. This paper is focused upon using a parallel programming algorithm than the previous alignment algorithms.

Keywords DNA · Genetics · Parallel algorithms · Thread

1 Introduction

DNA—Deoxyribonucleic Acid—replicating acts as main constituent of chromosomes existing in living entities. The genetic information carrier is DNA. The two strands of DNA are special double helix structure like a spiral ladder. The rings of the ladder are formed from the chemical groups as bases [1]. Bases combinations are

N. Singh
School of CS, University of Petroleum and Energy Studies, Dehradun, Utrakhand, India
e-mail: niharika@ddn.upes.ac.in; niharika1519@gmail.com

G. Rajput
Department of CS&E, G.L. Bajaj Institute of Technology & Management, Greater Noida, Utter Pradesh, India
e-mail: gauravrajput31@gmail.com; gaurav.rajput@glbitm.org

Y. Dixit · A. Sehgal (✉)
Department of Informatics, UPES School of Computer Science, Dehradun, Utrakhand, India
e-mail: Sehgalaastha15@stu.upes.ac.in

Y. Dixit
e-mail: dixityash15@stu.upes.ac.in

© Springer Nature Singapore Pte Ltd. 2020
V. Bhateja et al. (eds.), *Intelligent Computing and Communication*,
Advances in Intelligent Systems and Computing 1034,
https://doi.org/10.1007/978-981-15-1084-7_79

formed such that both the strands supplement each other. The resultant of the specific sequence establishes the genetic information major insight for DP. Additionally, the ideal answers for the subproblems add to the ideal arrangement of the given issue [2].

Clarify how will utilize dynamic programming for succession arrangement by utilizing DNA. By taking two groupings and look at them character by character, by setting them in two unique lines. Characters which will match will be placed in the same column and will be considered as a match, and those which do not match will be considered as a mismatch [3]. There can also be a case that there are character and a gap (−) in the sequences, which will also be considered as a mismatch. We can also use the gap to align the sequences perfectly. The characters that will match will be given a score of +1, the characters that will not match will be given a score of −1 and if we have a gap against a character it will be given a score of −2 and will be said to be Gap Penalty manner. One method of finding the fitness score is by arranging them in all the possible ways and finding out the best one from it, but the number of alignments is exponential which will result in a slow algorithm. Therefore, dynamic programming is a faster algorithm; it divides the problem into smaller instances and then solves it [4].

Giving two sequences Seq1 and Seq2 instead of determining the similarity between sequences as a whole, dynamic programming allows construction of the solution by determining all parallels between random prefixes of the two sequences. The algorithm begins with undersized prefixes and uses formerly computed results to elucidate the solution for larger prefixes [5].

Suppose we take an example of the following two sequences: ABCDGLJ and ABDGLJ, we will align them in the following manner:

ABCDGLJ
AB-CDGLJ
Fitness score: $6 * (1) + 1 * (-2) = 4$.

2 Methodology

Utilize dynamic programming for succession arrangement by utilizing DNA, take two successions and look at them character by character, by setting them in two distinct lines [6]. Characters which will match will be put in a similar section and will be considered as a match, and those which do not match will be considered as a crisscross. There can also be a case that there are a character and a gap (−) in the sequences, which will also be considered as a mismatch. We can also use the gap to align the sequences perfectly. The characters that will match will be given a score of +1, the characters that will not match will be given a score of −1 and if we have a gap against a character it will be given a score of −2 and will be said to be a Gap Penalty [7]. The main reasons of allotting these scores are to find a final score known as the Fitness score. Fitness score is nothing but the way we will arrange the two sequences in the best possible manner. One method of finding the fitness score is by arranging them in all the possible ways and finding out the best one from it, but the number of alignments is exponential which will result in a slow algorithm. So, dynamic programming is a faster alignment algorithm, it divides the problem into smaller instances and then solves it. Giving two sequences Seq(a) and Seq(b) are optimally aligned by following the directions of the arrows which are achieved by the Trace Back method. If the direction of the arrow is diagonal, we write both the characters of both the sequences. If, the arrow points towards some other direction we place a gap instead of that character [8].

3 Literature Review

Some of the algorithms/methods that have been used in the biological field for sequence alignment are.

3.1 *Combinatorial Extension*

The calculation includes a combinatorial expansion (CE) (shown in Fig. 1) of an arrangement way characterized by adjusted piece sets (AFPs) as opposed to the more ordinary systems utilizing dynamic programming and Monte Carlo advancement. AFPs, as the name proposes, are sets of parts, one from every protein, which give structure similitude. AFPs depend on nearby geometry, as opposed to worldwide highlights, for example, introduction of auxiliary structures and general topology. Mixes of AFPs that speak to conceivable persistent arrangement ways are specifically stretched out or disposed of in this manner prompting a solitary ideal arrangement. The calculation is quick and exact in finding an ideal structure arrangement and thus appropriate for database examining and investigation of substantial protein families. The strategy has been tried and contrasted and comes about because of Dali and

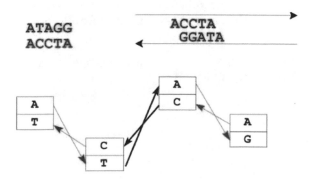

Fig. 1 Example of combinatorial extension

VAST utilizing a delegate test of comparable structures. A few new basic likenesses not distinguished by these different techniques are accounted for. This algorithm has a drawback that, this algorithm is good for comparing similar structures but when it comes to distinct structures, the final fragment may not be good [9].

3.2 Hirschberg Algorithm

In this algorithm Dynamic Programming (shown in Fig. 2) is used to find the calculation between the strings h1 and h2. The time required to complete this process is $O(|h1| * |h2|)$. An ideal arrangement which shows a genuine succession of tasks altering h1 into h2 can be done from the separation framework "n" utilizing the space complexity $O(|h1| * |h2|)$. This algorithm specifies that an ideal arrangement can be done in $O(|h1| * |h2|)$ time and just $O(|h2|)$ space utilizing double recursion [10]. There is a possibility that either there is no character or a single character. We have to cut string h1 in the center to shape h1a and h1b and to locate the relating spot to cut string h2 into h2a and h2b. Note, h2a and h2b may have very extraordinary dimension. The arrangement issue is then calculated iteratively on h1a and h2a and on h1b and h2b. With the help of the past column, one line of "l" can be found. This permits the last line of "l" to be computed in $O(|h2|)$ space. By including a limit push m($|h1| + 1$), and a limit section l($|h2| + 1$), the DPA can likewise be keep running backward from l($|h1| + 1, |h2| + 1$) to l(1, 1). Note that D(rev(h1), rev(h2)) = D(h1, h2). This permits the DPA to be run the two advances also in reverse [8]. It moves in forward direction on h1a and on all of h2 with a stage of $+1$ and in switch on s1b and all of h2 with a stage of -1. An ideal arrangement of h1 and h2 could be apportioned amongst s1a and s1b. This would give h2a and h2b. The alter separations of s1a and all conceivable h2a's are contained in the aftereffect of looking at s1a and h2. Essentially, the alter separations of h1b and all conceivable h2b's are contained in the consequence of looking at h1b and h2. The last or base column for the forward count and last or best line for the turnaround figuring are found in $O(|h2|)$ space. Including relating components of these two lines together gives the aggregate cost of altering 1 into h2 for every conceivable method for partitioning h2 into h2a and h2b. Picking a base aggregate gives an ideal h2a and h2b. The sub-issues of adjusting h1a to h2a and h1b with h2b are fathomed iteratively [11]. This method has a drawback

Fig. 2 Example of Hirschberg algorithm

```
(AGTACGCA, TATGC)
        /                              \
  (AGTA, TA)                       (CGCA, TGC)
   /      \                         /       \
(AG, )   (TA, TA)             (CG, TG)      (CA, C)
          /   \               /    \
      (T,T)  (A,A)        (C,T)   (G,G)
```

that the smaller sized alphabets in the sequences require more matches. Also, the DNA database is clustered with noncoding sequences [12].

3.3 Probalign Algorithm

Probalign takes as info unaligned protein or nucleic corrosive arrangements in FASTA design. eProbalign checks the dataset to ensure it acclimates with IUPAC nucleotide and amino corrosive one-letter shortened forms. Blank area between deposits/nucleotides in the groupings is stripped and the cleaned arrangements are passed on to the lining framework. The client can determine hole open, hole augmentation, and thermodynamic temperature parameters on the eProbalign input page. The info page gives a concise depiction of the parameters (help connection) and connections to the independent Probalign code with distribution and datasets. The three Probalign parameters on the information page are utilized for registering the segment work dynamic programming networks from which the back probabilities are inferred. This is the same as registering an arrangement of (problematic) pair wise arrangements (for each match of groupings in the information) and after that evaluating pair wise back probabilities by straightforward tallying. The thermodynamic temperature controls the degree to which problematic arrangements are considered. For instance, all conceivable problematic arrangements would be considered at vast temperature, while just the absolute best would be utilized at a temperature of zero. The relative hole parameters are utilized for the pair wise arrangements. In this way, Probalign registers the maximal expected exactness arrangement from the back probabilities similarly that Probcons does [13].

3.4 SSAP Method

The successive structure arrangement program (SSAP) technique uses a technique named as twofold programming to represent structure in terms of particle to molecule-based vectors in space. In the basic methods the Alpha Carbons are used for the basic buildup techniques but in the SSAP method the beta carbons are used in the place of alpha carbon [14]. A combination of grids is then developed containing the vector contrasts between neighbors for each match of buildups for which vectors were built. Dynamic programming connected to each subsequent lattice decides a progression of ideal neighborhood arrangements which are then summed into an "outline" grid to which dynamic writing computer programs are connected again to decide the general basic arrangement [15]. By and large, SSAP scores over 80 are related with exceedingly comparable structures. Scores in the vicinity of 70 and 80 demonstrate a comparable overlap with minor varieties. Structures yielding a score in the vicinity of 60 and 70 do not for the most part contain a similar overlap; however, more often than not has a place with a similar protein class with regular basic themes [16].

4 Proposed Work

One way to deal with register closeness between two arrangements is to produce every conceivable arrangement and pick the best one. Be that as it may, the quantity of arrangements between two successions is exponential and this will bring about a moderate calculation in this way, Dynamic Programming is utilized as a method to create speedier arrangement calculation. Dynamic Programming tries to take care of an occasion of the issue by utilizing as of now figured answers for littler occurrences of a similar issue. Giving two arrangements Seq(a) and Seq(b) as both of these are compared with each other and similarities and dissimilarities are find out according to the starting of the sequences and subproblems solutions are taken over with the help of dynamic programming (Fig. 3).

Let M = size of Seq1 and N = size of Seq2, the computation is arranged into an (N + 1) × (M + 1) array where entry (j, i) contains similarity between Seq2[1 ... j] and Seq1[1 ... i]. The algorithm computes the value for entry (j, i) by looking at just three previous entries:

(J − 1, i − 1) Diagonal Cell to (j, i)
(J − 1, i) Above Cell to entry (j, i) (J, i − 1) Left Cell to entry
J − 1, i
J, i
J − 1, i − 1
j, i − 1.

The most extreme estimation of the score of the arrangement situated in the cell (N − 1, M − 1) and the calculation will follow back from this cell to the primary passage cell (1, 1) to deliver the subsequent arrangement. In the event that the estimation of the cell (j, i) has been registered utilizing the estimation of the slanting cell, the arrangement will contain the Seq2 [h] and Seq1. On the off chance that the esteem has been registered utilizing the left cell, the arrangement will contain Seq1 [k] and a space ('−') in Seq2 [j]. There are three steps to perform the following:

Fig. 3 Formula for matrix generation

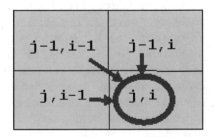

A. Filling the scoring matrix: By filling up the matrix by arranging the two sequences that are one in a row and another one in a column, respectively. By taking three values that are

1. Match $= +1$
2. Mismatch $= -1$
3. Gap $= -1$.

Now, the first row and first column will be filled by adding the gap values and the position of a particular point will be the maximum of three values that are of the beside position, diagonal position, and bottom position. In the bottom and beside value the gap value will be added and in the diagonal if there is a match then match value will be added otherwise the mismatch value. And then we will take the maximum of the three values and also the direction of that value.

B. TrBack: By traversing back from the most optimal value to the initial value that is zero. In this example the most optimal value is 3, so we will traverse from 3 to 0.

C. Alignment: Based upon the scoring matrix, by aligning the two sequences, align them from right to left, first take characters (C, C) with the diagonal direction, so both the characters will be placed one below the other. Next, having (A, A) with the upward direction, which means that will place a gap instead of a character and on the second place, will place that character to which the arrow is pointing. This means if there is a arrow in upward or sideward direction we will place a gap instead of a character at one position. By following this process we will get the best way in which the two sequences can be aligned (shown in Fig. 4).

S2 →

	-	A	C	T	A	T	A	G	A	C
-	0	-1	-2	-3	-4	-5	-6	-7	-8	-9
A	-1	1	0	-1	-2	-3	-4	-5	-6	-7
C	-2	0	2	1	0	-1	-2	-3	-4	-5
A	-3	-1	1	1	2	1	0	-1	-2	-3
G	-4	-2	0	0	1	1	0	1	0	-1
A	-5	-3	-1	-1	1	0	2	1	2	1
G	-6	-4	-2	-2	0	0	1	3	2	1
T	-7	-5	-3	-1	-1	1	0	2	2	1
A	-8	-6	-4	-2	0	0	2	1	3	2
A	-9	-7	-5	-3	-1	-1	1	1	2	2
C	-10	-8	-6	-4	-2	-2	0	0	1	3

S1 ↓

Optimal Alignment   A C - A G A G T A A C
                     A C T A T A G - A - C

Fig. 4 Scoring matrix

Fig. 5 Comparative analysis of Needleman–Wunsch algorithm versus proposed algorithm

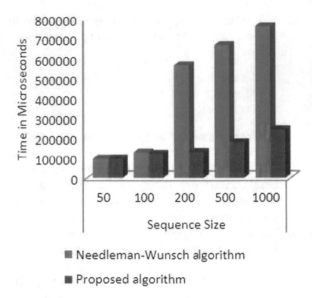

5 Result and Analysis

By taking the time taken by the Needleman–Wunsch calculation, and result calculated by the proposed methodology, the results are under in the same domain (shown in Fig. 5). The after effects of our tests are shown underneath:

As can be seen, the proposed approach works considerably speedier than the Needleman–Wunsch calculation when the grouping size is obviously enormous, i.e., atleast 200.

6 Conclusion

In this paper, introduced a parallel dynamic programming approach, to reduce the time of the DNA sequence alignment process. Toward the end, it might be specified that the subject opens up a wide extent of investigative examination with a view to investigate better change, assuming any. The creators recommend the accompanying territories of further research:

- Experimenting with genuine formation
- Comparing our approach, with different methodologies, to misuse their reciprocal qualities
- Exploiting more elevated amounts of parallelism
- Making a more keen decision of the cut-purposes of the DNA arrangements
- Applying these strategies to comparative issues, similar to the DNA multi-succession arrangement issue.

References

1. Burges, C.J.C.: Simplified support vector decision rules. In: Proceedings of the 13th International Conference on Machine Learning, Morgan Kaufmann, San Francisco, pp. 71–77 (1996)
2. Vapnik, V.: Statistical Learning Theory. Wiley, New York (1998)
3. Kushilevitz, E., Ostrovsky, R., Rabani, Y.: Efficient search for approximate nearest neighbor in high dimensional spaces. SIAM J. Comput. **30**, 457–474 (2000)
4. Cover, T.M., Hart, P.E.: Nearest neighbor pattern classification. IEEE Trans. Inf. Theory **13**, 21–27 (1967)
5. Breipohl, A.M.: Electricity price forecasting models. In: 2002 IEEE Power Engineering Society Winter Meeting. Conference Proceedings (Cat. No.02CH37309), New York, NY, USA, pp. 963–966 vol. 2 (2002)
6. Ramsay, B., Wang, A.J.: A neural network based estimator for electricity spot-pricing with particular reference to weekend and public holidays. ELSEVIER Neurocomp. **23**, 47–57 (1998)
7. Szkuta, B.R., Sanabria, L.A., Dillon, T.S.: Electricity price short-term forecasting using artificial neural networks; Hong, Y.-Y., Hsiao, C.-Y.: Locational marginal price forecasting in deregulated electricity markets using artificial intelligence. IEE Proc. Gen
8. Zhang, L., Luh, P.B., Kasiviswanathan, K.: Energy clearing price prediction and confidence interval estimation with cascaded neural networks. IEEE Trans. Power Syst. **18**(1) (2003)
9. Kumar, H., Singh, A.K.: An optimal replenishment policy for non instantaneous deteriorating items with stock dependent, price decreasing demand and partial backlogging
10. Chan, C.J.S.: Development of a profit maximization unit commitment program. MSc Dissertation, UMIST
11. Lim, B.I., Kim, S.-R., Kim, S.: The effect of the energy price increase on the energy poverty in Korea. Indian J. Sci. Technol. **8**, 790 (2015). https://doi.org/10.17485/ijst/2015/v8i8/69319
12. Jaggi, C.K.: An optimal replenishment policy for non-instantaneous deteriorating items with price dependent demand and time-varying holding cost. Int. Sci. J. Sci. Eng. Technol. **17**, 100–106 (2014)
13. Prabavathi, M., Gnanadass, R.: Energy bidding strategies for restructured electricity market. Int. J. Electr. Power Energy Syst. **64**, 956–966 (2015). https://doi.org/10.1016/j.ijepes.2014.08.018
14. Catalão, J.P.S., Mariano, S.J.P.S., Mendes, V.M.F., Ferreira, L.A.F.M.: Short-term electricity prices forecasting in a competitive market: a neural network approach. Electr. Power Syst. Res. **77**(10), 1297–1304 (2007)
15. Zhang, G., Eddy Patuwo, B., Hu, Y., Michael: Forecasting with artificial neural networks: the state of the art. Int. J. Forecast. **14**(1), 35–62 (1998)
16. Yonaba, H., Anctil, F., Fortin, V.: Comparing sigmoid transfer functions for neural network multistep ahead streamflow forecasting. J. Hydrol. Eng. **15**(4) (2010)

Data Mining Techniques for Videos Subscribers of Google YouTube

Rahul Deo Sah, Neelamadhab Padhy, Raja Ram Dutta
and Asit Kumar Mohapatra

Abstract As the access to computerized correspondence turns out to be progressively basic around the globe, it ends up less demanding to trade thoughts between societies, crossing once-restrictive geographic separations and national limits. Online video is especially interesting as a potential vector for social correspondence. As a visual resource, video offers the possibility to cross semantic and education hindrances. Video resources are additionally ready to catch social encounters, for example, move and music, which are hard to pass on through content-based media. Information mining is an expansive region that incorporates strategies from a few fields including AI, insights, design acknowledgment, man-made consciousness, and database frameworks, for the investigation of substantial volumes of information. There have been countless mining calculations attached in these fields to perform distinctive information examination tasks. Google give most intriguing thing and significant thing which is valued by each one whose most youthful to most established one in the time of present-day era. Google gives that thing is. YouTube is where sound and video has been seen by the clients. Loads of audio and recordings materials are on YouTube. Every client need to watch their fascinating sound and recordings material they visit. YouTube is where there is no bar of age or sexual orientation. Each one inquiries their very own decision. They can download or transfer the sound and recordings material. Presently in current age channels like Zee TV, Sony Entertainment, DD News, and so on their sound and recordings are transferred to YouTube and the greater part of clients watch their most loved channels and they can see sound

R. D. Sah (✉) · A. K. Mohapatra
Dr. Shyama Prasad Mukherjee University, Ranchi, India
e-mail: rahuldeosah@gmail.com

A. K. Mohapatra
e-mail: asitkm77@rediffmail.com

N. Padhy
Giet University, Gunupur, India
e-mail: dr.neelamadhab@gmail.com

R. R. Dutta
Birla Institute of Technology, Deoghar, India
e-mail: rajaramdutta@bitmesera.ac.in

© Springer Nature Singapore Pte Ltd. 2020
V. Bhateja et al. (eds.), *Intelligent Computing and Communication*,
Advances in Intelligent Systems and Computing 1034,
https://doi.org/10.1007/978-981-15-1084-7_80

819

recordings materials. In this paper, we need to dissected Video Subscribers or channel Subscribers with various information mining classifiers and there calculations. In the information mining procedures a bunches of grouping strategies are accessible. In this paper select managed and unsupervised characterization techniques. Bayes in calculations, Support Vector Machine, Deep learning, and Random Forest. In these order strategies accomplish best exactness rate out of four three 100% which are Navies Bayes, Deep Learning, and Support Vector Machine and rest Random Forest technique accomplish 95% in 4 min 48 s runtime yet in the three characterization strategies accomplish their objective at 4, 10, and 1 min 29 s.

Keywords Data mining · Feature extraction · Data cleaning · Classification methods

1 Introduction

Information Mining or "the proficient disclosure of important, on-evident data from a huge accumulation of data" [1] has an objective to find learning out of information and present it in a structure that is effectively fathomable to people. Learning recognition in databases is an exact procedure comprising of various unmistakable advances [2]. Information mining is the establishment step, which results in the revelation of obscure however accommodating learning from tremendous databases. A formal meaning of knowledge revelation in databases is given as pursues: "Information mining, or learning disclosure, is the PC helped procedure of burrowing through and breaking down tremendous arrangements of information and after that extricating the significance of the information. Information mining devices anticipate practices and future patterns, enabling organizations to make proactive, learning driven choices [3]. Data mining ability gives a buyer inclining way to deal with new and obscure examples in the information. The uncovered learning can be utilized by the human services directors to advance the prevalence of administration. Information mining is a wide zone that coordinates strategies from a few fields including AI, measurements, design acknowledgment, man-made brainpower, and database frameworks, for the examination of substantial volumes of information. It is normally utilized in a wide scope of profiling rehearses, for example, promoting, reconnaissance, and extortion identification, medicinal and logical disclosure. People have been physically extricating examples from information for quite a long time, however the expanding volume of information in current occasions has called for increasingly robotized approaches. As informational collections have developed in size and multifaceted nature, direct hands-on information examination has progressively been increased with circuitous, programmed information preparing. This has been helped by different revelations in software engineering, for example, neural systems, bunching, hereditary calculations, choice trees, and bolster vector machines. Information mining is the way toward applying these strategies to information with the expectation of revealing concealed examples. Information mining might be characterized as "the investigation and examination, via programmed or self-loader implies, of huge amounts of information so as to find significant examples and guidelines" [4]. Thus, it might be viewed as mining learning from a lot of information since it includes learning extraction, just as information/design investigation [5]. Online video, a universal, visual, and exceedingly shareable medium is appropriate to intersection geographic, social, and semantic obstructions. Inclining recordings specifically, by uprightness of achieving an expansive number of watchers in a limited capacity to focus time, are incredible as both influencers and pointers of global correspondence streams. Be that as it may, are new correspondence innovations really being utilized to share thoughts all around, or would they say they are essentially reflecting previous social channels? What is more, how do social, political, and geographic variables impact global correspondence? By dissecting utilization information from computerized correspondences stages, we can start to respond to these inquiries. In this paper, we center around slanting information from the YouTube video sharing stage to look at the global utilization of Online video.

2 Literature Survey

The proposed structure shows the examination to foresee sexual introduction of individuals from the checked copy of their handwriting styles [1]. The proposed system relies upon isolating the course of action of features from making tries out of male and female writers and getting ready classifiers to make sense of how to isolate between the two. Pictures are isolated using Otsu thresholding estimation. The going with features have been considered. Forming properties like tendency, ebb, and stream, surface and hazard are evaluated by enlisting neighborhood and overall features. Classification (innocent Bayes, SVM (Support Vector Machine, Adaboost, and Random Forest to develop a decision display well-being of foreseeing one get-together falls into the other. Support vector machine arranges the photos on the test dataset). Crossbreed classifiers are used to enlist hyperplane with a most outrageous edge. The count for the yield of a given SVM. In the direct word, a SVM plan tries to amass a decision display fit for foreseeing one grouping falls into the other. Reinforce vector machine arranges the photos on the test dataset [2, 3]. SVM is used to figure hyperplane with the best edge. The estimation for the yield of a given SVM. In the essential word, a SVM game plan attempts to create a decision demonstrate fit for foreseeing one characterization falls into the other. Half and half classify the photos on the test dataset. SVM is used to enroll hyperplane with a most extraordinary edge. The going with features have been considered. Forming characteristics like tendency, twist, surface, and commitment are evaluated by preparing neighborhood and overall features. Impressive research has just analyzed the survey propensities for YouTube clients. The notoriety of YouTube recordings have been appeared to pursue a non-Zipf, control law conveyance with a cutoff (Cha et al. [6]) with recordings' actives lives following a Pareto conveyance. Brodersen et al. [7] found that YouTube recordings are exceedingly nearby, with half of the recordings having 70% of their perspectives from a solitary country. The spread of recordings has been appeared to happen in stages, with social sharing, membership, and inquiry driving notoriety at beginning times while non-social, unified direct drive fame in later stages (Brodersen et al. 7). Late research has additionally taken a gander at how advanced innovation can be utilized to delineate comprehend the worldwide culture. Utilizing a substantial global email corpus, explored connections between global affinities and monetary, political, and social components. They see comparative examples of national and phonetic segregation in different countries where government oversight of the Internet is definitely not a huge factor. Their work proposes that the transmission of thoughts crosswise over societies is testing due to the lack of common cultural spaces, and we believe YouTube may be one of these spaces.

3 Proposed Structure for Data Mining Modeling

Today, special workplaces encourage accommodating affiliations information utilizing institutionalized investment funds data framework; as the structure contains a goliath degree of information, used to expel tied down data for making the sharp helpful end. The standard focal point of this examination is to make shrewdly. To build up this structure, for the demand of helpful applications client download in a minute and there total rating, for example, we have taken open dataset from open space The information mining demand approaches, viz., Naïve Bayes, Support Vector Machine, Deep Learning, and Random Forest are utilized… [8, 9] (Fig. 1).

In this paper organized the dataset in a given stepwise data mining structure modeling:

(i) Retrieving the Data (ii) Preprocessing of data (iii) Replacing the Missing Values (iv) Reordering the attributes (v) filtering examples (vi) sampling of data (vii) splitting the data (viii) handling the text values (ix) applying the tv for validation (x) applying tv on score (xi) Multiplication for training purposes (xii) sample for feature extraction (xiii) Go to Automatic Feature extraction (xiv) applying feature selection (xv) applying for set of feature selection (xvi) targeting the four algorithm like Naïve Bayes, Deep Learning, Random Forest and Support Vector Machine (xvii) validation for multiple training data set (xviii) Generates the batch (xix) Multiplying the training (xx) Make different simulator for different model (xxi) Multiply the model (xxii) Explain the prediction of different model (xxiii) Average of performance models.

4 Methodology

Deep Learning: Deep adapting (otherwise called profound organized learning or progressive learning) is a piece of a more extensive group of AI techniques dependent

(a)

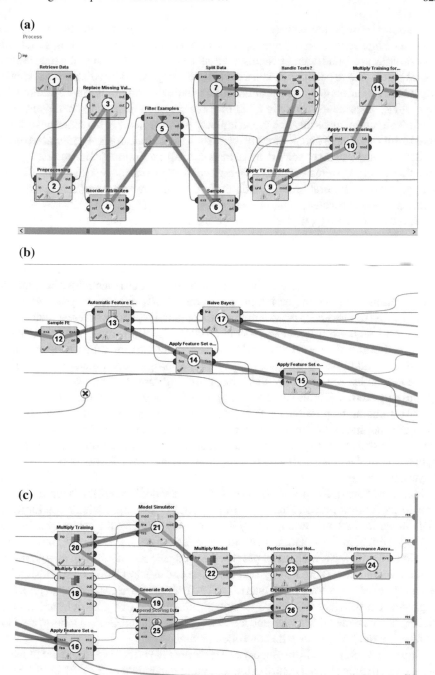

(b)

(c)

Fig. 1 **a** Structure for data mining modeling. **b** Structure for data mining modeling. **c** Structure for data mining modeling

on learning information portrayals, rather than errand explicit calculations. Learning can be managed, semi-regulated, or unsupervised. Profound learning structures, for example, profound neural systems, profound conviction systems, and intermittent neural systems have been connected to fields including PC vision, discourse acknowledgment, regular language preparing, sound acknowledgment, informal community separating, machine interpretation, bioinformatics, tranquilize plan, therapeutic picture investigation, material review, and tabletop game projects, where they have delivered results similar to and at times better than human specialists. Profound learning models are ambiguously roused by data preparing and correspondence designs in organic sensory systems yet have different contrasts from the auxiliary and utilitarian properties of natural minds (particularly human cerebrums), which make them contrary with neuroscience smidgens of proof.

Naïve Bayes: In AI, [10] Naïve Bayes classifiers are a group of basic "probabilistic classifiers" in view of applying Bayes' hypothesis with solid (credulous) freedom suspicions between the highlights. Naïve Bayes has been contemplated widely since the 1960s. It was presented (however not under that name) into the recovery network in the mid 1960s and remains a mainstream (pattern) technique for content classification, the issue of passing judgment on records as having a place with one classification or the other, (for example, spam or genuine, sports or legislative issues, and so on.) with word frequencies as the highlights. With fitting pre-preparing, it is focused in this area with further developed techniques including support vector machines. It additionally discovers application in programmed therapeutic finding. Naïve Bayes classifiers are very adaptable, requiring various parameters direct in the quantity of factors (highlights/indicators) in a learning issue. Greatest probability preparing should be possible by assessing a shut structure articulation, which takes straight time, instead of by costly iterative estimate as utilized for some different sorts of classifiers. Every one of these names references the utilization of Bayes' hypothesis in the classifier's choice principle, yet innocent Bayes is not (really) a Bayesian strategy.

Random Forest: Random Arbitrary [10] choice woodlands right for choice trees' propensity for over fitting to their preparation set. The principal calculation for irregular choice timberlands was made by Tin Kam Ho utilizing the arbitrary subspace strategy, which, in Ho's plan, is an approach to actualize the "stochastic separation" way to deal with the arrangement proposed by Eugene Kleinberg. An augmentation of the calculation was created by Leo Breiman and Adele Cutler, who enlisted "Irregular Forests" as a trademark (starting at 2019, possessed by Minitab, Inc.). The expansion consolidates Breiman's "sacking" thought and irregular determination of highlights, presented first by Ho and later freely by Amit and Geman so as to develop a gathering of choice trees with controlled change.

Support Vector Machine: Supervised learning [10] is the AI assignment of learning a capacity that maps a contribution to a yield dependent on precedent info yield sets. It construes a capacity from named preparing information comprising of a lot of preparing precedents. In directed adapting, every model is a couple comprising of an info object (ordinarily a vector) and the ideal yield esteem (likewise called the supervisory flag). An ideal situation will take into consideration the calculation to effectively decide the class names for inconspicuous occurrences. This requires the taking in a calculation to sum up from the preparation information to concealed circumstances in a "sensible" manner (see inductive inclination).

5 Dataset

Using dataset from public domain for retrieving The dataset contains the main 5000 rankings of the YouTube channels by an organization named Socialblade. The information contains different data on the YouTube channels, for example, the Social blade channel rankings, the evaluations conceded by Socialblade, the YouTube channel name, the quantity of recordings transferred by the channel, all outnumber of supporters on the channel and the all outnumber of perspectives on all the video content by the channel.

Affirmations: The information could not have been made without the diligent work of the Socialblade group. They really did the diligent work of positioning and gathering all the fundamental measurements of the YouTube channels, I just scratched it utilizing Python. A major yell out and gratitude to the Socialblade group for making this conceivable.

Motivation: The dataset can be utilized to perform exploratory information examination and perceptions which can help uncover some conceivable connections and bits of knowledge about variables driving the YouTube channel rankings. In spite of the fact that the information outlines just some fundamental data about the YouTube stations, I am almost certain that this information can be useful to every one of the tenderfoots and amateurs of information science.

5.1 Modeling and Design

Given by dataset for retrieving the data for processing and replacing the missing data values. Reordering all attributes and filtering the data. Now for extraction the data for making samples and the different samples have their different parameters for inputting the value now the data for validation process and make useful model.

6 Results Analysis

Dataset for retrieving or loading the data for all general proceeding steps and replacing the missing values reading its columns, next for filtering the data model on cases with label value applying the model on case with missing for target value and supplying the data for validation set and make samples. For training giving automatics feature extraction multiply the multiple feature set extraction. Now in multiple

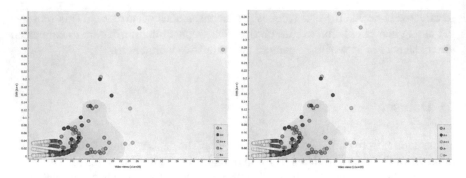

Fig. 2 Scattering plotting for random forest and support vector machine

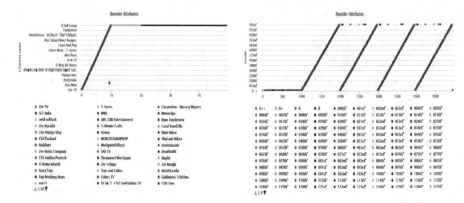

Fig. 3 Channels reorders attributes by grades of the channel and reorder attribute by videos view

algorithms like SVM, Deep Learning, Random Forest, and Naïve Bayes for multi training. Applying on resulting feature set for validated data. Now multiple training and multiple validations in applying on future set on scoring the data. Create model simulator and multiply the models. Finally make prediction for all cases and also calculate average performance for multiple methods (Figs. 2, 3, 4 and 5).

7 Conclusion

Data mining is the way toward applying these strategies to information with the expectation of revealing concealed examples. Subsequently, it might be viewed as mining information from a lot of information since it includes learning extraction, just as information/design examination Online video, a pervasive, visual, and very shareable medium, is appropriate to intersection geographic, social, and etymological boundaries. Inclining recordings specifically, by ethicalness of achieving countless

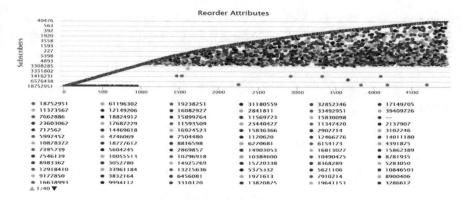

Fig. 4 Overall subscribers of given channels

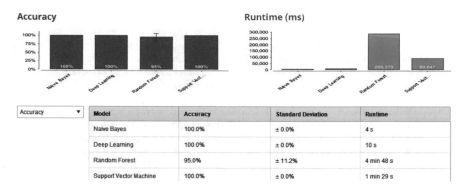

Model	Accuracy	Standard Deviation	Runtime
Naïve Bayes	100.0%	± 0.0%	4 s
Deep Learning	100.0%	± 0.0%	10 s
Random Forest	95.0%	± 11.2%	4 min 48 s
Support Vector Machine	100.0%	± 0.0%	1 min 29 s

Fig. 5 Overall performance of given by the models

in a limited ability to focus time, are groundbreaking as both influencers and pointers of global correspondence streams. Yet, are new correspondence advancements really being utilized to share thoughts all inclusive, or would they say they are just reflecting prior social channels? What is more, how do social, political, and geographic variables impact global correspondence? By investigating utilization information from computerized interchanges stages, we can start to respond to these inquiries. In this paper, we center around drifting information from the YouTube video sharing stage to inspect the worldwide utilization of Online video. In this paper select managed and unsupervised arrangement techniques. Bayesian calculations, Support Vector Machine, Deep learning, and Random Forest. In direct word classifiers (SVM, Deep Learning, Random Forest and Naïve Bayes) to develop a decision displays webbing for casting one get-together falls into the other. In other model data set passed in the different channels through find the methods. In these arrangement strategies accomplish best exactness rate out of four three achieve 100% which are Navies Bayes,

Deep Learning and Support Vector Machine and rest Random Forest strategy accomplish 95% running in 4 min 48 s runtime yet in the three characterization techniques accomplish their objective at 4,10, and 1 min 29 s.

References

1. Bhattacharyya, A.: On a measure of divergence between two statistical populations defined by their probability distribution. Bull. Calcutta Math. Soc. **35**, 99–109 (1943)
2. Bonacich, P.: Factoring and weighting approach to status scores and clique identification. J. Math. Soc. **2**, 112–120 (1972)
3. Borgatti, S.P., Halgin, D.S.: Analyzing Affiliation Networks, pp. 417–433. Sage (2011). Brieger, R.L.: The duality of persons and groups. Soc. Forces **53**(2), 181–190 (1974)
4. Han, J., Kamber, M.: Data Mining Concepts and Techniques (2001)
5. Chen, N., Lin, J., Hoi, S.C., Xiao, X., Zhang, B.: Ar-miner: mining informative reviews for developers from mobile app marketplace. In: Proceedings of the 36th International Conference on Software Engineering, pp. 767–778. ACM (2014)
6. Cha, M., Kwak, H., Rodriguez, P., Ahn, Y.-Y., Moon, S.: Analyzing the video popularity characteristics of large-scale user-generated content systems. IEEE/ACM Trans. Netw. (2009)
7. Brodersen, A., Scellato, S., Wattenhofer, M.: Youtube around the world: geographic popularity of videos. In: Proceedings of the 21st International Conference on world Wide Web. ACM (2012)
8. De Pauw, W., Jensen, E., Mitchell, N., Sevitsky, G., Vlissides, J., Yang, J.: Visualizing the execution of Java programs. In Software Visualization, pp. 151–162. Springer (2002)
9. Ghorashi, S.H., Ibrahim, R., Noekhah, S., Dastjerdi, N.S.: A frequent pattern mining algorithm for feature extraction of customer reviews. Int. J. Comput. Sci. Issues (IJCSI), 29–35 (2012)
10. Bala, S., et al.: Int. J. Comput. Sci. Mob. Comput. **3**(7), 960–967 (2014)

Prevalence of HCV in Shopian: Using Data Visualization (Data Mining)

Zubair Ahmad, Jitendra Seethlani and Sheikh Babar Gul

Abstract Data mining is an important subfield of Data Science and is also called knowledge discovery within a database. In this paper, we actually use data mining visualization techniques to analyze the huge database of District Shopian and assume prevalence rate of HCV in Thingific Area (Vehil) of District Shopian (J&K). Health Department of District Shopian uses different types of registers to maintain the data of patients. During survey, we go through the registers to collect data of symptomatic HCV patients analyze it by maintaining the disease trends. Data mining analytical tools were being used to detect prevalence of HCV infection distribution and also identify the reasons that contributed to such widespread of HCV in the area of Vehil

Keywords Descriptive analytics · Predictive analytics · Data visualization techniques hepatic C CASE STUDY

1 Introduction

Initially, we collect the data from District Surveillance Unit which is functioning under the control of chief medical officer Shopian and then, we analyze the existing data by using data mining techniques and maintain the disease trends of different diseases like Acute Diarrheal disease, Acute Respiratory infection, and Hepatitis C and B. On the basis of the disease trends, we devised a format of specific diseases and then distributed the said format to the different health facilities like Primary Health Centers, Community Health Centers, District Hospital, and Subcenters. During the

Z. Ahmad (✉) · J. Seethlani
Department of Computer Science, Sri Satya Sai University of Technology and Medical Science, Sehore 466001, MP, India
e-mail: zsheikh12@gmail.com

J. Seethlani
e-mail: dr.jsheetlani@gmail.com

S. B. Gul
Chief Medical Office, Block B, Mini Secretariat, Shopian 192303, J&K, India
e-mail: sheikhbabargull@gmail.com

© Springer Nature Singapore Pte Ltd. 2020
V. Bhateja et al. (eds.), *Intelligent Computing and Communication*,
Advances in Intelligent Systems and Computing 1034,
https://doi.org/10.1007/978-981-15-1084-7_81

course of collection of data from different health facilities/ zones of the district, data was received from PHC Vehil and while interpreting the data and giving due heed to the outcome of contemporary disease trends being maintained/ updated at the district level with special reference to different facilities/ zones of the area on monthly basis, a significant increase was observed in the hepatitis cases in comparison to the previous figures of PHC/Zone Vehil. In place to mention here that PHC Vehil is at a distance of 8 kms from district headquarter Shopian.

"According to National Centre for disease control, hepatitis C is a severe liver infection caused by the hepatitis C virus (HCV)", respectively. In comparison to national rates, in 2015, West Virginia reported the highest rate of HCV infection at 3.4 per 100,000 population in the United States (US). Since 2010, acute HCV has increased by 209%. HCV infection can cause acute (short-term) or chronic (lifelong) infection. Symptoms of HCV infections are analogous and can include malaise, anorexia, abdominal pain, jaundice, nausea, vomiting, diarrhea, and/or dark urine. Symptoms of HCV infection can take 14–180 days (2 weeks to 6 months) to display

2 Background

District Shopian comprises two Medical Blocks viz Shopian, and Keller with one (01) District Hospital, Two (02) CHC, Six (06) PHC's, Four (04) NTPHC Fifty-four (54) Subcenters, Five (05) MAC, and Two (02) Mobile MAC. The total number of villages is 231 with Eight (08) CD Blocks and Nine (09) Tehsils. The population of District Shopian is 2.66 lacs including 15% ST Population as per 2011 census, the total literacy rate of the district is 62.49%.The affected Vehil Zone is located in Block Shopian 10 kms away from District Headquarter (CMO's Office) and the said zone is having a population of 52000 souls, comprising 67 villages and catered by 16 Subcenters and one PHC (PHC Vehil).

Consequent upon the observance of increase in the hepatitis cases in Zone Vehil, in the very outset, we began to collect the data from all private labs in order to work out the possible outbreak of hepatitis B/C in the district and more than "30" positive cases came to surface, with majority of positives cases again belonging to zone Vehil. The team, accompanied by District Surveillance Unit Shopian and staff of PHC Vehil, conducted screening in village Vehil, having a population of "1958" souls, starting from the documented/ infected cases from the said village, followed by the close contacts of the said cases, and then to other inhabitants of the area at random, making a total of "600" cases that were screened by the team out of which 26 more cases of Hepatitis B/C were found positive. As per the findings conveyed by the Epidemiology Division Kashmir based on the screening made by the above-mentioned team, a team was again constituted at the District level and deputed to the area to look for the underlying causes for this epidemic. The key recommendations of the District team, indicating the common causes, are as follows (Fig. 1):

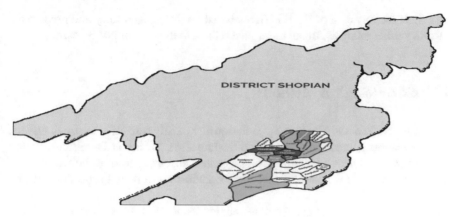

Fig. 1 Sophian district

1. The people of the area (male population) are mostly truck drivers, and labors who visit places outside Kashmir such as Jammu, Delhi, and Punjab) during each year which may be one of the causes of spread of infection;
2. In respect of the positive cases, there is a history of dental procedures/dental extractions;
3. There is a history of using a single shaving blade for different persons.

Kedziora et al. [1] discussed mechanisms involved in the development of chronic hepatitis C as potential targets of antiviral therapy. They have mainly focused on chronic hepatitis and conducted several experiments. Harder et al. [2] conducted case studies regarding hepatitis. They mainly focused adult person. Tahseen et al. [3] used the Neural Network techniques for hepatitis virus. They conducted automatic diagnosis system which able to extract the features. In [4] the author Larose discussed several data mining algorithms and investigated the medical database. Gibofsky et al. [5] discussed Epidemiology, pathophysiology, and diagnosis of rheumatoid arthritis. They have done systematic literature review and conclude Rheumatoid arthritis (RA) is one of the more common autoimmune disorders, affecting approximately 1% of the population worldwide.

3   Descriptive Analytics

Descriptive analysis or statistics does exactly what the name implies they "Describe", or summarize raw data and make it something that is interpretable by humans. They are analytics that describe the past. The past refers to any point of time that an event has occurred, whether it is 1 min ago, or 1 year ago. Descriptive analytics are useful because they allow us to learn from past behaviors, and understand how they might influence future outcomes. Descriptive Analysis is also called statistics. In this phase, we collect the raw data of HCV Patients based on their past history and prepare the data for further analysis. In our case, past represent the data of last 3 years.

4 Epidemiological Methods

i.   The people of the area (male population) are mostly fruit growers, truck drivers, and labors who visit places outside Kashmir such as Jammu, Delhi, and Punjab) during each year which may be one of the causes of spread of infection;
ii. In respect of the positive cases, there is a history of dental procedures/dental extractions.
iii. There is a history of using single shaving blade for different persons at a local barbershop.
iv. Positive cases have no signs and symptoms of jaundice and seemed apparently healthy;
v.   No history of dialysis/blood transfusion found;
vi. History of drug addiction could not be elicited;
vii. Pregnant ladies were also found to be infected;
viii. Age group from 12 to 70 years were found to be involved with majority of female cases.
ix. More than one person was found to be affected in some families.
x.   Mostly affected village is Vehil having more than 150 positive patients;
xi. Nearest Health facility is PHC Vehil.

5 Laboratory Method

Phase-I: Twenty (20) new cases of Hepatitis C were (07 Males and 13 females), with age group 24–70 years, diagnosed positive. Out of the said 20 cases, fourteen (14) cases are clustered in the same village in same Mohalla (Khan Mohalla and Malik Mohalla) (Fig. 2);

Phase II: Again, 450 cases (144 Males and 306 Females), with age group 15–70 years diagnosed positive via card method. These include "319'"cases diagnosed by the team constituted for the purpose under the banner of "SGMS Foundations" and remaining "131" reported from DH Lab. and Private Labs of District Shopian. **From among the above "470" hepatitis C positive cases, "300" cases are new while "170" cases are old/already detected ones**.

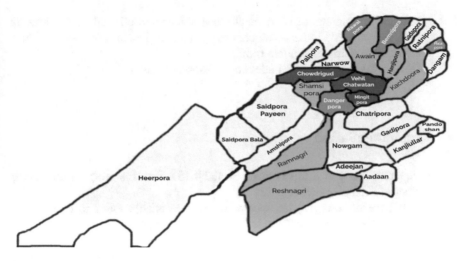

Fig. 2 Mohalla district

6 Action Taken

i.   Data collected on prescribed formats in respect of communicable diseases from all types of health facilities as well as from clinical Pvt. Establishments (Labs/Dental clinical, etc.) and subsequently interpreted at District/Divisional Level.

ii. Thorough screening held at different villages of zone Vehil, owing noticing/ observance of the significant increase in the Hepatitis positive cases in the zone as per the collected data.

iii. Awareness camps in order to import necessary education to the masses with special attention to different modes of transmission of the disease/ infection.

iv. The possible sources of spread of infection like unhygienic barbershops using a single blade for multiple people, unhygienic labs/medical shops and Dental clinics using unsterilized instruments, dressing materials, and following improper methods of biomedical waste management have been earmarked and sealed immediately.

v.   Directions also were given to health institutions and clinical establishments (Pvt. Owned) to keep the institutions/clinics properly fumigated to avoid any spread of infection on this score.

vi. Drug De-addictions programs conducted in zone Vehil under intimation to Civil/police administration.

vii. Viral Load/Genotype.

Viii. The close contacts of all the hepatitis B patients vaccinated by way of deputing special teams for the purposes.

ix. District/block level teams were put to the work of monitoring the spread of infection as well as preventive measures been taken.

   x.   Periodic inspections of Govt. health institutions and clinical establishments (Pvt. Labs/dental clinics conducted by the team of Doctors/Supervisory staff under supervision of Epidemiologist.

  xi.   Doctors/Supervisory staff under supervision of Epidemiologist.

7 Recommendation

a.   Adequate quantity of drugs is made available to suffice the needs of increasing number of hepatitis positive cases.

b.   Health Care persons require special training to enable them to combat the infection (its prevention, control, etc.) in the right perspective.

c.   Proper Screening of persons returning from the outside state may be done right at the entry points such as Airports, Railway stations, and Bus stands.

d.   Necessary Investigations like HIV, HBSAG, and HCV can be made mandatory before carrying out any medical/dental procedure.

e.   Special powers may be vested with health inspectors of Health Dept., to carry out inspections of all the clinical establishments (Govt/PVT) medical shops/Barbers, etc., and to impose on spot punishments Challah against them as and when required to achieve better results.

f.   Necessary funds may also be provided for IEC activities on this behalf.

g.   The services of one Gynecologist, Physician, and pediatrician with epidemiologist may also prove very much beneficial in the prevention and control of infection only after proper training in the field.

8 Screening

See Table 1.

9 Environmental Methods

As per the history of infected cases, the maximum patients have done dental procedures/extraction and some of them became the victim of single razor used by barber for different clients.

   Pertinent to mention here that drug addiction cannot be elicited.

Table 1 Village wise population screened in zone Vehil is mentioned

S. No.	Village name	Total population	Population screened
1.	Vehil	1958	602
2.	Mengipora	243	110
3.	Chowdergund	369	58
4.	Shamsipora	805	227
5.	Awind	881	141
6.	Ramnagri	3501	1392
7.	Reshnagri	1950	667
8.	Chetawan	386	230
9.	Hallapora	505	112
10.	Kachdoora	1363	464
11.	Nowgam	2112	203
12	Dangerpora	813	109
14.	Hanjipora	515	308
15.	Rawalpora	0f6	109
16.	Check kachdoora	1363	85
17.	Bemnipora	1054	244
	Total	18174	5061

HCV DETECTION AND TREATMENT FLOW:

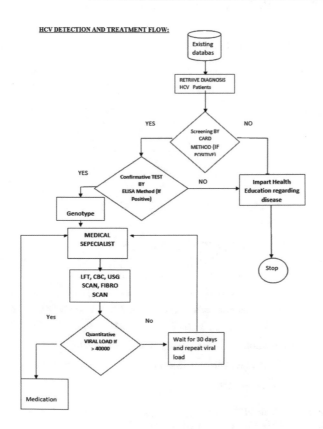

HCV Detection and Treatment Flow Model:
It is used to identify the causes of acute hepatitis C virus (HCV) infection and to
project HCV incidence cases. It begins with an antibody test. Antibodies to HCV
can be detected in blood usually within 2–3 months after virus infection test. HCV
antibody test is typically reported as positive and negative on the basis of antibodies.
Here in our flowchart, we use two types of methods to check presence of antibodies
in the blood. First one is card method; it is easy and very economical as compared to
second one. Second method is more reliable and also called confirmative test (ELISA
METHOD). After antigens examination if antigens are present, then we have to rule
out in which family virus belongs by doing genotype and then patients are referred
to a specialist for treatment and also check other complications. The most important
phase for the medication is the viral load also called RNA VIRAL load. Result of the
viral load and other investigations after comparing it with the international guidelines
of Hepatitis C (disease literature) helps the medical experts to put on the patient on
drug or not.

10 Statistical Analysis

10.1 Data Visualization Method

See Figs. 3, 4, 5, 6, 7 and 8.

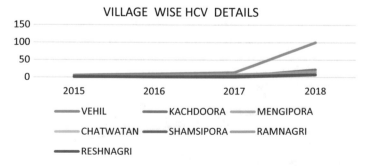

Fig. 3 Incidence of acute HCV cases by year of report, Shopian (VEHIL, RESHNAGRI,
CHOWDGERGUND, CHATWATAN, VEHIL), 2017–2018 ($n = 403$)

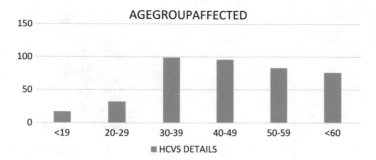

Fig. 4 Shows the distribution of acute HCV by age among cases reported between 2016 and 2018. The largest volume of acute HCV cases were the 30–39 years age group, followed by the 40–49 age group ($n = 403$)

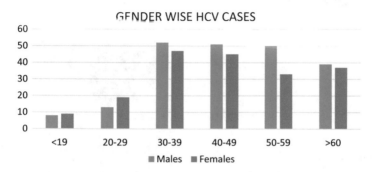

Fig. 5 Represents the incidence of acute HCV cases in District Shopian, by age and gender, in increments of 2 years. The highest incidence of acute HCV is among males aged 30–39, followed by females aged 30–39 ($n = 403$)

Fig. 6 Shows the gender distribution of chronic HCV infection in Shopian during 2017–2018. More cases were reported among males compared to female ($n = 403$)

Fig. 7 Distribution of acute
HCV cases by community,
Shopian, onset year
2016–2018 ($n = 403$)

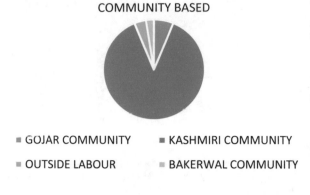

Fig. 8 Distribution of HCV
cases by OCCUPATION,
Shopian, onset year
2016–2018 ($n = 403$)

References

1. Kedziora, P., Figlerowicz, M., Formanowicz, P., Alejska, M., Jackowiak, P., Malinowska, N., Fratcza, A., Blazewicz, J., Figlerowicz, M.: Computational methods in diagnostics of chronic hepatitis C. Bull. Pol. Acad. Sci. Tech. Sci. **53**(3), 273–281 (2005)
2. Harder, A., Mehlhorn, H.: Comparative hepatitis: diseases caused by adult parasites or their distinct life cycle stages. In: Weber, O., Protzer, U. (eds.) Comparative Hepatitis. Birkhäuser Advances in Infectious Diseases, Birkhäuser Basel (2008). https://doi.org/10.1007/978-3-7643-8558-3_9
3. Jilani, T.A., Yasin, H., Yasin, M.M.: PCA-ANN for classification of hepatitis-C patients. Int. J. Comput. Appl. **14**(7) (2011)
4. Larose, D.T.: Data Mining Methods and Models. Wiley
5. Gibofsky, A.: Overview of epidemiology, pathophysiology, and diagnosis of rheumatoid arthritis. Am. J. Manag. Care **18**(13 Suppl), S295–302 (2012). PMID: 23327517

A New Approach for Matrix Completion Using Arrow Relation of FCA in Recommender Systems

G. Chemmalar Selvi and G. G. Lakshmi Priya

Abstract In this digital world, the Internet is the major source of rich amount of data where there are large numbers of choices available to the internet users. It requires the attention to scrutinize, itemize, and effectively project the essential information in order to violate the dispute of information overload. The information overload is an obstacle for many of the internet users. Recommender systems aim to provide solution to such obstacles by refining and seeking the large volume of exponentially growing information to provide the internet users with the personalized content and services. Here finding or predicting the user-item rating with the sparse data available in the rating matrix of the input data becomes a challenging task. This paper explores the problem of recommendation by presenting a FCA-based methodology which uses the mathematical tool in reducing the dimensionality of the user-item rating thereby determining the rating value of the unknown user with its corresponding item. The proposed method is simple and faster explained with an illustration to complete the user-item rating matrix from which the rating can be predicted by applying any of the recommendation algorithms.

Keywords Recommender systems · Formal concept analysis · Formal context · Matrix factorization · Cold-start problem

1 Introduction

With the growing volume of data, it is inexorable to deal with the problem of data representation. The data are represented by adopting many kinds of models like trees, graphs, dendrograms, etc. The Formal Concept Analysis (FCA) is one such method adopted in this paper to represent the input data as a formal context which is a tabular view of data with rows as objects and columns as attributes. The theory of FCA was

G. C. Selvi · G. G. L. Priya (✉)
Vellore Institute of Technology, Vellore, India
e-mail: lakshmipriya.gg@vit.ac.in

G. C. Selvi
e-mail: chemmalarselvi.g@vit.ac.in

© Springer Nature Singapore Pte Ltd. 2020
V. Bhateja et al. (eds.), *Intelligent Computing and Communication*,
Advances in Intelligent Systems and Computing 1034,
https://doi.org/10.1007/978-981-15-1084-7_82

given by Rudolf Wille in early 1980s which is the mathematical framework based on lattice theory [1] originated actually from the works of researchers [2, 3]. One of the properties of FCA is a method of exploratory data analysis which deals with clusters of formal concepts represented in the hierarchical structure from the data represented in the form of triplets called formal context [4]. It also acts as a powerful tool to deal with uncertainty and vagueness of data. To handle the uncertainty of data, the approach has been extensively explored such as fuzzy concept, interval–valued fuzzy concept, and rough set, approximate and triadic concept analysis. The literatures of these are examined with their independent mathematical background, methods, algorithms, and applications [5–9, 14].

FCA is an efficient tool for data analysis, knowledge representation, and knowledge management. FCA begins the data analysis by considering the incidence matrix which contains objects representing the rows, attributes representing the columns and the matrix value representing the relationship among the object and the attribute. It produces two kinds of output from the input data. One is the Concept Lattice. A concept lattice is a diagram representing a hierarchical structure of clusters of formal concept represented by constructing the line or Hasse diagram. It looks like a natural human-like concepts such as "Dog is an Animal", "Grasshopper is an insect." The second output of FCA is the collection of attribute implications. An attribute implication is true if there exists a valid dependency among the data such as "every number divisible by 3 and 4 is divisible by 6", "every individual with age more than 60 is retired."

In this paper, a new FCA-based methodology is presented to handle the problem of a recommender system. Recommender System is a type of information filtering system which aims at predicting the ratings or preferences of a user give to an item. The problem of predicting the user ratings for a new user or a new item who just enters the system domain is known as cold-start problem which is a challenging goal [10]. In this work, in place of matrix factorization [11], a new FCA-based methodology is proposed to complete the matrix by transforming the sparse matrix to dense matrix and predicting the user ratings. The input matrix is represented as an actual formal context and simplified it to a reduced formal context [12] to achieve the faster computation of matrix completion.

In the first section of the paper, the basic definitions of formal contexts, reduced formal context, formal concepts, and the concept lattice are discussed. In the second section, a new FCA-based methodology is proposed which aims at completing the formal context by reducing the actual formal context to reduced formal context. In the third section, an algorithm to reduce the actual formal context and to complete the actual formal context is presented. Also, this section discusses how the algorithm works by giving an example. Finally, this paper is concluded with the summary of the proposed FCA-based methodology which can be a possible solution in the application of recommender systems [10].

2 Related Mathematical Foundations

This section gives a brief review of some of the important terminologies used in the FCA to describe its definitions and meanings are discussed.

Definition 1 (*see* [4]) A formal context is a triplet represented as $C = (P,Q,R)$, where P and Q are two nonempty and finite sets, and R is a binary relation on P and Q.

Here, P and Q are the set of objects and set of attributes, respectively, where (p, q) $\in$ R indicates that the object p has the attribute q and (p, q) $\notin$ R indicate that the object p does not have the attribute q.

Definition 2 (*see* [4]) For every formal context (P, Q, R), the concept-forming operators $\uparrow: 2^P -> 2^Q$ and $\downarrow: 2^Q -> 2^P$ for every $A \subseteq P$ and $B \subseteq Q$ as:

$$A^\uparrow = \{q \in Q \mid for\ each\ p \in A : (p,q) \in R\}$$
$$B^\downarrow = \{p \in P \mid for\ each\ q \in B : (p,q) \in R\}$$

The operator $\uparrow$ assigns the subsets of Q to subsets of P. $A^\uparrow$ is the set of all attributes shared by all the objects in A. Analogously, the operator $\downarrow$ assigns the subsets of p to subsets of Q. $B^\downarrow$ is the set of all objects shared by all the attributes in B.

Definition 3 (*see* [4]) The formal concept is a pair (A, B) for any formal context (P, Q, R) such that $A \subseteq P$ and $B \subseteq Q$ where $A^\uparrow = B$ and $B^\downarrow = A$.

Here (A, B) is the formal concept if and only if set A consist of only those objects that are shared by all the attributes shared from the set B which is called extent of the concept. Also, set B consist of only those attributes that are shared from the set A which is called intent of the concept.

Thus, the concept-forming operators can be used to define formally the concept's extent as $Ext(P, Q, R) = \{ B^{\downarrow\uparrow} B \in Q\}$ and concept's intent as $Int(P, Q, R) = \{ A^{\uparrow\downarrow} A \in P\}$.

Definition 4 (*see* [4]) For the given formal context (P, Q, R), if (A_1, B_1) and (A_2, B_2) are the formal concepts then it holds a partial order relation $\leq$ such as: $(A_1, B_1) \leq (A_2, B_2)$ if and only if $A_1 \subseteq A_2$ and $B_2 \subseteq B_1$.

This partial order relation formally describes the subconcept—superconcept relationship which can be represented by drawing the line or Hasse diagram called concept lattice.

Definition 5 (*see* [4]) The concept lattice is defined as the hierarchical structure of partial order relation $\leq$ denoted by ß(ß, Q, R) such that $\{(A, B) \in 2^P\ X\ 2^Q \mid A^\uparrow = B$ and $B^\downarrow = A\}$.

Definition 6 (*see* [13]). The arrow relation for the formal context (P, Q, R) is defined as the relations $\rightarrow$, $\leftarrow$, $\updownarrow$ between the P and Q by

- $p \leftarrow q$ iff $(p, q) \notin R$ and if $\{p\}^\uparrow \subset \{p_1\}^\uparrow$ then $(p_1, q) \in R$.
- $p \rightarrow q$ iff $(p, q) \notin R$ and if $\{q\}^\downarrow \subset \{q_1\}^\downarrow$ then $(p, q_1) \in R$.
- $p \updownarrow q$ iff $p \leftarrow q$ and $p \rightarrow q$

The algorithm for constructing the reduced formal context from the actual formal context can be found in the paper [13]. The paper presented a new approach using arrow relation and irreducible formal concept to reduce the given formal context. In the following section, we present our FCA-based methodology which can be one such possible solution in solving the cold-start problem of recommender systems. This FCA-based methodology solves the matrix completion of the given user-item rating matrix by transforming the sparse rating matrix into a dense rating matrix.

3 Proposed Method

In this section, a FCA-based methodology for the recommender system is proposed. In the recommender system, a user-item rating matrix as an input where users are represented as rows and items are represented as columns. The rating value given by the user to item becomes the element of the cell. The FCA method is used for better and simple data representation as formal contexts.

The formal context is denoted as $M = (U, I, R)$ where U and I are two non-empty finite sets and R is the binary relation on $(U \times I)$. Hence, U and I are the collection of objects and collection of attributes, respectively, where $(u, i) \in R$ means that the user u has the relation with the item i and $(u, i) \notin R$ means that the user u do not have the relation with the item i. The relation R is represented by the rating value given by the user u to item i denoted as r_{ui} if R is defined by:

$$R = \{\forall(u, i | \exists r_{ui} \neq \emptyset) \in R)\}$$
$$R \neq \{\forall(u, i) | \exists! r_{ui} = \emptyset) \in R)\} \tag{1}$$

A formal context is said to be reduced if all the sets of objects and all the sets of attributes are irreducible. We apply the FCA-based arrow relation $(\rightarrow, \leftarrow, \updownarrow)$ on the actual formal context to transform it to the reduced actual context. The reduced formal context of $M = (U, I, R)$ is irreducible if no $u \in U$ and $i \in I$ are irreducible. This aims at removing the identical ratings between any two similar users or two similar items in the formal context which further helps in reducing the dimensionality of the rating matrix M.

The following is the proposed method:

Algorithm for Matrix Completion:

Input: User-Item Rating Matrix Z
Output: Predicting the unknown's in the Rating Matrix Z.

Method:

Step 1: Construct the user-item incidence matrix M where M = (U, I, R) of size u X i., given the rating matrix Z.

Step 2: Apply the FCA-based arrow relation $(\rightarrow, \leftarrow, \updownarrow)$ on M, between U and I by:

- $u \leftarrow i$ iff $(u, i) \notin R$ and if $\{u\}^{\uparrow} \subset \{u_1\}^{\uparrow}$ then $(u_1, i) \in R$.
- $u \rightarrow i$ iff $(u, i) \notin R$ and if $\{u\}^{\downarrow} \subset \{i_1\}^{\downarrow}$ then $(u, i_1) \in R$.
- $u \updownarrow i$ iff $u \leftarrow i$ and $u \rightarrow i$

Step 3: From the reduced formal context $M'' = (U', I', R_2)$, it is well observed that, there are no similar users or items.

Step 4: Calculate the rating value by walking through the cells where R does not exist in the reduced formal context as follows:

a. Put $r_{ij} = r_{u'i}$ iff $\exists\, u'_i$ in M''

b. Put $r_{ij} = \sum_I r_{u'i}$ iff $\exists\, \{u'_i\}$ in M''

Step 5: Transform completed reduced formal context M'' to actual formal context M.

Step 6: Repeat Step4 until the actual formal context M is completed.

4 Discussions

A sample dataset is considered representing the user-item rating matrix where there are 5 users, 4 items, rating scale from 1–5 and of size 5 × 4 dimensions (Fig. 1). Figure 1 shows that user U_1 had bought the item I_3 and given the rating of 2 scales and user U_5 had bought the item I_3 and given the rating scale of 4. The user U_4 has not bought the item $I_1, I_2,$ and I_4 for which the rating scale is unknown.

Example: The user-item rating matrix is denoted by the formal context M = (U_5, I_4, R) where binary relation R is defined by: (Refer Eq. 1)

$$R = \{\forall(u, i)|\exists r_{ui} \neq \emptyset) \in R)\}$$
$$R \neq \{\forall(u, i)|\exists! R_{ui} = \emptyset) \in R)\}$$

Fig. 1 User-item rating matrix

R	I_1	I_2	I_3	I_4
U_1	4	3	2	3
U_2	4	5		
U_3		3	2	3
U_4			2	
U_5		3	4	

Fig. 2 Formal context M of user-item rating matrix

R	I_1	I_2	I_3	I_4
U_1	X	X	X	X
U_2	X	X		
U_3		X	X	X
U_4			X	
U_5		X	X	

In Fig. 2, we construct the actual formal concept by defining the binary relation R between the set of users $U = \{ U_1, U_2, U_3, U_4, U_5 \}$ and set of items $I = \{I_1, I_2, I_3, I_4\}$ with the incidence matrix represented with the cross-mark (X) if there exist any ratings given by the user to the item.

From the formal context M, we apply the FCA-based arrow relation ($\rightarrow$, $\leftarrow$, $\updownarrow$) and illustrated in Fig. 3. Figure 3 shows the arrow-related exhibited by Fig. 2 with the relation R_1 (Refer Definition 6)

Now, the actual formal context is transformed into a reduced formal context and has removed similar group of users and similar group of items since its rating value can be easily achieved. The reduced formal context $M'' = ((U_2, U_3, U_5), (I_1, I_4), R_2)$ is shown in Fig. 4 of size 3×2 with the relation R_2.

Using Fig. 4, the rating r_{ui} of each user-item cell represented without cross-mark (X) is computed and the resultant and completed reduced formal context is shown as in Fig. 5.

The completed formal context M'' is composed into the actual formal context M and the rating r_{ij} of each user-item cell represented without cross-mark (X) are calculated and drawn in Fig. 6. Figure 6 shows the completion of the user-item rating matrix which is explained in Fig. 1.

R_1	I_1	I_2	I_3	I_4
U_1	X	X	X	X
U_2	X	X	$\leftarrow$	$\updownarrow$
U_3	$\updownarrow$	X	X	X
U_4		$\rightarrow$	X	
U_5	$\rightarrow$	X	X	$\updownarrow$

$(U_2 \leftarrow I_3)$ iff $(U_2, I_3) \notin R$
if $(U_2)^{\uparrow} \subset (U_1)^{\uparrow}$
then $(U_1, I_3) \in R$
$(U_5 \rightarrow I_1)$ iff $(U_5, I_1) \notin R$
if $(I_1)^{\uparrow} \subset (I_2)^{\uparrow}$
then $(U_5, I_2) \in R$

Fig. 3 Arrow relation of the formal context M

Fig. 4 Reduced formal context M''

R_2	I_1	I_4
U_2	X	
U_3		X
U_5		

Fig. 5 Completed reduced
formal context M″

R₂	I₁	I₄
U₂	4	4
U₃	3	3
U₅	3	3

Fig. 6 Completed formal
context M

R	I₁	I₂	I₃	I₄
U₁	4	3	2	3
U₂	4	5	2	3
U₃	4	3	2	3
U₄	4	3	2	3
U₅	4	3	4	3

In real-time, the recommender systems would be dealing with the millions of users and items where the dimension of the input rating matrix would be having heavy computation load. Thus, the above method presented would be one such possible solution to enhance and compute the rating prediction of any user of such a case.

5 Conclusion

The recommender system allows new opportunities to retrieve a variety of customized services from the Internet. The predominant issues of information overload to the internet users are solved by the evolution of information retrieval systems which gives the users to access the products and/or services which cannot be immediately used by the users. This paper has solved the problem of matrix completion in its way to predict the user ratings. It has presented the FCA-based methodology which is to best of the author's knowledge is the first attempt used to solve the completion of user-item rating matrix. The method was simple and faster which can be further enhanced to deal with the real-time amount of data and the result accuracy can be improvised.

References

1. Wille, R.: Restructuring lattice theory: an approach based on hierarchies of concepts. In: Rival, I. (eds.) Ordered Sets, pp. 445–470. Reidel, Dordrect–Boston (1982)
2. Barbut, M., Monjardet, B.: Ordre et classification, algèbre et combinatoire, vol. 1311. Hachette, Paris (1970)
3. Birkhoff, G.: Lattice Theory, vol. 25, 3rd edn. American Mathematical Society Coll Publication, Providence RI

4. Ganter, B., Wille, R.: Formal Concept Analysis: Mathematical Foundations. Springer, Berlin (1999)
5. Doerfel, S., Jaschke, R., Stumme, G.: Publication analysis of the formal concept analysis community. LNCS **7278,**77–95 (2012). Springer
6. Belohlavek, R., Vychodil, V.: Formal concept analysis and linguistic hedges. Int. J. Gen Syst. **41**(5), 503–532 (2012)
7. Ignatov, D.I., Gnatyshak1, D.V., Kuznetsov, S.O., Mirkin, B.G.: Triadic formal concept analysis and triclustering: searching for optimal patterns. Mach. Learn. **101**(1–3), 271–302 (2015). https://doi.org/10.1007/s10994-015-5487-y
8. Yao, Y.Y.: A comparative study of formal concept analysis and rough set theory in data analysis. LNAI **3066,** 59–66 (2004). Springer
9. Djouadi, Y.: Extended galois derivation operators for information retrieval based on fuzzy formal concept lattice. LNCS **6929,** 346–358 (2011). Springer
10. Isinkaye, F.O., Folajimi, Y.O., Ojokoh, B.A.: Recommendation systems: principles, methods and evaluation. Egypt. Inform. J. **16**(3), 261–273 (2015)
11. Koren, Y., Bell, R., Volinsky, C.: Matrix factorization techniques for recommender systems. Computer **8**, 30–37 (2009)
12. Burmeister, P., Holzer, R.: Treating incomplete knowledge in formal concept analysis. In: Formal Concept Analysis, pp. 114–126. Springer, Berlin (2005)
13. Belohlavek, R.: Introduction to formal concept analysis, vol. 47. Palacky University, Department of Computer Science, Olomouc (2008)
14. Zhang, W., Du, Y.J., Song, W.: Recommender system with formal concept analysis (2015)

Similarity Analysis of Residual Frames for Inter Frame Forgery Detection in Video

K. N. Sowmya, H. T. Basavaraju, B. H. Lohitashva, H. R. Chennamma
and V. N. Manjunath Aradhya

Abstract The credibility of digital videos obtained from various sources like surveillance cameras, smartphones, webcams used as evidence in courtrooms, medical world, and journalism needs to be verified for its authenticity and integrity to detect video forgery. The proposed forensic approach detects inter frame forgery due to frame duplication based on statistical correlation for the digital video. The statistical correlation in video content of the frames has been explored through textural features to identify frame duplication and is verified. Experimental results demonstrate accuracy in identifying frame replication.

Keywords Frame duplication · Frame replication · Video forgery · Inter frame forgery

1 Introduction

Today, wide employment of Surveillance cameras as part of security systems at homes, offices, commercial establishments, and public places generate a large amount of digital video data. The latest technological advancements in the digital world have made powerful editors freely available. These editors, when used with a spiteful objective to fabricate and falsify video data to masquerade or hide information representing truth are posing a challenge for the forensic research community. Anatomy of such forgeries is considered in video forensics to determine the integrity and authenticity of the digital video. Video authenticity is primarily concerned with determining the credentials of the given video as genuine or not. Example: Computer generated video or Natural video. The authenticity of the video can be established by determining the source camera used to capture the video sequence. Integrity of the digital

K. N. Sowmya (✉)
JSS Academy of Technical Education, Bengaluru, Karnataka, India

H. T. Basavaraju · B. H. Lohitashva · H. R. Chennamma · V. N. M. Aradhya
JSS Science & Technology University, Mysuru, Karnataka, India
e-mail: aradhya@sjce.ac.in

© Springer Nature Singapore Pte Ltd. 2020
V. Bhateja et al. (eds.), *Intelligent Computing and Communication*,
Advances in Intelligent Systems and Computing 1034,
https://doi.org/10.1007/978-981-15-1084-7_83

video primarily deals with, whether the given video has undergone any sort of modification upon its acquisition. Conformance of the originality of content for the video in question answers integrity issues.

Forensic analysis of the video can be active or passive in nature depending on the type of video. In an active approach, the embedded watermark can be fragile or semi fragile in nature. Successful retrieval of it establishes the integrity of the video when considered as evidence. Whenever there is a change in the original video content, embedded watermark gets disturbed and retrieval fails which indicates forgery. A digital signature helps to validate and authenticate Digital Video. Content-based digital signature helps to validate and authenticate digital video [1]. Content modification would quash the original digital signature that indicates digital forgery. Digital signature for the given video sequence remains unique and distinct for each video even though the source remains the same since it is based on visual content. The digital signature generated based on content can be embedded as watermark for authentication most of the time. The process of signature generation needs to use unique and distinct local/global features such that it is computationally infeasible to forge an original video by maintaining the same signature. Watermark and digital signature both need to be embedded and generated during the video creation or at the time of storage.

Passive forensics depends upon the naturally occurring properties or inherent fingerprint which is unique to the generated video due to its visual content, imaging device used and its characteristics. The statistical correlations of the original video remain consistent if the video is not forged. Disturbance in the correlations of a forged video is found for the specific features considered. Video forgery can happen at spatial level, temporal, or at spatiotemporal levels. Since video is comprised of group of pictures (GOP), usually forgery is found for at least a few numbers of frames depending on the frame rate and the content that need to be hidden or masked. Video forgery can be further classified into intra frame forgery and inter frame forgery. Intra frame forgery happens at pixel level or at selected object level through cloning, resampling, and upscale forgery. Inter frame forgery happens whenever there is spatiotemporal or temporal tampering at frame level due to cloning, additions, deletions, and frame shuffling to swindle truth.

2 Temporal Video Forgery

Temporal video forgery is commonly found in surveillance footage along with other types of digital videos. In these attacks, visual information captured, and stored by the surveillance camera is altered affecting the original recordings. It is executed on sequence of consecutive frames related by time 't'. Temporal video forgery attacks include Frame insertion, Frame removal, Frame reordering, Frame shuffling, and Frame cloning or replication. Temporal tampering can occur at scene level, shot level, or frame level.

2.1 Frame Addition

Frames from another video sequence having similar statistical properties are intentionally inserted at predetermined locations in a given video. The main aim here is to conceal the actual content and mislead through incorrect information. Gironia et al. [2] considered analysis of video to identify double encoding through sliding window of frames and VPF tool to recognize discontinuity in phase of a periodic function whenever frames were inserted or deleted. Inter frame forgery when the whole GOP is inserted or removed were not considered in their work.

2.2 Frame Deletion

Frames from the original video sequence are intentionally eliminated from a specific location or removed from different locations. Frame deletion-based attacks are usually carried out on surveillance videos to protect people involved in criminal activities by removing his/her presence. Chao et al. [3] identified inter frame forgery due to frame insertion and deletion through a discontinuity in optical flow sequence. Yuxing et al. [4] recognized inter frame forgery detection based on relative factor sequence (VFI) discontinuities after maximum sampling. It is based on generalized ESD test which follows approximate normal distribution. The relative sequence obtained varies for duplication and deletion.

2.3 Frame Shuffling

The sequence of events recorded through frames is jumbled or reordered such that the correct sequence is lost and wrong information is propagated. Yuting et al. [5] and Stamm et al. [6] have exploited the frame grouping strategy implemented by common video encoders to detect inter frame forgery detection.

2.4 Frame Averaging

An average frame between the set of specified frames is inserted in frame averaging attack and they appear to be more realistic along with original frames.

2.5   Frame Cloning/Replication

Statistical properties of the original videos are retained by cloning frames from the same video sequence. The occurrence of a specific event happening is kept in ambiguity through such approach. Guzin et al. [7] have proposed an inter frame forgery attack detection approach due to frame cloning and frame mirroring by considering binary features which are mirror invariant. PSNR and distance between consecutive source and duplicated frames are used to identify cloning. Jia et al. [8] have considered optical flow correlation and validations to detect frame duplication pairs.

3   Proposed Approach

The proposed statistical method clearly localizes inter frame forgery due to frame replication and the potential frames tampered are identified through frame similarity analysis. Textural-based entropy feature is considered that provide a low computational cost when compared to other methods. Fusion of statistical features like standard deviation and contrast is considered for statistical analysis along with entropy of residual frames to determine forgery as shown in Fig. 1.

The following sub section describes the steps adopted in our proposed inter frame forgery detection approach.

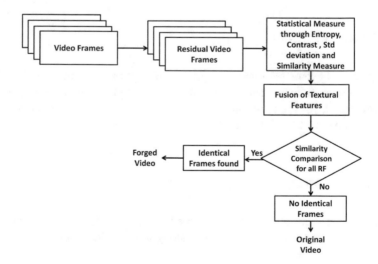

Fig. 1 Overview of the proposed inter frame forgery detection approach

3.1 Gray Level Thresholding

Initially, the frames obtained from the input video are converted to gray scale. The spatial variation in the gray frame is found with the help of second-order statistics through gray level co-occurrence matrix. Textural feature description can be obtained efficiently using these gray level co-occurrence matrices.

3.2 Residual Frames and Background Subtraction

The residual frames are obtained by successive subtraction between adjacent frames till end. Non static objects are identified through background subtraction. The textural features for the residual frames are obtained in the next stage for the residual frames of the gray level co-occurrence matrix as shown in Eq. (1). Textural features are considered since they do not have uniform intensity even if homogeneous regions exist.

$$RF_{i,j} = F_{i,j}^{t+1} - F_{i,j}^{t} \tag{1}$$

where $F_{i,j}$ represents the individual frame. 't' symbolize the frame number for all frames in the given video. $RF_{i,j}$ denote residual frames.

3.3 Entropy

Entropy of the frame identifies the mayhem of the residual frame and its value is used to measure the similarity of information content in the video frames. Entropy of a frame characterizes the textural property associated with the frame. Entropy is found by applying Shannon's approach as shown in Eq. (2). Similarity measurement is achieved by measuring the difference between each frame features versus all others.

$$RF_{Entropy} = \sum \left(NHc \times \log_2(NHc) \right) \tag{2}$$

where NH_c represents normalized histogram count for the residual frame.

3.4 Contrast

In general, the effect of brightness, contrast, saturation, and sharpness will vary depending on the content in each frame of the given video sequence. The surrounding environment influences the video content. Contrast is the disparity between maximum and minimum pixel intensities in the residual frame. It indicates brightness difference between objects or regions of the residual frame as shown in Eq. (3).

$$RF_{Contrast} = \sum_{i,j=0}^{n-1} RF_{ij}(i-j)^2 \qquad (3)$$

3.5 Standard Deviation

Standard deviation or the root mean square deviation associated with the residual frame helps us to locate the outlier residual frames deviating from normal. Abnormal patterns representing the suspected tampered frames can be bifurcated for further analysis.

3.6 Standard Minimum/Nearest Difference of Motion Residuals (Min$_{MR}$)

Find the extent of frame residuals for all frames for the given video. It is independent of total frames tampered, number of motion residual objects in a video sequence, illumination variation, recording device used, or compression codec adopted.

3.7 Fusion of Textural Features

In this stage, fusions of all the textural features are considered as shown in Eq. (4).

$$Fusion_{TF} = RF_{Entropy} + RF_{Contrast} + RF_{standard\ Deviation} + RF_{Min\_Residuals} \qquad (4)$$

$Fusion_{TF}$ contain the representation for the frames of the video. When the video frames are replicated due to forgery the fusion vector represents the original and forged frames. Frame similarity analysis helps us to identify the original and forged frames. Fusion of the textural feature vector provides uniqueness when compared to individual textural features when the video clip is original.

3.8   Similarity Analysis of Residual Frames

During this process, we compare the amalgamated feature representation $Fusion_{TF}$ for all residual frame values obtained. Whenever there is frame replication-based video forgery attack, identical frames can be located indicating forgery. Otherwise distinct values can be observed for all frames of the video.

Frames of the video are represented through a Video_Dictionary comprising key value pair. Frames form the keys and the values they contain represent the fusion of textural features extracted for each frame. In a Video_Dictionary each key is unique. The value held by the key is unique if not forged and will have repeated values when temporal forgery is done due to frame replication. Enumerating the key value pair for finding replicated value will help us to detect inter frame video forgery.

1   Experimental Results

We have conducted experiments to verify the efficacy of our approach to detect inter frame forgery attacks due to frame replication. The experimental videos considered have come from two different origins [9]. Video Gallery-1 videos have been taken from surveillance cameras (static or fixed camera) footages in the parking area. The video thus obtained has been broken into multiple clips of variable length. They have an aspect ratio of 1080 × 1920. These videos use H.264 codec. Video Gallery-2 videos have been captured manually from smartphone camera in a non static mode. All the videos in Video Gallery-2 have an aspect ratio of 720 × 1280 and are of variable length. The set of frames have been cloned at different position in these videos after the existing frames are deleted. The sample of few frames from VID_20190130_114022.mp4 has been shown in Fig. 2. It has 134 frames in the given video sequence. Original frames have been deleted and have been replaced by cloned frames from the same video. All the video sequence considered from Gallery 1 and Gallery 2 comprises a shot. Encoding of the video clip considered has undergone post processing operations such as quantization which leaves a unique fingerprint on the sequence itself. Our practical experiments reveal frame duplications whenever there is forgery and frame similarity in the sequence can be observed for forged frames as shown in Fig. 4. With the help of distance metric and Video_Dictionary value analysis, we can observe that frame 31–60 and frame 91–120 are identical in nature and either of them have been cloned. In a non forged video though they are homogeneous in nature identical frames cannot be seen in the results shown in Fig. 3. The length of the original video is maintained same in the forged video after forgery. Temporal forgery due to frame cloning can be clearly identified through the proposed approach.

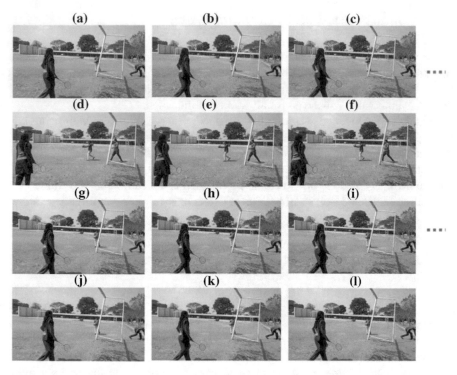

Fig. 2 **a–c** represents frames at position 30–32 and **d–f** represents frames at 90–92 from Video Gallery-2 for original video VID_20190130_114022.mp4. **g–i** represents frames at position 30–32 and **j–l** represents frames at 90–92 from Video Gallery-2 for forged video VID_20190130_114022.mp4

Fig. 3 Results of the Original video— VID_20190130_114022.mp4 from set 2

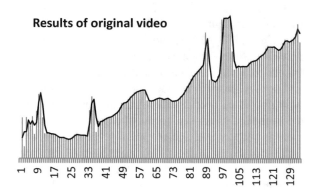

5   Conclusion

In this paper, the statistical approach adopted is passive in nature and has low computational complexity. It considers the statistical features of cloned frames in the

Fig. 4 Results of the forged video— VID_20190130_114022.mp4 from set 2

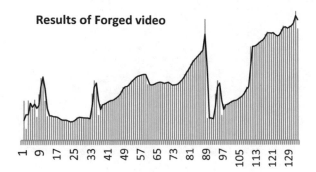

Results of Forged video

compromised frames to detect forgery. In future we would concentrate on all types of temporal forgery. Since no training samples or classifiers are used in our approach it is computationally inexpensive when compared to training-based approaches for video forgery detection.

Acknowledgements We wish to thank JSS MahaVidyapeetha, Karnataka, India for the constant support and encouragement in our work.

References

1. Sowmya, K.N., Chennamma, H.R., Rangarajan, L.: Video authentication using spatio temporal relationship for tampering detection. J. Inf. Secur. Appl. **41**, 159–169 (2018)
2. Gironi, A., Fontani, M., Bianchi, T., Piva, A., Barni, M.: A video forensic technique for detecting frame deletion and insertion. In: IEEE International Conference on Acoustics, Speech and Signal Processing (ICASSP), pp. 6226–6230 (2014). https://doi.org/10.1109/icassp.2014.6854801
3. Chao, J., Jiang, X., Sun, T.: A novel video inter-frame forgery model detection scheme based on optical flow consistency. In: Digital Forensics and Watermaking, pp. 267–281. Springer, Berlin (2013)
4. Wu, Y., Jiang, X., Sun, T., Wang, W.: Exposing video inter-frame forgery based on velocity field consistency. In: IEEE International Conference on Acoustics, Speech and Signal Processing (ICASSP), 2014, pp. 2674–2678 (2014)
5. Zhang, J., Su, Y., Zhang, M: Exposing digital video forgery by ghost shadow artifact. In: Proceedings of the First ACM workshop on Multimedia in forensics, pp. 49–54. ACM, 2009
6. Stamm, M.C., Lin, W.S., Liu, K.J.R.: Temporal forensics and anti-forensics for motion compensated video. IEEE Trans. Inf. Foren. Secur. **7**(4), 1315–1329 (2012)
7. Ulutas, G., Ustubioglu, B., Ulutas, M., Nabiyev, V.: Frame duplication/mirroring detection method with binary features. IET Image Proc. **11**(5), 333–342 (2017)
8. Jia, S., Zhengquan, X., Wang, H., Feng, C., Wang, T.: Coarse-to-fine copy-move forgery detection for video forensics. IEEE Access **6**, 25323–25335 (2018)
9. https://sites.google.com/view/hrchennamma/research-activities/jssstu-data-sets

Author Index

© Springer Nature Singapore Pte Ltd. 2020
V. Bhateja et al. (eds.), *Intelligent Computing and Communication*,
Advances in Intelligent Systems and Computing 1034,
https://doi.org/10.1007/978-981-15-1084-7